Dependent Data in Social Sciences Research

Mark Stemmler • Wolfgang Wiedermann •
Francis L. Huang
Editors

Dependent Data in Social Sciences Research

Forms, Issues, and Methods of Analysis

Second Edition

Springer

Editors
Mark Stemmler
Institute of Psychology
Friedrich-Alexander-University
of Erlangen-Nuremberg (FAU)
Erlangen, Germany

Wolfgang Wiedermann
Educational, School and Counseling
Psychology
University of Missouri
Columbia, MO, USA

Francis L. Huang
Educational, School and Counseling
Psychology
University of Missouri
Columbia, MO, USA

ISBN 978-3-031-56317-1 ISBN 978-3-031-56318-8 (eBook)
https://doi.org/10.1007/978-3-031-56318-8

This Springer imprint is published by the registered company Springer Nature Switzerland AG
The registered company address is: Gewerbestrasse 11, 6330 Cham, Switzerland

Preface

It has been now almost 10 years since the first edition of *Dependent Data in Social Sciences Research* was published. The book was a success (more than 23k downloads online). Now, we are proud to say that this new second edition reflects the new and exciting developments in the field of dependent and longitudinal data in the social sciences of the last decade. The new developments are in part due to the broad extension of the scope of the statisticians in the social sciences research to the Bayesian approach and to other disciplines like economics, physics or neuroscience.

All authors of this volume are leading experts in their field of applying or developing new statistical methods for dependent data scenarios. This second edition is again truly international, with authors from the USA, Europe and Asia. The book consists of 29 chapters and 7 sections. Therefore, the new second edition can be considered a compendium on the up-to-date and cutting-edge methods in the field. The seven sections are: (1) Growth Curve Modeling, Continuous Time Modeling and Dynamic Modeling, (2) Network Analysis and Causal Structure Learning, (3) Multilevel Analysis, (4) Longitudinal and Cross-sectional Dependent Categorical Data Analysis and Discrete Sequence Analysis, (5) Longitudinal Modeling and Estimation of Missing Data, (6) Item Response Modeling for Dependent Data and (7) Other Methods for the Analysis of Dependent Data.

Section I—Growth Curve Modeling, Continuous Time Modeling and Dynamic Modeling: Johan Oud starts out with a historical and philosophical background of continuous time modeling in the social sciences. He describes the introduction of continuous time modeling into social science by Simon and Coleman. (Jo)Han Oud also discusses the philosophical question whether change in continuous time is possible at all. Matt Miller and Paolo Ghisletta discuss the issue of time in latent growth curve modeling. Manuel Arnold, Pablo F. Cáncer, Eduardo Estrada, and Manuel Voelkle demonstrate model-based recursive partitioning as a powerful approach to analyzing heterogeneity between subjects. The *semtree* software package is a powerful tool for the recursive partitioning of structural equation models (SEM). The resulting SEM trees partition the sample into groups of similar individuals while identifying the most important predictors of group differences in the process. Mark Stemmler and Friedrich Lösel try to disentangle the bidirectional relationship between a child's social problem behavior and harsh parenting (here corporal punishment by the father). The child's problem behavior was rated by teachers from kindergarten and in school; fathers filled out a questionnaire to assess their use of harsh parenting. Continuous time modeling was applied to search for auto-regressive and cross-effects over time. In addition, an effect of a time-dependent variable, perceived sense of parenting competence, was explored. Hannes Meinlschmidt, Meike Sons and Mark Stemmler introduce a new approach to parameter estimation in continuous time modeling (ctsem) in the spirit of variational data assimilation or machine learning. A new algorithm is proposed; in addition, the authors complement the proposed algorithm with a practical example, comparing the results of this approach to those obtained via ctsem. Cédric Noel and Jang Schiltz introduce an extension of Nagin's finite mixture model to underlying Beta distributions. In addition, a new R package *trajeR*, which allows to calibrate the model, is introduced. Covid-19 related data during the first part of the pandemic is used to demonstrate the new model.

Section II—Network Analysis and Causal Structure Learning: Jonathan Park, Sy-Miin Chow and Peter C. M. Molenaar state that in our real world, variables and people seldom cooperate with our attempts to fit them into neat clusters. Therefore, Park et al. describe a fuzzy community detection method which permits variables to belong to multiple communities by varying degrees rather than restricting their membership to only one cluster. The authors also discuss how the application of fuzzy methodologies may inform the study of psychological symptoms and subjects. Takashi Nicholas Maeda, Yan Zeng and Shohei Shimizu are dealing with the central problem of science to elucidate the causal mechanisms underlying natural phenomena and human behavior. They present statistical methods for estimating causal relationships from observational data obtained from sources other than randomized experiments. Their new method goes beyond existing limits; in addition, the authors focus on hidden variable models, nonlinear models and mixed data models. Wolfgang Wiedermann and Bixi Zhang present and discuss linearizable non-linear models for which distributional and independence-based direction dependence measures to discern the direction of causation are applicable. Based on their simulation results, the authors suggest that direction of dependence properties of linearized models allow one to identify the direction of causation for a variety of non-linear functional forms. Furthermore, an empirical example is given and assumptions for model identification and extensions are discussed.

Li Wang, Shu-Min Liao and Daeyoung Kim propose an explorative approach to facilitate the understanding of dependence structures in longitudinal categorical data with ordinal outcome variables and categorical (nominal or ordinal) covariates. The authors propose an approach that utilizes a model-free association measure, the Scaled Checkerboard Copula Regression Association Measure (SCCRAM), developed for multivariate contingency tables with an ordinal response variable and a set of categorical predictors. Cody Ding presents a Bayesian network for discovering the causal structure in observational data. This approach is demonstrated using data on the effects of social adversity on posttraumatic stress and mood symptoms.

Section III—Multilevel Analysis: Simon Grund, Oliver Lüdtke and Alexander Robitzsch provide an overview of the different methods that have been recommended for handling missing data in multilevel analysis. They also present the features of multilevel data that need to be considered when these methods are used. Furthermore, the authors discuss what options the different methods provide for accommodating the structure and analysis of multilevel data. All applications are illustrated in a series of simulated examples. Yichi Zhang, Winnie Wing-Yee Tse and Mark Lai introduce different multilevel bootstrap procedures and described how and when to use them. Specifically, the authors review parametric, residual, wild, and cases bootstraps. They also present five ways of constructing confidence intervals (CIs) after obtaining a bootstrap sampling distribution: bootstrap normal CI, basic CI, studentized CI, percentile CI, and bias-corrected and accelerated CI. Bixi Zhang and Francis Huang suggest multilevel methods to deal with cross-classified data structures where observations are nested within two higher-level structures which in turn are not nested within each other. The authors introduce cross-classified random effects models (CCREMs) as an extension to standard multilevel modeling. This chapter also extends the use of a small sample adjustment approach of linear regression models with cluster-robust standard errors when used with cross-classified data structures and investigates its performance under various conditions. Ting Wang and Edgar Merkle describe a new self-normalization approach for score-based tests of mixed models, which addresses situations where there is dependence between scores. This differs from the traditional score-based tests, which require independence of scores. The authors first review traditional score-based tests and then propose a new, self-normalized statistic. A simulation study and the application of this new approach are demonstrated. Spyros Konstantopoulos, Wei Li and Bixi Zhang focus on a priori statistical power analysis in multilevel field experiments. Specifically, statistical power methods in two- and three-level cluster and block randomized designs are outlined. Altogether five experimental designs (two cluster and three block randomized) are defined.

Section IV—Longitudinal and Cross-sectional Dependent Categorical Data Analysis and Discrete Sequence Analysis: Jörg-Henrik Heine, Florian G. Hartmann and Christian Tarnai demonstrate the use of weighted configural frequency analysis with data from the German General Social Survey (ALL-BUS). Typical and atypical patterns of family educational triads from formal school-leaving qualifications are examined. In addition, the central question is addressed to what extent these patterns can be modeled in the longitudinal course of the birth cohorts of the respondents. From a methodological perspective, the influence of the design weights provided in the data, which compensate for the disproportionalities of two sub-samples and varying drawing probabilities of respondents the aid of functional CFA is suggested. Then Alexander von Eye and Wolfgang Wiedermann start a series of three chapters on configural frequency analysis (CFA). The first is dealing with CFA under multinormality. Thus far, CFA base models were specified either in terms of variable relations or in terms of functions that represent data generation processes. In this chapter, these two approaches are fused. Specifically, CFA base models are proposed in which variable relations are incorporated in tandem with cell probabilities that are based on the hypothesis of multinormality. The second chapter then extends the multinormality approach to incomplete tables. The authors introduce, new to the method of CFA, a joint application of CFA and methods for log-linear modeling of normally distributed variables in cross-classifications with structural zeros. This novel approach is illustrated in a real-world data example. The last chapter of this series deals with higher order CFA of groups of variables. A distinction is made between two groups of variables that, in standard CFA, are hypothesized to be completely independent of each other. An approach is presented in which first and higher order interactions among variables from the two variable groups are taken into account. Local deviations from such a model suggest that interactions exist higher than the ones taken into account. An interpretation of these deviations with the aid of functional CFA is suggested. Shu-Min Liao, Li Wang and Daeyoung Kim propose a graphical approach designed to explore and visualize the regression dependence structure in a multi-way contingency table with an ordinal dependent variable and a set of independent categorical (ordinal or nominal) variables. The proposed method is based on copula regressions developed for ordinal variables which enables prediction of the category of the ordinal dependent variable for any given combination of categories of independent variables in a model-free manner. A resampling technique to facilitate visualizing the uncertainty involved in the prediction and significance of the discovered regression associations is employed. Qianru Liang, Jimmy de la Torre, Mary E. Larimer and Eun-Young Mun present cognitive diagnostic modeling (CDM), which is an item-level analysis that accounts for attribute co-occurrences when characterizing attributes and classifying individuals attribute profiles. A study is introduced to demonstrate how intervention and gender affect transition probabilities from one state to another in a three-step latent transition CDM. The three-step latent transition CDM with covariates constitutes an appealing analytical tool for examining and explaining change in mental health symptoms with informative covariates.

Section V—Longitudinal Modeling and Estimation of Missing Data: Kristian Kleinke, Markus Fritsch, Mark Stemmler and Friedrich Lösel discuss strengths and weaknesses, and evaluates the robustness of three approaches for creating multiple imputations of missing data that do not necessarily rely on the assumption of independent, normal and homoscedastic errors: semi-parametric predictive mean matching, quantile regression-based multiple imputation and random forest-based multiple imputation. Simulations are displayed which contribute to establish practical guidelines regarding sample size requirements of the respective approaches. Kristian Kleinke and Jost Reinecke explain why the respective model that is used to create the imputations usually needs to reflect the assumed (and typically unknown) true data generating process, and if necessary, the mechanism that created the missing data patterns. The authors focus on nonlinear trajectories of the target variable over time and propose to use piecewise growth curve models as a relatively simple method to approximate a non-linear trajectory and to impute incomplete non-linear panel data. The results of a Monte Carlo simulation, where the data generating model was a piecewise growth curve model with two linear piecewise splines over 12 panel waves, reflecting first an increase in the target variable and a

decrease later on, are presented. Bo Ye, Recai Yucel and Donna Coffman inve: the impact of the imputation model on mediation analysis with clustered Specifically, the joint and variable-by-variable imputation models prior to a tilevel mediation analysis are considered. The authors also suggest theoretic analytical assessment of the bias under each imputation method. A compreh simulation study is conducted to understand the performance of imputation m in a repetitive sampling framework. Diane Losardo, Sy-Miin Chow, A.T. I Melissa Burkley and Edward Burkley discuss the development and use of pl missing designs in longitudinal panel studies. In their chapter, the authors pi several planned missingness designs for experience sampling/ecological mom assessment (EMA) studies and evaluate the statistical implications and trac involved in reducing the number of data points collected per person.

Section VI—Item Response Modeling for Dependent Data: Rainer Alexan icz, Linda Maurer, Anna Schultz and Marcus Mund introduce the Actor-P Interdependence Model (APIM) which is a complex regression-based mode turing the simultaneous effects within a dyad. The main focus of this cha to apply IRT models using the free and open software R in order to contras simple sum score-based path analysis and (2) two Structural Equation Mo approaches (one based on the covariance matrix for interval data and the based on the polychoric correlation for ordinal data) to (3) three Item Res Theory (IRT) models, the Partial Credit Model (PCM), the Generalized Credit Model (GPCM) and the Graded Response Model (GRM), each in and a multidimensional approach. Ferdinand Keller and Rainer Alexandrowic that for the evaluation of individual change during an intervention, several m have been proposed that are based on classical test theory, in particular, the r change index (RCI). The authors also deal with RCIs based on item response (IRT) which has been suggested to seem superior to CTT since IRT models varying standard errors over the trait. In the present chapter, a formal introd to the assessment of individual change based on these two approaches are giv addition, the two RCIs are compared using simulation and empirical studies.

Section: VII—Other Methods for the Analysis of Dependent Data: F Thomas discusses the issue of within-sample heterogeneity, which is a conc all matters of data collection and research considerations. This chapter intro finite mixtures as an approach to solving the problem of identifying unobs within-group heterogeneity.

Erlangen, Germany — Mark Sten
Columbia, MO, USA — Wolfgang Wieder
Columbia, MO, USA — Francis L. F
Winter, 2024

Acknowledgments

The first editor wants to express his gratefulness to Susanne and Quincy for their enduring support, comfort and love. The second editor is grateful to Anna and Linus for their endless support that made this edited volume possible. The third editor would like to thank Tracy, Billie and Lucas for their encouragement and humor.

In addition, we are very thankful to Faith Su (Associate Editor, Statistics) and Krishna(kumar) Pandurangan (Production Editor) of Springer Nature for their invaluable support and enduring efforts for making this endeavor possible.

The book is published in memoriam of John (Jack) J. McArdle (University of Southern California). Jack wrote the inaugurating chapter of the first edition. We were very sorry to learn that, unfortunately, Jack died in 2023; we were sorry to hear about his passing. He was a dear colleague, always friendly and helpful; we miss him a lot. We are thankful to his wife Carol Precott who wrote a small feature in his honor and provided a nice photo of him.

In Memoriam John J. McArdle, Jr. (1951–2022)

Jack McArdle was a brilliant scholar and a gifted teacher whose easygoing manner belied his keen intelligence and determination to find the right answer. He was best known for methodological developments in longitudinal analysis, particularly strategies for reducing bias from non-random missing data. Jack's interest in the structure of cognitive abilities over aging led to contributions in measurement and assessment, including equating scores from different cognitive measures across age groups (McArdle et al., 2009) and a random-interval retest experiment to investigate differences in reliability across age and abilities (McArdle & Woodcock, 1997). What may turn out to be Jack's most important substantive contribution was his research to improve cognitive measurement in the Health and Retirement Study (Fisher et al., 2013), now being used to track the epidemiology of cognitive impairment and decline in the US.

"Jackie" was born in New York City to Irish-American parents, the third child and oldest son of six siblings. During his early childhood, the family lived on the 13th floor of an apartment building in a subsidized housing complex in the Bronx. Rent was cheaper on the 13th floor and Jack recalled it being full of large families with kids having lots of fun. Perhaps this experience was the origin of his skepticism about numerology and other superstitions.

Jack was a natural athlete, excelling in competitive sports. He received an athletic scholarship to play baseball at the University of Notre Dame, but decided to attend Franklin and Marshall College, located in Pennsylvania and closer to home. He became a psychology major because his grades were too poor to continue in the pre-med program. After suffering a major knee injury while playing football, Jack became a more dedicated student, discovering mathematical psychology and computer programming.

When Jack was applying to doctoral programs, his father was ill and unemployed and the family was in severe financial trouble. Jack returned to New York to help his mother and younger sibs and to attend Hofstra University. There he worked full time at the computer center, loading tapes and operating the newly arrived mainframe, while still completing a PhD in 4 years. To cut down on commuting time, he often slept in his Volkswagen van in the school parking lot.

After receiving his doctorate in 1977, Jack moved to the University of Denver for a postdoctoral position with John Horn. His time in Colorado stretched to 6 years and became the foundation of multiple threads in Jack's career, including 35 years of continuous funding from the National Institute on Aging, part of the U.S. National Institutes of Health. His first grant funded the aggregation of data from longitudinal studies that collected Wechsler intelligence scores. Ultimately, Jack compiled more than 200,000 records of WAIS data, much of it obtained by visiting colleagues and photocopying the original test forms. Even Jack's students who did not participate in the copying and data entry remember the 30+ file cabinets holding the raw data.

A groundbreaking (but now largely forgotten) development in the early 1980s was RAM, Jack's 3-matrix simplification of path analysis (compared to the 8-matrix version in LISREL). He collaborated with Steve Boker on creating RAMPath (McArdle & Boker, 1990), the first software program to estimate structural models and display them graphically. Unlike many other portrayals of path models that were conceptual, Jack's diagrams conveyed exact algebraic expectations, including variable means.

Another happy development during this time for Jack and his wife Pamela was the arrival of daughter Cristin.

In 1984, Jack moved to the Department of Psychology at the University of Virginia (UVa), where he stayed until 2005. While at UVa, Jack founded the quantitative methods doctoral program, progressed to Full Professor and welcomed his two sons, Michael and Timothy. Among many valued colleagues at UVa was mentor and collaborator John Nesselroade. During these years, Jack was frequently in Washington DC, where his role as President of the Society for Multivariate Experimental Psychology led to involvement with the Federation of Brain and Behavioral Sciences and the Council of Scientific Society Presidents, organizations providing scientific input and advice to US government agencies and institutions.

In 2005, Jack was recruited to the University of Southern California (USC), where he headed the quantitative methods program in the Department of Psychology and joined his former mentor John Horn for his last years. Jack was accompanied in this move by his second wife, Carol (me).

Of his many awards, Jack was most proud of receiving the Jacob Cohen Award for Distinguished Contributions to Teaching and Mentoring, given by the American Psychological Association Division 5. Jack was a gifted teacher and his multivariate courses were standing room only. It did not seem to matter that his audio-visuals were sometimes perplexing, with overheads out of sequence or not matching what he was currently saying. Jack managed to convey complex concepts without sacrificing precision. He began with simple path diagrams, but soon those circles, squares and arrows progressed to matrix equations. He would say, "Imagine the circles are stacks of trash can lids and the squares are piles of pizza boxes." Jack gave confidence to students who were stats-phobic, leading many to discover an aptitude for quantitative analysis.

Jack's interest in college sports led to work with the National Collegiate Athletic Association (NCAA) studying racial inequality in the use of standardized tests. He headed the NCAA's Data Analysis Research Network for 20 years, leading the effort to employ empirical evidence to guide policy decisions about eligibility of student athletes for scholarships and sports participation (Petr & Paskus, 2019). Ironically, this involvement led to the most difficult period in Jack's professional life. As part of a lawsuit alleging racial bias by the NCAA, Jack's research came under scrutiny from a US Congressional subcommittee. All his correspondence and research files were subpoenaed, he spent dozens of hours in legal depositions and the accusation of racism left an enduring taint in the minds of some colleagues.

In actuality, Jack worked hard to promote diversity and inclusion. His graduate students and trainees came from all backgrounds and (unusually for the time) included many women. Many students, like Jack, came from economic disadvantage and educational institutions that were less known and less highly regarded. He valued motivation above other criteria.

Jack lived his statistical beliefs. Once, outside a convenience store, sons Mick and Tim asked for money to purchase $1 lottery tickets. Instead of lecturing them about the low probability of winning, Jack took out his wallet, removed two dollars and tossed them over his shoulder, saying, "It's better to throw your money in the gutter. That way, it might end up with someone who needs it."

In his early 60s, Jack began experiencing neurological symptoms, including difficulty with balance and handwriting. This was followed by trouble speaking spontaneously. Jack began reading his lectures from notes, a marked change from his prior loquacious self. In 2014, he was diagnosed with frontal-temporal dementia and eventually received a diagnosis of progressive suprabulbar palsy.

Jack had planned to retire to Hawai'i to write several books on longitudinal analysis. We had already purchased a lot on Hawai'i Island, where we spent visits camping in a converted school bus. Jack's diagnosis accelerated the timetable and in 2016 we relocated to Hawai'i. Over the following years, Jack gradually lost his mobility and became unable to speak more than a few words. He died at home in

2022. Until the end, he retained his interest in sports and current events as well as his sense of humor. Although Jack was not able to complete those books, he left an enduring legacy through his scholarship, mentoring and enthusiasm for life.

Hawi, Hawaii, USA
January, 2024

Carol Prescott

References

McArdle, J. J., Grimm, K., Hamagami, F., Bowles, R., & Meredith, W. (2009). Modeling life-span growth curves of cognition using longitudinal data with multiple samples and changing scales of measurement. *Psychological Methods*, 14, 126–149.

McArdle, J. J. & Woodcock, R. W. (1997). Expanding test retest designs to include developmental time lag components. *Psychological Methods, 2*, 403–435.

Fisher, G., McArdle, J. J., McCammon, R., Sonnega, A., & Weir, D. (2013). New Measures of Fluid Intelligence in the HRS, *HRS Documentation Report DR-027*. Survey Research Center, University of Michigan.

McArdle, J. J. & Boker, S. M. (1990). *RAMpath: A Computer Program for Automatic Path Diagrams*. Hillsdale, NJ: Lawrence Erlbaum Publishers.

Petr, T. A. & Paskus, T. S. (2019). Making the cut. How a quantitative psychologist changed college sports. In: Ferrer, E., Boker, S.M., & Grimm, K.J. (Eds.), *Longitudinal Multivariate Psychology*. New York, NY: Routledge Taylor & Francis Group.

Contents

About the Editors

Mark Stemmler, since 2011, is a Chair of Psychological Assessment, Quantitative Methods and Forensic Psychology at the Institute of Psychology at the Friedrich-Alexander University of Erlangen-Nuremberg (FAU), Germany, and an adjunct Professor at the College of Health and Human Development, Pennsylvania State University, USA. He accomplished his master's degree from the Technical University Berlin in 1989 and his PhD from the Pennsylvania State University in 1993. In 2002, he received his postdoctoral lecture qualification (Habilitation) from the FAU. From 2007 to 2011, he was a full Professor for Quantitative Methods at Bielefeld University, Germany. He currently serves on the editorial board of the *International Journal of Behavioural Development* (IJBD), *Methodology* and *Empirische Sonderpdagogik*. He published a Springer textbook on person-centered methods and configural frequency analysis. His research interests encompass psychological assessment of cognitive decline, program evaluation and methodology. He has worked on longitudinal studies in the USA and Germany.

Wolfgang Wiedermann is a Professor in the College of Education and Human Development at the University of Missouri-Columbia. He received his PhD in Quantitative Psychology from the University of Klagenfurt, Austria. His primary research interests include the development of methods for causal inference and causal structure learning in observational data, and methods for person-oriented research. He has edited books on new developments in statistical methods for dependent data (with Stemmler and von Eye) and on statistical methods for causal inference (with von Eye) and causal discovery (with Kim, Sungur and von Eye). In addition, he has co-authored books on configural frequency analysis and the general linear model (with von Eye). His work appears in leading quantitative methods journals, including *Psychological Methods*, *Multivariate Behavioral Research* and *Behavior Research Methods*, and top applied journals, including *Development and Psychopathology* and *Developmental Psychology*. He currently serves as an Associate Editor for *Behaviormetrika* and the *Journal for Person-Oriented Research*, and is an editorial board member of *School Psychology Review*.

Francis L. Huang is a Professor in the College of Education and Human Development at the University of Missouri (MU). He received his PhD from the University of Virginia and his MA from Teachers College, Columbia University. His research focuses on both methodological (e.g., analysis of clustered data) and substantive (e.g., school climate, bullying, disparities in disciplinary sanctions) areas of interest. He teaches courses on multilevel modeling, program evaluation and data management. He serves on the scientific review panel of the Institute of Education Sciences of the U.S. Department of Education, is an Associate Editor for *School Psychology Review*, and is on the editorial boards of *Prevention Science* and *School Psychology*. He has contributed to over 100 peer-reviewed articles and his research has been cited in outlets such as the *New York Times*, *The Washington Post* and *National Public Radio* (among others). Prior to joining MU, he was a Senior Scientist at the University of Virginia, has worked at the American Institutes for Research, was a management consultant and a high school teacher. He is chair of the School Community, Climate, and Culture Special Interest Group (SIG) of the American Educational Research Association (AERA).

Part I
Growth Curve Modeling, Continuous Time Modeling and Dynamic Modeling

Chapter 1
Continuous Time Modeling in the Social Sciences: History and Philosophical Background

Johan H. L. Oud

1.1 Introduction

While social research in the first decades after World War II was highly cross-sectionally orientated, the argument that causal claims need longitudinal data gained nevertheless widespread support. This led to a lot of longitudinal research and made especially the cross-lagged panel model (CLPM), in which the same subjects are measured repeatedly, very popular. Interest in continuous-time analysis emerged when it was realized that the discrete-time analyses of those data, by means of CLPM for example, are highly dependent on the time interval between the panel waves. This dependence makes comparison between the results of different studies impossible and the results within a single study difficult to interpret. A cross-lagged effect between X and Y might over a small interval, for example, be negative but change to positive over longer intervals. These discrete-time analysis problems disappear in a continuous-time analysis of the same data.

Starting from an overview of the methodological situation in psychology and sociology just after World War II, this contribution then depicts the birth of continuous time modeling in social science in the hands of Herbert Simon and James Coleman. Next, the addition of random error to the differential equation in the form of the Wiener process by Tuma and Hannan is followed. Further development in the two directions of structural equation modeling (SEM) and state space modeling (SSM) is described with the detour in structural equation modeling via the indirect parameter estimation procedure. It is observed in detail how the two merged in presently available flexible software, which, in addition to the choice of SEM or SSM, allows arbitrary time intervals within and between subjects and thus missing

J. H. L. Oud (✉)
Radboud University, Nijmegen, The Netherlands
e-mail: han.oud@ru.nl

M. Stemmler et al. (eds.), *Dependent Data in Social Sciences Research*,
https://doi.org/10.1007/978-3-031-56318-8_1

values at arbitrary places. Several philosophical issues related to continuous time modeling are discussed, culminating in the question whether change in continuous time is possible at all, a problem raised already 2500 years ago in the paradoxes of Zeno of Elea, which got a new impetus by the theory of quantum mechanics.

1.2 Social Science Methodology After World War II

The efforts of social research and methodology in the first decades after World War II concentrated on the analysis of cross-sectional data. This was the case for both psychology and sociology, which, apart from the cross-sectional orientation, had not yet much in common in methodology. Central topics in psychology were factor analysis, measurement theory, and experimental design. Factor analysis, invented by Spearman (1904) and hailed by many psychologists as the means to uncover the hidden factors of mind, initially followed the cross-sectional between-subjects procedure. This means that data were collected from different subjects, measurements thus repeated across subjects instead of within subjects, and assumed independent between subjects. Cattell et al. (1947) in the so-called P-technique attempted to apply the cross-sectional factor analysis procedure on longitudinal within-subject data, in fact on repeated measurements of one and the same subject. Because this does not account for the dependence of the data on the lagged measurements in the subject, P-technique is clearly fallacious in this respect, as was recognized repeatedly in the literature and later on by Cattell himself (Cattell, 1963). Even though some results of the P-technique may not deviate too much from the correct results (Molenaar & Nesselroade, 2009), there is little reason to use or recommend the use of the P-technique after appropriate procedures to take care of the dependence have become widely available.

The highly cross-sectional character of psychological measurement theory in the past and continued in more recent publications becomes immediately evident by inspecting and comparing the influential books of Torgerson (1958), Lord and Novick (1968), and van der Linden (2016). Longitudinal approaches are totally lacking. The interest in experimental research was and is shared by psychology with biology and several other disciplines. Of course, in many experimental designs, two or more measurements on the same subjects are required, for example, in the basic "pretest-posttest control group" design as well as in the so-called "repeated measurements" designs. However, the aim of repeated measurements is not an accurate description of development across time during and after treatment but, as remarked by Winer in his well-known textbook (Winer, 1962, p. 261), to let each subject serve as its own control and eliminate differences between subjects from the experimental error. Hereby, the assumption of compound symmetry is often made, which implies equal variances across time and measurements far apart in time correlating just as highly as the ones closer to each other. As confirmed also by Winer, compound symmetry over more than two points in time is a highly restrictive assumption. Real time series, especially in psychology, seldom conform to this.

Hypothesis testing problems as a result of the compound symmetry assumption can be avoided by multivariate analysis of variance (Winer, 1962, p. 282), but again, this does not result in any informative description of development, neither in discrete time nor in continuous time. Explicitly modeling the error structure across time was not yet done in this early period.

As said above, in this early period, the topics discussed in sociological methodology were different from those in psychology, although not less cross-sectional in character. The book by Blalock entitled "Causal Inferences in Nonexperimental Research" (Blalock, 1964) summarizes aptly what was going on in sociology. The interest was if and when causal conclusions could be drawn from relationships between variables in nonexperimental research. Although longitudinal data were not excluded, what one generally had in mind was cross-sectional research, variables from the same point in time. This can be exemplified by Lazarsfeld's very influential elaboration technique, in short called the Lazarsfeld method (Lazarsfeld, 1955), and going back to an address he gave as early as 1946 at a meeting of the American Sociological Society. By adding to an existing relationship between two variables a third test variable, and recomputing for the categories of the test variable the relationship, what can happen are four things:

(1) The relationship does not change in the categories of the test variable.
(2a) The relationship vanishes in the categories; the test variable (really or conceptually) precedes the other variables in time.
(2b) The relationship vanishes in the categories; the test variable (really or conceptually) comes in time between the other two variables,
(3) The relationship becomes different in different categories.

In case (1), the causal claim is considered corroborated. Case (2a) Lazarsfeld calls *explanation*: the causal claim is not justified, the original relationship is spurious, and the test variable a confounding variable, which explains the spurious relationship. Case (2b) Lazarsfeld calls *interpretation*: the causal relationship is real, but the test variable is an intervening variable (in more recent publications called mediator), which interprets how the causality works. Case (3) finally Lazarsfeld calls *specification* (elsewhere called interaction): again the causality is real, but its sign or strength differs in different categories.

Whereas Lazarsfeld's elaboration technique was formulated in terms of categorical variables, another very popular procedure in sociology at that time, path analysis, developed by geneticist Wright (1934), used continuous variables. The path diagram of path analysis initiated the tradition in social research to formulate one's hypothetical causal model by means of a graph, connecting variables (in squares or circles) by arrows to indicate the causal direction. For example, Lazarsfeld's explanation and interpretation would, respectively, get the following arrows: $X \leftarrow T \rightarrow Y$ and $X \rightarrow T \rightarrow Y$ (X hypothetical cause, Y hypothetical effect and T the test variable). Drawing arrows is easy. By building path diagrams for data that are in fact cross-sectional, one is easily tempted to specify instantaneous feedback relations, that is, direct or indirect reciprocal causal effects between variables that in fact do not differ in time. By means of these instantaneous reciprocities, however, social methodology entered a new and insidious area of statistics, the area of simultaneous equation modeling in econometrics (see, e.g., Goldberger, 1964; Johnston, 1972; Theil, 1971).

A simultaneous equation model can be put in structural or reduced form, and the structural form can take the less problematic recursive or more problematic nonrecursive form. The problems with regard to identification and estimation of the instantaneous feedback coefficients belong to the problematic nonrecursive domain. The problems were considered so serious by econometricians that some opted for estimating only the reduced form of the model (e.g., Liu, 1960), which is least problematic and can be handled by simple least-squares techniques. Others (e.g., Strotz & Wold, 1960; Wold & Jureen, 1953; Wold, 1960) rejected nonrecursive models with instantaneous feedback effects as being unrealistic and demanded recursive reformulation or at least recursive interpretation of the model. In his article introducing Granger-causality, Granger (1969) suggested that so-called instantaneous causality, "in which knowledge of the current values of a series helps in predicting the current value of a second series," may be an artifact of missing lagged data of the series and so many cases of instantaneous feedback would disappear if the economic variables would be recorded at more frequent intervals.

The most rigorous approach is avoiding all instantaneous effects by changing from cross-sectional to longitudinal data and replacing all instantaneous feedback relations by lagged ones. That is, letting dependent variables not depend on other contemporaneous variables but on the corresponding lagged ones. In a sense, Lazarsfeld did this for dichotomous variables in his famous "sixteen-fold table" by concentrating on the change cells in the table when analyzing the direction and strength of causation (Lazarsfeld, 1948, 1978; Lipset et al., 1954). The method of the sixteen-fold table was later extended to continuous variables and thus became the precursor of the well-known cross-lagged panel design. The term "cross-lagged panel" was already coined in 1963 by Campbell and Stanley (1963), but it took some more years to convince the majority of social scientists of the necessity of longitudinal data in nonexperimental causal research. As a special case of a recursive model, the cross-lagged panel model has the advantage that, just as the recursive model, it is automatically identified and can be estimated by simple least squares, which, for normally distributed error components, is then maximum likelihood (Johnston, 1972, pp. 369 & 378). However, this advantage is lost again when the variables are latent, as will be discussed in the next section.

1.3 Karl Jöreskog and the State Space Model Entering the Social Science Scene

The contribution to social science by Jöreskog, a former student of Herman Wold in Uppsala, has been enormous in methodological as well as statistical respect. Methodological, because he brought together two until then almost totally separated fields, factor analysis in psychology and simultaneous equation modeling in econometrics and sociology. He combined them by the formulation of an extremely general model, consisting of a measurement and an explanatory part. The measurement part covered most types of factor models and the explanatory part the simultaneous equation model forms: nonrecursive, recursive, cross-lagged panel, multivariate regression models, etc. The general model got the name structural equation model (SEM) and became extremely popular under this name both within and outside of social science. The abbreviation SEM is except for the model also used for the modeling approach. The most important feature of SEM is that the explanatory part instead of in terms of observed variables may be formulated in terms of latent variables, defined by the factor model in the measurement part. Not less important is that the measurement error variances of observed variables may be estimated and that the measurement errors at different time points may be specified to be correlated across time. This was shown by Jöreskog in several of the applications mentioned below, in which he also explained in detail how these important variances and correlations may be identified.

While in many fields maximum likelihood is the ideal by which estimators are evaluated, for a long time the authors in simultaneous equation modeling as well as in factor analysis complained about the computational problems in finding the maximum likelihood solution. Theil (1971, p. 525) simply said: "This estimator is only rarely used because it has to be solved from a very complicated set of estimation equations," while Goldberger (1964, p. 386) and Johnston (1972, p. 398) made similar remarks. According to Lawley and Maxwell (1971, p. vii) in their book about factor analysis as a statistical method: "The outstanding difficulty of applying the maximum likelihood method of estimation to factor analysis was the lack of satisfactory methods of obtaining numerical solutions. In the intervening period this difficulty has been almost entirely removed ... with a paper by K.G. Jöreskog ... in which he used a numerical method that was superior to any that had previously been employed in factor analysis." In 1972–1973 Jöreskog presented for the first time his general model, the maximum likelihood procedure for simultaneous estimation of measurement and explanatory part as a whole, and the program Linear Structural Relationships (LISREL) (Jöreskog & van Thillo, 1972) to do the computation (Jöreskog, 1973). One of the applications was Klein's model I, the very well-known and often estimated simultaneous equation model in econometrics. A more elaborate version of the 1973 chapter appeared in 1977 with several applications of recursive and nonrecursive models with and without measurement errors (Jöreskog, 1977). Also in 1977, a chapter especially devoted to longitudinal models was published (Jöreskog & Sörbom, 1977). The longitudinal SEM analyses follow a combined between-subjects/within-subject procedure with the dependencies between the repeated measurements within subjects accounted for in the model and the repetitions across subjects assumed to be independent.

State space modeling, being at the core of systems and control theory (see, e.g., Caines, 1988; Kalman, 1960; Kwakernaak & Sivan, 1972; Zadeh & Desoer, 1963), has quite different roots than SEM. It is also a highly statistical discipline. Different from other branches of statistics that are mainly or at least partly based on the between-subjects approach and thus profit from the independence assumption, state space modeling is almost exclusively reliant on single-subject research and therefore on within-subject dependent data. The distinguishing feature of a state space model was formulated by Kalman in his famous paper about what became later known as the Kalman filter, one of the main results in control theory and used, for example, in the historic 1969 Apollo XI mission to the moon. The state contains all "one has to know about the past behavior of the system in order to predict its future behavior" (Kalman, 1960, p. 39). The state is the memory of the system that makes the model perfectly Markov. If one wants to predict or control the future of a system, knowledge of the current state of the system (typically obtained by means of the Kalman filter) and its future input should be sufficient. If the state does not contain enough information to do this, then the number of latent state variables has to be extended until it does. This is analogous to cross-sectional factor analysis, where enough factors have to be specified to account for the common variance in the observed variables. Of course, the information summarized and kept in the cross-sectional case will be less than in the longitudinal state space modeling case.

Nevertheless, the similarity between factors and state variables, both being latent variables, was reason for several authors to study in detail the relations between SEM and state space modeling (SSM): MacCallum and Ashby (1986); Otter (1986); Oud et al. (1986), and later Chow et al. (2010). All of them except Otter confirmed SEM's multi-subject and SSM's single-subject orientation and agreed also that SEM is general enough to accommodate and estimate the SSM model. Otter commented only on the LISREL estimation procedure, which he considered a special case of the SSM procedure. But by first deriving the reduced form from the structural form and next recognizing the reduced form as a version of SSM, Otter did not oppose the generality of SEM as a model. While Oud in his dissertation about state space modeling in social science research included already a longitudinal application of SEM by means of LISREL (Oud, 1978), he and his colleagues were stimulated to formulate and estimate an SSM model as a SEM model for a multi-subject sample (Oud et al., 1986, 1990). By using the resulting SEM parameter estimates in the Kalman filter, they made this important result of control theory accessible for SEM and SEM users. In fact, the Kalman filter was used by them as a longitudinal factor score estimation procedure, and the relationships between the Kalman filter and two well-known cross-sectional factor score estimators, regression and Bartlett estimator (Lawley & Maxwell, 1971), were explained. The connection made by them between mainly cross-sectionally oriented SEM and longitudinally single-subject oriented SSM will also play an important role in the history of continuous time modeling.

In contrast to most cross-sectionally orientated measurement procedures in psychology, which use only present observed information to estimate a person's present latent properties and abilities, longitudinal factor score estimation by means of the Kalman filter combines present and past observed information into an optimal estimate of present factor scores. By cumulating all information of the individual subject across time and modeling as well as estimating its latent factor scores for all individual time points, the state space model and Kalman filter do in fact what Cattell's P-technique aimed at but could not do in a statistically correct way. In a later article by Oud et al. (1999), in addition to the Kalman filter, the Kalman smoother (Rauch et al., 1965) was explained and applied to educational data. The Kalman smoother uses also observed information at later points in time to improve estimates at previous points in time.

How does the within-subject model estimation procedure work in SSM, and how does it differ from SEM? This is discussed in detail by Oud (2004) and by Oud and Singer (2008). It turns out that the Kalman filter is also central in the most popular procedure to estimate the SSM model, the prediction error decomposition procedure. In this procedure, for all time points for which data are available in the subject, the Kalman filter estimates stepwise latent state values. Via the measurement equation, these are transformed into predicted data, which, next subtracted from the observed data, form predicted errors (also called "innovations"). A crucial property of the Kalman filter is that it allows to decompose the likelihood for these individual prediction errors and to maximize the likelihood accordingly. In SEM, no state estimation for individual time points takes place, the model is handled as a whole over the whole time range, and the data covariance or augmented moment matrix is directly compared to the model implied version, and maximization takes place by directly modifying the model parameters. If the number of subjects $N > 1$ or greater than the number of time points T, $N > T$, as was initially required by some SEM programs, the SEM procedure becomes a combined between-subjects/within-subject procedure. It is important to note, however, that, because it operates directly on the model as a whole, the within-subject part of SEM works quite differently from its stepwise counterpart in SSM.

Because it was proposed as an alternative to Cattell's P-technique and therefore indirectly to SSM as a procedure to analyze within-subject data (Molenaar, 1985), some comment should be dedicated to the Toeplitz matrix as an input matrix of the SEM program. The Toeplitz matrix is the result of taking overlapping stretches from the same time series of a single subject and handling them in fact as different subjects in the between-subjects part of the SEM procedure. So, as in the case of Cattell's P-technique, dependent within-subject data are analyzed as independent between-subjects data. The violation of the independence assumption is in a sense more serious here than in the P-technique, because in the latter, the stretches are different parts of the time series, whereas in the case of the Toeplitz matrix, the overlapping stretches partly contain exactly the same data. Clearly, the Toeplitz matrix procedure cannot produce maximum likelihood estimates, and it is therefore characterized as a pseudo-maximum likelihood estimator (Hamaker et al., 2003; Chow et al., 2010; Molenaar & Nesselroade, 1998). On the basis of extensive simulation research, its use could not be recommended for time series analysis, neither for small T (Hamaker et al., 2002, p. 365) nor for $T > N$ (Hamaker et al., 2003, p. 361). In the past, its use could perhaps be motivated by the fact that $T \geq N$ was not allowed by some SEM programs but by the availability of SSM and Kalman filter as well as SEM programs without such restrictions, there is not much reason left for its use in longitudinal analysis or $N = 1$ research. Additional problems in its application are that stretches can be chosen in different lengths (window sizes), which give different results without a clue as to the most appropriate size.

In addition, because stretches are shifted in position with regard to one another and so data of different time points are combined in the same variable, only time series with equal time intervals can meaningfully be analyzed.

1.4 Stochastic Linear Time-Invariant Continuous-Time State Space Model in Social Science

Continuous time modeling has deep roots in history. It goes back to Newton (1643–1727) and Leibniz (1646–1716), the inventors of differential and integral calculus. Calculus became the standard tool for analysis in physical science. Newton's laws of motion relate the position, speed, and acceleration of physical bodies by means of differential equations. Not less than two and a half centuries after the birth of calculus, statistician Maurice Bartlett realized that the discrete time nature of the observations in economics does not reflect any lack of continuity in the underlying series. "Thus theoretically it should often prove more fundamental to eliminate this imposed artificiality. An unemployment index does not cease to exist between readings, nor does Yule's pendulum cease to swing" (Bartlett, 1946, p. 31). Six years later, Herbert Simon, political scientist, computer scientist, psychologist, and Nobel Prize winner in economics, introduced the use of differential equations into social science (Simon, 1952). He did this by formalizing Homans' theory about the human group (Homans, 1950), relating the variables interaction, friendship, and activity by means of differential equations.

In 1968, the eminent sociologist James Coleman gave a first systematic and rather comprising contribution (Coleman, 1968), explaining how sociologists should take the step from cross-sectional to longitudinal research by means of calculus. He considered Newton's idea to compute a specific value for the change in a variable at a particular point in time, that is, at the point where the time interval becomes "vanishing small," "the only concept we have for systematically studying change" (Coleman, 1968, pp. 430–431). One year later, Hubert Blalock devoted more than half of his influential book on theory construction in sociology (Blalock, 1969) to the application of differential equations. This was followed by a series of interesting applications in the book by Doreian and Hummon (1976), who explicitly mentioned Coleman's 1968 paper as a direct intellectual forebear of their work. However, the differential equations used by Coleman and others in those early attempts in social science suffered from one important limitation. They did not allow for random error in the equations, and so, in spite of different kinds of estimation procedures for their parameters, in essence, they explained only deterministic behavior.

Continuous time modeling with random error in the form of the Wiener process was introduced into social science by Tuma and Hannan (1984). The Wiener process is also called Brownian motion after the Scottish botanist Robert Brown (1773–1858), who was the first to describe the random walk behavior of particles in a fluid. In 1928, mathematician and philosopher Norbert Wiener (1894–1964) succeeded to give a rigorous mathematical definition to the Brownian motion. We follow the common practice to write the Wiener process $\mathbf{W}(t)$ in honor of Wiener by capital letter $\mathbf{W}$. We meet the Wiener process in the so-called white noise $\frac{d\mathbf{W}(t)}{dt}$ of the continuous-time state space model, which consists of two equations,

$$\frac{d\mathbf{x}(t)}{dt} = \mathbf{A}\mathbf{x}(t) + \mathbf{B}\mathbf{u}(t) + \boldsymbol{\gamma} + \mathbf{G}\frac{d\mathbf{W}(t)}{dt}, \tag{1.1}$$

$$\mathbf{y}_{t_i} = \mathbf{C}\mathbf{x}_{t_i} + \mathbf{D}\mathbf{u}_{t_i} + \boldsymbol{\kappa} + \mathbf{v}_{t_i}, \tag{1.2}$$

dynamic state equation (1.1) and static measurement or read-out equation (1.2). The equations contain

$\mathbf{x}(t)$	state vector at t
$\mathbf{u}(t)$	exogenous vector at t
$\boldsymbol{\gamma}$	dynamic trait vector
$\mathbf{W}(t)$	Wiener process vector at t
$\mathbf{y}_{t_i}$	observed outcome vector at t_i
$\boldsymbol{\kappa}$	static trait vector
$\mathbf{v}_{t_i}$	measurement error vector at t_i
$\mathbf{A}$	drift matrix
$\mathbf{B}$	exogenous effect matrix
$\mathbf{G}$	diffusion matrix
$\mathbf{C}$	loading matrix
$\mathbf{D}$	feedthrough matrix

Both equations contain at the right-hand side in addition to an error term, which is the white noise pre-multiplied by the diffusion matrix in (1.1) and the measurement error in (1.2), a state term, an exogenous term, and a trait term. The discretely observed outcome variables in $\mathbf{y}_{t_i}$, $i \in Z$, measure the continuously developing latent state variables in $\mathbf{x}(t)$, $t \in R$, at discrete time point t_i, $\mathbf{x}_{t_i} = \mathbf{x}(t_i)$. Discrete measurements are also picked up from the continuously flowing exogenous variables: $\mathbf{u}_{t_i} = \mathbf{u}(t_i)$. However, while the values of the exogenous variables are taken for granted as coming from outside, the development of the state variables gets an explanation inside of the model by the matrices with names taken from meteorology, drift for the development of the state mean across time, and diffusion for the state dispersion development. In contrast to fixed exogenous variables, state and trait variables are random with trait variables standing for parts of state variables that are constant across time but differ between subjects. Because the measurement equation specified is discrete, the model as a whole is often called continuous-discrete state space model (e.g., Jazwinski, 1970; Singer, 1995).

White noise $\frac{d\mathbf{W}(t)}{dt}$ looks rather complicated but can be seen as the continuous-time analogue of the discrete time random walk $\mathbf{w}_{t_i} = \mathbf{w}_{t_{i-1}} + \mathbf{e}_{t_{i-1}}$ or $\mathbf{w}_{t_i} - \mathbf{w}_{t_{i-1}} = \mathbf{e}_{t_{i-1}}$, divided by the time interval $\Delta t_i = t_i - t_{i-1}$, that is, $\frac{\Delta \mathbf{w}_{t_i}}{\Delta t_i}$. The Wiener process can indeed be defined as the limit of such a random walk process. The fame of the Wiener process is undoubtedly due to two peculiar facts, which both stem from the erratic behavior of the sample paths as a result of the independence requirement (all increments $\mathbf{W}(t) - \mathbf{W}(s)$ for $s < t$ assumed independent across continuous time). First, as proven by Wiener, sample paths of the Wiener process are almost nowhere differentiable in the classical sense, and therefore, white noise cannot be defined as a derivative in the classical sense. It is easy to see that the value of $\frac{\Delta \mathbf{w}_{t_i}}{\Delta t_i}$ will not be finite anymore, if we let $\Delta t_i \rightarrow 0$ ($\frac{1}{\Delta t_i} \rightarrow \infty$) but do not at the same time let the correlation between $\mathbf{w}_{t_i}$ and $\mathbf{w}_{t_{i-1}}$ go to 1 (drop independence) (Priestley, 1981, pp. 158–159). Because of non-differentiability, Eq. (1.1) is often written multiplied by $\mathrm{d}t$ on both sides. Second, because of the unbounded variation of the erratic sample paths, which keep randomly changing as $s \rightarrow t$, the integrals with respect to the Wiener process cannot be defined as normal Riemann-Stieltjes integrals. Wiener was the first to define the stochastic integral in a mathematically sound way (Wiener stochastic integral), which was generalized in 1944 by the Japanese mathematician Itô (Itô stochastic integral). When the integrand is a deterministic function, the Itô integral reduces to the Wiener integral (Kuo, 2006, p. 37).

One of the reasons for the slow spread of continuous time modeling in social and other sciences could have been the difficulty in defining and handling random error. Independence in discrete time with nothing existing or happening between successive time points is easily defined. It is in a sense closing the eyes between throwing a dice two times. However, infinite repetition of independence across continuous time with no gaps between time points turned out to be a more difficult concept. Nevertheless, the mathematical problems are solved and need not be a reason for concern for the research practitioner. See, for example, Arnold (1974) and more recent works by Itô student Kuo (1996, 2006).

The trait variables in $\boldsymbol{\gamma}$ and $\boldsymbol{\kappa}$ need some comment, not only because they are less well-known in the state space model literature but also because of the radical difference between the two kinds. The trait variables $\boldsymbol{\kappa}$ in measurement equation (1.2) were specified under the name "random measurement bias" by Goodrich and Caines (1979), and their model was used again by Shumway and Stoffer (2000). If, for example, the observed variable is weight, one could think about systematic differences in the weight scales used by different subjects in the subject sample or in the case of achievement scores about teachers using different norms in the evaluation of student exams. These relatively simple constant measurement deviations differ fundamentally from the trait variables $\boldsymbol{\gamma}$ in the dynamic state equation (1.1). Trait variables $\boldsymbol{\gamma}$ were introduced by Hamerle et al. (1993) under the name "individual specific components." The presence of individual subject mean curves in the model solves two paradoxical situations in longitudinal research. The first is regression toward the mean. This, for instance, occurs in cross-sectional and longitudinal factor score estimation procedures. In a pure state model ($\boldsymbol{\gamma} = \mathbf{0}$), regression is toward the group mean developmental curve. The paradoxical situation is that the same subject, if part of a low functioning group, regresses toward the low mean curve and, if part of a higher functioning group, regression is toward the high mean curve. So, the same subject would get two different measurement results depending on the group he or she would be made a member of. This problem is solved in a state-trait model, where regression is toward the subject's own mean curve, which is in essence independent of the group mean curve. Another problem of the pure state model is that the subject specific mean curves are forced to coincide with the group mean curve. It means that the subject data show relatively big and enduring deviations from the group mean curve and the model tends to become unstable. In a stable state-trait model, the state converges to the subject specific mean curve, which is prevented in a pure state model. In the state-trait model, stability is restored by properly accounting for the trait variance. The state-trait model thus distinguishes trait variance (unobserved heterogeneity between subjects) clearly from stability and thus avoids confusion between both. In addition, because the trait variance is subtracted from the error variance, it can lead to a much more efficient estimation of the other parameters.

The discrete-time state space model replaces differential equation (1.1) by difference equation (1.3)

$$\mathbf{x}_{t_i} = \mathbf{A}*_{t_i,t_{i-1}}\mathbf{x}_{t_{i-1}} + \mathbf{B}*_{t_i,t_{i-1}}\mathbf{u}_{t_{i-1}} + \boldsymbol{\gamma}*_{t_i,t_{i-1}} + \mathbf{e}_{t_i\,t_{i-1}}, \tag{1.3}$$

$$\mathbf{y}_{t_i} = \mathbf{C}\mathbf{x}_{t_i} + \mathbf{D}\mathbf{u}_{t_i} + \boldsymbol{\kappa} + \mathbf{v}_{t_i}, \tag{1.4}$$

but keeps in (1.4) the same measurement equation as in the continuous-time model. Different discrete-time studies may entertain different time intervals $\Delta t_i = t_i - t_{i-1}$ between time points, and even within studies the time intervals may vary. This is the main reason why the continuous-time model or at least reporting results in terms of the continuous-time model is to be preferred to those of the discrete-time model. Both models are linear, but the relationship between both, between $\mathbf{A}$ and $\mathbf{A}*_{\Delta t_i}$, $\mathbf{B}$ and $\mathbf{B}*_{\Delta t_i}$, and $\boldsymbol{\gamma}$ and $\boldsymbol{\gamma}*_{\Delta t_i}$, is highly nonlinear. For this reason, direct comparison of different discrete-time models, that is, $\mathbf{A}*_{\Delta t_i}$, $\mathbf{B}*_{\Delta t_i}$, and $\boldsymbol{\gamma}*_{\Delta t_i}$ from models with different Δt_i does not make sense. The compelling conclusion is that instead of the difference equation (analyzed, e.g., in the popular cross-lagged panel model), the underlying differential equation in a continuous-time model should be formulated, studied, and interpreted. Oud and Delsing (2010) explain in detail the paradoxical differences appearing in the comparison of the continuous-time with the discrete-time model and of discrete-time models with one another. They show that even when different researchers choose the same constant time interval Δt between their measurement time points, this might not help in the comparison and interpretation of discrete-time results and may lead to deceptive conclusions.

To estimate the continuous-time model parameters, approximate and exact methods have been proposed in the past. Exact methods are based on the so-called exact discrete model (EDM), which Bergstrom (1988) claimed to have developed in 1961–1962. The EDM is the discrete-time model that exactly matches the solution of the differential equation in the continuous-time model at all discrete time points for all discretization intervals. The EDM is thus the discrete-time model but with the parameter matrices of difference equation (1.3) replaced by the solution matrices of differential equation (1.1), as follows:

$$
\begin{aligned}
\mathbf{A}*_{\Delta t_i} &= \mathrm{e}^{\mathbf{A}\Delta t_i},\\
\mathbf{B}*_{\Delta t_i} &= \mathbf{A}^{-1}(\mathrm{e}^{\mathbf{A}\Delta t_i} - \mathbf{I})\mathbf{B},\\
\boldsymbol{\gamma}*_{\Delta t_i} &= \mathbf{A}^{-1}(\mathrm{e}^{\mathbf{A}\Delta t_i} - \mathbf{I})\boldsymbol{\gamma},\\
&\mathbf{e}_{\Delta t_i} \sim N(\mathbf{0}, \mathbf{Q}*_{\Delta t_i}) \quad \text{with}\\
\mathbf{Q}*_{\Delta t_i} &= \mathrm{irow}[\mathbf{A}_{\#}^{-1}(\mathrm{e}^{\mathbf{A}_{\#}\Delta t_i} - \mathbf{I})\mathrm{row}\,\mathbf{Q}]\\
&\text{for } \mathbf{Q} = \mathbf{G}\mathbf{G}' \text{ and } \mathbf{A}_{\#} = \mathbf{A} \otimes \mathbf{I} + \mathbf{I} \otimes \mathbf{A}.
\end{aligned}
\tag{1.5}
$$

Here $\otimes$ is the Kronecker product (postmultiplying each element of the former matrix by the latter matrix), "row" the rowvec operation (putting the elements of a matrix row-wise in a column vector), and "irow" the inverse operation (putting the elements back into the matrix again). The nonlinearity is caused by the matrix exponential $\mathrm{e}^{\mathbf{A}}$ or $\mathrm{e}^{\mathbf{A}_{\#}}$ (Moler & Van Loan, 1978, 2003), occurring in each of the EDM matrices.

Some remarks should be made with regard to the EDM. First, because of the time-invariant character of the continuous-time model, the subscript t_i, t_{i-1} in the discrete-time model may be replaced by $\Delta t_i = t_i - t_{i-1}$. What matters in the continuous-time model are not the historic time points but only the length of the intervals. The subsubscript i is kept, because the length of the intervals within and between studies may differ and lead to different EDM parameter matrices. There is only one underlying continuous-time model but for each a confusing multitude of derived EDMs. A second important point regards the exogenous term $\mathbf{A}^{-1}[\mathrm{e}^{\mathbf{A}\Delta t_i} - \mathbf{I}]\mathbf{B}\mathbf{u}_{t_{i-1}}$. This looks similar to the trait term $\mathbf{A}^{-1}[\mathrm{e}^{\mathbf{A}\Delta t_i} - \mathbf{I}]\boldsymbol{\gamma}$ but differs in one crucial respect. In continuous time, the value of $\boldsymbol{\gamma}$ is constant across time, but the time path of the exogenous variables is not explained by the model and so unknown within each interval $[t_{i-1}, t_i]$. In EDM (1.5), it is assumed that $\mathbf{u}_{t_{i-1}}$ keeps the same value over the interval, so that $\mathbf{u}(t)$ follows a step function, but this need not be realistic. This problem got serious attention from Singer (1990), Hamerle et al. (1991), and Oud and Jansen (2000). Instead of the step function, which assumes that the exogenous variables just at the accidental measurement time points $\ldots, t_{i-1}, t_i, t_{i+1} \ldots$ show sudden increases or decreases, a piecewise linear function could be judged a better approximation of what happens in the interval. This means that we have two vector values to characterize the exogenous paths in each interval: $\mathbf{u}_{t_{i-1}}$ and $\mathbf{b}_{\Delta t_i} = \frac{\mathbf{u}_{t_i} - \mathbf{u}_{t_{i-1}}}{\Delta t_i}$. This leads to a new term in Eq. (1.3), $\mathbf{C}*_{t_i, t_{i-1}}\mathbf{b}_{\Delta t_i}$, and a new corresponding parameter matrix in EDM (1.5),

$$
\mathbf{C}*_{\Delta t_i} = \mathbf{A}^{-1}[\mathbf{B}*_{\Delta t_i} - \Delta t_i\mathbf{I}]\mathbf{B}, \tag{1.6}
$$

but not to any change in the continuous-time model. $\mathbf{B}*$ and $\mathbf{C}*$ are just different ways in which $\mathbf{B}$ impacts discrete time. Oud and Jansen (2000) generalized the piecewise linear approximation to piecewise polynomial approximation, implying more different ways in which **B** may impact discrete time.

The crucial role of drift matrix **A** should be mentioned. **A** has a big impact in every EDM matrix and plays a central role in the continuous-time model and its interpretation. Important properties of the model can be found by computing the eigenvalues of **A**. The eigenvalues first distinguish oscillatory from non-oscillatory movements. Complex eigenvalues imply oscillatory movement. The eigenvalues also define (asymptotic) stability or damping as stability is usually called in the case of oscillatory movements. Stability means that the model has an equilibrium state and that all movements of the state converge for $t \to \infty$ to the equilibrium state. The model is stable, if and only if all eigenvalues of **A** have negative real part. The equilibrium value $E[\mathbf{x}(t)] = E[\mathbf{x}(t_0)]$ or $\boldsymbol{\mu}_{\mathbf{x}(t)} = \boldsymbol{\mu}_{\mathbf{x}(t_0)}$ in the case of stability and constant exogenous input $\mathbf{u}(t) = \mathbf{u}(t_0) = \mathbf{u}_c$ is

$$\boldsymbol{\mu}_{\mathbf{x}(t)} = \boldsymbol{\mu}_{\mathbf{x}(t_0)} = -\mathbf{A}^{-1}\mathbf{B}\mathbf{u}_c. \tag{1.7}$$

For $t \to \infty$, as a result of stability, $\mathrm{e}^{\mathbf{A}(t-t_0)} \to \mathbf{0}$ and $\mathbf{A}^{-1}(\mathrm{e}^{\mathbf{A}(t-t_0)} - \mathbf{I})\mathbf{B}\mathbf{u}_c \to -\mathbf{A}^{-1}\mathbf{B}\mathbf{u}_c$. Because of commuting property $\mathbf{A}^{-1}\mathrm{e}^{\mathbf{A}(t-t_0)} = \mathrm{e}^{\mathbf{A}(t-t_0)}\mathbf{A}^{-1}$, choosing initial mean $\boldsymbol{\mu}_{\mathbf{x}(t_0)} = -\mathbf{A}^{-1}\mathbf{B}\mathbf{u}_c$ leaves $\boldsymbol{\mu}_{\mathbf{x}(t)}$ unchanged.

If we assume the system to be stationary, in addition to the state mean, the state covariance matrix is assumed to be in equilibrium, $\boldsymbol{\Phi}_{\mathbf{x}(t)} = \boldsymbol{\Phi}_{\mathbf{x}(t_0)}$, leading to equilibrium value

$$\boldsymbol{\Phi}_{\mathbf{x}(t)} = \boldsymbol{\Phi}_{\mathbf{x}(t_0)} = \mathbf{A}^{-1}\boldsymbol{\Phi}_{\gamma}\mathbf{A}'^{-1} - \mathrm{irow}[\mathbf{A}_{\#}^{-1}\mathrm{row}\,\mathbf{Q}]\,. \tag{1.8}$$

The novelty of the stationarity assumption is that the initial parameters in initial mean $\boldsymbol{\mu}_{\mathbf{x}(t_0)}$ and initial covariance matrix $\boldsymbol{\Phi}_{\mathbf{x}(t_0)}$ are totally defined in terms of other parameters as is clearly seen from (1.7) and (1.8). It means, in fact, that the initial parameters disappear and the total number of parameters to be estimated is considerably reduced. Although attractive, the stationarity assumption is quite restrictive and can be unrealistic in practice.

A final remark refers to the time scale of the model. As a main argument for using continuous time modeling, the possibility of comparing results between different time intervals was mentioned, within and between studies. For the comparison between different studies, it is important to put the results first on the same time scale. Changing the time scale unit, $t' = ct$ and thus $\Delta t_i{}' = c\Delta t_i$ (e.g., going from years in one study to months in another study, $t' = 12t$), indeed affects the parameter matrices but in a quite simple way, which in general does not require any re-estimation of the parameters. If one wants to compare one's own differential equation results (**A**, **B**, $\boldsymbol{\gamma}$, **G** in Eq. (1.1)) with those of a study that used time scale $t' = ct$, divide one's own results first by c. Next, by applying the EDM matrices (see (1.5)), also the discrete-time results can be compared.

1.5 Some Philosophical Issues in Structural Equation and State Space Modeling

A first issue is which of the two, SEM or SSM, is the more general one and how generality should be evaluated in this context. As mentioned above, the question was discussed by several authors. As SEM is able to estimate an SSM model and thereby contains the SSM model as a special case, no other conclusion can be drawn than that SEM in this respect is indeed more general than SSM. MacCallum and Ashby (1986) were most outspoken in this opinion and were convinced that SSM could benefit a lot from SEM. Still more general than the LISREL formulation of SEM is the reticular action model (RAM) formulation by McArdle and McDonald (1984). Somewhat paradoxically this enhancement of generality is achieved by putting four of the eight LISREL matrices (endogenous effect matrix $\boldsymbol{B}$, exogenous effect matrix $\boldsymbol{\Gamma}$, endogenous loading matrix $\boldsymbol{\Lambda}_y$, and exogenous loading matrix $\boldsymbol{\Lambda}_x$) in a single so-called asymmetric coefficient matrix $\mathbf{E}$ and the four remaining covariance matrices ($\boldsymbol{\Psi}$, $\boldsymbol{\Phi}$, $\boldsymbol{\Theta}_\epsilon$ and $\boldsymbol{\Theta}_\delta$) in a single so-called symmetric coefficient matrix $\mathbf{S}$. RAM has only one equation

$$\mathbf{v} = \mathbf{E}\mathbf{v} + \mathbf{u}, \tag{1.9}$$

with all observed and latent variables in single vector $\mathbf{v} = [\mathbf{y}' \ \mathbf{x}' \ \boldsymbol{\eta}' \ \boldsymbol{\xi}']'$ and all error variables in single vector $\mathbf{u} = [\boldsymbol{\epsilon}' \ \boldsymbol{\delta}' \ \boldsymbol{\zeta}' \ \boldsymbol{\xi}']'$.

The greater generality of RAM is caused by the fact that $\mathbf{E}$ contains many more coefficients than the total number of coefficients in the four matrices entering $\mathbf{E}$, and from the mathematical perspective, each of those coefficients could be specified to be nonzero. The same is true for $\mathbf{S}$. However, many of those extra coefficients of RAM in comparison to LISREL and of SEM in comparison to SSM do not make sense from a causal perspective. Examples are the coefficients that make the state variables $\mathbf{x}_{t_i}$ in $\boldsymbol{\eta}$ dependent on state variables $\mathbf{x}_{t_j}$ with $j > i$ in $\boldsymbol{\eta}$. It happens often in SEM analyses that the fit of the model is improved by freeing coefficients that let the present be influenced by the future. It is clear, however, that these mathematically acceptable coefficients (in the context of SEM) should nevertheless causally be rejected. Something else that can be done in SEM is letting $\mathbf{x}_{t_i}$ be influenced not only by $\mathbf{x}_{t_{i-1}}$ but also by earlier states $\mathbf{x}_{t_{i-j}}$ for $j > 1$, which makes the model non-Markov. As can easily be seen from (1.1)–(1.2) and (1.3)–(1.4), SSM has the two basic properties of non-anticipating (no restrictive effects on the past from the future) and Markov (present state contains all necessary information to predict the future state). The Markov property of the state, making the relevant past of the system available at each point in time, is crucial for optimal control. Within the constraints of these properties, SSM is extremely general. It is a first-order model, but as will be explained in the sequel, by extension of the state dimension, it is able to cover arbitrary higher-order ARMA models and in the continuous-time case CARMA models as well. An ARMA model makes a variable dependent on its own lagged values as well as on the error and its lagged values (Box & Jenkins, 1970); a

CARMA model does similarly in continuous-time (Oud et al., 2018), higher-order means with lags further back in time. In the so-called trivial extension of the state, the state is constructed such that all past states $\mathbf{x}_{t_{i-j}}$ for $j > 1$ are simply added to a later state at t_{i-1}. Because this can lead to an unduly big state, techniques are available in control theory to reduce its dimension to construct the so-called minimal state (Willems, 1975, p. 23–24). By entering the SSM model as part of SEM, all properties and techniques of SSM become available for SEM. So, while MacCallum & Ashby (1986) concluded that SSM could benefit a lot from SEM, it turns out that SEM could and still can benefit a lot from SSM.

We saw that the presence of coefficients in SEM that let variables later in time influence variables earlier in time, although mathematically correct in the context of SEM, must be and generally is causally rejected. The question is what to do with variables exactly equal in time. Many philosophical issues circle around simultaneity or instantaneity. The Greek philosopher Zeno (c. 490–c. 430 BCE) formulated the arrow paradox, which is about the possibility or impossibility of an arrow to have speed or change at an indivisible specific instance. Zeno concluded that no motion is possible. The arrow cannot move to where it is not, because an indivisible moment does not offer time to go there, and it cannot move to where it is, because it is already there. As discussed above, Coleman (1968, pp. 430–431) considered the concept of change at a specific point in time (implying the existence of variables with different values at the same time point) essential for studying change. However, by translating "specific point in time" immediately as "vanishing small" difference in time, he gave in fact the mathematical solution of the paradox and confirmed his causal acceptance of it. The differential quotient is defined as a limit, a value that can be approximated arbitrarily close but is, together with its specific point in time, never actually reached. In a sense, the solution avoids the crux of the paradox by letting the state approach the limit arbitrarily close but just not permitting exact equality. In spite of much debate, the solution was causally accepted by most mathematicians and philosophers of science. Similarly, white noise and the stochastic integral have been given mathematically correct definitions, which are in general causally accepted. Instantaneous relationships lead nevertheless often to debate and are only accepted hesitatingly or even not at all.

An interesting example is the continuous-time model (1.1)–(1.2) with or without throughput $\mathbf{D}$ (Willems, 1975). In case $\mathbf{D} = \mathbf{0}$, exogenous input $\mathbf{u}(t)$ appears only in the state equation (1.1). Because the state equation is a differential equation, the state $\mathbf{x}(t)$ is said to lag $\mathbf{u}(t)$ infinitesimally, there is no throughput, and the model is called proper or strongly causal. It is by the additional instantaneous effect from the exogenous input on the outcome $\mathbf{y}_{t_i}$ in measurement equation (1.2) why the model with $\mathbf{D} \neq \mathbf{0}$ is called improper or weakly causal in control theory. Although weakly, the qualification "causal" is not withheld from the model with $\mathbf{D} \neq \mathbf{0}$. This rather tolerant use of the term "causal" in a case of true instantaneity seems motivated by the measurement relationship. We have seen that the occurrence of trait variables $\boldsymbol{\kappa}$ in the measurement equation was called "random measurement bias" by Goodrich and Caines (1979). It may be no surprise that the measurement equation is also given the opportunity to compensate for time-varying measurement biases described by

$\mathbf{u}_{t_i}$. One could think about the decreasing quality of the weight scale influencing the measurement of the weight but not the weight itself. In Kalman filtering (Lewis, 1986; Oud et al., 1999), the measurement update or corrector $\hat{\mathbf{x}}(t)$ corrects the time update or predictor $\hat{\mathbf{x}}(t-)$ of the state. We have here the paradoxical situation that $t-$ and t refer to exactly the same point in time. The Kalman smoother adds even a third procedure to measure one and the same state at one and the same point in time, $\hat{\mathbf{x}}^s(t)$. Because the smoother uses also future information, in a sense, it is not only allowed to be in violation of strict causality but also of non-anticipation.

Another instantaneous relationship with interesting causal connections is the discrete time random walk. In social science, it gets most often the formulation

$$\mathbf{w}_{t_i} = \mathbf{w}_{t_{i-1}} + \mathbf{e}_{t_i} \tag{1.10}$$

with indeed strict instantaneity between $\mathbf{w}_{t_i}$ and $\mathbf{e}_{t_i}$. This is understandable, because $\mathbf{e}_{t_i}$ is simply what remains after subtracting the known previous step $\mathbf{w}_{t_{i-1}}$. However,

$$\mathbf{w}_{t_i} = \mathbf{w}_{t_{i-1}} + \mathbf{e}_{t_{i-1}} \tag{1.11}$$

could be considered more appropriate, if the remaining part $\mathbf{e}_{t_{i-1}}$ is taken to stand for the unknown causal influences on the system, which need some time to operate and to affect the system. Formulation (1.11) is more popular in control theory and, for instance, in simulation studies, where the remaining error component has actually to be fed into the system. It is interesting to see that the forward or instantaneous representation $\mathbf{e}_{t_i}$ and the backward or lagging representation $\mathbf{e}_{t_{i-1}}$ in spite of the totally different names and interpretations would be observationally equivalent and result in equally fitting models. From a causal standpoint, though, the backward representation is not less problematic than the forward representation, since it anticipates effects that in true time will happen only later in the interval $[t-1, t]$. The ambiguous representation in discrete time of the behavior between t and $t-1$ disappears in continuous time. It is no surprise that the Wiener process as continuous-time analogue of the discrete time random walk can be specified as a limit of the latter (Kuo, 2006, pp. 4–5) .

Because of its consequences for research in practice, the next issue is best described as the lagged and instantaneous effects dilemma. Cross-sectional research offers only the possibility to analyze instantaneous relationships. Longitudinal research with repeated measurements of the same variables offers by means of the cross-lagged panel model the possibility to replace all instantaneous relationships by the corresponding lagged relationships. But by maintaining in the cross-lagged panel model one or more of the instantaneous relationships, the famous simultaneous equations model arises. The estimation and interpretation of simultaneous equations systems has led to long and heated debates in econometrics, especially in the case of instantaneous feedback relationships, where the model is called nonrecursive or interdependent. The problem is that causation takes time but that the true causation interval could be smaller than the observation interval, in which case the cross-lagged relationships would not catch the true ones, but the instantaneous ones might be helpful and be closer to the true ones. This led to Fisher's very influential interpretation of simultaneous models "as limiting approximations to underlying non-simultaneous models" (Fisher, 1969, p. 495) and to the lagged and instantaneous effects dilemma, described by Oud and Delsing (2010).

Leaving out for clarity the error and exogenous components, (1.12) is the structural form of the simultaneous equations model and (1.13) the reduced form:

$$\mathbf{x}_{t_i} = \mathbf{A}_{ins}\mathbf{x}_{t_i} + \mathbf{A}_{lag}\mathbf{x}_{t_{i-1}}, \tag{1.12}$$

$$\mathbf{x}_{t_i} = (\mathbf{I} - \mathbf{A}_{ins})^{-1}\mathbf{A}_{lag}\mathbf{x}_{t_{i-1}}. \tag{1.13}$$

The dilemma is which effects to choose, the lagged ones in $\mathbf{A}_{lag}$ or the instantaneous ones in $\mathbf{A}_{ins}$. Estimated simultaneously, the result will be highly dependent on the length of the observation interval $\Delta t_i = t_i - t_{i-1}$. In general, a longer interval will lead to higher instantaneous effects in comparison to the lagged ones. It is clear that both should be taken in consideration, but it is not known in discrete time how to constrain the two types to find the true underlying effects. It was (Bergstrom, 1966) who gave a continuous-time interpretation to the simultaneous equations model by putting simple linear constraints on $\mathbf{A}_{lag}$ and $\mathbf{A}_{ins}$ and so defining an approximation $\mathbf{A}_{\dagger}$ of drift matrix $\mathbf{A}$:

$$\mathbf{A}_{ins} = \frac{1}{2}\mathbf{A}_{\dagger}\Delta t_i \tag{1.14}$$

$$\mathbf{A}_{lag} = \mathbf{I} + \frac{1}{2}\mathbf{A}_{\dagger}\Delta t_i, \tag{1.15}$$

leading to the following approximation: $\mathbf{A}_{\dagger\Delta t_i}$ of the exact $\mathbf{A}*_{\Delta t_i}$

$$\begin{aligned}
\mathbf{A}*_{\Delta t_i} &= \mathrm{e}^{\mathbf{A}\Delta t_i} = \textstyle\sum_{k=0}^{\infty}(\mathbf{A}\Delta t_i)^k/k! \\
&= \mathbf{I} + \mathbf{A}\Delta t_i + \tfrac{1}{2}\mathbf{A}^2\Delta t_i^2 + \tfrac{1}{6}\mathbf{A}^3\Delta t_i^3 + \tfrac{1}{24}\mathbf{A}^4\Delta t_i^4 + \ldots \\
&\qquad\qquad\qquad\qquad\qquad\qquad\qquad\qquad \text{(exact } \mathbf{A}\text{)}, \\
\mathbf{A}_{\dagger\Delta t_i} &= (\mathbf{I} - \mathbf{A}_{ins})^{-1}\mathbf{A}_{lag} = (\mathbf{I} - \tfrac{1}{2}\mathbf{A}_{\dagger}\Delta t_i)^{-1}(\mathbf{I} + \tfrac{1}{2}\mathbf{A}_{\dagger}\Delta t_i) \\
&= \mathbf{I} + \mathbf{A}_{\dagger}\Delta t_i + \tfrac{1}{2}\mathbf{A}_{\dagger}^2\Delta t_i^2 + \tfrac{1}{4}\mathbf{A}_{\dagger}^3\Delta t_i^3 + \tfrac{1}{8}\mathbf{A}_{\dagger}^4\Delta t_i^4 + \ldots \\
&\qquad\qquad\qquad\qquad\qquad\qquad\qquad \text{(approximation } \mathbf{A}_{\dagger}\text{)}.
\end{aligned} \tag{1.16}$$

The weights in the approximation ($\frac{1}{2}, \frac{1}{4}, \frac{1}{8}, \ldots$) of Bergstrom's approximate discrete model ADM, which can be shown to be based on a trapezoid approximation of the integral (Gard, 1988; Oud & Delsing, 2010), are only seen to decrease less quickly than in the exact series ($\frac{1}{2}, \frac{1}{6}, \frac{1}{24}, \ldots$) of his EDM. In a simulation study with different estimation procedures, Oud (2007) concluded that the quality of the ADM procedure in terms of root mean squared error (RMSE) was hardly lower than of the

EDM. So, by offering $\mathbf{A}_{\dagger}$ instead of both $\mathbf{A}_{lag}$ and $\mathbf{A}_{ins}$, Bergstrom's ADM gives a reasonable solution in continuous time to the lagged and instantaneous effects dilemma. Of course, it is even better to use one of the exact but nonlinearly based estimation procedures of the EDM, to be discussed in the sequel. An interesting point about the ADM is the diagonal elements of instantaneous matrix $\mathbf{A}_{ins}$ in the structural form. As self-loop coefficients, these are almost always specified to be zero in structural equation modeling. However, in the ADM, this can in general not be done, because it would deprive $\mathbf{A}_{\dagger}$, the approximated drift matrix, of the possibility to have diagonal elements unequal to zero as, for example, negative values in a stable model.

In the issue just discussed, the weakly causal status of the instantaneous effects was discussed and could be enhanced so to speak by combining them with the lagged effects into the continuous-time model of the ADM. The present issue is somewhat similar with regard to the instantaneous relationships in cross-sectional research. It assesses, in accordance with Coleman (1968, p. 444), whether and what, on the basis of cross-sectional research, can be said about the underlying continuous-time model. As is also often done in econometrics to make simultaneity plausible (Fisher, 1969), Coleman starts with the assumption of an (stable) equilibrium state: $\frac{d\mathbf{x}(t)}{dt} = \mathbf{0}$. Applying this to our state equation (1.1), while writing the error part as $\mathbf{w}(t) = \mathbf{G}\frac{d\mathbf{W}(t)}{dt}$, we get first

$$\mathbf{0} = \mathbf{A}\mathbf{x}(t) + \boldsymbol{\gamma} + \mathbf{B}\mathbf{u}(t) + \mathbf{w}(t) \tag{1.17}$$

and next by adding $\mathbf{x}(t)$ to both sides we get the (cross-sectional) structural form

$$\mathbf{x}(t) = (\mathbf{I} + \mathbf{A})\mathbf{x}(t) + \mathbf{B}\mathbf{u}(t) + \boldsymbol{\gamma} + \mathbf{w}(t)\ , \tag{1.18}$$

with associated reduced form

$$\mathbf{x}(t) = -\mathbf{A}^{-1}\mathbf{B}\mathbf{u}(t) + \mathbf{A}^{-1}[\boldsymbol{\gamma} + \mathbf{w}(t)]\ . \tag{1.19}$$

Thus, Coleman considered a cross-sectional model to be an approximation of a continuous-time dynamic model and tried to use cross-sectional data to estimate the parameters in matrices **A** and **B** of the differential equation model. For two reasons, however, the approximation is clearly inferior to the ADM. First, the assumption of a stable equilibrium for all people in the population is a very strong one, which seldom applies in practice. Second, for identification reasons, the self-loop coefficients or diagonals in the matrix $\mathbf{I} + \mathbf{A}$ must be fixed in a cross-sectional model, which is customarily done at 0 and therefore in **A** at -1. Although this negative feedback value of -1 is consistent with the assumption of the subjects being in a stable equilibrium state, it is a highly restrictive specification, since infinitely many other negative values would also be consistent with this assumption and cross-sectional data do not give any information about the true value. Observe that the diagonals of the structural form matrix $\mathbf{A}_{ins}$ in (1.12) are usually fixed at 0, which is even less realistic as approximations of the diagonals in **A** and all the more reason to apply the ADM. Coleman's attempt to get from cross-sectional data information about the underlying continuous-time model once again makes clear that the value of cross-sectional data in assessing dynamic causal relationships is extremely limited.

1.6 Indirect Parameter Estimation Procedure Criticized and Replaced

In 1990, when Singer (1990) started publishing his work on continuous time modeling, the stochastic linear time-invariant differential equation was already known in social science, thanks to an extensive discussion by Tuma and Hannan (1984) in their book "Social dynamics: Models and Methods." Tuma and Hannan also paid attention to parameter estimation, based on contributions from statistics and econometrics: A.W. Phillips (1957); Bergstrom (1966), P.C.B. Phillips (1972, 1974), Wymer (1972); Sargan (1974) and Robinson (1976). In fact, Tuma and Hannan described the indirect estimation procedure and were already aware of some of the problems associated with the indirect procedure. The indirect procedure consists in first estimating for equal time intervals $\Delta t_i = \Delta t$ the discrete-time model ($\mathbf{A}*_{\Delta t}$, $\mathbf{B}*_{\Delta t}$, and $\mathbf{Q}*_{\Delta t}$ in (1.3)), taken as a simple linear regression model, and next independently from the estimation procedure performing the nonlinear derivation of the continuous-time matrices (first $\mathbf{A}$ from $\mathbf{A}*_{\Delta t}$ and then $\mathbf{B}$ and $\mathbf{G}$ from $\mathbf{B}*_{\Delta t}$, and $\mathbf{Q}*_{\Delta t}$) as in (1.5). One problem, mentioned by Tuma and Hannan, is that the resulting statistical information such as standard errors refers to the discrete EDM parameters and not to the parameters in the composing continuous-time matrices such as $\mathbf{A}$ or $\mathbf{B}$. Also, because of the nonlinear relation between discrete and continuous-time parameters, constraints such as zero restrictions cannot be put on the continuous-time parameters. As an alternative, Tuma and Hannan propose the use of Bergstrom's ADM, which estimates and allows constraints on the parameters in the approximation $\mathbf{A}_{\dagger}$ of $\mathbf{A}$. But $\mathbf{A}_{\dagger}$, not being $\mathbf{A}$, has the drawback of the bias.

One of the few publications on continuous time modeling in the period 1984–1990 in social science is by Arminger (1986). Arminger also used the indirect procedure but added a measurement equation to the observed differential equation considered by Tuma and Hannan and estimated the combined model with LISREL. So, for the first time, a stochastic differential equation model with latent state variables was estimated by LISREL. He considered his treatment of the stochastic disturbances in the differential equation an improvement in comparison to earlier ad hoc analyses but acknowledged the work of Tuma and Hannan in this respect. The chapter contains an extensive application of the proposed continuous-time procedure on the Wheaton data (Wheaton, Muthén, Alwin and Summers, 1977), a popular dataset in LISREL.

In a series of publications from 1990 onward, Singer and colleagues (Singer, 1990; Hamerle et al., 1991; Singer, 1992a,b; Hamerle et al., 1993; Singer, 1993, 1995) heavily criticized the indirect procedure and its application by means of LISREL. They offered a suitable alternative in the form of a worked-out maximum likelihood procedure, implemented in the software package Linear Structural Differential Equations (LSDE) by Singer (1991). The main problem of the indirect procedure, as already mentioned by Tuma and Hannan, is the impossibility to put constraints on the continuous-time parameters during estimation. In a sense, it is contradictory to recommend the use of continuous time modeling because of its independence from the interval and applying the indirect procedure, which requires the chosen time intervals between the data to be equal. A correct procedure should allow the discrete parameters to be different in successive discrete-time intervals but at the same time the underlying continuous-time parameters to be equal. However, these simple equality constraints between continuous-time parameters in intervals of different sizes cannot be implemented in the indirect procedure. Neither can the piecewise linear function be used to describe the exogenous path: **B** in (1.6) and in (1.5) should be equal within each interval Δt_i, but there is no possibility to constrain them to be equal.

Next, Singer and colleagues also criticized the second step in the indirect procedure. The function relating **A** and $\mathbf{A}*_{\Delta t}$ is not one-to-one, which leads to an identification problem. Adding $\pm k\, 2\pi\, \Delta t^{-1}$ for arbitrary integer k to conjugate eigenvalue pairs $\lambda_{1,2} = a \pm bj$ (j imaginary $\sqrt{-1}$) of **A** does not change $\mathbf{A}*_{\Delta t}$. Each k leads to a different **A** but to the same $\mathbf{A}*_{\Delta t} = e^{\mathbf{A}\Delta t}$, and thus an identification problem is implied (see Phillips, 1973; Hamerle et al., 1991). Complex conjugate eigenvalue pairs imply oscillatory movement. The identification problem is made visible by increasing the frequency of the oscillating curve through the data points and observing that in the case of equal observation intervals, the curve with higher frequency can be forced to pass through exactly the same data points. This identification problem only arises in the case of complex eigenvalues and therefore not for a unidimensional state. Even in the multidimensional case, the problem is limited in the sense that there is only a finite number of integers k for which "aliasing" $\mathbf{A}_k$ leads to a real $\mathbf{G}_k$ and thus to a positive definite $\mathbf{G}_k\mathbf{G}_k'$ (Hansen & Sargent, 1983). The size of the finite set of observationally equivalent aliases additionally depends on the observation interval Δt, a smaller Δt leading to less aliases and a sufficiently small Δt to identification (Bergstrom, 1988, p. 379).

The identification problem rarely occurs in the case of unequal observation intervals (Oud & Jansen, 2000, p. 205) and has been proven by Hamerle et al. (1991, p. 213) to be solved in case piecewise linear functions are used to describe the exogenous paths. However, for both of these cases and for preventing directly the occurrence of aliases, nonlinear constraints need to be imposed directly on the continuous-time parameters during estimation. Singer's LSDE, a program for maximum likelihood (ML) estimation of the continuous-time state space model on the basis of panel data, made this possible. Basically, for estimating the state space model parameters in the $N = 1$ case, two procedures existed: the prediction error decomposition procedure, using the Kalman filter, and the EM decomposition procedure, using the Kalman smoother (Oud & Singer, 2008). These procedures were both generalized to $N > 1$: the prediction error decomposition procedure by Goodrich and Caines (1979) and the EM decomposition procedure by Singer (1990, 1993). Because both procedures have the basic SSM property of recursiveness, going from time point to time point and taking all relevant information by means of filter techniques from one time point to the next, Singer's procedure may be characterized as the direct SSM procedure to estimate the $N > 1$ continuous-time state space model. A version with time dependent parameters was published in 1998 (Singer, 1998).

In the course of the 1990s, LISREL got competitors, for instance, EQS (Bentler, 1994), AMOS (Arbuckle & Wothke, 1999), Mplus (Muthén & Muthén, 1999), and Mx (Neale, 1997). Most of them kept to linear constraints, but, for example, Mx offered nonlinear constraints, making it possible to compute and constrain the elements of $\mathbf{A}$ in the crucial matrix exponential $e^{\mathbf{A}\Delta t_i}$. Using Mx in this way, Oud and Jansen (2000) were able to replace the heavily criticized indirect SEM procedure by the direct one. Oud and Jansen covered not only time-invariant parameters as in (1.1)–(1.2) but allowed all parameters to be time-varying according to arbitrary polynomial schemes.

So, from 2000 onward, two quite different direct procedures were available for estimating the $N > 1$ continuous-time state space model: the direct SSM procedure and the direct SEM procedure. Oud (2004) and Oud and Singer (2008) explicated in detail the differences between the two. The main difference lies in the recursiveness of the SSM procedure. From time point to time point, the data are "prewhitened," transformed via filtering into independence by subtracting all relevant past information, and included as independent "innovations" in the loglikelihood equation. It means that the SEM loglikelihood decomposes in the SSM procedure into the sum of innovations and the SEM covariance or (augmented) moment matrix and other matrices collapse into small matrices of the individual time intervals. In the SEM procedure, no filtering takes place, and the state variables are derived out of the loglikelihood equation, leaving only direct functions of the model parameter matrices and the data.

The necessity of handling big matrices covering the whole time span of the model can be seen as a drawback of the SEM procedure but has also the advantage that all possible parameters from the initial time point until the final one are available and can be related via constraints. An important example is observed variables in social science, which consist of a common part, explained by the latent state variables, a random error part, but in general also of a specific part that is unique for the specific variable and will of course highly correlate across time. Such specific covariances across time, which are nonexistent in the SSM procedure, can be specified to be nonzero in the SEM procedure and constrained to be equal across time. In principle, these specific covariances could, under the name of "colored noise," be specified via additional state variables in the SSM procedure (Kwakernaak & Sivan, 1972, p. 356). However, not needing the detour of the state variables, specification via SEM is much easier and much more flexible. Crucial here is Kalman's definition of the state, which should contain all one has to know about the past behavior of the system in order to predict its future behavior. SEM is able to let the specific remaining parts, which cannot be subsumed under the common part and are not relevant to predict the future, just correlate across time without impacting the rest of the model.

As both direct procedures pretended to be exact and to give the ML solution, they should, in case of exactly the same model, give the same solution. Because of considerable technical differences, it was considered worthwhile to find out in simulation studies whether results indeed coincided. Two simulation studies were performed: Oud (2007), Oud and Singer (2008). In both, data were generated by the differential oscillator, which played such an important role in the criticism of the indirect procedure. Oud (2007) added also two approximate procedures to his study: Bergstrom's ADM procedure and Boker's multivariate latent differential equation (MLDE) procedure (Boker et al., 2004). As explained above, the ADM approximates $\mathbf{A}$ by taking a linearized version $\mathbf{A}_{\dagger}$, arising from constraining the structural form of the simultaneous equation model. The MLDE linearizes from the start by taking discrete versions of the derivatives $\frac{\mathrm{d}y(t)}{\mathrm{d}t}$ and $\frac{d^2y(t)}{\mathrm{d}t^2}$ and putting them in the middle of the observed time range. In essence, the MLDE reduces the longitudinal model to a cross-sectional model with three variables $y(t)$, $\frac{\mathrm{d}y(t)}{\mathrm{d}t}$, and $\frac{\mathrm{d}^2y(t)}{\mathrm{d}t^2}$, approximated at $t = t_m$. The conclusion for the direct procedures was that they indeed resulted in the same estimates, not only with regard to the parameter estimates but also with regard to the standard errors, which both were computed quite differently in both procedures. The extremely small differences in bias and RMSE occurring at a few places in the results were clearly caused by differences in computing precision. The two approximate procedures in Oud (2007) lagged considerably behind the exact procedures in quality of estimation results, with the ADM having a tendency to compensate for the biases by smaller standard errors of the parameter estimates.

1.7 Some Recent Developments

Toward a Flexible Direct SEM Procedure

Until 2011, the application of the direct SEM procedure, although possible by Mx (Neale, 1997), was less user-friendly, because the whole input script had to be rewritten for each new number of state and observed variables or for different observation intervals. In 2011, with the publication of OpenMx (Boker et al., 2011), the work became a lot easier. OpenMx is an open-source, R-based (R Development Core Team, 2011), freely available SEM program. Voelkle et al. (2012) and Voelkle and Oud (2013) applied OpenMx. In the first of these publications, the popular diagonalization method for computing the matrix exponential was still applied,

$$\mathrm{e}^{\mathbf{A}\Delta t_i} = \mathrm{e}^{\mathbf{MVM}^{-1}\Delta t_i} = \mathbf{M}\mathrm{e}^{\mathbf{V}\Delta t_i}\mathbf{M}^{-1}, \tag{1.20}$$

with $\mathbf{A}$ factorized in eigenvector matrix $\mathbf{M}$ and diagonal eigenvalue matrix $\mathbf{V}$. When analyzing oscillating processes as in Voelkle and Oud (2013), $\mathbf{V}$ contains complex eigenvalues. OpenMx allowed indeed the computation of complex eigenvalues but not to work with these in subsequent operations. Earlier, Oud (2007) and Oud and Singer (2008) solved the problem by expressing the complex values in terms of sine and cosine functions (Euler's formula: $e^{\pm bj} = \cos b \pm j \sin b$), which could in the univariate case indeed be handled by Mx and OpenMx. However, because it is quite cumbersome to extend this to the multivariate case, Voelkle and Oud (2013) tackled the problem quite differently by using Singer's oversampling method (Singer, 2012).

$$\begin{aligned} e^{\mathbf{A}\Delta t} &= \lim_{D\to\infty} \prod_{d=0}^{D-1} (\mathbf{I} + \mathbf{A}\delta_d) \\ &= \lim_{D\to\infty} \prod_{d=0}^{D-1} (\mathbf{I} - \frac{1}{2}\mathbf{A}\delta_d)^{-1}(\mathbf{I} + \frac{1}{2}\mathbf{A}\delta_d) \quad \text{for} \quad \delta_d = \frac{\Delta t}{D}. \end{aligned} \tag{1.21}$$

Oversampling is a general method to compute the matrix exponential without any need for extra measures to be taken in the case of complex eigenvalues. The idea is to work with intervals δ_d that are smaller than actually observed intervals Δt and to compensate for the approximation by letting instead $D \to \infty$. For convenience, the intervals δ_d in (1.21) are chosen equal, but that is not a necessity. In the rewrite in (1.21), one recognizes the ADM approximation $(\mathbf{I} - \frac{1}{2}\mathbf{A}\delta_d)^{-1}(\mathbf{I} + \frac{1}{2}\mathbf{A}\delta_d)$, which turned out to converge much more rapidly than $\mathbf{I} + \mathbf{A}\delta_d$. In the experience of Voelkle and Oud (2013) about $D = 30$ oversampling intervals suffice, which for many models is exact up to the precision limit of the SEM program. Singer (2012) recommended at least 20 oversampling intervals.

Continuous time modeling made the use of different observation intervals possible, but until 2011, it had to be assumed that all subjects obeyed the same, possibly unequal, intervals. It was a major step forward, made possible by OpenMx and implemented in the R-based OpenMx program CT-SEM developed by Voelkle and Oud (2013), that each subject was given the freedom of its own pattern of intervals. In fact, all intervals between and within subjects might be chosen differently. All time points and intervals should henceforth be written with two subscripts: $t_{k,i}$ and $\Delta t_{k,i}$, $k = 1, \ldots, N$ and $i = 1, \ldots, T_k$. The possibility of individually varying intervals has improved the flexibility of the procedure enormously. Monte Carlo simulations in Voelkle and Oud (2013) showed that the choice of individually varying intervals had no negative influence on the quality of the estimates. In fact, the simulations with oscillators proved that the application of individually varying time intervals was helpful in identifying the model, even if the mean interval over individuals did not meet the Nyquist-Shannon criterion. The Nyquist-Shannon criterion requires the sampling rate to be higher than twice the frequency of the signal in order to be able to reconstruct this signal from discrete-time observations.

The flexibility is all the greater, if one realizes that allowing individually varying intervals is equivalent with allowing between-subjects missing data. This is worked out in Oud and Voelkle (2014), which concludes that missing values indeed do not exist in a continuous-time analysis. Instead of avoiding "missings," the goal should be to infer as effectively as possible from all available observation occasions the underlying continuous-time process. Then it may be advantageous to deliberately collect data at varying time intervals between as well as within subjects. In turned out that by collecting data at arbitrary time points across the whole time scale, so creating "missings" on purpose, much more information was gathered about the course of the curves, and the true developmental curves in continuous time could be reproduced much more precisely. The article explained also why the phantom-variable missing value method, which remains within the discrete time framework, is of limited applicability and how the full information maximum likelihood (FIML) approach of OpenMx is effective in cases of only partly missing instruments in an individual at a specific time point.

Since the start of SEM, the conviction prevailed that SEM is only for large samples and cannot be performed for $N < T$ and especially not in the $N = 1$ case. For example, LISREL refused to start the analysis in the $N < T$ case, gave a warning, and first increased the values on the diagonal of the data covariance matrix to get this matrix positive definite before continuing. As made clear by Hamaker et al. (2003), Singer (2010), and Voelkle et al. (2012), however, the long-standing suggestion that a SEM analysis cannot be performed because of a non-positive definite covariance matrix $\mathbf{S}$ is wrong. Nothing in the likelihood function requires the sample covariance matrix $\mathbf{S}$ to be positive definite. It is the model implied covariance matrix Σ that should be positive definite, but this depends only on the chosen model and in no way on the sample size N. The incorrect requirement of a positive definite $\mathbf{S}$ for maximum likelihood estimation (Jöreskog & Sörbom, 1996, p. 24) may be explained by the modified likelihood function employed in LISREL and other SEM programs. This contains the quantity $\log|\mathbf{S}|$ (see Equation (3) in Voelkle et al., 2012), and thus these programs cannot handle $N = 1$ and N smaller than the number of columns/rows in $\mathbf{S}$. By employing the basic "raw data" or "full information" maximum likelihood function (RML or FIML), OpenMx avoids this problem.

In 2016, a new version of OpenMx became available under the name Extended OpenMx (Neale et al., 2016) and in 2017 a new version of CT-SEM under the name ctsem (Driver et al., 2017). Extended OpenMx implemented in addition to other new methodologies state space modeling by means of the Kalman filter. While CT-SEM interfaced with OpenMx, ctsem does so with Extended OpenMx. In this way, ctsem was able to combine and fully integrate the formerly quite separated direct SEM and direct SSM procedures. By means of the keywords "mxRAM" (SEM) and "Kalman" (SSM), one can in ctsem easily switch between the SEM and SSM procedures. Under both procedures, N may range from large to 1. So, the traditional distinction and opposition between the $N = 1$ SSM procedure and the large N SEM procedure does not exist anymore (see, e.g., Chow et al., 2010). Computing time depends most importantly on the number of observation time points T. For small T and small or no variation in observation intervals, SEM takes less time than SSM. SEM takes increasingly more time for larger T. This is caused by the recursive operation of SSM on relatively small-sized matrices, while in SEM, the size of most matrices increases by a multiple of the increase in T.

The SEM procedure with $N = 1$ has been applied extensively in an article about continuous-time autoregressive moving-average (CARMA) models by Oud et al. (2018). Except for a small simulation study in Singer (2012), the direct SEM continuous-time procedure had not been applied before on $N = 1$ data, in particular not on empirical data. In the article, also a subject-group-reproducibility test was designed and performed, benefiting from the fact that in the same analysis, each arbitrary individual subject from the total sample could be compared to the rest of the sample. The subject-group-reproducibility test, made possible in SEM by the $N = 1$ SEM analysis procedure, is a means to find out whether the individual subject model coincides with the entire group model and can be helpful in this way in the idiographic versus nomothetic debate in psychology.

Hierarchical Bayesian Analysis

Bayesian continuous-time analysis was introduced into social science by Oravecz et al. (2009, 2011, 2016). In their publications, they first discussed the Ornstein-Uhlenbeck process, which is the solution of the univariate version of differential equation (1.1), itself called Langevin equation. Next, in view of the example to be presented, they discussed a version with two state variables. Then, for describing and explaining interindividual differences between subjects in the sample, a hierarchical Bayesian extension of the Ornstein-Uhlenbeck process was considered. All parameters in the model were allowed to vary over individuals (thus not only the individual means, called trait variables $\boldsymbol{\gamma}$ in (1.1)). The analysis is hierarchical, because information is used from two levels, the individual level and the level of the sample as a whole. Bayesian parameter estimation was done by sampling from the posterior density using a Markov chain Monte Carlo (MCMC) algorithm. For all parameters, a uniform prior was assumed. A software program to do the analysis and the sampling from the joint posterior was written in MATLAB with subroutines in C++. The example was about the trajectories made by people in the two-dimensional core affect space: activation and pleasantness. The data were collected with experience sampling (respondents choosing themselves reporting time points), a clear example of individually varying time intervals, and the model described the two trajectories together with the cross-correlation. All subject-specific parameters were regressed on the well-known Big Five personality dimensions. As admitted by the authors (Oravecz et al., 2011, p. 486), a rather unrealistic feature of their model is that the drift matrix is assumed to be symmetric, meaning that the effect from activation on pleasantness and from pleasantness on activation is assumed to be equal.

A similar hierarchical Bayesian approach with no restrictions on the lengths of the time series, on the lengths of the intraindividual and interindividual observation intervals, or on the number of subjects was chosen by Driver and Voelkle (2018). Because all parameters are allowed to differ between subjects (also in the drift matrix), the estimation of N totally different individual subject models would be possible in one computer run. To inform the prior distribution, the authors emphasized that the estimation of subject-specific parameters is enhanced by using data from other subjects as well. It means that reasonable estimates could be obtained with less data than would be necessary in single-subject research and that it helps avoid the typical convergence problems with low numbers of subjects. Instead of a single set of parameters or multiple sets of independent parameters, one estimates population distributions for model parameters. And in contrast to hierarchical frequentist approaches, in which population distributions are only estimated, in Bayesian approaches, typically simultaneously subject-level parameters are sampled from the population distribution, "with the population distribution essentially serving as a prior distribution for the subject level parameters" (Driver & Voelkle, 2018, p. 776). Still priors are specified in hierarchical Bayesian models, but these are very vague expectations—so-called hyperpriors—regarding possible population distributions. Just as in the approach of Oravecz et al. (2009, 2011, 2016), in addition to the subject-level parameters and covariances between subject-level parameters, the estimation of relations between covariates and all subject-level parameters is possible.

The hierarchical Bayesian approach explained in Driver and Voelkle (2018) is implemented as an extension to ctsem (Driver et al., 2017), using the recursive state space formulation and relying on Stan software (Carpenter et al., 2017). The main functions of the extension are the ctModel and ctStanFit functions. The ctModel function allows the user to specify the continuous-time model, while the ctStanFit function translates the specification from ctModel into a model in the Stan language.

Time-Varying Parameters

The parameters of differential equation (1.1) are time-invariant, leading to solution matrices in the EDM (1.5) that do not depend on time t itself but only on the time interval Δt, the effect being that equal time intervals $\Delta t_i = \Delta t_j$, although on different places on the time scale, nevertheless give exactly the same results. Time invariance may not always be appropriate. Causal effects may change with age, for example. When children are learning to read, decoding speed may have a bigger influence on reading comprehension than at a later age. Three publications explored the use of time-varying parameters in the continuous-time model: Singer (1998), Oud and Jansen (2000), Chen et al. (2018). The first two started from the general time-varying differential equation (1.22), which replaces time-invariant **A**, **B**, and **G** in (1.1) by time-varying **A**(t), **B**(t), and **G**(t):

$$\frac{\mathrm{d}\mathbf{x}(t)}{\mathrm{d}t} = \mathbf{A}(t)\mathbf{x}(t) + \boldsymbol{\gamma} + \mathbf{B}(t)\mathbf{u}(t) + \mathbf{G}(t)\frac{\mathrm{d}\mathbf{W}(t)}{\mathrm{d}t} \tag{1.22}$$

with solution

$$\mathbf{x}(t) = \mathbf{\Phi}(t, t_0)\mathbf{x}(t_0) + \int_{t_0}^{t} \mathbf{\Phi}(t, s)\boldsymbol{\gamma}\mathrm{d}s + \int_{t_0}^{t} \mathbf{\Phi}(t, s)\mathbf{B}(s)\mathbf{u}(s)\mathrm{d}s$$
$$+ \int_{t_0}^{t} \mathbf{\Phi}(t, s)\mathbf{G}(s)\mathrm{d}\mathbf{W}(s),$$
$$\mathrm{cov}[\int_{t_0}^{t} \mathbf{\Phi}(t, s)\mathbf{G}(s)\mathrm{d}\mathbf{W}(s)] = \int_{t_0}^{t} \mathbf{\Phi}(t, s)\mathbf{G}(s)\mathbf{G}'(s)\mathbf{\Phi}'(t, s)\mathrm{d}s\,. \tag{1.23}$$

The so-called state transition matrix $\mathbf{\Phi}(t, t_0)$ is defined by $\frac{\mathrm{d}\mathbf{\Phi}(t,t_0)}{\mathrm{d}t} = \mathbf{A}(t)\mathbf{\Phi}(t, t_0)$, $\mathbf{\Phi}(t_0, t_0) = \mathbf{I}$. It solves the deterministic homogeneous part $\frac{\mathrm{d}\mathbf{x}(t)}{\mathrm{d}t} = \mathbf{A}(t)\mathbf{x}(t)$ in Eq. (1.22), $\mathbf{x}(t) = \mathbf{\Phi}(t, t_0)\mathbf{x}(t_0)$, and leads for commuting $\mathbf{A}(t)$ and $\int_{t_0}^{t} \mathbf{A}(s)\mathrm{d}s$ to $\mathbf{x}(t) = \mathrm{e}^{\int_{t_0}^{t} \mathbf{A}(s)\mathrm{d}s}\mathbf{x}(t_0)$ (Zadeh & Desoer, 1963, p. 340). In the time-invariant case $\mathrm{e}^{\int_{t_0}^{t} \mathbf{A}(s)\mathrm{d}s} = \mathrm{e}^{\mathbf{A}(t-t_0)}$ follows as well as the EDM parameter matrices in (1.5). Trait $\boldsymbol{\gamma}$ and exogenous input $\mathbf{u}(t)$ are not specified in Singer (1998), but they are in most of Singer's other publications and will be specified here too.

In the time-varying case, Singer (1998) kept the time path of the matrices $\mathbf{A}(t)$, $\mathbf{B}(t)$, and $\mathbf{G}(t)$ arbitrary but replaced the exact matrices of the EDM by approximate integrals of the simple Euler-Maruyama type (Singer, 1998, pp. 88–89):

$$\mathbf{x}_{t_{i+1}} = (\mathbf{I} + \mathbf{A}_{t_i}\delta_{t_i})\mathbf{x}_{t_i} + \boldsymbol{\gamma}\delta_{t_i} + \mathbf{B}_{t_i}\mathbf{u}_{t_i}\delta_{t_i} + \mathbf{G}_{t_i}\delta\mathbf{w}_{t_i}\,. \tag{1.24}$$

Although the approximations look rather coarse with no direct impact from $\mathbf{A}(t)$ on $\boldsymbol{\gamma}$, $\mathbf{B}(t)$, and $\mathbf{G}(t)$, the scheme (1.24) generates the exact discrete model in the limit $\delta \to 0$ (Singer, 1998, p. 89). When letting $\delta \to 0$ in this time-varying case, because of the possibly non-commuting property of $\mathbf{A}(t)$, one should, in the multiplication of successive $(\mathbf{I} + \mathbf{A}_{t_i}\delta_{t_i})$, later $(\mathbf{I} + \mathbf{A}_{t_j}\delta_{t_j})$, $j > i$, put before earlier ones. For estimating the model, Singer used a direct SSM procedure, similar to the one in the LSDE program (Singer, 1991) and treated in detail in Hamerle et al. (1993). An interesting point made by Singer on the basis of his simulations in this time-varying case is that pairs of eigenvalues can first be real, then suddenly jump to the complex plane, and finally return to the real line again.

While Singer kept the time-varying matrices $\mathbf{A}(t)$, $\mathbf{B}(t)$, and $\mathbf{G}(t)$ arbitrary and used approximations of the integrals, Oud and Jansen (2000) let the matrices vary continuously over the integration interval according to general polynomial schemes,

$$\mathbf{A}(t) = \sum_{i=0}^{n_a}(t - t_0)^i\mathbf{A}_i, \quad \mathbf{B}(t) = \sum_{i=0}^{n_b}(t - t_0)^i\mathbf{B}_i, \quad \mathbf{G}(t) = \sum_{i=0}^{n_g}(t - t_0)^i\mathbf{G}_i, \tag{1.25}$$

and tried to remain as close as possible to the exact discrete model. This was totally successful for $\mathbf{B}(t)$ and $\mathbf{G}(t)$ under the conditions of $\mathbf{A}(t)$ being time-invariant, $\mathbf{A}(t) = \mathbf{A}$, and $\mathbf{A}$ having distinct nonzero eigenvalues and nonzero sums of eigenvalue pairs. The condition of nonzero sums of eigenvalues pairs is implied by but less restrictive than asymptotic stability (all eigenvalues negative). Under the stated conditions, by using the formula for integration by parts, the exact integrals could be derived. However, time-varying drift matrix $\mathbf{A}(t) = \sum_{i=0}^{n_a}(t - t_0)^i \mathbf{A}_i$ turned out to be much more difficult to handle. Under the condition of all pairs of matrices $\mathbf{A}_i$ commuting, in which case $\mathbf{A}(t)$ and $\int_{t_0}^{t} \mathbf{A}(s)\mathrm{d}s$ commute, fast converging approximations could be derived. If this condition could not be satisfied or imposed, Singer's approximation method in terms of repeated $(\mathbf{I} + \mathbf{A}_{t_i}\delta_{t_i})$ was advised. For all cases, time-invariant, time-varying, exact, and approximate, Oud and Jansen (2000) used the direct SEM procedure, implemented by scripts in OpenMx (Neale, 1997).

The way time variance is handled by Chen et al. (2018) is by adding unknown latent state variables to the state and estimating them together with their extra parameters. The approach is comparable to the way colored noise is handled in SSM (Kwakernaak & Sivan, 1972). There the problem is correlation between the error components of existing variables in the model, and in order to get rid of the problem, extra variables are added and modeled in the state. In this case, the problem is the time variance of existing parameters that is modeled by extra variables in the state. The authors illustrate their approach with the well-known damped harmonic oscillator (Voelkle & Oud, 2013; Boker et al., 2004; Steele & Ferrer, 2011), which is a special case of the state space model (1.1)–(1.2). The extra variables are called "time-varying parameters" (TVPs). Two models for the TVP are proposed, one with a drift and diffusion parameter (O-U model) and one with only a diffusion parameter (stochastic noise model), which in fact is a special case of the O-U model. Because the extended model becomes nonlinear as a result of the interaction between the original and added variables, a direct SSM procedure, using the continuous-discrete extended Kalman filter (CDEKF), is applied as implemented in dynr. The dynr package, interfacing with R and C, handles a broad class of linear and nonlinear discrete- and continuous-time models, especially with regime switching (Ou et al., 2019).

Mediation Analysis

Just after World War II, Paul Lazarsfeld introduced mediation under the name of interpretation as a research method. It has since resulted in a huge amount of literature. The reason is that for many causal effects, $X \rightarrow Y$, after they have been found, the next question is how they work, $X \rightarrow M \rightarrow Y$? For example, in the case of problem behavior in children because of a bad education by the parents, the psychologist might want to know if there are any variables M in between, which could be influenced to diminish the behavior. It is interesting to find out whether and what progress 70 years of methodology has brought to the topic and what contribution continuous time modeling could make.

First, cross-sectional research was used to analyze mediation. In the course of many years, many sociologists and psychologists performed path analyses on simultaneously collected variables X, M, and Y to do two consecutive analyses: first to check whether an effect could be found between X and Y and, next, whether, in a combined analysis of all three variables together, the effect between X and Y totally or partly disappeared in favor of effects between X and M and between M and Y. The combination of the latter two effects was called indirect effect and the remaining part in the total effect between X and Y direct effect. The use of cross-sectional analyses got fundamental criticism from Maxwell and Cole (2007). Interestingly, it was via Lazarsfeld's "sixteenfold table" as a precursor of the cross-lagged panel model (CLPM) that many researchers were convinced to collect two versions of the three variables, X_1, M_1, Y_1 and X_2, M_2, Y_2 at two consecutive time points in time and to do the mediation analysis by CLPM. Later, it became even standard to collect three sets X_1, M_1, Y_1 and X_2, M_2, Y_2, as well as X_3, M_3, Y_3 at three time points to make the comparison between effect $X_1 \rightarrow Y_3$ and the effects $X_1 \rightarrow M_2$ and $M_2 \rightarrow Y_3$ possible. This made it possible to follow the mediation process through time with M also mediating in time, instead of analyzing mediation effects $X_1 \rightarrow M_2$ and $M_1 \rightarrow Y_2$ only at two time points.

However, as explained by Deboeck and Preacher (2016) in their informative article about mediation analysis, the application of the CLPM did not lead to less confusion. The reasons for confusion were partly the same as for discrete-time analyses in general. Numerous discrete-time studies have shown that results are dependent on the time lag used in the study with the same underlying continuous-time effect, leading to quite different effects in discrete-time studies. Several authors inside and outside the field of mediation analysis, such as Gollob and Reichardt (1987, 1991), Maxwell et al. (2011), and others, were well aware of this, but until the contribution of Deboeck and Preacher (2016), no attempt was made in mediation to make the results independent of the time lag by continuous-time analysis. Confusion is perhaps a big issue in mediation analysis, because the interplay between the effects of three variables simultaneously may lead to very unexpected and contradictory results in the autoregressive and cross-lagged effect functions at different time lags. Oud (2007, p. 97) and Deboeck and Preacher (2016, pp. 66–67) show, for example, that an effect may be negative for small lags and then for larger lags change to positive. The maximum and/or the minimum values of cross-lagged effects, which can easily be found in a continuous-time analysis, are almost always located at different lags. So, the advice to plan a study with the time lag catching the maximum effect is easy (see, e.g., Hoyle & Kenny, 1999, p. 218) but difficult to realize when no single lag exists at which all effects maximize. All of these problems and seemingly contradictory results disappear, however, when the analysis is done in continuous time.

How the continuous-time mediation analysis should be done is explained in detail by Deboeck and Preacher (2016) and repeated here. We start with a continuous-time analysis on the basis of longitudinal data with one, two, or more

Table 1.1 Full drift matrix $\hat{\mathbf{A}}$, full autoregression matrix $e^{\hat{\mathbf{A}}\Delta t}$, drift matrix $\hat{\mathbf{A}}_{ind}$ for indirect effect, autoregression matrix $e^{\hat{\mathbf{A}}_{ind}\Delta t}$ for indirect effect, drift matrix $\hat{\mathbf{A}}_{dir}$ for direct effect, autoregression matrix $e^{\hat{\mathbf{A}}_{dir}\Delta t}$ for direct effect ($\Delta t = 1$)

$$\begin{array}{c} \\ X \\ M \\ Y \end{array}\underset{\hat{\mathbf{A}}}{\begin{array}{ccc} X & M & Y \\ \hline -0.357 & 0 & 0 \\ 0.771 & -0.511 & 0 \\ -0.450 & 0.729 & -0.693 \end{array}} \quad \underset{e^{\hat{\mathbf{A}}\Delta t}}{\begin{array}{ccc} X & M & Y \\ \hline 0.700 & 0 & 0 \\ 0.500 & 0.600 & 0 \\ -0.100 & 0.400 & 0.500 \end{array}}$$

$$\begin{array}{c} \\ X \\ M \\ Y \end{array}\underset{\hat{\mathbf{A}}_{ind}}{\begin{array}{ccc} X & M & Y \\ \hline -0.357 & 0 & 0 \\ 0.771 & -0.511 & 0 \\ 0 & 0.729 & -0.693 \end{array}} \quad \underset{e^{\hat{\mathbf{A}}_{ind}\Delta t}}{\begin{array}{ccc} X & M & Y \\ \hline 0.700 & 0 & 0 \\ 0.500 & 0.600 & 0 \\ 0.167 & 0.400 & 0.500 \end{array}}$$

$$\begin{array}{c} \\ X \\ M \\ Y \end{array}\underset{\hat{\mathbf{A}}_{dir}}{\begin{array}{ccc} X & M & Y \\ \hline -0.357 & 0 & 0 \\ 0 & -0.511 & 0 \\ -0.450 & 0 & -0.693 \end{array}} \quad \underset{e^{\hat{\mathbf{A}}_{dir}\Delta t}}{\begin{array}{ccc} X & M & Y \\ \hline 0.700 & 0 & 0 \\ 0 & 0.600 & 0 \\ -0.267 & 0 & 0.500 \end{array}}$$

specific time intervals Δt. Suppose the estimated drift matrix $\hat{\mathbf{A}}$ is as in Table 1.1. The autoregression and cross-lagged effect functions for all intervals Δt in continuous time are then computed by the matrix exponential $e^{\hat{\mathbf{A}}\Delta t}$, diagonals giving the autoregression functions and off-diagonals the cross-lagged effect functions. These functions, which are mixture effects across time resulting from the pure continuous-time effects in the drift matrix, give a complete picture of how the autoregressions and cross-lagged effects develop across time and thus what values would be found in discrete-time analyses of the respective lags Δt. Table 1.1 shows that the total effect $X \rightarrow Y$ at lag or time interval $\Delta t = 1$ would be -0.100. To find the indirect effect function, we equal the corresponding effect $X \rightarrow Y$ in the drift matrix (-0.450) to 0 but keep the mediating effects $X \rightarrow M$ (0.771) and $M \rightarrow Y$ (0.729). The indirect effect at $\Delta t = 1$ in $e^{\hat{\mathbf{A}}_{ind}\Delta t}$ turns out to be 0.167. To find the direct effect function, we equal the mediating effects to 0 but keep the direct effect in the drift matrix. The direct effect at $\Delta t = 1$ in $e^{\hat{\mathbf{A}}_{dir}\Delta t}$ turns out to be -0.267. Adding for $\Delta t = 1$ the indirect effect of 0.167 to the direct effect of -0.267, we get the total effect of -0.100, as it should. It needs to be emphasized that Table 1.1 gives the values only for $\Delta t = 1$, but the continuous-time analysis gives the direct effect, indirect effect, and total effect functions over the whole continuous-time span from $\Delta t = 0$ until an arbitrary value Δt in the future. Not less important is the fact that the (nonlinear) direct and indirect effect curves need not be parallel, meaning that the difference between direct and indirect effect might increase and/or decrease across time and even change sign. So, a much more informative answer can be given than the simple yes-or-no decision in the discrete-time analysis.

1.8 Does Continuous Time Modeling Make Sense?

As highlighted earlier, continuous time modeling tries to describe something that seems actually impossible, namely, change at an indivisible point in time. Zeno of Elea concluded indeed that such motion is impossible and if everything is motionless at every instant and time is entirely composed of instants, then motion as such is impossible. The point he made becomes perhaps still more clear by two of his other paradoxes: the dichotomy paradox and the paradox of Achilles and the tortoise. If Achilles wants to catch up with the tortoise, he must first reach the spot where the tortoise was at the start. But if he reaches that spot, the tortoise has already reached another spot, which Achilles must reach next, etc. This goes on unto infinity, proving that Achilles will never reach the tortoise. The dichotomy paradox is perhaps the most paradoxical of Zeno's paradoxes. It says that a person, Achilles or the swift-footed huntress in antiquity Atalanta, who wants to reach some point at a distance can never start. If he or she wants to cover the distance, first, half of the distance must be covered, but for covering half of the distance, first, a quarter must be covered, one-eighth, one-sixteenth, etc. Conclusion: he or she can never start, let alone cover the whole distance.

What to do with Zeno's devastating criticism of motion and especially motion in continuous time. It took around 2300 years before the paradoxes seemed successfully solved by means of Augustin-Louis Cauchy's limit concept, which he developed to lend a firm basis to calculus. Both the paradox of Achilles and the tortoise and the dichotomy paradox for a long time were considered solved by limit

$$\lim_{n\to\infty}\sum_{k=0}^{n-1} sv^k = s\frac{1}{1-v} \tag{1.26}$$

of sum $\sum_{k=0}^{n-1} sv^k = s + sv + sv^2 + sv^3 + \ldots + sv^{n-1}$. Multiply the sum by $1 - v$ to find $1 - v^n$ or

$$s + sv + sv^2 + sv^3 + \ldots + sv^{n-1} = s\frac{1-v^n}{1-v} \tag{1.27}$$

which for all values $v < 1$ becomes for $n \to \infty$ equal to limit (1.26). Details of the solution are given by Theunissen and Oud (2021). In the case of Achilles and the tortoise, s is the head start of the tortoise and v the speed of the tortoise as a fraction of the speed of Achilles. If the tortoise got, for example, head start $s = 100$ meters and has $v = 0.1$ of the speed of Achilles, Achilles catches up with the tortoise at the distance of $s\frac{1}{1-v} = 100\frac{1}{1-0.1} = 111.1111\ldots$ meters. In the dichotomy paradox, $v = 1/2$, and for $s = d/2$, each arbitrary distance d is reached by limit $s\frac{1}{1-v} = (d/2)\frac{1}{1-1/2} = d$ in successive steps $d/2, d/4, d/8, \ldots$.

However, to get at the limit in both solutions, an infinite number of steps has to be taken. This entails the important philosophical question, whether time and space are not only mathematically but also physically infinitely divisible. Zeno made his point by rejecting the infinite divisibility of time and space. It appears that, at least with regard to the divisibility of time and space, Zeno was right. According to quantum theory, space and time consist of discrete units with the smallest sizes being, respectively, Planck length ($\approx 1.616 \times 10^{-35}$ meters) and Planck second ($\approx 5.391 \times 10^{-44}$ seconds). The Planck length is the distance that light travels in one Planck second. As the radius of a hydrogen atom is "only" 2.5×10^{-11} meters, this clarifies how minuscule Planck length and second actually are. The hydrogen atom turns out to be in relation to the Planck length about what the universe is to us.

If space and time are discrete indeed and continuous spacetime has to be replaced by so-called "granular" spacetime, we must conclude that the paradox solution (1.26) does not give a perfect representation of reality, nor can continuous time modeling claim to do that. Efforts have been made over the last 70 years to adapt the continuous algebraic rules to granular spacetime (Crouse & Skufca, 2019). Theunissen and Oud (2021) present an adaptation of paradox solution (1.26) to these new rules, a discrete granular solution. They make clear that no inconsistency exists between the solution in continuous spacetime and the discrete granular solution. Granular spacetime can be viewed as a restriction on continuous spacetime, saving only what has the smallest possible size in the universe ("hodon" for smallest possible space unit and "chronon" for smallest possible time unit). The discrete granular solution and the adapted rules converge for increasing numbers of units, distances in spacetime, to the continuous spacetime solution and methods. Theunissen and Oud give a method to compute the distances for which continuous spacetime methods can still be considered valid. In practice, adapting continuous-time methods turns out to be only necessary for extremely small distances, approaching the Planck length and second. For all other distances in spacetime and in many other areas as, for example, social science, traditional continuous-time methods continue to be the appropriate way of data analysis, although their fundamentally approximate character should be recognized. One of the interesting application fields and success stories of continuous-time methods is in finance (Merton, 1990), where the continuous-time curves are the approximations of the combined discrete decisions taken by millions of people. It proves that the application of continuous-time methods makes sense not only in fields traditionally seen as continuous but has also a role in approximating frequently and rapidly succeeding discrete events.

1.9 Conclusion

Starting from the methodological situation in psychology and sociology just after World War II, the birth of continuous time modeling in social science in the hands of Herbert Simon and James Coleman is depicted. Next, the addition of random error to the differential equation in the form of the Wiener process by Tuma and Hannan is followed. Further development in two directions of structural equation modeling (SEM) and state space modeling (SSM) is described with the detour via the indirect parameter estimation procedure in structural equation modeling. It is observed in detail how the two came together in the extremely flexible software package ctsem, which, in addition to the choice of SEM ("mxRAM") or SSM ("Kalman"), allows arbitrary time intervals within and between subjects and thus missing values at arbitrary places. Finally, several philosophical issues related to continuous time modeling are discussed, culminating in the question whether change in continuous time is possible at all, a problem raised already 2500 years ago in the paradoxes of Zeno of Elea, which got a new impetus by the theory of quantum mechanics. It is concluded that the question and its answers do not restrain the usefulness of continuous time modeling in social science but rather extend it to all cases of less frequently and less rapidly succeeding discrete events than those approaching the Planck second.

Of course, not all aspects of continuous time modeling in social science could be dealt with. One topic not discussed is nonlinearity (see, e.g., Singer, 2008, 2018; Chow et al., 2007, 2016). This is very much work in progress, and as remarked by the developers of dynr, a software package specializing in nonlinearity and regime switching, recent results are still subject to approximations and truncation errors (Ou et al., 2019). Not dealt with either are the continuous-time autoregressive latent trajectory (CALT) model by Delsing and Oud (2008) and the second-order differential equation model proposed in its place by Oud (2010). Both are continuous-time alternatives for the autoregressive latent trajectory (ALT) model, developed earlier by Curran and Bollen (2001) and Bollen and Curran (2004, 2006), but would require going into too much detail about the latent trajectory model.

Although continuous-time analysis indeed turns out to have some history in social science, it is much less developed than in some other fields, for example, in economics. At least two topics might be worthwhile to pay attention to in the future and include in software packages such as ctsem and dynr. The first is the distinction between stock and flow variables. As observed by Chambers et al. (2018), continuous-time models can take account of the behavior of variables during the observation interval, thus allowing "a proper distinction to be made in estimation between stock variables (measured at points in time) and flow variables (measures as integrals of a rate of flow over the observation period)" (p. 319). For example, wealth is a stock since it can be measured at a point in time, but income is a flow because measured over a period of time. The second topic refers to the measurement part of the model. The current versions of ctsem and dynr assume a linear measurement model with normally distributed errors and so do not allow specification of nonlinear item response models. However, Driver and Voelkle (2018) confirm that nonlinearities in the measurement model and link of ctsem are possible with some modifications. In the SEM option of ctsem, it must also be possible to specify more complicated measurement models encompassing several observation intervals or the whole time span of the model simultaneously.

References

Arbuckle, J., & Wothke, W. (1999). *AMOS 4.0 User's Guide*. Smallwaters.
Arminger, G. (1986). Linear stochastic differential equation models for panel data with unobserved variables. In N. B. Tuma (Ed.) *Sociological methodology* (pp. 187–212). Jossey-Bass.
Arnold, L. (1974). *Stochastic differential equations*. Wiley.
Bartlett, M. (1946). On the theoretical specification and sampling properties of autocorrelated time-series. *Journal of the Royal Statistical Society (Supplement), 7*, 27–41.
Bentler, P. (1994). *EQS 4.0*, BMDP Statistical Software, Los Angeles.
Bergstrom, A. (1966). Nonrecursive models as discrete approximations to systems of stochastic differential equations. *Econometrica, 34*, 173–182.
Bergstrom, A. (1988). The history of continuous-time econometric models. *Econometric Theory, 4*, 365–383.
Blalock, Jr., H. (1964). *Causal inferences in nonexperimental research*. University of North Carolina Press.
Blalock, Jr., H. (1969). *Theory construction: From verbal to mathematical formulations*. Prentice-Hall.
Boker, S., Neale, M., & Rausch, J. (2004). Latent differential equation modeling with multivariate multi-occasion indicators. In K. van Montfort, J. H. Oud & A. Satorra (Eds.), *Recent developments on structural equation models: Theory and applications* (pp. 151–174). Kluwer Academic Publishers.
Boker, S. M., Neale, M., Maes, H., Wilde, M., Spiegel, M., Brick, T., Spies, J., Estabrook, R., Kenny, S., Bates, T., Mehta, P., & Fox, J. (2011). OpenMx: An open source extended structural equation modeling framework. *Psychometrika, 76*, 306–317.
Bollen, K. A., & Curran, P. (2004). Autoregressive latent trajectory (ALT) models: a synthesis of two traditions. *Sociological Methods and Research, 32*, 336–383.
Bollen, K., & Curran, P. (2006). *Latent curve models: A structural equation perspective*. Wiley.
Box, G. E. P., & Jenkins, G. M. (1970). *Time series analysis: Forecasting and control*, Holden-Day.
Caines, P. (1988). *Linear stochastic systems*. Wiley.
Campbell, D. T., & Stanley, J. (1963). Experimental and quasi-experimental designs for research on teaching. In N. Gage (Ed.), *Handbook of research on teaching* (pp. 171–246). Rand McNally.
Carpenter, B., Gelman, A., Hoffman, M., Lee, D., Goodrich, B., Betancourt, M., Brubaker, M., Guo, J., Li, P., & Riddell, A. (2017). Stan: A probabilistic programming language. *Journal of Statistical Software, 76*, 1–32.
Cattell, R. (1963). The structuring of change by P-technique and incremental R-technique. In C. Harris (Ed.), *Problems in measuring change* (pp. 167–198). University of Wisconsin Press.
Cattell, R., Cattell, A., & Rhymer, R. (1947). P-technique demonstrated in determining psychophysical source traits in a normal individual. *Psychometrika, 12*, 267–288.
Chambers, M. J., McCrorie, J. R., & Thornton, M. A. (2018). Continuous time modelling based on an exact discrete time representation. In K. van Montfort, J. H. Oud & A. Satorra (Eds.), *Continuous time modeling in the behavioral and related sciences* (pp. 317–357). Springer.
Chen, M., Chow, S.-M., & Hunter, M. D. (2018). Stochastic differential equation models with time-varying parameters. In K. van Montfort, J. H. Oud & M. C. Voelkle (Eds.), *Continuous time modeling in the behavioral and related sciences* (pp. 151–174). Springer.

Chow, S.-M., Ferrer, E., & Nesselroade, J. R. (2007). An unscented Kalman filter approach to the estimation of nonlinear dynamical systems models. *Multivariate Behavioral Research, 42*, 283–321.

Chow, S.-M., Ho, M.-H., Hamaker, E. L., & Dolan, C. V. (2010). Equivalence and differences between structural equation modeling and state-space modeling techniques. *Structural Equation Modeling, 17*, 303–332.

Chow, S.-M., Lu, Z., Sherwood, A., & Zhu, H. (2016). Fitting nonlinear ordinary differential equation models with random effects and unknown initial conditions using the stochastic approximation expectation-maximization (SAEM) algorithm. *Psychometrika, 81*, 102–134.

Coleman, J. C. (1968). The mathematical study of change. In H. Blalock, Jr. & A. Blalock (Eds.), *Methodology in social research* (pp. 428–478). McGraw-Hill.

Crouse, D., & Skufca, J. (2019). On the nature of discrete spacetime, Part I: The distance formula, relativistic time dilation and length contraction in discrete space-time. *Logique & Analyse, 246*, 177–223.

Curran, P., & Bollen, K. (2001). The best of two worlds: Combining autoregressive and latent curve models. In L. M. Collins & A. G. Sayer (Eds.), *New methods for the analysis of change* (pp. 105–135). American Psychological Association.

Deboeck, P. R., & Preacher, K. J. (2016). No need to be discrete: A method for continuous time mediation analysis. *Structural Equation Modeling: A Multidisciplinary Journal, 23*, 61–75.

Delsing, M., & Oud, J. H. (2008). Analyzing reciprocal relationships by the continuous-time autoregressive latent trajectory model. *Statistica Neerlandica, 62*, 58–82.

Doreian, P., & Hummon, N. P. (1976). *Modeling social processes*. Elsevier.

Driver, C. C., Oud, J. H., & Voelkle, M. C. (2017). Continuous time structural equation modelling with R package ctsem. *Journal of Statistical Software, 77*, 1–35.

Driver, C. C., & Voelkle, M. C. (2018). Hierarchical Bayesian continuous time dynamic modeling. *Psychological Methods, 23*, 774–749.

Fisher, F. M. (1969). Causation and specification in econometric theory and econometrics. *Synthese, 20*, 489–500.

Gard, T. C. (1988). *Introduction to stochastic differential equations*. Marcel Dekker.

Goldberger, A. S. (1964). *Econometric theory*. Wiley.

Gollob, H., & Reichardt, C. (1987). Taking account of time lags in causal models. *Child Development, 58*, 80–92.

Gollob, H., & Reichardt, C. (1991). Interpreting and estimating indirect effects assuming time lags really matter. In L. Collins & J. Horn (Eds.), *Best methods for analysis of change: Recent advances, unanswered questions, future directions* (pp. 243–259). American Psychological Association.

Goodrich, R. L., & Caines, P. E. (1979). Linear system identification from non-stationary cross-sectional data. *IEEE Transactions on Automatic Control, 24*, 403–411.

Granger, C. (1969). Investigating causal relations by econometric models and cross-spectral methods. *Econometrica, 37*, 424–438.

Hamaker, E. L., Dolan, C. V., & Molenaar, P. C. (2002). On the nature of SEM estimates of ARMA parameters. *Structural Equation Modeling, 9*, 347–368.

Hamaker, E. L., Dolan, C. V., & Molenaar, P. C. (2003). ARMA based SEM when the number of time points T exceeds the number of cases N: Raw data maximum likelihood. *Structural Equation Modeling, 10*, 352–379.

Hamerle, A., Nagl, W., & Singer, H. (1991). Problems with the estimation of stochastic differential equations using structural equations models. *Journal of Mathematical Sociology, 16*, 201–220.

Hamerle, A., Singer, H., & Nagl, W. (1993). Identification and estimation of continuous time dynamic systems with exogenous variables using panel data. *Econometric Theory, 9*, 283–295.

Hansen, L., & Sargent, T. J. (1983). The dimensionality of the aliasing problem. *Econometrica, 51*, 377–388.

Homans, G. (1950). *The human group*. Harcourt, Brace & World.

Hoyle, R. H., & Kenny, D. A. (1999). Sample size, reliability, and tests of statistical mediation. In R. Hoyle (Ed.), *Statistical strategies for small sample research* (pp. 243–259). Sage.

Jazwinski, A. H. (1970). *Stochastic processes and filtering theory*. Academic.
Johnston, J. (1972). *Econometric methods*. McGraw-Hill.
Jöreskog, K. (1973). A general method for estimating a linear structural equation system. In A. Goldberger & O. Duncan (Eds.), *Structural equation models in the social sciences* (pp. 85–112). Seminar Press.
Jöreskog, K. (1977). Structural equation models in the social sciences: Specification, estimation and testing. In P. Krishnaiah (Ed.), *Applications of statistics* (pp. 265–287). North Holland.
Jöreskog, K., & Sörbom, D. (1977). Statistical models and methods for analysis of longitudinal data. In A. Goldberger & O. Duncan (Eds.), *Structural equation models in the social sciences* (pp. 85–112). North Holland.
Jöreskog, K. G., & Sörbom, D. (1996). *LISREL 8: User's reference guide*. Scientific Software International.
Jöreskog, K. G., & van Thillo, M. (1972). *LISREL, A general computer program for estimating a linear structural equation system involving multiple indicators of unmeasured variables*. Educational Testing Service.
Kalman, R. (1960). A new approach to linear filtering and prediction problems. *Journal of Basic Engineering (Trans. ASME, ser. D), 82*, 35–45.
Kuo, H.-H. (1996). *White noise distribution theory*. CRC Press.
Kuo, H.-H. (2006). *Introduction to stochastic integration*. Springer.
Kwakernaak, H., & Sivan, R. (1972). *Linear optimal control systems*. Wiley.
Lawley, D., & Maxwell, A. (1971). *Factor analysis as a statistical method*. Butterworths.
Lazarsfeld, P. F. (1948). The use of panels in social research. *Proceedings of the American Philosophical Society, 92*, 405–410.
Lazarsfeld, P. F. (1955). Interpretation of statistical relations as a research operation. In P. F. Lazarsfeld & M. Rosenberg (Eds.), *The language of social research* (pp. 115–125). The Free Press.
Lazarsfeld, P. F. (1978). Some episodes in the history of panel analysis. In D. B. Kandel (Ed.), *Longitudinal research on drug abuse: Empirical findings and methodological issues*. Hemisphere Press.
Lewis, F. L. (1986). *Optimal estimation with an introduction to stochastic control theory*. Wiley.
Lipset, S., Lazarsfeld, P., Barton, A., & Linz, J. (1954). The psychology of voting: An analysis of political behavior. In G. Lindzey (Ed.), *Handbook of social psychology* (pp. 1124–1174). Addison-Wesley.
Liu, T.-C. (1960). Underidentification, structural estimation and forecasting. *Econometrica, 28*, 855–865.
Lord, F. M., & Novick, M. R. (1968). *Statistical theories of mental test scores*, Addison-Wesley.
MacCallum, R., & Ashby, F. G. (1986). Relationships between linear systems theory and covariance structure modeling. *Journal of Mathematical Psychology, 30*, 395–418.
Maxwell, S., & Cole, D. (2007). Bias in cross-sectional analyses of longitudinal mediation. *Psychological Methods, 12*, 23–44.
Maxwell, S., Cole, D., & Mitchell, M. (2011). Bias in cross-sectional analyses of longitudinal mediation: Partial and complete mediation under an autoregressive model. *Multivariate Behavioral Research, 46*, 816–841.
McArdle, J. J., & McDonald, R. P. (1984). Some algebraic properties of the reticular action model for moment structures. *British Journal of Mathematical and Statistical Psychology, 37*, 234–251.
Merton, R. (1990). *Continuous time finance*. Blackwell.
Molenaar, P. C. (1985). A dynamic factor model for the analysis of multivariate time series. *Psychometrika, 50*, 201–218.
Molenaar, P. C., & Nesselroade, J. R. (1998). A comparison of pseudo-maximum likelihood and asymptotically distribution-free dynamic factor analysis parameter estimation in fitting covariance structure models to block-Toeplitz matrices representing single-subject multivariate time-series. *Multivariate Behavioral Research, 33*, 313–342.

Molenaar, P. C., & Nesselroade, J. R. (2009). The recoverability of P-technique factor analysis. *Multivariate Behavioral Research, 44*, 130–141.

Moler, C., & Van Loan, C. (1978). Nineteen dubious ways to compute the exponential of a matrix. *Siam Review, 20*, 3–49.

Moler, C., & Van Loan, C. (2003). Nineteen dubious ways to compute the exponential of a matrix, twenty-five years later. *Siam Review, 45*, 3–49.

Muthén, L., & Muthén, B. (1999). *Mplus, the comprehensive modeling program for applied researchers user's guide*. Muthen & Muthen.

Neale, M. C. (1997). *Mx: Statistical modeling* (4th ed.). Department of Psychiatry.

Neale, M. C., Hunter, M. D., Pritikin, J. N., Zahery, M., Brick, T. R., Kirkpatrick, R. M., Estabrook, R., Bates, T., Maes, H., & Boker, S. M. (2016). OpenMx 2.0: Extended structural equation and statistical modeling. *Psychometrika, 81*, 535–549.

Oravecz, Z., Tuerlinckx, F., & Vandekerckhove, J. (2009). A hierarchical Ornstein-Uhlenbeck model for continuous repeated measurement data. *Psychometrika, 74*, 395–418.

Oravecz, Z., Tuerlinckx, F., & Vandekerckhove, J. (2011). A hierarchical latent stochastic differential equation model for affective dynamics. *Psychological Methods, 16*, 468–490.

Oravecz, Z., Tuerlinckx, F., & Vandekerckhove, J. (2016). Bayesian data analysis with the bivariate hierarchical Ornstein-Uhlenbeck process model. *Multivariate Behavioral Research, 51*, 106–119.

Otter, P. W. (1986). Dynamic structural systems under indirect observation: Identifiability and estimation aspects from a system theoretic perspective. *Psychometrika, 51*, 415–428.

Ou, L., Hunter, M. D., & Chow, S.-M. (2019). What's for dynr: A package for linear and nonlinear dynamic modeling in R. *The R Journal, 11*, 1–20.

Oud, J. H. (1978). *Systeem-methodologie in sociaal-wetenschappelijk onderzoek [Systems methodology in social science research], Doctoral dissertation*, Alfa, Nijmegen.

Oud, J. H. (2004). SEM state space modeling of panel data in discrete and continuous time and its relationship to traditional state space modeling. In K. van Montfort, J. H. Oud & A. Satorra (Eds.), *Recent developments on structural equation models: Theory and applications* (pp. 13–40). Kluwer Academic Publishers.

Oud, J. H. (2007). Comparison of four procedures to estimate the damped linear differential oscillator for panel data. In K. van Montfort, J. H. Oud & A. Satorra (Eds.), *Longitudinal models in the behavioral and related sciences* (pp. 19–39). Lawrence Erlbaum Associates.

Oud, J. H. (2010). Second-order stochastic differential equation model as an alternative for the ALT and CALT models. *Advances in Statistical Analysis, 94*, 203–215.

Oud, J. H., Delsing, M. J. (2010). Continuous time modeling of panel data by means of SEM. In K. van Montfort, J. H. Oud & A. Satorra (Eds.), *Longitudinal research with latent variables* (pp. 201–244). Kluwer Academic Publishers.

Oud, J. H., & Jansen, R. A. (2000). Continuous time state space modeling of panel data by means of SEM. *Psychometrika, 65*, 199–2015.

Oud, J. H., Jansen, R. A., van Leeuwe, J. F., Aarnoutse, C. A., & Voeten, M. J. (1999). Monitoring pupil development by means of the Kalman filter and smoother based upon SEM state space modeling. *Learning and Individual Differences, 11*, 121–136.

Oud, J. H., & Singer, H. (2008). Continuous time modeling of panel data: SEM versus filter techniques. *Statistica Neerlandica, 62*, 4–28.

Oud, J. H., van den Bercken, J. H., & Essers, R. (1986). Longitudinal factor score estimation using the Kalman filter. *Kwantitatieve Methoden, 20*, 109–129.

Oud, J. H., van den Bercken, J. H., & Essers, R. (1990). Longitudinal factor score estimation using the Kalman filter. *Applied Psychological Measurement, 14*, 395–418.

Oud, J. H., & Voelkle, M. C. (2014). Do missing values exist? Incomplete data handling in cross-national longitudinal studies by means of continuous time modeling. *Quality & Quantity, 48*, 3271–3288.

Oud, J. H., Voelkle, M. C., & Driver, C. C. (2018). SEM based CARMA time series modeling for arbitrary N. *Multivariate Behavioral Research, 53*, 36–56.

Phillips, A. (1957). The estimation of parameters in systems of stochastic differential equations. *Biometrika, 46*, 67–76.
Phillips, P. C. (1972). The structural estimation of a stochastic differential equation system. *Econometrica, 40*, 1021–1042.
Phillips, P. C. (1973). The problem of identification in finite parameter continuous time models. *Journal of Econometrics, 1*, 351–362.
Phillips, P. C. (1974). The estimation of some continuous time models. *Econometrica, 42*, 803–824.
Priestley, M. (1981). *Spectral analysis and time series*. Academic Press.
R Development Core Team (2011). *R: A language and environment for statistical computing*, R Foundation for Statistical Computing, Vienna.
Rauch, H., Tung, F., & Striebel, C. (1965). Maximum likelihood estimates of linear dynamic systems. *AIAA Journal, 3*, 1445–1450.
Robinson, P. (1976). The estimation of linear differential equations with constant coefficients. *Econometrica, 44*, 751–764.
Sargan, J. (1974). Some discrete approximations to continuous stochastic models. *Journal of the Royal Statistical Society B, 36*, 74–90.
Shumway, R. H., & Stoffer, D. S. (2000). *Time series analysis and its applications*. Springer.
Simon, H. (1952). A formal theory of interaction in small groups. *American Sociological Review, 17*, 202–211.
Singer, H. (1990). *Parameterschätzung in zeitkontinuierlichen dynamischen Systemen [Parameter estimation in continuous time dynamic systems]*, Hartung-Gorre, Konstanz.
Singer, H. (1991). *LSDE - A program package for the simulation, graphical display, optimal filtering and maximum likelihood estimation of linear stochastic differential equations: User's guide*, Author, Meersburg.
Singer, H. (1992a). The aliasing phenomenon in visual terms. *Journal of Mathematical Sociology, 17*, 39–49.
Singer, H. (1992b). *Zeitkontinuierliche Dynamische Systeme[Continuous time dynamical systems]*, Campus, Frankfurt.
Singer, H. (1993). Continuous-time dynamical systems with sampled data, errors of measurement and unobserved components. *Journal of Time Series Analysis, 5*, 527–545.
Singer, H. (1995). Analytical score function for irregularly sampled continuous time stochastic processes with control variables and missing values. *Economic Theory, 11*, 721–735.
Singer, H. (1998). Continuous panel models with time dependent parameters. *Journal of Mathematical Sociology, 23*, 77–98.
Singer, H. (2008). Nonlinear continuous time modeling approaches in panel research. *Statistica Neerlandica, 62*, 29–57.
Singer, H. (2010). SEM modeling with singular moment matrices Part I: ML-estimation of time series. *Journal of Mathematical Sociology, 34*, 301–320.
Singer, H. (2012). SEM modeling with singular moment matrices. Part II: ML-estimation of sampled stochastic differential equations. *Journal of Mathematical Sociology, 36*, 22–43.
Singer, H. (2018). Langevin and Kalman importance sampling for nonlinear continuous-discrete state-space models. In K. van Montfort, J. H. Oud & M. C. Voelkle (Eds.), *Continuous time modeling in the behavioral and related sciences* (pp. 389–435). Springer.
Spearman, C. (1904). "General Intelligence", objectively determined and measured. *American Journal of Psychology, 5*, 201–293.
Steele, J., & Ferrer, E. (2011). Latent differential equation modeling of self-regulatory and coregulatory affective processes. *Multivariate Behavioral Research, 46*, 956–985.
Strotz, R. H., & Wold, H. O. (1960). Recursive vs. nonrecursive systems: An attempt at synthesis. *Econometrica, 28*, 417–427.
Theil, H. (1971). *Principles of econometrics*. Wiley.
Theunissen, T., & Oud, J. H. (2021). Achilles and the tortoise 2500 years after Zeno. *Logique & Analyse, 255*, 245–265.
Torgerson, W. S. (1958). *Theory and methods of scaling*. Wiley.

Tuma, N. B., & Hannan, M. T. (1984). *Social dynamics: Methods and models*. Academic.
van der Linden, W. (ed.) (2016). *Handbook of item response theory, Vol. 1: Models*. Chapman & Hall.
Voelkle, M., & Oud, J. H. (2013). Continuous time modeling with individually varying time intervals for oscillating and non-oscillating processes. *British Journal of Mathematical and Statistical Psychology, 66*, 103–126.
Voelkle, M., Oud, J. H., Davidov, E., & Schmidt, P. (2012). An SEM approach to continuous time modeling of panel data: Relating authoritarianism and anomia. *Psychological Methods, 17*, 176–192.
Voelkle, M., Oud, J. H., von Oertzen, T., & Lindenberger, U. (2012). Maximum likelihood dynamic factor modeling for arbitrary N and T using SEM. *Structural Equation Modeling: A Multidisciplinary Journal, 19*, 329–350.
Wheaton, B., Muthén, B., Alwin, D., & Summers, G. (1977). Assessing reliability and stability in panel models. In D. Heise (Ed.), *Sociological methodology* (pp. 84–136). Jossey-Bass.
Willems, J. C. (1975). Minimal realization in state space form from input/output data. In H. Buffart & J. H. Oud (Eds.), *Systeemleer* (pp. 22–61). Stenfert Kroese.
Winer, B. (1962). *Statistical principles in experimental design*. McGraw-Hill.
Wold, H. (1960). A generalization of causal chain models. *Econometrica, 28*, 443–463.
Wold, H., & Jureen, L. (1953). *Demand analysis*. Wiley.
Wright, S. (1934). The method of path coefficients. *Annals of Mathematical Statistics, 5*, 161–215.
Wymer, P. C. (1972). Econometric estimation of some stochastic differential equation systems. *Econometrica, 40*, 565–578.
Zadeh, L., & Desoer, C. (1963). *Linear system theory: The state space approach*. McGraw-Hill.

Chapter 2
Time in Latent Growth Curve Models

Matt L. Miller and Paolo Ghisletta

Time is what keeps everything from happening at once.

—Ray Cummings

2.1 Time in Social Science Research

A Brief Introduction to Time as a Variable

The social sciences seek to understand the human experience. This experience unfolds across time, so it is important for the social sciences to examine longitudinal data (Baltes & Nesselroade, 1979; Lindenberger et al., 2011). In such studies, time provides a metric to the progression of these events, though it does not cause the observed changes. The causes of any change may be studied along with the outcome, such as when symptoms of depression are studied in groups treated with an antidepressant versus a placebo, a time-invariant covariate. Another example is when memory change is studied in parallel with changes in the volume of brain regions associated with memory (Miller et al., 2021); that is, the cause of difference also changes over time, a time-varying covariate (Grimm et al., 2017). On the other hand, changes in outcome are sometimes simply measured as they change across time without considering causes, perhaps to abstract a phenomenon's change without the distraction of other phenomena. The choice of method should be guided

M. L. Miller (✉)
REACH Institute, Arizona State University, Tempe, AZ, USA
e-mail: mlmill47@asu.edu

P. Ghisletta
Faculty of Psychology and Educational Sciences, University of Geneva, Geneva, Switzerland

Faculty of Psychology, UniDistance Suisse, Brig, Switzerland

Swiss National Centre of Competence in Research LIVES, University of Geneva, Geneva, Switzerland
e-mail: paolo.ghisletta@unige.ch

M. Stemmler et al. (eds.), *Dependent Data in Social Sciences Research*,
https://doi.org/10.1007/978-3-031-56318-8_2

by the research question under study, but to simplify discussion in this chapter, we will confine ourselves to the latter type of study, in which a single phenomenon is examined.[1]

One of the main differences in the treatment of time itself in longitudinal data analysis is whether time is measured continuously or discretely. For example, if it is December 31, 2022, Alice was born on January 1, 2000, and Bob was born on December 31, 2000, we could say they both are 22 years old, treating time discretely, or we could say that Alice is 22.99726... years old and Bob is 22.00000... years old, treating time continuously. If we were studying reading skills, this would matter more if we were studying them at an age when reading ability changes rapidly versus very little. As a practical matter, we always round measurements in our research to some extent whether due to instrumental accuracy, the numeric precision of our computers, or the number of digits we are willing to type.

2.2 Outline of the Latent Growth Curve Model

Two popular, related methods for analyzing longitudinal data are multilevel modeling (MLM) and structural equation modeling (SEM) (McArdle, 1986; Meredith & Tisak, 1990; Singer & Willett, 2003), both often referred to as latent growth curve modeling (LGCM). LGCMs are generally used for "big N, small T" studies, in which the number of participants is large, but the number of measurement occasions (usually called "waves") is relatively small, as compared to techniques such as time series analysis, more commonly used for "small N, big T" studies.[2] LGCMs in MLM and SEM, though often conceived of as different techniques, in most cases are equivalent models of change (Curran, 2003). The SEM framework is attractive to many researchers because of its intuitive interpretation and application, including the use of easy-to-read path diagrams (Grimm et al., 2017). Thus, for the sake of simplicity, we will confine most of our discussion in the remainder of this chapter to LGCM in the SEM framework. However, the handling of time is equally important in MLM, and the considerations we discuss herein are equally applicable.[3] Readers interested in information about using MLM beyond this discussion can find excellent guidance in resources that cover both methods (e.g., Singer & Willett, 2003; Grimm et al., 2017).

---

[1] When research is focused on time-varying covariates, it is particularly appropriate to consider dynamic models of change (Voelkle et al., 2018). Such models are outside the scope of this discussion, though the metric of time is equally important in such models.

[2] Because we separate the concepts of measurement occasion, or *wave*, from time and its coding, we use W to represent a number of occasions in the remainder of this chapter.

[3] It is often assumed that because MLM handles time rather naturally, research applying this MLM must treat time rigorously. A careful analysis of unpublished data from a review of such studies (Ghisletta et al., 2020) revealed that this is not always the case. We recommend careful attention to the general points discussed herein no matter what method is used to model longitudinal data.

Overall Model

So as to focus on time, throughout this chapter we omit some very useful elements common in longitudinal models, such as covariates and multiple groups. Though we omit them from our example models, the treatment of time is at least as important with these extra elements as it is in simplified models. So we can use common vocabulary and variable names, we will describe one of the simplest cases: a single outcome measure, y, linear change, and participants measured across a total of W occasions, numbered $w = \{1, 2, \ldots W\}$.

The coding of time is the central point of this chapter, so we must deal with it as a more complex quantity than a simple repetition of w. Thus, we will designate the coding of time as t. The time coding for individual i at occasion w will be designated as t_{iw}. While it may sometimes be the case that $t_{iw} = w - 1$ (i.e., the intercept occurs during the first wave, time increments by one at each successive wave, and every participant is measured at equal intervals of time), this is by no means universal. However, for our simple example, it is easiest to start thinking about it as a case in which the occasions are equally spaced in time, and we can thus have $t_{iw} = \{0, 1, 2 \ldots W - 1\}$ for all i.

Given this setup, the simple LGCM (Fig. 2.1) is[4]

$$y_{iw} = \alpha_i + t_{iw}\beta_i + \epsilon_{iw} \tag{2.1}$$

with y_{iw} the outcome for individual i at occasion w, α_i the intercept (the overall level on y at occasion $w = 1$, thus at $t_{iw} = 0$), t_{iw} the time interval from the first wave to the measurement occasion w, β_i the growth factor, representing the change in y per unit time, and ϵ_{iw} the residual, with

$$\alpha_i = \mu_\alpha + \eta_{\alpha i} \tag{2.2}$$

$$\beta_i = \mu_\beta + \eta_{\beta i} \tag{2.3}$$

where μ_α and μ_β are the fixed effects (i.e., the average values across the sample) for the intercept and growth factor, respectively. The random effects (representing individual variations around the means) for intercept and the growth factor, $\eta_{\alpha i}$ and $\eta_{\beta i}$, respectively, are then centered at 0 and distributed

$$\begin{bmatrix} \eta_{\alpha i} \\ \eta_{\beta i} \end{bmatrix} \sim \text{MVNorm}\left(\begin{bmatrix} 0 \\ 0 \end{bmatrix}, \begin{bmatrix} \sigma_\alpha^2 & \sigma_{\alpha\beta} \\ \sigma_{\alpha\beta} & \sigma_\beta^2 \end{bmatrix} \right) \tag{2.4}$$

[4] We provide formulas in the style that is usually encountered in the MLM framework instead of the matrix algebra notation often used in SEM for simplicity's sake; it is mathematically equivalent for most of the models we present.

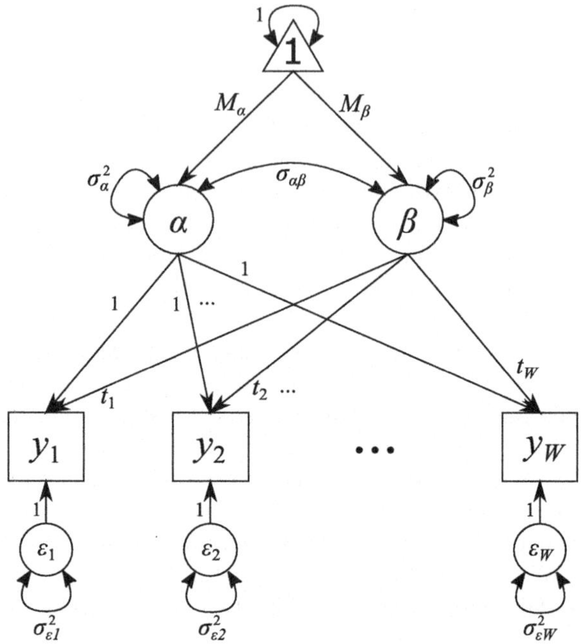

Fig. 2.1 Squares represent manifest variables, circles latent variables, and the triangle the constant. One-headed arrows are regression weights or factor loadings, and two-headed arrows are variances or covariances. Arrows with numbers are not estimated (i.e., they are fixed parameters), and arrows with names are estimated parameters except for t_w, which are usually fixed

with the residual term in Eq. (2.1) distributed

$$\epsilon_{iw} \sim \text{Norm}(0, \sigma^2_{\epsilon w}) \,. \tag{2.5}$$

The relationship between the outcome y_{iw} and time is given in Eq. (2.1), the level-1 equation. How this relationship varies across individuals is given in Eqs. (2.2) and (2.3), the level-2 equations. We can rewrite the model as a single equation:

$$y_{iw} = (\mu_\alpha + \eta_{\alpha i}) + t_w(\mu_\beta + \eta_{\beta i}) + \epsilon_{iw} \,. \tag{2.6}$$

Note that we have now removed the subscript i from t_w. Though the most general form of the LGCM allows each participant to have unique times of measurement, in SEM a common assumption is that individuals are measured at the same intervals between occasions. The term $(\mu_\alpha + \eta_{\alpha i})$ represents the *intercept*, the value when all other predictors are zero. In our simple linear model, where $t_1 = 0$, this is the initial value. Its two parts, μ_α and $\eta_{\alpha i}$, represent the sample's mean intercept and the amount by which individual i differs from that mean. Equation (2.4) says that $\eta_{\alpha i}$ has a mean of zero and a standard deviation of σ_α, the amount that individuals vary in their intercept. This distribution also tells us that individuals' deviations from the mean intercept covary with the deviations from the mean slope ($\sigma_{\alpha\beta}$).

The term $t_w(\mu_\beta + \eta_{\beta i})$ represents the size of change over time. The basis vector t_w represents the measurement of time. The combined sum $(\mu_\beta + \eta_{\beta i})$, then, is how large the change is per unit of time *as expressed in the basis vector*. With our simple example, in which $t_w = w - 1 = \{0, 1, 2, 3, \ldots, W\}$, this is the amount of change between waves. Because this simple interpretation is so common, this term is often called the *slope*. However, in many cases it is not actually a linear slope but a measure of the size of change in a more complicated trajectory or even the size of just one component of an overall trajectory. Therefore, we call this a *growth factor*. Similar to the intercept, μ_β is the mean size of growth and $\eta_{\beta i}$ is the amount by which individual i differs from the mean. Equation (2.4) shows that $\eta_{\beta i}$ has a mean of zero and a standard deviation of σ_β, which is the amount by which individuals vary in their change per unit time, along with the covariance with the intercept $\sigma_{\alpha\beta}$, which denotes the linear association between the intercept and the growth factor.

Finally, ϵ_{iw} is the unexplained variance or the *residual*, the part of the outcome y that cannot be explained by the growth components α and β of our model. Its distribution indicates that $\sigma_{\epsilon w}$ is the standard deviation of the unexplained part of the outcome at wave w. The model *without* the residual term is called the *predicted* value. However, another source of residual variance is possible: If the model of change is incorrect, for example, we have specified a linear trajectory when the change actually has a curved shape, $\sigma_{\epsilon w}$ will be inflated for some occasions by the amount that the modeled trajectory is different from the true shape; in such a case, all other parameters may be biased as well and may not have useful interpretations.

The Basis Vector

Though we discuss other parameters, the key feature that we will be dealing with is the values of t_w, which together we refer to as the *basis vector*.[5] We refer to individual elements of the basis vector as *basis coefficients*; they determine how much a growth factor changes the measurement at the corresponding time point. So, if the basis coefficient is zero, the growth factor has no influence, and at that wave only the intercept and residual determine the outcome. In our example, the coefficients then proceed as $\{0, 1, 2, \ldots\}$ because at each time point, the growth contributes an additional unit amount to the measurement, so if the value of the growth factor were 3, the expected measurement would start at the intercept, then increase by three units, then increase by another three units (six total), and so forth.

So the basis vector represents landmarks on the timeline that we use to define how other parameters in the model are interpreted. To keep everything separate, we will introduce one more bit of notation, κ, which we will use to represent clock or calendar time. The point at which $\kappa = 0$, called the *epoch*, is arbitrary since we will mostly be dealing with differences in κ, but for convenience we will set it to the

[5] This term comes from linear algebra and was used by Meredith and Tisak (1990), in their seminal work on LGCM. Though it is not usually referred to by this name in MLM, the concepts we discuss apply equally to the values used to represent time in that framework.

Table 2.1 Enumerations of time in this chapter

Symbol	Name	Definition
w	Wave number	The measurement occasion, numbered from 1 to W, the total number of waves
κ_w	Clock time	The time indicated on the clock or date on the calendar at measurement occasion w
κ_0	Clock time intercept	The time indicated on the clock at the intended model intercept
t_w	Basis coefficient	The value used in the model to represent wave w on the y-axis

Note: i is added when individuals have any differences in timing; see text for details

clock time that is intended to be the intercept and denote it as κ_0 (note that as we have defined it, w cannot be zero, so this should not be confused with clock time at "wave 0"). Thus, if the intercept occurs at the first occasion, $\kappa_1 = \kappa_0$. If participants have unique intercepts, such as their date of birth when using age as the metric of time, we add a subscript i to the intercept, κ_{0i}. We have now introduced several ways of enumerating data collection occasions, so for reference we summarize them in Table 2.1. When individuals vary in the times their data were collected within occasions, we add the subscript i to κ and t (κ_{iw} and t_{iw}) to indicate that each individual has their own value for occasion w.

The intercept need not be the first occasion, and in fact there does not need to be a zero in the vector if the intercept is outside the period in which data were collected[6] or between data collection occasions. The sizes of the basis coefficients set the scale of the estimates: We can set them so the parameters reflect change per day, per decade, or any other metric. Beyond these two simple manipulations, we can also affect the shape of the trajectory by giving the vector values that are not simple multiples of the data collection times, such as when data are collected at $\kappa = \{0, 1, 2, 3\}$, but we set the basis vector to $t = \{0, 1, 4, 9\}$. The decisions we make about the vector's exact values will directly affect the interpretation of almost every other parameter in the model via dictating the shape of change of our model. One of the strengths of SEM is that, though the basis vector is often specified *a priori* with fixed values specifying a particular shape (a *fixed-basis* model), the shape can also be estimated from the data by allowing most of the basis coefficients to be free parameters (a *latent-basis* model).

A graphical analogy of how the basis vector represents time will be useful for further explanation. Imagine that we have collected reading skills data for a sample of children at ages of 10, 11, 13, 15, and 17. We draw the hypothesized trajectory of the data on graph paper (Fig. 2.2, top left), with the x-axis representing time (in terms of k) and the y-axis representing the outcome's magnitude. The scalings of

[6] We often recommend that the value zero is situated somewhere in the period of data collection to facilitate parameters' interpretation and to discourage extrapolation beyond the time period that was studied (Aiken & West, 1991).

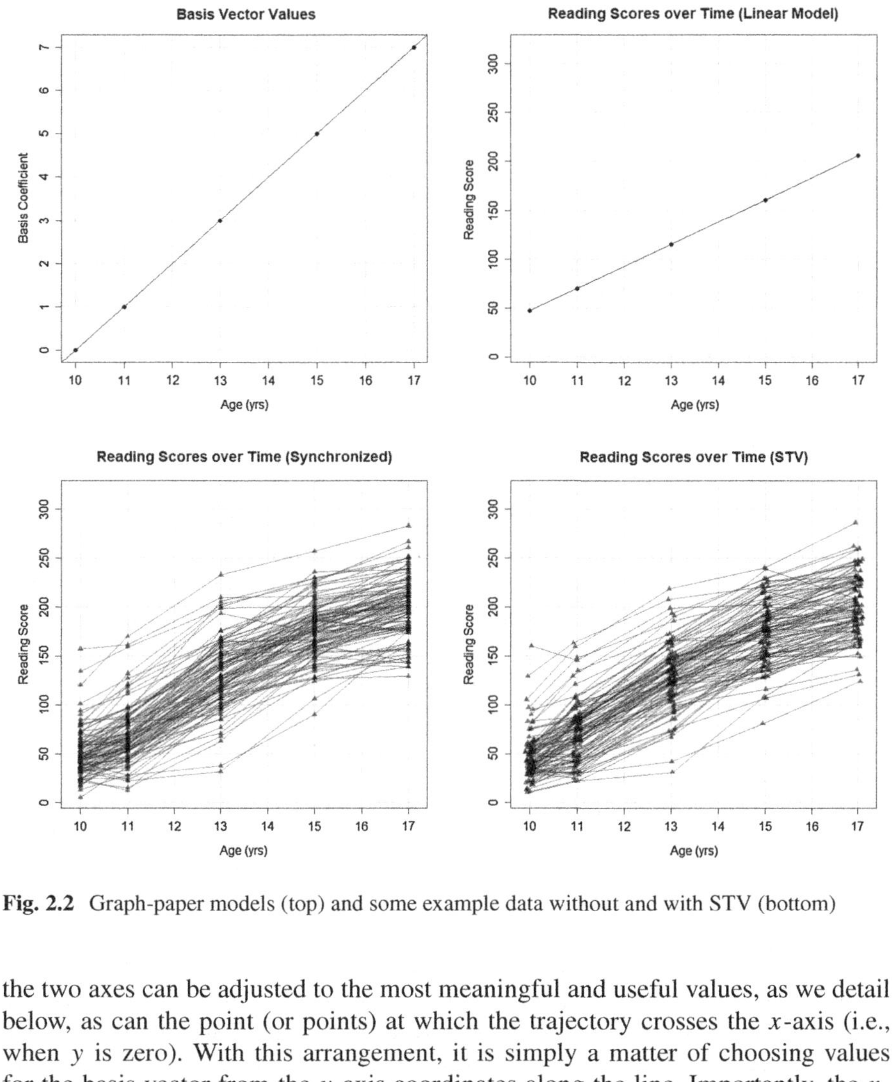

Fig. 2.2 Graph-paper models (top) and some example data without and with STV (bottom)

the two axes can be adjusted to the most meaningful and useful values, as we detail below, as can the point (or points) at which the trajectory crosses the x-axis (i.e., when y is zero). With this arrangement, it is simply a matter of choosing values for the basis vector from the y-axis coordinates along the line. Importantly, the x-axis coordinates that you choose along the line should correspond to the times at which the data were collected. Time is thereby represented by the x coordinates that correspond to the y coordinates chosen for the basis vector(s).[7] Note also that the values from the y-axis should always be on the same scale across occasions so that the associated growth factor has a consistent meaning across waves. Whenever we use this analogy, we will call it the *graph-paper model*. Given the linear model

[7] In MLM, this usually amounts to having time directly enter the equation so that $t_{iw} = \kappa_{iw}$ or a mathematical transformation of κ, such as its square.

scaling and intercept depicted in the top left of Fig. 2.2, if the estimate of the intercept is 47 and the estimate of the growth factor is 22.7, the model would indicate the mean trajectory shown in the upper right of Fig. 2.2.

Effects of Intercept Location and Scaling of Growth Factor Basis

Given that the LGCM is mainly applied to longitudinal data to address hypotheses related to change, we discuss briefly two model specification features encoded within the basis vector that affect estimates of change-related parameters. The principle effects of these features occur in the sample mean (μ_β) and the variance of the growth factor (σ^2_β) and in the correlation between the intercept and the growth factor ($\sigma_{\alpha\beta}$).

First, the location of the intercept, that is, its centering (i.e., where the trajectory crosses the y axis in the graph-paper model), directly affects the estimates of the change-related parameters. By centering the intercept we mean choosing a particular time value for which the basis vector has value 0. This is often chosen to match the beginning of the study, at the first wave ($w = 0$), but certainly need not be. For instance, when the time operationalization is calendar time ($t = \kappa$), then the value 0 of the basis vector may have a particular meaning. If the substantive question is about change over chronological age, the point $t = 0$ would mean that the intercept is centered at age 0, an age (probably birth or conception) that many substantive hypotheses do not address. Just as in ordinary least squares regression, in the LGCM the advice to center predictors so as to obtain parameter estimates defined in the data space is useful (Aiken & West, 1991). Hence, age is often centered on a meaningful value, which could be invariant across individuals (e.g., at age 60 years) or idiosyncratic (e.g., at the average value of each individual, Wainer, 2000; Mehta & West, 2000, which in the language of MLM is equivalent to the group-centering approach, Bryk & Raudenbush, 1987). Centering the intercept affects not only the intercept-related parameters (its mean μ_α and variance σ^2_α) but also the latent correlation $\sigma_{\alpha\beta}$, which has been shown to vary in its estimate from negative, to zero, to positive (Biesanz et al., 2004; Rogosa & Willett, 1985; Rovine & Molenaar, 1998). Hence, whether individuals with a high general level of performance (i.e., high intercept score) are likely to increase (i.e., $\sigma_{\alpha\beta} > 0$ or decrease (i.e., $\sigma_{\alpha\beta} < 0$) over the course of the study depends on the location of the intercept.

Second, the scaling of the growth factor, which is operationalized by the magnitude of its loadings (i.e., the basis vector), also alters the change-related parameters. Here, also, basic rules from regression apply: The most straightforward interpretation of a slope is, everything else held constant, the change in the predicted value of the outcome for a one-unit change of that slope's predictor (Aiken & West, 1991). Imagine that our study of reading skills was conducted over a single year instead of from ages 10 to 17 and that the children were measured in five equidistant waves. One could choose to scale the growth factor with conveniently rounded values of $\kappa_w = 0, 0.25, 0.5, 0.75, 1$, meaning that change is defined over 1 year.

Or, change could be defined over months, by specifying $\kappa_w = 0, 3, 6, 9, 12$, or even days ($\kappa_w = 0, 91, 182, 273, 364$). The three scalings are fully collinear, so the LGCMs estimated by them are fully equivalent. Yet, parameters' interpretation will vary greatly (change over 1 year vs. 1 month vs. 1 day) as will parameter estimates. Following standard statistical expectation rules, the means are scaled by the multiplying factor, whereas the variances by its square. For instance, if $\mu_\beta = 24$ and $\sigma_\beta = 144$ when time is scaled in years, then the analogous estimates would be 2 and 1 when scaled in months. Note that the p-values of these parameters will not change, but their standard errors will be scaled accordingly. For computational reasons, we advise avoiding scalings that are very large or very small (Kline, 2016).

2.3 Sampling-Time Variation (STV)

In our discussion of SEM so far, we have generally treated participants measured at the same wave, w, as if they were measured at the same times, κ_w, or the same time intervals, $\kappa_w - \kappa_{w-1}$. In our example of children's reading skills, such data could look like the lower left of Fig. 2.2, with each measurement occasion having all participants' times horizontally aligned. However, it is often not possible to collect data at the same intervals for all participants. This is why measurement occasions are often referred to as *waves*—participants are not measured simultaneously, but instead within the confines of a given interval. When this occurs, the true measurement times need the subscript by individual, κ_{iw}, from our original LGCM (Eq. (2.1)). This is very likely to happen in "big N, small T" studies, given that the more numerous the participants, the higher the chances that the planned schedule for data collection interferes with participants' own schedule. Consequently, participants will be assessed at different intervals, both between (participant A is on a different schedule than participant B) and within individuals (the same participant is assessed at different intervals). This phenomenon is called *sampling-time variation* (STV). Such data would then resemble the lower right of Fig. 2.2, with a variation between participants' x-axis value in each wave; the figure shows a variation of ± 6 weeks from the intended interval. When researchers analyze such data with models that treat these varying intervals as if they were equal, STV can lead to biased parameter estimates, reduced statistical efficiency, and mischaracterized measurement reliability (Coulombe et al., 2015; Miller & Ferrer, 2017).

Introduction to STV

Consider a researcher who collects data on children every 2 weeks, only on weekdays (represented as white squares in Fig. 2.3). The interval of interest to the researcher is 14 days, but if there are many children in the study, it is unlikely that

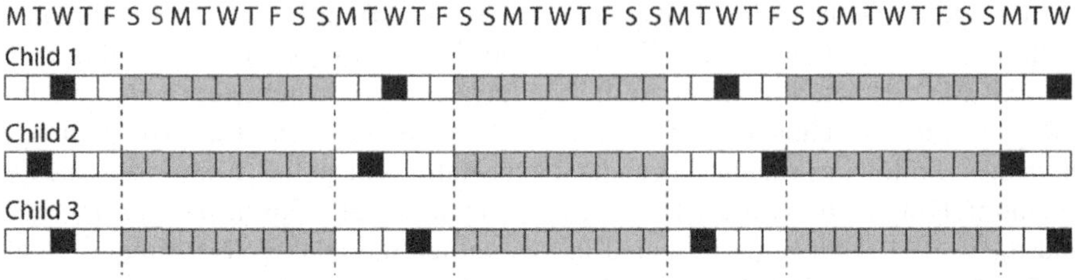

Fig. 2.3 Hypothetical data collection example. Gray boxes indicate days that are not part of data collection intervals (the off week and the preceding and following weekends), white boxes weekdays available for data collection, and black boxes days on which data are assessed on the hypothetical children

all participants can be assessed on the same day at every wave. In fact, a pattern like that shown in Fig. 2.3 could realistically result (black squares represent assessment days). Child 1 is regularly assessed on Wednesdays every 14 days, but child 2 has a longer interval of 17 days between the second and the third assessment (differing intervals between participants). Child 2 also has a shorter interval of 10 days between the third and the fourth assessment (differing intervals within participant). Child 3 has some variation in assessment dates but is generally somewhat closer to the desired interval of 14 days than child 2. The sampling times can have different distributions: If participants are as likely to be assessed before as they are after the planned assessment day, then the distribution will be symmetrical. On the other hand, if the researcher attempts to schedule assessment only at the desired interval or later, but never earlier, this will lead to a distribution that is strongly right-skewed. The magnitude of variations from the desired interval also depends on choices made by the researcher, such as how long before or after the desired date of assessment is acceptable. Singer and Willett (2003) have also noted that deviations from the desired interval tend to increase over the course of a study, a phenomenon they refer to as "occasion creep." In the lower right panel of Fig. 2.2, this would be represented if the distribution of horizontal misalignments grew wider at later occasions. All of these represent different types of STV.

Within most longitudinal studies, there is a particular interval that is initially intended by the researchers; in the preceding example, that was 14 days. We call this the *desired interval*; the time points that correspond to measurements at the desired intervals are the *intended times*. We refer to the distribution of deviations from the desired intervals as the *STV distribution* of the study. When referred to by itself, we mean the distribution of all deviations across all individuals or the distribution of *interindividual STV*. When we mean the distribution of deviations within a single participant, we call it the *intraindividual STV* distribution. In most studies, the possible maximum size of STV distribution will be limited by the maximum allowed deviation from the desired interval. Because this is reflected in the width of the wave when visualized as in Fig. 2.3, we call this the *wave width* and adopt *occasion creep* from Singer and Willett (2003) to describe when the wave width grows at later waves in a study. The time between collection waves is called *interwave width* (IWW); the larger the ratio of wave width to IWW is, the larger the variability in STV distribution, and thus the greater the potential consequences of STV. When all times for a given wave are treated as being equal—that is, when all t_w are equal even when κ_{iw} varies for a given w—we call this *time binning*. Treating time as discrete instead of continuous is one form of time binning. In the graph-paper model, this amounts to taking many different values on the x-axis and clustering them all onto a limited set of unique values. In the example above, this is equivalent to pretending that all four children were assessed exactly on every other Wednesday, thereby implying no STV. Many didactic works on LGCM in the SEM framework present models assuming that waves have been time-binned to a single value per wave for the sake of convenience and simplicity (e.g., Singer & Willett, 2003; Bollen & Curran, 2006). However, SEM does not require time binning data with STV, as we illustrate below.

Sources of STV

Studies with STV can be put into three broad categories based on its source and handling: participant and research design characteristics, data availability constraints, and theoretical motivation. The first of these, characteristics of participants and research design, is probably the most common. Real-world research involves working with participants whose personal circumstances may make regular scheduling difficult or impossible, an issue that increases as the study's N grows larger. Issues with participant retention and scheduling are pervasive; an entire literature exists on this topic (e.g., Abshire et al., 2017; Ferris et al., 2021), in particular for special populations (e.g., Davies et al., 2014; Nicholson et al., 2011). But it is not just participants who have time constraints—some data collection efforts are too large for any reasonably sized research team to collect simultaneously and so are often forced to schedule participants across a range of times. For these reasons, researchers frequently design studies so there is a range of dates around waves' desired intervals for collecting data. Since it is often impossible to maintain the desired interval within successive waves (indeed, occasion creep makes it difficult to even maintain constant wave widths), this design necessity leads to STV. Fortunately, recognizing this before data are collected in many cases will allow researchers to avoid time binning such data by simply recording the exact time of data collection across the study and applying the methods described below.

The second category of STV encompasses studies that actually have the same underlying causes of STV as the first type. Many publicly available datasets now censor exact data collection dates and exact participant birthdates to protect privacy. This is because of concerns about reidentification, in which the intended anonymization of data is overcome by matching minimal amounts of data to public records and big data sources (Rocher et al., 2019). Thus, this category differs from the first in that the exact time of data collection (or the exact time relative to birthdate) is not available and cannot be recovered for analysis. Thus, data collection intervals are necessarily rounded rather than provided as a more precise value representing within-month or even within-year variations. It is recognized that such a rough approach to intervals may be practically inconsequential in some portions of the lifespan, but in others, such as early childhood, rounding measurement intervals likely leads to gross misestimation of the underlying developmental process (Baltes & Nesselroade, 1979). It is also possible that ethics concerns may even prevent the original researchers from recording timing information beyond a certain resolution, in which case no one can use exact timing information for data analysis. In such cases, time binning is a necessary consequence of ethical obligations for participants' privacy. When working with this kind of data, the effects of time binning should be mitigated as described below since it cannot be avoided altogether.

The third category of STV arises when the research hypothesis entails discrete changes in discrete time intervals. For example, if an educational researcher believes that some outcome depends on completion of a full grade level, with no expected effects of partial completion, then binning time by grade is theoretically justified. However, if research includes hypotheses in which the outcome is influenced by the number of hours spent reading or by chronological age or in which the grade effect may be moderated by chronological age, this would suggest continuous change, and thus time binning would not be justifiable. When studies include only hypotheses with discrete time outcomes, it is unnecessary to apply any of the methods to account for STV or to mitigate time binning. It seems likely, however, that in many fields requiring longitudinal data (e.g., development, aging), processes in which such hypotheses are justifiable are not very common.

Effects and Importance of Time Binning with STV

When data are time-binned, STV enters the model as variance that cannot be explained by the model of change (Miller & Ferrer, 2017). If you think of the graph-paper model and imagine moving the x axis location back and forth by a small amount (lower left of Fig. 2.2), the deviations of the y value from the value at the implied times represent this excess variance. The distribution of this variance depends on the STV distribution and the shape of the trajectory around the implied time. If the trajectory is linear in the neighborhood of the implied time and STV is distributed normally and centered on the implied time, then the extra variance will be normally distributed and centered at 0 and will enter simply as increased residual variance at that time point. Greater residual variance in an LGCM implies reduced measurement reliability, which, among others, reduces the power to detect variance in change (Hertzog et al., 2006). Other situations, such as when the STV is not centered on the implied time or is skewed or the trajectory is nonlinear around the implied time, will produce excess variance distributions that may have more wide-reaching impacts.

In all cases, even when STV is centered and normally distributed, we expect a decrease in statistical efficiency due to time binning. This is because there is an upper limit on statistical efficiency, the Cramér–Rao bound, which is less than or equal to the inverse of the Fisher information of the data (Shao, 2003; Cramér, 1946; Rao, 1945). The Fisher information is itself inversely proportional to variance for normal distributions (Casella & Berger, 2001). Because the most common estimation methods for both MLM and SEM asymptotically achieve the Cramér–Rao bound (Daniels, 1961), anything that increases residual variance, such as time binning, will reduce relative efficiency in some model component. This is important not only because we typically want our estimates to be as close as possible to the population value (i.e., unbiased), but we also want them to vary as little as possible if we were to repeat the study (i.e., efficient). Indeed, maximal statistical efficiency is a key assumption for power and thus sample size estimation for future studies.

Despite this theoretical expectation, at least one study (Coulombe et al., 2015) reports having found no change in statistical efficiency, although they, too, concluded that STV leads to overestimated residual variance. Given these theoretically unexpected results, which are also at conflict with unpublished results we have obtained, we conducted a simulation study of statistical efficiency of time-binned data with STV. Data were simulated in R from the model given in Eqs. (2.1)–(2.5) with $W = 5$ measurement occasions. Parameters for the model were taken from typical values found in a comprehensive review of recent LGCM literature in the social sciences (Miller et al., 2023).[8] A sample size of $N = 500$ was simulated with both simultaneous measurement occasions and occasions simulating occasion creep with a triangular distribution skewed to the right. This model of STV, with participants providing data either at the time intended by the researchers or later, is a common assumption in the STV literature (Aydin et al., 2014; Coulombe et al., 2015). Though hardly ever studied, this simulation also models occasion creep by making these delays cumulative. If the intended interval were 30 days, at wave 1 a typical realization of this STV simulation has most participants (30%) measured on the intended day or 1 day later (26%), with a maximum delay of 6 days (the left panel of Fig. 2.4). At each successive wave, the STV distribution begins to approximate a right-skewed normal distribution (the center panel of Fig. 2.4) as it accumulates deviations from the intended intervals, in line with the central limit theorem. By the last occasion, the STV distribution looks very close to normal, though most participants arrive for measurement over a week later than initially intended (the right panel of Fig. 2.4). These new results therefore extend our earlier results (Miller & Ferrer, 2017) by modeling another common pattern of STV. These data were fit to the matching linear LGCM across 5000 replications, and the parameter bias and efficiency was assessed relative to results from the simultaneously measured data (i.e., with zero STV).

We found that, under these conditions, the estimate of the variance of the growth factor (σ_β) was 4.6% less efficient with time-binned STV data than with

[8] $\mu_\alpha = 21, \mu_\beta = 7, \sigma_\alpha = 100, \sigma_\beta = 3, \sigma_{\alpha\beta} = 4.33$ ($\rho = .25$), and ϵ varying over w to maintain a nominal $R^2 = 0.7$ for all observed variables.

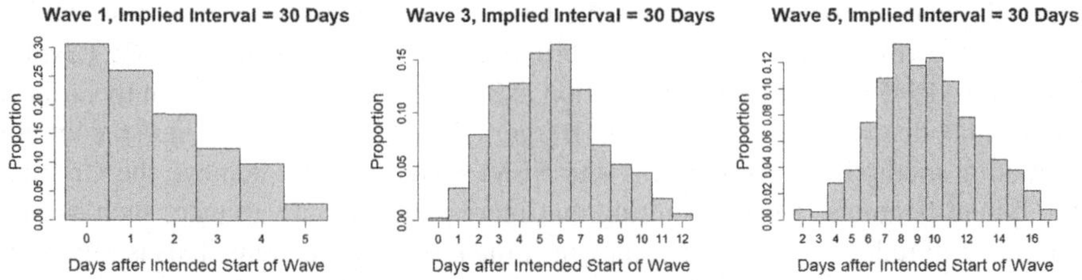

Fig. 2.4 Typical realization of simulated STV, if intended interval is 30 days, at waves 1, 3, and 5

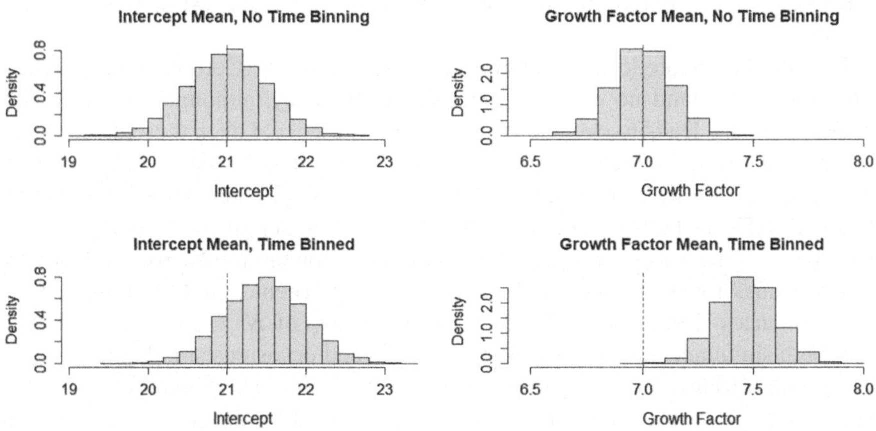

Fig. 2.5 Comparison of distribution of estimates across simulations for μ_α (left) and μ_β (right) for non-time-binned and time-binned data. Dashed vertical line represents the true value

simultaneous data; the estimate of the residual at wave 1 ($\sigma_{\epsilon 1}$) was 1.6% less efficient, and the estimate of the mean of the intercept was 0.6% less efficient. All other estimates had efficiency reductions less than 0.5% or no reduction but were not all unbiased. The estimate of the mean of the intercept (μ_α) was inflated by 2.2% on average (in terms of relative bias; Fig. 2.5, left, compare top and bottom) and the estimate of the mean of the growth factor (μ_β) by 6.7% (Fig. 2.5, right, compare top and bottom). All other estimates were biased by less than 1%. The most concerning finding was the effect on parameter coverage, that is, the proportion of cases in which the 95% confidence interval contained the true value of the parameter, which by definition should be 95% of the replications. However, for the mean of the intercept, the coverage was only 84%; as the graph of the parameter estimates suggests (Fig. 2.5, bottom left), the effect on the mean of the growth factor's coverage was much more severe, with only 6% coverage of the simulated value. Thus, we found bias in estimation, loss in efficiency, and loss of coverage for key LGCM parameters. In particular, any statistical conclusion about change would be contaminated by time binning's effects on the growth factor.

Given previous findings of upward biased residuals (Miller & Ferrer, 2017) or more pervasive patterns of bias (Aydin et al., 2014; Sterba, 2014; Coulombe et al., 2015; Estrada & Ferrer, 2019) and conflicting reports of effects on parameter efficiency (Coulombe et al., 2015), it should be clear that the effects of time binning STV data vary based on STV distribution, experimental design, and other factors. These effects are difficult to predict, at best. Thus, it is clear that time binning should be avoided for LGCMs whenever possible. Yet, rare are the textbooks on LGCM that explicitly discuss STV and its potentially deleterious effects, though remedies exist.

Handling STV in LGCMs

Explicit Approaches

Fortunately, STV can be accounted for as long as it is taken into consideration, preferably from the beginning of data collection. The first step is to record the time of data collection with appropriate precision, whenever possible. This will prevent analysts from sole reliance on an imprecise and/or rounded time variable because nothing better is available. What "appropriate precision" means depends primarily on the intended interval and the rate of change of the outcome, as well as the precision with which the outcome can be measured and magnitude of change that is meaningful—in other words, it is as much a methodological as a substantive matter. As a general rule of thumb, consider retaining precision to either at least 5% of the intended interval or the time in which the outcome is expected to change by no more than 1% or 2%, whichever is more precise. One should avoid any rounding that discards time periods in which the outcome could change more than by a small proportion of what would be a meaningful amount in the intended interval. These rules of thumb are based on our experience with longitudinal data; the exact amount of rounding that is acceptable is an open empirical question that will undoubtedly vary based on the factors we have enumerated and possibly others. This step may result in vastly heterogeneous time values across participants as a result of highly individualized assessment schedules. In extreme scenarios, each participant may have entirely unique times of assessment.

Second, the analyses must use the precise (unrounded) values of time, rather than those of a convenient rounded variable. In the case of the MLM, using individual time information is very simple, as long as it is available. One simply uses the time difference from the intercept as the value for t_w instead of a proxy such as occasion (w). That is, $t_{iw} = \kappa_{iw} - \kappa_0$. MLM software has been computationally optimized to account for sparse matrices, that is, matrices that present few nonzero numbers (corresponding to assessment times) and many zeros (corresponding to nonassessment times) (Bates, 2004).

In SEM, the process is slightly more complicated because of the way that SEMs are mathematically represented for estimation (for details, see Grimm et al., 2017, p. 81–82). One must use what is known as *definition variables*, which replace the

t_w shown in Fig. 2.1 with participant-specific loadings t_{iw} (McArdle & Hamagami, 1996; Mehta & Neale, 2005; Neale et al., 2006). That is, each participant's loadings on the growth factor will have the values of that participant's time of assessment. In the extreme case scenario where each participant has his/her own time of assessment values, each participant will thus have a unique basis vector. However, two major issues hamper the use of definition variables. First, they are not widely implemented in commonly used software; to our knowledge, it is only available in two popular packages at the time of this writing: Mplus (Muthén & Muthén, 2017) (with the `TSCORES` syntax) and OpenMx (Neale et al., 2016). Therefore, there is at least one proprietary and one open-source solution available for researchers with STV data. Second, computation time with highly idiosyncratic values for definition variables is usually very high, transforming even a simpler linear LGCM into a minutes- or hours-long computer job. Clearly, the sparse matrix optimization algorithms implemented in MLM software are not available in SEM software.

Coding Examples for SEMs

For our examples, we use data generated to match our children's reading skills example for 520 participants assessed near ages 10, 11, 13, 15, and 17. Age of assessment includes STV of up to 6 weeks before or after each child's birthday (Fig. 2.2, lower right shows 100 random cases). We include coding samples for Mplus and OpenMx to fit linear, quadratic, and latent-basis models. We also include a model with nonlinear latent parameters, a Gompertz curve model of learning, as such nonlinear models may fit hypotheses better and provide more interpretable parameters (Ghisletta et al., 2020). The data and code can be found at https://osf.io/zmbq8/.

There are several ways to implement linear and quadratic models; we provide two alternatives. The code includes comments noting which performs better and which is more flexible (this is usually a trade-off). The comments also note any potential difficulties fitting the models to the example dataset, including when a specific realization does not provide an admissible solution. However, inadmissible solutions are generally due to the latent variable covariance matrix being nonpositive-definite; there are potential solutions to such problems, (e.g., McNeish & Bauer, 2020) that are beyond the scope of this chapter.

A final caution is warranted here for nonlinear models. These are estimated via Taylor series expansion approximation (Browne & Toit, 1991), which sums linear approximations of the function's shape at the given time points. Because the actual function curves away from the linear approximation at those time points, introducing STV moves data away from those times which could lead to biased estimates, though to what degree this may be a problem is still an open empirical question. This issue compounds already-recognized limitations of these models, such as problems with model convergence (Grimm et al., 2017, p. 304). Therefore, while very useful, nonlinear LGCMs in the SEM framework with STV data should be approached cautiously.

When You Must Use Time-Binned Data

As we have seen, with some datasets—archival or historical data and, in particular, data that have been anonymized to prevent reidentification—researchers simply will not be able to use precise time data in their models, since the data do not exist. In such cases, one must instead mitigate the effects of time binning. To conclude, we provide several suggestions for such circumstances.

Possibly the most important step to take is to center the modeled time points within each wave of the data; for example, if dates are expressed only as year, use the middle of the year as the modeled time point. This will mitigate effects due to skewed STV distributions. In particular, do not model the first wave as the earliest date and the last wave as the latest date possible in the data—at least one study (Estrada & Ferrer, 2019) has found widespread estimation bias due to the inconsistent STV distribution between waves. One should avoid setting constraints (other than nonnegativity) on residual variances, since the effects of STV are likely to appear in bias of these parameters; constraints may force the model to move the bias to another parameter of more substantive interest. This likewise applies to constraining residual variance to be equal across waves; occasion creep may lead to higher residual variance as waves proceed. Consider using a fixed-basis instead of latent-basis model when theory provides a reasonable trajectory shape, since data-driven techniques may be more susceptible to unmodeled sources of error. One should also be cautious in interpreting estimates of residual variance, since they are likely to be inflated and may be inefficient. In all cases, one should report that time binning was necessary despite the data having STV and that residual variance is therefore likely overestimated, power may be lower than prior estimates suggest, especially for the covariance of the intercept and growth factor, and the accuracy of estimates may be affected.

2.4 Conclusion

Research on STV is still in its nascency, and yet some convergence about its deleterious statistical effects have appeared in the literature. We believe there are at least two areas of research that deserve closer attention in relation to STV. First, the few studies on the effects of time binning have not included missing data, an unavoidable feature of "big N, small T" studies. Likewise, the many studies on missing data in SEM have not considered STV. Both phenomena create holes in an otherwise full matrix of data, and they are most likely related. Missing data enjoy the useful and popular classification of Rubin (1974), who contributed immensely to the understanding of their effects. Future research should study STV from the angle of missing data to consider the shared and unique properties of either feature of longitudinal data. Second, nearly nothing is known about the sensitivity and specificity of common SEM fit indices in the presence of STV. Only Aydin et al. (2014) and Miller and Ferrer (2017) addressed these questions. They concluded that only the chi square statistic, with its degrees of freedom, obtained by comparing the tested model to a fully saturated model, does, under certain conditions, reveal misfit due to imprecise analyses of STV. Other commonly used indices were insensitive to time binning, thereby useless to identify negative statistical consequences due to unmodeled STV. Future research ought to focus on fit indices that are capable of detecting misfit caused by time binning when STV variation warrants against it.

As alluded to before, we have found STV to be present also in applications of the LGCM in the MLM literature. A third area of research should investigate how STV affects change estimates obtained via the MLM statistical framework. Given that, under certain conditions, the LGCM is equivalent across both SEM and MLM frameworks (McArdle & Hamagami, 1996), we would expect general results about bias and efficiency to replicate. Yet, other results are bound to be specific, given the particular features of MLM, such as reduced computational burden in the presence of sparse data, which could mean that the handling of STV is more tractable within the MLM framework than in SEM. Second, commonly used fit indices are quite different across the two frameworks. Whereas in SEM there are many indices that reflect the overall quality of the model (e.g., Goodness of Fit Index, Jöreskog and Sörbom, 1988; root mean square error of approximation, Steiger and Lind, 1980, though as noted above, none of these seem to detect time binning), in MLM most indices are geared toward the comparison of alternative models (e.g., Akaike Information Criterion, AIC Akaike, 1973; Bayesian information criterion, Schwarz, 1978), in which case they are not informative about the overall adjustment of a model (Cantoni et al., 2021 reviewed existing MLM fit indices and found that only 6 of 25 allow evaluating the overall quality of a model, without comparing it to an alternative one). Thus, any work aimed at studying how goodness of fit indices are affected by STV, and whether they detect that time binning was unacceptably imprecise, will be for the most part specific to either framework. At this time, we cannot guess which, if any, of the two sets of fit (MLM or SEM specific) indices may show superior performance in the presence of STV.

With increasingly precise research questions being asked of increasingly sensitive measures in the social sciences, the effects of model misspecification that used to be considered ignorable, such as time binning, become more important to address. A first step is to inform, whenever possible, our statistical models about the STV present in our data. Since there are also cases in which working with time-binned data is unavoidable and because current methods of accounting for STV may not precisely account for it in some circumstances, such as when using definition variables in a nonlinear SEM, further research on the effects and mitigation of STV is also warranted.

References

Abshire, M., et al. (2017). Participant retention practices in longitudinal clinical research studies with high retention rates. *BMC Medical Research Methodology, 17*, 30. https://doi.org/10.1186/s12874-017-0310-z.

Aiken, L. S., & West, S. G. (1991). *Multiple regression: Testing and interpreting interaction*. Sage Publications.

Akaike, H. (1973). Information theory as an extension of the maximum likelihood principle. In B. N. Petrov & F. Csáki (Eds.), *2nd International Symposium on Information Theory* (pp. 267–281). Akadémiai Kiadó.

Aydin, B., Leite, W. L., & Algina, J. (2014). The consequences of ignoring variability in measurement occasions within data collection waves in latent growth models. *Multivariate Behavioral Research, 49*, 149–160. https://doi.org/10.1080/00273171.2014.887901.

Baltes, P. B., & Nesselroade, J. R. (1979). History and rationale of longitudinal research. In: J. R. Nesselroade & P. B. Baltes (Eds.), *Longitudinal research in the study of behavior and development* (pp. 1–39). Academic Press.

Bates, D. (2004). Sparse Matrix Representations of Linear Mixed Models. Technical Report. R Development Core Team. https://www.researchgate.net/publication/268413922_Sparse_Matrix_Representations_of_Linear_Mixed_Models.

Biesanz, J. C., et al. (2004). The role of coding time in estimating and interpreting growth curve models. *Psychological Methods, 9*(1), 30–52. https://doi.org/10.1037/1082-989X.9.1.30.

Bollen, K. A., & Curran, P. J. (2006). *Latent curve models: A structural equation perspective*. Wiley.

Browne, M. W., & du Toit, S. H. C. (1991). Models for learning data. In L. M. Collins & J. L. Horn (Eds.) *Best methods for the analysis of change* (pp. 47–68). American Psychological Association.

Bryk, A. S., & Raudenbush, S. W. (1987). Application of hierarchical linear models to assessing change. *Psychological Bulletin, 101*, 147–158. https://doi.org/10.1037/0033-2909.101.1.147.

Cantoni, E., Jacot, N., & Ghisletta, P. (2021). Review and comparison of measures of explained variation and model selection in linear mixed-effects models. *Econometrics and Statistics*. Advance online publication. https://doi.org/10.1016/j.ecosta.2021.05.005.

Casella, G., & Berger, R. L. (2001). *Statistical inference*, 2nd edn. Cengage Learning.

Coulombe, P., Selig, J. P., & Delaney, H. D. (2015). Ignoring individual differences in times of assessment in growth curve modeling. *International Journal of Behavioral Development, 40*(1), 76–86. https://doi.org/10.1177/0165025415577684.

Cramér, H. (1946). *Mathematical methods of statistics*. Princeton University Press.

Curran, P. J. (2003). Have multilevel models been structural equation models all along? *Multivariate Behavioral Research, 38*, 529–569. https://doi.org/10.1207/s15327906mbr3804_5.

Daniels, H. E. (1961). *The asymptotic efficiency of the maximum likelihood estimator*. In J. Neyman (Ed.) (pp. 151–163).

Davies, K. et al. (2014). Improving retention of very old participants in longitudinal research: Experiences from the Newcastle 85+ study. *PLoS ONE, 9*(10), e108370. https://doi.org/10.1371/journal.pone.0108370.

Estrada, E., & Ferrer, E. (2019). Studying developmental processes in accelerated cohort-sequential designs with discrete-and continuous-time latent change score models. *Psychological Methods, 24*, 708–734. https://doi.org/10.1037/met0000215.

Ferris, E. B., et al. (2021). Recruitment and retention strategies for community based longitudinal studies in diverse urban neighborhoods. *JMIR Formative Research, 5*(3), e18591. https://doi.org/10.2196/18591.

Ghisletta, P., et al. (2020). On the use of growth models to study normal cognitive aging. *International Journal of Behavioral Development, 44*, 88–96. https://doi.org/10.1177/0165025419851576.

Grimm, K. J., Ram, N., & Estabrook, R. (2017). *Growth modeling: Structural equation and multilevel modeling approaches*. Guilford Press.

Hertzog, C., et al. (2006). On the power of multivariate latent growth curve models to detect correlated change. *Psychological Methods, 11*, 244–252. https://doi.org/10.1037/1082-989X.11.3.244.

Jöreskog, K. G., & Sörbom, D. (1988). *LISREL 7: A guide to the program and applications*. SPSS.

Kline, R. B. (2016). *Principles and practice of structural equation modeling*, 4th. edn. Guilford Press.
Lindenberger, U., et al. (2011). Cross-sectional age variance extraction: What's change got to do with it? *Psychology and Aging, 26*(1), 34–47. https://doi.org/10.1037/a0020525.
McArdle, J. J. (1986). Latent variable growth within behavior genetic models. *Behavior Genetics, 16*, 163–200. https://doi.org/10.1007/BF01065485.
McArdle, J. J., & Hamagami, F. (1996). Multilevel models from a multiple group structural equation perspective. In G. A. Marcoulides & R. E. Schumacker (Eds.), *Advanced structural equation modeling* (pp. 89–124). Psychology Press.
McNeish, D., & Bauer, D. J. (2020). Reducing incidence of nonpositive definite covariance matrices in mixed effects models. *Multivariate Behavioral Research, 57*, 318–340. https://doi.org/10.1080/00273171.2020.1830019.
Mehta, P. D., & Neale, M. C. (2005). People are variables, too: Multilevel structural equations modeling. *Psychological Methods, 10*(3), 259–284. https://doi.org/10.1037/1082-989X.10.3.259.
Mehta, P. D., & West, S. G. (2000). Putting the individual back into individual growth curves. *Psychological Methods, 5*, 23–43. https://doi.org/10.1037/1082-989x.5.1.23.
Meredith, W., & Tisak, J. (1990). Latent curve analysis. *Psychometrika, 55*(1), 107–122. https://doi.org/10.1007/BF02294746.
Miller, M. L., & Ferrer, E. (2017). The effect of sampling-time variation on latent growth curve models. *Structural Equation Modeling: A Multidisciplinary Journal, 24*, 831–854. https://doi.org/10.1080/10705511.2017.1346476.
Miller, M. L., Ferrer, E., & Ghisletta, P. (2023). The current practice of latent growth curve modeling in the social and behavioral sciences: Observations and recommendations. https://doi.org/10.31234/osf.io/9c2ew, https://psyarxiv.com/9c2ew.
Miller, M. L., Ghisletta, P., et al. (2021). Changes in cerebral arterial pulsatility and hippocampal volume: A transcranial Doppler ultrasonography study. *Neurobiology of Aging, 108*, 110–121. https://doi.org/10.1016/j.neurobiolaging.2021.08.014.
Muthén, L. K., & Muthén, B. O. (2017). *Mplus user's guide*, 8th edn. Muthén & Muthén.
Neale, M. C., Boker, S. M., et al. (2006). *Mx: Statistical modeling*, 7th edn. Virginia Commonwealth University, Department of Psychiatry.
Neale, M. C., Hunter, M. D., et al. (2016). OpenMx 2.0: Extended structural equation and statistical modeling. *Psychometrika, 81*(2), 535–549. https://doi.org/10.1007/s11336-014-9435-8.
Nicholson, L. M., et al. (2011). Recruitment and retention strategies in longitudinal clinical studies with low-income populations. *Contemporary Clinical Trials, 32*(3), 353–362. https://doi.org/10.1016/j.cct.2011.01.007.
Rao, C. R. (1945). Information and accuracy attainable in the estimation of statistical parameters. *Bulletin of the Calcutta Mathematical Society, 37*(3), 81–91.
Rocher, L., Hendrickx, J. M., & de Montjoye, Y.-A. (2019). Estimating the success of re-identifications in incomplete datasets using generative models. *Nature Communications, 10*(1), 3069. https://doi.org/10.1038/s41467-019-10933-3.
Rogosa, D. R., & Willett, J. B. (1985). Understanding correlates of change by modeling individual differences in growth. *Psychometrika, 50*, 203–228. https://doi.org/10.1007/BF02294247.
Rovine, M. J., Molenaar, P. C. M. (1998). The covariance between level and shape in the latent growth curve model with estimated basis vector coefficients. *Methods of Psychological Research Online, 3*(2), 95–107.
Rubin, D. B. (1974). Estimating causal effects of treatments in randomized and nonrandomized studies. *Journal of Educational Psychology, 66*(5), 688–701. https://doi.org/10.1037/h0037350.
Schwarz, G. (1978). Estimating the dimension of a model. *Annals of Statistics, 6*(2), 461–464. https://doi.org/10.1214/aos/1176344136.
Shao, J. (2003). *Mathematical statistics*, 2nd edn. Springer Science+Business Media.
Singer, J. D., & Willett, J. B. (2003). *Applied longitudinal data analysis: Modeling change and event occurrence*. Oxford University Press, Inc.

Steiger, J. H., & Lind, J. M. (1980). Statistically based tests for the number of common factors. *Annual Meeting of the Psychometric Society*. Iowa City, IA.

Sterba, S. K. (2014). Fitting nonlinear latent growth curve models with individually varying time points. *Structural Equation Modeling: A Multidisciplinary Journal, 21*, 630–647. https://doi.org/10.1080/10705511.2014.919828.

Voelkle, M. C., et al. (2018). The role of time in the quest for understanding psychological mechanisms. *Multivariate Behavioral Research, 53*, 782–805. https://doi.org/10.1080/00273171.2018.1496813.

Wainer, H. (2000). The centercept: An estimable and meaningful regression parameter. *Psychological Science, 11*(5), 434–436. https://doi.org/10.1111/1467-9280.00284.

Chapter 3
Score-Guided Recursive Partitioning of Continuous-Time Structural Equation Models

Manuel Arnold, Pablo F. Cáncer, Eduardo Estrada, and Manuel C. Voelkle

3.1 Introduction

In this work, we combine continuous-time dynamic panel models (e.g., Oud & Jansen, 2000; Voelkle et al., 2012) with model-based recursive partitioning (e.g., Zeileis et al., 2008) and provide a tutorial using data from the Survey of Health, Ageing and Retirement in Europe (Börsch-Supan et al., 2013). Furthermore, we extend earlier work put forward by Brandmaier et al. (2018) on the recursive partitioning of continuous-time dynamic panel models by leveraging a family of score-based tests. These tests were recently used by Arnold et al. (2021b) to guide the recursive partitioning of structural equation models (SEMs; Bollen, 1989). The resulting approach offers a computationally efficient data-driven way to identify predictors of group differences in continuous-time dynamic panel model parameters.

Supplementary Information The online version contains supplementary material available at https://doi.org/10.1007/978-3-031-56318-8_3.

M. Arnold (✉) · M. C. Voelkle
Department of Psychology, Humboldt-Universität zu Berlin, Berlin, Germany
e-mail: manuel.arnold@hu-berlin.de; manuel.voelkle@hu-berlin.de

P. F. Cáncer
Department of Social Psychology and Methodology, Universidad Autónoma de Madrid, Madrid, Spain

Department of Psychology, Universidad Pontificia Comillas, Madrid, Spain

E. Estrada
Department of Social Psychology and Methodology, Universidad Autónoma de Madrid, Madrid, Spain
e-mail: eduardo.estrada@uam.es

M. Stemmler et al. (eds.), *Dependent Data in Social Sciences Research*,
https://doi.org/10.1007/978-3-031-56318-8_3

Panel studies, in which the same individuals are repeatedly observed at multiple time points, are among the most popular longitudinal designs in psychological research. Especially when experimental designs are infeasible, panel designs offer ways to investigate causal relationships from observational data. To this end, researchers typically rely on panel models such as the cross-lagged panel model or one of its many derivatives (see Usami et al., 2019; Zyphur et al., 2020a, b for an overview). Common features of these models are regression parameters that quantify reciprocal effects and temporal stability of repeatedly observed variables. In psychological research and social sciences, panel models are often specified and estimated within the SEM framework.

An important distinguishing feature of panel models is the treatment of time. Arguably, most panel models in psychological research are discrete-time (DT) models. These models consider time implicitly by leveraging the temporal ordering inherent to longitudinal data and directly estimate the relationship between a measurement and the following one. Importantly, DT models require the time interval between all measurements to be the same. More specifically, DT models yield an average estimate of the regression strength between consecutive time points. These averaged effects are useful when all time intervals are constant between measurements but may be incorrect for some or even all measurement occasions when the time intervals differ. For instance, consider a panel study where the first three waves of data were collected with a time interval of 1 year. Additionally, a follow-up measurement was gathered 3 years after wave three. Fitting a DT panel model on this particular data would lead to averaged regression parameters that underestimate the strength of the relation between the first three waves and overestimate the relation between wave three and the follow-up wave. A more detailed discussion about the effects of unequally spaced measurement occasions is provided by Oud and Delsing (2010) and Voelkle et al. (2012).

Continuous-time (CT) models (Oud & Jansen, 2000; Voelkle et al., 2012) overcome the limitations of DT models and account for differences in time intervals through the use of stochastic differential equations. In a nutshell, CT models do not directly estimate the relations between consecutive measurements. Instead, they employ non-linear functions that take CT parameters and the time interval as arguments, where the CT parameters represent the time-independent regression strength, which is then damped or amplified by the time interval. Another important advantage of the CT model stems from the time-independent nature of the CT parameters, which can be used to interpolate or extrapolate the strength of relations between variables for any time interval. Thus, CT models facilitate the comparison of results from studies with different sampling intervals. Until recently, one possible hurdle that prevented the more widespread use of CT models is that they are mathematically more intricate than DT models. However, rapid development in different software packages, such as the **ctsem** package (Driver et al., 2017; Driver & Voelkle, 2018) for the statistical software language R (R Core Team, 2022), has made CT modeling much more accessible.

Already more than 20 years ago, Oud and Jansen (2000) put forward a CT model with mixed effects that contained so-called trait variables, which accounted for stable differences between individuals (see also Hamaker et al., 2015 for a discussion of trait variables for DT models). Moreover, Oud and Jansen (2000) also discussed the inclusion of covariates (which they called input variables) to accommodate different mean trajectories. More recently, Driver and Voelkle (2018) presented a hierarchical Bayesian approach to estimate CT models, allowing for individual variation in all types of model parameters. Although these mixed-effects and hierarchical Bayesian approaches account for individual differences, detecting different kinds of heterogeneity and finding the most important predictor variables remains challenging, particularly in large data sets.

Usually, panel data does not only consist of the data that is of core interest when fitting the model but also various other covariates, such as demographic variables, socio-economic variables, or other background information. Many of these covariates may explain individual differences, and their effects can be explicitly tested by including them in the model. However, hypotheses about predictors of heterogeneity are often vague. So, how should a researcher proceed if there are many potentially relevant covariates? Which covariates should be included in the model, and which role (e.g., moderators, predictors of mean differences, interaction terms) should these covariates play?

One answer to these questions is model-based recursive partitioning (Strobl et al., 2009; Zeileis et al., 2008), which was originally introduced in machine learning. Model-based recursive partitioning is a data-driven way to hierarchically split data into homogeneous groups by recursively selecting predictors of group differences from a potentially large set of candidate variables. The resulting segmented models can be visualized and effectively illustrate the split decisions made by the algorithm. Because these visualizations often resemble a tree-like shape, recursive partitioning is also known as the decision tree paradigm. In the psychological literature, recursive partitioning of SEMs is referred to as SEM trees and is almost exclusively performed with the R package **semtree** (Brandmaier et al., 2013).

Brandmaier et al. (2018) already combined CT models and SEM trees. However, their approach was limited in terms of practicality due to an excessive runtime and suboptimally implemented covariate testing. More particularly, they relied on a likelihood-ratio-based procedure, originally suggested by Brandmaier et al. (2013), that involved different ad hoc adjustments to correct for multiple testing. As Arnold et al. (2021b) pointed out, this implementation either favored the selection of certain variable types over others or severely lacked statistical power to detect heterogeneity. To solve these issues, Arnold et al. (2021b) replaced the ad hoc corrections with the correct distribution of the likelihood-ratio tests, which allows an unbiased covariate selection based on likelihood ratios. More importantly, they also implemented an alternative covariate testing procedure based on a family of score-based tests, following the unified model-based recursive partitioning framework put forward by Zeileis et al. (2008). Score-based tests (Merkle et al., 2014; Merkle & Zeileis, 2013) monitor fluctuations in case-wise derivatives of the log-likelihood function to detect parameter heterogeneity. Compared to likelihood-ratio tests, score-based tests require no model estimation to evaluate covariates, which makes them very appealing for recursive partitioning. Arnold et al. (2021b) implemented score-guided covariate testing in the **semtree** package. Moreover, they showed that SEM trees guided by score-based tests typically compute in a fraction of the time needed by likelihood-ratio-guided trees and offer unbiased covariate selection.

In the present chapter, we make use of these advantages and present score-guided continuous-time structural equation model (CTSEM) trees. The remainder is structured as follows: Sections 3.2 and 3.3 give an overview of CT models and score-guided model-based recursive partitioning, respectively. Section 3.4 illustrates the score-guided CTSEM trees by exploring the dynamics of life satisfaction between married couples using data from the SHARE study. We conclude this chapter by discussing the recommendations and limitations of our approach in Sect. 3.5.

3.2 Continuous-Time Models

The following section outlines the SEM specification of CT models, focusing on the interpretation of the different CT model parameters and software implementations. More thorough introductions are provided by Oud and Delsing (2010) and Voelkle et al. (2012).

The CT models discussed in this work are closely related to the more well-known DT models, such as the cross-lagged panel model. Here, we denote the time interval of individual i between two consecutive measurements that occur at t_j and t_{j-1} with $\Delta t_{i,j}$. The index j denotes the measurement occasion, and t the time point of the measurement. Following Voelkle et al. (2012), we can express the DT model for individual i at the time point t_j as a function of CT parameters and the time interval:

$$\begin{aligned}\mathbf{y}\left(t_{i,j}\right) = e^{\mathbf{A}\Delta t_{i,j}}\mathbf{y}\left(t_{i,j} - \Delta t_{i,j}\right) + \left(\mathbf{I} - e^{\mathbf{A}\Delta t_{i,j}}\right)\boldsymbol{\xi}_i + \mathbf{A}^{-1}\left(e^{\mathbf{A}\Delta t_{i,j}} - \mathbf{I}\right)\mathbf{b} \\ + \mathbf{w}\left(\Delta t_{i,j}\right)\end{aligned} \tag{3.1}$$

In the above equation, $\mathbf{y}(t_{i,j})$ is the measurement of m variables of primary scientific interest from individual i assessed at time point t_j. $e^{(\bullet)}$ represents the matrix exponential. The $m \times m$ matrix $\mathbf{A}$ denotes the so-called drift matrix that characterizes the relationships among the variables of interest. Auto effects are on the main diagonal of the drift matrix, and cross effects are placed on the off-diagonal positions. $\mathbf{I}$ is a $m \times m$ identity matrix. The m-variate random vector $\boldsymbol{\xi}_i$ indicates the temporally stable deviation of individual i from the group means, and its elements are often referred to as random intercepts or trait variables. The random vector $\boldsymbol{\xi}_i$ is assumed to be multivariate, normally distributed with mean of zero and covariance matrix $\boldsymbol{\Phi}_{\boldsymbol{\xi}}$. The m-variate vector $\mathbf{b}$ contains the CT intercepts. Finally, $\mathbf{w}(\Delta t_{i,j})$ is the DT prediction error term. It is a random variable and assumed to be multivariate normally distributed with mean zero and DT covariance matrix

$$\text{irow}\left[\mathbf{A}_{\#}^{-1}\left(e^{\mathbf{A}_{\#}\Delta t_{i,j}}-\mathbf{I}\right)\text{row}\left(\mathbf{Q}\right)\right], \tag{3.2}$$

where $\mathbf{A}_{\#} = \mathbf{A} \otimes \mathbf{I} + \mathbf{I} \otimes \mathbf{A}$ and $\otimes$ symbolizes the Kronecker product. The symmetric matrix $\mathbf{Q}$ of dimensionality $m \times m$ contains the so-called diffusion parameters that describe the covariance of the prediction errors in CT. The operator row puts the elements of the diffusion matrix row-wise into a column vector, while irow performs the inverse operation and stacks the elements back into a matrix.

For readers unfamiliar with CT models, Eq. (3.1) may look daunting at first, and the link to DT models might not be very apparent. To illustrate the close relationship, let us assume that the time intervals across all individuals and measurements are exactly one, that is, $\Delta t_{i,j} = 1 \ \forall \ i, j$. Moreover, let us assume we fitted a bivariate model (where two variables of interest were measured repeatedly) and estimated the following values for the drift and intercept parameters:

$$\mathbf{A} = \begin{bmatrix} -0.713 & 0.203 \\ 0.203 & -0.713 \end{bmatrix}, \quad \mathbf{b} = \begin{bmatrix} 0 \\ 0 \end{bmatrix} \tag{3.3}$$

By plugging the estimates and the time interval into Eq. (3.1), we obtain the following results (note that we applied some rounding):

$$\mathbf{y}\left(t_{i,j}\right) = \begin{bmatrix} 0.5 & 0.1 \\ 0.1 & 0.5 \end{bmatrix} \mathbf{y}\left(t_{i,j}-1\right) + \begin{bmatrix} 0.5 & 0.9 \\ 0.9 & 0.5 \end{bmatrix} \boldsymbol{\xi}_i + \mathbf{w}(1) \tag{3.4}$$

The model in Eq. (3.4) resembles a conventional cross-lagged panel model with random intercepts much more closely. In fact, since the time intervals are constant, we could just as well have estimated a DT model and would have obtained exactly the same parameters as in Eq. (3.4). However, with the help of the CT parameters in Eq. (3.3), we can calculate the corresponding DT parameters not only for equally spaced time intervals but also for any arbitrary time intervals. For example, solving the matrix exponential for a time interval of 3 instead of 1 yields the following DT autoregressive and cross-lagged coefficients:

$$e^{\begin{bmatrix} -0.713 & 0.203 \\ 0.203 & -0.713 \end{bmatrix} \cdot 3} = \begin{bmatrix} 0.14 & 0.076 \\ 0.076 & 0.14 \end{bmatrix} \tag{3.5}$$

By comparing the DT autoregressive and cross-lagged parameters in Eqs. (3.4) and (3.5), we find that the longer time interval of 3 led to considerably smaller autoregressive and cross-lagged effects, indicating that the strength of relationships shrinks as more time passes. In summary, parameters provided by CT models can be transformed into DT parameters for any given time interval. This relationship not only makes CT parameters interpretable in a familiar way but also makes CT models highly useful as they allow to inter- or extrapolate to time intervals of interest that need not be observed.

The CT model introduced in Eq. (3.1) can be extended in various ways. A straightforward extension, which is a central part of most SEMs, is the inclusion of a measurement part that allows to include latent variables and accounts for measurement errors in the observed variables. The measurement part is defined as

$$\widetilde{\mathbf{y}}\left(t_{i,j}\right) = \mathbf{\Lambda}\boldsymbol{\eta}\left(t_{i,j}\right) + \boldsymbol{\varepsilon}\left(t_{i,j}\right), \tag{3.6}$$

where $\widetilde{\mathbf{y}}\left(t_{i,j}\right)$ is a k-variate vector with the manifest or directly observed variables measured at the time point $t_{i,j}$. The manifest variables are related to the latent factors contained in the m-variate vector $\boldsymbol{\eta}(t_{i,j})$, weighted by the factor loadings in the $k \times m$ matrix $\mathbf{\Lambda}$, plus the corresponding measurement errors in the k-variate vector $\boldsymbol{\varepsilon}(t_{i,j})$. The measurement errors are assumed to be independent and identically multivariate normally distributed with mean zero and $k \times k$ covariance matrix $\mathbf{\Psi}$. If the CT model includes a measurement part, the latent variables $\boldsymbol{\eta}$ in Eq. (3.6) replace the manifest variables $\mathbf{y}$ in Eq. (3.1). While cross-sectional SEMs typically require each latent variable to be measured by multiple manifest indicators, panel models can separate measurement errors from dynamic variability using a single indicator (cf. Schuurman et al., 2015). In these cases, the factor loading matrix $\mathbf{\Lambda}$ simply reduces to an identity matrix of order m. Other model extensions involve accounting for different trajectories between individuals by adding exogenous predictor variables to the model (cf. Oud & Delsing, 2010; Oud & Jansen, 2000). Due to a lack of space, we will not discuss predictor variables in the context of this chapter. Instead, we will turn to model-based recursive partitioning as an alternative in the next section.

Specification and estimation of CT models are usually performed with specialized software packages such as the **ctsemOMX** (Driver et al., 2017) and the **ctsem** (Driver & Voelkle, 2018) R packages. The **ctsemOMX** package uses a mixed-effects approach and interfaces the **OpenMx** R package (Neale et al., 2016) for model estimation. In contrast, the **ctsem** package is based on the Stan software (Carpenter et al., 2017) and supersedes **ctsemOMX** in large parts. Among other things, **ctsem** offers a fully hierarchical model formulation that allows all model parameters to differ across individuals and provides diverse estimation options, including maximum likelihood, maximum a posteriori, and Bayesian estimation based on Hamiltonian Monte Carlo sampling. Despite these enhancements, we chose to rely on the **ctsemOMX** package for model estimation since it outputs OpenMx-type models, which are supported by the **semtree** package we used for model-based recursive partitioning.

3.3 SEM Trees

Recursive Partitioning

SEM trees are an instance of model-based recursive partitioning, which is a variant of recursive partitioning used to approach prediction and classification problems. Recursive partitioning was first pioneered in the seminal work on automated interaction detection (AID) by Morgan and Sonquist (1963) and was later popularized in the form of the classification and regression trees (CART; Breiman et al., 2017) and the C4.5 (Quinlan, 1986, 1993) algorithms. Recursive partitioning offers a simple nonparametric approach to regression problems by recursively selecting predictor variables that split a sample into individuals with similar values of a response variable. A key feature of this technique is that it produces decision trees that are easy to visualize and interpret.

In the context of recursive partitioning, decision trees are tree-shaped structures depicting the split decisions of a tree made to partition the sample hierarchically. For the sake of simplicity, we will limit this exposition to binary decision trees because these are the types of decision trees produced by the SEM tree algorithm. Binary decision trees consist of nodes, where each node has either no successor (or children) and is called a leaf or has exactly two successors and is called an inner node. The first node of the tree (which has no parent) is called the root node and is typically placed at the top of the tree. Each inner node of a tree corresponds to a split decision. Each split decision involves a covariate which is used to determine what subsets will be associated with its children.

The fact that recursive partitioning often leads to large and thus hard-to-interpret trees (e.g., Chan & Loh, 2004) motivated research into model-based recursive partitioning. Model-based recursive partitioning incorporates parametric models into the nodes of the trees and aims to identify groups of individuals with similar values of the model parameters. Important early algorithms are M5 (Quinlan, 1993), QUEST (Loh & Shih, 1997), CRUISE (Kim & Loh, 2001), GUIDE (Loh, 2002), and LOTUS (Chan & Loh, 2004). Zeileis et al. (2008) proposed a general framework for model-based recursive partitioning that overcomes problems due to a biased variable selection (i.e., favoring the selection of variables with many unique values over the selection of variables with few unique values) and an extensive runtime. Their approach uses so-called structural change tests (e.g., Zeileis, 2005; Zeileis & Hornik, 2007) that monitor the fluctuations in individual contributions to the fitting function (also known as scores) to test potential covariates (see Zeileis & Hornik, 2007). The framework of Zeileis et al. (2008) was implemented in the R packages **party** (Zeileis et al., 2008) and **partykit** (Hothorn & Zeileis, 2015) that provide a wide array of tools for recursive partitioning.

SEM trees and the associated R package **semtree** put forward by Brandmaier et al. (2013) offer specialized tools for the recursive partitioning of SEMs. Brandmaier et al. described SEM trees as a hybrid approach that combines the confirmatory SEM framework with the exploratory decision tree paradigm. Thus, SEM trees allow users to specify relationships between observed and latent variables based on prior hypotheses while simultaneously exploring heterogeneity. Initially, SEM trees used a covariate testing procedure built upon likelihood ratios and applied various correction procedures to prevent a biased variable selection. However, Arnold et al. (2021b) demonstrated that the correction measures were either insufficient or diminished the statistical power to detect heterogeneity. As a remedy, Arnold et al. presented score-guided SEM trees following the general model-based recursive partitioning framework by Zeileis et al. (2008). Score-guided SEM trees offer unbiased variable selection and high statistical power and compute much more quickly than the original SEM trees. In addition to the score-guided trees, Arnold et al. (2021b) also implemented a new unbiased likelihood-ratio-based covariate testing procedure.

The Score-Guided SEM Tree Algorithm

The score-guided SEM tree algorithm can be described in four steps:

1. Specify a template SEM.
2. Fit the template SEM on all observations in the current node.
3. Assess whether the SEM parameters are constant or vary with respect to a covariate.
4. Choose the covariate that is associated with the largest group difference. If the group differences reach statistical significance (given a predefined significance level), split the sample associated with the current node into two subsamples, and repeat the procedure beginning with Step 2 in both child nodes. Otherwise, terminate.

We briefly outline the above steps in the following, and note some particularities for fitting score-guided CTSEM trees with the **semtree** package. A general and more detailed account of score-guided SEM trees is given by Arnold et al. (2021b).

Step 1: Template Model

The first step of fitting a SEM tree is the specification of a template SEM. The template model represents the confirmatory part of a SEM tree and reflects hypotheses about the research question. It determines which relationships among observed variables and latent constructs are to be estimated. The template model can be specified with general-purpose SEM packages such as **laavan** (Rosseel, 2012) or **OpenMx**. CT models can be specified with the **ctsemOMx** package.

By default, all free parameters contribute to split decisions and are estimated in every node of the tree. However, this behavior can be adjusted if desired. **semtree** offers the option to specify so-called focus parameters. By declaring certain parameters as focus parameters, the SEM tree will only consider heterogeneity in these parameters when testing covariates. Thus, focus parameters are useful in situations where heterogeneity in some parameters is more relevant than in others. Another option is the use of global parameter constraints that set the values of some freely estimated parameters to equality. An estimate of a globally constrained parameter is obtained once using the complete sample and then used in all nodes of the tree. Thus, global constraints can reduce the complexity of a tree, facilitating estimation and interpretation. However, erroneously constraining a parameter to equality that is actually different across groups will lead to a severely misspecified tree (cf. Arnold et al., 2021b).

Step 2: Model Estimation

Model estimation within SEM trees follows the standard estimation techniques described in the SEM literature (e.g., Bollen, 1989; Kline, 2016). In fact, the **semtree** package interfaces to the widely used **lavaan** or **OpenMx** packages for model estimation, depending on which software was used to specify the template model. In the case of CT models, which were specified with the help of the **ctsemOMX** package, model estimation is performed using **OpenMx** by minimizing the following log-likelihood function for multivariate normally distributed data:

$$\sum_{i=1}^{n} \ln L\left(\boldsymbol{\theta}; \mathcal{Y}_{\mathbf{i}}, \mathbf{u}_i\right) = -\frac{1}{2}\sum_{i=1}^{n} \left\{ \begin{matrix} \left[\mathcal{Y}_{\mathbf{i}} - \boldsymbol{\mu}\left(\boldsymbol{\theta}, \mathbf{u}_i\right)\right]^{\mathrm{T}} \boldsymbol{\Sigma}(\boldsymbol{\theta}, \mathbf{u}_i)^{-1} \left[\mathcal{Y}_{\mathbf{i}} - \boldsymbol{\mu}\left(\boldsymbol{\theta}, \mathbf{u}_i\right)\right] + \\ \ln\left[\left|\boldsymbol{\Sigma}\left(\boldsymbol{\theta}, \mathbf{u}_i\right)\right|\right] + o_i k \ln\left(2\pi\right) \end{matrix} \right\} \tag{3.7}$$

Here, $\boldsymbol{\theta}$ is an l-variate vector containing the parameters of the CT model. $\mathcal{Y}_{\mathbf{i}}$ represents the repeated measurements of the manifest variables $\tilde{\mathbf{y}}$ of individual i that are stacked into a vector of size $o_i \cdot k$, where o_i symbolizes the number of measurements of individual i. $\mathbf{u}_i$ is o_i-variate vector with the measurement occasions of individual i. $\boldsymbol{\mu}(\boldsymbol{\theta}, \mathbf{u}_i)$ and $\boldsymbol{\Sigma}(\boldsymbol{\theta}, \mathbf{u}_i)$ denote the $o_i \cdot k$-variate model-implied mean vector and the $o_i \cdot k \times o_i \cdot k$ model-implied covariance matrix of the repeated measurements $\mathcal{Y}_{\mathbf{i}}$, respectively. Note that the model-implied means and covariance are functions of the model parameters as well as the time intervals between measurements. Thus, the model-implied means and covariances will vary when individuals are measured with different time intervals (cf. Oud & Jansen, 2000). Finally, n represents the number of individuals in the current node. Maximizing Eq. (3.7) yields the maximum likelihood parameter estimates $\hat{\boldsymbol{\theta}}$.

Step 3: Covariate Testing

Score-guided SEM trees apply a family of score-based tests to detect heterogeneity. These tests originated in econometrics, better known as structural change tests (e.g., Zeileis, 2005; Zeileis & Hornik, 2007), and were popularized in psychometrics in the past decade (Merkle et al., 2014; e.g., Merkle & Zeileis, 2013; Wang et al., 2014). More specifically, score-based tests are used to test the following null hypothesis of invariant parameters

$$H_0: \quad \boldsymbol{\theta}_i = \boldsymbol{\theta}_0, \quad i = 1, \ldots, n, \tag{3.8}$$

where $\boldsymbol{\theta}_i$ denotes a vector of model parameters that potentially vary between n individuals of the (sub-) sample associated with the current node. In other words, rejecting the null hypothesis implies that the individuals in the current node do not follow an identical distribution and cannot be represented by a single set of parameter values $\boldsymbol{\theta}_0$.

Score-based tests analyze the individual or casewise derivatives of the log-likelihood with respect to the model parameters:

$$\mathbf{s}\left(\hat{\boldsymbol{\theta}}; \mathcal{Y}_i, \mathbf{u}_i\right) = \left[\frac{\partial \ \ln L\left(\hat{\boldsymbol{\theta}}; \mathcal{Y}_i, \mathbf{u}_i\right)}{\partial \hat{\theta}_1}, \ldots, \frac{\partial \ \ln L\left(\hat{\boldsymbol{\theta}}; \mathcal{Y}_i, \mathbf{u}_i\right)}{\partial \hat{\theta}_l}\right]^{\mathrm{T}} \tag{3.9}$$

These derivatives, also called (maximum likelihood) scores, capture the sensitivity of the log-likelihood function associated with individual i to infinitesimal changes in the model parameters. They indicate how well the model parameters represent the corresponding individuals. Intuitively speaking, the scores are similar to residuals in regression models, where individuals closer to the regression line are associated with smaller residuals than individuals further away. Likewise, small absolute score values indicate a good fit between an individual and the model parameters, whereas large absolute scores generally point to misfit. In homogeneous samples, the scores fluctuate randomly around zero, much like residuals of a correctly specified regression model. However, when the sample is heterogeneous, the variability in the scores is not entirely random anymore but is also due to a systematic part that reflects group differences. For example, let us assume that there is a group difference between young and old individuals. This group difference would then result in scores that tend to be negative for younger individuals and positive for older individuals or vice versa.

In practice, the scores can either be obtained analytically or numerically. The implementation of Arnold et al. (2021b) computes the scores analytically using a fast and efficient method put forward by von Oertzen and Brick (2014). Unfortunately, this method assumes that the SEM is linear in parameters, which is not the case for CT models due to the matrix exponential function and the parameter constraints in Eq. (3.2). To solve this issue, we implemented a new method into the **semtree** package that numerically approximates the derivatives of the matrix exponential function and derives all other parts analytically. We found that this hybrid approach was much faster than a purely numerical approach using **OpenMx**'s `imxRowGradients` function.

To construct a test statistic to test the null hypothesis in Eq. (3.8), we first order the scores according to the covariate whose effect is being tested and then cumulate the sorted scores. The resulting l-variate cumulative score process is formally defined as

$$\mathbf{CSP}\left(\hat{\boldsymbol{\theta}}; s\right) = \frac{1}{\sqrt{n}}\mathbf{I}\left(\hat{\boldsymbol{\theta}}\right)^{-1/2}\sum_{h=1}^{s}\mathbf{S}\left(\hat{\boldsymbol{\theta}}; \mathcal{Y}_h, \mathbf{u}_h\right), \tag{3.10}$$

where the index s denotes the number of sorted individuals entering the equation and the h selects the individuals until $h = s$. For example, suppose we want to assess parameter heterogeneity with respect to the covariate age. For $s = 1$, only the scores of the youngest individual would be used to compute the cumulative score process. For $s = 2$, the process is calculated by summing the scores of the youngest and second youngest individuals and so forth. The $l \times l$ matrix $\mathbf{I}\left(\hat{\boldsymbol{\theta}}\right)^{-1/2}$ is the root of the inverse of the Fisher information matrix evaluated at the parameter estimates. This matrix decorrelates the scores of the l parameter estimates.

Hjort and Koning (2002) show that under the assumption of homogeneous or invariant parameters, the fluctuation in the cumulative score process follows a Brownian bridge. This result allows us to calculate p-values and critical values for a formal test of parameter homogeneity by comparing the observed cumulative score process to its theoretical limiting distribution. Multiple test statistics for different levels of measurements of the covariates have been proposed. For example, Merkle and Zeileis (2013) suggest test statistics for continuous covariates, and Merkle et al. (2014) put forward further statistics for ordinal and categorical covariates. Detailed instructions on how these test statistics can be applied in practice are given by Wang et al. (2014). Although no single test statistic will always work best in any situation, we limit this presentation to the maximum Lagrange multiplier (maxLM) test statistic, which is currently the default approach in the **semtree** package for the evaluation of continuous covariates.

The maxLM test statistic is given by

$$maxLM = \max_{s=\underline{s},\ldots,\overline{s}}\left\{\left[\frac{s}{n}\left(1-\frac{s}{n}\right)\right]^{-1}\sum_{r=1}^{l}\mathbf{CSP}\left(\hat{\boldsymbol{\theta}}; s\right)_r^2\right\}, \tag{3.11}$$

where $\mathbf{CSP}\left(\hat{\boldsymbol{\theta}}; s\right)_r$ refers to the r-th parameter of the cumulative score process derived from the scores of the sorted first s-th individuals. The statistic selects the maximum of the sums of the squared scores of all parameters. In other words, the statistic searches for the point in the cumulative score process with the most pronounced heterogeneity and disregards heterogeneity in other points. Additionally, maxLM contains a scaling term $\left[\frac{s}{n}\left(1-\frac{s}{n}\right)\right]^{-1}$, which gives equal weight to all individuals. A disadvantage of this scaling term is that one needs to specify an interval $\left[\underline{s}, \overline{s}\right]$ of the covariate values considered, meaning that parameter differences outside these boundaries are not taken into account. For additional information on the maxLM and alternative test statistics, see Merkle and Zeileis (2013).

In addition to obtaining p-values and critical values, the maxLM test statistic also gives grounds to locate split points in covariates by selecting the value of the covariate that is associated with the maximum. Importantly, split point selection is one main difference between score-guided SEM trees and the general model-based recursive partitioning framework by Zeileis et al. (2008), who recommend an exhaustive split search. While an exhaustive split search will provide the most accurate split points, the maxLM-based approach used to grow score-guided SEM trees is computationally less demanding and performed well in a simulation study, as reported by Arnold et al. (2021b).

Step 4: Covariate Selection

The covariate testing procedure outlined in Step 3 is performed for every covariate provided to the SEM tree. The tree picks the covariate associated with the smallest p-value, and if this p-value is smaller than a pre-specified value, the sample in the current node is partitioned. Then, the tree continues to split the child nodes recursively until it finds no more significant splits. Besides hypothesis tests, the **semtree** package also offers additional stopping criteria. For instance, users can specify a minimum number of individuals per node.

The evaluation of multiple covariates increases the risk of a type I error. The **semtree** package allows the user to apply the Bonferroni procedure to adjust p-values. However, given many covariates, the Bonferroni procedure will drastically reduce the tree's power and may produce sparse trees. Therefore, we recommend working with unadjusted p-values and being mindful that the number of covariates may increase the number of splits.

3.4 Demonstration of Score-Guided CTSEM Trees

To highlight the usage and utility of score-guided CTSEM trees, we predicted heterogeneity in the dynamics of self-reported quality of life of married couples in older age, using data from the SHARE study (Börsch-Supan et al., 2013). Supplementary material, including the R code with the analyses and instructions for data preparation, can be found in the Electronic Supplementary Material 3.1.

SHARE Subsample

The SHARE is a large pan-European panel study monitoring public health and socio-economic living conditions. Since 2004, the SHARE has been administered yearly, and by 2022, it will include data from roughly 140,000 individuals aged 50 or older. For demonstration purposes, we selected a subsample of 1037 married, opposite-sex couples that participated in six waves. The time intervals between measurements varied substantially, warranting the use of a CT model. The mean interval was 2.57 years with a standard deviation of 0.88 years. Our primary outcome of interest was the self-reported quality of life, measured with a 12-item version of the longer control, autonomy, self-actualization, and pleasure (CASP) scale originally consisting of 19 items (Hyde et al., 2003). We aggregated the scores from the 12 items into a single mean. Furthermore, we used a total of ten different covariates to explore heterogeneity in the quality of life. A summary of all covariates is given in Table 3.1. For ease of demonstration, we used only complete cases. However, SEM trees generally allow missing values in the data used for model estimation, although it requires complete data for the covariates.

Table 3.1 Covariates used in the SEM tree

Covariate	Description	Variably type
`age_diffabs`	Absolute age differences between spouses	Continuous
`nchild`	Number of children	Count
`hhsize_mn`	Number of people living in the households	Count
`isced_diffabs`	Absolute education level difference between spouses	Ordinal (7 levels)
`politics_diffabs`	Absolute difference in political views from left- to right-leaning between spouses	Continuous
`spousenet_F`	Is the male spouse part of the female spouse's social network?	Dummy
`spousenet_M`	Is the female spouse part of the male spouse's social network?	Dummy
`snsatis_diffabs`	The absolute difference between female and male spouses' satisfaction with their social networks	Continuous
`propfriends_F`	The proportion of friends in the female spouse's social network, excluding acquaintances and family members	Proportion
`propfriends_M`	The proportion of friends in the male spouse's social network, excluding acquaintances and family members	Proportion

Note: The values of all covariates except `isced_diffabs`, `politics_diffabs`, and `snsatis_diffabs` were taken from the first wave. `isced_diffabs` indicated the mode and `politics_diffabs` and `snsatis_diffabs` the mean across all waves. `propfriends_M` and `propfriends_F` were measured as the number of friends divided by the total members of one's social network

CT Model

The first step of fitting a CTSEM tree involves the specification and estimation of a template model. We modeled the dynamics in the self-reported quality of life between spouses with a bivariate CT model, where one process represented the life satisfaction of the female spouse and the other life satisfaction of the male spouse. In this model, the cross-lagged effects describe how changes in the life satisfaction of one spouse influence the life satisfaction of the other spouse over time.

We started the search for an appropriate template model by fitting a bivariate CT model containing random intercepts and measurement error variance. Unfortunately, this model did not converge. The estimation problems were not surprising since they are also frequent for the DT counterpart of our model, which is the stable autoregressive trait and state model (Kenny & Zautra, 2002). Therefore, we ignored measurement errors for the sake of convergence. The resulting model is similar to the model in Eq. (3.1).

We specified the template model using the following `ctModel` function call from the **ctsemOMx** package:

```
m <- ctModel(manifestNames = c("CASP_M", "CASP_F"),
                                   latentNames = c("M", "F"),
            Tpoints = 6, LAMBDA = diag(2),
            CINT = matrix(c("cint_M", "cint_F"), nrow = 2, ncol = 1),
            MANIFESTMEANS = matrix(0, nrow = 2, ncol = 1),
            MANIFESTVAR = matrix(0, nrow = 2, ncol = 2),
            TRAITVAR = "auto")
```

The `manifestNames` argument requires the name of the observed variables as they are called in the data set. Additional names for the latent processes can be set via the `latentNames` argument. The argument `Tpoints` requires the number of measurement occasions as an input. The remaining arguments are used to specify the parameters of the CT models, which is done by providing a single character (e.g., `"auto"`) or a matrix to `ctModel`. If this matrix contains characters (e.g., `"cint_M"`), the corresponding parameters are estimated freely, while numeric values (e.g., 0) fix the corresponding parameter to these values. In the above function call, we requested the estimation of CT intercepts via the CINT arguments. We fixed the means and variances of the manifest variables to zero via the `MANIFESTMEANS` and `MANIFESTVAR` arguments, respectively. Additionally, we requested the estimation of random intercepts of the latent processes by setting the TRAITVAR argument to `"auto."` Also, not explicitly specified, `ctModel` automatically adds drift and diffusion parameters to the model.

Next, we estimated the specified model using the ctFit function:

```
fit <- ctFit(dat = df, ctmodelobj = m, stationary = "all",
            retryattempts = 25)
```

We provided the `ctFit` function with our data set named `df` in the wide format using the `dat` argument (see Driver et al., 2017, for instructions on preparing the data). The CT model `m` that we specified in the previous step is supplied to `ctFit` via

Table 3.2 Parameter estimates of the template model

Parameter	Value	Matrix	SE	Parameter	Value	Matrix	SE
drift_M	−0.762	DRIFT	0.056	cint_M	−0.013	CINT	0.093
drift_F_M	0.166	DRIFT	0.055	cint_F	0.000	CINT	0.106
drift_M_F	0.096	DRIFT	0.051	traitvar_M	15.717	TRAITVAR	0.826
drift_F	−0.820	DRIFT	0.063	traitvar_F_M	12.409	TRAITVAR	0.768
diff_M	18.098	DIFFUSION	1.012	traitvar_F	19.397	TRAITVAR	0.986
diff_F_M	4.409	DIFFUSION	0.758				
diff_F	19.189	DIFFUSION	1.149				

Note: The parameter labels were taken directly from the output provided by **ctsemOMX**. `drift` drift parameter, `diff` diffusion parameter, `cint` CT intercept, `traitvar` variance/covariance parameter of the random intercepts. For drift parameters, `M` and `F` indicate the auto-effects of male and female spouses. `F_M` marks a cross-effect of the male spouse on the female spouse, and `M_F` is the cross-effect of the female spouse on the male spouse. For the variance/covariance parameters, `M` and `F` denote variances and `F_M` marks covariances

the `ctmodelobj` argument. The two remaining arguments are optional, and we used them in order to stabilize model estimation within the CTSEM tree. `stationary` set to `"all"` means that the initial variance and mean parameters were not estimated but constrained to ensure stationary variance for this process, avoiding potential overfitting issues. The argument `retryattempts` determines the number of estimation attempts, which we increased from 5 to 25.

Table 3.2 presents the estimated parameter values of the template model. Using the `plot` function, we can visualize the strength of the autoregressive and the standardized cross-lagged effects over time, as depicted in Fig. 3.1. Standardization of the cross-lagged effects makes them comparable and is achieved through multiplication by the ratios of the asymptotic variance of the repeatedly measured variables (Oud, 2002). Apparently, there is little to no cross-lagged effect over time. In other words, self-reported quality of life seems to be rather independent between married couples in the complete sample.

CTSEM Trees

To investigate whether there are subsamples with a more pronounced coupling between wife and male spouse, we fit the CTSEM tree in the following. Tree construction is performed by passing the template model, the data set, a list of the names of the covariates, and some options to the `semtree` function of the **semtree** package. We will first discuss the different options and meta-parameters guiding the tree-creation process.

Most options and meta-parameters of a SEM tree can be set by creating a control object via the `semtree.control` function. The `method` argument allows users to choose between different covariate testing procedures explained by Brandmaier et al. (2013) and Arnold et al. (2021b). To make use of the score-guided covariate test-

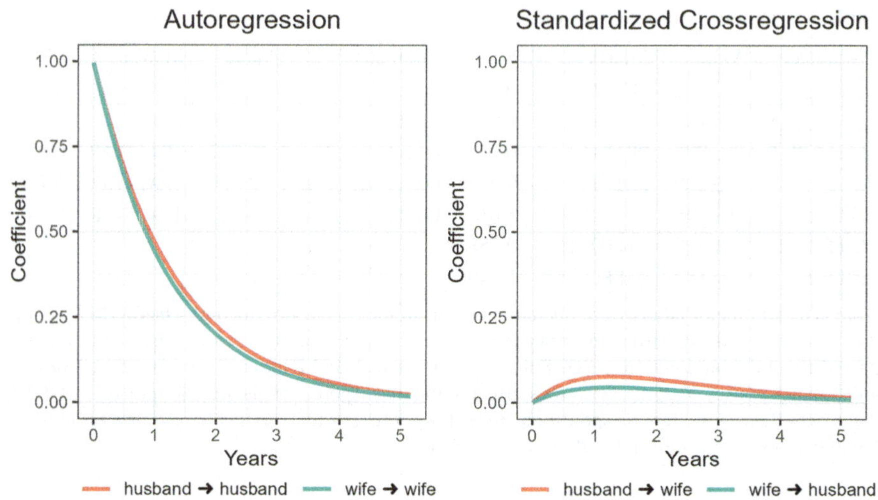

Fig. 3.1 Autoregressive and standardized cross-lagged effects as functions of the time interval between measurements

ing approach, we set `method` to "`score`." Other arguments of `semtree.control` can be used to limit the growth of the tree. The `min.N` argument sets the minimum of cases required to be in a node to continue splitting. Similarly, the `min.bucket` argument determines the minimum sample size of a node. `min.N` and `min.bucket` are very useful arguments to avoid estimating trees where the models in the leaf nodes are fitted on too few individuals and become hard to interpret. The argument `max.depth` limits the depth of the tree, that is, the length of the longest path from a root to a leaf. Finally, setting `bonferroni` to `TRUE` applies a Bonferroni correction to adjust the *p*-values at each split to control for multiple testing. Using a Bonferroni correction guarantees that the number of unnecessary splits (or type I errors) will not exceed 5% of the total splits on average. However, given many covariates, the tree will become increasingly conservative and will likely stop splitting prematurely. Instead of limiting tree growth in advance, users can also reduce the maximum depth of a fitted tree by using the `prune` function. In the present demonstration, we used the following `semtree.control` object:

```
ctrl <- semtree.control(method = "score", max.depth = 2,
                        min.bucket = 100)
```

Other options can be selected similarly as with the `semtree.control` function via the `semtree.constraints` function. `semtree.constraints` allows users to declare focus parameters (`focus.parameters` arguments) and set parameters to equality across the tree (`global.invariance` argument). Especially the use of focus parameters may prove helpful as heterogeneity in parameters that are less important to one's research interest can be ignored. However, for the sake of brevity, we will not use the `semtree.constraints` function in this demonstration.

Finally, we invoke the `semtree` function to fit the CTSEM tree:

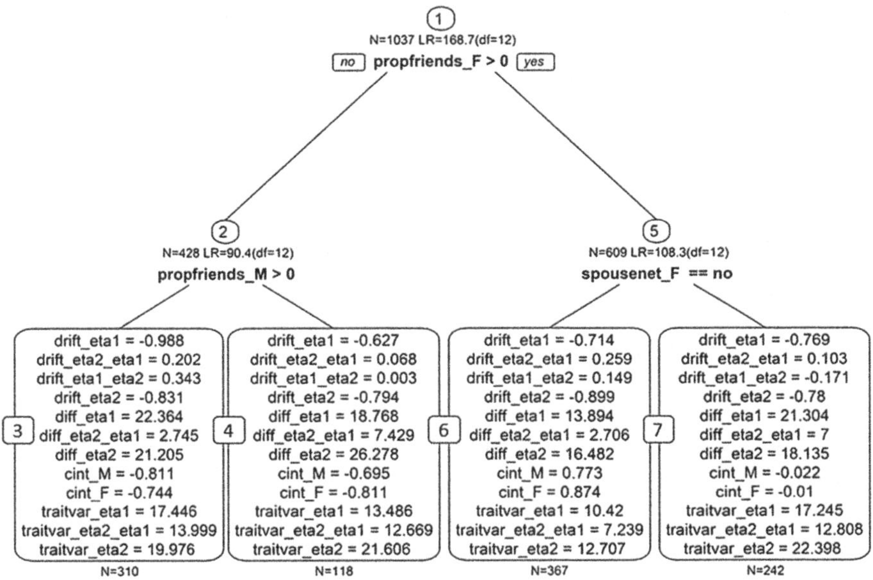

Fig. 3.2 CTSEM tree modeling self-reported quality of life in married couples

```
tree <- semtree(model = fit, data = df_all, control = ctrl,
                predictors = covariates)
```

We provided the `semtree` with the template model using the model argument. Under `data`, we supplied the tree with the data used to fit the CT model plus all covariates. The `predictor` argument requires a character vector with all the names of the covariates as in the data frame `df_all`. Finally, we provided `semtree` with the `semtree.control` object `ctrl`.

The resulting SEM tree can be either printed to the R console or visualized with the `plot` function, as depicted in Fig. 3.2. We found that the tree selected three covariates with information about the couples' social networks to split the sample. The covariates `propfriends_F` and `propfriends_M` measured the proportions of friends in the social circle of the female spouse and male spouse, respectively. The covariate `spousenet_F` was a dummy variable indicating whether the male spouse belonged to the female spouse's social network. At the root (Node 1), the tree chose to split the sample into 428 couples where no friends were part of the female spouse's social network (Node 2) and 609 couples where the female's spouse social network included friends (Node 5). Conditional on the first split, the subsample with couples with no friends in the female spouse's social network (Node 2) was further split into 310 couples where also the male spouse's social network included no friends (Node 3) and 118 couples with friends in the male spouse's social network (Node 4). Finally, the subsample with couples where the female spouse's social network included friends (Node 5) was further split into 367 couples where the male spouse was part of the female spouse's social network (Node 6) and 242 couples

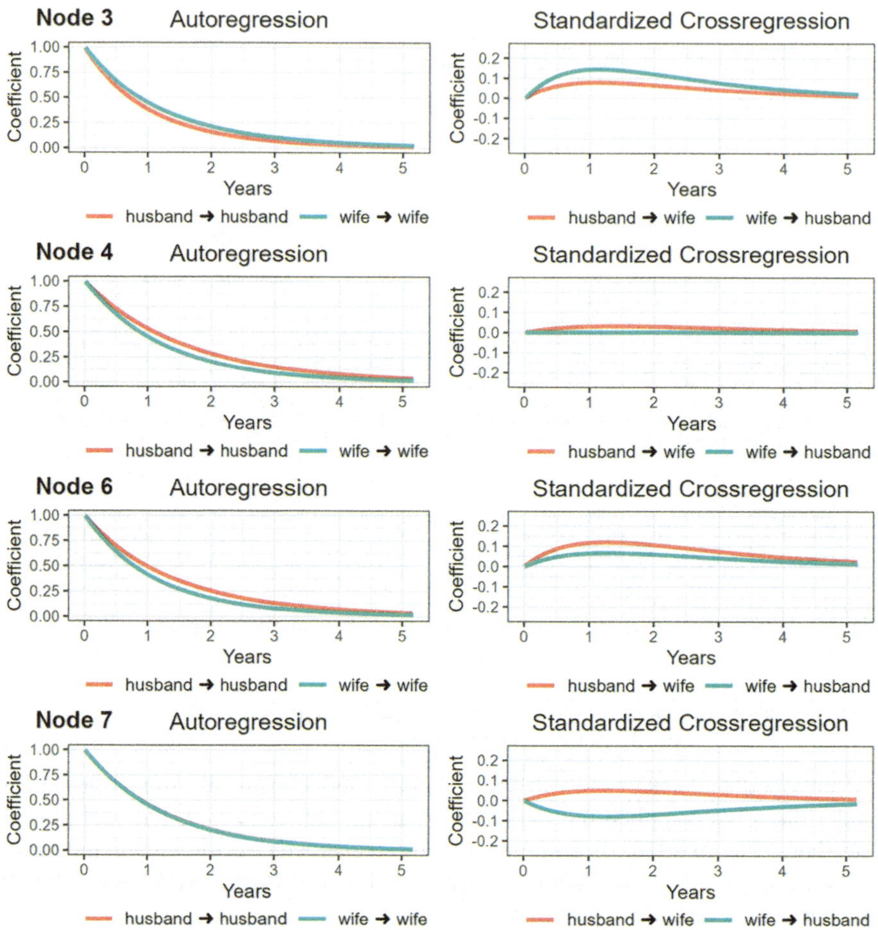

Fig. 3.3 Autoregressive and cross-lagged effects as functions of the time interval between measurements separated for the leaf nodes. *Note*: The scale limits of the *y*-axis differ between plots visualizing autoregression and crossregression

where the male spouse was not (Node 7). The tree stopped after these splits because we limited the maximum depth of the tree to two. However, Nodes 3, 6, and 7 possibly contained enough couples to give grounds for further splits, which could identify other important predictors of differences in the self-reported quality of life.

Figure 3.2 also shows the parameter values of the CT models associated with the leaf nodes. In R, the parameters of a tree can also be assessed via the `coef` and `parameter` functions. While `coef` outputs a list with the parameters of all leafs, `parameter` outputs a table with parameter estimates of all nodes of the tree.

Figure 3.3 visualizes the strength of the autoregressive and cross-lagged effects over time, separated for all subsamples associated with the leafs. As can be seen

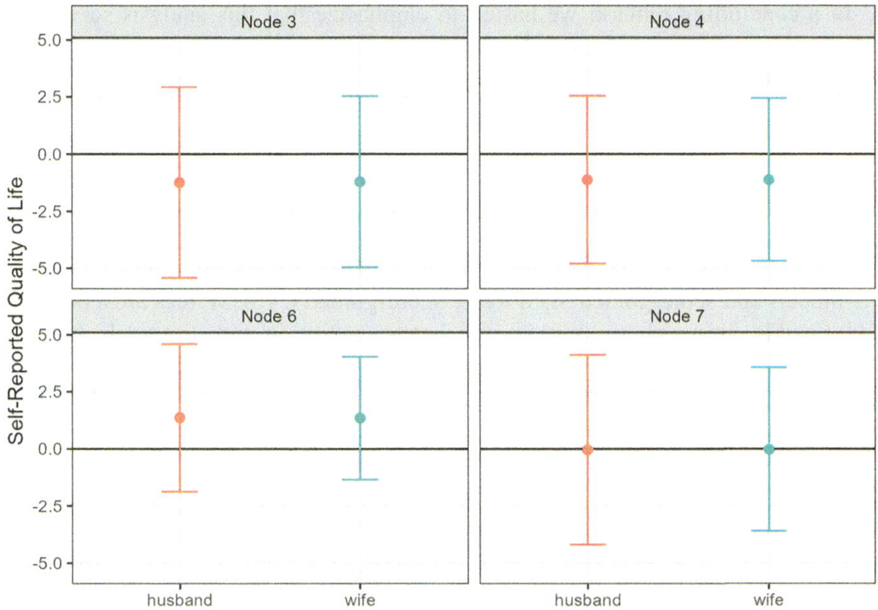

Fig. 3.4 Group means and individual differences separated for the leaf nodes. *Note*: The dots mark the means, and the error bars indicate one positive and negative standard deviation of the random effects, approximately covering 68% of the individual deviations from the group average

from the figure, there was hardly any visible heterogeneity in the autoregressive effects across subsamples. However, the cross-effects differed slightly. While couples associated with Node 4, in which friends were only part of the male spouse's social network, followed dynamically independent courses, couples in the other subsamples affected each other in their quality of life ratings. The strength of the cross-effects generally peaked after 1 year. In couples with no friends in the social circle (Node 3), a positive deviation from the female spouse's expected ratings predicted a following positive deviation from the male spouse's expected rating. This pattern was reversed in couples where the female spouse's social circle included friends and the male spouse (Node 6). Interestingly, in couples where the female spouse's social circle included friends but not the male spouse (Node 7), positive deviations from the female spouse's ratings predicted a decrease in the self-reported quality of life by the male spouse.

Besides differences in the dynamics, we also found heterogeneity in the mean of self-reported quality of life, as shown in Fig. 3.4. Individuals living in marriages where there were no friends in the female spouse's social network (Nodes 3 and 4) were, on average, unhappier than couples where the social network of the female spouse included friends (Nodes 6 and 7). Interestingly, the highest quality of life was reported by couples where the female spouse's social network included not only friends but also the male spouse (Node 6).

In a concluding remark, we hasten to emphasize that this analysis served primarily illustrative purposes and that further research is needed before the resulting patterns can be interpreted substantially.

3.5 Discussion

In the present chapter, we presented scored-guided CTSEM trees, which combine CT models and score-guided SEM trees. Score-guided CTSEM trees are a powerful and flexible approach to account for heterogeneity in time intervals between measurement occasions and heterogeneity in parameters of CT models. Score-guided CTSEM trees address the problem of variable selection by identifying important covariates that predict group differences from a potentially large set of candidate variables and provide interpretable output, such as a visualization of the fitted decision tree. Our approach advances the pioneering work by Brandmaier et al. (2018), who put forward likelihood-based recursive partitioning of CT models. However, the earlier approach used a suboptimal covariate testing procedure that led to biased trees and was held back by an excessive runtime (see Arnold et al., 2021a). Our score-guided CTSEM trees overcome these obstacles, making recursive partitioning of CT models possible even on regular desktop computers or notebooks.

We want to use the remainder of this chapter to point out some weaknesses of our implementation of score-guided CTSEM trees and point toward future developments. We will start with some well-known general problems of tree methods (see Strobl et al., 2009 for more details): First, trees are often unstable; that is, even small changes in the data may produce a very different tree. For instance, the more or less arbitrary choice between two covariates of almost equal importance (or two equally important cutpoints on the same covariate) propagates through the tree, potentially leading to very different structures. Because of this instability, predictions of single trees often show high variability. Second, trees cannot find conditional effects of two covariates when these variables do not affect the outcome unconditionally (in other words, when there is an interaction but no main effects). Finally, although trees can, in theory, approximate any non-linear associations between covariates and outcomes with a set of dichotomous splits, they are inefficient in estimating truly linear relationships between a covariate and the outcome. Fortunately, all these problems are mitigated with random forests (Breiman, 2001). Random forests are ensemble methods consisting of multiple trees (usually up to several thousand) fitted on resampled data. In the context of SEMs, random forests were popularized by Brandmaier et al. (2016). Although random forests usually outperform trees in terms of predictive power (e.g., Fernández-Delgado et al., 2014), they are harder to visualize and interpret than trees. Thus, researchers derive metrics such as variable importance, which ranks the impact of each covariate on the outcomes. Because random forests are based on trees and are typically computationally intensive, they will profit dramatically from the score-guided CTSEM tree approach described in this chapter.

Score-guided CTSEM trees and forests are not always the most appropriate tool to explore heterogeneity. First and foremost, the performance of trees and forests depends on the covariates available. The tree will overlook even substantial subgroup differences if none of these covariates are related to heterogeneity in the CT model parameters. In these situations, researchers may resort to finite mixture models (e.g., Jedidi et al., 1997; Lubke & Muthén, 2005) as an alternative approach that can detect heterogeneity without relying on covariates. If the question of how a covariate affects model parameters is of primary interest, the information about the relationships provided by trees is often too sparse. In this case, researchers may find the individual parameter contribution framework (Arnold et al., 2020, 2021a) more fruitful than recursive partitioning. Individual parameter contribution regression allows modeling individual differences in model parameters as functions of covariates and thus can estimate much more fine-grained relationships between covariates and outcomes than trees.

There are several promising avenues for future research and improvements. On the CT modeling side, the performance and flexibility of the score-guided CTSEM trees could be improved by implementing a new interface to the **ctsem** package. Our implementation is based on the older **ctsemOMX** package since both **ctsemOMX** and **semtree** interface to the **OpenMx** package for model estimation. However, **ctsemOMX** is superseded in terms of functionality by the newer **ctsem** package. Using **ctsem** instead of **ctsemOMX** would open up the use of Bayesian methods and optimizers from the Stan software. On the score-guided tree side, the testing of covariates could be improved in small samples. Our score-guided approach is based on the model-based recursive partitioning framework by Zeileis et al. (2008). Recently, Huth et al. (2022) proposed a new permutation approach that estimates the sampling distribution of the score-based test statistic through resampling the dataset. This permutation approach avoids distributional assumptions leading to improved power but comes at the cost of increased runtime. Finally, the use of score-guided CTSEM forests should be systematically investigated as forests shore up some crucial weaknesses of the trees. Besides these limitations and room for improvement, we are convinced that score-guided CTSEM trees have great potential for the investigation of heterogeneity in panel data.

References

Arnold, M., Oberski, D. L., Brandmaier, A. M., & Voelkle, M. C. (2020). Identifying heterogeneity in dynamic panel models with individual parameter contribution regression. *Structural Equation Modeling: A Multidisciplinary Journal, 27*(4), 613–628. https://doi.org/10.1080/10705511.2019.1667240

Arnold, M., Brandmaier, A. M., & Voelkle, M. C. (2021a). Predicting differences in model parameters with individual parameter contribution regression using the R package ipcr. *Psych, 3*(3), 360–385. https://doi.org/10.3390/psych3030027

Arnold, M., Voelkle, M. C., & Brandmaier, A. M. (2021b). Score-guided structural equation model trees. *Frontiers in Psychology, 11*, 564403. https://doi.org/10.3389/fpsyg.2020.564403

Bollen, K. A. (1989). *Structural equations with latent variables*. Wiley. https://doi.org/10.1002/9781118619179

Börsch-Supan, A., Brandt, M., Hunkler, C., Kneip, T., Korbmacher, J., Malter, F., Schaan, B., Stuck, S., & Zuber, S. (2013). Data resource profile: The survey of health, ageing and retirement in Europe (SHARE). *International Journal of Epidemiology, 42*(4), 992–1001. https://doi.org/10.1093/ije/dyt088

Brandmaier, A. M., von Oertzen, T., McArdle, J. J., & Lindenberger, U. (2013). Structural equation model trees. *Psychological Methods, 18*(1), 71–86. https://doi.org/10.1037/a0030001

Brandmaier, A. M., Prindle, J. J., McArdle, J. J., & Lindenberger, U. (2016). Theory-guided exploration with structural equation model forests. *Psychological Methods, 21*(4), 566–582. https://doi.org/10.1037/met0000090

Brandmaier, A. M., Driver, C. C., & Voelkle, M. C. (2018). Recursive partitioning in continuous time analysis. In K. van Montfort, J. H. Oud, & M. C. Voelkle (Eds.), *Continuous time modeling in the behavioral and related sciences* (pp. 273–298). Springer International Publishing.

Breiman, L. (2001). Random forests. *Machine Learning, 45*(1), 5–32. https://doi.org/10.1023/A:1010933404324

Breiman, L., Friedman, J. H., Olshen, R. A., & Stone, C. J. (2017). *Classification and regression trees*. Routledge. https://doi.org/10.1201/9781315139470

Carpenter, B., Gelman, A., Hoffman, M. D., Lee, D., Goodrich, B., Betancourt, M., Brubaker, M., Guo, J., Li, P., & Riddell, A. (2017). Stan: A probabilistic programming language. *Journal of Statistical Software, 76*(1), 1–32. https://doi.org/10.18637/jss.v076.i01

Chan, K.-Y., & Loh, W.-Y. (2004). LOTUS: An algorithm for building accurate and comprehensible logistic regression trees. *Journal of Computational and Graphical Statistics, 13*(4), 826–852. https://doi.org/10.1198/106186004X13064

Driver, C. C., & Voelkle, M. C. (2018). Hierarchical Bayesian continuous time dynamic modeling. *Psychological Methods, 23*(4), 774–799. https://doi.org/10.1037/met0000168

Driver, C. C., Oud, J. H. L., & Voelkle, M. C. (2017). Continuous time structural equation modeling with R package ctsem. *Journal of Statistical Software, 77*(5), 1–35. https://doi.org/10.18637/jss.v077.i05

Fernández-Delgado, M., Cernadas, E., Barro, S., & Amorim, D. (2014). Do we need hundreds of classifiers to solve real world classification problems? *Journal of Machine Learning Research, 15*, 3133–3181.

Hamaker, E. L., Kuiper, R. M., & Grasman, R. P. P. P. (2015). A critique of the cross-lagged panel model. *Psychological Methods, 20*(1), 102–116. https://doi.org/10.1037/a0038889

Hjort, N. L., & Koning, A. (2002). Tests for constancy of model parameters over time. *Journal of Nonparametric Statistics, 14*(1–2), 113–132. https://doi.org/10.1080/10485250211394

Hothorn, T., & Zeileis, A. (2015). partykit: A modular toolkit for recursive partytioning in R. *Journal of Machine Learning Research, 16*(118), 3905–3909.

Huth, K. B. S., Waldorp, L. J., Luigjes, J., Goudriaan, A. E., van Holst, R. J., & Marsman, M. (2022). A note on the structural change test in highly parameterized psychometric models. *Psychometrika, 87*(3), 1064–1080. https://doi.org/10.1007/s11336-021-09834-6

Hyde, M., Wiggins, R. D., Higgs, P., & Blane, D. B. (2003). A measure of quality of life in early old age: The theory, development and properties of a needs satisfaction model (CASP-19). *Aging & Mental Health, 7*(3), 186–194. https://doi.org/10.1080/1360786031000101157

Jedidi, K., Jagpal, H. S., & DeSarbo, W. S. (1997). Finite-mixture structural equation models for response-based segmentation and unobserved heterogeneity. *Marketing Science, 16*(1), 39–59. https://doi.org/10.1287/mksc.16.1.39

Kenny, D. A., & Zautra, A. (2002). Trait-state models for longitudinal data. In L. M. Collins & A. G. Sayer (Eds.), *Decade of behavior: 2002: 1. New methods for the analysis of change* (pp. 243–263). American Psychological Association.

Kim, H., & Loh, W.-Y. (2001). Classification trees with unbiased multiway splits. *Journal of the American Statistical Association, 96*(454), 589–604. https://doi.org/10.1198/016214501753168271

Kline, R. B. (2016). *Principles and practice of structural equation modeling* (Methodology in the social sciences: 2016) (Vol. 3, 4th ed.). The Guilford Press.
Loh, W.-Y. (2002). Regression trees with unbiased variable selection and interaction detection. *Statistica Sinica, 12*(2), 361–368.
Loh, W.-Y., & Shih, Y.-S. (1997). Split selection methods for classification trees. *Statistica Sinica, 7*(4), 815–840. http://www.jstor.org/stable/24306157
Lubke, G. H., & Muthén, B. (2005). Investigating population heterogeneity with factor mixture models. *Psychological Methods, 10*(1), 21–39. https://doi.org/10.1037/1082-989X.10.1.21
Merkle, E. C., & Zeileis, A. (2013). Tests of measurement invariance without subgroups: A generalization of classical methods. *Psychometrika, 78*(1), 59–82. https://doi.org/10.1007/s11336-012-9302-4
Merkle, E. C., Fan, J., & Zeileis, A. (2014). Testing for measurement invariance with respect to an ordinal variable. *Psychometrika, 79*(4), 569–584. https://doi.org/10.1007/s11336-013-9376-7
Morgan, J. N., & Sonquist, J. A. (1963). Problems in the analysis of survey data, and a proposal. *Journal of the American Statistical Association, 58*(302), 415–434. https://doi.org/10.1080/01621459.1963.10500855
Neale, M. C., Hunter, M. D., Pritikin, J. N., Zahery, M., Brick, T. R., Kirkpatrick, R. M., Estabrook, R., Bates, T. C., Maes, H. H., & Boker, S. M. (2016). Openmx 2.0: Extended structural equation and statistical modeling. *Psychometrika, 81*(2), 535–549. https://doi.org/10.1007/s11336-014-9435-8
Oud, J. H. L. (2002). Continuous time modeling of the cross-lagged panel design. *Kwantitatieve Methoden, 69*, 1–26.
Oud, J. H. L., & Delsing, M. J. M. H. (2010). Continuous time modeling of panel data by means of SEM. In K. van Montfort, J. H. Oud, & A. Satorra (Eds.), *Longitudinal research with latent variables* (pp. 201–244). Springer. https://doi.org/10.1007/978-3-642-11760-2_7
Oud, J. H. L., & Jansen, R. A. R. G. (2000). Continuous time state space modeling of panel data by means of sem. *Psychometrika, 65*(2), 199–215. https://doi.org/10.1007/BF02294374
Quinlan, J. R. (1986). Induction of decision trees. *Machine Learning, 1*(1), 81–106. https://doi.org/10.1007/BF00116251
Quinlan, J. R. (1993). *C4.5: Programs for machine learning* (The Morgan Kaufmann Series in Machine Learning). Morgan Kaufmann Publishers.
R Core Team. (2022). *R: A language and environment for statistical computing* [Computer software]. R Foundation for Statistical Computing, Vienna, Austria.
Rosseel, Y. (2012). lavaan: An R package for structural equation modeling. *Journal of Statistical Software, 48*(2), 1–36. https://doi.org/10.18637/jss.v048.i02
Schuurman, N. K., Houtveen, J. H., & Hamaker, E. L. (2015). Incorporating measurement error in $n = 1$ psychological autoregressive modeling. *Frontiers in Psychology, 6*, 1038. https://doi.org/10.3389/fpsyg.2015.01038
Strobl, C., Malley, J., & Tutz, G. (2009). An introduction to recursive partitioning: Rationale, application, and characteristics of classification and regression trees, bagging, and random forests. *Psychological Methods, 14*(4), 323–348. https://doi.org/10.1037/a0016973
Usami, S., Murayama, K., & Hamaker, E. L. (2019). A unified framework of longitudinal models to examine reciprocal relations. *Psychological Methods, 24*(5), 637–657. https://doi.org/10.1037/met0000210
Voelkle, M. C., Oud, J. H. L., Davidov, E., & Schmidt, P. (2012). An SEM approach to continuous time modeling of panel data: Relating authoritarianism and anomia. *Psychological Methods, 17*(2), 176–192. https://doi.org/10.1037/a0027543
von Oertzen, T., & Brick, T. R. (2014). Efficient Hessian computation using sparse matrix derivatives in RAM notation. *Behavior Research Methods, 46*(2), 385–395. https://doi.org/10.3758/s13428-013-0384-4
Wang, T., Merkle, E. C., & Zeileis, A. (2014). Score-based tests of measurement invariance: Use in practice. *Frontiers in Psychology, 5*, 438. https://doi.org/10.3389/fpsyg.2014.00438

Zeileis, A. (2005). A unified approach to structural change tests based on ML scores, F statistics, and OLS residuals. *Econometric Reviews, 24*(4), 445–466. https://doi.org/10.1080/07474930500406053

Zeileis, A., & Hornik, K. (2007). Generalized M-fluctuation tests for parameter instability. *Statistica Neerlandica, 61*(4), 488–508. https://doi.org/10.1111/j.1467-9574.2007.00371.x

Zeileis, A., Hothorn, T., & Hornik, K. (2008). Model-based recursive partitioning. *Journal of Computational and Graphical Statistics, 17*(2), 492–514. https://doi.org/10.1198/106186008X319331

Zyphur, M. J., Allison, P. D., Tay, L., Voelkle, M. C., Preacher, K. J., Zhang, Z., Hamaker, E. L., Shamsollahi, A., Pierides, D. C., Koval, P., & Diener, E. (2020a). From data to causes I: Building a general cross-lagged panel model (GCLM). *Organizational Research Methods, 23*(4), 651–687. https://doi.org/10.1177/1094428119847278

Zyphur, M. J., Voelkle, M. C., Tay, L., Allison, P. D., Preacher, K. J., Zhang, Z., Hamaker, E. L., Shamsollahi, A., Pierides, D. C., Koval, P., & Diener, E. (2020b). From data to causes II: Comparing approaches to panel data analysis. *Organizational Research Methods, 23*(4), 688–716. https://doi.org/10.1177/1094428119847280

Chapter 4
Studying the Interaction Between Harsh Parenting and the Child's Social Behavior Problems over Time Using Continuous Time Modeling

Mark Stemmler and Friedrich Lösel

4.1 Introduction

In recent years, the number of longitudinal data projects in the social sciences has increased, and at the same time, new statistical methods have emerged that enable researchers to investigate how variables evolve, vary, and relate to one another over time (cf. Stemmler et al., 2015). The latest statistical developments are marked by the continuous time modeling in the behavioral and related sciences (cf. van Montfort et al., 2018). *Continuous time* (CT) models are extensions of the discrete time (DT) models. In DT models the flow of time is encompassed in discrete steps, while time is viewed as a continuous variable in CT models. Although CT models can offer solutions for and conceptual and practical problems associated with the DT approach (Oud & Delsing, 2010; Voelkle et al., 2012), CT models are not yet widely used (Ryan et al., 2018). This is because of two reasons. One is that the statistical methods and resulting findings are not easily understood, and the other reason is that the application of CT models requires specialized software.

Supplementary Information The online version contains supplementary material available at https://doi.org/10.1007/978-3-031-56318-8_4.

M. Stemmler (✉)
Institute of Psychology, Friedrich-Alexander-University of Erlangen-Nuremberg (FAU), Erlangen, Germany
e-mail: mark.stemmmler@fau.de

F. Lösel
Friedrich-Alexander-University (FAU) Erlangen-Nuremberg, Institute of Psychology, Erlangen, Germany

Cambridge University, Institute of Criminology, Cambridge, UK

M. Stemmler et al. (eds.), *Dependent Data in Social Sciences Research*,
https://doi.org/10.1007/978-3-031-56318-8_4

Although prevalence data show a decrease of parental use of corporal punishment (i.e., spanking or hitting a child) in Western societies, it is still a frequent method of disciplining children. There exists a rather extensive empirical literature on the association between corporal punishment and children's adjustment (Gershoff, 2002, 2010). Whereas some authors state a generally negative impact on children (e.g., Straus, 2010), others view it as less problematic when it is occasionally and mildly used to stop inadequate child behavior in an intact parent-child relationship (e.g., Baumrind et al., 2010). Against this partially controversial background, this chapter analyzes data from a rather large longitudinal study, the Erlangen-Nuremberg Development and Prevention Study (ENDPS, Lösel et al., 2013), to investigate potential bidirectional and reciprocal influences between the child's social problem behavior and harsh parental discipline over time. Starting with a small model CT, we tested the covarying processes between the above mentioned variables over time. Then successively, time-independent and time-dependent variables were added, resulting in a complex analysis to model the interactive dynamics between the child's problem behavior and fathers' harsh discipline. In two further steps, the model was extended. First, two time-independent variables, the child's intelligence, measured in kindergarten, and the child's difficult temperament, rated by the parents, were added into the model. Finally, perceived sense of parental competence by fathers was included as a time-dependent variable. In the electronic supplement, the data set and the R scripts are provided. The following paragraphs give a brief introduction to DT with CT models.

4.2 Discrete Versus Continuous Time Modeling

The Discrete Time Framework

DT models are longitudinal models, usually based on cross-lagged panels with autoregression and cross-effects, in which the data generating process takes place in discrete moments in time. Usually, a set of at least two or more variables are measured repeatedly over time. The measures show a serial dependency which is based on auto-correlations. This dependency can be described in autoregressive equations, where the relationship between the values of the observed variables depends on the consecutive ordering of the measurements. The most basic DT model is a first-order Vector Auto-Regressive (VAR) model (cf. Ryan et al., 2018). A set of V variables are measured at *N* different measurement points. In the VAR model, y_τ is the outcome variable, here a V $\times$ 1 column vector of variables measured at occasion τ (for $\tau = 2, \ldots, N$); τ is limited to positive integers; e.g., there is no 0.25th measurement point. y_τ is regressed on $y_{\tau-1}$ in a regression-like equation (see Eq. 4.1):

$$y_T = c + \Phi y_{T-1} + \epsilon_T \tag{4.1}$$

where Φ represents the V × V matrix with autoregressive and cross-lagged coefficients that regress y_{τ} on $y_{\tau-1}$. The V × 1 column vector ϵ_{τ} represents the variable-specific error terms. The errors are normally distributed with a mean of zero and the V × V variance-covariance matrix ψ; c represents the V × 1 column vector of the intercepts. In DT models the statistical assumption is that there are equally spaced measurement points $\Delta t_i = 1$, an assumption which is often not met in reality. Oud and Delsing (2010) explained that two autoregressive effects of the same value cannot be compared, if in one study $\Delta t_i = 2$ and in another study $\Delta t_i = 1$. Therefore, that interpretation of an auto-effect depends on the chosen observation interval Δt_i.

The goal of our study was to find out (1) whether negative parental behavior like corporal punishment evokes or increases the child's problem behavior over time or whether the child's problematic behavior "leads" to the use of negative parental behavior. (2) Secondly, we wanted to know at what time in the development of the child there is a peak or maximum in use of corporal punishment and problematic behavior. (3) Thirdly, we addressed other variables which might have an impact on the bidirectional relationship between the parent and the child. We picked two time-independent (i.e., stable) variables, intelligence and the child's temperament. (4) At last, we looked at a possible moderating effect of time-depending variables. Here, we choose the meta-cognition of parents' behavior, especially fathers' perceived parental sense of competence (cf. Deboeck & Preacher, 2016; Meinlschmidt et al., 2024, this volume). Those research questions would not have been possible to answer satisfactorily and completely within the traditional discrete framework.

The Continuous Time Framework

Instead of discrete observations, the Continuous Time (CT; cf. van Montfort et al., 2018) framework looks at developmental processes as functions of the continuous time t (Oud, 2024, this volume). To model the constant dynamics which can happen at any moment in time, the mathematical class of differential equations is applied. For instance, take a first-order Stochastic Differential Equation (SDE) model, where the position of a V-dimensional system at any point in time $y_{(t)}$ changes relative to a baseline position μ. A similar approach can be found in the area of field vectors in physics (Oud & Delsing, 2010). Humans can be seen as dynamic systems which change constantly over time; in addition, they strive for a non-stressed state of equilibrium. A perfect picture for the dynamics being modeled by the CT framework is a person who takes a pain killer because he or she suffers from a headache. The continuous reciprocal dynamics between the efficacy of the drug and the felt pain can be adequately modeled within the CT framework. The strive for the end point of an equilibrium is the relief of any pain (cf. Ryan et al., 2018).

The major difference between CT and DT models can be found in the matrix exponential

$$A_{\Delta_{ti}} = e^{A\Delta_{ti}} \tag{4.2}$$

$A_{\Delta_{ti}}$ is the discrete time autoregression matrix based on the time interval $\Delta_{ti} = t_i - t_{i-1}$ for (i = 1, 2, ...), and A is the so-called drift matrix. In the CT framework, the drift matrix is the continuous time autoregression matrix multiplied by time interval of the exponent (see Eq. 4.2).

Next, we present an example (from our analyses) of the discrete time *autoregression matrix* (left: $\Delta_{ti} = 1$) and the corresponding continuous time *drift matrix* (right):

$$A_{\Delta_{ti}} = 1 : \begin{pmatrix} 0.28 \; 0.49 \\ 0.15 \; 0.20 \end{pmatrix}; \quad A : \begin{pmatrix} -2.39 & 3.20 \\ 1.01 & -2.36 \end{pmatrix} \tag{4.3}$$

The most striking differences can be found in the diagonals of the drift matrix. Here all auto-effects are negative (called eigenvalues), and they are all positive in the autoregression matrix. In continuous time modeling, the exact values of $y(t_i) = A_{\Delta_{ti}} y(t_{i-1})$ depend on the evolved time multiplied by the exponential matrix of the drift matrix. This matrix is calculated by means of differential equations (Oud, 2024, this volume).

$$\frac{dy(t)}{dt} = Ay(t) \tag{4.4}$$

In Eq. (4.4), the initial value is a fixed assumed baseline value, usually y_0, because the left side of the equation is seen as a "vector of velocities, describing in what direction and with what magnitude the system will move an instant later in time" (Ryan et al., 2018, p. 30). The full multivariate stochastic differential regression-like equation with an intercept and error term looks like this:

$$\frac{dy(t)}{dt} = b + Ay(t) + G\frac{dW(t)}{dt} \tag{4.5}$$

b contains the continuous time intercepts and $G\frac{dW(t)}{dt}$ the continuous time error process. W(t) represents the so-called Wiener process which is in case of an autoregression (AR) of 1.0 a random walk in time and in case of AR = 0 is representing independence or noise (Voelkle et al., 2012). W(t) also includes the diffusion matrix; a significant diffusion matrix signalizes increased intra-person variation.

To sum up this brief introduction to the DT and CT framework, the following benefits of the CT framework can be listed: (1) CT models can deal with unequal time intervals between measurement points which is the case more often than not.

(2) DT models are not generalizable beyond the time intervals specified in the data. The drift matrix can handle any time intervals (of course, the models assume sufficient data points to calculate the dynamics over time). (3) CT models can also calculate how the cross-lagged effects are changing depending on the time interval between measurement points; one can therefore estimate a maximum effect (Deboeck & Preacher, 2016).

4.3 Method

Sample and Participation Rates

The data was taken from the Erlangen-Nuremberg Development and Prevention Study (ENDPS; Lösel & Stemmler, 2012a). Starting in kindergarten, ENDPS combined a prospective-longitudinal study on the origin and development of problem behavior in children with the evaluation of a child- and parent-oriented prevention program called EFFEKT (Lösel et al., 2009). The original sample of the core study consisted of 675 kindergarten children (336 boys, 339 girls) from 609 families. Of the parents, 98.9% were biological mothers and 96.3% biological fathers. The sample was largely representative of young families living in Erlangen and Nuremberg (North Bavaria). According to an index of the socioeconomic status (SES; Geißler, 1994) which included income, education, profession, and housing conditions, 13.3% of the families were lower class, 32.3% lower middle class, 30.6% middle class, 15.4% upper middle class, and 3.0% upper class. Approximately 86% of the parents were married at Time 1. The ENDPS contains seven waves (T_1 thru T_7) of data collection that were carried out annually in the first 3 years and later at larger intervals. The average age of the children at each data assessment was 5.53 years (T_1), 6.61 years (T_2), 7.70 years (T_3), 9.93 years (T_5), 11.41 years (T_6), and 14.77 years (T_7). The retention rates varied over time, but we even had data from about 90% in the last wave.

Measures

Numerous data from different sources were collected in the Erlangen-Nuremberg study. They range from interviews with parents and observations during home visits to psychological tests, physiological measurements, and medical data on the children, as well as assessments in kindergarten, elementary school, and secondary school.

(Social) Problem Behavior To measure the extent of the children's behavior problems, we used the kindergarten educators and school teacher's version of the

Social Behavior Questionnaire (PSBQ; cf. Tremblay et al., 1992) which we adapted into German. In the PSBQ, the items were rated on a 3-point scale (from "not true" to "almost always true"), indicating the extent to which the child shows certain behavioral problems. The overall problem scale consists of 36 items. We used the subscale *Social Behavior Problems* (11 items; $\alpha = 0.88$). It consists of items on *Physical Aggression* (e.g., "It bites or hits other Children") and on *Destruction of Things/Delinquency* (e.g., "It destroys things that belong to others" or "It steals outside of home"). The ratings of kindergarten educators and school teachers were z-standardized to make them comparable.

Parenting Behavior The fathers' parenting behavior was assessed using a German translation of the Alabama Parenting Questionnaire (APQ; Shelton et al., 1996). The questionnaire consists of 42 items covering different facets of parenting behavior, like monitoring and supervision, parental involvement, positive reinforcement, and corporal punishment. Each of the items is rated on a five-point scale ranging from 1 ("never") to 5 ("always"). Scale scores are obtained by calculating the means of the respective items. According to our research question, in this paper the *Corporal Punishment* scale for fathers was used. This scale contains three items such as "You spank your child with your hand when he/she has done something wrong". Internal consistency of the German scale adaptation was $\alpha = 0.44$ for *Corporal Punishment* (Lösel et al., 2003). It was similar as in other German applications of this short scale (Reichle & Franiek, 2009).

Child's Intelligence The child's *Intelligence* was assessed using the German version of the Kaufman-Assessment Battery for Children (K-ABC; Melchers & Preuß, 1994). The German K-ABC was scaled on a German speaking sample of $N = 3098$ from Germany, Austria and Switzerland. The K-ABC measures non-education and education dependent cognitive abilities. We used the mean of the two aspects of intelligence. The split-half reliabilities for children of kindergarten age were reported between $r = 0.86$ and $r = 0.93$ (Melchers & Preuß, 1994).

Temperament The child's t*emperament* was rated by the parents who were interviewed during a home visit (Stemmler & Lösel, 2010). The parent present at the interview, rated whether their child was on a 5-point rating scale with higher scores representing a more difficult temperament (i.e. negative mood, low regularity and adaptability).

Perceived Parental Sense of Competence The *Parental Sense of Competence Scale* (PSOC; Johnston & Mash, 1989) was used to assess certain cognitions about the parents' self-perception in the parenting role. The total questionnaire consists of 16 items, which are rated on a scale from "strongly disagree" (1) to "strongly agree" (6). We selected the items for the assessment of sense of competence in parenting (e.g., "Problems that a father faces can usually be solved"). We used only father's ratings and the internal consistency of the respective scale was $\alpha = 0.70$ (cf. Lösel et al., 2003).

Statistical Analyses

For the analyses below, we first used both five time points, i.e., T_1, T_2 T_3 T_5 and T_6 and then four time points, i.e. T_1, T_2 T_3 and T_5. The selection was based on the requirement, that all the variables in our multi-informant study were assessed at each measurement point. At T_4 and T_6 not all of the repeated measures were available for the analyses and T_7 was a pure self-rating assessment by the study children. The measures included in the analyses were *Problem Behavior* rated by teachers (kindergarten educators and school teachers), use of *Corporal Punishment* and *Perceived Parental Self-Efficacy* both rated by fathers. The K-ABC *Intelligence* test was conducted by the research team in the kindergarten (T_1); and the parents rated their child's *Temperament* also at T_1.

We ran a number of continuous time models using the R-package *ctsem* (Driver et al., 2017). We started with the most parsimonious model, that is, the two covarying variables *Problem Behavior* and *Corporal Punishment.* We then added two time-independent variables, that is, *Intelligence* and *Temperament* and *Perceived Sense of Competence*.

In Fig. 4.1, one can see the model for the first analyses, that is, the time varying effects of (Social) *Problem Behavior* and *Corporal Punishment.* There are auto-regressive as well as cross-effects over time. The unexplained variance of the latent constructs *w* may also correlate between the constructs.

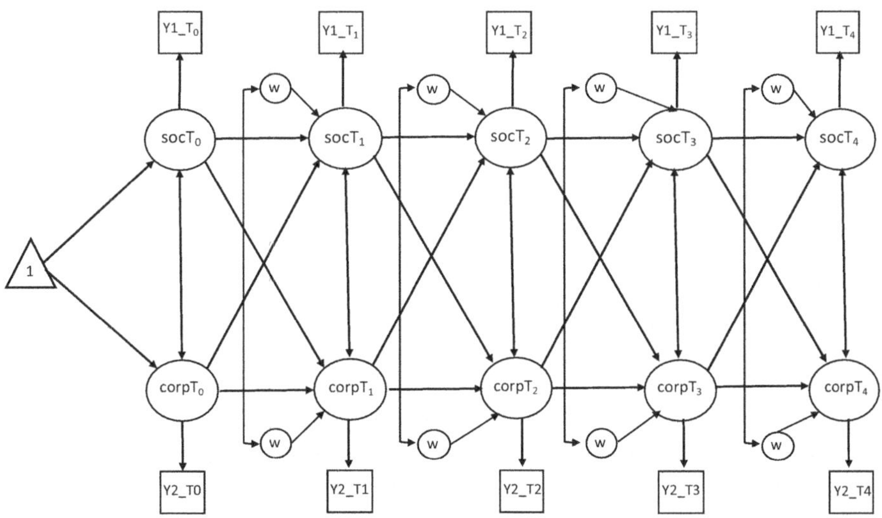

Fig. 4.1 Continuous time model with two variables and four measured time points. *Note*: In ctsem, measurement point T_0 is considered to be the assumed baseline measure, which is not measured. "socT_1" is the rating of the child's (social) problem behavior at Time 1; "corpT_1" is the father's rating of the use of corporal punishment at Time 1

The individuals in the model were observed at four occasions on four manifest variables (they are ordered by the variable). Arrows between the same variables describe auto-regressive effects; the cross-lagged arrows represent the cross-effects between the correlating variables. Ctsem calculates the $T-1$ time intervals between the measurement occasions; in our data, the time intervals were sometimes longer than 1 year.

Results

Bivariate Correlations

First, we looked at the bivariate correlations of the involved variables across time. The correlations of *Problem Behavior* at T_1 across time were $r = 0.64^*$ (T_2), $r = 0.58^*$ (T_3), $r = 0.23^*$ (T_5), and $r = 0.25^*$ (T_6); n varied from $n = 525$ to 669. The within-time correlations were lower at T_5 and T_6. However, the child's problem behavior is relatively stable across time, especially within consistent raters. The correlations of father's *Corporal Punishment* at T_1 across time were $r = 0.66^*$ (T_2), $r = 0.58^*$ (T_3), $r = 0.50^*$ (T_5), and $r = 0.16^*$ (T_6); n varied from $N = 447$ to 592. The correlations suggest a high consistency over time until the child was hitting adolescence. The correlations of father's *Perceived Parental Competence* at T_1 across time were $r = 0.57^*$ (T_2), $r = 0.58^*$ (T_3), and $r = 0.48^*$ (T_5); sample size varied from $N = 445$ to 594. This construct seems to be even more stable over time than *Problem Behavior* and *Corporal Punishment.*

The *between-construct* correlations revealed the following results between *Problem Behavior* and *Corporal Punishment*: $r = 0.21^*$ (T_1), $r = 0.17^*$ (T_2), $r = 0.18^*$ (T_3), $r = 0.18^*$ (T_5), and $r = 0.09^*$ (T_6); n varied from $N = 456$ to 591. These were small associations until early adolescence.

The negative correlations between *Intelligence* at T_1 with *Problem Behavior (PB)* and *Corporal Punishment (CP)* across time were $r = -0.08$ (PB; T_1), $r = -0.17^*$ (CP; T_1), $r = -0.03$ (PB; T_2), $r = -0.13^*$ (CP; T_2), $r = -0.03$ (PB; T_3), $r = -0.18^*$ (CP; T_3), $r = -0.12^*$ (PB; T_5), $r = -0.15^*$ (CP; T_5), and $r = -0.14^*$ (PB; T_6), $r = 0.01$ (CP; T_6); sample size varied from $N = 447$ to 594. There was no significant association between intelligence and problem behavior in kindergarten; it turns to a significant small effect in school. In sum, intelligence correlated low and negatively with the two constructs, indicating that higher intelligence goes along with less social problem behavior in children and less harsh punishment by fathers. Father's corporal punishment correlated significantly and stably with intelligence across time before adolescence.

The correlations between *Temperament* at T_1 with *Problem Behavior (PB)* and *Corporal Punishment (CP)* across time were $r = 0.37^*$ (PB; T_1), $r = 0.11^*$ (CP; T_1), $r = 0.29^*$ (PB; T_2), $r = 0.10^*$ (CP; T_2), $r = 0.20^*$ (PB; T_3), $r = 0.11^*$ (CP; T_3), $r = 0.07$ (PB; T_5), $r = 0.09$ (CP; T_5), and $r = -0.12^*$ (PB; T_6), $r = 0.06$ (CP; T_6);

sample size varied from $N = 455$ to 591. Temperament correlated positively with the two time-independent constructs, indicating that a more difficult temperament leads to more social problem behavior in children and to more harsh punishment by fathers. There is no association between temperament and corporal punishment in adolescence. The child's problem behavior correlated significantly and stably with temperament across time until adolescence (with a dip at early adolescence).

Continuous Time Model with Two Variables

The results from the first CT model are presented in Table 4.1. For all CT models in this chapter, the involved variables were z-standardized, and the observed variances were fixed to zero. The auto-effects of the two variables *Problem Behavior (PB)* and *Corporal Punishment (CP)* are almost equally strong and significant (PB: -2.40^{*} and CP: -2.37^{*}). It should be remembered that auto-effect eigenvalues have to be negative, representing a velocity back toward an equilibrium. However, the cross-effects revealed that the influence of PB onto CP is much greater than vice versa (PB: 3.21^{*} and CP: 1.01^{*}). One can also see that there is a significant bivariate effect. PB influences CP and CP has a reciprocal effect of PB. The maximum point

Table 4.1 Continuous time parameter: child's social problem behavior (rated by teachers) and corporal punishment (father's self-report)

	Continuous time parameter			
	Problem behavior		Corporal punishment	
Parameter	Estimate	[95% CI][a]	Estimate	[95% CI][a]
Drift matrix (A)				
Auto-effects	−2.40*	[−4.28; −1.53]	−2.37*	[−2.37; −1.39]
Cross-effects	3.21*	[1.82; 6.08]	1.01*	[0.38; 2.46]
Diffusion matrix (Q)				
q	2.13*	[1.51; 3.59]	3.98*	[2.51; 5.79]
$qJ = qS$	−2.33*	[−5.00; −1.07]		
Baseline (T0)				
M_{t0}	0.00	[−0.07; 0.07]	0.00	[−0.06; 0.08]
var_{t0}	1.00*,[b]	[0.90; 1.11]	1.00*	[0.89; 1.11]
cov_{t0}	0.75*	[0.66; 0.85]		
Model indices	−2*LL*	*df*	*N*	
	13,468.51	51	675	

Note: Continuous time parameter estimates for auto-effects are always negative but correspond to positive effects in discrete time

$p < 0.05$

[a]Confidence intervals for many parameters in CT models may not be symmetric around the point estimate. If no upper or lower bound of the confidence interval was provided in the ctsem output, the confidence interval was computed by multiplying the standard error of the estimate by 1.96 (see Driver et al., 2017, p. 17). Formula for calculating the discrete time parameters $A_{\Delta ti} = e^{A\Delta ti}$

[b]Manifest variances were fixed to zero; only z-standardized variables were used

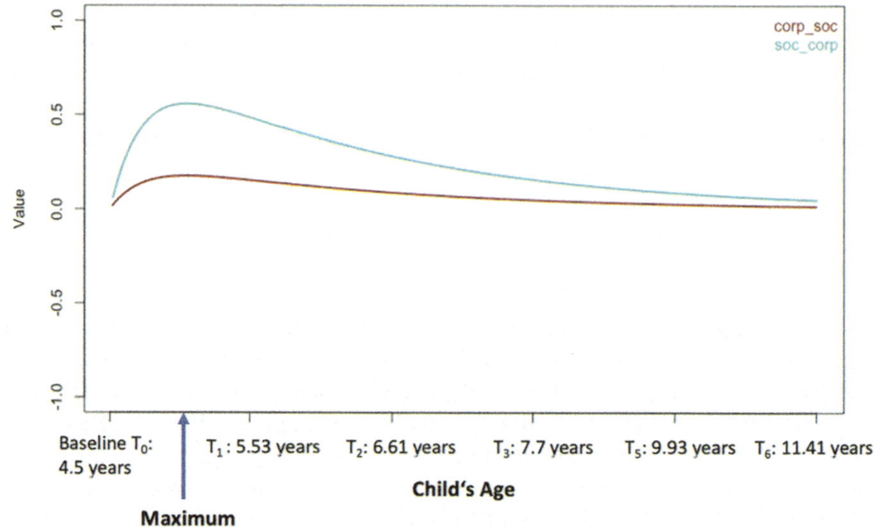

Fig. 4.2 Cross-effects of social problem behavior and corporal punishment. *Note*: In ctsem, measurement point T_0 is considered to be the assumed baseline measure, which is not measured. "soc_corp" represent cross-effects from social problem behavior to corporal punishment; "corp_soc" represent cross-effects from corporal punishment to social problem behavior

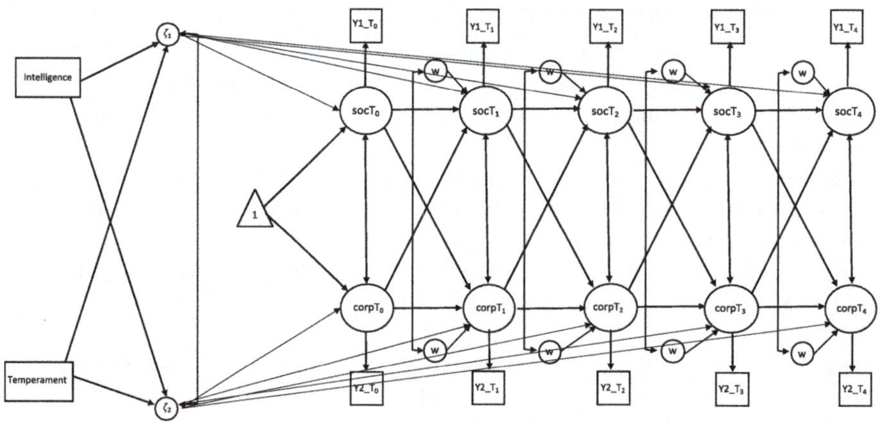

Fig. 4.3 Continuous time model with two variables and two independent time variables measured at four time points. *Note*: "$socT_1$" is the rating of the child's (social) problem behavior at Time 1; "$corpT_1$" is the father's rating of the use of corporal punishment at Time 1. "Intelligence" is the child's intelligence measured in kindergarten, and "Temperament" is the parental rating of the child's (difficult) temperament

of the strongest effect of PB on to Corp is between ages 4.5 and 5.5 years (see Fig. 4.2). The diffusion matrix showed significant effects indicating that there is still unobserved intra-person heterogeneity in the error variance. Therefore, we searched for more explanatory variables to be included in the model.

Continuous Time Model with Two Time-Independent Variables

For the second CT model, two time-independent variables, *Intelligence* and *Temperament*, were added (see Fig. 4.3). The influence of the two time-independent, and therefore not changing, variables is valid across all time points (see arrows starting from $\zeta_{1,2}$). The statistical results of the model are shown in Table 4.2. For all CT models in this chapter, the involved variables were z-standardized, and the observed variances were fixed to zero. The auto-regression effects (PB: -3.46^* and CP: -3.49^*) as well as the cross-effects came out even a little stronger (PB: 5.10^* and CP: 1.61^*) suggesting that the two time-independent variables set more explainable variance free. Intelligence had a significant impact on PB (0.02^*) and CP (-0.02^*). There was a positive relationship between intelligence and PB and a negative one with CP; however, the effects were small. Temperament had no significant impact on PB (−0.11), but an increasing and significant effect on CP (0.21^*), indicating that

Table 4.2 Continuous time parameter: child's social problem behavior (rated by teachers) and corporal punishment (father's self-report) including time-independent variables

	Continuous time parameter			
	Problem behavior		Corporal punishment	
Parameter	Estimate	[95% CI][a]	Estimate	[95% CI][a]
Drift matrix (A)				
Auto-effects	−3.46*	[−4.20; −2.92]	−3.49*	[−4.62; −2.72]
Cross-effects	5.10*	[4.45; 6.02]	1.61*	[1.12; 2.34]
Time-independent variables[b]				
Intelligence	0.02*	[0.01; 0.02]	−0.02*	[−0.02; −0.02]
Temperament	−0.11	[−0.33; 0.11]	0.21*	[0.05; 0.37]
Diffusion matrix (Q)				
q	2.75*	[2.17; 3.52]	5.70*	[2.50; 7.05]
$qJ = qS$	−3.96*	[−4.90; −3.31]		
Baseline (T0)				
M_{t0}	0.57*	[0.01; 1.12]	0.74*	[0.20; 1.26]
var_{t0}	0.95*,[c]	[0.85; 0.97]	0.95*	[0.85; 0.97]
cov_{t0}	0.70*	[0.62; 0.78]		
Model indices	−2*LL*	*df*	*N*	
	19,979.56	67	675	

Note: Continuous time parameter estimates for auto-effects are always negative but correspond to positive effects in discrete time

$p < 0.05$

[a]Confidence intervals for many parameters in CT models may not be symmetric around the point estimate. If no upper or lower bound of the confidence interval was provided in the ctsem output, the confidence interval was computed by multiplying the standard error of the estimate by 1.96 (see Driver et al., 2017, p. 17)

[b](Difficult) temperament: higher values represent more difficult temperament. Formula for calculating the discrete time parameters $A_{\Delta ti} = e^{A\Delta ti}$

[c]Manifest variances were fixed to zero; only standardized variables were used

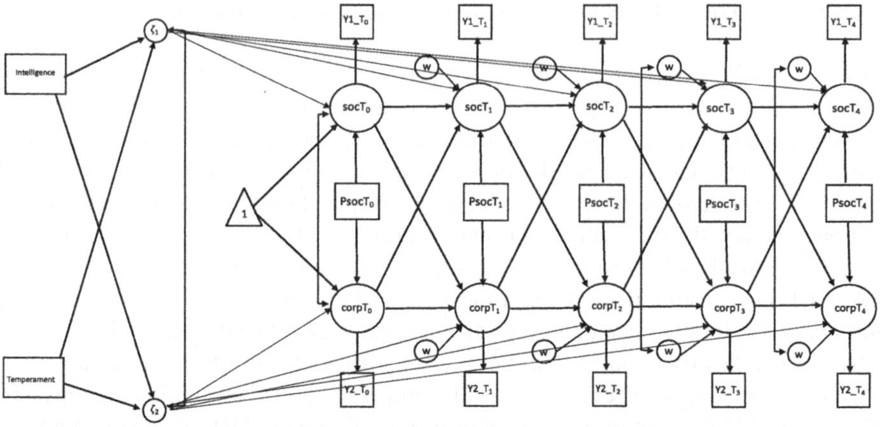

Fig. 4.4 Continuous time model with two variables, two time-independent variables and one time-dependent variable measured at four time points. *Note*: "socT$_1$" is the rating of the child's (social) problem behavior at Time 1; "corpT$_1$" is the father's rating of the use of corporal punishment at Time 1. The time-independent variables are "Intelligence" which is the child's intelligence measured in kindergarten and "Temperament" which is the parental rating of the child's (difficult) temperament. The time-dependent variable is the father's perceived sense of competence ("PsocT$_1$" measures at Time 1)

a difficult temperament increases fathers' harsh punishment. Again, the diffusion matrix showed significant error variance.

Continuous Time Model with Two Time-Independent Variables and One Additional Time-Independent Variable

For the last CT model, one other time-dependent variable, *Perceived Sense of Competence* (Psoc) rated by the father, was added (Fig. 4.4). At each time point, there was a direct influence from Psoc on the covarying variable PB and CP, while the two time-independent variables were still in the model. Again, all variables were z-standardized. The observed variances were not fixed to zero because with fixed variances the output generated positive eigenvalues that is an indicator that the CT model did not converge satisfactorily. Now, only four measurement points were included in the analysis. The statistical results of the model can be taken from Table 4.3. The addition of Psoc to the model had a valuable impact on the effects; it explained significant intra-person variability in the error or diffusion matrix. The error variance of CP (11.83) and the error covariance with PB (0.95) were no longer significant. Psoc had a significant and negative effect on PB (-0.07^*) and CP (-0.07^*) over time. Now PB still had a small, but negative significant effect on CP (-0.21^*).

Furthermore, intelligence showed a significant and negative effect on PB (-0.004^*) but no longer a significant influence on CP (-0.02). Temperament

Table 4.3 Continuous time parameter: child's social problem behavior (rated by teachers), corporal punishment (father's self-report), perceived sense of competence (father's self-report) as time-dependent variable and intelligence and (difficult) temperament as time-independent variables

	Continuous time parameter			
	Problem behavior		Corporal punishment	
Parameter	Estimate	[95% CI][a]	Estimate	[95% CI][a]
Drift matrix (A)				
Auto-effects	−0.27*	[−2.07; −0.27]	−13.82	[−62.60; 34.98]
Cross-effects	−0.05	[−2.22; 2.13]	10.93	[−28.68; 50.54]
Time-independent variables[b]				
Intelligence	−0.004*	[−0.004; −0.004]	−0.02	[−0.03; 0.07]
Temperament	0.05*	[0.06; 0.09]	−0.11	[−0.90; 0.71]
Time-dependent variable				
Parental competence	−0.07*	[−0.10; −0.04]	−0.08*	[−0.12; −0.03]
Diffusion matrix (Q)				
q	0.40*	[0.24; 0.56]	11.83	[−28.94; 52.60]
$qJ = qS$	0.95	[−1.60; 3.50]		
Baseline (T0)				
M_{t0}	0.87*	[0.26; 0.87]	0.91*	[0.34; 0.91]
var_{t0}	0.60*,[c]	[0.50; 0.71]	0.83*	[0.71; 0.96]
cov_{t0}	0.70*	[0.62; 0.80]		
Model indices	−2LL	*df*	*N*	
	22,582.98	78	675	

Note: Continuous time parameter estimates for auto-effects are always negative but correspond to positive effects in discrete time
$p < 0.05$
[a]Confidence intervals for many parameters in CT models may not be symmetric around the point estimate. If no upper or lower bound of the confidence interval was provided in the ctsem output, the confidence interval was computed by multiplying the standard error of the estimate by 1.96 (see Driver et al., 2017, p. 17)
[b](Difficult) temperament: higher values represent more difficult temperament. Formula for calculating the discrete time parameters $A_{\Delta ti} = e^{A\Delta ti}$
[c]Only standardized variables were used

now showed a significant and positive impact on PB (0.05*) but no significant impact on CP (−0.11).

4.4 Discussion

Three different CT models (from simple to more complex) were applied. It turned out that the autoregressive effects of corporal punishment and social problem behavior were almost identical for the two variables, suggesting that they both have a significant momentum or strive toward an equilibrium or stable system (Deboeck & Preacher, 2016). "This means that if the system takes on a position away from equilibrium (e.g., due to a random shock from the stochastic part of the model on either Down or Tired), the system will inevitably return to its equilibrium position over time. It is for this reason that the equilibrium position or fixed point in stable systems is also described as the attractor point, and stable systems are described as equilibrium-reverting" (Ryan et al., 2018). We started out with the research question whether corporal punishment "causes" social problem in the child or vice versa. Using CT modeling we showed that there is a significant bidirectional reciprocal relationship over time. It turned out that the effect of problem behavior on corporal punishment is greater than vice versa. Moreover, we found that the maximum impact of problem behavior was before the school entry at about 5 years. At the end of childhood, this influence tends to diminish. This result is in line with findings from a nationally representative sample of US households with children (Finkelhor et al., 2019). Here, the maximum age of spanking children of both gender was between age 4 and 6. This is also the age when children were most aggressive to their parents (Ulman & Straus, 2003). Children at this age widen their action space and can create more situations of conflict (partially similar to the so-called Terrible Twos in early childhood). The maximum point or the peak influence of corporal punishment in the interaction with the child is an indication that the child's defiant behavior is retroactive to the fathers. One can assume that the father's use of harsh discipline might be a somewhat helpless effort in response to the child's problem behavior, because a parental sense of competence in education leads to less corporal punishment and vice versa (Patterson, 1982). This was supported by other findings in the ENDPS data, where fathers' self-perceived problematic parent-child interactions correlated with a broader range of disciplinary techniques and more inconsistency in parenting (Haupt et al., 2014). In the same vain, our research also showed a substantial connection between negative parenting and antisocial behavior (Beelmann et al., 2006). Furthermore, positive treatment effects from our combined child and parent training (EFFEKT) resulted in a significant lowering of the child's social problem behavior and at the same time in a decrease in negative parenting like inconsistent discipline (Stemmler et al., 2007). Other studies suggest that corporal punishment is not only a replicated risk factor for children's antisocial behavior and other problem, but is also partially a reaction to aggression of the child (e.g., Farrington, 2005; Ulman & Straus, 2003). Of course, this does not justify any severe corporal punishment as a parenting practice. In fact, it indicates helplessness that can lead to various forms of child abuse and neglect (Bender & Lösel, 2015). On a positive note, the rate of use of corporal punishment declined over recent decades in Western societies, and more than 50% of the children have not experienced any spanking in the last year (see Finkelhor et al., 2019). As harsh parenting is detrimental to the child's social problem behavior, healthy development preventive measures against the use of corporal punishment should focus on increasing parental use of positive parenting as a protective influence (Lösel & Farrington, 2012). Accordingly, reviews and meta-analyses showed significant desirable effects of family-oriented prevention programs on child behavior (Farrington et al., 2017; Lösel & Bender, 2017). Similar mean positive effects were found for programs in German-speaking countries (Weiss et al., 2022). However, worldwide, more long-term evaluations of family-oriented prevention are needed and these may benefit from the statistical approaches presented in this article.

Our results regarding a (difficult) temperament were inconsistent. In the second CT model with two covarying effects and two time-independent variables (i.e., intelligence and temperament), temperament revealed no significant impact on problem behavior but on corporal punishment (a somewhat surprising increasing effect). In the third and most complex model with parenting sense of competence as an additional time-dependent variable, temperament displayed a significant and positive effect on problem behavior (here, the effect on corporal punishment lost its significance). The literature on developmental psychology stresses the issue that difficult temperament increases the child's externalizing behavior (Paulussen-Hoogeboom et al., 2008). In a study on Dutch families, positive parenting behavior was able to moderate the negative effects between a difficult temperament and the child's problem behavior (Karreman et al., 2010).

The results concerning the child's intelligence were also partially inconsistent across the CT models. In the second CT model, intelligence significantly increased problem behavior and reduced corporal punishment (although non-significantly). In the third and final model, intelligence had a significant reducing effect on problem behavior and on corporal punishment. This is in line with other studies where intelligence had a buffering effect on externalizing behavior and on family adversity (Flouri et al., 2015; Ttofi et al., 2016). Because the third and most complex CT model showed a better fit (reduction in error variance) and because the statistical results were in line with the literature, we consider those as the most appropriate findings (see also Nisbett et al., 2012).

The partially rather small effects and some inconsistencies may have been due to characteristics of our sample. The area of Erlangen-Nuremberg has a population in relatively good socioeconomic and educational framing conditions. Accordingly, high scores of corporal punishment and child behavior problems were not frequent. In addition, our sample contained not only males but also females, who typically exhibit less externalizing problems in childhood (Lösel & Stemmler, 2012 b; Moffitt et al., 2001; Stemmler & Lösel, 2012). These factors may have reduced the variance and the number of extreme cases that would have contributed to stronger effects. As in other fields of research on problem behavior, more replications with different samples and methods are needed (e.g., Lösel, 2018). However, our analyses still showed the advantages of CT models over DT. In CT models no equal spacing is required. In our ENDPS data, only the first three measurement points had an equal spacing of about 1 year; thereafter for T_5 and T_6, the spacing was longer than 1 year. With the CT model, the maximum peak of the reciprocal influence between social problem behavior and corporal punishment was apparent. We recommend the use of CT models although the statistical methods are not easy to apply, but the use of the ctsem R-package is straightforward (Driver et al., 2017; see also Driver, 2022). Therefore, we recommend more prospective longitudinal studies using the CT framework to disentangle the interplay of risk, promotive, and protective factors in the child's social development (Lösel & Bender, 2003; Lösel & Farrington, 2012).

Acknowledgement Many thanks to Dr. Meike Sons for her help in setting up the ctsem R-scripts and on other statistical issues.

References

Baumrind, D., Larzelere, R. E., & Owens, E. B. (2010). Effects of preschool parents' power assertive patterns and practices on adolescent development. *Parenting: Science and Practice, 10*(3), 157–201.

Beelmann, A., Lösel, F., Stemmler, M., & Jaursch, S. (2006). Beurteilung von sozialen Verhaltensproblemen und Erziehungsschwierigkeiten im Vorschulalter [Assessment of social behavioral problems and parental difficulties at preschool age]. *Diagnostica, 52*(4), 189–198.

Bender, D., & Lösel, F. (2015). Risikofaktoren, Schutzfaktoren und Resilienz bei Misshandlung und Vernachlässigung [Risk factors, protective factors, and resilience in child abuse and neglect]. In U. T. Egle, P. Joraschky, A. Lampe, I. Seiffge-Krenke, & M. Cierpka (Eds.), *Sexueller Missbrauch, Misshandlung, Vernachlässigung [Sexual abuse, physical abuse, and neglect]* (4th ed., pp. 77–103). Schattauer.

Deboeck, P. R., & Preacher, K. J. (2016). No need to be discrete: A method for continuous time mediation analysis. *Structural Equation Modeling: A Multidisciplinary Journal, 23*(1), 61–75.

Driver, C. C. (2022). *Inference with cross-lagged effects – Problems in time*. OSF Preprints.

Driver, C. C., Oud, J. H., & Voelkle, M. C. (2017). Continuous time structural equation modeling with the R package ctsem. *Journal of Statistical Software, 77*, 1–35.

Farrington, D. P. (2005). Childhood origins of antisocial behavior. *Clinical Psychology & Psychotherapy, 12*, 177–190.

Farrington, D. P., Gaffney, H., Lösel, F., & Ttofi, M. (2017). Systematic reviews of the effectiveness of developmental prevention programs in reducing delinquency, aggression, and bullying. *Aggression and Violent Behavior, 33*, 91–106.

Finkelhor, D., Turner, H., Wormuth, B. K., Vanderminden, J., & Hamby, S. (2019). Corporal punishment: Current rates from a national survey. *Journal of Child and Family Studies, 28*, 1991–1997. https://doi.org/10.1007/s10826-019-01426-4

Flouri, E., Midouhas, E., & Joshi, H. (2015). Family and neighbourhood risk and children's problem behaviour: The moderating role of intelligence. *Intelligence, 53*, 33–42.

Geißler, R. (Ed.). (1994). *Soziale Schichtung und Lebenschancen in Deutschland [Social strata and life opportunities in Germany]*. Enke.

Gershoff, E. T. (2002). Corporal punishment by parents and associated child behaviors and experiences: A meta-analytic and theoretical review. *Psychological Bulletin, 128*, 539–579.

Gershoff, E. T. (2010). More harm than good: A summary of scientific research on the intended and unintended effects of corporal punishment on children. *Law & Contemporary Problems, 73*, 31–56.

Haupt, H., Lösel, F. & Stemmler M. (2014). Quantile regression analysis and other alternatives to ordinary least squares regression: A methodological comparison on corporal punishment. *Methodology: European Journal of Research Methods for the Behavioral and Social Sciences, 10*(3), 81–91. https://doi.org/10.1027/1614-2241/a000077

Johnston, C., & Mash, E. J. (1989). A measure of parenting satisfaction and efficacy. *Journal of Clinical Child Psychology, 18*, 167–175.

Karreman, A., de Haas, S., van Tuijl, C., van Aken, M. A. G., & Deković, M. (2010). Relations among temperament, parenting and problem behavior in young children. *Infant Behavior and Development, 33*(1), 39–49.

Lösel, F. (2018). Evidence comes by replication, but needs differentiation: The reproducibility issue in science and its relevance for criminology. *Journal of Experimental Criminology, 14*, 257–278. https://doi.org/10.1007/s11292-017.9297-z

Lösel, F., & Bender, D. (2003). Resilience and protective factors. In D. Farrington & J. Coid (Eds.), *Early prevention of adult antisocial behaviour* (pp. 130–204). Cambridge University Press.
Lösel, F., & Bender, D. (2017). Parenting and family-oriented programs for the prevention of child behavior problems: What the evidence tells us. *Zeitschrift für Familienforschung [Journal of Family Research, SI], 11*, 217–239.
Lösel, F., & Farrington, D. (2012). Direct protective and buffering protective factors in the development of youth violence. *American Journal of Preventive Medicine, 43*(Suppl. 2), 8–23.
Lösel, F., & Stemmler, M. (2012a). Preventing child behavior problems in the Erlangen-Nuremberg Development and Prevention Study: Results from preschool to secondary school age. *International Journal of Conflict and Violence, 6*, 214–224.
Lösel, F., & Stemmler, M. (2012b). Continuity and patterns of externalizing and internalizing behavior problems in girls: A variable- and person-oriented study from preschool to youth age. *Psychological Test and Assessment Modeling, 54*, 307–319.
Lösel, F., Beelmann, A., Jaursch, S., Scherer, S., Stemmler, M., & Wallner, S. (2003). *Skalen zur Messung elterlichen Erziehungsverhaltens bei Vorschul- und Grundschülern [Scales for measuring parental parenting behavior in preschool and primary school children]*. Universität Erlangen-Nürnberg.
Lösel, F., Stemmler, M., Jaursch, S., & Beelmann, A. (2009). Universal prevention of antisocial development: Short- and long-term effects of a child and parent-oriented program. *Monatsschrift für Kriminologie und Strafrechtsreform [Monthly Journal for Criminology and Criminal Law Reform], 92*, 289–308.
Lösel, F., Stemmler, M., & Bender, D. (2013). Long-term evaluation of a bimodal universal prevention program: Effects on antisocial development from kindergarten to adolescence. *Journal of Experimental Criminology, 9*(4), 429–449.
Meinlschmidt, H., Sons, M., & Stemmler, M. (2024). A new (machine learning) algorithm to estimate continuous time dynamic models. In M. Stemmler, W. Wiedermann, & F. Huang (Eds.), *Dependent data in the social sciences research: Forms, issues and methods of analysis* (2nd ed.). Springer Publishing Company.
Melchers, P., & Preuß, U. (1994). *K-ABC: Kaufman-assessment battery for children* (German Version). Swets.
Moffitt, T. E., Caspi, A., Rutter, M., & Silva, P. A. (2001). *Sex differences in antisocial behaviour: Conduct disorder, delinquency, and violence in the Dunedin longitudinal study*. Cambridge University Press.
Nisbett, R. E., Aronson, J., Blair, C., Dickens, W., Flynn, J., Halpern, D. F., & Turkheimer, E. (2012). Intelligence: New findings and theoretical developments. *American Psychologist, 67*(2), 130–159. https://doi.org/10.1037/a0026699
Oud, J. H. L. (2024). Continuous time modeling in the social sciences: History and philosophical background. In M. Stemmler, W. Wiedermann, & F. Huang (Eds.), *Dependent data in the social sciences research: Forms, issues and methods of analysis* (2nd ed.). Springer Publishing Company.
Oud, J. H. L., & Delsing, M. J. M. H. (2010). Continuous time modeling of panel data by means of SEM. In K. van Montefort, J. Oud, & A. Satorra (Eds.), *Longitudinal research with latent variables* (pp. 201–244). Springer.
Patterson, G. R. (1982). *Coercive family process*. Castalia.
Paulussen-Hoogeboom, M. C., Stams, G. J. J., Hermanns, J. M., Peetsma, T. T., & van den Wittenboer, G. L. (2008). Parenting style as a mediator between children's negative emotionality and problematic behavior in early childhood. *The Journal of Genetic Psychology, 169*(3), 209–226.
Reichle, B., & Franiek, S. (2009). Erziehungsstil aus Elternsicht: Deutsche erweiterte Version des Alabama Parenting Questionnaire für Grundschulkinder (DEAPQ-EL-GS) [Parenting style from the parents' perspective: German extended version of the Alabama Parenting Questionnaire for Elementary School Children (DEAPQ-EL-GS)]. *Zeitschrift für Entwicklungspsychologie und Pädagogische Psychologie [Journal of Developmental Psychology and Educational Psychology], 41*(1), 12–25. https://doi.org/10.1026/0049-8637.41.1.12

Ryan, O., Kuiper, R. M., & Hamaker, E. L. (2018). A continuous time approach to intensive longitudinal data: What, why and how. In K. V. Montfort, J. H. L. Oud, & M. C. Voelkle (Eds.), *Continuous time modeling in the behavioral and related sciences* (pp. 27–54). Springer.

Shelton, K. K., Frick, P. J., & Wotton, J. (1996). Assessment of parenting practices in families of elementary school-age children. *Journal of Clinical Child Psychology, 25*, 317–329.

Stemmler, M., & Lösel, F. (2010). Different patterns of boys' externalizing behaviour and their relation to risk factors: A longitudinal study of preschool children. *Bulletin de la Société des Sciences Médicales du Grand Duché de Luxembourg, 1*, 53–67.

Stemmler, M., & Lösel, F. (2012). The stability of externalizing behavior in boys from preschool age to adolescence: A person-oriented analysis. *Psychological Test and Assessment Modeling, 54*, 195–207.

Stemmler, M., Beelmann, A., Jaursch, S., & Lösel, F. (2007). Improving the family environment to prevent child behavior problems: A study on the parent training of the EFFEKT program. *International Journal of Hygiene and Environmental Health, 210*, 563–570.

Stemmler, M., Eye, A. V., & Wiedermann, W. (2015). Dependent data in social sciences research. *Springer Proceedings in Mathematics & Statistics, 145*, 2015.

Straus, M. A. (2010). Criminogenic effects of corporal punishment by parents. In M. Herzog-Evans & I. Dran-Rivette (Eds.), *Transnational criminology manual* (pp. 373–390). Wolf Legal Publishing.

Tremblay, R. E., Vitaro, F., Gagnon, C., Piché, C., & Royer, N. (1992). A prosocial scale for the preschool behavior questionnaire: Concurrent and predictive correlates. *International Journal of Behavioral Development, 15*, 227–245.

Ttofi, M. M., Farrington, D. P., Piquero, A. R., Lösel, F., DeLisi, M., & Murray, J. (2016). Intelligence as a protective factor against offending: A meta-analytic review of prospective longitudinal studies. *Journal of Criminal Justice, 45*, 4–18.

Ulman, A., & Straus, M. A. (2003). Violence by children against mothers in relation to violence between parents and corporal punishment by parents. *Journal of Comparative Family Studies, 34*, 41–60.

Van Montfort, K., Oud, J. H., & Voelkle, M. C. (Eds.). (2018). *Continuous time modeling in the behavioral and related sciences*. Springer International Publishing.

Voelkle, M., Oud, J. H. L., Davidov, E., & Schmidt, P. (2012). An SEM approach to continuous time modeling of panel data: Relating authoritarianism and anomia. *Psychological Methods, 17*, 176–192.

Weiss, M., Schmucker, M., & Lösel, F. (2022). Meta-analysis on the effects of family-oriented prevention in German-speaking: A contribution to cross-cultural generalization. *International Journal of Developmental Science, 16*, 3–18. https://doi.org/10.3233/DEV-210320

Chapter 5
A Variational Approach to Continuous Time Dynamic Models

Hannes Meinlschmidt, Meike Sons, and Mark Stemmler

5.1 Introduction

Continuous time (CT) models are a well-established concept to extract information about the evolution of psychological processes from longitudinal data as it arises, e.g., in long-term and large-scale panel studies (Boker, 2002; Oud & Delsing, 2010; Voelkle et al., 2012; Ryan et al., 2018). The basic idea is that the moods, feelings, and generally, the behavior of sentient beings evolve in a continuous manner in time, so that a model that describes (a particular part of) these traits should naturally also be of continuous time form. With this point of view, change and interaction occurs continuously, and data collection in studies is merely an evaluation of the state of the traits under consideration at a given point in time. A conceptual model is illustrated in Fig. 5.1. CT models are complemented by discrete time (DT) models which are the "classical" viewpoint taken by many researchers in the past, in which the temporal evolution of the traits is modeled to occur from one time point to the next in a discrete manner. This represents the point of view of change occurring when we measure it.

H. Meinlschmidt (✉)
FAU Erlangen-Nürnberg, Department of Mathematics, Chair for Dynamics, Control, Machine Learning and Numerics (AvH-Professorship), Erlangen, Germany
e-mail: meinlschmidt@math.fau.de

M. Sons
Bayerisches Landesamt für Gesundheit und Lebensmittelsicherheit, Arbeitsmedizinisches Institut für Schulen (AMIS-Bayern), Bamberg, Germany
e-mail: meike.sons@lgl.bayern.de

M. Stemmler
Institute of Psychology, Friedrich-Alexander-University of Erlangen-Nuremberg (FAU), Erlangen, Germany
e-mail: mark.stemmler@fau.de

M. Stemmler et al. (eds.), *Dependent Data in Social Sciences Research*,
https://doi.org/10.1007/978-3-031-56318-8_5

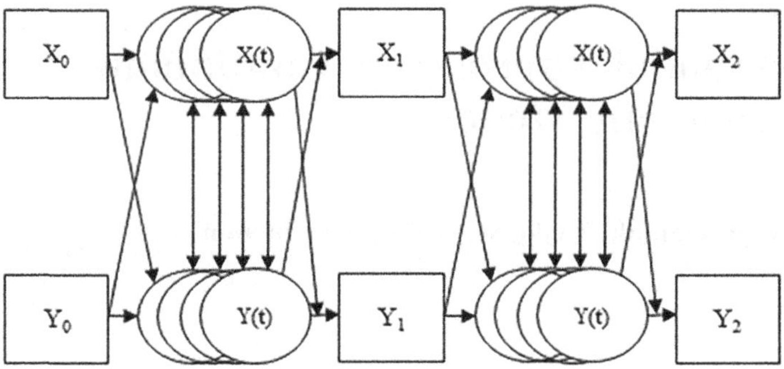

Fig. 5.1 Conceptual model. A continuous time perspective on a cross-lagged panel model. The first three of the several time points for the two-process continuous time model are illustrated. Occasions at which no observations are made are represented by latent variables (circles). The processes influence one another continuously over time

CT models offer a few practical advantages over DT models which are all related to the fundamental difference of CT models being structurally independent of the underlying time intervals used in data collection; for example, different lengths of time intervals between panel evaluations are naturally incorporated into CT models, whereas for DT models, particular care has to be taken in order to not obtain results which depend explicitly on the time intervals in data collection (Voelkle & Oud, 2013). This in particular also concerns the case where data may be missing for individuals at certain collection time points (Oud & Voelkle, 2014). In this sense, CT models are ideally suited to handle big and complex, imperfect data like panel data (Oud & Jansen, 2000). Also, the study of dynamic relationships between variables such as cross-lagged panel effects is easily possible within the framework of CT models. Examining how cross-effects evolve and vary as a function of the time interval between evaluations leads to a more dynamic view on the underlying processes and avoids oversimplification (Dormann et al., 2019). We do however not go into a detailed comparison between CT and DT models here but only refer to, e.g., (Ryan et al., 2018; Voelkle et al., 2012) and the references therein for a comprehensive exposition.

In fact, we rather take the CT modeling approach for longitudinal data for granted but propose a different way of deriving and calculating the actual model parameters which disposes of the connection to DT models nearly completely. Indeed, a most common approach (Oud & Jansen, 2000; Oud et al., 2008; Voelkle et al., 2012) to CT modeling of panel data as we want to consider goes via a Structural Equation Model (SEM) and the Exact Discrete Model (EDM), although there of course also exist other approaches such as filter methods (e.g., Oud et al. (2008)). In the SEM approach, one supposes that the underlying model is of continuous time and sets up a large SEM by relating the CT model at time evaluation points with an associated DT model, the EDM. More precisely, if $\mathbf{x}$ represents the, say, V-dimensional vector of quantities of interest, evolving over time, and $t_0, t_1, t_2, \ldots$ are the time points of data collection, then a linear DT model in state-space form would be given by the difference equation

$$\mathbf{x}(t_i) = \mathbf{A}(\Delta t_i)\mathbf{x}(t_{i-1}) + \mathbf{b}(\Delta t_i) + \mathbf{w}(\Delta t_i), \tag{5.1}$$

where $\Delta t_i = t_i - t_{i-1}$ is the time interval between data collection time points, $\mathbf{A}(\Delta t_i)$ is the $V \times \mathrm{V}$ transition matrix from time point t_{i-1} to t_i, and $\mathbf{b}(\Delta t_i)$ and $\mathbf{w}(\Delta t_i)$ are the V-dimensional intercept vector and random innovations depending on Δt_i (Voelkle et al., 2012, eq. (13)). The sought for quantity here is $\mathbf{A}(\Delta t_i)$. The associated linear CT model would be described by the continuous time (stochastic) differential equation[1]

$$\dot{\mathbf{x}}(t) = A\mathbf{x}(t) + b + G\frac{\mathrm{d}W(t)}{\mathrm{d}t} \tag{5.2}$$

with the time derivative $\dot{\mathbf{x}}$, the *drift matrix* A ($V \times V$), the continuous time intercept b, and a continuous time error process (random walk, Wiener process) induced by the $V \times V$ lower triangle matrix G (Voelkle et al., 2012, (14)). Relating the explicit solution to (5.2) with the DT model (5.1), one obtains the EDM

$$\mathbf{x}(t_i) = e^{A\Delta t_i}\mathbf{x}(t_{i-1}) + A^{-1}\big[e^{A\Delta t_i} - I\big]b + \mathbf{w}(\Delta t_i),$$

with the matrix exponential $e^{A\cdot}$ and the $V \times V$ identity matrix I and the covariance

$$\operatorname{cov}\mathbf{w}(\Delta t_i) = \int_{t_{i-1}}^{t_i} e^{A(t_i-s)}GG^{\top}e^{A^{\top}(t_i-s)}\,\mathrm{d}s.$$

In particular, we have $\mathbf{A}(\Delta t_i) = e^{A\Delta t_i}$, which is the fundamental link between $\mathbf{A}$ and the drift matrix A estimated in a CT model. The foregoing identities for $\mathbf{x}(t_i)$ are then used to set up an SEM whose parameters are highly nonlinear in the sought for quantities, most obviously A which is modulated by the matrix exponential. The SEM parameters are fitted to the given data by, e.g., a maximum-likelihood estimator (Voelkle et al., 2012, p. 184). Nowadays there are sophisticated packages which automatically take care of all the mathematical setup such as `ctsem` (Driver et al., 2017; Driver & Voelkle, 2018) for R. Still, the SEM approach fundamentally links back to the discrete time setup in the way it is set up and fitted.

We next propose a purely continuous time, dynamical systems-type approach to CT modeling. This conforms to the following point of view for CT models as

[1] We stick with boldface notation $\mathbf{x}$ for the *trajectories*; otherwise, matrices and vectors related to the DT model are written in boldface, too, whereas the corresponding quantities in the CT model are in regular font.

in (Voelkle et al., 2012, p. 179):

> One can also conceive of such a continuous time model as a dynamical system. At each point in time, the system has a specific configuration, but time itself never acts as an explanatory variable in the model. Instead, the model itself is an explanatory model. Arguably, this is often a more realistic view of the world compared with models that include time explicitly as a predictor (e.g., HLMs or LGMs). For example, when using an LGM to study learning, we typically do not assume that time "causes" learning, although we mathematically model it this way.

This is done in the next section. We follow up with a preliminary comparison between the results obtained by `ctsem` and the proposed algorithm, respectively, for a practical example relating satisfaction with health and satisfaction with work from the German Socio-Economic Panel (SOEP, Sozio-oekonomisches Panel (SOEP)).

5.2 Machine Learning-Type Approach to Continuous Time Modeling

We next describe a direct way of estimating the continuous time matrix A in a CT model given by (5.2) by a direct mathematical optimization problem in terms of a variational problem. The idea follows the basic paradigm of continuous time modeling by assuming that the behavior of a subject in the respective dataset follows a predetermined pattern over time whose properties can be captured by a linear continuous time differential equation such as (5.2) above. There will be no reformulation in terms of the DT values in an SEM or the likes, but we use the way how the individual subjects' trajectories depend on A *directly*. This will prove to work in a very direct and mathematically fully tractable manner.

To make the idea clear, we consider only the simplest version of (5.2) and do not include intercepts or stochastic noise at first. We come back to these extensions later. That is, the CT model is now given by

$$\dot{\mathbf{x}}(t) = A\mathbf{x}(t). \tag{5.3}$$

For the sake of exposition, we further explain the approach first for one individual subject $\mathbf{z}$. Say, we have $(k + 1)$ evaluations of panel data $z_0, z_1, \ldots, z_k$ for this subject (vectors of a fixed length), each obtained at the respective times $t_0, t_1, \ldots, t_k$ which we assume to be in increasing order. The solution to the linear continuous time differential equation (5.3) for $\mathbf{z}$ starting from the initial value z_0 at time t_0 is given by $\mathbf{z}(t) = e^{A(t-t_0)}z_0$, using the matrix exponential function. We can imagine $t \mapsto \mathbf{z}(t)$ as the trajectory of the initial starting point z_0 induced by the linear differential equation with matrix A. Thus, under the foregoing assumption, we would expect that the panel data points z_i each lie on that trajectory, so $z_i = \mathbf{z}(t_i) = e^{A(t_i-t_0)}z_0$ for every i. Of course, due to random factors, we cannot expect this equality to hold precisely, but only in an approximating sense. From this point of view, we are looking for a matrix A such that the discrepancy between the panel data

$z_1, z_2, \ldots, z_k$ and the trajectory evaluations $e^{A(t_1-t_0)}z_0, e^{A(t_2-t_0)}z_0, \ldots, e^{A(t_k-t_0)}z_0$ is minimal, for instance, in a least-squares sense. That means that A should be a solution to the minimization problem

$$\min_A \sum_{i=1}^{k} \left\| e^{A(t_i-t_0)}z_0 - z_i \right\|_2^2, \tag{5.4}$$

where $\|\cdot\|_2$ is the standard Euclidean norm. (We do not include $i = 0$ in the minimization problem by construction: The trajectory $t \mapsto z(t)$ always starts at z_0 at time t_0.) In an optimal world without perturbations or random effects, we could hope to be able to determine A exactly such that the foregoing minimization problem (5.4) has an optimal value of 0. But in reality, there will be discrepancies between z_i and the trajectory point $\mathbf{z}(t_i) = e^{A(t_i-t_0)}z_0$. Moreover, (5.4) is a nonconvex optimization problem such that we can only expect a *local* solution instead of a global one. But still, generally, by (locally) solving (5.4), we determine A such that the trajectory evaluations $\mathbf{z}(t_1), \mathbf{z}(t_2), \ldots, \mathbf{z}(t_k)$ fit the observed data $z_1, z_2, \ldots, z_k$ in a (locally) *best possible way* with the given data. This is the basis of a *variational formulation* of continuous time modeling of cross-lagged panel data. From this point of view, we do not require any implicit or explicit statistical assumptions about probability distributions of random elements or perturbations in the given panel data. Mathematically, this approach is similar to variational techniques in so-called data assimilation (Asch et al., 2016).

Moreover, if we view this task as one of the fitting values given by a *model function*, here represented by the parameters in A (its coefficients) and solving the linear differential equation (5.3), to a given set of *measurements*, then we can interpret this approach exactly as a form of machine learning. Note that A only indirectly creates the model values: We use it to obtain time-continuous trajectories in dependence on A, which are then fitted to the observed data or measurements $z_1, z_2, \ldots, z_k$ at the time instants t_i.

The above description was given for one single subject. Next we formulate the abovementioned approach for a full panel set. Suppose that we have N total subjects and for each subject, indexed by j, we have a number of $(k_j + 1)$ data points $z_0^j, z_1^j, \ldots, z_{k_j}^j$ associated with time points $t_0^j, t_1^j, \ldots, t_{k_j}^j$ which we again supposed to be ordered in an increasing fashion for each j. (The time points of evaluation are allowed to vary with each subject j.) Then the associated full minimization problem becomes

$$\min_A \sum_{j=1}^{N} \omega_j \sum_{i=1}^{k_j} \left\| \mathbf{z}^j(t_i^j) - z_i^j \right\|_2^2, \tag{5.5}$$

where each function $\mathbf{z}^j$ is the solution to the time-continuous linear differential equation (5.3) for $\mathbf{z}^j$ with initial value z_0^j at time t_0^j, so $\mathbf{z}^j(t_0^j) = z_0^j$. We could specify that this means that $\mathbf{z}^j$ is given by

$$\mathbf{z}^j(t) = e^{A(t-t_0^j)} z_0^j$$

and insert this formula in (5.5) to get the analog to (5.4). It will however be convenient for later considerations to retain the abstract description for z^j as the solution to (5.3) with initial value $\mathbf{z}^j(t_0^j) = z_0^j$. In the full formulation (5.5), we moreover have included a weighting parameter $\omega_j \geq 0$ for each subject. This parameter could be used, for example, to normalize each outer summand in (5.5) (indexed by j) with respect to the number of subject-specific data points k_j, which we could do by choosing $\omega_j = 1/k_j$ for this case. However, for example, it may also be reasonable to choose these parameters ω_j as 1 for each j in which case a subject with a higher number of data points would indirectly be considered more important to be well fitted than subjects with a lower number of data points. (In principle, it would also be possible to have a parameter ω_i^j for every individual data point. This could be used if extensive prior knowledge about individual data points was available.)

Intercepts and Trait Variables So far, the optimization approach to CT modeling of cross-lagged panel data was described for the simplest case of the underlying differential equation (5.3). It is also possible to include continuous time intercepts b (possibly with a selector $b = Bu$ for subgroups of subjects) and continuous time trait variables κ^j (for each subject j) in these considerations in a very straightforward way. We refer to Oud and Delsing (2010) for a comprehensive overview of these aspects. The modification would be to keep the minimization problem (5.5), but now $\mathbf{z}^j$ is defined to be the solution to

$$\dot{\mathbf{x}}(t) = A\mathbf{x}(t) + b + \kappa^j \tag{5.6}$$

with initial value $\mathbf{z}(t_0^j) = z_0^j$, and in addition to minimizing with respect to A in (5.5), we also minimize with respect to b (or u in $b = Bu$) and the collection $\kappa^1, \kappa^2, \ldots, \kappa^N$. Regarding the subgroup selectors $b = Bu$, it would be interesting to consider recursive partitioning techniques which were very recently also transferred to continuous time models (Arnold et al., 2023). We note that also Eq. (5.6) admits a unique solution satisfying the given initial value, and there is an explicit formula (the *variation of constants* or *Duhamel* formula) which we could use if so desired.

Stochastic Differential Equation Formulation It seems natural to include also a stochastic error term as in the CT model (5.2) in the linear differential equations (5.3) or (5.6) and then also to optimize with respect to the matrix G. Then $\bar{\mathbf{z}}^j$ is the solution to the continuous time *stochastic* differential equation (5.2) with initial value $\bar{\mathbf{z}}(t_0^j) = z_0^j$. (We use the bar in $\bar{\mathbf{z}}^j$ as a visual cue that $\bar{\mathbf{z}}^j$ is supposed to be a solution to the *stochastic* differential equation.) The trajectories $\bar{\mathbf{z}}^j$ are now random variables since they arise as solutions to a stochastic differential equation. Accordingly, we cannot optimize the trajectories directly but only stochastic indicators such as moments. We choose the expected value, and the associated minimization problem then becomes

$$\min_{A,G} \sum_{j=1}^{N} \omega_j \sum_{i=1}^{k_j} \mathbb{E}\Big[\|\bar{\mathbf{z}}^j(t_i^j) - z_i^j\|_2^2 \Big]. \tag{5.7}$$

However, since (5.2) is a *linear* stochastic differential equation, we know that the relationship between the white noise induced by G and the solutions $\bar{\mathbf{z}}^j$ is linear. Using this and a few more properties of the Wiener process W when expanding the square in (5.7), it turns out that (5.7) is equivalent to

$$\min_{A,G} \sum_{j=1}^{N} \omega_j \sum_{i=1}^{k_j} \left[\|\mathbf{z}^j(t_i^j) - z_i^j\|_2^2 + \int_{t_0^j}^{t_i^j} \|e^{A(t_i^j - s)} G\|_{\mathrm{Fr}}^2 \, \mathrm{d}s \right], \tag{5.8}$$

where each $\mathbf{z}^j$ is again the solution to the *deterministic* (so, *nonstochastic*) linear differential equation (5.6) with initial value $\mathbf{z}(t_0^j) = z_0^j$. Here, $\|\cdot\|_{\mathrm{Fr}}$ denotes the Frobenius norm of a matrix. We derive this in the Appendix. But in (5.8), we see that any optimal pair for the variables A and G will necessarily be of the form $(\bar{A}, 0)$, since G *only* occurs in a sum of nonnegative terms in the objective function to be minimized, so it will always be optimal to choose $G = 0$. In this sense, we postulate that we can always consider the deterministic problem setup (5.5), which is exactly the resulting problem with $G = 0$, in order to determine the dynamics of the problem, that is, A and possible trait vectors, without loss of generality. The situation would change if G was given and fixed and thus not subject to minimization in (5.7), but it is not clear how one would construct G from given data. Vice versa, given or having optimized for the dynamics, we expect that one could devise a second-stage optimization problem to estimate G or $GG^T = Q$ from the data. We leave this for future work.

Measurement Equation and Error There is also a straightforward extension of the method described so far to also include a measurement equation and associated measurement error. Then, we assume that the trajectories, $\mathbf{z}^j$ or $\bar{\mathbf{z}}^j$, respectively, are in fact *latent*, and we only directly observe a, say, linear function of them at the time points t_i^j, induced by a matrix C, with an error w_i^j and an offset d^j:

$$y_i^j = C\mathbf{z}^j(t_i^j) + d^j + w_i^j.$$

Then we would replace $\mathbf{z}^j(t_i^j)$ in (5.5) and (5.8) by y_i^j. Since this is an affine-linear function in $\mathbf{z}^j(t_i^j)$, this modification gives rise to only marginal mathematical changes in the approach. In particular, assuming that the expected value of the measurement errors is zero and that they are uncorrelated with respect to the random error in the stochastic differential equation (5.2), the derivation of (5.8), and the conclusion that it suffices to consider the deterministic counterpart of the respective formulation, stays correct.

Particularities of This Approach The approach as laid out so far seems promising to be very effective mainly due to three reasons.

First, minimizing (5.5) with respect to the drift matrix A will certainly involve solving the differential equation (5.6) for each $\mathbf{z}^j$ many times, at least as often as the current value of the function to be minimized needs to be evaluated. For a large number N of trajectories, this will amount to a huge amount of solves. On the other hand, numerical methods for differential equations are extremely well researched, and highly efficient solvers are ubiquitous in nearly every programming language which we can make use of, in particular, because, typically, the dimension V of the state vectors will not be large at all.

Second, while (5.5) is highly nonlinear in the optimization variable A, the latter's number of variables V^2 will rather be very low for standards in numerical optimization. Similarly to the situation as for the differential equation, this allows to use the full width of available theory for nonlinear optimization problems and associated efficient solvers. If we include subject-specific trait variables, the number of optimization variables increases quite drastically by $V \times N$; however, while the overall dynamics induced by A are coupled across all subjects, subject-specific traits only influence one particular trajectory, effectively leading to a certain decoupling between optimization for each subject-specific trait. (For the same reason, we do not expect problems regarding over-fitting, neither in this regard nor without subject-specific traits, since then the number of degrees of freedom V^2 is very small compared to the number of subjects.)

Still, there is the caveat that if we denote the function to be minimized in (5.5)—the so-called *objective function*—by $f(A)$, then we will require at least the gradient $\nabla f(A)$ of f to use more efficient optimization algorithms. That is, we need to understand and derive for which directions H (this is a matrix) the objective function f will infinitesimally decrease when going from A toward $A + H$. This requires either some more or less involved mathematics to derive an analytic expression for the gradient $\nabla f(A)$, or usage of *automatic differentiation* tools, which are ubiquitous nowadays, to derive the gradient automatically on a machine level. The latter could however become a bottleneck in performance, such that the former is preferable in general, if feasible. Having an explicit description of the gradient $\nabla f(A)$ also allows to optimize the efficiency with which it is calculated. Note that there are also results on an explicit description of the gradients for the maximum-likelihood estimation which is done in `ctsem` (Singer, 1990), but it seems that these have been mostly superseded by automatic differentiation tools these days. We also mention that the particular form of the objective function to be minimized in (5.5) or (5.8), respectively, is susceptible to so-called stochastic optimization algorithms such as the nowadays ever present SGD (Stochastic Gradient Descent) which could provide further performance speedups.

Third, and most importantly, there is room for generalization. Indeed, the whole approach does not at all rely on an explicit formula for the solution to the differential equation (5.6). It is thus straightforward to generalize the whole approach to, e.g., time-dependent drift matrices $A(t)$. Here, the amount of parameters to be optimized would increase drastically; formally, it would be uncountable, but practically, it would become $V^2 \times$ (the amount of time steps used in a numerical discretization of the differential equation).

Recently, regularization for continuous time models, leading to sparsity in the parameters, that is, less complex models, has been considered by Orzek, Arnold, and Völkle in Orzek et al. (2023); Orzek and Voelkle (2023). Extending the minimization problems (5.5) or (5.8) with such regularizations is very natural and could be considered without further ado. (In the present formulation such a regularization approach would only make sense with trait variables and intercepts, since the parameters in A are already few enough.) This would require to use adapted optimization algorithms to actually solve the optimization problems, but these are well researched and very efficient nowadays. We refer to the references mentioned before.

Going even further, one could switch to *nonlinear* models of the form

$$\dot{\mathbf{x}}(t) = F_\theta(\mathbf{x}(t))$$

with a nonlinear function parametrized by a vector of parameters θ such as, for example, a Neural Network. In this case, a stochastic differential equation formulation would also be much more difficult to handle. While such a generalization would not at all be straightforward, it nevertheless seems very promising to investigate. We leave this for future work.

Finally, for full disclosure, we point out again that the present approach is inherently deterministic so far. In particular, by design, there will be no confidence intervals for the estimated parameters, and we do not yet estimate the diffusion matrix Q; only the actual dynamics induced by A will be estimated.

Empirical Example

Work, pressures, strains, and stresses within the workplace, which occupy most people's waking time, have been identified as being a potentially important health factor (Faragher et al., 2013; Judge & Watanabe, 1993). Numerous theories now exist, developed from a wide range of perspectives, postulating a direct link between organizational/workplace stress and personal well-being. Meta-analytic results show that an increase in work satisfaction would be associated with improved health satisfaction. Furthermore, there is evidence for a reciprocal effect from health satisfaction to job satisfaction (Faragher et al., 2013; Judge & Watanabe, 1993). We tested cross-lagged effects between satisfaction with health and satisfaction with work using Continuous Time Structural Equation Modeling (`ctsem`). While many questions might be asked using this approach, the questions we will address here are the following: What are the continuous temporal dynamics of satisfaction with health and satisfaction with work? How much variation does there exist in such dynamics? And, in particular, how do the results obtained using a `ctsem` approach compare with those yielded by the above variational algorithm, and what are possible implications for model fitting?

Sample In this example, we used data from the German Socio-Economic Panel (SOEP, Sozio-oekonomisches Panel (SOEP)), an ongoing survey of German households and persons which has been conducted annually since 1984. More precisely, we chose the measurements from 1984 to 2021.

Measures We are interested in the association between satisfaction with health and satisfaction with work. In the SOEP participants were asked to rate their satisfaction with various aspects of their lives on an 11-point scale, ranging from 0 (totally unhappy) to 10 (totally happy). The question in this regard is as follows: "How satisfied are you today with the following areas of your life?" The chosen items are "With your health" and "With your work."

Model We used two different approaches to obtain a model for the foregoing aspects in the given data.

First, we fitted an SEM-based continuous time (CT) cross-lagged panel model (Dormann et al., 2019; Driver et al., 2017; Voelkle et al., 2012; Voelkle & Oud, 2013) to our data. We fit the model using the open-source software `R` (version 4.2.2; `R` Core Team (2022)) and the package `ctsem` (version 3.7.2, Driver et al. (2017); Driver & Voelkle (2018)), which interfaces to `OpenMx` (Neale et al., 2016) via `ctsemOMX` (version 1.0.4). The precise setup of the model in `R` can be found in the Appendix.

Second, we used the variational approach to Continuous Time Modeling as it was explained before. The code used to generate the results was written from scratch in `python` and can be found on `github` (Meinlschmidt, 2023).

Since this is a first glimpse and a trial run for the variational type approach to continuous time modeling, we do not consider time-continuous intercepts or individual traits in the empirical example.

Results

`ctsem` The results of the continuous time models estimated by `ctsem` are summarized in Table 5.1. Estimates of the auto- and cross-effects for both processes are displayed in the drift matrix which resulted to

$$A_{\text{ctsem}} = \begin{pmatrix} -0.457024 & 0.262225 \\ 0.376226 & -0.5775 \end{pmatrix}.$$

The time-continuous auto-effects on the diagonal of the drift matrix reflect the processes' temporal stability; negative values for auto-effects are expected in the

Table 5.1 Continuous time parameter estimates: satisfaction with health and satisfaction with work

	Satisfaction Health (OV)		Satisfaction Work (OV)	
Parameter	Estimate	[95% CI][a]	Estimate	[95% CI][a]
Drift matrix (A)				
Auto-effects	−0.457*	[−0.462; −0.452]	−0.578*	[−0.583; −0.572]
Cross-effects	0.262*	[0.258; 0.267]	0.376*	[0.371; 0.382]
Diffusion matrix (Q)				
q	4.105*	[4.082; 4.129]	5.399*	[5.367; 5.432]
$qJ = qS$	0.388*	[0.384; 0.408]		
Baseline ($t0$)				
M_{t0}	2.468*	[2.453; 2.484]	2.307*	[2.29; 2.324]
var_{t0}	4.42*	[4.375; 4.467]	5.272*	[5.218; 5.328]
cov_{t0}	1.462*	[1.425; 1.5]		
Model indices				
−2LL	3,496,829			
Degrees of freedom	2837			

[a]Confidence intervals for many parameters in CT models may not be symmetric around the point estimate

*$p < 0.05$

Note: Continuous time parameter estimates for auto-effects are always negative but correspond to positive effects in discrete time

continuous time approach. Looking at the derived auto-effects, so the diagonal entries of the matrix $\mathbf{A}_{\text{ctsem}}(\Delta t) = e^{A_{\text{ctsem}}\Delta t}$ for $0 < \Delta t < 37$ years (see Fig. 5.2), the average within-person stability of satisfaction with health for a discrete time interval of 1 year ($\Delta t = 1$) was 0.66. As expected, stability coefficients decreased with increasing time. The stability of satisfaction with work was lower with 0.59 for a discrete time interval of 1 year ($\Delta t = 1$) and asymptotically approaches zero with increasing time. Of primary interest for our research question are the cross-effects of the drift matrix, which reflect the inter-relatedness of the processes (e.g., an increase in X leads to a decrease/increase in Y). We analyze the significance of the effects using 95% likelihood-based confidence intervals. The drift coefficients make it possible to estimate the auto-regressive and cross-effects for any chosen time interval Δt between 0 and 37 years. In order to predict the relationships between variables over a given time period, the continuous time parameters are used to calculate the discrete time parameters (Dormann et al., 2019). The cross-effects, so the non-diagonal entries of $\mathbf{A}_{\text{ctsem}}(\Delta t) = e^{A_{\text{ctsem}}\Delta t}$ as a function of the time interval between observations ($0 < \Delta t < 37$ years) are shown in Fig. 5.3. Regarding the relationship between satisfaction with health and satisfaction with work, we found that both the positive effect of satisfaction with health on satisfaction with work (coef. = 0.24 for $\Delta t = 1$ year) and the positive effect of satisfaction with work on satisfaction with health (coef. = 0.16 for $\Delta t = 1$ year) reach their respective maxima at a time length between 2 and 3 years and decreases for longer times, although

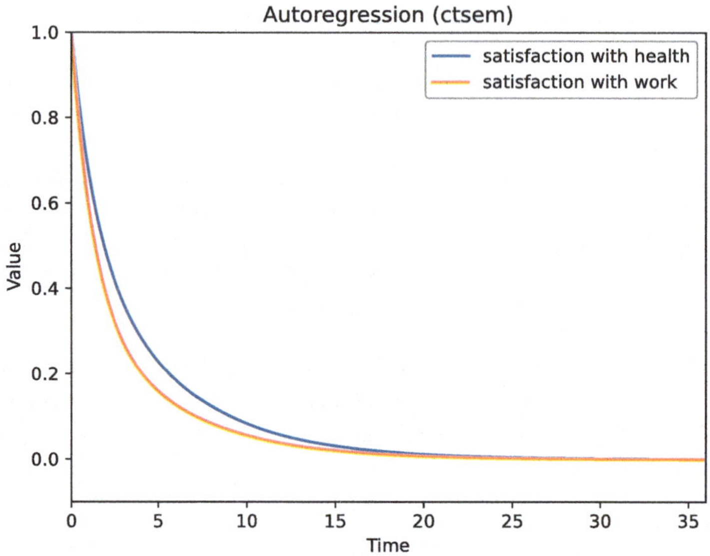

Fig. 5.2 Auto-effects derived from the continuous time model for satisfaction with health and satisfaction with work. The plot shows the auto-effects on the diagonal of $\mathbf{A}_{\text{ctsem}}(\Delta t)$ as a function of a time interval $0 < \Delta t < 37$ years. The stability of satisfaction with health (blue) and the stability of satisfaction with work (orange)

the maximum is decisively smaller than for the positive effect of satisfaction with health on satisfaction with work. (Indeed the maxima must occur at the same time for mathematical reasons related to the matrix exponential.)

Variational Approach We next present results for the SOEP panel data with regard to the relationship between satisfaction with health and satisfaction with work, as described earlier.

Before we present the actual results, we comment on the computations and methodology. We have used the normalization $\omega_j = 1/k_j$. The minimization was solved by a Quasilinear Newton Method (BFGS) which was supplied with the objective function to be minimized as in (5.5) and its gradient with respect to the parameters in A. The code for these calculations was written in Python whose SciPy package offers the aforementioned BFGS optimization method. For the initialization of A, we used the matrix which was obtained from `ctsem` as in Table 5.1.

The two-dimensional data points were set up such that the first variable corresponds to "satisfaction with health" and, accordingly, the second one corresponds to "satisfaction with work." Then the optimization result, so the matrix A

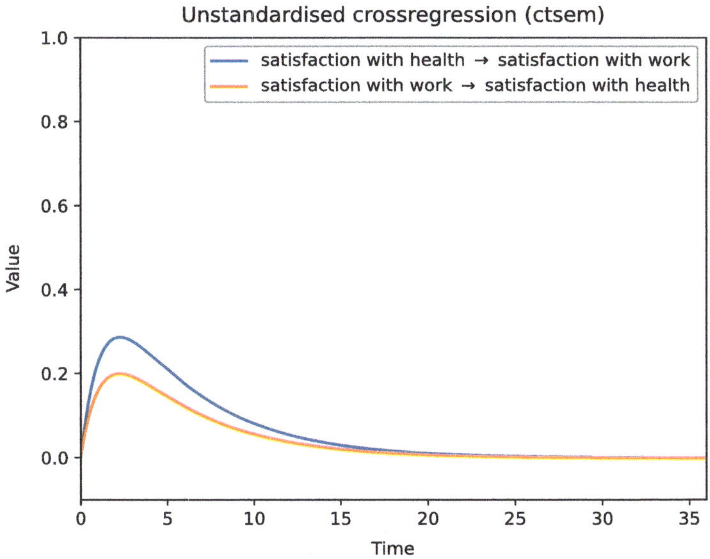

Fig. 5.3 Cross-effects derived from the continuous time model for satisfaction with health and satisfaction with work. The plot shows the cross-lagged parameters on the off-diagonal of $\mathbf{A}_{\text{ctsem}}(\Delta t)$ as a function of a time interval of $0 < \Delta t < 37$ years. Satisfaction with health predicting subsequent changes in satisfaction with work (blue) and changes in satisfaction with work predicting subsequent changes in satisfaction with health (orange)

minimizing—at least locally—problem (5.5) was found to be

$$A_{\text{learn}} = \begin{pmatrix} -0.258455 & 0.220716 \\ 0.386343 & -0.49353 \end{pmatrix}.$$

Let us note that, in order to validate our expectation that the possible phenomenon of over-fitting should not occur due to the low number $V^2 = 4$ of degrees of freedom in this example, the algorithm was run several times with a reduced dataset, where only 50% or 25% of the participants in the SOEP study were chosen randomly and used to optimize for A_{learn}, and the resulting matrix A_{learn} was indeed stable.

Looking at the matrix values for the discrete time drift matrix as derived from the continuous time one, $\mathbf{A}_{\text{learn}}(\Delta t) = e^{A_{\text{learn}}\Delta t}$ as depicted in Figs. 5.4 (auto-effects) and 5.5 (cross-effects), we find the expected behavior decreasing values for the auto-effects. The average within-person stability of satisfaction with health for a discrete time interval of 1 year ($\Delta t = 1$) was 0.8, whereas the stability of satisfaction with work was lower with 0.63 for the same discrete time interval. The cross-effects again exhibit a maximum after which they decline; however, this time the maximum occurs at about $\Delta t = 4$ years. Again, the positive influence of satisfaction with health on satisfaction with a coefficient of 0.27 at $\Delta t = 1$ year is more substantial than the reverse one with a coefficient of 0.15 at $\Delta t = 1$ year, as seen in the graphs.

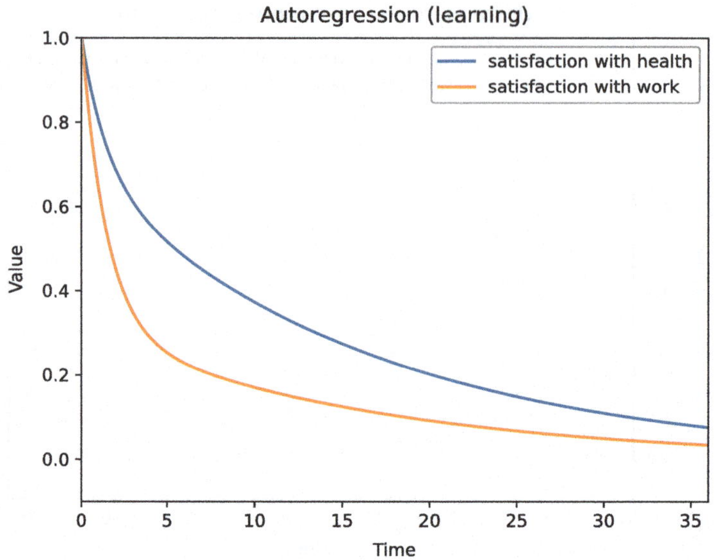

Fig. 5.4 Auto-effects derived from the continuous time model for satisfaction with health and satisfaction with work via the machine learning approach. The plot shows the auto-effects on the diagonal of $\mathbf{A}_{\text{learn}}(\Delta t)$ as a function of a time interval $0 < \Delta t < 37$ years. The stability of satisfaction with health (blue) and the stability of satisfaction with work (orange)

Discussion

We next compare the different results A_{ctsem} and A_{learn} for the time-continuous drift matrices and associated consequences for the time-discrete drift matrices $\mathbf{A}_{\text{ctsem}}$ and $\mathbf{A}_{\text{learn}}$.

Firstly, we remark that if f denotes the function to be minimized in the optimization problem (5.5), then $f(A_{\text{learn}})$ is 12% less (so: better) than $f(A_{\text{ctsem}})$ which is a substantial but not completely outrageous difference. Note that A_{ctsem} was the starting point for the numerical optimization algorithm used to solve the minimization problem (5.5) and the algorithm moved away to find A_{learn}, so A_{ctsem} is not a local minimum for (5.5). Thus, A_{ctsem} is a reasonable first approximation of a matrix minimizing (5.5), but not an actual optimal solution. Of course this comparison is a bit unfair since `ctsem` actually estimates—read: optimizes for—more parameters, in particular, the diffusion matrix $Q = GG^T$. However, if our main interest is in finding a matrix A such that the present panel data is best described by the dynamics induced by the associated linear differential equation, then it makes sense to compare the performances of A_{learn}, so A_{ctsem} only with regard to the optimization problem (5.5).

Comparing the actual parameters in A_{ctsem} and A_{learn}, we observe that the non-diagonal entries in each matrix are reasonably well matched, whereas the diagonal entries differ more substantially. However, the qualitative properties of a matrix

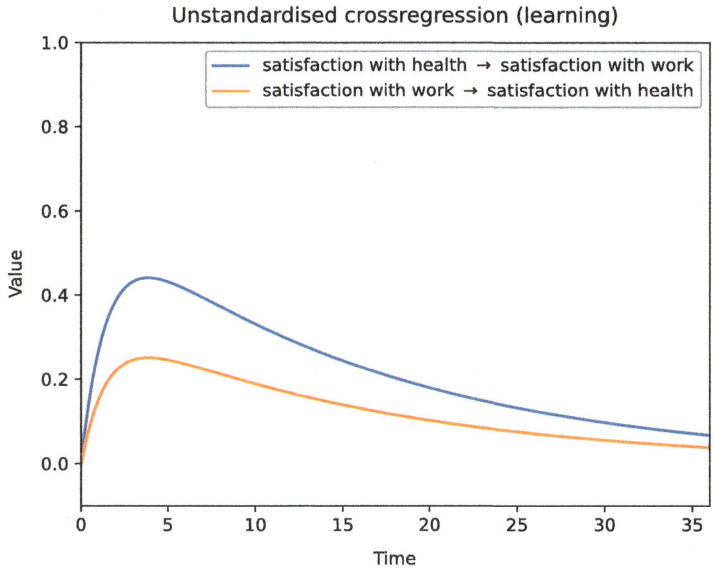

Fig. 5.5 Cross-effects derived from the continuous time model for satisfaction with health and satisfaction with work via the machine learning approach. The plot shows the cross-lagged parameters on the off-diagonal of $\mathbf{A}_{\text{learn}}(\Delta t)$ as a function of a time interval of $0 < \Delta t < 37$ years. Satisfaction with health predicting subsequent changes in satisfaction with work (blue) and changes in satisfaction with work predicting subsequent changes in satisfaction with health (orange)

A are better represented by its spectral properties, so by its eigenvalues $\lambda(A)$ and eigenvectors $v(A)$. Here we find (rounded and ordered)

$$\lambda(A_{\text{ctsem}}) = (-0.19, -0.84), \quad v(A_{\text{ctsem}}) = \left(\begin{pmatrix} 0.71 \\ 0.7 \end{pmatrix}, \begin{pmatrix} -0.57 \\ 0.82 \end{pmatrix} \right)$$

and

$$\lambda(A_{\text{learn}}) = (-0.06, -0.69), \quad v(A_{\text{learn}}) = \left(\begin{pmatrix} 0.74 \\ 0.67 \end{pmatrix}, \begin{pmatrix} -0.45 \\ 0.89 \end{pmatrix} \right).$$

While the eigenvalues are again rather different, it becomes apparent that the eigenvectors, so the principal directions of either A along which the solution to the differential equation (5.3) will evolve, are well matched. We can also observe that in Fig. 5.6 where the trajectories of a few randomly selected individuals in the panel study under the linear differential equation (5.3) are shown for $A = A_{\text{ctsem}}$ and $A = A_{\text{learn}}$, respectively. For $A = A_{\text{ctsem}}$, the dashed red lines indicate the eigenvectors $v(A_{\text{ctsem}})$ scaled by their respective (inverted) eigenvalue to show how the trajectory evolution aligns with the eigenvectors, making them a most important quantity of interest in this context.

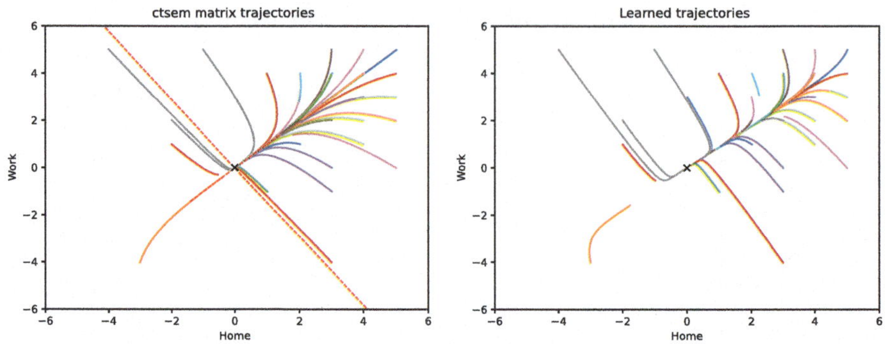

Fig. 5.6 Trajectories of a few randomly selected individuals in the panel study under the linear differential equation (5.3) for $A = A_{\text{ctsem}}$ (left) and $A = A_{\text{learn}}$ (right). For $A = A_{\text{ctsem}}$, the dashed red lines indicate the eigenvectors $v(A_{\text{ctsem}})$ scaled by their respective inverted eigenvalue

With respect to the questions of study for the SOEP data, so the influence of satisfaction with health on satisfaction with work and vice versa, we again refer to Figs. 5.2, 5.3, 5.4 and 5.5. It is apparent that the decay in the auto-effects as Δt increases is much slower for $\mathbf{A}_{\text{learn}}(\Delta t)$ than in $\mathbf{A}_{\text{ctsem}}(\Delta t)$; the mathematical reason behind this is that the eigenvalues $\lambda(A_{\text{learn}})$ of A_{learn} are less negative than those of A_{ctsem} in $\lambda(A_{\text{ctsem}})$. This means that the eigenvalues $\lambda(\mathbf{A}_{\text{learn}}(\Delta t = 1)) = e^{\lambda(A_{\text{learn}})}$ of $\mathbf{A}_{\text{learn}}(\Delta t = 1)$ are closer to 1 than those of $\mathbf{A}_{\text{ctsem}}$. On the other hand, the cross-effects exhibit quite similar behavior for both results, with the respective peaks being a bit higher and also for larger values of Δ for A_{learn} or $\mathbf{A}_{\text{learn}}(\Delta t)$, respectively, obtained from the learning approach.

We conclude that *qualitatively*, the results A_{ctsem} and A_{learn} of the different approaches agree, while *quantitatively*, they yield different conclusions.

Appendix

`ctsem` *Settings*

The following R snippet was used to perform the `ctsem` model estimation:

Program Code

```
data <- read.csv("soep_panel_file.csv")

data_selection <- data.frame(
    data$pid,
    data$syear,
    data$plh0171,
    data$plh0173
)
```

```
names(data_selection) <- c("pid","syear","hsat","wsat")

library("ctsem")
library("ctsemOMX")

wide <- ctLongToWide(
    datalong = data_selection,
    id = "pid",
    time = "syear",
    manifestNames = c("hsat", "wsat")
)

wide_int <- ctIntervalise(
    datawide = wide,
    Tpoints = 37,
    n.manifest = 2,
    manifestNames = c("hsat", "wsat"),
    individualRelativeTime = TRUE,
    imputedefs = FALSE
)

model <- ctModel(
    type = 'omx',
    n.latent = 2,
    n.manifest = 2,
    Tpoints = 37,
    manifestNames = c("hsat", "wsat"),
    latentNames = c("hsat", "wsat"),
    LAMBDA = diag(2),
    MANIFESTMEANS = matrix(data=0, nrow=2, ncol=1),
    MANIFESTVAR = matrix(data=0, nrow=2, ncol=2),
    DRIFT ="auto",
    CINT = matrix(data=0, byrow=T, nrow=2, ncol=1),
    DIFFUSION = "auto",
    TRAITVAR = NULL,
    T0TRAITEFFECT = NULL,
    MANIFESTTRAITVAR = NULL,
    startValues = NULL
)

set.seed(1)

model_fit <- ctFit(
    dat = wide_int,
    ctmodelobj = model,
    transformedParams = T
)

summary(model_fit, verbose=T)
```

Reformulation of the Variational Stochastic Approach

Let $\bar{\mathbf{x}}(t)$ be the solution to the linear stochastic differential equation (5.2) with initial value $\bar{\mathbf{x}}(t_0) = x_0$. Then $\bar{\mathbf{x}}(t)$ is given by (Arnold (1974, Cor. 8.2.4))

$$\bar{\mathbf{x}}(t) = e^{A(t-t_0)}x_0 + \int_{t_0}^{t} e^{A(t-s)}b\,\mathrm{d}s + \int_{t_0}^{t} e^{A(t-s)}G\,\mathrm{d}W(s). \tag{5.9}$$

Set further

$$\mathbf{x}(t) = e^{A(t-t_0)}x_0 + \int_{t_0}^{t} e^{A(t-s)} b \, \mathrm{d}s,$$

which is the solution to the nonstochastic linear differential equation (5.6) with initial value $\mathbf{x}(t_0) = x_0$ and with $\kappa^j = 0$, and let some time $T > t_0$ and datum x_T be given. Then, expanding the square,

$$\|\bar{\mathbf{x}}(T) - x_T\|_2^2 = \|\mathbf{x}(T) - x_T\|_2^2 + 2\Big(\mathbf{x}(T) - x_T, \int_{t_0}^{T} e^{A(T-s)} G \, \mathrm{d}W(s)\Big) + \Big\|\int_{t_0}^{T} e^{A(T-s)} G \, \mathrm{d}W(s)\Big\|_2^2,$$

where in the second term on the right, we mean the inner product. We next consider the expectation of the foregoing equation. From Arnold (1974, Thm. 4.4.14), the expectation of the stochastic Itô integral in (5.9) is zero, that is,

$$\mathbb{E}\Big[\int_{t_0}^{T} e^{A(T-s)} G \, \mathrm{d}W(s)\Big] = 0,$$

and we have the *Itô isometry*

$$\mathbb{E}\Big\|\int_{t_0}^{T} e^{A(T-s)} G \, \mathrm{d}W(s)\Big\|_2^2 = \int_{t_0}^{T} \mathbb{E}\,\|e^{A(T-s)}G\|_{\mathrm{Fr}}^2 \, \mathrm{d}s = \int_{t_0}^{T} \|e^{A(T-s)}G\|_{\mathrm{Fr}}^2 \, \mathrm{d}s.$$

Accordingly,

$$\mathbb{E}\,\|\bar{\mathbf{x}}(T) - x_T\|_2^2 = \|\mathbf{x}(T) - x_T\|_2^2 + \int_{t_0}^{T} \|e^{A(T-s)}G\|_{\mathrm{Fr}}^2 \, \mathrm{d}s.$$

Every summand in (5.7) is precisely of the foregoing form, so the last identity yields (5.8).

References

Arnold, L. (1974). *Stochastic differential equations: Theory and applications.* Dover Publications.

Arnold, M., Cáncer, P. F., Estrada, E., & Voelkle, M. (2023). Score-guided recursive partitioning of continuous-timestructural equation models (Chapter 3, this volume). In M. Stemmler et al. (Eds.), *Dependent data in social sciences research*. Springer.

Asch, M., Bocquet, M., & Nodet, M. (2016). *Data assimilation: Methods, algorithms, and applications.* SIAM. https://doi.org/10.1137/1.9781611974546.

Boker, S. M. (2002). Consequences of continuity: The hunt for intrinsic properties within parameters of dynamics in psychological processes. *Multivariate Behavioral Research, 37*, 405–422. https://doi.org/10.1207/S15327906MBR3703_5.

Dormann, C., Guthier, C., & Cortina, J. M. (2019). Introducing continuous time meta-analysis (CoTiMA). *Organizational Research Methods, 23*(4), 620–650. https://doi.org/10.1177/1094428119847277.

Driver, C. C., Oud, J., & Voelkle, M. (2017). Continuous time structural equation modeling with R package `ctsem`. *Journal of Statistical Software, 77*, 1–35. https://doi.org/10.18637/jss.v077.i05.

Driver C. C., & Voelkle, M. C. (2018). Hierarchical bayesian continuous time dynamic modeling. *Psychological Methods, 23*(4), 774–799. https://doi.org/10.1037/met0000168.

Faragher, E. B., Cass, M., & Cooper, C. L. (2013). The relationship between job satisfaction and health: A meta-analysis. In C. L. Cooper (Ed.), *From stress to wellbeing* (Vol. 1, pp. 254–271). Palgrave Macmillan. https://doi.org/10.1057/9781137310651_12.

Judge, T. A., & Watanabe, S. (1993). Another look at the job satisfaction-life satisfaction relationship. *Journal of Applied Psychology, 78*, 939–948. https://doi.org/10.1037/0021-9010.78.6.939.

Meinlschmidt, H. (2023). https://github.com/hannesmeinlschmidt/cautious-carnival.

Neale, M. C., Hunter, M. D., Pritikin, J. N., et al. (2016). OpenMx 2.0: Extended structural equation and statistical modeling. *Psychometrika, 81*, 535–549. https://doi.org/10.1007/s11336-014-9435-8.

Orzek, J. H., Arnold, M., & Voelkle, M. C. (2023). Striving for sparsity: On exact and approximate solutions in regularized structural equation models. *Structural Equation Modeling: A Multidisciplinary Journal*, 1–18. https://doi.org/10.1080/10705511.2023.2189070.

Orzek, J. H., & Voelkle, M. C. (2023). Regularized continuous time structural equation models: A network perspective. *Psychological Methods*. https://doi.org/10.1037/met0000550.

Oud, J. H. L., & Delsing, M. J. M. H. (2010). Continuous time modeling of panel data by means of SEM. In K. van Montfort, J. H. L. Oud, & A. Satorra (Eds.), *Longitudinal research with latent variables* (pp. 201–244). Springer. https://doi.org/10.1007/978-3-642-11760-2_7.

Oud, J. H. L., & Jansen, R. A. R. G. (2000). Continuous time state space modeling of panel data by means of SEM. *Psychometrika, 65*, 199–215. https://doi.org/10.1007/BF02294374.

Oud, J. H. L., & Singer, H. (2008). Continuous time modeling of panel data: SEM versus filter techniques. *Statistica Neerlandica, 62*, 4–28. https://doi.org/10.1111/j.1467-9574.2007.00376.x.

Oud, J. H. L., & Voelkle, M. C. (2014). Do missing values exist? Incomplete data handling in cross-national longitudinal studies by means of continuous time modeling. *Quality & Quantity, 48*, 3271–3288. https://doi.org/10.1007/s11135-013-9955-9.

R Core Team (2022). R: *A language and environment for statistical computing*. R Foundation for Statistical Computing. https://www.R-project.org/.

Ryan, O., Kuiper, R. M., & Hamaker, E. L. (2018). A continuous time approach to intensive longitudinal data: What, why and how? In K. van Montfort, J. H. L. Oud, & M. C. Voelkle (Eds.), *Continuous time modeling in the behavioral and related sciences* (pp. 27–54). Springer. https://doi.org/10.1007/978-3-319-77219-6_2.

Singer, H. (1990). *Parameterschätzung in zeitkontinuerlichen dynamischen Systemen*. Konstanzer Dissertationen–282.

Sozio-oekonomisches Panel (SOEP), Daten der Jahre 1984–2018. https://doi.org/10.5684/soep-core.v35.

Voelkle, M. C., Oud, J. H. L., Davidov, E., & Schmidt, P. (2012). An SEM approach to continuous time modeling of panel data: Relating authoritarianism and anomia. *Psychological Methods, 17*(2), 176–192. Erratum in: Psychological Methods (2012) 17(3), 384. https://doi.org/10.1037/a0027543.

Voelkle, M. C., & Oud, J. H. L. (2013). Continuous time modelling with individually varying time intervals for oscillating and non-oscillating processes. *British Journal of Mathematical and Statistical Psychology, 66*, 103–126. https://doi.org/10.1111/j.2044-8317.2012.02043.x.

Chapter 6
Finite Mixture Models for an Underlying Beta Distribution with an Application to COVID-19 Data

Cédric Noel and Jang Schiltz

6.1 Introduction

Finite mixture models in the sense of Nagin (2005) are fuzzy logic cluster analysis models for time series. Starting from a sample of trajectories, the aim is to detect a number of subgroups of the sample, so that subjects in the same group exhibit quite similar data trajectories, whereas two subjects from two different groups have trajectories that differ in some sense. These models are part of a larger strand of models that analyze latent evolutions in longitudinal data.

Model-based approaches to trajectory analysis comprise latent class growth analysis, growth mixture modeling, and group-based trajectory models (van der Nest et al., 2020). The objective of all these approaches is to capture information about interindividual differences in intraindividual change over time (Nesselroade, 1991).

There are a host of model-based techniques for analyzing outcome development over time (see Verbeken et al., 2014). According to van der Nest et al. (2020), they can be divided into models comprising the estimation of one latent class, such as growth curve models (Laursen & Hopf, 2006), or more than one latent class, such as latent class growth analysis (Berlin et al., 2014), examples of which include growth mixture models (Muthén & Shedden, 1999) and group-based trajectory models (Nagin, 2005). Quite similar models have been developed independently in bioinformatics (see Nadif & Role, 2021).

C. Noel · J. Schiltz (✉)
Department of Finance, University of Luxembourg, Luxembourg, Luxembourg
e-mail: jang.schiltz@uni.lu

M. Stemmler et al. (eds.), *Dependent Data in Social Sciences Research*,
https://doi.org/10.1007/978-3-031-56318-8_6

Growth curve models are not categorizing subjects, but try to explain the relationship between explanatory variables and the trajectories of interest as a whole. Most modern econometric techniques fall into this category (see Greene, 2018).

In latent class growth analysis, the relationship between the explanatory variables and the trajectories of interest is group specific. Growth mixture modeling, introduced by Muthén and Shedden (1999), is a very suitable framework to handle the issue of unobserved heterogeneity. In growth mixture modeling, there are random effects influencing all the parameters defining the typical group trajectories for all subjects in the sample. This has the advantage that fewer groups are required to specify a satisfactory model, but the groups are not always clearly defined, and group cross-over effects may exist. Group-based trajectory modeling, also called finite mixture models (Nagin, 2005) on the other hand, has no random effects to capture individual differences in a continuous way. All individual deviations from the typical trajectory of a group are treated as residual errors. It was originally introduced by Nagin and Land (1993) and is actually specifically designed to detect the presence of distinct subgroups among a set of trajectories. Thus, it represents an interesting compromise between analysis around a single mean trajectory and case studies (Von Eye & Bergman, 2003). Compared to subjective classification methods, the nonparametric mixed model has the advantage of providing a formal framework for testing the existence of distinct groups of trajectories. Furthermore, in finite mixture models, the error variance is supposed to remain constant over all groups and the complete timeline. This implies that far fewer parameters need to be estimated, and these kinds of models remain useful for small samples. Schiltz (2015) introduced a generalization of this model in which the error variance remains constant inside a given group but is allowed to differ across groups, which is a far more realistic assumption in a lot of applications. While the conceptual aim of the analysis is to identify clusters of individuals with similar trajectories, the model's estimated parameters are not the result of a classical cluster analysis but of maximum likelihood estimation (Nagin, 2005). Let's also point out that finite mixture models are actually fuzzy logic cluster analysis of time series models and should not be confounded with finite mixture models in the sense of McLachlan and Peel (2000), which just refer to mixtures of probability distributions.

This chapter extends finite mixture models to the case of an underlying Beta distribution and presents the **R** package **trajeR**, which allows to calibrate the model. The package can be seen as an extension of the **SAS** procedure **Proc Traj** developed by Jones et al. (2001), Jones and Nagin (2007) and their **Stata** version (Jones & Nagin, 2012), to the generalized finite mixture model, with the additional advantage that, as an **R** package, it is an open-source software and can be adapted to the needs of the users.

There are a hundreds of research papers that have been published about the COVID-19 pandemic over the last 2 years, and an excellent overview about the different strands of research thematics and the corresponding papers can be found in Spiegelhalter and Masters (2021). The use of a cluster analysis methodology in this context is however much rarer. Chandu (2020) analyzes 89 countries by means of a K-means clustering algorithm and identifies two groups of countries, showing that a high COVID-19 case fatality rate, higher proportion of positive test results, higher percentage of GDP spent as public health expenditure, and a greater percentage of elderly people are related. Zarikas et al. (2020) use hierarchical cluster analysis to analyze active cases, active cases per population, and active cases per population and per area based on Johns Hopkins epidemiological data, and they identify four different shapes in the evolution rate of COVID-19. Alvarez et al. (2020) use hierarchical cluster analysis methods on time series with the aim of identifying groups of countries with a similar spread of the coronavirus. They show that there are groups of countries with differentiated contagion dynamics and conclude that the actions taken by the countries, the speed at which they were taken, and the number of tests carried out may explain part of the differences in these dynamics. Rahman et al. (2020) propose a data-driven dynamic clustering framework for moderating the adverse economic impact of COVID-19 flare-up. They model lockdown as a clustering problem and design a dynamic clustering algorithm for localized lockdown by taking into account the pandemic, economic, and mobility aspects. James and Menzies (2020) propose a hierarchical and K-means cluster-based method to analyze COVID-19 pandemic cases and death counts. They study the change in both quantities over time and identify anomalous countries in the progression from COVID-19 cases to deaths. This analysis can aid in highlighting the most and least significant public policies in minimizing a country's COVID-19 mortality rate. Kinnunen et al. (2021) identify groups of countries with similar COVID strategies on the one hand and groups with similar performance success on the other hand and construct a composite COVID mitigation index for comparative purposes, thus implying how to redesign the strategic policies. They use dynamic clustering with a Gaussian mixture model in the sense of McLachlan on a 2-month period in order to achieve this.

In the second part of this chapter, we use a finite mixture model to analyze COVID-19-related data during the first part of the pandemic. We identify a classification of the world into five groups of countries with respect to the evolution of the contamination rate and describe the five groups as well as their typical contamination trajectories. We also show that the median population age is the main predictor of group membership. We do however not see any sign of efficiency of the sanitary measures taken by the different countries against the propagation of the virus.

The remainder of this chapter is structured as follows. In Sect. 6.2, we present the generalized finite mixture model for an underlying Beta distribution. Section 6.3 introduces our R package **trajeR**, and Sect. 6.4 shows how the model introduced in Sect. 6.2 can be calibrated with the help of **trajeR** by means of an example with simulated data. In Sect. 6.5, we present the example with real-world data about the recent COVID-19 pandemic. Section 6.6 finally contains the conclusion with an outlook on further research.

6.2 The Generalized Finite Mixture Model for an Underlying Beta Distribution

Finite Mixture Models

In group-based trajectory modeling, we consider a population of size N divided into K latent classes. The assignment of the individuals into classes is based on the degree of similarity of the developmental trajectories.

More precisely, consider a time-varying variable of interest Y defined in a population Ω of size N. Let $Y_i = y_{i_1}, \cdots, y_{i_T}$ be T measures of the variable Y, taken at times $t_1, \cdots, t_T$ for subject number i.

The aim of the analysis is to divide the population into K sub-populations $G_1, \ldots G_K$, which are homogeneous in the sense that two subjects in the same group have similar trajectories for the variable of interest Y and two subjects in different groups have quite different trajectories for the variable of interest Y.

Let $P^k(Y_i)$ be the probability of Y_i given membership in group G_k and $P(Y_i)$ the unconditional probability of observing the realization Y_i of Y. Furthermore, for a given group G_k, we suppose conditional independence for the sequential realizations of the elements y_{i_t} over the T periods of measurements. Then,

$$P(Y_i) = \sum_{k=1}^{K} P\left(G_i = k\right) P^k(Y_i). \tag{6.1}$$

By definition of a finite mixture model (Nagin, 2005), the density f of Y is given by

$$f(y_i; \psi) = \sum_{k=1}^{K} \pi_k g_k(y_i; \Theta_k). \tag{6.2}$$

The role of the parameters Θ_k is to describe the shape of the trajectories in group k.

This model supposes no structural between-subject variability within a class; hence, the error variance is assumed to be constant inside a given class. Moreover, the group size $\pi_k > 0$ denotes the probability of a given subject to belong to group number k, and thus

$$\sum_{k=1}^{K} \pi_k = 1.$$

Since in practice it is difficult to constraint the π_k to be numbers between 0 and 1, we link the π_k to a set of parameters $\theta_1, \ldots, \theta_K$ such that

$$\pi_k = \frac{e^{\theta_k}}{\sum_{k=1}^{K} e^{\theta_k}}.$$

The model depends thus on the parameter set $\psi = (K, \theta_1, \cdots, \theta_{K-1}, \Theta_1, \cdots, \Theta_K)$.

There are two interesting generalizations of this basic finite mixture model. First, we can test if group membership of the individuals is influenced by a static set of R risk variables $X = (X_1 \cdots X_R)$, which are typically measured before the start of the trajectories Y. Second, we can investigate the relationship between the trajectories Y and a time-dependent covariate W, which is independent of X. Let's point out that this is basically regression analysis, so there is no automatic proof of causality here. We test if there is a significant relationship between Y and W and can only conclude that W influences the trajectories of Y if we get the causal status of the relationship by other means.

Combining these two extensions, the conditional density of Y given X and W is given by

$$f(y_i|x_i, w_i) = \sum_{k=1}^{K} \left(P(G_i = k|X_i = x_i) \prod_{t=1}^{T} P\left(Y_{i_t} = y_{i_t}|X_i = x_i, W_i = w_i, G_i = k\right) \right), \tag{6.3}$$

which can be written as

$$f(y_i|x_i, w_i) = \sum_{k=1}^{K} \left(\sum_{j=1}^{R} \frac{e^{x_i^j \theta_k^j}}{1 + e^{x_i^j \theta_k^j}} \prod_{t=1}^{T} P\left(Y_{i_t} = y_{i_t}|W_i = w_i, G_i = k\right) \right). \tag{6.4}$$

Nagin (2005) defined this model for underlying logit, (censored) normal and zero inflated Poisson distributions. In this chapter, we extend the model to an underlying Beta distribution.

The Beta Distribution

The Beta distribution is quite useful for modeling percentages and proportions, since it takes values between 0 and 1. Its density depends on the two positive shape parameters α and β. The primary advantage of this distribution is the flexibility of the shape of its density, as can be seen in Fig. 6.1.

This flexibility allows to capture a lot of non-normal distributions within the interval [0; 1], which should help resolve the problem that common Gaussian

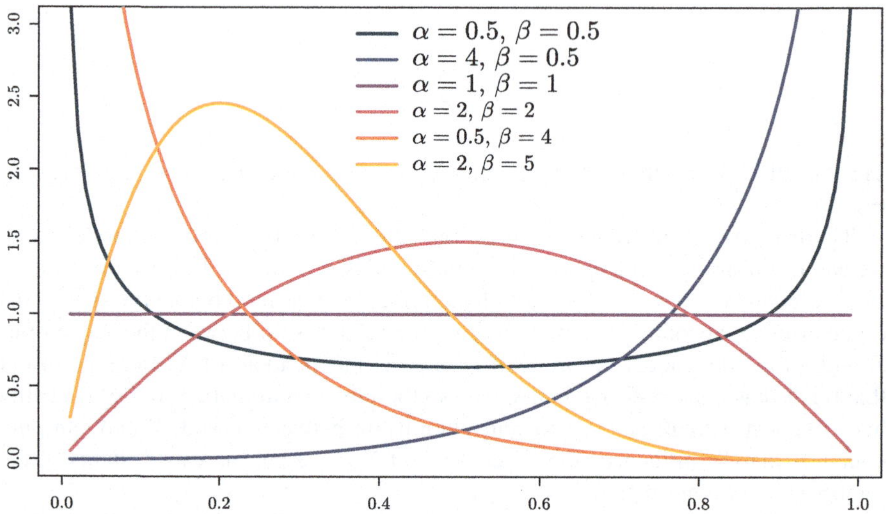

Fig. 6.1 Example of different shapes of Beta density for some parameters

mixture models may extract too many latent groups due to non-normality of the outcome as highlighted by Bauer and Curran (2003).

In a regression context, another parametrization of the density is commonly used (Ferrari & Cribari-Neto, 2004). Let Y be a random variable following a Beta distribution with mean μ. Consider the parameter ϕ defined by

$$var(Y) = \frac{\mu(1-\mu)}{1+\phi}.$$

ϕ can be interpreted as a precision parameter, in the sense that a large value of ϕ implies a small variance of Y. The density f of Y can be written as

$$f(y; \mu; \phi) = \frac{\Gamma(\phi)}{\Gamma(\mu\phi)\Gamma((1-\mu)\phi)} y^{\mu\phi-1}(1-y)^{(1-\mu)\phi-1},$$

where $0 < \mu < 1$ and $\phi > 0$.

Finite Mixture Models for the Beta Distribution

As mentioned in the last paragraph, Beta distributions only have values between 0 and 1, which might seem quite reductive. But any continuous variable Y with bounds (a, b) can of course be transformed to the standard unit interval using $Y^* = (Y-a)/(b-a)$, and so the use of Beta distributions could actually be meaningful for most datasets.

For the finite mixture model with underlying Beta distribution, called here the BETA model, we consider a latent variable y_{it}^* such that

$$y_{it}^* = f(a_{it}; \beta_k, \delta_k) + \varepsilon_{it} = \beta_k A_{it} + \delta_k W_t + \varepsilon_{it}, \tag{6.5}$$

where $\varepsilon_{it} \sim \mathcal{N}(0, \sigma_k)$, $A_{it} = (1, a_{it}, a_{it}^2, \cdots, a_{it}^{n_\beta - 1})^t$, $W_t = (w_{i1}, \cdots, w_{in_\delta})^t$, $\beta_k = (\beta_{k1}, \cdots, \beta_{kn_\beta})$, and $\delta_k = (\delta_{k1}, \cdots, \delta_{kn_\delta})$. n_δ denotes the dimension of the covariate W and n_β the number of measurements of Y for every individual. Here, and in the rest of this chapter, we take into account that in some applications, data are actually given as a function of the age a_{it} of the subjects at the time of measurement.

The usual assumption is that $y_{it} = 1$ if $y_{it}^* > 0$ and $y_{it} = 0$ if $y_{it}^* \leq 0$.

The density of y_{it} conditional to membership in group G_k can then be written as

$$g_k(y_{it}; \mu_{k,it}, \phi_{k,it}) = \frac{\Gamma(\phi_{k,it})}{\Gamma(\mu_{k,it}\phi_{k,it})\Gamma((1-\mu_{k,it})\phi_{k,it})} y_{it}^{\mu_{k,it}\phi_{k,it}-1}(1-y_{it})^{(1-\mu_{k,it})\phi_{k,it}-1},$$

with

$$\mu_{k,it} = \frac{e^{\beta_k A_{it} + \delta_k W_{it}}}{1 + e^{\beta_k A_{it} + \delta_k W_{it}}} \text{ and } \phi_{k,it} = \zeta_k A_{it}, \tag{6.6}$$

where $\zeta_k = (\zeta_{k1}, \cdots, \zeta_{kn_\zeta})$ is a vector defining the polynomial modeling the precision parameter ϕ_k, which can vary over time.

One difficulty with the Beta distribution is that the density may converge to infinite in the neighborhood of 0 and 1 for some values of the parameters. If the data y contain zeros or ones, it is advisable to transform the raw data y into $(y \cdot (n-1) + 0.5)/n$, where n is the sample size (Smithson & Verkuilen, 2006).

To fit the model, we use the quasi-Newton method BFGS, which calculates an iterative approximation of the inverse of the Hessian matrix to maximize the log-likelihood of the data, which is given by

$$l = \sum_{i=1}^{n} \log \left(\sum_{k=1}^{K} \sum_{j=1}^{R} \frac{e^{x_i^j \theta_k^j}}{1 + e^{x_i^j \theta_k^j}} \prod_{t=1}^{T} \frac{\Gamma(\phi_{k,it})}{\Gamma(\mu_{k,it}\phi_{k,it})\Gamma((1-\mu_{k,it})\phi_{k,it})} \right.$$
$$\left. \times\, y_{it}^{\mu_{k,it}\phi_{k,it}-1}(1-y_{it})^{(1-\mu_{k,it})\phi_{k,it}-1} \right). \tag{6.7}$$

6.3 The R Package trajeR

Finite mixture models can be calibrated by means of a maximum likelihood approach (Symons, 1981) or the EM algorithm (Dempster et al., 1977). Jones and

Nagin (2007) have developed a **SAS** procedure **Proc Traj** to calibrate finite mixture models, as well as a **Stata** version (Jones & Nagin, 2012). Our **R** package **trajeR** allows to calibrate finite mixture models for all commonly used densities (Noel & Schiltz, 2022). The proofs of all algorithms inside this package can be found in Noel's PhD thesis (Noel, 2024). For an underlying Beta distribution, **trajeR** maximizes the log-likelihood by direct optimization of its derivative. The package **trajeR** is written in **R** and **C++**. The linkage between **R** and **C++** is achieved through the packages **Rcpp** (Eddelbuettel & François, 2011) and **RcppArmadillo** (Eddelbuettel & Sanderson, 2014).

Missing values are supposed to be missing at random, following Rubin's missing data mechanisms (Rubin, 1976).

The package **trajeR** is built around the core function `trajeR`, which fits the model and computes its parameters for given degrees of the polynomial trajectories in the different groups. The function signature for `trajeR` is

```
R> trajeR(Y, A, Risk = NULL, TCOV = NULL, degre, degre.phi = 0,
+         Model, Method = "L",
+         ssigma = FALSE, ymax = max(Y) + 1, ymin = min(Y) - 1,
+         hessian = TRUE, itermax = 100, paraminit = NULL,
+         ProbIRLS = TRUE, refgr = 1, + fct = NULL, diffct = NULL,
+         nbvar = NULL, nls.lmiter = 50)
```

Some of these arguments are mandatory, while others are optional.

The mandatory arguments are the main data matrices `Y`, `A`, as well as `degre`, `degre.phi`, `Model`, and `Method`.

Here, `Y` is the matrix containing the values of the variable of interest, and `A` is the matrix containing the age or time variable. In most applications, this matrix just contains times of measurement that are the same for each individual in the sample, implying that all lines of the matrix `A` are equal, but this is not necessarily the case. `A` can, for instance, contain the age of the different individuals at the times of measurement, which is generally different for each individual in the sample.

`degre` is a vector indicating the degree of the polynomials describing the typical trajectories in the different groups. Implicitly, the dimension of this vector also indicates the number of groups into which we want to divide the population, so there is no need for a separate argument for the number of groups.

`degre.phi` is a vector indicating the degree of the polynomials describing the precision parameters in the different groups.

`Model` is a string defining the underlying distribution used in the model. The possible choices are LOGIT for the Logistic Regression Mixture Model, CNORM for the Censored Normal Mixture Model, ZIP for the Zero Inflated Poisson Mixture model, and BETA for the BETA model.

`Method`, finally, is a string to decide which algorithm is used for estimating the model parameters. In case of the BETA model, only `L` for direct optimization is possible.

The optional arguments are `Risk`, `TCOV`, `degre.nu`, `ssigma`, `ymax`, `ymin`, `hessian`, `itermax`, `paraminit`, `ProbIRLS`, `refgr`, `fct`, `diffct`, `nls.lmiter`, `ng.nl`, and `nbvar`.

`Risk` is a data matrix that contains the values of the covariate X modifying the group membership probability. By default, there is no such variable, and `Risk` is a one-column matrix with value 1.

`ProbIRLS` allows to decide which method is used to compute the predictor probabilities. If its value is TRUE (default setting), we use the IRLS method, and if it is FALSE, we use the optimization method.

`TCOV` is an optional data matrix containing a time-dependent covariate W that influences the trajectories themselves. By default, its value is NULL.

To ensure the identifiability of the parameters of the predictor, we have to fix a reference group. This can be done by the `refgr` command. Its default value is 1.

`hessian` indicates if we want to calculate the Hessian matrix, the default value is FALSE. If the method used is direct optimization, the Hessian matrix is computed by inverting the Fisher Information Matrix.

`itermax` gives the maximal number of iterations for the `optim` function or for the EM algorithm.

The choice of the initial parameters is very important in optimization problems. We can specify these initial parameters by `paraminit`. By default, `trajeR` calculates the initial value based on the range or the standard deviation of the data (for details, see Noel, 2024).

The output of `trajeR` for the BETA model is an object of class `Trajectory.BETA`, as described in following sections.

6.4 An Example with Simulated Data

We use the simulated dataset `BETA\_data01` that comes with installing the package **trajeR** from the CRAN repository. The dataset consists of a simulated two-group solution, based on the BETA model presented in paragraph 2. The sample consists of 500 trajectories with 6 time points each, the values of the variable Y being between 0 and 1. We take respective group sizes $\pi_1 = 0.378$ and $\pi_2 = 0.622$. In group one, the typical trajectory is a polynomial of degree 2 with parameters $\beta_1 = (-6, 3.7, -0.5)$, whereas in group 2, the typical trajectory is a polynomial of degree 1 with parameters $\beta_2 = (3.8, -1)$.

The variable of interest Y_i is contained in `data[,2:7]`; the time variable A_i in `data[,8:13]`; the time-dependent covariate W, which could, for instance, indicate the presence of a characteristics of the individual, in `data[,15:20]`; and a covariate X influencing group membership probability in `data[,14]`. Hence:

- `data[,2:7]` is a matrix with values between 0 and 1.
- `data[,8:13]` is a matrix with timepoints from 1 to 6.
- `data[,14]` is a column with real numbers.
- `data[,15:20]` is a matrix with real numbers.

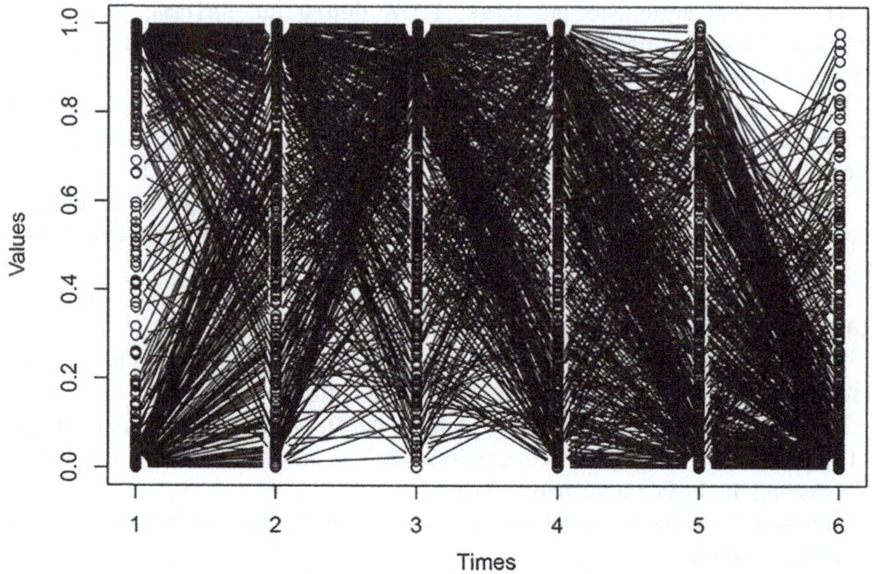

Fig. 6.2 Plot of the individual trajectories

We first plot the trajectories of Y to get a first impression of our sample (|Fig. 6.2).

To fit the model, we specify the number of groups to two, ng=2, and define the degree of the polynomial shape of the trajectories for each of the two groups. Here we choose a line and a quadratic polynomial, so degree is the vector (1, 2).

We specify `hessian=TRUE` to ask the computation of the Hessian matrix.

In case of a Beta distribution, the computations can be influenced by the starting values of the algorithm. By default, the starting value for the dispersion parameter is set to 5 for each group, which can give rise to numerical errors. So we use the `paraminit` parameter to manually set more sensible values.

Finally, to use the likelihood method, which is the only available method for the BETA model, we call `trajeR` with option `Method ="L"`. So, for this example, the code to launch the software is

```
> paraminit = c(0, -1.12, 0, 0, 1.15 , 0, 0, 0, 0, 0)
> solL = trajeR(Y = data[,2:7], A = data[,8:13],
+              param=paraminit,
+              degre=c(2,1), degre.phi = c(1,1),
+              Model = "BETA", Method = "L", hessian = T)
```

The basic results are contained in the object solL.

```
> solL

## trajeR with 2 groups with trajectories of degrees 2 and 1.
## Model : Beta
## Method : Likelihood
##
##   group   Parameter   Estimate   Std. Error   T for H0:    Prob>|T|
##                                                param.=0
## -------------------------------------------------------------------
##     mean
##       1   Intercept   -4.98565      0.17979   -27.72978           0
##              Linear    3.21808       0.1122    28.68113           0
##           Quadratic   -0.43902       0.0153    -28.6953           0
##     var.
##       1   Intercept    0.93348      0.10147      9.1999           0
##              Linear   -0.08262       0.0251    -3.29207     0.00101
##
##     mean
##       2   Intercept    3.41974      0.07685    44.49832           0
##              Linear   -0.87554      0.01926   -45.46406           0
##     var.
##       2   Intercept    1.30877      0.08325    15.72096           0
##              Linear   -0.19443      0.02042    -9.52225           0
## -------------------------------------------------------------------
##       1         pi1    0.37889      0.02176           0           0
##       2         pi2    0.62111      0.02176     22.7122           0
## -------------------------------------------------------------------
## Likelihood : 3188.759
```

We find quite good estimations for all parameters. We also see that the precision parameter Φ of the underlying Beta distribution is constant in both groups, which means that the variability of the Beta distribution is not time dependent in this example.

Next, we add a covariate X with the aim of analyzing if it influences the group membership probabilities. By default, the first group is the reference group, meaning that the potential influence of the covariate is compared to the first group. This setting can be changed with the argument `refgr`. The code to launch **trajeR** thus becomes

```
> paraminit = c(solL$theta[2],0,
+               solL$beta,
+               solL$phi)
> solLRisk = trajeR(Y = data[,2:7], A = data[,8:13], Risk = data[,14],
+                   param=paraminit,
+                   degre=c(2,1), degre.phi = c(1,1),
+                   Model = "BETA", Method = "L", hessian = T)
```

We get the following results:

```
> solLRisk

## trajeR with 2 groups with trajectories of degrees 2 and 1.
## Model : Beta
## Method : Likelihood
##
##   group   Parameter   Estimate   Std. Error   T for H0:   Prob>|T|
##                                               param.=0
## -------------------------------------------------------------------
##    mean
##       1   Intercept   -4.96716     0.1871    -26.54745          0
##              Linear    3.20966     0.11866    27.04812          0
##           Quadratic    -0.4382     0.01629   -26.90758          0
##    var.
##       1   Intercept    0.91587     0.10342     8.85578          0
##              Linear   -0.07829     0.02587    -3.02569     0.0025
##
##    mean
##       2   Intercept    3.41974     0.07942    43.05827          0
##              Linear   -0.87558     0.01992   -43.96029          0
##    var.
##       2   Intercept    1.30819     0.08369    15.63067          0
##              Linear   -0.19445     0.02041    -9.52701          0
## -------------------------------------------------------------------
##       1    Baseline          0          NA          NA         NA
##
##       2   Intercept    0.64936     0.06268    10.36058          0
##                   X   -0.29799     0.12577     -2.3694    0.01788
##
---------------------------------------------------------------------
## Likelihood : 3261.767
```

Here, instead of the estimations of the group sizes, we get the changes of the second group with respect to the first group, which is our baseline group. The estimate of the *X* parameter indicates the size of group 2 (0.649), and the estimate of the *X* parameters tells us that we can be 98% sure that an increase of 1 in the *X* variable diminishes the probability of membership in group 2 by 30%. Hence, in this example, the covariate *X* has a very determining influence on group membership.

Finally, we add a time-dependent covariate and test if it influences the shape of the trajectories directly. Of course, we actually only test if there is a significant relationship between the covariate and the trajectories. The causality part needs to come from other sources. We introduce these possible effects by using the option TCOV in the command `trajeR`.

```
> paraminit = c(0, -1.12, 0, 0, 1.15 , 0, 0, 0, 0, 0),TCOV = data[,15:20],
> solLTCOV = trajeR(Y = data[,2:7], A = data[,8:13],
+              param=paraminit,
+              degre=c(2,1), degre.phi = c(1,1),
+              Model = "BETA", Method = "L", hessian = T)
```

We get the following results:

```
> solLTCOV

## trajeR with 2 groups with trajectories of degrees 2 and 1.
## Model : Beta
## Method : Likelihood
##
##   group   Parameter   Estimate   Std. Error   T for H0:   Prob>|T|
##                                                param.=0
## -------------------------------------------------------------------
##    mean
##      1  Intercept   -5.01739      0.18071   -27.76494          0
##            Linear     3.2314      0.11271    28.67047          0
##         Quadratic   -0.44015      0.01538   -28.61454          0
##             TCOV1    0.02612      0.01182     2.21005    0.02718
##    var.
##      1  Intercept    0.95954      0.10311     9.30624          0
##            Linear    -0.0882       0.0256    -3.44491    0.00058
##
##    mean
##      2  Intercept    3.41911      0.07772    43.99116          0
##            Linear   -0.87459      0.01941   -45.06335          0
##             TCOV1   -0.02487      0.00897    -2.77147    0.00561
##    var.
##      2  Intercept    1.32971      0.08561    15.53248          0
##            Linear   -0.19918      0.02103    -9.46955          0
## -------------------------------------------------------------------
##      1        pi1    0.37874       0.0224           0          0
##      2        pi2    0.62126       0.0224    22.09257          0
## -------------------------------------------------------------------
## Likelihood : 3195.311
```

The estimations of the parameter TCOV1 show that there is a significant relationship between the covariate and the typical trajectories for both groups. Indeed, an increase of 1 of the covariate induces an increase of 0.026 of the typical trajectory of group 1 and a decrease of 0.025 of the typical trajectory of group 2.

There are various possibilities to present the different results graphically. The basic graph consists in plotting the typical trajectories of the groups, which can be done with the command `plotrajeR` included in our **trajeR** package, applied to an object of class `Trajectory`. By default, colors are gray scale, but we can specify any color we want (Fig. 6.3).

```
> plotrajeR(solLTCOV, col =vcol)
```

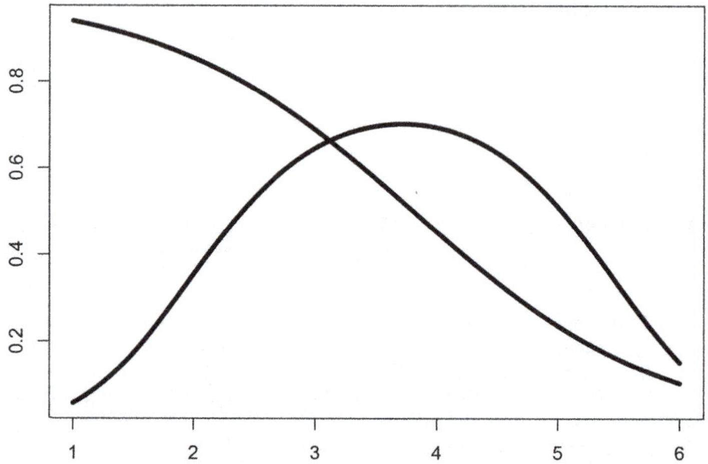

Fig. 6.3 Typical trajectories for the two groups

But the graph looks much nicer if we add the initial longitudinal data. In order to achieve this, we have to specify the variables Y and A in the function `plot()`. Here, we use the colors pink and light blue for the individual trajectories in the two groups and purple and dark blue for the typical trajectories (Fig. 6.4).

```
> # Defining the colours
> trans = "70"
> col1 = "#034569"
> col1.1 = paste0("#64AAD0", trans)
> col2 = "#750062"
> col2.1 = paste0("#D962C7", trans)
> cols1 = c(col1.1, col2.1)
> cols2 = c(col1, col2)
> vcol = c(cols1, cols2)
>
> plotrajeR(solLRisk, Y = data[,2:7], A = data[,8:13], col = vcol)
```

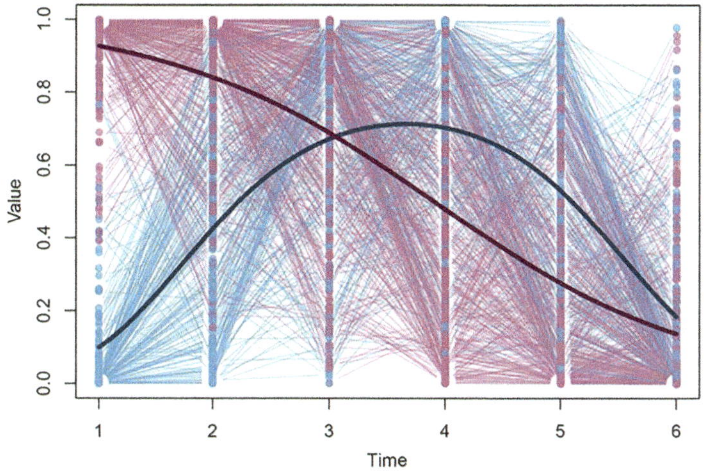

Fig. 6.4 Individual trajectories together with the two typical group trajectories

We can add the average of the data points to the plot. For each group and each time point, the mean of the data is calculated and added to the plot. Usually, the means remain of course quite close to the typical group trajectories (Fig. 6.5).

We use the argument `mean` =TRUE.

```
> # colour's defintion
> trans = "70"
> col1 = "#034569"
> col1.1 = paste0("#64AAD0", trans)
> col2 = "#750062"
> col2.1 = paste0("#D962C7", trans)
> cols1 = c(col1.1, col2.1)
> cols2 = c(col1, col2)
> vcol = c(cols1, cols2)
>
> plotrajeR(solLRisk, Y = data[,2:7], A = data[,8:13],
+       col = vcol, mean = TRUE, alpha = 0.3)
```

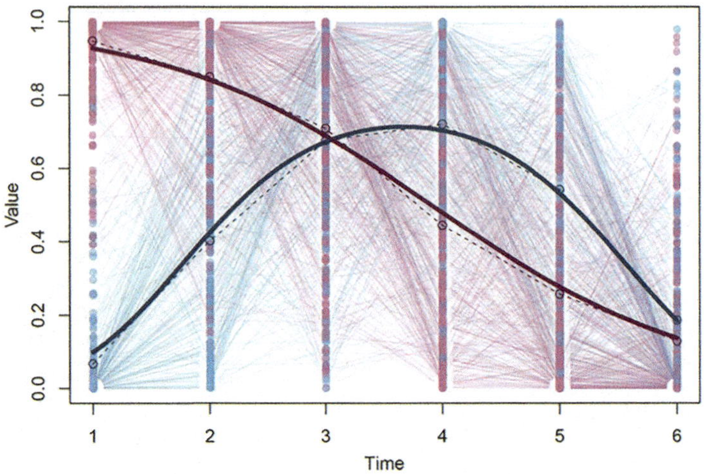

Fig. 6.5 Individual trajectories together with the two typical group trajectories (full lines) and the mean trajectories of each group (dashed lines)

Finally, if we want to show the impact of a particular value of the time covariate on the trajectory, we can add this to the plot by `plotcov` option.

The full line is the trajectory with the time covariate matrix, and the dashed one shows the impact on this trajectory of a particular value. In this case, we see what happens when the covariate has constant values −5 and 5 during the entire time interval (Fig. 6.6).

```
> plotrajeR(solLTCOV, col = vcol,
+  plotcov = c(-5, -5, -5, -5, -5, -5, 5, 5, 5, 5, 5, 5))
```

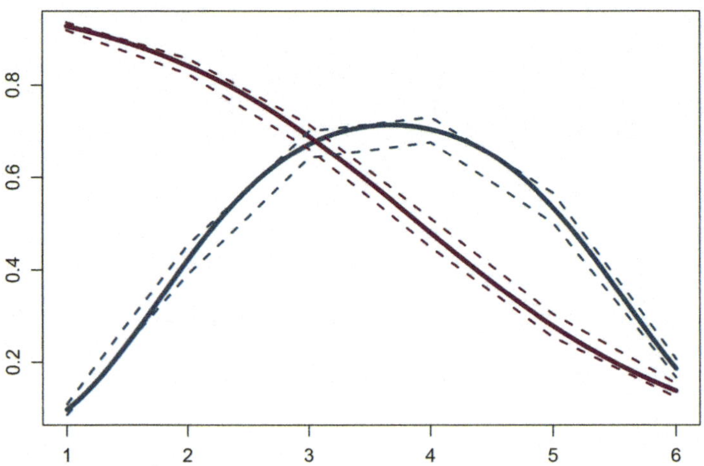

Fig. 6.6 Typical group trajectories without (full lines) and with (dashed lines) influence of two particular values of the covariate

6.5 An Application to COVID-19 Data

Data

We use daily data from 191 countries obtained from "Our World In Data" Hasell et al. (2020).

This dataset is built from information supplied by the European Centre for Disease Prevention and Control, government reports, Oxford COVID-19 Government Response Tracker, World Bank, United Nations Statistics Division, and Eurostat. Our main variable of interest is the rate of contamination with the COVID-19 virus. Moreover, we consider the covariates' new cases, population size (in million inhabitants), total cases per million people, median age of the population, population density, number of inhabitants over 65 (in million inhabitants), government response stringency index, GDP per capita, extreme poverty index, cardiovascular death rate, diabetes prevalence rate, index of handwashing facilities, rate of hospital beds per thousand inhabitants, life expectancy, and index of human development.

Finally, we use a stringency index, a composite measure of the nine government actions school closures, workplace closures, cancelation of public events, restrictions on public gatherings, closures of public transport, stay-at-home requirements, public information campaigns, restrictions on internal movements, and international travel controls.

First we need to do some data cleaning, replacing, for instance, some missing values for the median age, population density, population over 65, GDP per capita, and life expectancy variables.

Alvarez et al. (2020), who use the same dataset, have created time series by considering a period of 100 days after the confirmation of the tenth COVID-19 case. We choose another approach. We transform all variables into monthly data and obtain thus a 16-period panel, starting in January 2020 and finishing in April 2021. This allows to smooth out periodicity problems that occur with daily data, for instance, because on weekends, some countries do not publish any COVID-19-related data. Moreover, for a lot of smaller countries, it is very difficult to determine the exact beginning of the epidemic. Since we interpret missing data as missing at random, and not as 0, this approach is more convenient anyway.

Using the notations of the previous paragraph, we hence observe the rate of contamination Y_{it} for country i at time t, for $n = 190$ countries at $T = 16$ time points. $Y_i = (Y_{i1}, \cdots, Y_{i16})$ gives the time series for country i. Figure 6.7 shows the trajectories of the contamination rate for all 190 countries in our sample.

Model Selection

To decide on the number of groups in our model, we use the BIC and AIC criteria together with the metric developed by Kass and Wasserman (1995). Indeed, let p_k

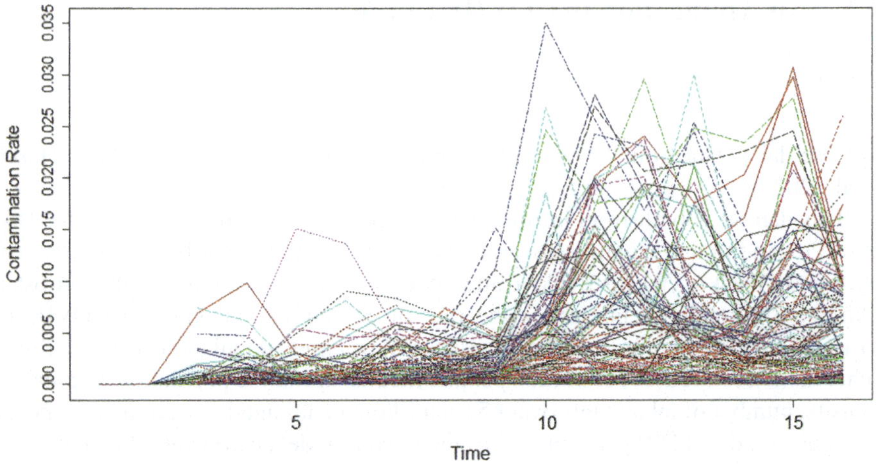

Fig. 6.7 Contamination rates for all countries

be the probability that a model with k groups is the correct model. They show that p_k can be approximated by

$$p_k \approx \frac{e^{BIC_k - BIC_{max}}}{\sum_k e^{BIC_k - BIC_{max}}}.$$

To start with, we decide to model the typical trajectories with polynomials of degree 3 to be able to take into account two inflection points and to use polynomials of degree 2 to model the time dependency of the precision parameters zeta of the underlying Beta distribution.

Optimization methods can heavily depend upon the choice of the starting values of the different parameters. In case of a bad choice, the algorithm may not converge at all or converge to a local extremum. Unfortunately, there is no established method to choose these starting points. We chose a three-step approach to solve this issue. In a first step, we chose k horizontal lines as starting values, i.e., take the parameters $\beta_k = (\beta_0, 0, 0, \dots)$ and $\zeta_k = (\zeta_0, 0, 0, \dots)$ for $k = 2, \dots, 10$. Every time, we check if the model we found is well defined by computing its average posterior probability, i.e., the average posterior probability of membership for each group for those individuals that were assigned to it. In an ideal situation, the assignment probability for each individual would be 1, and the average posterior probability (AvePP) would be 1 too. Nagin (2005) considers that it should be at least 0.7 for all groups, so we use that criteria too. We also remove models that contain empty groups.

In a second step time, we try some affine functions to take into account the growth of the data during the time period. The first two steps exhibit mostly two-group and some three-group solutions. In a third step, we then use the parameters of the

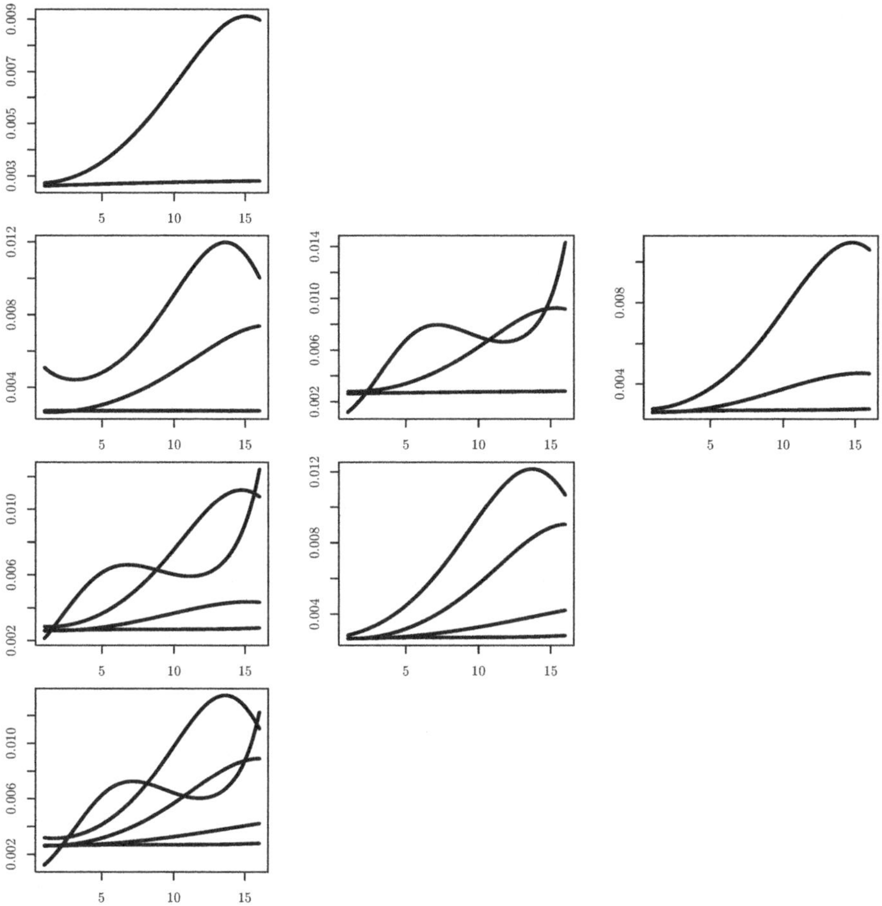

Fig. 6.8 Typical trajectories of the different possible models

different groups found in the previous steps as starting values and also find four- and five-group solutions. The typical trajectories of all seven possible models with less than ten groups are presented in Fig. 6.8.

The BIC and AIC, as well as Kass and Wasserman's p_k, are presented in Table 6.1.

The lowest values of BIC and AIC are obtained for the model with five groups with BIC $= -15558.41$ and AIC $= -31241.46$. The Kass and Wasserman criterion also clearly indicates that we should retain the five-group solution.

Looking at the parameter estimates for the selected five-group solution, we see that actually in group 5, the first- and second-degree parameters of ζ are nonsignificant, meaning that for this group, the precision parameter Φ is constant over time. We take this into account by rerunning the model with a constant ζ for group 5 and get the final model parameters presented in Table 6.2.

Table 6.1 Model selection criteria

Number of groups	AIC	BIC	Prob
2	−29851.99	−14902.64	0.00000
3	−30341.00	−15142.28	0.00000
3	−29945.96	−14936.64	0.00000
3	−30777.14	−15352.23	0.00000
4	−30839.69	−15370.52	0.00000
4	−31192.78	−15547.06	0.00001
5	−31241.46	−15558.41	0.99999

Table 6.2 Parameters of the final model

Param.	sd	Test	Param.	sd	Test	Param.	sd	Test
Beta 1			Zeta 1			Pi 1		
−5.902	0.018	0.000	14.648	0.309	0.000	0.302	0.034	0.000
−0.052	0.013	0.000	−1.253	0.072	0.000			
0.020	0.002	0.000	0.045	0.004	0.000			
−0.001	0.000	0.000						
Beta 2			Zeta 2			Pi 2		
−5.927	0.003	0.000	20.026	0.280	0.000	0.183	0.028	0.000
−0.015	0.003	0.000	−1.818	0.088	0.000			
0.005	0.001	0.000	0.066	0.005	0.000			
0.000	0.000	0.005						
Beta 3			Zeta 3			Pi 3		
−5.659	0.133	0.000	9.454	0.374	0.000	0.159	0.028	0.000
−0.119	0.066	0.071	−0.600	0.099	0.000			
0.040	0.009	0.000	0.022	0.005	0.000			
−0.002	0.000	0.000						
Beta 4			Zeta 4			Pi 4		
−5.962	0.011	0.000	13.212	0.360	0.000	0.319	0.034	0.000
0.018	0.005	0.000	0.007	0.085	0.934			
−0.002	0.001	0.000	−0.008	0.004	0.061			
0.000	0.000	0.000						
Beta 5			Zeta 5			Pi 5		
−7.511	0.370	0.000	7.422	0.146	0.000	0.036	0.014	0.008
0.911	0.143	0.000						
−0.102	0.016	0.000						
0.004	0.001	0.000						

Figure 6.9 shows the evolution of the precision parameters over time. We see that the trajectory of the parameters ϕ for group 5 is a horizontal line, and so, for this group, the precision is constant over time, while for most other groups, the trajectories become more homogeneous over time.

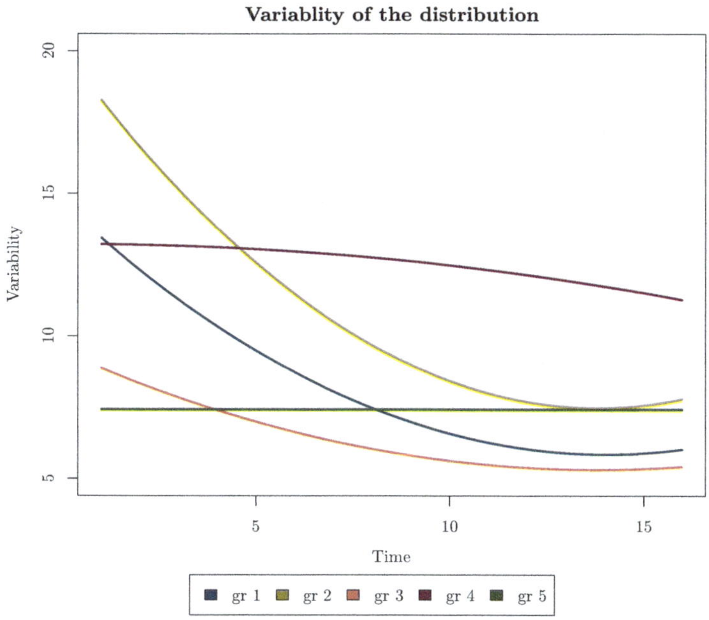

Fig. 6.9 Variation of the precision parameters over time

Description of the Groups

The final model exhibits 4 main groups with 57, 35, 30, and 61 countries, respectively, and 1 small group with 7 countries. Group membership of all countries in the sample can be found in Table 6.6 in the appendix and can also be seen on the world map in Fig. 6.10.

Figure 6.11 shows the typical evolution of the contamination rate in the five groups, together with the initial trajectories of the 191 countries.

Group 1 contains countries with a quite fast and regularly increasing contamination rate, while the countries in group 2 exhibit a much smaller growth of contamination rate. In the countries of group 3, the contamination rate grows even faster than in group 1, till the end of 2020, and starts declining at the beginning of 2021. Group 4 contains all countries that declare a contamination rate that stays very close to zero over the whole time interval in our study. The countries in group 5 have quite a particular behavior with two inflection points. This group contains countries from South America like Chile, Peru, or Brazil where the second increase of the contamination rate can be explained by the appearance of a new variant, as described in Hojo de Souza et al. (2021), but also countries from the Middle East like Kuwait, Qatar, and Oman (Table 6.3).

Table 6.6 shows the principal characteristics of the countries in the different groups.

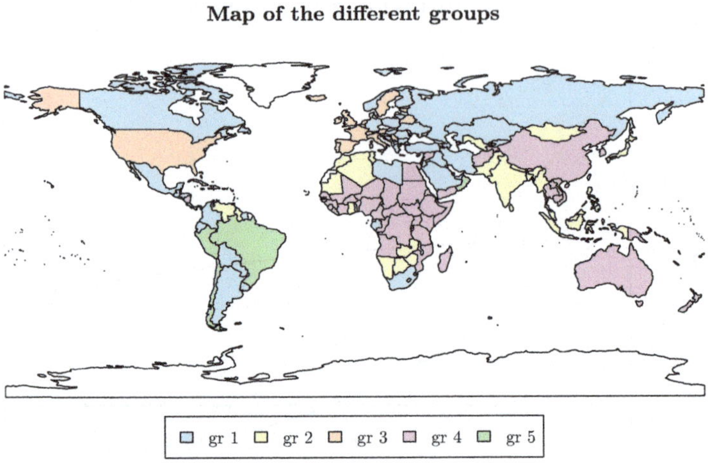

Fig. 6.10 World map with the geographic distribution of the five groups

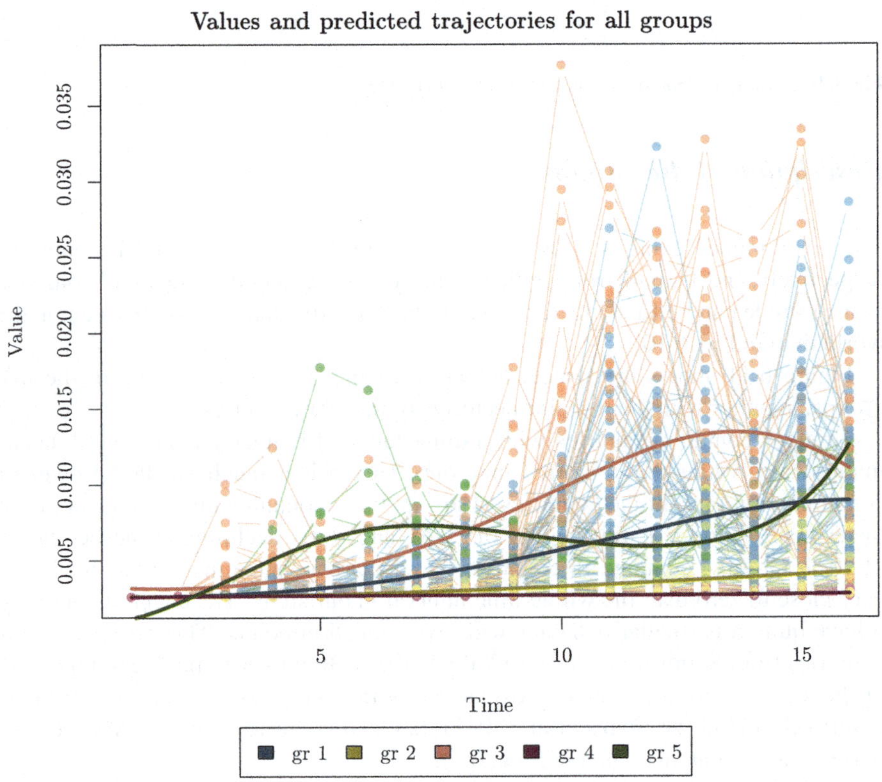

Fig. 6.11 Evolution of the contamination rate for the five groups and all countries of the sample

Table 6.3 Means and standard deviations of the descriptive variables for each group

Variables	Group 1 mean (sd)	Group 2 mean (sd)	Group 3 mean (sd)	Group 4 mean (sd)	Group 5 mean (sd)
Population size	19.97 (31.3)	76.41 (235.85)	23.38 (61.44)	48.49 (184.5)	39.64 (77.13)
Median age	33.67 (7.76)	28.24 (6.88)	41.67 (5.29)	23.49 (7.51)	32.13 (2.19)
Population density	122.67 (220.76)	207.31 (242.47)	1148.75 (3729.09)	126.64 (153.86)	286.19 (524.35)
Aged 65 and older	10.54 (6.22)	6.75 (4.69)	17.08 (5.38)	4.83 (3.3)	5.27 (3.7)
Aged 70 and older	6.87 (4.25)	4.16 (3.21)	10.76 (3.43)	2.81 (2.13)	3.23 (2.34)
GDP per capita	20583.62 (14511.5)	10267.22 (8325.13)	50138.17 (41050.34)	7655.52 (12468.09)	40673.95 (38570.57)
Extreme poverty	3.63 (6.68)	12.96 (16.69)	0.81 (0.67)	31.28 (24.1)	2.73 (1.24)
Cardiovascular death rate	259.68 (110.33)	285.53 (113.24)	172.99 (108)	306.11 (121.94)	161.7 (56.6)
Diabetes prevalence	8.35 (3.3)	7.75 (3.63)	6.93 (2.84)	7.72 (5.59)	10.95 (4.08)
Handwashing facilities	72.84 (22.95)	55.53 (28.91)	95.88 (2.6)	32.5 (27)	96.6 (1.13)
Hospital beds	3.63 (2.18)	2.47 (2.5)	4.45 (2.67)	1.87 (2.1)	1.79 (0.38)
Life expectancy	75.2 (4.56)	70.73 (6.42)	80.97 (3.16)	67.18 (7.41)	77.9 (1.95)
Human development index	0.79 (0.09)	0.68 (0.1)	0.9 (0.05)	0.59 (0.14)	0.8 (0.04)
Stringency index	64.75 (16.98)	64.1 (15.74)	58.01 (18.17)	49.9 (16.22)	68.72 (22.02)

We observe that the contamination rate is higher in the groups with a higher median age or with a higher proportion of old people. Age variables are indeed highest in groups 3 and 1 and lowest in group 4. We also see that countries in groups with the largest increase of contamination rate have the highest life expectancy and GDP per capita and the lowest poverty rate. Moreover, group 3 has by far the highest population density, whereas there is no visible trend linking population density to the contamination rate in the other four groups.

On the other hand, we see that the average stringency index is almost the same for all five groups, so at a first glance, there does not seem to be a link between the contamination rate and the stringency index.

Carrillo-Larco and Castillo-Cara (2020) use a K-means clustering algorithm on cross-sectional data and find a division of the countries into clusters that are quite similar to our groups. They use some selected diseases, socioeconomic status, air pollution, and health system as descriptive variables but do of course not get any typical evolution trajectories, since they perform a static analysis.

In the following paragraph, we will formally test which of the descriptive variables have a significant influence on group membership.

Predictors of Trajectory Group Membership

We test several covariates to understand if some of them influence the group membership probabilities. We use median age, population density, proportion of people elder than 65 older, life expectancy, and the mean of the stringency index over the observed time period as risk measures X that potentially influence group membership in the generalized finite mixture model. Since that part of the model is essentially equivalent to a multinomial logit model, we have to use one of the groups as comparison group and test then for all covariates if and, if yes, how much they influence the membership probabilities for the other groups. We choose group 4 as comparison group. Table 6.4 reports the results.

The median age affects membership to groups 1 ($p = 0.02445$), 2 ($p = 0.088$), and especially 5 (p=0), in the sense that countries with an older population have a higher chance to belong to these groups. The high influence of the median age on the membership probability of group 5 is somewhat counterbalanced by the fact that a high proportion of people older than 65 has a highly significant $p = 0.002$ but negative influence on group membership in that group.

Figure 6.12 shows the boxplots of the median age for all five groups of countries.

We see that the median of the median age covariant is by far highest in group 5 and actually nearly all countries in group 5 have a relatively high median age. More generally, we detect a clear link between a high median age of a population and membership in a group with a high contamination rate.

Finally, we see that a high value of the stringency index has a highly significant positive influence on group membership in all four groups. Accordingly, a high value of the stringency index also decreases the probability of membership in the

Table 6.4 Predictors of group membership

	Group 1			Group 2		
	Estimate	Std. error	Prob>|T|	Estimate	Std. error	Prob>|T|
intercept	−16.812	4.681	0	−4.805	3.422	0.16
Median age	0.193	0.086	**0.024**	0.172	0.101	**0.088**
Population density	−0.003	0.002	0.093	0.000	0.001	0.869
Aged 65 older	−0.021	0.132	0.871	−0.060	0.126	0.631
Life expectancy	0.073	0.080	0.364	−0.073	0.071	0.304
Stringency index	0.112	0.023	**0**	0.092	0.023	**0**
	Group 3			Group 5		
	Estimate	Std. error	Prob>|T|	Estimate	Std. error	Prob>|T|
intercept	−67.733	19.400	0	−73.689	23.469	0.002
Median age	0.129	0.158	0.412	0.418	0.205	**0.041**
Population density	0.000	0.001	0.784	0.000	0.001	0.926
Aged 65 older	0.109	0.178	0.542	−0.640	0.206	**0.002**
Life expectancy	0.646	0.223	**0.004**	0.646	0.283	**0.023**
Stringency index	0.185	0.054	**0.001**	0.228	0.075	**0.002**

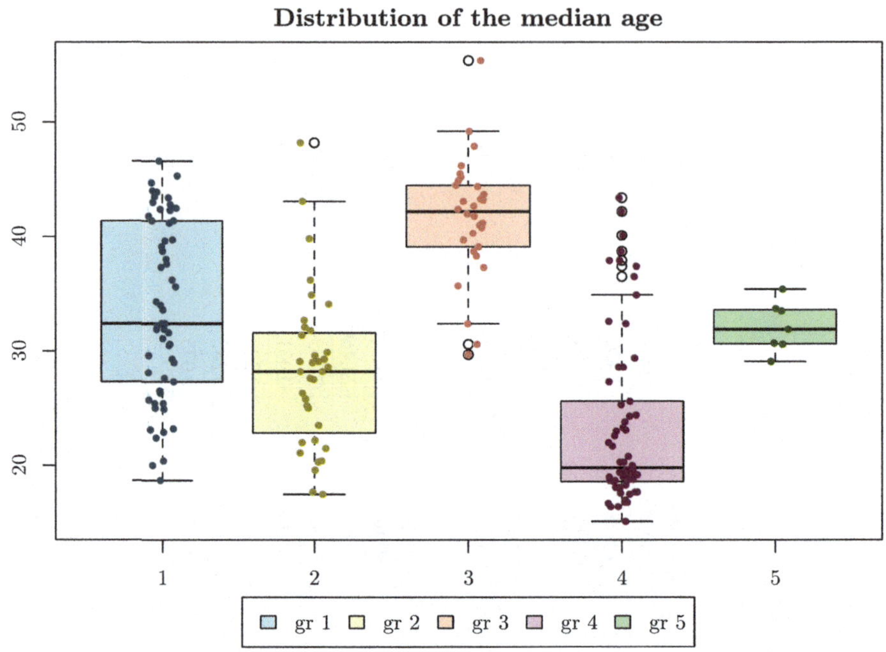

Fig. 6.12 Boxplots of the median age for all five groups

baseline group. Now, this results seem of course surprising. But we actually did something here that we are not supposed to do. In generalized finite mixture models, the covariates X that potentially influence group membership have to be measured before the beginning of the trajectories Y. This is not the case for the stringency index in our example. The government measures that are summarized in that index have not been taken before the pandemic but during the first part of the outbreak. So, it is obvious that the stringency index cannot influence group memberships here. What our result shows is actually rather the contrary. The governments of countries in groups with a higher contamination rate have usually taken harsher measures, which explains that the coefficients in groups 1, 2, 3, and 5 are significant and strictly positive with respect to group 4, which contains the countries not affected a lot by the pandemic during the time of our study.

On the other hand, these four significant theta coefficients are quite close one to each other. If we want to understand if they differ significantly among themselves, we can use a χ^2-based test of multiple contrasts (Nagin, 2005). The degree of freedom of this test equals the number of equality constraints being tested or the number of different coefficients being tested minus 1. Here, the degree of freedom is 3, since we perform pairwise equality tests between the four theta coefficients. More precisely, we perform the test

$$H_0 : \theta_2^{st} = \theta_3^{st} \text{ and } \theta_2^{st} = \theta_4^{st} \text{ and } \theta_2^{st} = \theta_5^{st} \tag{6.8}$$

$$H_1 : \theta_2^{st} \neq \theta_3^{st} \text{ or } \theta_2^{st} \neq \theta_4^{st} \text{ or } \theta_2^{st} \neq \theta_5^{st} \tag{6.9}$$

If we use the notations $\theta = \left(\theta_2^{st}, \theta_3^{st}, \theta_4^{st}, \theta_5^{st}\right)'$, $q = (0, 0, 0)'$ and

$$H = \begin{pmatrix} 1 & -1 & 0 & 0 \\ 1 & 0 & -1 & 0 \\ 1 & 0 & 0 & -1 \end{pmatrix},$$

our hypotheses can be written

$$H_0 : H\theta = q \tag{6.10}$$

$$H_1 : H\theta \neq 0. \tag{6.11}$$

The χ^2 distance is then defined by

$$\chi^2 = \left(H\hat{\theta} - q\right)' \left(HV_{\hat{\theta}}H'\right)^{-1} \left(H\hat{\theta} - q\right), \tag{6.12}$$

where $\hat{\theta}$ denotes the counterpart vector of the coefficients and $V_{\hat{\theta}}$ is the matrix variance/covariance of the parameter estimates in $\hat{\theta}$.

We obtain $\chi^2 = 5.62$ with 3 degrees of freedom, which is far short of significance. So, whereas we have shown that groups 1, 2, 3, and 5 have a significantly higher stringency index than group 1, we do not find any significant pairwise difference between these four groups.

We have done these pairwise comparison tests by hand, since currently, the **R** package **trajeR** does not allow to perform these automatically. This is foreseen in a future extension of the package together with other tools of more detailed description of the groups obtained by the cluster analysis.

Stringency as Time-Dependent Covariate

The generalized finite mixture model allows to test if time-dependent covariates have a significant relationship with the shape of the typical trajectories in the different groups. Here it makes sense to analyze if the different measures the governments took against the spread of the pandemic were successful. In other words, we are interested in testing if the typical contamination rate trajectories were flattened in case of a high stringency index.

Thus, here W_{it} denotes the stringency index at time t, for $1 \leq t \leq 16$, for country i. For this application, we had to remove countries without values for the

Table 6.5 Parameters of the final model with time dependent covariates

Param.	sd	Test	Param.	sd	Test	Param.	sd	Test	Param.	sd	Test
Beta 1			Phi 1			Delta 1			Prob. 1		
−5.843	0.026	0.000	14.337	0.317	0.000	0.001	0.000	0.001	0.328	0.039	0.00
−0.120	0.024	0.000	−1.164	0.076	0.000						
0.029	0.004	0.000	0.040	0.004	0.000						
−0.001	0.000	0.000									
Beta 2			Phi 2			Delta 2			Prob. 2		
−5.927	0.003	0.000	19.866	0.570	0.000	0.000	0.000	0.955	0.175	0.030	0.00
−0.014	0.004	0.000	−1.710	0.125	0.000						
0.005	0.001	0.000	0.061	0.006	0.000						
0.000	0.000	0.001									
Beta 3			Phi 3			Delta 3			Prob. 3		
−5.602	0.117	0.000	9.624	0.369	0.000	0.010	0.001	0.000	0.156	0.030	0.00
−0.421	0.070	0.000	−0.521	0.097	0.000						
0.076	0.009	0.000	0.016	0.005	0.003						
−0.003	0.000	0.000									
Beta 4			Phi 4			Delta 4			Prob. 4		
−5.972	0.012	0.000	12.887	0.372	0.000	0.000	0.000	0.955	0.301	0.035	0.00
0.012	0.005	0.018	0.148	0.085	0.082						
−0.001	0.001	0.043	−0.015	0.004	0.000						
0.000	0.000	0.027									
Beta 5			Phi 5			Delta 5			Prob. 5		
−7.304	0.366	0.000	7.384	0.137	0.000	0.004	0.001	0.004	0.040	0.016	0.01
0.701	0.147	0.000									
−0.078	0.017	0.000									
0.003	0.001	0.000									

stringency, index[1] and in case of isolated missing data, we completed the variable by linear interpolation.

Table 6.5 shows the parameter estimations of the model. Their interpretation is the same as before, except for the delta coefficients, which illustrate the relationship between the stringency index and the different typical group trajectories.

For groups 1, 3, and 5, i.e., groups with a high contamination rate, delta parameters are significant and positive, although very small, while for the other two groups, they are not significant. Here, again, the small but significant relationship between the delta parameters and the typical trajectories mostly indicates that the countries with a high contamination rate took slightly stricter measures against the spread of the virus than countries that were less affected by the virus. But conversely, this also shows that we cannot see any sign of efficiency of the sanitary measures taken by the different countries against the propagation of the virus during the first part of the pandemic.

6.6 Conclusion

In this chapter, we presented an extension of finite mixture models to Beta distributed data, the so-called BETA model, and showed how to calibrate this kind of models with our **R** package **trajeR**. Further extensions of the package will allow to handle multi-trajectory models, as well as incorporate different strategies to handle missing data, which is quite important in some research areas like medicine, where missing data often cannot be assumed to happen randomly (there are, for instance, patient dropouts, due to the death of patients).

In a second part of this chapter, we used the BETA model to analyze COVID-19-related data during the first part of the pandemic. We identified a classification of the world into five groups of countries with respect to the evolution of the contamination rate and described the five groups as well as their typical contamination trajectories. We also showed that the median population age is the main predictor of group membership. We did however not see any sign of efficiency of the sanitary measures taken by the different countries against the propagation of the virus. This is in accordance with the results of the meta analysis of Herby et al. (2022), who also conclude that the lockdowns did not have a large effect on the mortality rate.

In further works, it would be interesting to include other variables in these studies, for example, the presence of variants. It would also be interesting to perform a deeper study of the possible effect of the different measures represented in the stringency index by means of dual trajectory modeling. Finally, it would

[1] Antigua and Barbuda, Armenia, Comoros, Equatorial Guinea, Grenada, Guinea-Bissau, Liechtenstein, Maldives, Marshall Islands, Micronesia (country), Montenegro, North Macedonia, Saint Kitts and Nevis, Saint Lucia, Saint Vincent and the Grenadines, Samoa, Sao Tome and Principe, Vatican.

be interesting to study the second wave of the pandemic and also include the vaccination rate in the study.

Appendix

See Table 6.6.

Table 6.6 Countries belonging to each group

Group 1	Albania, Argentina, Azerbaijan, Bahamas, Belize, Bolivia, Bosnia and Herzegovina, Bulgaria, Canada, Cape Verde, Colombia, Costa Rica, Croatia, Cyprus, Denmark, Djibouti, Dominican Republic, Ecuador, Equatorial Guinea, Finland, Gabon, Georgia, Germany, Greece, Guatemala, Honduras, Hungary, Iran, Iraq, Jordan, Kazakhstan, Kosovo, Kyrgyzstan, Latvia, Lebanon, Libya, Lithuania, Malta, Mexico, Moldova, North Macedonia, Norway, Palestine, Paraguay, Poland, Romania, Russia, Sao Tome and Principe, Saudi Arabia, Seychelles, Slovakia, South Africa, Suriname, Turkey, Ukraine, United Arab Emirates, Uruguay
Group 2	Algeria, Antigua and Barbuda, Bangladesh, Barbados, Botswana, Comoros, Cuba, El Salvador, Eswatini, Gambia, Ghana, Guyana, India, Indonesia, Jamaica, Japan, Lesotho, Malaysia, Mauritania, Mongolia, Morocco, Myanmar, Namibia, Nepal, Pakistan, Philippines, Saint Lucia, Saint Vincent and the Grenadines, Sri Lanka, Trinidad and Tobago, Tunisia, Uzbekistan, Venezuela, Zambia, Zimbabwe
Group 3	Andorra, Armenia, Austria, Bahrain, Belarus, Belgium, Czechia, Estonia, France, Iceland, Ireland, Israel, Italy, Liechtenstein, Luxembourg, Monaco, Montenegro, Netherlands, Panama, Portugal, San Marino, Serbia, Singapore, Slovenia, Spain, Sweden, Switzerland, United Kingdom, United States, Vatican
Group 4	Afghanistan, Angola, Australia, Benin, Bhutan, Brunei, Burkina Faso, Burundi, Cambodia, Cameroon, Central African Republic, Chad, China, Congo, Cote d'Ivoire, Democratic Republic of Congo, Dominica, Egypt, Eritrea, Ethiopia, Fiji, Grenada, Guinea, Guinea-Bissau, Haiti, Kenya, Laos, Liberia, Madagascar, Malawi, Mali, Marshall Islands, Mauritius, Micronesia (country), Mozambique, New Zealand, Nicaragua, Niger, Nigeria, Papua New Guinea, Rwanda, Saint Kitts and Nevis, Samoa, Senegal, Sierra Leone, Solomon Islands, Somalia, South Korea, South Sudan, Sudan, Syria, Taiwan, Tajikistan, Tanzania, Thailand, Timor, Togo, Uganda, Vanuatu, Vietnam, Yemen
Group 5	Brazil, Chile, Kuwait, Maldives, Oman, Peru, Qatar

References

Alvarez, E., Brida, J.G., & Limas, E. (2020). *Comparisons of COVID-19 dynamics in the different countries of the World using Time-Series clustering.* https://doi.org/10.1101/2020.08.18.20177261

Bauer, J. D., & Curran, P. J. (2003). Distributional assumptions of growth mixture models: implications for overextraction of latent trajectory classes. *Psychological Methods, 8*(3), 338–363.

Berlin, K. S., Parra, G. R., & Williams, N. A. (2014). An introduction to latent variable mixture modeling. *Journal of Pediatric Psychology, 39*(2), 174–203.

Carrillo-Larco, R. M., & Castillo-Cara, M. (2020). Using country-level variables to classify countries according to the number of confirmed COVID-19 cases: An unsupervised machine learning approach. *Welcome Open Research, 5*, 56.

Chandu, V. C. (2020). Identification of spatial variations in covid-19 epidemiological data using k-means clustering algorithm: a global perspective. medRxiv preprint. https://doi.org/10.1101/2020.06.03.20121194

Dempster, A. P., Lair N. M., & Rubin, D. B. (1977). Maximum likelihood from incomplete data via the EM algorithm. *Journal of the royal statistical society. Series B, 39*(1), 1–38.

Eddelbuettel, D., & François, R. (2011). Rcpp: Seamless R and C++ integration. *Journal of Statistical Software, 40*(8), 1–18.

Eddelbuettel, D., & Sanderson, C. (2014). RcppArmadillo: Accelerating R with high-performance C++ linear algebra. *Computational Statistics and Data Analysis, 71*, 1054–1063.

Ferrari, S. L. P., & Cribari-Neto, F. (2004). Beta regression for modelling rates and proportions. *Journal of Applied Statistics, 31(7)*, 799–815.

Greene, W. H. (2018). *Econometric analysis* (8th ed.). Pearson.

Hasell, J. Mathieu, E. Beltekian, D. Macdonald, B. Giattino, C., Ortiz-Ospina, E. Roser, M., & Ritchie, H. (2020). A cross-country database of COVID-19 testing. *Scientific Data, 7*, 345.

Herby, J. Jonung, L., & Hanke, S. H. (2022). *A literature review and meta-analysis of the effects of lockdowns on COVID-19 mortality*. Studies in Applied Economics (Vol. 200). John Hopkins Institute for Applied Economics.

Hojo de Souza, F. S. Satchiko Hojo-Souza, N. Maciel da Silva C., & Guidoni, D. L. (2021). Second wave of COVID-19 in Brazil: Younger at higher risk. *European Journal of Epidemiology, 36*, 441–443.

James, N., & Menzies, M. (2020). Cluster-based dual evolution for multivariate time series: Analyzing COVID-19. *Chaos, 30*, 061108.

Jones, B. L., & Nagin, D. S. (2007). Advances in group-based trajectory modeling and an SAS procedure for estimating them. *Sociological Methods & Research, 35*(4), 542–571.

Jones, B. L., & Nagin, D. S. (2012). A stata plugin for estimating group-based trajectory models. Heinz College Research Working Paper.

Jones, B. L. Nagin, D. S., & Roeder, K. (2001). A SAS procedure based on mixture models for estimating developmental trajectories. *Sociological Methods & Research, 29(3)*, 374–393.

Kass, R. E., & Wasserman, L. (1995). A reference Bayesian test for nested hypotheses and its relationship to the Schwarz criterion. *Journal of the American Statistical Association, 90*(431), 928–934.

Kinnunen, J. Georgescu, I. Hosseini, Z., & Androniceanu, A. M. (2021). Dynamic indexing and clustering of government strategies to mitigate Covid-19. *Entrepreneurial Business and Economics Review, 9*(2).

Laursen, B. P., & Hopf, E. (2006). Person-centered and variable centered approaches to longitudinal data. *Merrill-Palmer Quarterly, 52*(3), 377–389.

McLachlan, G., & Peel, D. (2000). *Finite mixture models*. Wiley.

Muthén, B. O., & Shedden, K. (1999). Finite mixture modeling with mixture outcomes using the EM algorithm. *Biometrics, 55*, 463–469.

Nadif, M., & Role, F. (2021). Unsupervised and self-supervised deep learning approaches for biomedical text mining. *Briefings in Bioinformatics, 22*(2), 1592–1603.

Nagin, D. S. (2005). *Group-based modeling of development*. Harvard University Press.

Nagin, D. S., & Land, K. C. (1993). Age, criminal careers and population heterogeneity: Specification and estimation of a nonparametric, mixed Poisson model. *Criminology, 31*, 327–362.

Nesselroade, J. R. (1991). Interindividual differences in intraindividual change. In L. A. Collins & J. L. Horn (Eds.), *Best methods for the analysis of change* (pp. 92–106). American Psychological Association.

Noel, C. (2024). *On a generalisation of Nagin's finite mixture model*. PhD Thesis. University of Luxembourg. Luxembourg.

Noel, C., & Schiltz, J. (2022) *trajeR, an R package for cluster analysis of time series*. Working Paper. University of Luxembourg. Luxembourg.

Rahman, M. A., Zaman, N., Asyhari, A. T., Al-Turjman, F., Alam Bhuiyan, M. Z., & Zolkipli M. F. (2020). Data-driven dynamic clustering framework for mitigating the adverse economic impact of Covid-19 lockdown practices. *Sustainable Cities and Society, 62*, 102372.

Rubin, D. B. (1976). Inference and Missing Data. *Biometrika, 63*(3), 581–592.

Schiltz, J. (2015). A generalization of Nagin's finite mixture model. In: M. Stemmler, A. Von Eye & W. Wiedermann (Eds.), *Dependent data in social sciences research* (pp. 107–126). Springer.

Smithson, M., & Verkuilen, J. (2006). A better lemon squeezer? Maximum-likelihood regression with beta-distributed dependent variables. *Psychological Methods, 11*(1), 54–71.

Spiegelhalter, D., & Masters, A. (2021). *Covid by numbers*. Penguin.

Symons, M. J. (1981). Clustering criteria and multivariate normal mixtures. *Biometrics, 37*(1), 35–43.

van der Nest, G. Lima Passos, V., Candel, M. J. J. M., & van Breukelen, G. J. P. (2020). An overview of mixture modelling for latent evolutions in longitudinal data: Modelling approaches, fit statistics and software. *Advances in Life Course Research, 43*, 100323.

Verbeken, G. Fieuws, S., & Molenberghs, G. (2014). The analysis of multivariate longitudinal data: A review. *Statistical Methods in Medical Research, 2014* (1), 42–59.

von Eye, A., & Bergman, L. R. (2003). Research strategies in developmental psychopathology: Dimensional identity and the person-oriented approach. *Development and Psychopathology, 15*, 553–580.

Zarikas, V. Poulopoulos, S. G., Gareiou, Z., & Zervas, E. (2020). Clustering analysis of countries using the covid-19 cases dataset. *Data in Brief, 31*, 105787.

Part II
Network Analysis and Causal Structure Learning

Chapter 7
What the Fuzz!? Leveraging Ambiguity in Dynamic Network Models

Jonathan J. Park, Sy-Miin Chow, and Peter C. M. Molenaar

Part of the scientific endeavor revolves around understanding the fundamentals of the world around us. Typically, this process involves the reduction of a complex set phenomenon into descriptive categories that have utility in describing these systems in simpler and heuristic ways; however, our attempts to discretize nature are not always beneficial or consistent. For example, chemistry texts frequently describe and disagree on the boundary between polar covalent bonds and ionic bonds with various resources listing the boundary at varying values of electronegativity (see Fig. 7.1; Sproul, 2001). This results in a large number of bonds whose classification is ambiguously polar covalent *or* ionic dependent on the source used. Similarly, the study of psychopathology in the last century has been dominated by attempts to qualify various psychological disorders into distinct categories comprised of unique symptoms; however, research has consistently indicated a high degree of overlap between various psychopathologies and their symptom profiles (Borsboom & Cramer, 2013; Isvoranu et al., 2022). Formally, we argue that many of the boundaries in contemporary research are "fuzzy." The goal of this chapter is to introduce social and behavioral scientists to the broader concept of fuzziness in community detection algorithms and how we may be able to leverage ambiguity when making decisions regarding boundaries between variables.

Research reported in this manuscript was supported by the National Institutes of Health under grants U24AA027684 and UL1 TR002014; and National Science Foundation grant IGE-1806874

J. J. Park (✉)
Department of Psychology, The University of California, Davis, Davis, CA, USA
e-mail: imJPark@UCDavis.edu

S.-M. Chow · P. C. M. Molenaar
Department of Human Development and Family Studies, The Pennsylvania State University, University Park, PA, USA
e-mail: symiin@psu.edu; pxm21@psu.edu

M. Stemmler et al. (eds.), *Dependent Data in Social Sciences Research*,
https://doi.org/10.1007/978-3-031-56318-8_7

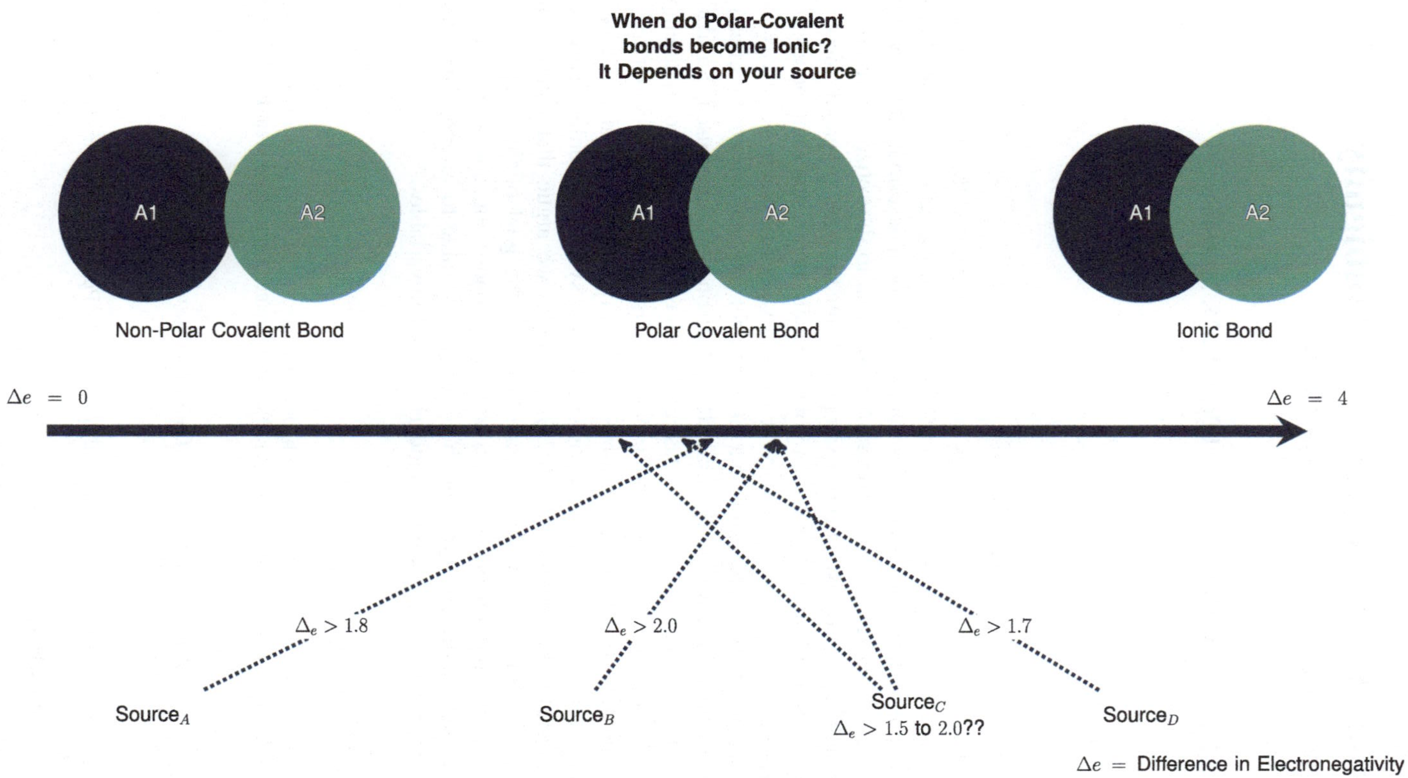

Fig. 7.1 Inconsistent decision rules for the boundary between polar covalent and ionic bonds from various sources

Psychopathology
The study of psychopathology has recently seen a paradigm shift, with increased focus on the observed intercorrelations among variables in lieu of latent variable approaches (Borsboom & Cramer, 2013). This shift—dubbed the network approach—has resulted in a number of novel methods for tackling the study of psychopathology. One of these approaches, community detection, has seen widespread adoption in the network approach to psychopathology and broader psychological and behavioral networks.

Historically, latent variable models have been widely applied in the study of psychopathology. These approaches posit that an underlying disorder (e.g., major depression) is the root common cause for the correlations between indicator variables (i.e., symptoms: anhedonia, loss of appetite, loss of concentration). In contrast, the network approach to psychopathology suggests that the common correlation among a set of symptoms is—in and of itself—the manifestation of the psychological disorder (Borsboom & Cramer, 2013). This conceptualization of symptom-level data has also been used to help explain the great deal of correlation and overlap between symptoms across different psychological disorders with the boundaries between disorders being described as "fuzzy" in the literature (Borsboom & Cramer, 2013; Cramer et al., 2010). However, despite this acknowledgement regarding the fuzziness between psychological disorders, attempts to identify discrete communities of symptoms have still been the most common approaches (e.g., Goekoop & Goekoop, 2014; Groen et al., 2019; Imperiale et al., 2021; Golino & Epskamp, 2017). While these have provided valuable insight regarding the structure and stability of symptom networks in relation to broader diagnostic criteria, relatively few papers in the psychological and behavioral sciences have attempted to separate psychological disorders by their symptom profiles while acknowledging their fuzzy boundaries (see Castro et al., 2019, 2018; Blanken et al., 2018, for examples).

An additional focal point in the application of psychological networks is the concept of "bridges" (Castro et al., 2018; Blanken et al., 2018; Cramer et al., 2010; Fried et al., 2017). Broadly, bridges are any form of node in a network which may increase connectivity and the likelihood of contagion—the activation of one disorder by another through common/shared symptoms—from one community of nodes to another (Jones et al., 2021; Cramer et al., 2010). The identification of these bridges has been posited as a major avenue of importance for network psychopathology for identifying targets for intervention to prevent contagion. Prior work by Jones et al. (2021) proposed measures for quantifying and defining bridge centrality measures which have shown a strong ability to identify bridge symptoms in psychopathology networks; however, the metrics themselves assume knowledge of the underlying community membership of the symptoms themselves which may not be available or known to researchers. Thus, in some instances, an exploratory procedure for identifying bridges may be desirable. In addition, given that bridges are theorized as variables that exist between communities or strongly relate two communities together, they themselves may exhibit fuzzy or unclear membership in community detection solutions (Blanken et al., 2018).

The current work will introduce community detection approaches and describe the differences between discrete community detection methods with their fuzzy alternatives. We will provide simulated and empirical illustrations which highlight the key differences between discrete and fuzzy community detection solutions and discuss their potential implications in the analysis of subjects when applying these methods to both longitudinal and cross-sectional data.

The Adjacency Matrix and Discrete Community Detection

Prior to conducting k-means—or any community detection algorithm—one must define their metric for distance between nodes within a network. Here, we discuss the adjacency matrix, $\mathbf{A}$, as

$$\mathbf{A} = \begin{bmatrix} a_{11} & a_{12} & \cdots \\ \vdots & \ddots & \\ a_{K1} & & a_{KK} \end{bmatrix} \tag{7.1}$$

where any cell a_{ij}, $i \neq j$, represents the distance between the ith and jth subjects. The adjacency matrix is ultimately the heart and soul of community detection in our endeavor to teach computers to find communities from data and is thus the key for addressing any specific hypotheses regarding our data. Minor differences in how we define our metric of distance can have marked impacts on the results we may yield from even the same set of analyses. Thus, it is of critical importance that we consider *how* the adjacency matrix is determined for any given analysis.

Community detection has its origins in statistical physics and broader graph theory with the goals of identifying clusters or communities of nodes which are more common to one another than they are to nodes which belong to other communities. In everyday life, our understanding of communities is relatively clear; we may have a clear understanding of where our families begin or end or the boundaries of the cities in which we live. However, teaching a computer algorithm to identify the differences between objects involves translating what we inherently know and take for granted into quantifiable tasks which a computer can execute. Below, we introduce two concepts, discrete community detection and a simple fuzzy alternative. Following that, we will provide some illustrations regarding how these may be applied to data for useful insights.

In this section, we discuss a relatively simple discrete community detection algorithm: k-means clustering. Formally, k-means clustering operates by assigning all subjects to k-random partitions and calculating their centroids (e.g., means) and attempting to minimize the following objective function:

$$J(X) = \sum_{j=1}^{C} \sum_{i=1}^{n_j} \left\| \mathbf{x}_i - \bar{\mathbf{x}}_j \right\|^2 \tag{7.2}$$

where C is the total number of communities to partition, n_j is the number of subjects in the jth community, and $\left\| \mathbf{x}_i - \bar{\mathbf{x}}_j \right\|$ is the Euclidean distance between individual i's values on a vector of variables, $\mathbf{x}$, and the jth community's means, $\bar{\mathbf{x}}_j$. Ultimately, this results in the within-cluster variance being minimized. Upon random assignment and initial calculation of the centroids, individual points are moved into clusters which reduce the within-cluster variance further. Once every case has been shuffled, the values for the centroids are recalculated, and the procedure is repeated iteratively until some convergence criterion is met.

However, one may see from the formulation of the k-means algorithm that individuals belong to singular communities with no chance for overlap, whereas this chapter began with the goal of loosening this restriction. Thus, we introduce the fuzzy extension of the k-means algorithm: fuzzy c-means (FCM; Bezdek, 1974).

Fuzzy Community Detection
Fuzzy community detection is a subfield of the broader domain of cluster analysis. As noted previously, community detection in the behavioral sciences has typically focused on the grouping of subjects in a person-by-person adjacency matrix where discrete and orthogonal groups are identified (e.g., Gates et al., 2017; Goekoop and Goekoop, 2014). Fuzzy clustering offers researchers the flexibility of estimating the likelihood of membership for subjects across multiple groups. For example, instead of discretely categorizing an individual as being in Group A or B, a fuzzy cluster analysis would indicate the likelihood that an individual belongs to Group A and the separate likelihood that they belong to Group B.

Fuzzy clustering has found success in bioinformatics where it has been used to identify clusters of genes which coactivate with one another and connect multiple physiological systems which do not discretely belong to one group or another (Valafar, 2002). This issue of genetic markers mirrors the issues currently faced in the study of psychological networks where we attempt to identify symptoms which interact with one another and belong to multiple psychopathologies. Classically, fuzzy clustering has been conducted using fuzzy c-means which is similar to the discrete k-means clustering algorithm.

In contrast to mixture models (Morin & Wang, 2016), which assume that the communities, often denoted as *classes*, are latent variables with their own parametric distributions (e.g., categorical distributions), fuzzy clustering—as a direct extension of k-means and related clustering approaches—involves relatively little distributional assumptions. In brief, fuzzy clustering utilizes a fuzzifier operator on degree (denoted herein as μ_{ij}), calculated as the distance between member i and each assigned community centroid relative to its distance from other communities. In other words, degree is a summary statistics that describes an individual's strength of association with the jth community that can be readily calculated from empirical data. Unlike other mixture modeling alternatives, the fuzzifier operator, as described in detail below, does not require distributional assumptions on community membership but rather may be seen as a heuristic approach to allow for uncertainty in individuals' community membership. This is a contrast to mixture approaches that rely heavily on distributional assumptions on the latent classes and other variables within a class to allow for use of algorithms such as the expectation–maximization algorithm for model estimation purposes (Chow et al., 2015). In the strict sense,

k-means clustering on the Euclidean distances of raw data can been shown to produce results identical to the Gaussian Mixture Model with an identity covariance matrix (Roweis & Ghahramani, 1999). Likewise, under specific parameterizations, the clustering results of fuzzy c-means can produce identical results to Gaussian Mixtures at set or fixed values of the fuzzifier (Ichihashi et al., 2001). However, these restrictions may not necessarily apply when unique definitions of "distance" are applied such as when distances are defined as structural similarities between individual dynamic models. Furthermore, the equivalence between these models exists under specific conditions which—once altered—result in deviations from any possible Gaussian Mixture Modeling result (Ichihashi et al., 2001).

Specifically, fuzzy clustering draws from fuzzy logic and allows for cluster membership to be determined based on degree weights rather than a dichotomous (Yes/No) classification. While similar in many respects to k-means, fuzzy c-means differs in some key ways. Fuzzy c-means minimizes the following objective function:

$$J(U, X) = \sum_{j=1}^{C} \sum_{i=1}^{N} \mu_{ij}^{m} \left\| \mathbf{x}_i - \bar{\mathbf{x}}_j \right\|^2 \tag{7.3}$$

where all terms are as defined previously, except for the addition of μ_{ij}^{m}. Here, μ_{ij}^{m} is an element of the larger $\mathbf{U}$ membership degree matrix of dimension $N \times C$ with N being the total sample size, and C, the total number of communities for a given value of m where $m > 1$—is typically referred to as the *fuzzifier* and determines the extent of fuzziness of the community assignment and is used as a *label* to indicate that the element μ_{ij} is calculated for that specific fuzzifier value. An element of $\mathbf{U}$ conceptualizes how closely the ith subject is to the jth community's centroid relative to other communities. Formally, μ_{ij}^{m} may be calculated for any arbitrary value of m as

$$\mu_{ij}^{m} = \frac{1}{\sum_{k=1}^{C} \left(\frac{\left\| \mathbf{x}_i - \bar{\mathbf{x}}_j \right\|}{\left\| \mathbf{x}_i - \bar{\mathbf{x}}_k \right\|} \right)^{\frac{2}{m-1}}} \tag{7.4}$$

where we see that μ_{ij}^{m} is inversely related to the distance between a data point and its assigned cluster and is bounded from 0 and 1 where the row sum of $\mu_{i,}$ for any ith individual—the degree to which the ith subject belongs to any community sums to 1. A membership degree closer to 1 then indicates that a subject is closer—and thus more likely—to be a member of an assigned community relative to another. Furthermore, we can see that as the fuzzifier approaches a value of 1, values will become binary resulting in the discrete, k-means solution.

The calculation of the centroids is then reworked due to the introduction of fuzziness. In particular, the centroids are now weighted by the degree of membership and can be expressed as

$$\bar{\mathbf{x}}_j = \frac{\sum_{i=1}^{N} \mu_{ij}^m \mathbf{x}_i}{\sum_{i=1}^{N} \mu_{ij}^m}, \forall j = 1, 2, \ldots, C \tag{7.5}$$

where the centroid of the jth cluster is the mean of all data points weighted by their degree of membership within that specific cluster. FCM has limitations—as with k-means—where the number of clusters must be determined usually as dictated by theory or by some cluster selection criteria; some of which we will describe in the forthcoming section. We will review several cluster quality indices that can be used to help guide researchers in making decisions on the number of communities to enumerate from community detection results.

Cluster Quality Indices The quality of a set of partitions in community detection can be assessed in multiple ways depending on the knowledge a researcher has beforehand. For example, given prior knowledge of underlying group memberships, the Hubert–Arabie Adjusted Rand Index (referred to henceforth as the ARI; Hubert and Arabie, 1985) may be used (Lane et al., 2019). The ARI typically takes values ranging from 0 to 1 with 1 indicating a perfect match between the "true" (or reference) and estimated group membership. The ARI is calculated as

$$ARI_{HA} = \frac{\binom{N}{2}(a+d) - [(a+b)(a+c) + (c+d)(b+d)]}{\binom{N}{2}^2 - [(a+b)(a+c) + (c+d)(b+d)]} \tag{7.6}$$

where N is the number of subjects, a represents the number of true positives (i.e., hits), b indicates the number of false negative classifications (i.e., pairs classified as separate when they actually belonged together), c is the number of false positive classifications, and d is the number of true negative classifications. In evaluation, excellent recovery is considered $ARI_{HA} > 0.90$, good recovery is $ARI_{HA} > 0.80$, moderate recovery is $ARI_{HA} > 0.65$, and values below $ARI_{HA} = 0.65$ are considered poor (Lane et al., 2019; Steinley, 2004). Generally, the ARI was designed to compare the number of correct classification cases (relative to the true or reference membership) relative to the incorrect ones, correcting for correct decisions by chance.

Additional metrics may be used to describe the quality of a fuzzy cluster solution in terms of how clean the "cuts" are separating groups from one another. This can be accomplished in multiple ways including with the partition coefficient (PC; Bezdek, 1974) and the partition entropy (PE; Bezdek, 1975). In addition, some quality indices exist for both hard and fuzzy community detection such as the silhouette index and its fuzzy extension (Campello & Hruschka, 2006). We begin with the PE. Given the membership degree matrix from Eq. 7.4, the PC is computed as

$$PC = \frac{1}{N} \sum_{j=1}^{C} \sum_{i=1}^{N} \mu_{ij}^{m\,2} \tag{7.7}$$

where better PCs are bound between $\frac{1}{C}$ and 1 with PCs closer to a value of 1 indicating "harder" communities and values closer to $\frac{1}{C}$ indicating fuzzier communities (Sheng Li, 2011). Since every membership degree is bound between 0 and 1, squaring each membership degree has the effect of penalizing subjects whose membership degrees are low for every community than if a subject belongs very strongly to a single partition. While the PC is an indicator of how clearly separated a set of clusters may be, this may still not reflect the underlying structure of the data (see Trauwaert, 1988; Bezdek, 2013, for more discussion). For instance, the separation between groups in reality may truly be fuzzy. Thus, defining a "hard" solution as the optimum—as in the case of an "ideal" PC—may not be best in practice.

Similarly, the PE, computed as

$$PE = -\frac{1}{N}\sum_{j=1}^{C}\sum_{i=1}^{N}\mu_{ij}^{m}\log\left(\mu_{ij}^{m}\right) \tag{7.8}$$

has ranges between 0 and 1 where values closer to 0.00 are typically considered higher quality partitions with 0.00 indicating a perfect partition; more formally, that all subjects were perfectly placed into a single community with no overlap. PE is similar to the PC in that it assesses cluster solutions by how strongly individuals are clustered into one group relative to multiple groups with minimal connections between communities with slightly different interpretations. With PC being described as characterizing the degree of certainty in assigning N subjects to one of the C communities. Alternatively, PE is better conceptualized as the average fuzzy entropy within the membership degree matrix (see Bezdek, 1975).

Finally, the silhouette index (SI) and its fuzzy form (FSI) are measures of how close an individual data point is from the closest out-of-community neighbor. Beginning with the SI, if we define the a_{ij} as the average distance between the ith subject and members of the jth community it was assigned to, let d_{ij} represent the average distance of a subject to another community's members, and finally define b_{ik} as the minimum of the set d_{ik} where $k \neq j$ (i.e., the average distance from the ith subject to the closest community it was not assigned to), we can derive the silhouette index of the ith subject as

$$s_i = \frac{b_{ik} - a_{ij}}{\max\{a_{ij}, b_{ik}\}} \tag{7.9}$$

where the denominator is a normalization term that selects whichever value is greater: the average distance between a subject and its assigned neighbors or the average distance between a subject and the members of the closest neighboring community. The full silhouette index for the entire sample is then just the arithmetic mean of the individual silhouettes:

$$SI = \frac{1}{N} \sum_{i=1}^{N} s_i \tag{7.10}$$

The fuzzy extension proposed by Campello and Hruschka, 2006 leverages the membership degree matrix, **U**, described in Eq. 7.4 and is written as

$$FSI = \frac{\sum_{i=1}^{N} (\mu_{ij}^{m} - \mu_{ik}^{m})^{\alpha} s_i}{\sum_{i=1}^{N} (\mu_{ij}^{m} - \mu_{ik}^{m})^{\alpha}} \tag{7.11}$$

where α is an optional scaling coefficient that is typically set to a value of 1 (Campello & Hruschka, 2006). The FSI is a transformation of the s_i for each subject by a factor of the difference between the first (j) and second (k) largest membership degrees. This results in essentially a weighted average of the SI by the difference between the two strongest membership degrees a subject has for any two communities. Higher values of the SI and FSI are indicative of clearer separation between communities and are compared across the number of enumerated groups.

Overall, both the PC and PE are measures of how "hard" a partition of communities are and will generally agree with one another due to a formal relation between the two metrics (see Bezdek, 1975). Furthermore, the silhouette indices are a measure of the amount of separation between communities. All of the measures have been historically used to assess cluster validity by comparing values obtained across the number of communities enumerated and selecting for the partition which produces the highest PC, SI, and SF and the lowest PE (Bezdek, 2013). However, much debate and discussion exists regarding whether an ideal fuzzy solution should be based on whether it is a good hard solution (Bezdek, 2013). In the following illustrations, the PE, PC, silhouette indices, and the ARI were used to assist in making decisions regarding the number of communities where appropriate to enumerate by comparing the values across the number of communities enumerated. In instances where the results of the fit indices clashed with expectations, we follow up with more thorough explanations below.

7.1 Simulated Illustration

We now proceed with showing how discrete and fuzzy community detection methods may yield discrepant results in a simulated illustration. We highlight the key differences between fuzzy and discrete community detection methods when applied to dynamic networks by simulating synthetic dynamic networks from two overlapping communities which evolve over time. Both methods are then applied to the simulated data. The illustration highlights the advantages and limitations of each approach and provides insights into the relative importance of considering fuzzy or discrete community structure in different types of dynamical processes. We begin by discussing the construction of an adjacency matrix when modeling dynamic networks.

Given that **A** is a symmetric matrix that defines how similar two subjects are, there are a host of ways for us to take the values from a dynamic model to define "similarity." Take—for instance—that we decide to fit a VAR model to our data in the form of lag order 1:

$$\eta_{(t)} = \mu + \mathbf{\Phi}\eta_{(t-1)} + \zeta_t \tag{7.12}$$

where η is a p-variate vector of scores, μ is a p-variate vector of intercepts, $\mathbf{\Phi}$ is a $p \times p$ lagged regression matrix, and ζ_t is a p-variate residual vector assumed to have zero means and covariance matrix, $\mathbf{\Psi}$.

Several parameters in the VAR(1) equation could be utilized when attempting to identify unique communities of individuals. For instance, subjects may differ in their mean levels across variables. In which case, μ may be of interest to us. In addition, if one is interested in the dynamics and if groups of individuals may change similarly to one another through time, the $\mathbf{\Phi}$ matrix may be worth investigating. Other researchers have also leveraged the residual covariance matrices or otherwise extracted information from them (e.g., the graphical VAR) to further model contemporaneous effects (Gates et al., 2017; Epskamp et al., 2018).

However, in addition to leveraging different elements of the VAR(1) model, we must also consider how we transform the information contained in the coefficient matrices to our definition of "distance" in the adjacency matrix. We may refer to past work on clustering VAR-based models for how they operationally defined their adjacency matrices. For instance, the subgrouping option in the group iterative multiple model estimation (S-GIMME; Gates et al., 2017) procedure counts the number of statistically significant coefficients that are similar between any pair of subjects. If Subject X and Subject Y both share a connection, ϕ_{13}, that is both statistically significant and identical in sign (i.e., $\pm$), then it is added into the cell, a_{xy}, of the adjacency matrix. Other methods like the subgrouping option for chain graphical VAR (scgVAR) models additionally count the common statistically nonsignificant—or null (e.g., $\phi_{13} = 0.00$), as well as the common significant paths. In a separate domain of work by Bulteel and colleagues (2016), VAR models have also been clustered based on the Euclidean distances between the elements of the $\mathbf{\Phi}$ between any pair of subjects. Thus, while some algorithms prioritize commonality in the raw structure, others may prioritize the relative distance between coefficient matrices themselves. These nuanced differences may lead to various effects on the clustered solution. For instance, the alternating least squares VAR (ALS VAR; Bulteel et al., 2016)—in an effort to maximally separate the generated clusters—tends to avoid allowing subgroups to share common paths with one another. In this way, the ALS VAR is able to detect differences between groups based on effect sizes (i.e., $\phi_{13} = 0.21$ vs. $\phi_{13} = 0.63$). In contrast, the S-GIMME and scgVAR approaches do not have this particular strength, but rather, rely on significance testing results to put individuals into discrete clusters.

The definition of our adjacency matrix becomes all the more complex when we integrate the role of fuzziness into its construction. We may need to further consider what makes a subject "fuzzy"? When constructing adjacency matrices in a manner similar to S-GIMME or scgVAR models, it may be a subject with many structural similarities to multiple groups and could be indicative of a subject that is functionally between two "states," if we define a state as a common (sub)group model. In the following illustrations, we will use the raw structural counts similar to S-GIMME's approach for clustering VAR models by following these steps:

1. Estimate VAR(1) models on each subject in the dataset.
2. Prune all nonsignificant paths from the resulting $\mathbf{\Phi}_1$ matrices.
3. Construct an adjacency matrix, $\mathbf{A}$, where each cell, a_{ij}, is a count of the number of elements of $\mathbf{\Phi}$ that the ith and jth subjects share in common.
   - Commonality is defined as the same nonzero edge with the same polarity (e.g., two participants both shared a statistically significant entry in $\mathbf{\Phi}$ with the same sign).
4. Apply k-means and fuzzy c-means to the resulting adjacency matrix to extract solutions.

Data Simulation The goal of our simulated illustration is to highlight the differences that may arise when discrete and fuzzy community detection methods are applied to dynamic processes. Thus, we address two conditions: (1) how well do these methods perform when no clear communities exist and (2) how well do these algorithms perform when fuzzy subjects exist in the data?

To address these goals, we simulated two VAR(1) models comprised of ten variables each. The two discrete subgroups shared all ten autoregressive effects and were separated by either zero or nine cross-regressions. The zero unique cross-regression condition will allow us to address the first condition, while the nine unique effects would allow the community detection methods to delineate the groups from one another. Fuzzy subjects were then simulated from a 50% mixture distribution such that they shared 50% of the characteristics of the two distinct subgroups. The resulting networks for the typical fuzzy subjects as well as the two distinct subgroups can be found in Fig. 7.2. We simulated a total of $N = 30$ subjects, where $n_{g1} = 10$, $n_{g2} = 10$, and ten subjects came from a mixture of the two data-generating VAR(1) models resulting in two discrete groups with ten "fuzzy" subjects for subsequent analysis.

Simulation Results Following data generation, VAR(1) models were estimated for each subject, and an adjacency matrix was constructed using the steps detailed in the previous section. Following this, we applied both the discrete k-means and the fuzzy c-means algorithms to the adjacency matrix comprised of structural path counts. The results of the hard and fuzzy clustering algorithms on the simulated dynamic networks are presented in Fig. 7.3.

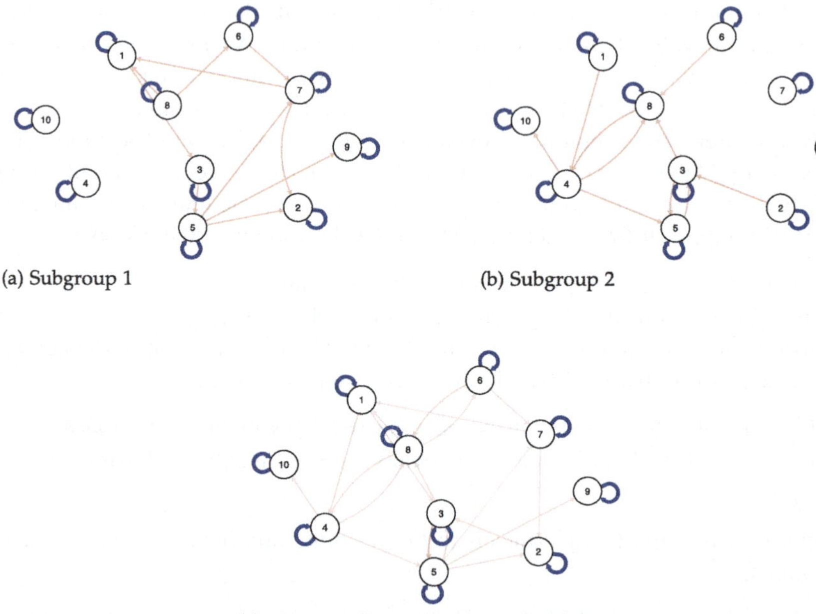

(a) Subgroup 1

(b) Subgroup 2

(c) Average Dynamic Network of Fuzzy Subjects

Fig. 7.2 Visual depiction of the simulated dynamic network structures. In (**a**), we see the dynamic network structure for the first subgroup, (**b**) shows the second subgroup's dynamic network structure, and (**c**) is the average estimated dynamic network structure of the fuzzy subjects

Our first aim was to address how well discrete and fuzzy community detection methods perform when faced with essentially no community structure. We found that—for our illustration—fuzzy methods performed well in *not* committing to a membership assignment in contrast to the hard clustering methods which essentially must enumerate the number of groups requested. When enumerating a two- and three-community solution out of the discrete clustering methods, we found essentially random assignment for both conditions as seen in the top two panels of Fig. 7.3a. When assessing the SI, we found that the three-community solution was preferred. The PC and PE were not used in this case as they are metrics for fuzzy partition quality only. In contrast, the bottom panel of Fig. 7.3a) shows that the membership degrees produced by the two-community fuzzy clustering method rarely deviated away from 0.50 for the fuzzy subjects. This indicated that the fuzzy subjects existed perfectly between either community (i.e., $\mu_{ij} = \frac{1}{C} = \frac{1}{2}$). Interestingly, the three-community solution was preferred by the PC, PE, and the FSI in the fuzzy case despite the fact that the subjects were drawn from a mixture of the two true populations. We followed these results by applying a threshold of 0.70 to the membership degrees (i.e., **U**), similar to that used by Bezdek, 1974 for the two-class solution and found that all of the simulated fuzzy subjects were below the threshold; thus, the two-community solution served multiple purposes:

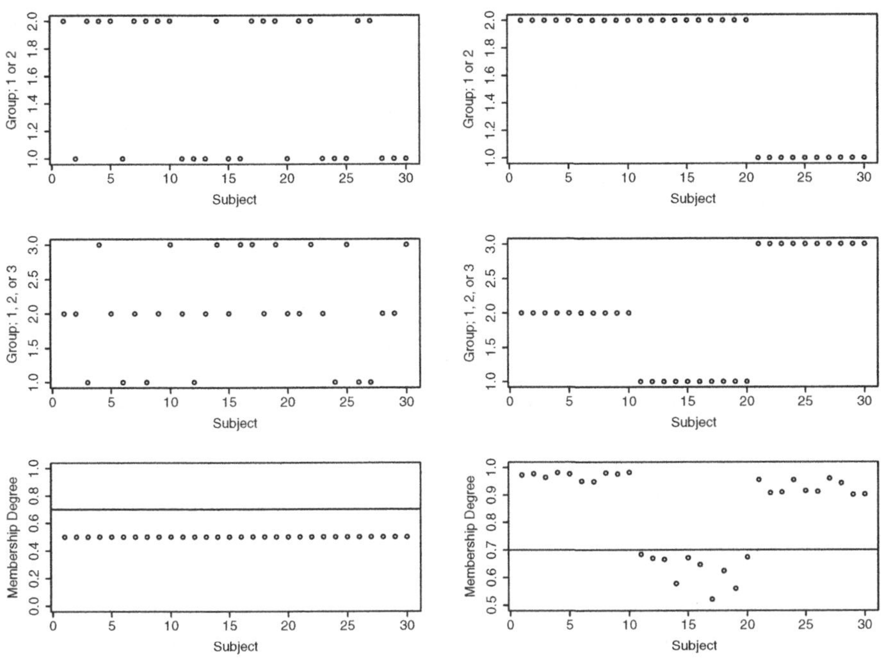

(a) No simulated subgroups (b) Two simulated subgroups

Fig. 7.3 Cluster solutions for (**a**) no subgroups and (**b**) two subgroups with fuzzy subjects. Top two rows represent two- and three-enumerated subgroups in the hard clustering solution where the Y-axis is the group membership assignment. The bottom panel indicates the fuzzy clustering solution where the Y-axis indicates the membership degree

(1) it correctly separated subjects that belonged to distinct communities, and (2) it correctly identified the fuzzy subjects when a threshold was applied without the need to enumerate a third group.

In our second aim, we sought to elucidate how discrete and fuzzy methods handle ambiguous cases. The results from the illustration suggest that the fuzzy community detection algorithm generally enumerated fewer groups than the discrete clustering algorithm. In addition, it placed smaller membership degree weights (i.e., μ_{ij}) on fuzzy subjects as illustrated in the bottom panel of Fig. 7.3b. In contrast, the discrete community detection algorithm tended to group the fuzzy subjects into one of the two dominant communities depending on the random seed used; however, when a third group was enumerated, both the discrete and fuzzy community detection methods placed all fuzzy subjects into a third category as seen in the top two panels of Fig. 7.3b. Even though such a three-cluster solution reduced the fuzzy subjects' impact in biasing the estimated structures of the two dominant communities, this did create new questions concerning the appropriateness of viewing these subjects as constituting their own distinct community or if their fuzziness is not sufficient for enumerating an additional group.

To summarize, our simulated illustration highlighted some key differences between the discrete and fuzzy methods for community detection when fitted to adjacency matrices constructed from VAR(1) dynamic networks. First, we found that—in the presence of significant interindividual variability—fuzzy c-means provided relatively ambiguous membership degree weights for all subjects ($\sim$0.50). In contrast, the discrete clustering method identified as many groups as we enumerated—a limitation of the function itself—and resorted to randomly forcing subjects into communities. It should be noted that this shortcoming can be addressed by assessing quality indices such as the partition coefficient, partition entropy, or other related measures. Second, we also found differences in the performance between discrete and fuzzy methods when applied to a simulated condition where two true groups existed along with subjects drawn from a mixture of the two groups. We found that the fuzzy approach correctly identified subjects as fuzzy and assigned relatively low weights to their membership degrees. In contrast, the discrete method would force fuzzy subjects into one category or another and correctly separated out the fuzzy subjects when tasked with enumerating a third community.

Empirical Illustration

In this section, we apply both discrete and fuzzy community detection methods to an empirical dataset borrowed from De Vos et al., 2017. The dataset contains $N = 54$ subjects who either have Major Depressive Disorder (MDD) or were pair-matched controls ($n_{dep} = 27$; $n_{con} = 27$). All subjects were assessed on 14-affect items and were asked to rate their level (e.g., happiness) since the previous measurement occasion. Subjects filled out surveys over the course of a month (30 days) at an interval of three occasions per day resulting in 90 possible assessments per subject. Data were preprocessed using code provided by the authors to ensure our results are closely matched to their own. More information regarding the study and data structure may be found in the work by De Vos et al. 2017. For the purposes of this illustration, two communities were extracted from the community detection algorithms and assessed based on their characteristics.

Following the procedure outlined previously, we began by fitting a VAR(1) model to each subject in the dataset using all 14-affect variables. An adjacency matrix was then constructed where each cell, a_{ij} where $i \neq j$ indicated the number of common, nonzero edges the ith and jth subjects shared in common with one another in their matrices. Following this, both k-means and c-means community detection algorithms were applied to the adjacency matrix to identify communities.

Results from the empirical illustration are presenting in Fig. 7.4. Notably, Fig. 7.4a reveals the two-group solution supplied by the k-means algorithm. Broadly, we discovered that subject group assignments did not align well with clinical group membership; however, closer inspection of the adjacency matrix revealed that many subjects—even within the same community—shared almost no common paths with one another. Similarly, the two-community solution from fuzzy c-means depicted in Fig. 7.4b revealed no true structure when a threshold of 0.70 was applied to the membership degrees following precedent set by Bezdek, 1974. This result aligns with the observed adjacency matrix which suggests that many subjects share few—

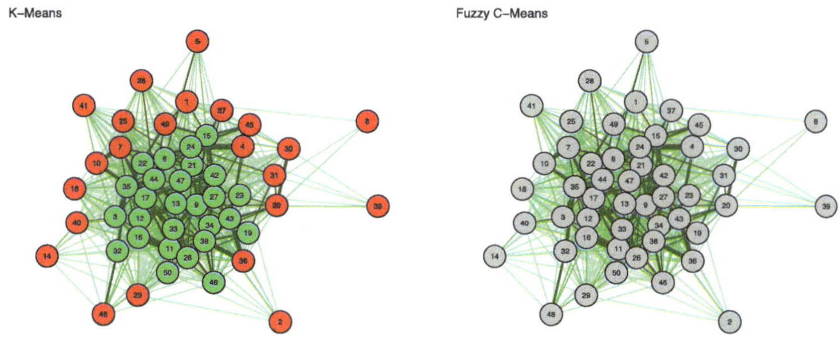

(a) k-means clustering results on de Vos dataset (b) Fuzzy c-means clustering results on de Vos dataset

Fig. 7.4 Community detection solutions for (**a**) k-means where subjects with identical colors (red/green) are assigned to the same subgroup. (**b**) is the results from fuzzy c-means. Gray subjects are those whose membership degree was below an arbitrary threshold of 0.70 signaling some degree of ambiguity in the cluster assignment

if any—common structural paths with one another and similarly both our empirical illustration and the results from De Vos et al., 2017 that noted the large degree of individual differences in the data. Thus, fuzzy community detection may be valuable for identifying and diagnosing the degree of within-sample heterogeneity in dynamics in addition to identifying fuzzy subjects.

Clustering Means Structures

A subsequent set of analyses were conducted on the cross-sectional data to test the performance of k- and c-means clustering algorithms when applied to cross-sectional networks. Using data provided by De Vos et al., 2017, we calculated the within-subjects means for each subject on all 14 variables and created an adjacency matrix where each cell was defined as the Euclidean distance between any pair of subjects multivariate means.

Figure 7.5 highlights the key results from applying both k- and c-means community detection methods on the cross-sectional means. Specifically, Fig. 7.5a, b highlights the clustering results from fitting a two- and three-group solution to the k-means algorithm. Notably, a significant portion of the subjects in the “middle” of happiness and depression becomes reclassified into a third group. A comparison based on the SI index indicated that the two-cluster partition was favorable due to the index having a higher value in the two-group solution compared to the three-group solution. However, this resulted in a fairly wide dispersion of subjects across the depressed continuum as shown in Fig. 7.5a. In relation to the clinical diagnoses, we found an $ARI = 0.38$ which equates to approximately 10 subjects of the total 54 being misclassified indicating relatively poor agreement with the clinical memberships.

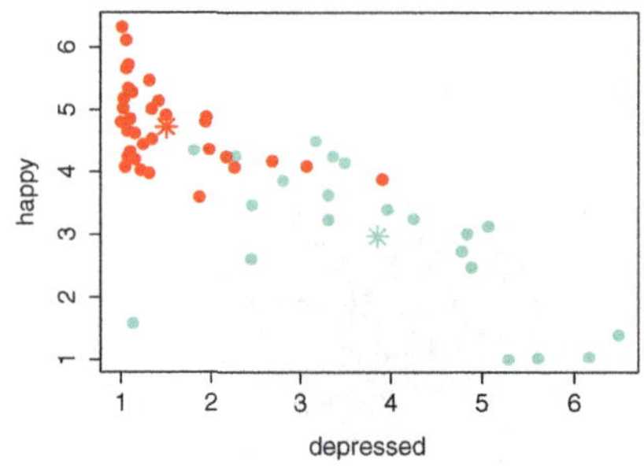

(a) *k*-means with 2 communities enumerated

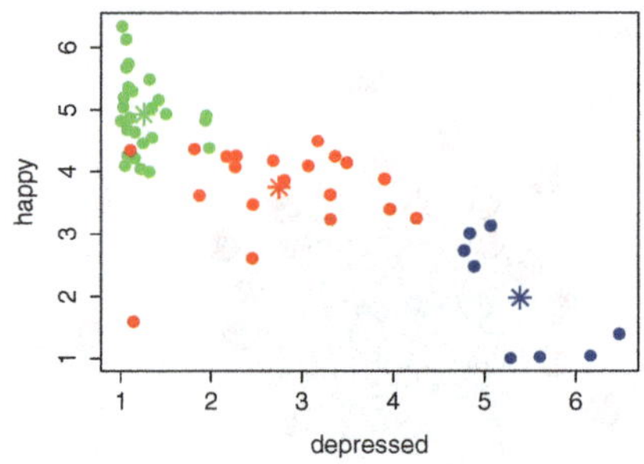

(b) *k*-means with 3 communities enumerated

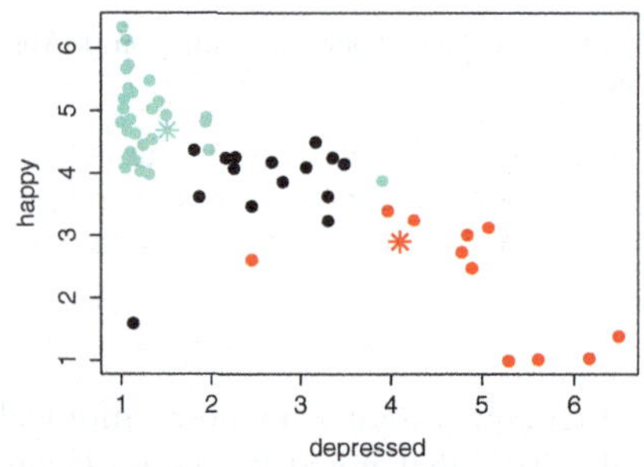

(c) Fuzzy *c*-means with 2 communities enumerated

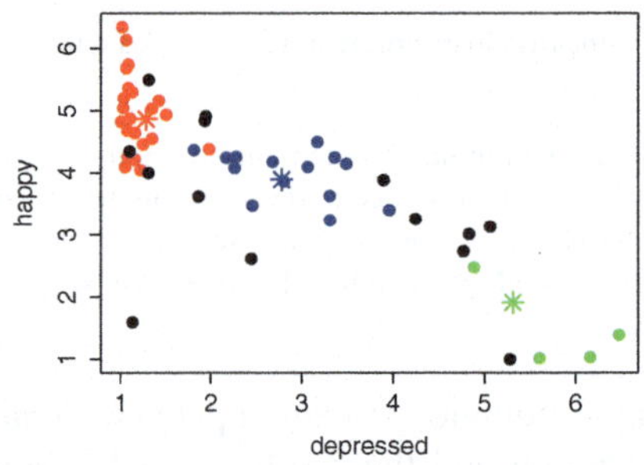

(d) Fuzzy *c*-means with 3 communities enumerated

Fig. 7.5 Community detection results for both k-means (**a** and **b**) and c-means (**c** and **d**) on 2 of the 14 dimensions analyzed. Different coloration indicates separate group membership, whereas black subjects denote those with a membership degree below 0.70 signaling some degree of ambiguity in the cluster assignment

Similarly, in Fig. 7.5c, d, a two- and three-group solution was extracted using fuzzy c-means, and the two-group solution was deemed favorable based on the PC, PE, and FSI. Inspection of Fig. 7.5c indicated that most fuzzy subjects—those with a membership degree less than 0.70—were relatively average on both happiness and depression in comparison to the other two identified communities. This group of fuzzy subjects ($n_{fz} = 15$) is comprised primarily of individuals with major depressive disorder (MDD; $n = 13$). This finding may be relevant as the sample itself is comprised on subjects undergoing outpatient treatment for their MDD and thus may be indicative of a positive prognosis. Furthermore, the number of subject incorrectly classified has decreased from 10 subjects to 5 (1 control and 4 MDD) in relation to the clinical memberships. In this instance, the ARI was not used as there was no clear way to relate the subjects that were deemed to be below threshold to a "true"/reference group (see Eq. 7.6).

In summary, our empirical illustration highlighted key differences between the results of discrete and fuzzy community detection when applied to both individual dynamics and sample-level data. With regard to the dynamics, we found that discrete clustering methods—due to their very nature—forced individuals into communities with other subjects even when they had very few—or no—similarities in their dynamics as illustrated in Fig. 7.4. Furthermore, the fuzzy community detection approaches—when a threshold was applied—were less likely confidently placed a subject into any community. The lack of confidence in the membership degrees may indicate that the individuals exhibit a significant degree of heterogeneity in their dynamics over time—a finding similar to that of De Vos et al., 2017. Thus, fuzzy community detection methods may serve as an avenue for identifying communities of homogeneous subjects who share similar dynamic processes—as with standard community detection—but may also assist in identifying and diagnosing the degree to which individual differences exist in a given sample. Furthermore, our cross-sectional analyses illustrated that the results of discrete and fuzzy community detection methods can significantly agree with one another; however, it also highlights how fuzzy methods may highlight more nuanced cases on the boundaries of communities which may be different enough to warrant individual investigation such as the subjects with clinical diagnoses of MDD who were given membership degrees below our threshold value.

7.2 Discussion

Fuzzy community detection methods are an underutilized tool in the psychological and behavioral sciences. Their ability to quantify the ambiguity in cluster assignments may prove to be a valuable tool in the search for clarity in psychological symptom networks in both cross-sectional and longitudinal research applications. The current chapter sought to introduce behavioral scientists to the concept of fuzziness and how ambiguity in community assignments may yield more nuanced information than discrete approaches.

We illustrated how fuzzy methods may be robust when no communities are present in the data and are capable—in simulated conditions—of identifying fuzzy cases in contrast to discrete methods. These results could suggest that fuzzy community detection methods—with appropriate thresholding—could be used to identify cases that exist *between* communities of variables or otherwise connect them together. This follows in the vein of bridges in networks and could be used to identify variables (i.e., symptoms or subjects) that are equally related to multiple communities.

Our empirical results highlighted two sets of analyses to illustrate the differences that may arise when applying fuzzy and discrete methods to intensive longitudinal data and cross-sectional data. The findings showed important differences between the two methods. Specifically, our analysis on the intensive longitudinal data highlighted the lack of confidence in the group membership assignments produced by the fuzzy community detection approach paralleling that many subjects shared few—if any—dynamics in common with one another. In contrast and—by its very nature—the discrete community detection method forced subjects into groups that contained subjects with no similarities with one another. Thus, these methods could prove to be a vital tool for diagnosing the degree to which individual differences are present in a given dataset and highlight individuals who may be confidently grouped with one another for group-level modeling while allowing the flexibility to identify "fuzzy" subjects that may be better targeted with individual modeling. Furthermore, our results on the mean-level data indicated the potential for fuzzy methods to clarify the boundaries between groups of individuals while simultaneously highlighting individuals who may exist "between" two or more confidently grouped sets of subjects. These "bridge subjects" may be individuals who exist at an intersection between multiple groups and may prove to be interesting targets for further study. These speculations are interesting targets in need of further study and should be addressed in future research. In particular, it is critical that the validity of the fuzzy subjects identified in this—and other—study be verified and compared to real and tangible outcomes to ensure that fuzziness captured in our example is meaningful or an artifact of noise in the data.

While similar studies have applied fuzzy-related concepts and algorithms (e.g., Blanken et al., 2018; Castro et al., 2018, 2019), greater use of membership degrees may allow greater flexibility if the threshold for "confidence" can be tuned in the future depending on the structure of the data. Past work primarily used arbitrarily derived cutoffs (e.g., Bezdek, 1974, 2013; Gunderson, 1978). Future research could expand upon our simulated illustrations and provide clear guidelines on how and if thresholding based on the membership degree weights is a valid approach for identifying and classifying fuzzy cases in a dataset. Furthermore, the fuzzy community detection approach discussed here could be compared against other algorithms and methods (e.g., mixture modeling) and extended to allow for greater complexity and a data-driven approach to determine the optimal number of communities in the dataset rather than relying on user input which may be error prone.

Additionally, our use of fuzzy c-means as the fuzzy community detection algorithm is somewhat limited. Recent developments in the past decade have addressed some of the limitations with fuzzy c-means by creating fuzzy alternatives to more sophisticated algorithms such as the fuzzy WalkTrap and fuzzy Newman–Girvan approaches. These extensions allow for an automated search for the optimal number of clusters without *a priori* specification and could be implemented in future studies to assess how these more complex fuzzy algorithms perform in the context of identifying fuzzy subjects and variables. Following these developments, more work could be done to integrate fuzzy methods into existing algorithms that cluster on individual dynamics (e.g., Gates et al., 2017; Park et al., 2022; Bulteel et al., 2016) which could allow for the identification of not only groups of relatively homogeneous subjects but also subjects that may require individual modeling to obtain clearer estimates of their dynamics without bias from significantly different subjects.

As a tool, fuzzy methods can bring a wealth of information to social and behavioral researchers and has a large degree of potential for future investigations. From outlier control to extracting information from ambiguity, the benefits to employing fuzzy methods are far removed from its namesake and are quite clear indeed.

References

Bezdek, J. C. (1974). Cluster validity with fuzzy sets. *Journal of Cybernetics, 3*, 58–73.

Bezdek, J. C. (1975). Mathematical models for systematics and taxonomy. In *Proceedings of the 8th International Conference on Numerical Taxonomy, 1975*.

Bezdek, J. C. (2013). *Pattern recognition with fuzzy objective function algorithms*. Springer Science & Business Media.

Blanken, T. F., Deserno, M. K., Dalege, J., Borsboom, D., Blanken, P., Kerkhof, G. A., & Cramer, A. O. (2018). The role of stabilizing and communicating symptoms given overlapping communities in psychopathology networks. *Scientific reports, 8*(1), 1–8.

Borsboom, D., & Cramer, A. O. (2013). Network analysis: An integrative approach to the structure of psychopathology. *Annual Review of Clinical Psychology, 9*, 91–121.

Bulteel, K., Tuerlinckx, F., Brose, A., & Ceulemans, E. (2016). Clustering vector autoregressive models: Capturing qualitative differences in within-person dynamics. *Frontiers in Psychology, 7*, 1540.

Campello, R. J., & Hruschka, E. R. (2006). A fuzzy extension of the silhouette width criterion for cluster analysis. *Fuzzy Sets and Systems, 157*(21), 2858–2875.

Castro, D., Ferreira, F., de Castro, I., Rodrigues, A. R., Correia, M., Ribeiro, J., & Ferreira, T. B. (2019). The differential role of central and bridge symptoms in deactivating psychopathological networks. *Frontiers in Psychology, 10*, 2448.

Castro, D., Ferreira, F., Mendes, A. S., & Ferreira, T. B. (2018). Bridges between bipolar and borderline personality disorders: Clarifying comorbidity through the analysis of the complex network of connections between symptoms. *The Psychologist: Practice & Research Journal, 1*(1), 45.

Chow, S.-M., Witkiewitz, K., Grasman, R., & Maisto, S. A. (2015). The cusp catastrophe model as cross-sectional and longitudinal mixture structural equation models. *Psychological Methods, 20*(1), 142.

Cramer, A. O., Waldorp, L. J., Van Der Maas, H. L., & Borsboom, D. (2010). Comorbidity: A network perspective. *Behavioral and Brain Sciences, 33*(2–3), 137–150.

De Vos, S.,Wardenaar, K. J., Bos, E. H.,Wit, E. C., Bouwmans, M. E., & De Jonge, P. (2017). An investigation of emotion dynamics in major depressive disorder patients and healthy persons using sparse longitudinal networks. *PLoS One, 12*(6), e0178586.

Epskamp, S., Waldorp, L. J., Mõttus, R., & Borsboom, D. (2018). The Gaussian graphical model in cross-sectional and time-series data. *Multivariate Behavioral Research, 53*(4), 453–480.

Fried, E. I., van Borkulo, C. D., Cramer, A. O., Boschloo, L., Schoevers, R. A., & Borsboom, D. (2017). Mental disorders as networks of problems: A review of recent insights. *Social Psychiatry and Psychiatric Epidemiology, 52*, 1–10.

Gates, K. M., Lane, S. T., Varangis, E., Giovanello, K., & Guiskewicz, K. (2017). Unsupervised classification during time-series model building. *Multivariate Behavioral Research, 52*(2), 129–148.

Goekoop, R., & Goekoop, J. G. (2014). A network view on psychiatric disorders: Network clusters of symptoms as elementary syndromes of psychopathology. *PloS One, 9*(11), e112734.

Golino, H. F., & Epskamp, S. (2017). Exploratory graph analysis: A new approach for estimating the number of dimensions in psychological research. *PloS One, 12*(6), e0174035.

Groen, R. N., Wichers, M., Wigman, J. T., & Hartman, C. A. (2019). Specificity of psychopathology across levels of severity: A transdiagnostic network analysis. *Scientific Reports, 9*(1), 18298.

Gunderson, R. (1978). Application of fuzzy isodata algorithms to star tracker pointing systems. *IFAC Proceedings Volumes, 11*(1), 1319–1323.

Hubert, L., & Arabie, P. (1985). Comparing partitions. *Journal of Classification, 2*(1), 193–218.

Ichihashi, H., Miyagishi, K., & Honda, K. (2001). Fuzzy c-means clustering with regularization by K-L information. In *10th IEEE International Conference on Fuzzy Systems.(Cat. No. 01CH37297)* (pp. 924–927).

Imperiale, M. N., Lieb, R., Calkins, M. E., & Meinlschmidt, G. (2021). Multimorbidity networks of mental disorder symptom domains across psychopathology severity levels in community youth. *Journal of Psychiatric Research, 141*, 267–275.

Isvoranu, A.-M., Epskamp, S., Waldorp, L., & Borsboom, D. (2022). *Network psychometrics with R: A guide for behavioral and social scientists*. Routledge.

Jones, P. J., Ma, R., & McNally, R. J. (2021). Bridge centrality: A network approach to understanding comorbidity. *Multivariate Behavioral Research, 56*(2), 353–367.

Lane, S. T., Gates, K. M., Pike, H. K., Beltz, A. M., & Wright, A. G. (2019). Uncovering general, shared, and unique temporal patterns in ambulatory assessment data. *Psychological Methods, 24*(1), 54.

Morin, A. J., & Wang, J. C. (2016). *A gentle introduction to mixture modeling using physical fitness performance data*. John Wiley & Sons.

Park, J. J., Chow, S.-M., Epskamp, S., & Molenaar, P. (2022). Subgrouping with chain graphical var models. https://doi.org/10.31234/osf.io/u3ve8.

Roweis, S., & Ghahramani, Z. (1999). A unifying review of linear Gaussian models. *Neural Computation, 11*(2), 305–345.

Sheng Li, C. (2011). The improved partition coefficient. *Procedia Engineering, 24*, 534–538.

Sproul, G. (2001). Electronegativity and bond type: Predicting bond type. *Journal of Chemical Education, 78*(3), 387.

Steinley, D. (2004). Properties of the Hubert-arable adjusted rand index. *Psychological Methods, 9*(3), 386.

Trauwaert, E. (1988). On the meaning of Dunn's partition coefficient for fuzzy clusters. *Fuzzy Sets and Systems, 25*(2), 217–242.

Valafar, F. (2002). Pattern recognition techniques in microarray data analysis: A survey. *Annals of the New York Academy of Sciences, 980*(1), 41–64.

Chapter 8
Causal Discovery with Hidden Variables Based on Non-Gaussianity and Nonlinearity

Takashi Nicholas Maeda, Yan Zeng, and Shohei Shimizu

8.1 Introduction

Statistical causal inference (Pearl, 2000; Imbens & Rubin, 2015) is a methodology that combines domain knowledge and data to support decision-making based on understanding causal mechanisms. Recently, an interdisciplinary field at the intersection of machine learning and statistical causal inference has begun to arise within the overall field of artificial intelligence (AI) (Schölkopf et al., 2021; Kaddour et al., 2022). The reason is that causal knowledge helps analyze and improve the reliability, explainability, and fairness of machine learning methods.

Traditional machine learning methods are unable to infer causality. Let us provide an example to illustrate the difference between machine learning and statistical causal inference. A well-known paper (Messerli, 2012) shows a strong correlation between the consumption of chocolate and the number of Nobel laureates in 23 countries. Using machine learning, we can ask how many Nobel laureates are likely to be found in a specific country given the chocolate consumption or how much chocolate is likely to be consumed given the number of Nobel laureates. The

T. N. Maeda (✉)
Computer Centre, Gakushuin University, Tokyo, Japan

Center for Advanced Intelligence Project, RIKEN, Tokyo, Japan
e-mail: takashi.maeda@gakushuin.ac.jp

Y. Zeng
Department of Computer Science and Technology, Tsinghua University, Beijing, China

Center for Advanced Intelligence Project, RIKEN, Tokyo, Japan

S. Shimizu
Faculty of Data Science, Shiga University, Shiga, Japan

Center for Advanced Intelligence Project, RIKEN, Tokyo, Japan
e-mail: shohei-shimizu@biwako.shiga-u.ac.jp

M. Stemmler et al. (eds.), *Dependent Data in Social Sciences Research*,
https://doi.org/10.1007/978-3-031-56318-8_8

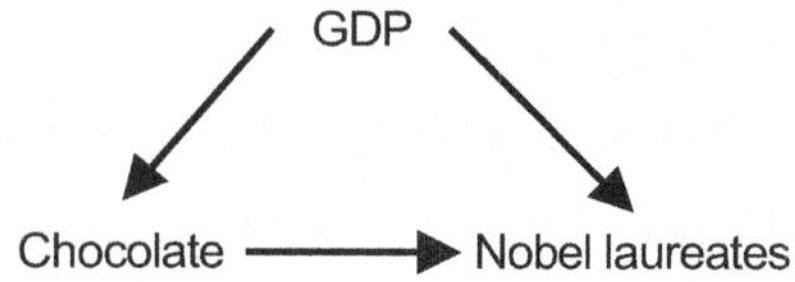

Fig. 8.1 Candidate causal graph of three variables, chocolate consumption, number of Nobel laureates, and GDP

objective is to estimate how many Nobel laureates are likely to be found in a given country. However, even if we find that the more chocolate countries consume, the more Nobel laureates they have, that does not tell us whether increasing chocolate consumption would increase the number of Nobel laureates. To answer this kind of question, we need to use statistical causal inference.

A typical procedure for performing statistical causal inference is described as follows. For example, suppose we want to estimate how the number of Nobel laureates would increase if we increased chocolate consumption, i.e., the intervention effect of chocolate consumption on the number of Nobel laureates. We first draw a causal graph that represents the causal structure related to the number of Nobel laureates and chocolate consumption using domain knowledge, such as the causal graph in Fig. 8.1. Based on the causal graph, the theory of statistical causal inference (Pearl, 1995; Shpitser & Pearl, 2008) is used to derive which variables should be considered for adjustment. Then, those variables are used for adjustment to estimate the intervention effect. In the example in Fig. 8.1, GDP, which is the common cause of chocolate consumption and the number of Nobel laureates, is the variable for which adjustment is necessary. To estimate the intervention effect, we take the expectation of Nobel laureates with respect to GDP after obtaining its conditional expectation conditioned on GDP and chocolate consumption. The conditional expectation is estimated by conducting a regression analysis with the number of Nobel laureates as the objective variable and GDP and chocolate consumption as the explanatory variables, Thus, as in this example, causal graphs should be provided as an input to statistical causal inference procedures.

In addition to intervention effects, various other analyses can be performed if causal graphs can be drawn. Several examples are given below.

1. Explainability of machine learning models: A causal concept called probabilities of causation (Pearl, 1999) is used to improve the explainability of machine learning models (Galhotra et al., 2021). For example, suppose that the prediction result is 0 (failure) when in fact the value of a binary feature is 0. In cases such as this, a type of probability of causation, probability of sufficiency, is defined as the probability that the prediction result would be 1 (success) if the value of the feature were set to be 1 given the prediction result and the value of the feature. If this probability is large, then it is considered sufficient for the feature to be 1 in order for the predicted outcome to be 1 (success). Probabilities of causation are counterfactual concepts that examine what would happen if something different from what actually happened were to happen. It has been shown (Kuroki & Cai, 2008) that structural information contained in causal graphs is useful to obtain the tighter bound of probabilities of causation.

2. Fairness of machine learning models: As the social implementation of machine learning advances, there is an increasing demand to create fair machine learning models. Many definitions of fairness have been proposed using the language of causal inference (Plecko & Bareinboim, 2022; Kaddour et al., 2022). Such definitions make it possible to determine which explanatory variables (features) could be chosen to create fair machine learning models (Kusner et al., 2017), for example. Roughly speaking, we consider a machine learning model fair concerning a variable (e.g., gender or race) if its predictions would not change if we intervened on the variable. This sort of fairness analysis can also be conducted using the idea of causal mediation analysis (Nabi et al., 2022; Wiedermann & Li, 2020).
3. Transfer learning: The idea of statistical causal inference has been shown to enable the automatic selection of adjustments that should be made to construct a machine learning model that is also effective in a different domain, by representing assumptions of transfer learning or domain adaptation in machine learning using a causal graph (Zhang et al., 2020).
4. Analysis of prediction mechanisms: Statistical causal inference can be used to analyze the prediction mechanisms of machine learning models. For example, one can examine how changing the value of an explanatory variable would change the prediction of a machine learning model (Blöbaum & Shimizu, 2017). In various scientific fields, including life sciences, we try to understand a phenomenon by examining what would happen if we intervened on variables related to the phenomenon. Similarly, in understanding machine learning models, it would be useful to examine what would happen if we intervened on machine learning models using the framework of statistical causal inference.

Thus, statistical causal inference using causal graphs is essential in improving the explainability, fairness, and performance needed for reliable AI. To perform statistical causal inference, a causal graph needs to be drawn by the analyst, but it often is the case that insufficient domain knowledge is available for this purpose. Then, a causal structure search methodology that uses data to infer a causal graph, i.e., causal discovery, is useful. There are two major approaches to causal discovery.

The first is a non-parametric approach that makes no specific assumptions about the functional form of causal relationships or distributional form in a causal model (Spirtes et al., 1993). In this case, the only clue with which to infer the causal graph from data is the conditional independence between the observed variables. Although the assumptions are indeed less restrictive, in most cases, only part of the structure of the causal graph can be identified. For example, the direction of causality often cannot be inferred. Note that, in principle, no assumptions are made regarding the functional form or distributions, but in practice, in many implementations, linearity is often assumed for the functional forms and Gaussian distributions for the distribution of continuous variables.

The second approach (Shimizu, 2014; Zhang & Hyvärinen, 2016) makes some assumptions about the functional forms and distributions. This approach originated from the need to alleviate the difficulty that the first approach is often unable to infer the direction of causality. That is, less restrictive assumptions are desirable, but the estimation results are often unhelpful because in most cases, the methods in the first approach are unable to produce useful results.

The first major result to address this difficulty is the LiNGAM (Shimizu et al., 2006), which assumes linearity, acyclicity, absence of hidden common causes, and non-Gaussian continuous error distributions. Linearity is often assumed in the methods and implementations of the aforementioned non-parametric approach and is also assumed in the LiNGAM. It has been shown theoretically that the causal graph is identifiable in the LiNGAM model, for any distribution that is non-Gaussian and continuous. In addition to the advantage in identifiability over conventional methods, LiNGAM has the advantage that it can theoretically be shown that the probability of correct estimation increases as the sample size increases (Genin, 2021). This property is important in guaranteeing that the confidence regions computed are reasonable.

The idea of the LiNGAM has been extended to nonlinear cases (Hoyer et al., 2009; Zhang & Chan, 2006; Zhang & Hyvärinen, 2009; Peters et al., 2014), such as showing that the same idea can uniquely identify models with nonlinear additive errors by examining independence between error terms. An extensive and updated list of references is available at https://www.shimizulab.org/lingam/lingampapers. Such an approach based on models with some restrictions on functional forms also has the advantage that it is easier to deal with individual-level causation (Pearl, 2000) in addition to population-level causation than approaches that makes no assumptions (Kiritoshi et al., 2021). Further, it has been found that variables are uniquely identifiable if they follow certain discrete distributions, for example, if they follow a Poisson or negative binomial distribution (Park &Park, 2019). See Wiedermann and Zhang (2023), a chapter of this book, as well for related topics.

Combining these causal discovery approaches and domain knowledge enables statistical causal inference to be performed by setting up a causal graph with the help of data and causal structure search methods, even if it is difficult to do so with domain knowledge alone. However, many methodological challenges remain regarding applications of causal discovery. For example, major challenges include managing the existence of unobserved common causes and handling mixed cases of discrete and continuous variables.

If there is a hidden common cause between variables whose causal relations are to be examined, the problem of inferring causal graphs becomes more difficult. This is an issue not only in causal discovery but also in causal inference in general. When such unobserved common causes exist, the classical non-parametric approach can warn about the possibility of unobserved common causes among variables (Spirtes et al., 1993), but, in principle, cannot proceed any further with inference. Conversely, approaches that make some assumptions about the functional or distribution form, such as LiNGAM, not only warn but may also be able to estimate the causal relations among variables in the presence of unobserved common causes (Hoyer et al., 2008; Salehkaleybar et al., 2020; Genin, 2021). However, even in cases where estimation is possible in principle, efficient estimation methods are still under development to avoid problems involving local optima and small sample sizes. Note that although temporal information as background knowledge is useful for estimating causal directions, the problem of hidden common causes remains.

Most existing methods and software implementations assume that all variables are either discrete or continuous and cannot handle cases involving a mix of both discrete and continuous variables. In particular, when using statistical causal inference concerning the explainability and fairness of machine learning models, it is quite difficult to avoid handling the case where both types of variables are mixed. In the last few years, several studies have dealt with such mixed cases (Wei et al., 2018; Zeng et al., 2022; Li et al., 2022; Handhayani & Cussens, 2020; Andrews et al., 2019). More research in this direction would be necessary to build a general theory and develop efficient algorithms. It would be also useful to customize general-purpose methods so that they are designed to satisfy specific needs for each application domain.

8.2 Non-Gaussian and Nonlinear Models for Causal Discovery

We first introduce basic models with no hidden variables in the LiNGAM approach and discuss some extensions. This model assumes linearity, acyclicity, absence of hidden common causes, and non-Gaussian continuous error variables. Mathematically, the model for observed variables x_i $(i = 1, \ldots, p)$ is written as

$$x_i = \sum_{j \in \mathrm{pa}(x_i)} b_{ij} x_j + e_i \qquad (i = 1, \ldots, p), \tag{8.1}$$

where $\mathrm{pa}(x_i)$ is the set of parents of observed variable x_i. Error variables e_i $(i = 1, \ldots, p)$ are independent and follow non-Gaussian continuous distributions. This independence between error variables follows from the assumption that there are no unobserved or hidden common causes.

The LiNGAM of Eq. (8.1) is identifiable (Shimizu et al., 2006). This means that the coefficients b_{ij} can be estimated uniquely based on the distribution of the observed variables. Drawing a causal graph based on the zero and non-zero pattern of the coefficients is feasible. If the coefficient b_{ij} is non-zero (zero), a directed edge from the variable x_j to x_i is (not) drawn.

This basic linear model has been extended in various ways (Shimizu, 2014, 2022; Peters et al., 2017). Among these extensions, two nonlinear extensions have been studied extensively. One is the additive nonlinear model (ANM) (Hoyer et al., 2009; Peters et al., 2014), in which the structural equations are additive in noises, i.e., error variables.

$$x_i = f_i\left(\mathrm{pa}(x_i)\right) + e_i \qquad (i = 1, \ldots, p). \tag{8.2}$$

The other is the post-nonlinear causal model (PNL) (Zhang & Chan, 2006; Zhang & Hyvärinen, 2009)

$$x_i = \sum_{j \in \mathrm{pa}(x_i)} f_{ij}^{(2)} \left(f_{ij}^{(1)}(x_j) + e_i \right) \qquad (i = 1, \ldots, p), \tag{8.3}$$

where $f_{ij}^{(1)}$ represents the nonlinear relation of x_i and their parents and $f_{ij}^{(2)}$ can be interpreted as the measurement distortion. Both of the models are identifiable under certain reasonable conditions (Hoyer et al., 2009; Zhang & Hyvärinen, 2009; Peters et al., 2014). Estimation is possible by finding a model that maximizes the independence between the error variables (Hoyer et al., 2009; Peters et al., 2014; Zhang & Chan, 2006; Uemura et al., 2022) such as LiNGAM (Shimizu et al., 2006, 2011; Hyvärinen & Smith, 2013).

We offer a Python code package (Ikeuchi et al., 2022) available at https://github.com/cdt15/lingam. Herein we present a brief description of methods available in this package. Our package implements two major algorithms for the basic LiNGAM in Eq. (8.1), i.e., the original LiNGAM discovery algorithm based on independent component analysis (ICA) (Shimizu et al., 2006) and a direct method called DirectLiNGAM (Shimizu et al., 2011) with an information-theoretic independence measure (Hyvärinen & Smith, 2013). The package further offers utilities including computing bootstrap probabilities of directed paths and edges (Komatsu et al., 2010) and incorporating prior knowledge of topological causal orders in the estimation by DirectLiNGAM. Two extensions of the basic LiNGAMs are available for multi-group analysis (Shimizu, 2012; Kadowaki et al., 2013).

In addition, the package offers two time-series extensions of the basic LiNGAM. VAR-LiNGAM (Hyvärinen et al., 2010) combines it with vector autoregressive models (VAR), and VARMA-LiNGAM (Kawahara et al., 2011) does the same with vector autoregressive moving average models (VARMA). The estimation is performed by first computing the residuals of VAR and VARMA and then applying either ICA-based LiNGAM or DirectLiNGAM.

Further, our package can perform a nonlinear causal discovery based on regression with subsequent independence test (RESIT) (Peters et al., 2014) assuming the additive noise model in Eq. (8.2). Python code for estimating the post-nonlinear causal model in Eq. (8.3) (Uemura et al., 2022) is available at another GitHub repository https://github.com/rafcc/abpnl.

Another important extension is LiNGAM with hidden common causes or latent factors (Hoyer et al., 2008; Shimizu et al., 2009). We implemented the repetitive causal discovery (RCD) algorithm (Maeda & Shimizu, 2020) and the CAM-UV (causal additive models with unobserved variables) algorithm (Maeda & Shimizu, 2021) for hidden common cause cases. We also implemented the multi-domain LiNGAM algorithm for latent factors (MD-LiNA) (Zeng et al., 2021). The package also provides a mixed-data extension of the basic LiNGAM: linear mixed (LiM) causal discovery algorithm. LiM extends LiNGAM to handle mixed data that consists of both continuous and discrete variables (Zeng et al., 2022).

In the following Sects. 8.3 and 8.4, we review recent advances on hidden variable models and mixed data models in the LiNGAM framework using assumptions on functional forms.

8.3 Causal Models with Unobserved Variables

Using a causal discovery method that assumes the absence of unobserved variables leads to incorrect causal estimation when the data are affected by unobserved variables. Many functional causal model-based approaches have assumed the absence of unobserved variables, such as LiNGAM and additive noise models (ANMs). However, in many cases, assuming the absence of unobserved variables is often unrealistic. This section introduces two methods for discovering causal relationships from data affected by unobserved variables. The repetitive causal discovery (RCD) algorithm (Maeda & Shimizu, 2020) is for linear cases, and the CAM-UV (causal additive models with unobserved variables) algorithm (Maeda & Shimizu, 2021) is for nonlinear cases. In regard to the former method, we introduce the improved RCD (I-RCD) algorithm (Maeda, 2022), which is an improved version of RCD.

RCD assumes that the relationship between each pair of (observed or unobserved) variables is linear and that the external influence of each (observed or unobserved) variable is non-Gaussian. In addition, we assume that (observed or unobserved) data are generated from a process represented graphically by a directed acyclic graph (DAG). The generation model is formulated using Eq. 8.4.

$$x_i = \sum_j b_{ij} x_j + \sum_k \lambda_{ik} f_k + e_i \tag{8.4}$$

where x_i denotes an observed variable, b_{ij} is the causal strength from x_j to x_i, f_k denotes a latent confounder, λ_{ik} denotes the causal strength from f_k to x_i, and e_i is an external effect. The external effect e_i and the latent confounder f_k are assumed to follow non-Gaussian continuous-valued distributions with zero mean and nonzero variance and are mutually independent. The zero/nonzero pattern of b_{ij} and λ_{ik} corresponds to the absence/existence pattern of directed edges. Latent confounders f_k are assumed to be mutually independent. In a matrix form, the model is described as Eq. 8.5:

$$\mathbf{x} = \mathbf{B}\mathbf{x} + \mathbf{\Lambda}\mathbf{f} + \mathbf{e} \tag{8.5}$$

where the connection strength matrices $\mathbf{B}$ and $\mathbf{\Lambda}$ collect b_{ij} and λ_{ik} and the vectors $\mathbf{x}$, $\mathbf{f}$, and $\mathbf{e}$ collect x_i, f_k, and e_i.

RCD has two goals. First, it extracts the pairs of observed variables that are affected by the same latent confounders. This is formulated by $\mathbf{C}$ whose element c_{ij} is defined by Eq. 8.6:

$$c_{ij} = \begin{cases} 0 & (\text{if } \forall k, \lambda_{ik} = 0 \vee \lambda_{jk} = 0) \\ 1 & (\text{otherwise}) \end{cases} \tag{8.6}$$

Element c_{ij} equals 0 when there is no latent confounder affecting variables x_i and x_j. Element c_{ij} equals 1 when variables x_i and x_j are affected by the same latent confounders.

The second goal is to estimate the absence/existence of the causal relations between the observed variables that do not have the same latent confounder. This is defined by a matrix **P** whose element p_{ij} is expressed by Eq. 8.7:

$$p_{ij} = \begin{cases} 0 & (\text{if } b_{ij} = 0 \text{ or } c_{ij} = 1) \\ 1 & (\text{otherwise}) \end{cases} \tag{8.7}$$

$p_{ij} = 0$ when $c_{ij} = 1$ because we do not aim to identify the causal direction between the observed variables that are affected by the same latent confounders.

Finally, RCD outputs a causal graph where a bidirected arrow indicates the pair of variables that have the same latent confounders and a directed arrow indicates the causal direction of a pair of variables that are not affected by the same latent confounder. For example, assume that using the data generation model shown in Fig. 8.2a, the final goal is to draw a causal diagram shown in Fig. 8.2b, where variables f_1 and f_2 are latent confounders and variables A–H are observed variables.

In the presence of latent confounders, it is not possible to determine the causal order of the observed variables as in methods based on existing causal function models. Therefore, the RCD algorithm (1) infers the direct and indirect causes of each observed variable, (2) determines the direct cause of each variable, and finally (3) determines variable pairs with unobserved common causes. The first process is shown in Algorithm 1. The RCD initializes M_i to the empty set for each $x_i \subset X$. The RCD algorithm repeats causal inference between variables in each $U \subset X$ with $(l + 1)$ elements. Number l is initialized to 1; if there is no change in M, l is increased by 1. If there is a change in M, l is set to 1. If l exceeds n, the iteration terminates. Variable *changed* has information on whether or not there is a change in M in the iteration. The second process of RCD is to find the direct causes of each variable. RCD creates set P_i, which is the set of all the parents of x_i. When x_j is an ancestor but not a parent of x_i, the effect of x_j on x_i is mediated through

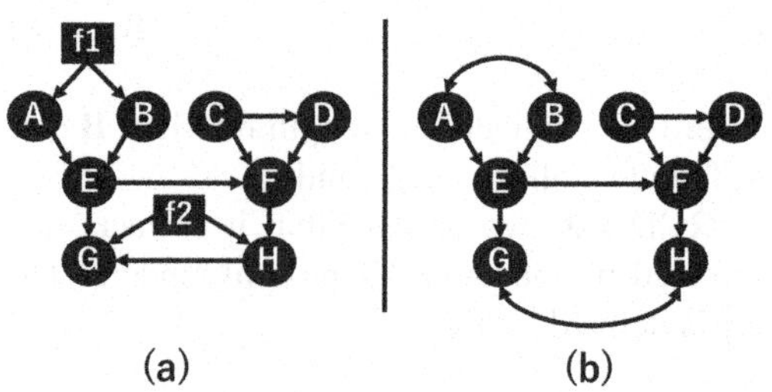

Fig. 8.2 (**a**) Data generation model (f_1 and f_2 are latent confounders). (**b**) Causal graph that RCD produces. A bidirected arrow indicates that two variables are affected by the same latent confounders

Algorithm 1 The I-RCD algorithm: identification of ancestors

Require: X: the set of observed variables, α_C: the alpha level for Pearson's correlation, α_I: the alpha level for independence test, α_S: the alpha level for Shapiro-Wilk test, n: the maximum number of explanatory variables.
Ensure: M: the sequence $\{M_i\}$ where M_i is a set of ancestors of x_i.

```
for all i do
    M_i ← ∅
end for
l ← 1
while l ≤ n do
    changed ← FALSE
    for all U ⊆ X; (|U| = l + 1) do
        H_U ← ⋂_{x_j ∈ U} M_j
        S ← ∅
        for all x_j ∈ U do
            y_j ← the residual obtained by regression of x_j on H_U
            t_j ← the p-value of Shapiro-Wilk test of y_j
        end for
        if ∀t_k < α_S then
            for all x_i ∈ U do
                if M_i ∩ U = ∅ then
                    for all x_j ∈ U \ {x_i} do
                        c_ij ← the p-value of linear correlation between y_i and y_j
                    end for
                    if ∀c_ij < α_C then
                        s_i^U ← the residual obtained by regression of y_i on {U \ {x_i}}
                        for all x_j ∈ U \ {x_i} do
                            h_ij ← the p-value of the HSIC between s_i^U and y_j
                        end for
                        if ∀h_ij > α_I then
                            S ← S ∪ {x_i}
                        end if
                    end if
                end if
            end for
            if |S| = 1 then
                for all x_i ∈ S do
                    M_i ← M_i ∪ (U \ {x_i})
                end for
                changed ← TRUE
            end if
        end if
    end for
    if changed = TRUE then
        l ← 1
    else
        l ← l + 1
    end if
end while
return M
```

$M_i \setminus \{x_j\}$. Therefore, $x_j \perp\!\!\!\perp x_i \mid M_i \setminus \{x_j\}$. RCD creates P_i where each $x_j \in P_i$ satisfies $x_j \in M_i$ and that x_j is dependent of $M_i \setminus \{x_j\}$. The third process of RCD is to identify pairs of variables that have the same latent confounders. RCD infers that two variables are influenced by the same confounders if they are linearly correlated even after all parental influences are removed. RCD identifies x_i and x_j as a variable pair having the same latent confounder when the residuals of x_i and x_j obtained by the regression on P_i and P_j, respectively, are mutually dependent.

CAM-UV is for nonlinear data, while RCD is for analyzing linear data. It is an extension of causal additive models (CAMs) (Bühlmann et al., 2014) to incorporate unobserved variables. CAMs assume that the structural equations are additive in the variables and error terms. CAM-UV assumes the following: Let $X = \{x_i\}$ denote the set of observed variables, $Y = \{y_i\}$ the set of unobserved variables, and $V = \{v_i\}$ the set of all the observed and unobserved variables ($V = X \cup Y$). We assume the data generation model is formulated as

$$v_i = z_i + w_i + n_i, \quad z_i = \sum_{x_j \in P_i} f_j^{(i)}(x_j), \quad w_i = \sum_{y_k \in Q_i} f_k^{(i)}(y_k) \tag{8.8}$$

where z_i is the sum of the direct effects of observed variables on v_i, w_i is the sum of the direct effects of unobserved variables on v_i, $f_j^{(i)}$ is a nonlinear function, P_i is the set of observed direct causes of v_i, Q_i is the set of unobserved direct causes of v_i, and n_i is the external effect on v_i. We assume that all the external effects are mutually independent. We also assume that the causal structure of the observed and unobserved variables forms a DAG. In addition, we assume that all the causal functions and the external effects in CAM-UV satisfy the following condition: If variables v_i and v_j have terms involving functions of the same external effect n_k, then v_i and v_j are mutually dependent (i.e., $(n_k \not\perp\!\!\!\perp v_i) \wedge (n_k \not\perp\!\!\!\perp v_j) \Rightarrow (v_i \not\perp\!\!\!\perp v_j)$).

While RCD cannot identify the causal relationship between observed variables having the same latent confounders, CAM-UV cannot identify the causal relationship between observed variables having an unobserved causal path (UCP) or an unobserved backdoor path (UBP) that are defined as below. The explanatory chart for the definitions is shown in Fig. 8.3.

Definition 1 A directed path from an observed variable to another is called a *causal path* (CP). A CP from x_j to x_i is called an *unobserved causal path* (UCP) if it ends with the directed edge connecting x_i and its unobserved direct cause (i.e., $x_j \to \cdots \to y_m \to x_i$ where y_m is an unobserved direct cause of x_i). □

Definition 2 An undirected path between x_i and x_j is called a *backdoor path* (BP) if it consists of the two directed paths from a common ancestor of x_i and x_j to x_i and x_j (i.e., $x_i \leftarrow \cdots \leftarrow v_k \to \cdots \to x_j$, where v_k is the common ancestor). A BP between x_i and x_j is called an *unobserved backdoor path* (UBP) if it starts with the edge connecting x_i and its unobserved direct cause and ends with the edge

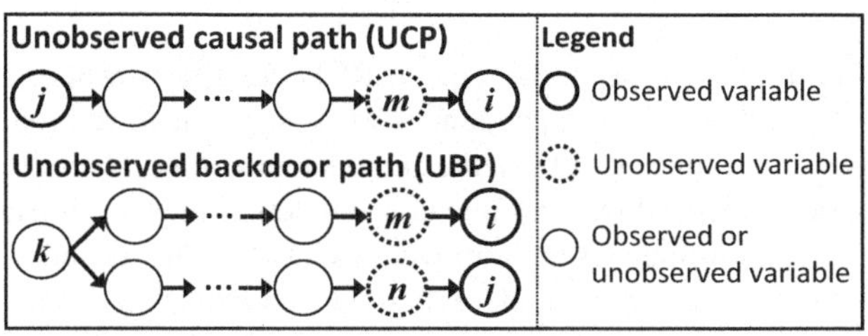

Fig. 8.3 Definitions of an unobserved causal path (UCP) and an unobserved backdoor path (UBP)

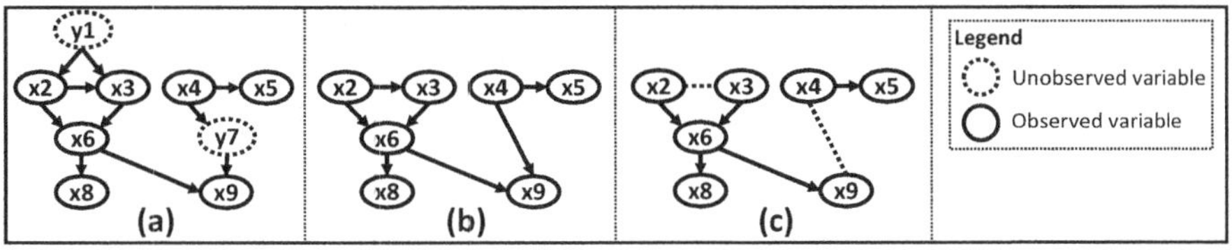

Fig. 8.4 (**a**) Data generation process. (**b**) True causal graph of observed variables. (**c**) Causal graph inferred by the method: Each dashed edge indicates that the causal relationship cannot be determined by the method

connecting x_j and its unobserved direct cause (i.e., $x_i \leftarrow y_m \leftarrow \cdots \leftarrow v_k \rightarrow \cdots \rightarrow y_n \rightarrow x_j$, where v_k is the common ancestor and y_m and y_n are the unobserved direct causes of x_i and x_j, respectively). The undirected path $x_i \leftarrow y_k \rightarrow x_j$ is also a UBP, as v_k, y_m, and y_n can be the same variable. □

The CAM-UV algorithm outputs causal graphs in which directed edges indicate direct causal relationships and dashed undirected edges indicate variable pairs that have a UCP or a UBP. Assume that the data generation process is as shown in Fig. 8.4a, in which y_1 and y_7 are unobserved variables and the other nodes indicate observed variables. Ideally, the causal graph shown in Fig. 8.4b should be recovered. However, the goal is to recover the causal graph shown in Fig. 8.4c where the dashed undirected edges between x_2 and x_3 and between x_4 and x_9 indicate that their causal relationships cannot be identified.

The CAM-UV algorithm first determines the directed edges. The detailed procedure is listed in Algorithm 2, which consists of two steps. The method first extracts the candidates of the parents of each observed variable, and then it determines the parents of each observed variable. The notations G_1 and G_2 in Algorithms 2 and 3 indicate GAM regression functions. Those functions perform differently in different lines and different iterations.

The first step of Algorithm 2 involves each M_i collecting observed variables that are not descendants of x_i, which we call the candidates of the parents of x_i. The method first initializes each M_i to an empty set. Then it repeats finding a sink for each K that satisfies $K \subseteq X$ and $|K| = t$. That is, each K is a set consisting of t observed variables. The value of t starts at 2. It is incremented by 1 when each M_i remains unchanged through an iteration, and it is updated to 2 when at least one M_i changes during the iteration. When the method determines that x_b is a sink of K, it updates M_b by adding each variable in $K \setminus \{x_b\}$ to M_b. The iteration ends when t exceeds d, which is a hyperparameter that is set as the maximum number of $|K|$. The purpose of d is to reduce the computation time, and it should be set according to the sparsity of the causal relationships.

To find a sink for each K, the method first finds the most endogenous variable x_b in K. Such a x_b maximizes the independence between $(x_b - G_1(M_b \cup K \setminus \{x_b\}))$ and $(\{x_j - G_2(M_j) | x_j \in K \setminus \{x_b\}\})$. We use the p-value of the Hilbert-Schmidt

Algorithm 2 The CAM-UV algorithm: determination of directed edges

Require: i.i.d samples of a p-dimensional distribution on $\{x_1, \cdots, x_p\}$ X, maximum number of variables to examine causality for each step d, significance level for independence test α.
Ensure: the sets of the parents $\{M_1, \cdots, M_p\}$.

\# PHASE 1: Extracting the candidates of the parents of each variable.
for all $i = 1$ **to** p **do**
  Initialize $M_i \leftarrow \emptyset$.
end for
Initialize $t \leftarrow 2$.
while $t \leq d$ **do**
  Initialize $noChange \leftarrow$ True.
  for all $K \in \{K | K \subseteq X, |K| = t\}$ **do**
    \# Finding the most endogenous variable x_b in K
    $x_b \leftarrow \arg\max_{x_i \in K} \widehat{\text{p-HSIC}}(x_i - G_1(M_i \cup K \setminus \{x_i\}), \{x_j - G_2(M_j) | x_j \in K \setminus \{x_i\}\})$
    \# Computing the independence between the residuals
    $e \leftarrow \widehat{\text{p-HSIC}}(x_b - G_1(M_b \cup K \setminus \{x_b\}), \{x_j - G_2(M_j) | x_j \in K \setminus \{x_b\}\})$
    $h \leftarrow \max_{x_j \in K \setminus \{x_b\}} \widehat{\text{p-HSIC}}(x_b - G_1(M_b), x_j - G_2(M_j))$
    \# Checking whether x_b is really a sink of K
    if $(\alpha < e) \wedge (\alpha > h)$ **then**
      \# When x_b is a sink of K, add each variable in $K \setminus \{x_b\}$ to M_b
      $M_b \leftarrow M_b \cup (K \setminus \{x_b\})$
      $noChange \leftarrow$ False
    end if
  end for
  \# If each M_i remains unchanged, increment t by one. If not, substitute 2 for t.
  if $noChange =$ True **then**
    $t \leftarrow t + 1$
  else
    $t \leftarrow 2$
  end if
end while
\# PHASE 2: Determining the parents of each variable.
for $i = 1$ **to** p **do**
  for all $x_j \in M_i$ **do**
    \# Checking whether x_j is parent of x_i
    if $\alpha < \widehat{\text{p} - \text{HSIC}}(x_i - G_1(M_i \setminus \{x_j\}), x_j - G_2(M_j))$ **then**
      \# When x_j is not a parent, remove it from M_i $M_i \leftarrow M_i \setminus \{x_j\}$
    end if
  end for
end for
return $\{M_1, \cdots, M_p\}$

Independence Criteria (HSIC) (Gretton et al., 2008) for measuring independence, and we also use the GAM regression method proposed by (Wood, 2004).

In the second step, the algorithm determines the parents of each observed variable. If $x_j \in M_i$ satisfies $x_i - G_1(M_i \setminus \{x_j\}) \perp\!\!\!\perp x_j - G_2(M_j)$, it is not a parent of x_i. Therefore, the method removes each x_j satisfying the above equation from M_i and defines the variables remaining in M_i as the parents of x_i. The reason why the variables remaining in M_i are the parents of x_i is as follows. Each directed path from each x_j in M_i to x_i is blocked by the parents of x_i that are included in M_i (i.e. $M_i \cap P_i$). If x_j is not a parent of x_i, all the directed paths from x_j to x_i are blocked by $(M_i \cap P_i \setminus \{x_j\})$. Then, $(x_i - G_1(M_i \setminus \{x_j\}))$ and $(x_j - G_2(M_j))$ are mutually independent. If x_j is a parent of x_i, there is a direct causal effect $x_j \rightarrow x_i$, and it is not blocked by $(M_i \cap P_i \setminus \{x_j\})$. Then, $(x_i - G_1(M_i \setminus \{x_j\}))$ and $(x_j - G_2(M_j))$ cannot be mutually independent. Therefore, the variables remaining in M_i are parents of x_i.

Algorithm 3 The CAM-UV algorithm: determination of undirected dashed edges

Require: i.i.d samples of a p-dimensional distribution on $\{x_1, \cdots, x_p\}$ X, series of the sets of the parents $\{M_1, \cdots, M_p\}$, significance level for independence test α.
Ensure: set of variable pairs having a UCP or UBP C.

```
C ← ∅.
for all i, j ((1 ≤ i, j ≤ p) ∧ (i ≠ j)) do
    if (x_i ∉ M_j) ∧ (x_j ∉ M_i) then
        if α > p−HSIC(x_i − G_1(M_i), x_j − G_2(M_j)) then
            C ← C ∪ {{x_i, x_j}}
        end if
    end if
end for
return C
```

After determining the direct causal relationships, the method determines variable pairs having UBPs or UCPs (i.e., variable pairs connected with dashed undirected edges). The detailed procedure is listed in Algorithm 3. If the residual of x_i regressed on M_i and that of x_j regressed on M_j are mutually dependent, there is a UCP or UBP between them. Therefore, the method connects x_i and x_j with a dashed undirected edge.

The following are possible future developments regarding these methods. First, the RCD algorithm and the CAM-UV algorithm should be extended for models involving discrete variables such as categorical and binary variables. In many datasets, it is rare that all variables take continuous values, so the method proposed in this study should be improved to accommodate them. Second, methods should be developed to extract variables that are directly causally related to or influenced by the same unobserved variable only for a specific variable among all observed variables. The RCD algorithm is very computationally time-consuming. Therefore, if one is interested in a particular variable, it may be possible to reduce the computation time by developing a method that concentrates on a particular variable and extracts causal relationships with the surrounding variables. The third is an extension of the RCD algorithm to infer causal graphs from data generated from cyclic causal structures. The RCD algorithm removes the influence from the ancestors of both variables when estimating the causal relationship between two continuous variables. This suggests that the method proposed in this study has the potential to be extended to causal discovery from data generated from cyclic causal structures. Finally, the fourth necessary extension is a way to set the alpha level threshold used to determine independence in the RCD algorithm and the CAM-UV algorithm to the optimal one for the data. The optimal values of this element of the algorithms are expected to vary depending on the number of data and the probability distribution that each variable follows. Currently, it is necessary to manually find the values of the arguments that will output the most plausible causal graph. An algorithm and theory to rationally determine the set value of this argument is needed.

8.4 Causal Discovery for Observed Variables and Latent Factors

In this section, we first introduce a linear mixed (LiM) causal discovery algorithm for observed variables, which handles mixed data that consist of both continuous and discrete variables. It achieves identification of causal structures between such observed mixed variables. In many scenarios, the underlying causal relations are usually embedded between latent variables (or factors) that cannot be directly measured, other than through observed variables. As such, we finally introduce a multi-domain LiNGAM algorithm for latent factors that discovers causal structures among latent factors from observed data.

Linear Mixed Causal Model

Most causal discovery methods and software implementations assume that all variables are either discrete or continuous and cannot handle cases involving a mix of both discrete and continuous variables. In many real-world applications, the collected data is often a mixture of both continuous and discrete variables. Taking the datasets in the UC Irvine Machine Learning Repository,[1] for instance, *Boston Housing Dataset* both has discrete variables, e.g., CHA (if tract bounds river or not), and continuous variables, e.g., TAX (full-value property-tax rate per 10,000 dollars), IND (proportion of non-retail business acres per town), etc. The *Abalone Dataset* both has discrete variables, e.g., sex (male, female or infant), and has continuous variables, e.g., height (longest shell measurement), etc.

When encountering such mixed data, one may ignore the discrete variables and apply the methods for continuous variables to estimate the partial causal network or utilize a discretization strategy to discretize the continuous variables, so that they can use those methods for discrete data (Monti & Cooper, 1998; Chen et al., 2017). These methods attempt to convert different mixed data types into the same type, which may easily lose information, inducing non-negligible estimation errors. Existing causal discovery algorithms for mixed data, by and large, can be categorized into two classes: constraint-based and score-based ones. Constraint-based methods take the best of conditional independence to infer the causal graph from mixed data, including Cui et al. (2016), Sedgewick et al. (2019), Tsagris et al. (2018), Handhayani and Cussens (2020), Li et al. (2022). They cannot always guarantee identifiability and are sensitive to samples or sample composition. Score-based methods for mixed data optimize a likelihood score with mixed data's characteristics and the commonly used greedy equivalence search framework. Such methods include Li and Shimizu (2018), Huang et al. (2018), Andrews et al. (2019),

[1] The website of the UC Irvine Machine Learning is https://archive-beta.ics.uci.edu/.

Wei et al. (2018). Specifically, Li and Shimizu (2018) employed the score function from LiNGAM and the logistic regression model, Huang et al. (2018) from the regression model in the reproducing kernel Hilbert space (RKHS), and Andrews et al. (2019) from the degenerate distributions. However, they do not provide the model's identifiability results and may return Markov equivalence classes. Wei et al. (2018) developed a mixed causal model and prove the identifiability when there are bivariate mixed variables. However, in their bivariate identifiability, they assume the noises of continuous variables follow the Laplace distributions and that the intercepts for discrete variables are zeros. In addition, such a bivariate identifiability cannot handle multivariate cases theoretically as the multivariate data usually exists in many applications.

Thus, herein we introduce a linear structural causal model that consists of both continuous and discrete variables, along with its estimation method for the causal structure (Zeng et al., 2022).[2] Suppose we are given p observed random variables, including discrete and continuous ones, i.e., $X = \{x_1, \ldots, x_p\}$. Since a categorical variable with T categories can be regarded as $(T - 1)$ binary variables, we assume that each discrete variable is binary (Wei et al., 2018). Further, we use the following assumptions from Li and Shimizu (2018):

A1. Observed variables x_i $(i = 1, \ldots, p)$ form a directed acyclic graph (DAG).

A2. The value assigned to each continuous variable x_i is a linear function of its parent variables denoted by $\mathrm{pa}(x_i)$ plus a non-Gaussian error term e_i, that is,

$$x_i = e_i + c_i + \sum_{j \in \mathrm{pa}(x_i)} b_{ij} x_j, \quad e_i \sim Non - Gaussian(\cdot), \tag{8.9}$$

where the error terms e_i are continuous random variables with non-Gaussian densities, and the error variables e_i are independent of each other. Coefficients b_{ij} and intercepts c_i are constants.

A3. For each discrete variable x_i, its value equals 1 if the linear function of its parent variables $\mathrm{pa}(x_i)$ plus a Logistic error term e_i is larger than 0; otherwise, its value equals 0. That is,

$$x_i = \begin{cases} 1, & e_i + c_i + \sum_{j \in \mathrm{pa}(x_i)} b_{ij} x_j > 0 \\ 0, & \text{otherwise} \end{cases}, \quad e_i \sim Logistic(0, 1), \tag{8.10}$$

where the error terms e_i are identical to those in Eq. (8.9) but follow the logistic distribution.

If a causal model for mixed data satisfies assumptions A1–A3, then this SCM is called a linear mixed causal model, abbreviated as LiM. Our model can be rewritten

[2] For sufficient identifiability conditions in bivariate as well as multivariate cases, please refer to Zeng et al. (2022).

as:

$$x_i = f_i(\mathrm{pa}(x_i), e_i), \quad e_i \sim P(e_i), \tag{8.11}$$

where $f_i \in \mathcal{F}$ and $P(e_i) \in \mathcal{P}$.

$$\mathcal{F} = \{f^{con}, f^{dis} | f^{con}(x, e) = bx + e + c, f^{dis}(x, e) = \begin{cases} 1, & bx + e + c > 0 \\ 0, & \text{otherwise} \end{cases}\},$$

is a set of two functions which work on continuous and discrete variables, respectively, where x, e, b, and c are a random variable of X, error term, coefficient, and intercept, respectively. $\mathcal{P} = \{P^{con}, P^{dis}\}$ denotes the set of probabilistic distributions for continuous and discrete variables.

To uncover the causal structure for the LiM model from mixed data, we propose an integrated hybrid score-based learning method, as illustrated in Algorithm 4. First, it performs global updates, estimating the connection strengths matrix $\hat{\mathbf{B}}$ in one step with continuous optimization techniques. It then performs a local update to search over the skeleton space, estimating one candidate DAG with one changing edge at each iteration in a combinatorial optimization manner.

Regarding the global search phase, we seek to solve the following continuous optimization problem:

$$\begin{aligned} \min_{\mathbf{B}} \quad & \mathcal{L}(\mathbf{B}) + \Lambda||\mathbf{B}||_1 \\ \text{subject to} \quad & h(\mathbf{B}) = 0, \end{aligned} \tag{8.12}$$

where $\mathbf{B}$ is the adjacency matrix, $h(\mathbf{B})$ is an acyclicity constraint that ensures that $\mathbf{B}$ is a DAG (Zheng et al., 2018), Λ is a regularization parameter, and $||\cdot||_1$ is an l_1 sparsity regularization. $\mathcal{L}(\mathbf{B})$ is based on the negative log-likelihood of the data, i.e.,

$$\mathcal{L}(\mathbf{B}) = -\log(P(\mathbf{X})) \tag{8.13}$$

$$= -\sum_{t}^{n}\sum_{i}^{p} z_i\{x_{i,t}\log[\sigma(\mathbf{B})] + (1 - x_{i,t})\log[1 - \sigma(\mathbf{B})]\} + \tag{8.14}$$

$$(1 - z_i)\log p_i(x_{i,t} - \sum_{k \in \mathrm{pa}(x_i)} b_{ik}x_{k,t}), \tag{8.15}$$

where $x_{i,t}$ is the tth sample of the ith variable x_i and n is the sample size. P_d and P_c denote the probability distribution of discrete and continuous variables, respectively. z_i is an indicator variable, where $z_i = 1$ if x_i is discrete while $z_i = 0$ otherwise. $\sigma(\mathbf{B}) = \frac{1}{1+e^{-(c_i+\sum_{l \in \mathrm{pa}(x_i)} b_{il}x_{l,t})}}$, while p_i is the density function of the non-Gaussian error terms e_i of continuous variables. In our method, though we specified p_i to

Algorithm 4 LiM algorithm

Require: Data $\boldsymbol{X}$; indicator vector Z; threshold ϵ; tolerance parameter ω.
Ensure: Connection strengths matrix $\mathbf{B}^*$.
Phase I: Global search
1: Optimize Eq. (8.12) to obtain $\hat{\mathbf{B}}$ using QPM with the tolerance parameter ω,
2: Rule out edges whose connection strengths are below ϵ: $\hat{\mathbf{B}} = \hat{\mathbf{B}} \circ \mathbf{0}(\hat{b}_{ij} < \epsilon)$.
Phase II: Local search
3: Initiate a temporary minimum log-likelihood as $\mathcal{L}_{tmp}(\mathbf{B}^*) = \mathcal{L}(\hat{\mathbf{B}})$.
4: **while** $\mathbf{B} \in \text{Ske}(\hat{\mathbf{B}})$ **and** $h(\mathbf{B}) < \omega$ **do**
5: Compute the negative log-likelihood $\mathcal{L}(\mathbf{B})$ for $\mathbf{B}$.
6: **if** $\mathcal{L}(\mathbf{B}) < \mathcal{L}_{tmp}(\mathbf{B}^*)$ **then**
7: $\mathcal{L}_{tmp}(\mathbf{B}^*) = \mathcal{L}(\mathbf{B})$.
8: **end if**
9: **end while**

be the density function of Laplace distributions, we can use other density functions as well. Following the optimization procedure in Zheng et al. (2018); Zeng et al. (2021), we leverage the quadratic penalty method (QPM) to estimate **B**, converting Eq. (8.12) into an unconstrained function:

$$\min_{\mathbf{B}} \mathcal{S}(\mathbf{B}), \tag{8.16}$$

where $\mathcal{S} = \mathcal{L}(\mathbf{B}) + \Lambda||\mathbf{B}||_1 + \frac{\rho}{2}h(\mathbf{B})^2$ is the quadratic penalty function and ρ is a regularization parameter. Then we utilize the L-BFGS-B (Byrd et al., 1995) to solve Eq. (8.16). Due to the machine precision, it is well-known that it is hard for the estimated b_{ij} to receive absolute zeros if such pairs have no edges (Zheng et al., 2018). Hence, we give a small fixed threshold ϵ to rule out those whose estimated effects are lower than ϵ.

Regarding the local search phase, to mitigate the local optima issue, we tackle further the following combinatorial optimization problem:

$$\mathbf{B}^* = \arg \min_{\mathbf{B} \in \text{Ske}(\hat{\mathbf{B}}), h(\mathbf{B}) < \omega} \mathcal{L}(\mathbf{B}), \tag{8.17}$$

where $\text{Ske}(\hat{\mathbf{B}})$ represents a set where the containing DAGs entail the same skeleton as $\hat{\mathbf{B}}$ and ω is a tolerance parameter. Compared with the traditional approaches that search over the discrete space of DAGs with the full graph, we perform our structure search over the narrowed space of DAGs within the estimated skeleton, which possesses an advantage in computational efficiency.

There are several restrictions for the LiM model. First, LiM is identifiable under causal sufficiency, whereas in many realistic applications, causal sufficiency may not hold, and there may exist latent variables or confounders in the underlying causal graphs. Second, the linearity assumption may decrease the learning effect and limit the applicability. It is essential to generalize the identifiability to cases where there are nonlinear relationships. Third, there may also exist other types of discrete variables, e.g., multi-categorical variables or ordinal variables. Ignoring ordinal information may have adverse effects on model estimation (von Eye & Wiedermann, 2018). Thus, it is desirable to extend our method to cover general data types with scalability.

LiNGAM for Latent Factors

Most existing causal discovery methods focus on identifying causal structures between observed variables, while in many scenarios, the observed ones may not be the underlying causal variables but are generated by latent variables or factors that are causally related (Shimizu et al., 2009; Xie et al., 2020). The underlying causal relations are embedded between latent factors that cannot be directly measured. For instance, researchers in psychology may study the causal relations between latent factors, e.g., anxiety, depression, and coping ability, which can only be measured by questionnaires (Silva et al., 2006; Galbraith et al., 2002); educational researchers may investigate the effects of organization (role ambiguity, role conflict, classroom climate, superior support, etc.) and personality (self-esteem, external locus of control) on three facets of burnout in full-time elementary teachers (Byrne, 2013).

To identify the causal structure among latent factors, existing approaches could be categorized into covariance-based and non-Gaussianity-based ones. Covariance-based methods employ the covariance structure of data alone, e.g., BuildPureClusters algorithm (Silva et al., 2006) or FindOneFactorClusters algorithm (Kummerfeld & Ramsey, 2016), to ascertain the number of latent factors as well as the structure of latent factors. However, these algorithms can only output structures up to the Markov equivalence class for latent factors. Non-Gaussianity-based methods address this indistinguishable identification problem by taking advantage of the non-Gaussianity of data with some assumptions about the functional forms and distributions. Specifically, Shimizu et al. (2009) leveraged the non-Gaussianity of data and firstly achieved identification of a unique causal structure between latent factors based on LiNGAM. Recently, to avoid the local optima of the NICA, Cai et al. (2019) designed the so-called Triad constraints, while Xie et al. (2020) developed the GIN condition. Both proposed a two-phase method to learn the structure or causal orderings among latent factors. Further, to identify the latent hierarchical structures where the children of latent variables can be latent as well, Xie et al. (2022) took the best of the GIN condition, while Huang et al. (2022) leveraged rank constraints.

Note that all the abovementioned methods focus on data that originates from the same domain, i.e., single-domain data. However, in many real-world applications, data is often collected under distinct conditions and may originate from different domains, resulting in distinct distributions and/or various causal effects. For instance, in neuroinformatics, functional magnetic resonance imaging (fMRI) signals are frequently extracted from multiple subjects or over time (Smith et al., 2011);

in biology, a particular disease is measured by distinct medical equipment (Dhir & Lee, 2020), etc. Most methods allow the existence of latent variables or confounders to handle multi-domain data, whereas there exist only a few methods to learn the causal structure among latent factors with only observed data. It is desirable to perform causal discovery from multi-domain data to uncover the structure among latent factors.

To this end, we introduce multi-domain linear non-Gaussian acyclic models for latent factors (MD-LiNA) to represent the causal mechanism of latent factors, which tackles not only single-domain but also multi-domain data (Zeng et al., 2021). Suppose we have data from M domains. Let $\boldsymbol{x}^{(m)}$ and $\boldsymbol{f}^{(m)}$ $(m = 1, \ldots, M)$ be the random vectors that collect p_m observed variables and q_m latent factors in domain m, respectively. The total number of observed variables for all domains is $p = \sum_{m=1}^{M} p_m$, while that of latent factors is $q = \sum_{m=1}^{M} q_m$. We focus on linear models,

$$\begin{aligned} \boldsymbol{f}^{(m)} &= \boldsymbol{B}^{(m)} \boldsymbol{f}^{(m)} + \boldsymbol{\varepsilon}^{(m)}, \\ \boldsymbol{x}^{(m)} &= \boldsymbol{G}^{(m)} \boldsymbol{f}^{(m)} + \boldsymbol{e}^{(m)}, \end{aligned} \tag{8.18}$$

where $\boldsymbol{\varepsilon}^{(m)}$ and $\boldsymbol{e}^{(m)}$ are random vectors that collect external influences and errors, respectively, and are independent of each other. $\boldsymbol{B}^{(m)}$ is a matrix that collects causal effects $b_{ij}^{(m)}$ between $f_i^{(m)}$ and $f_j^{(m)}$, while $\boldsymbol{G}^{(m)}$ collects factor loadings $g_{ij}^{(m)}$ between $f_j^{(m)}$ and $x_i^{(m)}$. $f_i^{(m)}$ is assumed to have zero means and unit variances. Note that in a specific domain m, data is generated with the same $b_{ij}^{(m)}$ and $g_{ij}^{(m)}$. $\varepsilon_i^{(m)}$ and $e_i^{(m)}$ are sampled independently from the identical distributions. In different domains, $\boldsymbol{B}^{(m_1)}$ $(\boldsymbol{G}^{(m_1)})$ and $\boldsymbol{B}^{(m_2)}$ $(\boldsymbol{G}^{(m_2)})$ may be different, but they have a shared causal structure.

To integrate the multi-domain data effectively, we leverage an idea from simple coding representation (Shimodaira, 2016), wherein an observation is represented as an augmented one with p dimensions, where only p_m dimensions come from the original domain, while the other $(p - p_m)$ dimensions are padded with zeros. Any augmented data are expressed with a bar, e.g., $\bar{\boldsymbol{x}}$ and $\bar{\boldsymbol{f}}$. With such representations, we obtain,

$$\begin{aligned} \bar{\boldsymbol{f}} &= \bar{\boldsymbol{B}} \bar{\boldsymbol{f}} + \bar{\boldsymbol{\varepsilon}}, \\ \bar{\boldsymbol{x}} &= \bar{\boldsymbol{G}} \bar{\boldsymbol{f}} + \bar{\boldsymbol{e}}, \end{aligned} \tag{8.19}$$

where $\bar{\boldsymbol{B}} = \mathrm{Diag}(\boldsymbol{B}^{(1)}, \ldots, \boldsymbol{B}^{(M)})$, $\bar{\boldsymbol{G}} = \mathrm{Diag}(\boldsymbol{G}^{(1)}, \ldots, \boldsymbol{G}^{(M)})$, and $\bar{\boldsymbol{\varepsilon}}$ and $\bar{\boldsymbol{e}}$ are independent of each other. To guarantee factors $\boldsymbol{f}$ in different domains are represented by the same concepts (factors) of interest, we introduce factors of interest $\tilde{\boldsymbol{f}}$, which are embedded as different concepts with causal relations. As depicted in Fig. 8.5, suppose $\bar{\boldsymbol{f}}$ are linearly generated by $\tilde{\boldsymbol{f}}$,

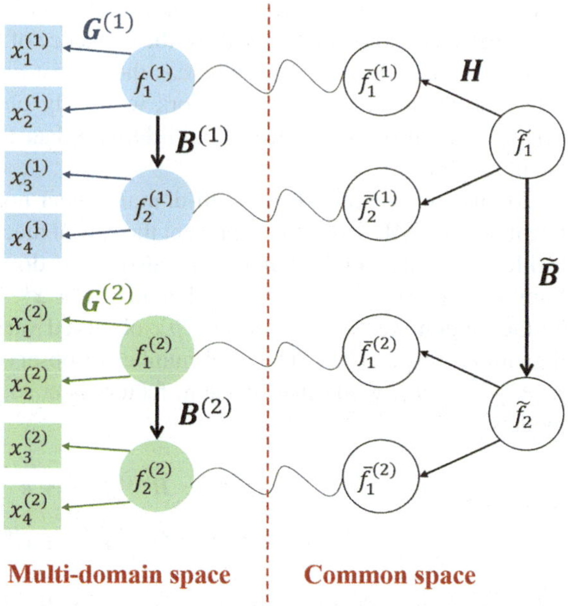

Fig. 8.5 An MD-LiNA model. Variables in the same color (light blue and light green) are in the same domain. Observed variables $\boldsymbol{x}^{(1)}$ in domain (1) entail its latent factors $\boldsymbol{f}^{(1)}$. Augmented latent factors $\bar{\boldsymbol{f}}$ are obtained using the coding representation method, which are signified by the curved waved lines (These waved lines are used to emphasize the common space and can be simply replaced by directed edges). $\tilde{\boldsymbol{f}}$ are shared latent factors of interest, whose structure is shared by $\boldsymbol{f}^{(m)}$ from different domains.

$$\bar{\boldsymbol{f}} = \boldsymbol{H}\tilde{\boldsymbol{f}}, \tag{8.20}$$

where $\boldsymbol{H} \in \mathbb{R}^{q \times \tilde{q}} (\tilde{q} \leq q)$ is a transformation matrix and $\tilde{q}$ is the number of $\tilde{\boldsymbol{f}}$. To conclude, an MD-LiNA model satisfies the following assumptions:

A1. $\boldsymbol{f}^{(m)}$ are generated linearly from a directed acyclic graph (DAG) with non-Gaussian distributed external variables $\boldsymbol{\varepsilon}^{(m)}$, as in Eq. (8.18).

A2. $\boldsymbol{x}^{(m)}$ are generated linearly from $\boldsymbol{f}^{(m)}$ plus errors $\boldsymbol{e}^{(m)}$, as in Eq. (8.18). Note that errors could be either Gaussian or non-Gaussian.

A3. Each f_i has at least two pure measurement variables.[3]

A4. Each $\bar{f}_i^{(m)}$ is linearly generated by only one latent in $\tilde{\boldsymbol{f}}$, and each $\tilde{f}_i$ generates at least one latent in $\bar{\boldsymbol{f}}$, as in Eq. (8.20).

For detailed explanations about the model and identifiability, please refer to Zeng et al. (2021).

To uncover the causal structure between latent factors from multi-domain observed data, we exhibit a two-phase framework (measurement-model and structure-model phases) in Algorithm 5. Regarding Phase I, to learn the measurement models (relating the structure from $\boldsymbol{x}^{(m)}$ to $\boldsymbol{f}^{(m)}$), we use confirmatory factor analysis (CFA) (Reilly & O'Brien, 1996), after employing Triad (Cai et al.,

[3] Pure measurement variables are those that have only one single latent factor parent (Silva et al., 2006).

Algorithm 5 MD-LiNA algorithm

Require: Data $\boldsymbol{X}^{(1)}, \ldots, \boldsymbol{X}^{(M)}$; M.
Ensure: Factor loadings $\bar{\boldsymbol{G}}$ ($\boldsymbol{G}^{(1)}$); effects matrix $\tilde{\boldsymbol{B}}$ ($\boldsymbol{B}^{(1)}$).

Phase I: Measurement models

1: Find the number of latent factors q_m and locate latent factors for each domain m by Triad constraints;
2: Get augmented data $\bar{\boldsymbol{X}} = \text{Diag}(\boldsymbol{X}^{(1)}, \ldots, \boldsymbol{X}^{(M)})$ and estimate $\bar{\boldsymbol{G}}$ by CFA;
3: Estimate $\bar{\boldsymbol{f}} \doteq (\bar{\boldsymbol{G}}^T \bar{\boldsymbol{G}})^{-1} \bar{\boldsymbol{G}}^T \bar{\boldsymbol{X}}$;

Phase II: Structure models

4: **if** $M > 1$ **then**
5: Optimize Eq. (8.21) iteratively for $\boldsymbol{H}$ and $\tilde{\boldsymbol{B}}$ until convergence using QPM;
6: Update $\tilde{\boldsymbol{B}}$ with regard to $\boldsymbol{H}$;
7: **else**
8: Optimize Eq. (8.22) to get $\boldsymbol{B}^{(1)}$ with QPM;
9: **end if**

10: **return** $\bar{\boldsymbol{G}}$ and $\tilde{\boldsymbol{B}}$ ($M > 1$); or $\boldsymbol{G}^{(1)} = \bar{\boldsymbol{G}}$ and $\boldsymbol{B}^{(1)}$ ($M = 1$).

2019) to yield the structure between latent factors $\boldsymbol{f}^{(m)}$ and observed variables $\boldsymbol{x}^{(m)}$, i.e., $\bar{\boldsymbol{G}}$. Thereafter, we get $\bar{\boldsymbol{f}}$. Regarding Phase II, to learn the structure models (recording the causal relations among $\boldsymbol{f}^{(m)}$ or $\tilde{\boldsymbol{f}}$), when $M > 1$, we optimize with

$$\begin{aligned} &\min_{\tilde{\boldsymbol{B}}, \boldsymbol{H}} \bar{\mathcal{F}}(\tilde{\boldsymbol{B}}, \boldsymbol{H}), \quad \text{s.t.} \quad h(\tilde{\boldsymbol{B}}) = 0, \\ &\text{where} \quad \bar{\mathcal{F}}(\tilde{\boldsymbol{B}}, \boldsymbol{H}) = \mathcal{F}(\tilde{\boldsymbol{B}}, \boldsymbol{H}) + \mathcal{E}(\boldsymbol{H}) + \lambda_3 \|\boldsymbol{H}\|_{1*}, \end{aligned} \tag{8.21}$$

where $\tilde{\boldsymbol{B}}$ is a matrix that collects causal effects between shared latent factors of interest $\tilde{\boldsymbol{f}}$ and $h(\tilde{\boldsymbol{B}})$ is the acyclicity constraint (Zheng et al., 2018). $\mathcal{E}(\boldsymbol{H}) = \|\bar{\boldsymbol{f}} - \boldsymbol{H}\tilde{\boldsymbol{f}}\|^2$ is the reconstruction error of $\boldsymbol{H}$, which characterizes the dependence relations between $\bar{\boldsymbol{f}}$ and $\tilde{\boldsymbol{f}}$. $\| \cdot \|_{1*}$ represents the sparsity constraint, and $\mathcal{F}(\tilde{\boldsymbol{B}}, \boldsymbol{H})$ stands for the log-likelihood of MD-LiNA. When $M = 1$, we only need to estimate the factor loading $\boldsymbol{G}^{(1)} = \bar{\boldsymbol{G}}$ and the matrix $\boldsymbol{B}^{(1)}$, thus optimizing,

$$\begin{aligned} &\min_{\boldsymbol{B}^{(1)}} \mathcal{F}(\boldsymbol{B}^{(1)}, \hat{\boldsymbol{G}}^{(1)}), \quad \text{s.t.} \quad h(\boldsymbol{B}^{(1)}) = 0, \\ &\text{where} \quad \mathcal{F}(\boldsymbol{B}^{(1)}, \hat{\boldsymbol{G}}^{(1)}) = -\mathcal{L}(\boldsymbol{B}^{(1)}, \hat{\boldsymbol{G}}^{(1)}) + \lambda_1 \|\boldsymbol{B}^{(1)}\|_{1*} + \lambda_2 \|\boldsymbol{B}^{(1)}\|^2, \end{aligned} \tag{8.22}$$

where $\hat{\boldsymbol{G}}^{(1)}$ is the estimate of $\boldsymbol{G}^{(1)}$ and $\| \cdot \|^2$ is the ℓ_2 regularization. λ_1 and λ_2 are regularization parameters. $\mathcal{L}(\boldsymbol{B}^{(1)}, \hat{\boldsymbol{G}}^{(1)})$ stands for the log-likelihood function of LiNA that characterizes the independence relations between $\boldsymbol{\varepsilon}$. As for the optimization method, we leverage the quadratic penalty method (QPM), transforming constrained problems of Eqs. (8.21) and (8.22) into unconstrained ones for the estimation.

Future research directions in line with this approach (of identifying causal structures between latent factors and latent variables) include (i) allowing arbitrary distributions for external influences of latent factors, (ii) allowing nonlinear relationships (Xie et al., 2020), and (iii) allowing observed measurement variables to affect latent factors (Huang et al., 2022).

8.5 Conclusions

A bottleneck in automating statistical causal inference is that the causal graph must be known. Causal discovery aims to infer causal graphs from data as well as background knowledge. This would greatly advance the automation of statistical causal inference and would allow users to focus on considering which assumptions and domain knowledge should be accepted and employed, greatly reducing the hurdles to using causal inference.

Acknowledgments S.S. was partially supported by the grants ONR N00014-20-1-2501 and KAKENHI 20K11708.

References

Andrews, B., Ramsey, J., & Cooper, G. F. (2019). Learning high-dimensional directed acyclic graphs with mixed data-types. In *Proceedings of Machine Learning Research*, volume 104 of *Proceedings of Machine Learning Research* (pp. 4–21). PMLR.

Blöbaum, P., & Shimizu, S. (2017). Estimation of interventional effects of features on prediction. In *Proc. 2017 IEEE 27th International Workshop on Machine Learning for Signal Processing (MLSP)* (pp. 1–6). IEEE.

Bühlmann, P., Peters, J., & Ernest, J. (2014). CAM: Causal additive models, high-dimensional order search and penalized regression. *Annals of Statistics, 42*(6), 2526–2556.

Byrd, R. H., Lu, P., Nocedal, J., & Zhu, C. (1995). A limited memory algorithm for bound constrained optimization. *SIAM Journal on Scientific Computing, 16*(5), 1190–1208.

Byrne, B. M. (2013). *Structural equation modeling with Mplus: Basic concepts, applications, and programming*. Routledge.

Cai, R., Xie, F., Glymour, C., Hao, Z., & Zhang, K. (2019). Triad constraints for learning causal structure of latent variables. In *Advances in neural information processing systems* (Vol. 32).

Chen, Y.-C., Wheeler, T. A., & Kochenderfer, M. J. (2017). Learning discrete Bayesian networks from continuous data. *Journal of Artificial Intelligence Research, 59*, 103–132.

Cui, R., Groot, P., & Heskes, T. (2016). Copula PC algorithm for causal discovery from mixed data. In *Joint European Conference on Machine Learning and Knowledge Discovery in Databases* (pp. 377–392). Springer.

Dhir, A., & Lee, C. M. (2020). Integrating overlapping datasets using bivariate causal discovery. In *Proceedings of the AAAI Conference on Artificial Intelligence* (Vol. 34, pp. 3781–3790).

Galbraith, J. I., Moustaki, I., Bartholomew, D. J., & Steele, F. (2002). *The analysis and interpretation of multivariate data for social scientists*. Chapman and Hall/CRC.

Galhotra, S., Pradhan, R., & Salimi, B. (2021). Explaining black-box algorithms using probabilistic contrastive counterfactuals. In *Proceedings of the 2021 International Conference on Management of Data* (pp. 577–590).

Genin, K. (2021). Statistical undecidability in linear, non-gaussian causal models in the presence of latent confounders. In M. Ranzato, A. Beygelzimer, Y. Dauphin, P. Liang, & J. W. Vaughan (Eds.), *Advances in neural information processing systems* (Vol. 34, pp. 13564–13574). Curran Associates.

Gretton, A., Fukumizu, K., Teo, C. H., Song, L., Schölkopf, B., & Smola, A. J. (2008). A kernel statistical test of independence. In *Advances in neural information processing systems* (Vol. 20, pp. 585–592).

Handhayani, T., & Cussens, J. (2020). Kernel-based approach for learning causal graphs from mixed data. In *Proceedings of the 10th International Conference on Probabilistic Graphical Models*, volume 138 of *Proceedings of Machine Learning Research* (pp. 221–232). PMLR.

Hoyer, P. O., Janzing, D., Mooij, J., Peters, J., & Schölkopf, B. (2009). Nonlinear causal discovery with additive noise models. In *Advances in neural information processing systems* (Vol. 21, pp. 689–696). Curran Associates.

Hoyer, P. O., Shimizu, S., Kerminen, A., & Palviainen, M. (2008). Estimation of causal effects using linear non-Gaussian causal models with hidden variables. *International Journal of Approximate Reasoning, 49*(2), 362–378.

Huang, B., Low, C., Xie, F., Glymour, C., & Zhang, K. (2022). Latent hierarchical causal structure discovery with rank constraints. In *Advances in neural information processing systems.*

Huang, B., Zhang, K., Lin, Y., Schölkopf, B., & Glymour, C. (2018). Generalized score functions for causal discovery. In *Proceedings of the 24th ACM SIGKDD International Conference on Knowledge Discovery & Data Mining* (pp. 1551–1560).

Hyvärinen, A., & Smith, S. M. (2013). Pairwise likelihood ratios for estimation of non-Gaussian structural equation models. *Journal of Machine Learning Research, 14*, 111–152.

Hyvärinen, A., Zhang, K., Shimizu, S., & Hoyer, P. O. (2010). Estimation of a structural vector autoregressive model using non-Gaussianity. *Journal of Machine Learning Research, 11*, 1709–1731.

Ikeuchi, T., Ide, M., Zeng, Y., Maeda, T. N., & Shimizu, S. (2022). Python package for causal discovery based on LiNGAM. *Journal of Machine Learning Research, 22*, 1–8. In press.

Imbens, G. W., & Rubin, D. B. (2015). *Causal inference in statistics, social, and biomedical sciences.* Cambridge University Press.

Kaddour, J., Lynch, A., Liu, Q., Kusner, M. J., & Silva, R. (2022). Causal machine learning: A survey and open problems.

Kadowaki, K., Shimizu, S., & Washio, T. (2013). Estimation of causal structures in longitudinal data using non-Gaussianity. In *Proceedings of the 23rd IEEE International Workshop on Machine Learning for Signal Processing (MLSP2013)* (pp. 1–6).

Kawahara, Y., Shimizu, S., & Washio, T. (2011). Analyzing relationships among ARMA processes based on non-Gaussianity of external influences. *Neurocomputing, 74*(12–13), 2212–2221.

Kiritoshi, K., Izumitani, T., Koyama, K., Okawachi, T., Asahara, K., & Shimizu, S. (2021). Estimating individual-level optimal causal interventions combining causal models and machine learning models. In *Proceedings of The KDD'21 Workshop on Causal Discovery*, volume 150 of *Proceedings of Machine Learning Research* (pp. 55–77). PMLR.

Komatsu, Y., Shimizu, S., & Shimodaira, H. (2010). Assessing statistical reliability of LiNGAM via multiscale bootstrap. In *Proceedings of 20th International Conference on Artificial Neural Networks (ICANN2010)* (pp. 309–314). Springer.

Kummerfeld, E., & Ramsey, J. (2016). Causal clustering for 1-factor measurement models. In *Proceedings of the 22nd ACM SIGKDD International Conference on Knowledge Discovery and Data Mining* (pp. 1655–1664).

Kuroki, M., & Cai, Z. (2008). Formulating tightest bounds on causal effects in studies with unmeasured confounders. *Statistics in Medicine, 27*(30), 6597–6611.

Kusner, M. J., Loftus, J., Russell, C., & Silva, R. (2017). Counterfactual fairness. In *Advances in neural information processing systems*, (Vol. 30). Curran Associates.

Li, C., & Shimizu, S. (2018). Combining linear non-Gaussian acyclic model with logistic regression model for estimating causal structure from mixed continuous and discrete data. arXiv preprint arXiv:1802.05889.

Li, Y., Xia, R., Liu, C., & Sun, L. (2022). A hybrid causal structure learning algorithm for mixed-type data (Vol. 36, pp. 7435–7443).

Maeda, T. N. (2022). I-RCD: An improved algorithm of repetitive causal discovery from data with latent confounders. *Behaviormetrika, 49*(2), 329–341.

Maeda, T. N., & Shimizu, S. (2020). RCD: Repetitive causal discovery of linear non-Gaussian acyclic models with latent confounders. In *Proceedings of the 23rd International Conference on Artificial Intelligence and Statistics (AISTATS2010)*, volume 108 of *Proceedings of Machine Learning Research* (pp. 735–745). PMLR.

Maeda, T. N., & Shimizu, S. (2021). Causal additive models with unobserved variables. In *Proc. 37th Conference on Uncertainty in Artificial Intelligence (UAI2021)* (pp. 97–106). PMLR.

Messerli, F. H. (2012). Chocolate consumption, cognitive function, and Nobel laureates. *New England Journal of Medicine, 367*, 1562–1564.

Monti, S., & Cooper, G. F. (1998). A multivariate discretization method for learning Bayesian networks from mixed data. In *Proceedings of the 14th Conference on Uncertainty in Artificial Intelligence (UAI1998)* (pp. 404–413).

Nabi, R., Malinsky, D., & Shpitser, I. (2022). Optimal training of fair predictive models. In *Conference on Causal Learning and Reasoning* (pp. 594–617). PMLR.

Park, G., & Park, H. (2019). Identifiability of generalized hypergeometric distribution (GHD) directed acyclic graphical models. In *Proceedings of the 22nd International Conference on Artificial Intelligence and Statistics (AISTATS2019)* (pp. 158–166). PMLR.

Pearl, J. (1995). Causal diagrams for empirical research. *Biometrika, 82*(4), 669–688.

Pearl, J. (1999). Probabilities of causation: three counterfactual interpretations and their identification. *Synthese, 121*(1), 93–149.

Pearl, J. (2000). *Causality: Models, reasoning, and inference*. Cambridge University Press.

Peters, J., Janzing, D., & Schölkopf, B. (2017). *Elements of causal inference: Foundations and learning algorithms*. The MIT Press.

Peters, J., Mooij, J. M., Janzing, D., & Schölkopf, B. (2014). Causal discovery with continuous additive noise models. *Journal of Machine Learning Research, 15*, 2009–2053.

Plecko, D., & Bareinboim, E. (2022). Causal fairness analysis.

Reilly, T., & O'Brien, R. M. (1996). Identification of confirmatory factor analysis models of arbitrary complexity: The side-by-side rule. *Sociological Methods & Research, 24*(4), 473–491.

Salehkaleybar, S., Ghassami, A., Kiyavash, N., & Zhang, K. (2020). Learning linear non-Gaussian causal models in the presence of latent variables. *Journal of Machine Learning Research, 21*, 39–1.

Schölkopf, B., Locatello, F., Bauer, S., Ke, N. R., Kalchbrenner, N., Goyal, A., & Bengio, Y. (2021). Toward causal representation learning. *Proceedings of the IEEE*, 109(5):612–634.

Sedgewick, A. J., Buschur, K., Shi, I., Ramsey, J. D., Raghu, V. K., Manatakis, D. V., Zhang, Y., Bon, J., Chandra, D., Karoleski, C., et al. (2019). Mixed graphical models for integrative causal analysis with application to chronic lung disease diagnosis and prognosis. *Bioinformatics, 35*(7), 1204–1212.

Shimizu, S. (2012). Joint estimation of linear non-Gaussian acyclic models. *Neurocomputing, 81*, 104–107.

Shimizu, S. (2014). LiNGAM: Non-Gaussian methods for estimating causal structures. *Behaviormetrika, 41*(1), 65–98.

Shimizu, S. (2022). *Statistical causal discovery: LiNGAM approach*. Springer.

Shimizu, S., Hoyer, P. O., & Hyvärinen, A. (2009). Estimation of linear non-Gaussian acyclic models for latent factors. *Neurocomputing, 72*, 2024–2027.

Shimizu, S., Hoyer, P. O., Hyvärinen, A., & Kerminen, A. (2006). A linear non-Gaussian acyclic model for causal discovery. *Journal of Machine Learning Research, 7*, 2003–2030.

Shimizu, S., Inazumi, T., Sogawa, Y., Hyvärinen, A., Kawahara, Y., Washio, T., Hoyer, P. O., & Bollen, K. (2011). DirectLiNGAM: A direct method for learning a linear non-Gaussian structural equation model. *Journal of Machine Learning Research, 12*, 1225–1248.

Shimodaira, H. (2016). Cross-validation of matching correlation analysis by resampling matching weights. *Neural Networks, 75*, 126–140.

Shpitser, I., & Pearl, J. (2008). Complete identification methods for the causal hierarchy. *Journal of Machine Learning Research, 9*, 1941–1979.

Silva, R., Scheines, R., Glymour, C., & Spirtes, P. (2006). Learning the structure of linear latent variable models. *Journal of Machine Learning Research, 7*, 191–246.

Smith, S., Miller, K., Salimi-Khorshidi, G., Webster, M., Beckmann, C., Nichols, T., Ramsey, J., & Woolrich, M. (2011). Network modelling methods for FMRI. *NeuroImage, 54*(2), 875–891.

Spirtes, P., Glymour, C., & Scheines, R. (1993). *Causation, prediction, and search* (2nd ed.). Springer. MIT Press (2000).

Tsagris, M., Borboudakis, G., Lagani, V., & Tsamardinos, I. (2018). Constraint-based causal discovery with mixed data. *International Journal of Data Science and Analytics, 6*(1), 19–30.

Uemura, K., Takagi, T., Takayuki, K., Yoshida, H., & Shimizu, S. (2022). A multivariate causal discovery based on post-nonlinear model. In *Proceedings of the First Conference on Causal Learning and Reasoning*, volume 177 of *Proceedings of Machine Learning Research* (pp. 826–839). PMLR.

von Eye, A., & Wiedermann, W. (2018). Strengthening arguments based on scale levels? *Journal for Person-Oriented Research, 4*(1), 45.

Wei, W., Feng, L., & Liu, C. (2018). Mixed causal structure discovery with application to prescriptive pricing. In *Proceddings of the 27rd International Joint Conference on Artificial Intelligence (IJCAI2018)* (pp. 5126–5134).

Wiedermann, W., & Li, X. (2020). Confounder detection in linear mediation models: Performance of kernel-based tests of independence. *Behavior Research Methods, 52*(1), 342–359.

Wiedermann, W., & Zhang, B. (2023). Direction of dependence in non-linear models via linearization. In M. Stemmler, W. Wiedermann, & F. Huang (Eds.), *Dependent data in social sciences research: Forms, issues, and methods of analysis* (2nd ed.). Springer.

Wood, S. N. (2004). Stable and efficient multiple smoothing parameter estimation for generalized additive models. *Journal of the American Statistical Association, 99*(467), 673–686.

Xie, F., Cai, R., Huang, B., Glymour, C., Hao, Z., & Zhang, K. (2020). Generalized independent noise condition for estimating latent variable causal graphs. In *Advances in neural information processing systems* (Vol. 20).

Xie, F., Huang, B., Chen, Z., He, Y., Geng, Z., & Zhang, K. (2022). Identification of linear non-gaussian latent hierarchical structure. In *International Conference on Machine Learning* (pp. 24370–24387). PMLR.

Zeng, Y., Shimizu, S., Cai, R., Xie, F., Yamamoto, M., & Hao, Z. (2021). Causal discovery with multi-domain LiNGAM for latent factors. In *Proceedings of the 30th International Joint Conference on Artificial Intelligence (IJCAI2021)*.

Zeng, Y., Shimizu, S., Matsui, H., & Sun, F. (2022). Causal discovery for linear mixed data. In *Proceedings of the First Conference on Causal Learning and Reasoning (CLeaR2022)*, volume 177 of *Proceedings of Machine Learning Research* (pp. 994–1009). PMLR.

Zhang, K., & Chan, L.-W. (2006). ICA with sparse connections. In *Proceedings of the 7th International Conference on Intelligent Data Engineering and Automated Learning (IDEAL 2006)* (pp. 530–537).

Zhang, K., Gong, M., Stojanov, P., Huang, B., & Glymour, C. (2020). Domain adaptation as a problem of inference on graphical models. In *Advances in neural information processing systems* (Vol. 20).

Zhang, K., & Hyvärinen, A. (2009). On the identifiability of the post-nonlinear causal model. In *Proceedings of the 25th Conference on Uncertainty in Artificial Intelligence (UAI2009)* (pp. 647–655).

Zhang, K., & Hyvärinen, A. (2016). Nonlinear functional causal models for distinguishing causes form effect. In *Statistics and causality: Methods for applied empirical research*. Wiley & Sons.

Zheng, X., Aragam, B., Ravikumar, P. K., & Xing, E. P. (2018). DAGs with NO TEARS: Continuous optimization for structure learning. In *Advances in neural information processing systems* (Vol. 31). Curran Associates.

Chapter 9
Direction of Dependence in Non-linear Models via Linearization

Wolfgang Wiedermann and Bixi Zhang

9.1 Introduction

Methods of causal learning and direction of dependence to evaluate causal structures of variable relations in observational data have experienced rapid development (Peters et al., 2017; Shimizu, 2022; Wiedermann et al., 2021) and have been successfully applied in various research areas including, for example, genetics (Kim et al., 2008), neuroscience (Lee & Kim, 2019), epidemiology (Rosenström et al., 2012), education (Li et al., 2022), medicine (Zhang et al., 2022), and psychology (von Eye & DeShon, 2012; Wiedermann et al., 2020). Methods of causal discovery are usually concerned with learning the causal structure of observational data in exploratory multivariable research settings, i.e., data settings where all involved variables are potentially eligible as causes or effects. In contrast, methods of direction of dependence usually address confirmatory bivariate research settings in which one is interested in probing and validating a causal target model for two variables x and y against plausible alternatives while adjusting for potential covariates.

Statistical methods of Direction Dependence Analysis (DDA; Wiedermann & Li, 2018) and causal machine learning algorithms, such as the Linear Non-Gaussian Acyclic Model (LiNGAM; Shimizu et al., 2006) and extensions thereof (see, e.g., Hoyer et al., 2008b; Hyvärinen et al., 2010; Maeda & Shimizu, 2020; Shimizu et al., 2011), have in common that they assume linearity of the variable relation and utilize non-normality of variables to learn causal structures beyond Markov-equivalence.

W. Wiedermann (✉)
Educational, School and Counseling Psychology, University of Missouri, Columbia, MO, USA
e-mail: wiedermannw@missouri.edu

B. Zhang
The Graduate Center, City University of New York, New York, NY, USA

M. Stemmler et al. (eds.), *Dependent Data in Social Sciences Research*,
https://doi.org/10.1007/978-3-031-56318-8_9

The latter can be explained by the fact that linear models are completely symmetric under normality, that is, two causally competing models (e.g., one that treats x as the cause and y as the effect, or, in path notation $x \rightarrow y$, and the reverse causation model $y \rightarrow x$) explain the same amount of variability and will have the same likelihoods. Thus, under linearity and normality, no conclusions concerning the causal direction of effect can be made, echoing the widely held believe that "correlation $\neq$ causation."

Under non-normality, however, causally competing linear models become distinguishable which opens the door to probe the causal direction of effects in observational data. Here, early studies by Dodge and Rousson (2000, 2001), Muddapur (2003), and Dodge and Yadegari (2010) showed that higher than second moments of variables enable one to formulate asymmetric facets of the Pearson correlation, i.e., facets that contain information concerning cause and effect. These results have been extended to linear regression errors of competing causal models by Wiedermann et al. (2014), Wiedermann and Hagmann (2016), Wiedermann and von Eye (2015a, b), and Wiedermann (2015, 2018). In the area of causal machine learning, Shimizu et al. (2006) were among the first who proposed learning the causal structure of linear variable networks using distributional properties of variables that become accessible under non-normality. Further, Shimizu et al. (2011) showed that independence properties of predictors and errors also enable one to recover the underlying causal graph of a linear variable network (see also Wiedermann & von Eye, 2015b; Wiedermann et al., 2017).

In the area of causal machine learning, extensions of the LiNGAM algorithm to non-linear variable relations have been proposed by, e.g., Hoyer et al. (2008a), Zhang and Hyvärinen (2010, 2016), and Monti et al. (2020)—see also the related chapter by Maeda et al. (2024) in this volume. Here, Hoyer et al. (2008a) showed that asymmetry in the independence properties of predictors and errors holds for an additive noise model (ANM), $y = f(x) + \varepsilon$ with $f(.)$ being a non-linear function of x and ε denoting an independent, arbitrarily distributed, additive error term. Note that this extension does not require variables deviating from normality (except for the case where $f(.)$ is linear), implying that non-linearity of variable relations alone is sufficient to identify the direction of causation. Further, Zhang and Hyvärinen (2010) showed that similar statements hold for post non-linear (PNL) models of the form $y = f_2(f_1(x) + \varepsilon)$, where $f_1(.)$ and $f_2(.)$ are non-linear functions. Here, the second non-linear function $f_2(.)$ represents measurement distortion. The PNL contains the linear model as well as the ANM as special cases. ANMs and PNL models usually rely on asymmetry in the independence structure of the error and the putative cause (i.e., independence of cause and error holds in the correctly specified model and non-independence holds in the reverse causation model; cf. Peters et al., 2014) to test the direction of causation. Similar asymmetry features hold for the linear model under non-normality (for details see DDA's third component outlined below).

In contrast to the developments in causal machine learning, methods of direction of dependence have almost exclusively been discussed in the linear variable case assuming non-normality of variables. Therefore, the present chapter introduces principles of direction of dependence for non-linear models. In contrast to ANMs and PNL models, we focus on multiplicative error structures for two reasons: First, under non-linearity, error terms influence the predictor often in a non-additive form. Thus, multiplicative error structures may be better suited to capture model noise. Second, under multiplicative errors, methods of direction of dependence can become accessible through proper linearization, that is, in addition to asymmetry in the independence structure of cause and error, distributional asymmetry of observed variables and errors becomes testable for non-linear variable relations.

We start with introducing linearizable regression functions and discuss asymmetry properties of cause and effect in linearized models. Then, we present methods of direction of dependence for linearized models and report the results of a Monte Carlo simulation study to evaluate the performance of direction of dependence statistics to identify the direction of causation. An empirical example is presented from allometric research. The chapter closes with a discussion of the central assumptions required to guarantee unbiasedness of the presented approach and potential extensions.

9.2 Linearizable Regression Functions

Linearizable regression functions describe cases in which non-linear functions can be transformed such that the resulting regression function is additive and linear in the model parameters. Formally, linearizable regression functions can be expressed as

$$h(y) = f(x, \boldsymbol{\theta}) + \varepsilon$$

with y being the outcome, x denoting the predictor, $\boldsymbol{\theta}$ describing a vector of model parameters, h and f representing appropriate functions that are linear in the model parameters and monotone in their outcomes, and ε denoting random errors.

Various linearizable regression functions have been proposed (cf. Daniel & Wood, 1980). The *exponential model* and the *power model* constitute to prominent examples of linearizable non-linear functions. In both cases, observed variables can be transformed using the log-transformation to achieve linearity of the transformed variable relations.

The exponential model can be written as $y = \alpha e^{\beta x}$ with α and β denoting the regression coefficients (intercept and regression slope) and e being Euler's constant. Taking the natural logarithm of the outcome, i.e., $\widetilde{y} = h(y) = \log(y)$, one obtains the linearized equation $\widetilde{y} = \widetilde{\alpha} + \beta x$ (with $\widetilde{\alpha} = \log(\alpha)$). Thus, instead of estimating parameters using non-linear least squares, one can regress $\widetilde{y}$ on x (note that f constitutes the identify function $f(x) = x$ in this case) and use ordinary least squares (OLS) regression to estimate regression parameters (with $\alpha = e^{\widetilde{\alpha}}$). The power model (also known as the power law) constitutes another common linearizable non-linear function. In its original form, the power model can be written as $y = \alpha x^{\beta}$. To achieve linearity, again, the log-transformation is applied. However, instead of only transforming the outcome variable, both variables are log-transformed, i.e., h and f constitute the natural logarithm. Thus, after linearization, the power model can be expressed as $\widetilde{y} = \widetilde{\alpha} + \beta\widetilde{x}$ with $\widetilde{y} = \log(y)$, $\widetilde{x} = \log(x)$, and $\widetilde{\alpha} = \log(\alpha)$. In other words, regression constants α and β can be estimated through regressing $\widetilde{y}$ on $\widetilde{x}$ using OLS regression (with $\alpha = e^{\widetilde{\alpha}}$).

Considering the context of learning the causal structure of variable relations, it is important to emphasize that the linearizable exponential model is an example of a one-sided data transformation. That is, to achieve linearity, data transformation is only applied to the putative outcome. The putative predictor enters the model in non-transformed form. This poses challenges in the context of testing the direction of dependence. As shown in the next section, many statistical measures to learn the causal direction of dependence are constructed such that test statistics larger than zero point at one causal model and test statistics smaller than zero provide evidence for the competing causal model. For the exponential model, this implies that positive test statistics would support a model of the form $x \rightarrow \widetilde{y}$ and negative statistics point at the $\widetilde{y} \rightarrow x$. In the latter model, however, one regresses x on the log-transformed $\widetilde{y}$ which does not correspond to the causally reversed exponential model $y \rightarrow \widetilde{x}$. Therefore, to confirm a model of the form $x \rightarrow \widetilde{y}$, one has to show that $x \rightarrow \widetilde{y}$ finds support by DDA and causal properties of $y \rightarrow \widetilde{x}$ do not contradict this initial conclusion.

In contrast, linearization of the power model follows the tradition of transform-both-sides regression (Carroll & Ruppert, 1984) where the same transformation is applied to predictor and outcome. This simplifies the application of DDA because one directly compares $\widetilde{x} \rightarrow \widetilde{y}$ with the reverse causation model $\widetilde{y} \rightarrow \widetilde{x}$. Thus, positive DDA test statistics provide empirical evidence for $\widetilde{x} \rightarrow \widetilde{y}$, and negative test statistics suggest that $\widetilde{y} \rightarrow \widetilde{x}$ better approximates the underlying causal mechanism. For this reason, in the present chapter, we focus on power models and their linearization using the log-transformation.

Although we restrict our discussion of linearized DDA to the power model, one can expect that the presented approach finds application in various areas of the empirical sciences. For example, in biology, the power law is successfully applied in allometric scaling (West et al., 1997) and to explain cellular development (Azevedo & Leroi, 2001). In linguistics, the power model reappears in the form that the frequency of words in human languages is inversely proportional to their ranks (cf. Piantadosi, 2014, i.e., a power model with $\beta = -1$ which is also known as Zipf's law). In the social sciences, the power law has been suggested to predict the performance in any given learning task and the number of training sessions (Newell & Rosenbloom, 1981) and has successfully been applied in various domains such as problem solving (Blessing & Anderson, 1996), numerosity judgments (Palmeri, 1997), and social judgments (Smith et al., 1988). Thus, overall, one can expect that many non-linear variable relations can be adequately described by the power model. Furthermore, as described in the next section, linearization of the power model via the log-transformation opens the door to straightforwardly applying principles of direction dependence modeling (originally developed for linear models) to particular non-linear causal models.

9.3 Distinguishing Cause and Effect in Linearized Models

In the following section, we discuss properties of linearized regression models (specifically, linearization of the power model) with an emphasis on the distinction between cause and effect. We present statistics and inferential procedures that can retain the fundamental asymmetry of cause and effect inherent to any unidirectional causal mechanism. That is, a causal mechanism of the form $x \rightarrow y$ implies that changes in x will lead to changes in y; however, due to the unidirectional nature of the underlying mechanism, changes in y will not lead to any changes in x.

This asymmetry has consequences for the probability distributions of x and y. The probability distribution of the outcome y conditional on the hypothetical intervention $do(x) = x_i$ differs from the marginal distribution of y, i.e., $\Pr(y \mid do(x) = x_i) \neq \Pr(y)$ (with $do(.)$ representing an intervention operator; cf. Rubin, 1974; Pearl, 1995). In contrast, the probability distribution of x conditional on $do(y) = y_i$ will not differ from the marginal distribution of x, i.e., $\Pr(x \mid do(y) = y_i) = \Pr(x)$. In linear models, measures of dependence that incorporate information of higher moments of variables (which become accessible under non-normality) can help to retain this asymmetry (Dodge & Rousson, 2000, 2001; Shimizu et al., 2006, 2011; Wiedermann et al., 2021; Wiedermann & Li, 2018). In the present section, we argue that similar principles become accessible after linearization of non-linear variable relations.

As outlined in the previous section, we focus on the power model (with multiplicative error) as a widely applicable function to describe non-linear variable relations and the log-transformation as a straightforward tool for linearization. Let

$$y = \alpha_{yx} x^{\beta_{yx}} e^{\varepsilon_{yx}}$$

be the "true" causal model ($x \rightarrow y$) with α_{yx} being the regression constant, β_{yx} describing the causal effect of x on y, and $e^{\varepsilon_{yx}}$ representing a multiplicative error term assumed to be independent of x with expected value 1 and variance $\sigma^2_{\varepsilon_{yx}}$. The causally mis-specified model $y \rightarrow x$ can be written as

$$x = \alpha_{xy} y^{\beta_{xy}} e^{\varepsilon_{xy}}$$

with specifications similar to the "true" model with the exception that we do not impose any independence assumptions on the mis-specified error term. In linearized form, the two models can be expressed as

$$\widetilde{y} = \widetilde{\alpha}_{yx} + \beta_{yx}\widetilde{x} + \varepsilon_{yx}$$

and

$$\widetilde{x} = \widetilde{\alpha}_{xy} + \beta_{xy}\widetilde{y} + \varepsilon_{xy}.$$

with $\widetilde{y} = \log(y)$, $\widetilde{x} = \log(x)$, $\widetilde{\alpha}_{yx} = \log\left(\alpha_{yx}\right)$, and $\widetilde{\alpha}_{xy} = \log\left(\alpha_{xy}\right)$. Since the log-transformation is a monotonic transformation, we do not expect any structural change in the variable relation. Therefore, evaluating the status of $\widetilde{x}$ and $\widetilde{y}$ as cause or effect allows one to deduce statements about the underlying (non-linear) causal mechanism.

Throughout the chapter, we assume that the model $\widetilde{x} \rightarrow \widetilde{y}$ (and therefore $x \rightarrow y$) constitutes the "true" causal model. We impose the following assumptions on the linearized "true" model (for the causally mis-specified model, we do not make any distributional or independence assumptions about the errors aside from $\sigma^2_{\varepsilon_{xy}} > 0$):

(A1) $\widetilde{x}$ is a transformed continuous predictor variable (the cause) that deviates from the normal distribution and whose cause is exogenous.
(A2) ε_{yx} is an additive error term with zero mean and variance $\sigma^2_{\varepsilon_{yx}}$ that is independent of $\widetilde{x}$.

Note that assumption (A1) addresses distributional requirements of $\widetilde{x}$. When x is normal, for example, the transformed variable $\widetilde{x}$ will deviate from normality which implies that, in contrast to the standard linear case, direction of dependence analysis can be expected to be applicable even when observed variables are normally distributed. However, this does not mean that no distributional assumptions are made about x. In fact, instead of assuming that variables deviate from the normal distribution, one assumes that original variables do not follow a log-normal distribution.

To test the direction of dependence of the two competing models $\widetilde{x} \rightarrow \widetilde{y}$ and $\widetilde{y} \rightarrow \widetilde{x}$, we make use of the three components of the DDA framework: (1) differences in higher moments of observed variables, (2) differences in higher moments of regression residuals, (3) differences in the independence structure of predictors and errors of the competing models (for a detailed overview see Wiedermann and Li (2018) and Wiedermann (2022)).

Higher Moments of Observed Variables

After linearization of the *x*-*y* relation, the outcome $\widetilde{y}$ can be assumed to be generated by the sum of a (weighted) cause $\widetilde{x}$ and an independent error term ε_{yx}. Thus, the distribution of the outcome can be characterized as a convolution of the distributions of the cause and the independent error. Distributional implications of this convolution can be used to retain the asymmetry between cause and effect.

For this purpose, we make use of *ij*-th central product moments defined as higher powers of the Pearson correlation, $\text{cor}(\widetilde{x}, \widetilde{y})_{ij} = \text{cov}\,(\widetilde{x}, \widetilde{y})_{ij} / \left(\sigma^i_{\widetilde{x}}\sigma^j_{\widetilde{y}}\right)$ with $\text{cov}\,(\widetilde{x}, \widetilde{y})_{ij} = \mathbb{E}\left[(\widetilde{x} - \mathbb{E}\,[\widetilde{x}])^i(\widetilde{y} - \mathbb{E}\,[\widetilde{y}])^j\right]$ being the higher-order covariance. For

$i = j = 1$, for example, $\text{cor}(\widetilde{x}, \widetilde{y})_{11} = \rho_{\widetilde{x}\widetilde{y}}$ refers to the standard Pearson correlation coefficient of $\widetilde{x}$ and $\widetilde{y}$. Using $i = 3$ and $j = 0$, one arrives at a measure of the skewness of $\widetilde{x}$, i.e., $\text{cor}(\widetilde{x}, \widetilde{y})_{30} = \mathbb{E}\left[(\widetilde{x} - \mathbb{E}[\widetilde{x}])^3\right]/\sigma^3_{\widetilde{x}}$, and $i = 0$ and $j = 3$ leads to a measure of skewness of $\widetilde{y}$, i.e., $\text{cor}(\widetilde{x}, \widetilde{y})_{03} = \mathbb{E}\left[(\widetilde{y} - \mathbb{E}[\widetilde{y}])^3\right]/\sigma^3_{\widetilde{y}}$. Higher orders of the Pearson correlation can be used to formulate asymmetric facets of the Pearson correlation, that is, facets that preserve the status of the variables as cause and effect. The first asymmetric formulation takes the form $\rho^3_{\widetilde{x}\widetilde{y}} = \text{cor}(\widetilde{x}, \widetilde{y})_{03}/\text{cor}(\widetilde{x}, \widetilde{y})_{30}$ (Dodge & Rousson, 2000, 2001) which expresses the cube of the Pearson correlation as the ratio of the skewness of the outcome $\widetilde{y}$ and the skewness of the predictor $\widetilde{x}$. The second asymmetric facet utilizes two measures for the co-skewness of $\widetilde{x}$ and $\widetilde{y}$ (asymmetry measures that take into account the relation between the variables) and can be expressed as $\rho_{\widetilde{x}\widetilde{y}} = \text{cor}(\widetilde{x}, \widetilde{y})_{12}/\text{cor}(\widetilde{x}, \widetilde{y})_{21}$. Because the Pearson correlation $\rho_{\widetilde{x}\widetilde{y}}$ is bounded on the interval $[-1, 1]$, we can make use of the following two DDA statistics

$$\Delta(\gamma)_1 = \text{cor}(\widetilde{x}, \widetilde{y})^2_{30} - \text{cor}(\widetilde{x}, \widetilde{y})^2_{03} \tag{9.1}$$

$$\Delta(\gamma)_2 = \text{cor}(\widetilde{x}, \widetilde{y})^2_{21} - \text{cor}(\widetilde{x}, \widetilde{y})^2_{12}. \tag{9.2}$$

Here, the first measure relies on the skewnesses of $\widetilde{x}$ and $\widetilde{y}$, and one obtains $\Delta(\gamma)_1 = \left(1 - \rho^6_{\widetilde{x}\widetilde{y}}\right)\gamma^2_{\widetilde{x}} > 0$, under the model $\widetilde{x} \rightarrow \widetilde{y}$, and $\Delta(\gamma)_1 = \left(\rho^6_{\widetilde{x}\widetilde{y}} - 1\right)\gamma^2_{\widetilde{y}} < 0$, under the competing model $\widetilde{y} \rightarrow \widetilde{x}$. However, to apply this measure, one has to assume symmetry of the error term ε_{yx}. When this assumption does not hold, the second DDA measure can be used. Here, without making any distributional assumptions about ε_{yx}, one obtains $\Delta(\gamma)_2 = \rho^2_{\widetilde{x}\widetilde{y}}\left(1 - \rho^2_{\widetilde{x}\widetilde{y}}\right)\gamma^2_{\widetilde{x}} > 0$, for the model $\widetilde{x} \rightarrow \widetilde{y}$, and $\Delta(\gamma)_2 = \rho^2_{\widetilde{x}\widetilde{y}}\left(\rho^2_{\widetilde{x}\widetilde{y}} - 1\right)\gamma^2_{\widetilde{y}} < 0$, under the competing model $\widetilde{y} \rightarrow \widetilde{x}$.

As a third measure of higher moment properties of observed variables, the measure proposed by Hyvärinen and Smith (2013) can be employed. This measure serves as an approximation of the likelihood ratio of the two competing models which, in the present context, is given by

$$R = \frac{1}{N}\log L\left(\widetilde{x} \rightarrow \widetilde{y}\right) - \frac{1}{N}\log L\left(\widetilde{y} \rightarrow \widetilde{x}\right)$$

with $L\left(\widetilde{x} \rightarrow \widetilde{y}\right)$ and $L\left(\widetilde{y} \rightarrow \widetilde{x}\right)$ describing the likelihoods of the two candidate models. Under the model $\widetilde{x} \rightarrow \widetilde{y}$, one observed $R > 0$; for the competing model $\widetilde{y} \rightarrow \widetilde{x}$, one obtains $R < 0$. As a third-moment based approximation, Hyvärinen and Smith (2013) suggested

$$\hat{R} = \rho_{\widetilde{x}\widetilde{y}}\,\mathbb{E}\left[\widetilde{x}^2\widetilde{y} - \widetilde{x}\widetilde{y}^2\right] \tag{9.3}$$

which leads to $\hat{R} = \gamma_{\tilde{x}}\left(\rho^2_{\tilde{x}\tilde{y}} - \rho^3_{\tilde{x}\tilde{y}}\right) > 0$ for $\tilde{x} \rightarrow \tilde{y}$ and $\hat{R} = \gamma_{\tilde{y}}\left(\rho^3_{\tilde{x}\tilde{y}} - \rho^2_{\tilde{x}\tilde{y}}\right) < 0$ under $\tilde{y} \rightarrow \tilde{x}$. Note that $\tilde{x}$ and $\tilde{y}$ need to be multiplied by the signs of their respective skewnesses prior computation of $\hat{R}$, that is, $\text{sgn}\left(\gamma_{\tilde{x}}\right)\tilde{x}$ and $\text{sgn}\left(\gamma_{\tilde{y}}\right)\tilde{y}$, with $\text{sgn}\left(\gamma_{\tilde{x}}\right) = 1$ if $\gamma_{\tilde{x}} > 0$ and $\text{sgn}\left(\gamma_{\tilde{x}}\right) = -1$ if $\gamma_{\tilde{x}} < 0$ (similar definitions apply to $\text{sign}\left(\gamma_{\tilde{y}}\right)$), are used instead of $\tilde{x}$ and $\tilde{y}$ to compute $\hat{R}$.

Therefore, for all three difference measures in Eqs. (9.1, 9.2, and 9.3), the sign of the differences enables one to perform causal model selection. Positive difference statistics point at the model $\tilde{x} \rightarrow \tilde{y}$; negative difference statistics support the reverse causation model $\tilde{y} \rightarrow \tilde{x}$. To perform statistical inference, one can use bootstrap confidence intervals.

Higher Moments of Residuals

The second DDA component focuses on asymmetry properties of residuals obtained from competing regression models. Residuals carry information about the direction of dependence for two reasons: First, the Pearson correlation coefficient between observed variables can be expressed as the negative association between regression residuals. That is, one obtains $\text{cor}(\tilde{x}, \tilde{y}) = -\text{cor}(\varepsilon_{yx}, \varepsilon_{xy})$ (cf. Wiedermann & Hagmann, 2016). Second, the error term of the causally mis-specified model can be expressed as a function of the "true" predictor and the "true" error term. That is, in the standardized case, one obtains $\varepsilon_{xy} = \left(1 - \rho^2_{\tilde{x}\tilde{y}}\right)\tilde{x} - \rho_{\tilde{x}\tilde{y}}\varepsilon_{yx}$. Thus, one can expect that the causal structure and distributional features of $\tilde{x}$ and $\tilde{y}$ are, to some extent, reflected in residuals from causally competing models. Making use of higher-order correlations, two measures are straightforwardly available:

$$\Delta(\gamma_\varepsilon)_1 = \text{cor}\left(\varepsilon_{yx}, \varepsilon_{xy}\right)^2_{03} - \text{cor}\left(\varepsilon_{yx}, \varepsilon_{xy}\right)^2_{30} \tag{9.4}$$

$$\Delta(\gamma_\varepsilon)_2 = \text{cor}\left(\varepsilon_{yx}, \varepsilon_{xy}\right)^2_{21} - \text{cor}\left(\varepsilon_{yx}, \varepsilon_{xy}\right)^2_{12}. \tag{9.5}$$

In analogy to DDA measures that are based on observed variable distributions, the first residual measure given in (9.4) assumes symmetry of the "true" error term and quantifies differences in residual skewnesses of $\tilde{x} \rightarrow \tilde{y}$ and $\tilde{y} \rightarrow \tilde{x}$. Here, $\Delta(\gamma_\varepsilon)_1 > 0$ implies that $\tilde{x} \rightarrow \tilde{y}$ should be preferred over $\tilde{y} \rightarrow \tilde{x}$. In contrast, $\Delta(\gamma_\varepsilon)_1 < 0$ provides evidence for the model $\tilde{y} \rightarrow \tilde{x}$. For asymmetrically distributed "true" errors, the second residual measure can be applied. Here, $\Delta(\gamma_\varepsilon)_2 > 0$ can be expected to hold under the model $\tilde{x} \rightarrow \tilde{y}$ and $\Delta(\gamma_\varepsilon)_2 < 0$ under the competing model $\tilde{y} \rightarrow \tilde{x}$.

Further, a third measure is available if one applies the Hyvärinen-Smith co-skewness statistics on regression residuals, that is,

$$\hat{R}_\varepsilon = \text{cor}\left(\varepsilon_{yx}, \varepsilon_{xy}\right)\mathbb{E}\left[\varepsilon_{xy}\varepsilon^2_{yx} - \varepsilon^2_{xy}\,\varepsilon_{yx}\right]. \tag{9.6}$$

Here, one obtains $\hat{R}_\varepsilon > 0$ for the model $\widetilde{x} \to \widetilde{y}$ and $\hat{R}_\varepsilon < 0$ for the competing model $\widetilde{y} \to \widetilde{x}$. Note that, similar to the Hyvärinen-Smith measures applied on observed variables, the test statistic in (9.6) requires that the skewnesses of ε_{yx} and ε_{xy} have the same sign. Thus, prior to computing the test statistic, one needs to multiply the residuals by the sign of their respective skewnesses. For all residual-based DDA measures, bootstrapping can be applied to perform statistical inference.

Independence Properties

The last DDA component focuses on the independence structure between predictors and errors of the two competing models. Assuming that the functional form of the target model (here $\widetilde{x} \to \widetilde{y}$) is correctly specified, systematic non-independencies between predictors and errors emerge from the presence of hidden confounders and reverse causation. In particular the latter has consequences for selecting between two causally competing models. In case of a reverse causation bias, that is, when the model $\widetilde{y} \to \widetilde{x}$ instead of $\widetilde{x} \to \widetilde{y}$ is entertained, the error term of the causally mis-specified model can be written as $\varepsilon_{xy} = \left(1 - \rho^2_{\widetilde{x}\widetilde{y}}\right)\widetilde{x} - \rho_{\widetilde{x}\widetilde{y}}\varepsilon_{yx}$ (assuming that $\widetilde{x}$ and $\widetilde{y}$ have been standardized). Further, reconsidering the "true" causal mechanism, $\widetilde{y} = \rho_{\widetilde{x}\widetilde{y}}\widetilde{x} + \varepsilon_{yx}$, it follows that both terms are linear functions of the same variates (the "true" predictor $\widetilde{x}$ and the "true" error term ε_{yx}) and will, therefore, be stochastically non-independent. This can rigorously be proven by making use of the Darmois-Skitovich theorem (Darmois, 1953, Skitovich, 1953; for details see, for example, Shimizu et al. (2011) and Wiedermann et al. (2017)). In essence, the Darmois-Skitovich theorem states that when two linear functions of independent variables (z_j; $j = 1, \ldots, k$; $k \geq 2$), $v_1 = \sum_{j=1}^{k} \alpha_j z_j$ and $v_2 = \sum_{j=1}^{k} \beta_j z_j$ ($\alpha_j\beta_j \neq 0$ with α_j and β_j being constants), are stochastically independent, all random variates z_j ($j = 1, \ldots, k$) will follow a normal distribution. We can, therefore, conclude that v_1 and v_2 are stochastically non-independent when at least one common z_j (with $\alpha_j\beta_j \neq 0$) deviates from the normal distribution. The latter statement has implications for the task of selecting between two causally competing models. As outlined above, the "true" outcome and the error term obtained from the causally mis-specified model are linear functions of two common components whenever $\rho_{\widetilde{x}\widetilde{y}} \neq 0$. Therefore, it follows that $\widetilde{y}$ and ε_{xy} will be stochastically non-independent whenever at least one of the two components deviates from the normal distribution. Since $\widetilde{x}$ is non-normal by (A1), non-independence can be expected to occur for the causally mis-specified model. In contrast, independence is expected to hold for the correctly specified model by (A2), and one arrives at the following simple decision rules:

- When independence holds for $\widetilde{x} \to \widetilde{y}$ and independence does not hold for $\widetilde{y} \to \widetilde{x}$, then the causal effect is more likely to go from $\widetilde{x}$ to $\widetilde{y}$.

- When independence holds for $\widetilde{y} \rightarrow \widetilde{x}$ and independence does not hold for $\widetilde{x} \rightarrow \widetilde{y}$, then the causal effect is more likely to go from $\widetilde{y}$ to $\widetilde{x}$.

The above decision rules apply under any form of non-normality and serve as the basis for various causal discovery algorithms, e.g., DirectLiNGAM (Shimizu et al., 2011), RESIT (Peters et al., 2014), as well as non-linear additive noise models (Hoyer et al., 2008a) and post non-linear causal models (Zhang & Hyvärinen, 2010). Note that identifiability of the multiplicative power model based on independence properties of predictors and errors directly follows from the PNL causal model, because the multiplicative error model is a special case of a PNL model with $f_2(.) = \exp(.)$ and $f_1(x) = \log(x)$.

Furthermore, as outlined by Wiedermann and Li (2018) and Maeda and Shimizu (2020), independence testing of predictors and errors can also be used to detect the presence of influential hidden confounders. Here, a confounder is likely to be present when non-independence is observed for both causal models.

To test independence properties of predictors and errors, omnibus measures of independence, such as the Hilbert-Schmidt Independence Criterion (HSIC; Gretton et al., 2008) or Brownian distance correlation (Székely et al., 2007), can be used to detect stochastic non-independence of linearly uncorrelated variates (recall that OLS regression residuals will be uncorrelated with the model predictors by construction). Computationally simpler measures are available through considering non-linear correlations, that is, Pearson correlation tests performed on non-linearly transformed variates (Hyvärinen et al., 2001).

In addition to separately evaluating the independence structure using an omnibus independence test, difference statistics of the form

$$\Delta(T) = T\left(\widetilde{y}, \varepsilon_{xy}\right) - T\left(\widetilde{x}, \varepsilon_{yx}\right),$$

with $T(.)$ being an independence measure, can be applied. Here, $\Delta(T) > 0$ describes the case in which non-independence is more pronounced in the model $\widetilde{y} \rightarrow \widetilde{x}$, implying that $\widetilde{x} \rightarrow \widetilde{y}$ may constitute a better approximation of the data-generating mechanism. In contrast, $\Delta(T) < 0$ implies that non-independence occurs to a larger extent in the model $\widetilde{x} \rightarrow \widetilde{y}$, thus favoring a model of the form $\widetilde{y} \rightarrow \widetilde{x}$. Resampling techniques can be applied to construct confidence limits for the difference measure.

A Unified Framework

DDA integrates the presented asymmetry properties of observed variables and model errors into one unified framework (Wiedermann & von Eye, 2015a, b, 2024; Wiedermann & Li, 2018; Wiedermann et al., 2021). Although DDA incorporates asymmetry properties of third as well as fourth higher moments (Wiedermann et al., 2021), we focus on third higher moments with the understanding that similar principles can be applied with respect to fourth moments. Considering the three

DDA components, each causal model shows unique asymmetry patterns which enables one to make empirical statements about the flow of causation. Here, the model $\tilde{x} \rightarrow \tilde{y}$ finds empirical support when:

1. $\tilde{y}$ is closer to symmetry than $\tilde{x}$ (skewness-based measures can be used when the error term of $\tilde{x} \rightarrow \tilde{y}$ is symmetrically distributed; otherwise, co-skewness-based measures can be applied)
2. the model errors ε_{yx} are closer to symmetry than ε_{xy} (skewness-based measures can be used when one accepts the assumption of error symmetry in the target model; alternatively, co-skewness-based measures are available)
3. $\tilde{x}$ and ε_{yx} are independent and, at the same time, non-independence holds for $\tilde{y}$ and ε_{xy} (when non-independence holds in both models, a hidden confounder is likely to be present)

In the following section, we evaluate the performance of DDA measures for linearized regression models.

9.4 Monte Carlo Simulation

To evaluate the performance of DDA measures to distinguish competing (linearized) causal models, a simulation study was conducted in the R statistical programing environment (R Core Team, 2024). Data were generated according to the power model $y = \alpha_{yx}\, x^{\beta_{yx}}\, e^{\varepsilon_{yx}}$. The intercept α_{yx} was set to 1, and values for the causal effect β_{yx} were selected to cover five different scenarios, specifically, $\beta_{yx} = -0.25$, 0, 0.25, 1, and 2. The resulting functional relations are summarized in Fig. 9.1. Here, $\beta_{yx} = 0$ reflects the case of no causal relation, $\beta_{yx} = 1$ corresponds to the case of a linear causal relation (with multiplicative error), $\beta_{yx} = -0.25$ describes a concave downward relation (an exponent of -0.25 is considered typical for rates of maximal population growth, heartbeat, and cellular metabolism; West et al., 1997), $\beta_{yx} = 0.25$ reflects a convex upward relation (typical for, e.g., times of blood circulation and embryonic growth and development; West et al., 1997), and $\beta_{yx} = 2$ corresponds to a concave upward relation between variables.

The predictor (x) and the error term (ε_{yx}) were either sampled from the standard normal distribution (with a skewness of 0) or various gamma distributions with a scale parameter of 1 and shape parameters $2/\gamma^2$ to obtain skewnesses of $\gamma =$ 0.75, 1.5, and 2.25. Gamma-distributed predictors were divided by the theoretical standard deviation to rescale the population variance to one. Further, the gamma-distributed errors were standardized to exhibit a population mean of zero and a population standard deviation of 0.3. Sample sizes were set to $n = 500$ and 1000. The simulation design was fully crossed resulting in 5 (causal effect β_{yx}) $\times$ 4 (distribution of x) $\times$ 4 (distribution of ε) $\times$ 2 (sample size) $=$ 160 experimental conditions. Five hundred samples were generated under each simulation condition.

For each sample, predictor and outcome scores were log-transformed using $\tilde{x} = \log(x)$ and $\tilde{y} = \log(y)$. Next, the magnitude of non-linearity was eval-

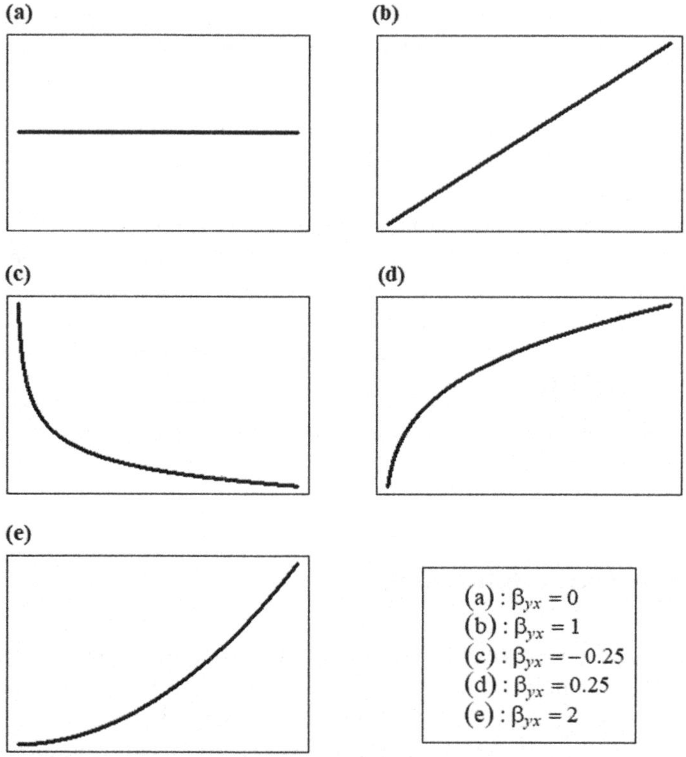

Fig. 9.1 Selected functions for causal effect assessment. (**a**) No relation, (**b**) linear relation, (**c**) concave downward relation, (**d**) convex upward relation, (**e**) concave upward relation

uated for raw and log-transformed data using the polynomial regression models $y = \beta_0 + \beta_1 x + \beta_2 x^2$ and $\tilde{y} = \beta_0 + \beta_1 \tilde{x} + \beta_2 \tilde{x}^2$. When the null hypothesis of $\beta_2 = 0$ is not rejected, one can conclude that the variable relation is sufficiently linear. Despite the structural mis-specification of the polynomial regression performed on raw scores (the polynomial model assumes an additive error), one would expect that the polynomial model shows smaller model-data discrepancies than the corresponding linear model. In the polynomial regression model performed on log-transformed variates, one would expect that the null hypothesis of $\beta_2 = 0$ is solely rejected by chance according to the nominal significance level of 5%.

In the last step, log-transformed variables were used to test direction dependence properties of $\tilde{x}$ and $\tilde{y}$ to discern statements about the causal relation of x and y. Specifically, co-skewness differences of (log-transformed) observed variables and competing regression residuals were used to test distributional implications of the underlying causal model. Ninety-five percent percentile bootstrap confidence intervals (CIs; based on 500 resamples) were used for statistical inference.

Further, separate gamma-approximated HSIC tests were applied to evaluate independence properties of the competing models $\tilde{x} \rightarrow \tilde{y}$ and $\tilde{y} \rightarrow \tilde{x}$. Calculations were performed using the R implementation of DDA provided by Wiedermann and Hirni (2022; cf. www.dda-project.com). Causal model selection was performed using the DDA decision guidelines presented in the previous section. That is, co-skewness differences (either based on observed variables or on regression residuals) larger than zero point at a causal model of the form $\tilde{x} \rightarrow \tilde{y}$, and co-skewness differences smaller than zero favor the competing model $\tilde{y} \rightarrow \tilde{x}$. Further, when the HSIC retains the null hypothesis of independence for $\tilde{x} \rightarrow \tilde{y}$ and, at the same time, rejects the null hypothesis for $\tilde{y} \rightarrow \tilde{x}$, the former is selected. Conversely, when independence is violated in the model $\tilde{x} \rightarrow \tilde{y}$ and satisfied in the model $\tilde{y} \rightarrow \tilde{x}$, the model $\tilde{y} \rightarrow \tilde{x}$ is selected.

9.5 Results

In the following section, we summarize the results obtained from the Monte Carlo simulation experiment. First, we focus on the case of no causal relation between x and y (i.e., $\beta_{yx} = 0$; cf. Fig. 9.1a). Due to absence of a causal effect, DDA measures performed on $\tilde{x}$ and $\tilde{y}$ are expected to select one of the two competing causal models solely by chance. Similarly, F-tests of the two estimated polynomial regression models are expected to indicate the presence of a quadratic effect by chance for both raw and log-transformed variables. Next, we focus on non-zero causal effect scenarios, that is, cases (b)–(e) in Fig. 9.1. Here, for each causal effect scenario, one expects that the power to select the correct model increases with sample size and magnitude of skewness of variables.

Type I Errors

Table 9.1 summarizes empirical Type I error rates for the main effects of the simulation study (i.e., aggregated across remaining conditions) for the five DDA measures and the F-tests of polynomial regressions to detect non-linearity of variable relations for raw and log-transformed data. Type I error rates of the DDA measures reflect the portion of selecting the correct model (columns labeled with $x \rightarrow y$) or causally mis-specified model (columns labelled with $y \rightarrow x$) despite a causal effect of zero. Overall, Type I error rates are close to zero for all three DDA measures regardless of sample size and skewness of x and ε_{yx}. However, the HSIC constitutes an exception with respect to the proportion of selecting the causally mis-specified model. Here, empirical rate of selecting $y \rightarrow x$ increases to 12.9% when the error distribution is heavily skewed (i.e., $\gamma_{\varepsilon} = 2.25$).

Overall, we can conclude that DDA measures are overly conservative in statistical decision-making. Similar results have been reported for DDA measures in case

Table 9.1 Type I error rates of DDA measures and F-tests to detect potential non-linearities

	HSIC		Co-skewness		RHS		Residual co-skewness		Residual RHS		Non-linearity	
	$x \to y$	$y \to x$	$x \to y$	$y \to x$	$x \to y$	$y \to x$	$x \to y$	$y \to x$	$x \to y$	$y \to x$	Raw	Log
Sample size												
500	0.008	0.042	0.001	0.004	0.001	0.004	0.000	0.000	0.001	0.001	0.046	0.052
1000	0.007	0.044	0.003	0.003	0.001	0.003	0.000	0.001	0.000	0.000	0.045	0.049
Skewness of x												
0	0.008	0.044	0.003	0.003	0.001	0.003	0.000	0.001	0.001	0.001	0.050	0.049
0.75	0.004	0.047	0.001	0.006	0.000	0.008	0.001	0.000	0.000	0.000	0.047	0.053
1.50	0.006	0.046	0.001	0.002	0.001	0.002	0.000	0.000	0.001	0.000	0.039	0.047
2.25	0.012	0.035	0.004	0.003	0.002	0.002	0.000	0.001	0.000	0.001	0.046	0.055
Skewness of ε_{yx}												
0	0.009	0.004	0.005	0.001	0.003	0.000	0.000	0.001	0.000	0.000	0.048	0.054
0.75	0.010	0.007	0.003	0.001	0.001	0.001	0.001	0.001	0.001	0.001	0.050	0.056
1.50	0.007	0.031	0.001	0.005	0.000	0.004	0.000	0.000	0.000	0.001	0.045	0.046
2.25	0.005	**0.129**	0.000	0.007	0.000	0.010	0.000	0.000	0.001	0.000	0.039	0.047

Note: RHS = Hyvärinen-Smith R; Inflated Type I error rates (rates above the upper robustness limit of 0.075) are marked bold

of linear variable relations (Wiedermann, 2021). Here, it is important to note that conservative Type I errors do not invalidate DDA test statistics. Instead, Type I errors close to zero imply reduced statistical power in case of a non-zero causal effect. Further, Type I error rates of the F-test to detect potential non-linearities (last two columns of Table 9.1) are close to the nominal significance level of 5% regardless of whether polynomial regression is conducted with raw or log-transformed data, as expected.

Statistical Power

We now focus on non-zero causal effects to determine the statistical power of DDA to detect the correct causal model based on log-transformed variables. Before we summarize power characteristics of the five DDA measures, we focus on the performance of polynomial regression F-tests to evaluate the presence of non-linearity before and after log-transformation. Table 9.2 gives the rejection rates for the null hypothesis of linearity for all non-zero causal scenarios and all considered predictor and error distributions (rates were computed across both levels of sample size). Columns labelled with "raw" give the results for untransformed x and y variables; columns labelled with "log" refer to results for log-transformed variables.

The proportion of significant results for raw data reflect the power of the F-test to detect deviations from linearity. For skewed predictors and curvilinear relations, power rates tend to be larger than 80%, in the majority of cases. Power rates tend to decrease with the skewness of the error term. Note that rejection rates of the F-test performed on raw data are also well above the 5% nominal significant level for linear x-y relations together with skewed variables (recall that the polynomial regression based on raw data is structurally mis-specified due to the presence of a multiplicative instead of an additive error). Most important, testing the presence of a curvilinear effect for log-transformed variables results in rejection rates close to the nominal significance level of 5% regardless of the functional relationship of original variables and distributions of predictors and errors. Thus, we can conclude that log-transformation successfully removed the non-linearity of the x-y relations.

We now turn to power results obtained for the five DDA measures. Figure 9.2 summarizes statistical power rates to identify the correct causal model as a function of the skewness of the predictor (γ_x) and the skewness of the "true" error ($\gamma_{\varepsilon_{yx}}$), for the considered causal effect scenarios and a sample size of $n = 500$. Figure 9.3 gives the corresponding simulation results for a sample size of $n = 1000$. Overall, power patterns were virtually identical for moderate and large samples. In the latter case, DDA measures have higher power to detect the true causal direction of effect for large samples, as expected. Further, rates of erroneously selecting the mis-specified model were close to zero across all simulation conditions, suggesting that the risk of selecting the reverse causal model instead of the correctly specified model is close to zero.

Table 9.2 Rejection rates of the *F*-test to evaluate the presence of non-linearity for raw and log-transformed data

		Linear relation ($\beta_{yx} = 1$)		Convex upward ($\beta_{yx} = 0.25$)		Concave downward ($\beta_{yx} = -0.25$)		Concave upward ($\beta_{yx} = 2$)	
γ_x	$\gamma_{\varepsilon_{yx}}$	Raw	Log	Raw	Log	Raw	Log	Raw	Log
0	0	0.088	0.045	0.103	0.040	0.323	0.055	0.996	0.049
0	0.75	0.081	0.045	0.107	0.050	0.249	0.047	0.986	0.061
0	1.50	0.084	0.070	0.084	0.055	0.208	0.054	0.969	0.043
0	2.25	0.067	0.052	0.068	0.041	0.149	0.044	0.914	0.045
0.75	0	0.229	0.050	0.371	0.060	0.705	0.052	1.000	0.043
0.75	0.75	0.206	0.054	0.306	0.052	0.593	0.044	1.000	0.048
0.75	1.50	0.169	0.045	0.254	0.047	0.462	0.047	0.999	0.048
0.75	2.25	0.151	0.043	0.191	0.052	0.363	0.054	0.998	0.056
1.50	0	0.456	0.059	0.975	0.042	1.000	0.049	1.000	0.054
1.50	0.75	0.458	0.051	0.950	0.066	1.000	0.047	1.000	0.042
1.50	1.50	0.410	0.055	0.906	0.042	1.000	0.057	1.000	0.060
1.50	2.25	0.364	0.045	0.850	0.046	0.992	0.046	1.000	0.053
2.25	0	0.612	0.060	0.998	0.056	1.000	0.053	1.000	0.045
2.25	0.75	0.579	0.045	0.997	0.060	1.000	0.043	1.000	0.054
2.25	1.50	0.553	0.041	0.987	0.058	1.000	0.035	1.000	0.050
2.25	2.25	0.535	0.047	0.979	0.067	0.997	0.053	1.000	0.047

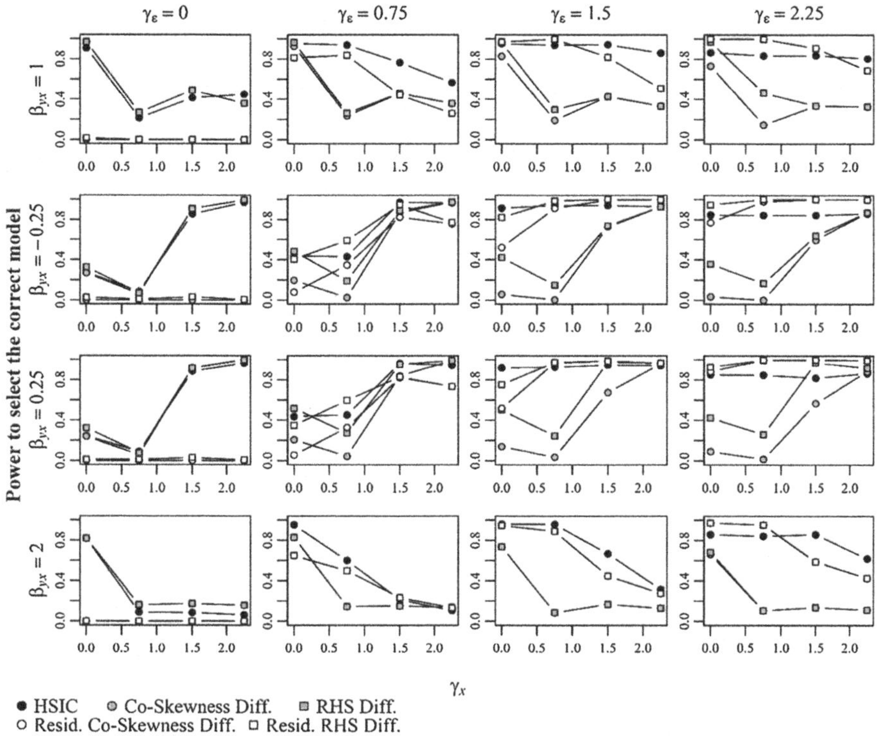

Fig. 9.2 Power to detect the correct causal model as a function of skewness of x, the skewness of ε_{yx}, and the four non-zero causal effect scenarios ($n = 500$)

Power of the HSIC tends to be larger than 80% for all considered causal effect scenarios when model errors are at least moderately skewed (i.e., $\gamma_{\varepsilon_{yx}} \geq 1.5$)—the case of a concave upward relation ($\beta_{yx} = 2$) constitutes an exception when error skewness is moderate ($\gamma_{\varepsilon_{yx}} = 1.5$) and the predictor is heavily skewed ($\gamma_x = 2.25$). For slightly skewed errors ($\gamma_{\varepsilon_{yx}} = 0.75$), the power of the HSIC is larger than 80% for the linear case, and power systematically increases with the skewness of the predictor when variable relations follow a concave downward or convex upward trend. In contrast, the power decreases with the skewness of the predictor when the causal effect manifests in a concave upward curve ($\beta_{yx} = 2$) and errors are normal or slightly skewed. When the errors follow a normal distribution (i.e., $\gamma_{\varepsilon_{yx}} = 0$), the power of the HSIC for linear ($\beta_{yx} = 1$), concave downward ($\beta_{yx} = -0.25$), and convex upward ($\beta_{yx} = 0.25$) variable relations tends to follow an S-shaped curve. Overall, one can expect that the HSIC possesses adequate power when errors are sufficiently non-normal. For concave upward relations, sufficient asymmetry of error and predictor distributions is required to guarantee sufficient statistical power.

Power curves for co-skewness differences slightly decrease with the skewness of the errors across all considered causal effect scenarios. In contrast, the Hyvärinen-

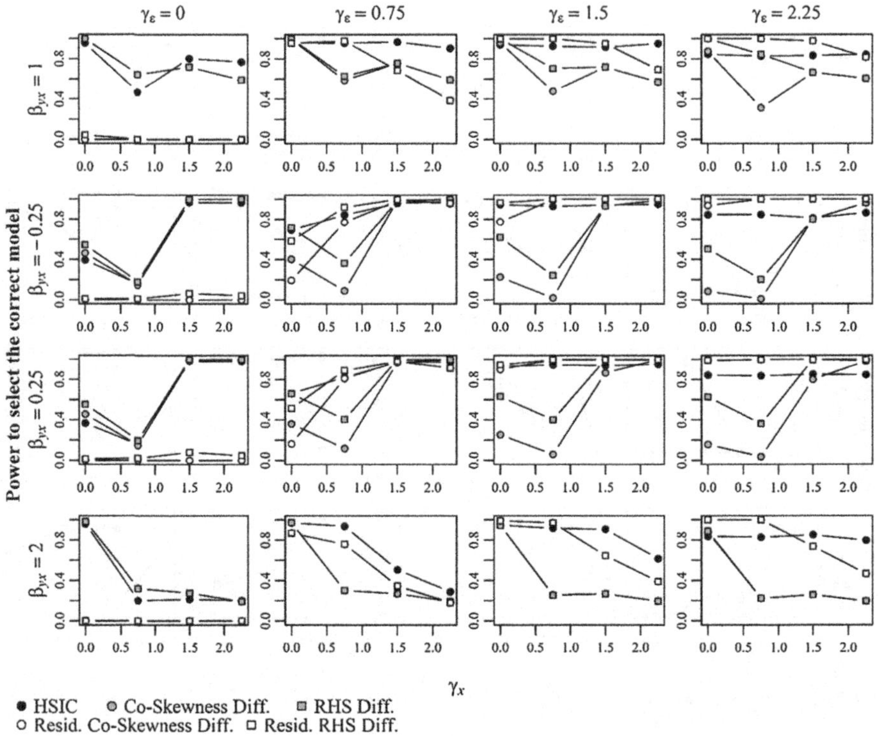

Fig. 9.3 Power to detect the correct causal model as a function of skewness of x, the skewness of ε_{yx}, and the four non-zero causal effect scenarios ($n = 1000$)

Smith co-skewness measure is less affected by the distribution of the error term. In general, the Hyvärinen-Smith measure has a power advantage for concave downward ($\beta_{yx} = -0.25$) and convex upward ($\beta_{yx} = 0.25$) relations for symmetric and slightly skewed predictors. For both DDA measures, the power to detect the true causal model tends to be larger for normal errors than for slightly skewed error terms. This can be explained by the fact that a log-transformed normal variate exhibits a higher skewness of $\tilde{x}$ than a log-transformed gamma-distributed variable with a skewness of 0.75. This power disadvantage disappears for moderately and highly skewed predictors as long as the variable relation can be described by either a linear ($\beta_{yx} = 1$), concave downward ($\beta_{yx} = -0.25$), or a convex upward ($\beta_{yx} = 0.25$) function. For concave upward relations ($\beta_{yx} = 2$), the power tends to decrease with the skewness of the predictor, independent of the level of skewness of the error term. Thus, overall, one can detect the correct causal model based on differences in co-skewnesses, regardless of the distribution of the errors, when variable relations follow concave downward or convex upward trends and predictors are at least moderately skewed ($\gamma_x \geq 1.5$).

Next, we focus on the power of differences in residual co-skewnesses. Power curves for residual co-skewness differences and the Hyvärinen-Smith measure applied on residuals of competing models are virtually identical, except for concave downward ($\beta_{yx} = -0.25$) and convex upward ($\beta_{yx} = 0.25$) relations, together with symmetric/slightly skewed predictors and slightly to moderately skewed errors. As expected, the statistical power is close to zero when the errors follow a normal distribution. Under a linear variable relation with skewed errors, the power to detect the correctly specified model is close to 100% when predictors are normal or slightly skewed. Skewness of the errors needs to be more pronounced to achieve acceptable power, when predictor skewness is moderate to high. For concave downward ($\beta_{yx} = -0.25$) and convex upward ($\beta_{yx} = 0.25$) relations, the power systematically increases with the asymmetry of the predictor when errors are slightly skewed ($\gamma_{\varepsilon_{yx}} = 0.75$); power can be expected to be above 80% regardless of the predictor skewness, when errors are moderately or highly skewed (i.e., $\gamma_{\varepsilon_{yx}} \geq 1.5$). For the considered concave upward trend ($\beta_{yx} = 2$), this DDA measure performs adequately when the skewness of the predictor is less pronounced.

Overall, we can conclude that DDA measures performed on log-transformed variables possess properties that enable one to determine the causal orientation of the transformed variable relation. The statistical power, however, depends on the sample size, the magnitude of deviations from Gaussianity, and the non-linear effect scenario considered. In particular the latter deserves closer attention. For almost all considered DDA measures, we observe that power curves for the concave upward case ($\beta_{yx} = 2$) differ from the remaining non-linear effect scenarios (the HSIC constitutes an exception for moderately to highly skewed variables). It is important to note that this should not be attributed to the different functional form (concave upward versus concave downward/convex upward), but to the magnitude of the non-linear effect and the magnitude of deviations of predictor and errors from the normal distribution. For smaller non-linear effects, larger deviations from normality of (raw) predictors and errors lead to larger distributional differences of log-transformed variates (regardless of the sign of the causal effect; for this reason, power patterns of DDA measures for $\beta_{yx} = -0.25$ and 0.25 are virtually identical). However, for larger non-linear effects, larger deviations from normality of (raw) predictors and errors can mitigate distributional differences of transformed variables. In other words, for highly non-linear relations, (raw) variables closer to the normal distribution are preferable.

Finally, to mimic the situation in which one starts the analysis of direction of dependence with the causally mis-specified model (here, erroneously using $y \rightarrow x$ as the target model instead of $x \rightarrow y$), we repeated the Monte Carlo simulation experiment and treated $\log(y)$ as the putative predictor and $\log(x)$ as the putative outcome. Overall, Type I error rates and power curves of DDA measures were virtually identical to the ones reported in Figs. 9.2 and 9.3, with the important difference that DDA now favors the alternative model, as expected. In other words, regardless of the input model, DDA, based on competing linearized regression models, can detect the underlying causal mechanism, confirming that prior knowledge about the causal mechanism is not required for causal model selection.

9.6 Real-World Data Example

In the following section, we present a real-world data example to illustrate the application of the presented approach. We use Crile and Quiring's (1940) classic data on brain and body weights of animals. These data have been used to deduce conclusions concerning the evolution of body and brain in "lower" and "higher" vertebrates (Jerison, 1973). The relation between brain weight and body weight of "lower" vertebrates (reptiles, amphibians, and fish) is assumed to follow a simple, constant brain/body ratio, indicating that selection pressure did not require major changes of the brain. In contrast, relative to the body, the brain of "higher" vertebrates (mammals and birds) increased due to selection pressure to adapt to the need of increased information processing and neural control (Jerison, 1969).

The brain-body weight relation is often modeled in the form of an allometric equation that follows the power law considered in the present approach. Thus, for "higher" vertebrates one would expect that the brain-body weight relation can be described via $y = \alpha x^{\beta}$ with y being the body weight and x being the brain weight. The parameter β is known as the coefficient of allometry, and α describes the coefficient of encephalization (cf. Bauchot et al., 1976). Because the allometric equation used to explain the relation between body and brain weight constitutes a linearizable function, the coefficients of allometry and encephalization can either be estimated via non-linear regression estimation or linear OLS estimation of log-transformed brain and body weights.

Dodge and Rousson (2000, 2016) used Crile and Quiring's (1940) data to evaluate the direction of dependence of the brain-body weight relation. Based on DDA measures that focus on observed variable distributions, the authors provided evidence that body weight is more likely to affect brain weight in "lower" vertebrates. For "higher" vertebrates, however, DDA measures suggested the opposite causal direction. That is, changes in brain weight are more likely to cause changes in body weight and not vice versa. von Eye and Wiedermann (2018) re-analyzed the data from a configural perspective using non-hierarchical log-linear models and also concluded that, for "higher" vertebrates, a causal model of the form brain → body is more likely to hold than the reverse causal model body → brain.

In the present re-analysis of direction dependence properties of the brain-body weight relation, we focus on 375 "higher" vertebrates (24.3% birds, 14.9% carnivores, 14.9% primates, 14.7% rodents, 16.0% ungulates, and 15.2% odd toed ungulates). In contrast to Dodge and Rousson (2000, 2016), we focus on all three components of DDA, i.e., higher moment properties of observed variables and residuals and independence properties of competing causal models. Figure 9.4 (left panel) suggests that the power model is well suited to describe the relation between

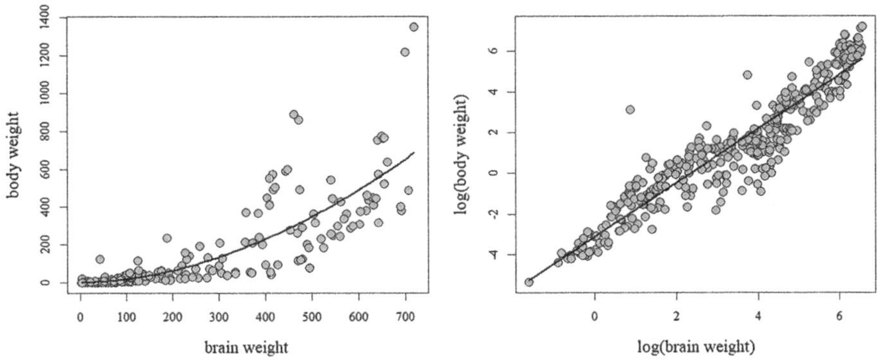

Fig. 9.4 Raw and log-transformed relation between body and brain weights of 375 animals classified as higher vertebrates

brain weight (in gram) and body weight (in kilo), confirming a positive (concave upward) relation between brain weight and body weight.

We now focus on the linearized regressions to evaluate properties of direction of dependence. Taking the natural logarithm of brain and body weights results in the linear variable relation given in Fig. 9.4 (right panel). The corresponding linear regression equation takes the form log(body) $= \widetilde{\alpha} + \beta$ log(brain) with $\widetilde{\alpha} = -3.13$ ($SE = 0.09, p < 0.001$) and $\beta = 1.33$ ($SE = 0.02, p < 0.001, R^2 = 0.89$)—indicating a strong linear relation. Before we inspect direction dependence properties of competing linearized regression models, we evaluate distributional requirements of DDA. As expected, the log-transformation reduces the degree of non-normality of both variables. Specifically, we observe a reduction in skewness from 1.42 (D'Agostino $z = 8.78, p < 0.001$) to -0.28 ($z = -2.19, p = 0.028$) for brain weights and from 3.05 (z = 13.56, p < 0.001) to 0.00 (z = 0.004, p = 0.997) for body weights. Although log-transformation eliminates asymmetry of the body weight distribution, we still observe significant deviations from symmetry for brain weights. Previous application of observed-variable based DDA by Dodge and Rousson (2000, 2016) suggests that these small deviations are sufficient to apply principles of direction of dependence.

To test the causal direction of the (log-transformed) brain-body relation, we use co-skewness difference measures for observed variables and regression residuals. Ninety-five percent bias-corrected and accelerated (BCa) bootstrap CIs were applied for statistical inference (based on 2000 resamples). Co-skewness differences significantly larger than zero indicate that brain → body is better able to describe the underlying causal mechanism than body → brain. In addition, we evaluated independence properties of the causally competing models using separate gamma-approximated HSIC tests and 95% BCa CIs for the difference in HSIC statistics (again, difference scores significantly larger than zero point at the target model brain → body).

The co-skewness difference measure based on observed variables is significantly larger than zero ($\Delta(\gamma)_2 = 0.025$, 95% CI = [0.001, 0.060]) indicating that brain → body is favored over body → brain which is in line with previous findings of Dodge and Rousson (2000, 2016). In addition, the Hyvärinen-Smith measure also supports the target model (RHS = 0.253, 95% CI = [0.231, 0.329]). Further, because the residuals of the target model are sufficiently symmetric, we can also make use of the difference measures based on skewnesses. Again, the difference measure is significantly larger than zero ($\Delta(\gamma)_1 = 0.078$, 95% CI = [0.001, 0.179]), thus supporting the model brain → body.

DDA results for competing model residuals, however, are inconclusive. Here, we observe non-significant co-skewness difference statistics of $\Delta(\gamma_\varepsilon)_2 = 0.002$, 95% CI = [−0.017, 0.052] and $RHS_\varepsilon = 0.010$, 95% CI = [−0.018, 0.133] which can be explained by the fact that residual distributions in both models are too close to normality; for brain → body we observe a residual skewness of −0.09 ($z = -0.76$, $p = 0.447$), and for the competing model body → brain, we obtain a residual skewness of 0.058 ($z = 0.464$, $p = 0.643$). Thus, as expected, co-skewnesses for regression residuals are close to zero, and no decision concerning the direction of effect can be made. Finally, separate HSIC tests suggest substantial non-independence in both models (i.e., HSIC(brain → body) = 3.05, $p < 0.001$; HSIC(body → brain) = 5.38, $p < 0.001$), indicating the presence of hidden common causes of body and brain weight.

However, the magnitude of non-independence is larger in the causally reversed model, suggesting that the model brain → body better fits the data than body → brain (Pollaris & Bontempi, 2020). An HSIC difference of Δ(HSIC) = 2.34 (95% BCa CI = [1.19, 3.55]) supports this hypothesis. Further, the Breusch-Pagan homoscedasticity test indicates no violations in the target model brain → body ($\chi^2(1) = 0.94$, $p = 0.332$) and the presence of (at least a weak form of) heteroscedasticity in the alternative model body → brain ($\chi^2(1) = 3.98$, $p = 0.046$). This asymmetry pattern can also be interpreted as evidence for the target model (Wiedermann et al., 2017).

Taken together, observed variable co-skewnesses as well as independence properties provide empirical support for the model that treats (log) brain weight as the cause and (log) body weight as the effect which is in line with previous analyses. Therefore, we conclude that an allometric model of the form body = α brain$^\beta$ better approximates the underlying causal mechanism than the reverse causation model.

9.7 Discussion

The present chapter introduced DDA in the context of linearizable non-linear variable relations. Statistics of direction of dependence were presented for cases in which the non-linear variable relation can be approximated by a power model with multiplicative error (as a special case of more general post non-linear causal models). In this case, applying the log-transformation for the tentative predictor and the tentative outcome leads to a linear model with additive error term. Therefore, the entire arsenal of DDA measures (so far restricted to the standard linear case) becomes accessible. Results of a simulation study confirmed that linearized DDA enables one to identify the direction of causation of non-linear variable relations. The performance of DDA to detect the "true" model is, however, affected by the magnitude of the non-linear relation and the magnitude of non-normality of observed variables. Further, in a real-world data example, the presented approach supported the theory that brain weight development is more likely to drive changes in body weight among higher vertebrates than vice versa.

Linearized DDA rests on a number of assumptions. First, for distributional DDA measures for observed variables and regression errors, the assumption of non-normality of variables (required for the standard linear case) is replaced with the assumption that untransformed variables deviate from the log-normal distribution. The reason for this is that under log-normality, log-transformed variables will follow a normal distribution and measures using higher than second moments can no longer be applied. In other words, distributional DDA measures can be used to distinguish causally competing models whenever transformed variables $\widetilde{x}$ and $\widetilde{y}$ sufficiently deviate from normality.

Second, all three DDA components assume independence of cause and error in the correctly specified model. Independence implies that no hidden confounders are present. When hidden confounders influence the causal relation of interest, additional assumptions about the hidden confounders are required to utilize distributional DDA measures. In the linear case, for example, Wiedermann (2022) recently showed that co-skewness measures can be used to obtain an unbiased estimate of the causal effect as long as the confounder (or the convolution of many hidden confounders) follows a symmetric distribution. Therefore, one may speculate that, under symmetric confounding, co-skewness-based measures are also able to retain cause-effect asymmetry to perform causal model selection in linearized models.

Furthermore, the third DDA component (testing independence of cause and error in the competing models) is known to indicate non-independence in both causal models in the linear confounded case (Wiedermann & Li, 2018; Maeda & Shimizu, 2020). This enables one to uniquely distinguish between the two unconfounded models ($x \rightarrow y$ and $y \rightarrow x$) and a third model in which substantive confounding is present. Whether similar properties can be expected for linearized models is material for future work.

Third, the presented approach assumes that linearization via the log-transformation results in a model that is linear in its parameters and additive. Thus, performance of DDA measures (note that independence testing can be applied under non-linearity but assumes additivity of errors) ultimately depends on how well the initial power model (with multiplicative error) approximates the true (unknown) causal mechanism. While the power model has repeatedly been shown to describe non-linear variable relations well across the empirical sciences, sensitivity analyses should be considered in practice to determine and contrast the fit of the power model against plausible alternatives.

The presented approach can be extended in various ways. For example, the present chapter discussed linearized DDA with an emphasis on third higher moments. In the standard linear case, however, DDA principles can easily be extended to fourth moments (i.e., kurtosis and co-kurtosis; cf. Dodge & Yadegari, 2010; Hyvärinen & Smith, 2013; Wiedermann, 2018). Thus, fourth moments-based DDA may also be useful for causal model selection in the linearized case. A systematic evaluation of the performance of fourth moment-based DDA measures is material for future studies.

Finally, in addition to the power model (and its linearization via log-transformation of both sides of the regression), various alternative linearizable regression functions exist. As illustrated above, the exponential model is linearizable through log-transformation of the outcome variable only. In this case, DDA model selection rules need to be adjusted for the two additional auxiliary regression models ($\widetilde{y} \rightarrow x$ and $\widetilde{x} \rightarrow y$) that emerge in distributional measures of direction of dependence (an example of such an adjustment is given by Li and Wiedermann (2020)).

References

Azevedo, R. B. R., & Leroi, A. M. (2001). A power law for cells. *Proceedings of the National Academy of Sciences, 98*(10), 5699–5704. https://doi.org/10.1073/pnas.091485998

Bauchot, R., Platel, R., & Ridet, J.-M. (1976). Brain-body weight relationships in Selachii. *Copeia, 1976*(2), 305. https://doi.org/10.2307/1443950

Blessing, S. B., & Anderson, J. R. (1996). How people learn to skip steps. *Journal of Experimental Psychology: Learning, Memory, and Cognition, 22*(3), 576–598. https://doi.org/10.1037/0278-7393.22.3.576

Carroll, R. J., & Ruppert, D. (1984). Power transformations when fitting theoretical models to data. *Journal of the American Statistical Association, 79*(386), 321–328. https://doi.org/10.1080/01621459.1984.10478052

Crile, G., & Quiring, D. P. (1940). A record of the body weight and certain organ and gland weights of 3690 animals. *Ohio Journal of Science, 40*(5), 219–259.

Daniel, C., & Wood, F. S. (1980). *Fitting equations to data*. Wiley.

Darmois, G. (1953). Analyse générale des liaisons stochastiques: Etude particulière de l'analyse factorielle linéaire [General analysis of stochastic links]. *Revue de l'Institut International de Statistique/Review of the International Statistical Institute, 21*(1/2), 2–8. https://doi.org/10.2307/1401511

Dodge, Y., & Rousson, V. (2000). Direction dependence in a regression line. *Communications in Statistics – Theory and Methods, 29*(9–10), 1957–1972. https://doi.org/10.1080/03610920008832589

Dodge, Y., & Rousson, V. (2001). On asymmetric properties of the correlation coefficient in the regression setting. *The American Statistician, 55*(1), 51–54. https://doi.org/10.1198/000313001300339932

Dodge, Y., & Rousson, V. (2016). Statistical inference for direction of dependence in linear models. In W. Wiedermann & A. von Eye (Eds.), *Statistics and causality: Methods for applied empirical research* (pp. 45–62). Wiley.

Dodge, Y., & Yadegari, I. (2010). On direction of dependence. *Metrika, 72*(1), 139–150. https://doi.org/10.1007/s00184-009-0273-0

Gretton, A., Fukumizu, K., Teo, C. H., Song, L., Schölkopf, B., & Smola, A. J. (2008). A kernel statistical test of independence. *Advances in Neural Information Processing Systems, 20*, 585–592.

Hoyer, P. O., Janzing, D., Mooij, J. M., Peters, J., & Schölkopf, B. (2008a). Nonlinear causal discovery with additive noise models. *Advances in Neural Information Processing Systems, 21*, 689–696.

Hoyer, P. O., Shimizu, S., Kerminen, A. J., & Palviainen, M. (2008b). Estimation of causal effects using linear non-Gaussian causal models with hidden variables. *International Journal of Approximate Reasoning, 49*(2), 362–378. https://doi.org/10.1016/j.ijar.2008.02.006

Hyvärinen, A., & Smith, S. M. (2013). Pairwise likelihood ratios for estimation of non-Gaussian structural equation models. *Journal of Machine Learning Research, 14*, 111–152.

Hyvärinen, A., Karhunen, J., & Oja, E. (2001). *Independent component analysis*. Wiley.

Hyvärinen, A., Zhang, K., Shimizu, S., & Hoyer, P. O. (2010). Estimation of a structural vector autoregression model using non-Gaussianity. *Journal of Machine Learning Research, 11*, 1709–1731.

Jerison, H. J. (1969). Brain evolution and dinosaur brains. *The American Naturalist, 103*(934), 575–588. https://doi.org/10.1086/282627

Jerison, H. J. (1973). *Evolution of the brain and intelligence*. Academic Press.

Kim, J.-M., Jung, Y.-S., Sungur, E. A., Han, K.-H., Park, C., & Sohn, I. (2008). A copula method for modeling directional dependence of genes. *BMC Bioinformatics, 9*(1), 225. https://doi.org/10.1186/1471-2105-9-225

Lee, N., & Kim, J.-M. (2019). Copula directional dependence for inference and statistical analysis of whole-brain connectivity from fMRI data. *Brain and Behavior, 9*(1), e01191. https://doi.org/10.1002/brb3.1191

Li, X., & Wiedermann, W. (2020). Conditional direction dependence analysis: Evaluating the causal direction of effects in linear models with interaction terms. *Multivariate Behavioral Research, 55*(5), 786–810. https://doi.org/10.1080/00273171.2019.1687276

Li, X., Bergin, C., & Olsen, A. A. (2022). Positive teacher-student relationships may lead to better teaching. *Learning and Instruction, 80*, 101581. https://doi.org/10.1016/j.learninstruc.2022.101581

Maeda, T. N., & Shimizu, S. (2020). Causal discovery of linear non-Gaussian acyclic models in the presence of latent confounders. *arXiv:2001.04197*. https://doi.org/10.48550/arXiv.2001.04197

Maeda, T. N., Zeng, Y., & Shimizu, S. (2024). Causal discovery with hidden variables based on non-Gaussianity and non-linearity. In M. Stemmler, W. Wiedermann, & F. L. Huang (Eds.), *Dependent data in social sciences research: Forms, issues, and methods of analysis* (2nd ed.). Springer.

Monti, R. P., Zhang, K., & Hyvärinen, A. (2020). Causal discovery with general non-linear relationships using non-linear ICA. *Proceedings of Machine Learning Research, 115*, 186–195.

Muddapur, M. (2003). On directional dependence in a regression line. *Communications in Statistics – Theory and Methods, 32*(10), 2053–2057. https://doi.org/10.1081/STA-120023266

Newell, A., & Rosenbloom, P. S. (1981). Mechanisms of skill acquisition and the law of practice. In J. R. Anderson (Ed.), *Cognitive skills and their acquisition* (pp. 1–55). Laurence Erlbaum Associates.

Palmeri, T. J. (1997). Exemplar similarity and the development of automaticity. *Journal of Experimental Psychology: Learning, Memory, and Cognition, 23*(2), 324–354. https://doi.org/10.1037/0278-7393.23.2.324

Pearl, J. (1995). Causal diagrams for empirical research. *Biometrika, 82*(4), 669–688. https://doi.org/10.1093/biomet/82.4.669

Peters, J., Mooij, D., Janzing, D., & Schölkopf, B. (2014). Causal discovery with continuous additive noise models. *Journal of Machine Learning Research, 15*, 2009–2053.

Peters, J., Janzing, D., & Schölkopf, B. (2017). *Elements of causal inference: Foundations and learning algorithms*. MIT Press.

Piantadosi, S. T. (2014). Zipf's word frequency law in natural language: A critical review and future directions. *Psychonomic Bulletin & Review, 21*(5), 1112–1130. https://doi.org/10.3758/s13423-014-0585-6

Pollaris, A., & Bontempi, G. (2020). Latent causation: An algorithm for pairs of correlated latent variables in linear non-Gaussian structural equation modeling. *BNAIC/BeNeLearn, 2020*, 209–223.

R Core Team. (2024). *R: A language and environment for statistical computing*. R Foundation for Statistical Computing. http://www.R-project.org/

Rosenström, T., Jokela, M., Puttonen, S., Hintsanen, M., Pulkki-Råback, L., Viikari, J. S., Raitakari, O. T., & Keltikangas-Järvinen, L. (2012). Pairwise measures of causal direction in the epidemiology of sleep problems and depression. *PLoS One, 7*(11), e50841. https://doi.org/10.1371/journal.pone.0050841

Rubin, D. B. (1974). Estimating causal effects of treatments in randomized and nonrandomized studies. *Journal of Educational Psychology, 66*(5), 688–701. https://doi.org/10.1037/h0037350

Shimizu, S. (2022). *Statistical causal discovery: LiNGAM approach*. Springer. https://doi.org/10.1007/978-4-431-55784-5

Shimizu, S., Hoyer, P. O., Hyvärinen, A., & Kerminen, A. (2006). A linear non-Gaussian acyclic model for causal discovery. *Journal of Machine Learning Research, 7*, 2003–2030.

Shimizu, S., Inazumi, T., Sogawa, Y., Hyvärinen, A., Kawahara, Y., Washio, T., Hoyer, P. O., & Bollen, K. (2011). DirectLiNGAM: A direct method for learning a linear non-Gaussian structural equation model. *Journal of Machine Learning Research, 12*, 1225–1248.

Skitovich, W. P. (1953). On a property of the normal distribution. *Doklady Akademii Nauk SSSR/Reports of the Academy of Sciences USSR, 89*, 217–219.

Smith, E. R., Branscombe, N. R., & Bormann, C. (1988). Generality of the effects of practice on social judgment tasks. *Journal of Personality and Social Psychology, 54*(3), 385–395. https://doi.org/10.1037/0022-3514.54.3.385

Székely, G. J., Rizzo, M. L., & Bakirov, N. K. (2007). Measuring and testing dependence by correlation of distances. *Annals of Statistics, 35*(6), 2769–2794. https://doi.org/10.1214/009053607000000505

von Eye, A., & DeShon, R. P. (2012). Directional dependence in developmental research. *International Journal of Behavioral Development, 36*(4), 303–312. https://doi.org/10.1177/0165025412439968

von Eye, A., & Wiedermann, W. (2018). Locating event-based causal effects: A configural perspective. *Integrative Psychological and Behavioral Science, 52*(2), 307–330. https://doi.org/10.1007/s12124-018-9423-0

West, G. B., Brown, J. H., & Enquist, B. J. (1997). A general model for the origin of allometric scaling laws in biology. *Science, 276*(5309), 122–126. https://doi.org/10.1126/science.276.5309.122

Wiedermann, W. (2015). Decisions concerning the direction of effects in linear regression models using fourth central moments. In M. Stemmler, A. von Eye, & W. Wiedermann (Eds.), *Dependent data in social sciences research: Forms, issues, and methods of analysis* (Vol. 145, pp. 149–169). Springer. https://doi.org/10.1007/978-3-319-20585-4_7

Wiedermann, W. (2018). A note on fourth moment-based direction dependence measures when regression errors are non normal. *Communications in Statistics – Theory and Methods, 47*(21), 5255–5264. https://doi.org/10.1080/03610926.2017.1388403

Wiedermann, W. (2021). Asymmetry properties of the partial correlation coefficient: Foundations for covariate adjustment in distribution-based direction dependence analysis. In W. Wiedermann, D. Kim, E. A. Sungur, & A. von Eye (Eds.), *Direction dependence in statistical modeling: Methods of analysis* (pp. 81–110). Wiley.

Wiedermann, W. (2022). Third moment-based causal inference. *Behaviormetrika, 49*(2), 303–328. https://doi.org/10.1007/s41237-021-00154-8

Wiedermann, W., & Hagmann, M. (2016). Asymmetric properties of the Pearson correlation coefficient: Correlation as the negative association between linear regression residuals. *Com-*

munications in Statistics – Theory and Methods, 45(21), 6263–6283. https://doi.org/10.1080/03610926.2014.960582

Wiedermann, W., & Hirni, M. (2022). *Direction dependence analysis in R*. www.ddaproject.com

Wiedermann, W., & Li, X. (2018). Direction dependence analysis: A framework to test the direction of effects in linear models with an implementation in SPSS. *Behavior Research Methods, 50*(4), 1581–1601. https://doi.org/10.3758/s13428-018-1031-x

Wiedermann, W., & von Eye, A. (2015a). Direction of effects in mediation analysis. *Psychological Methods, 20*(2), 221–244. https://doi.org/10.1037/met0000027

Wiedermann, W., & von Eye, A. (2015b). Direction-dependence analysis: A confirmatory approach for testing directional theories. *International Journal of Behavioral Development, 39*(6), 570–580. https://doi.org/10.1177/0165025415582056

Wiedermann, W., & von Eye, A. (2024). *Direction dependence analysis: Foundations and statistical methods*. Cambridge University Press.

Wiedermann, W., Hagmann, M., & von Eye, A. (2014). Significance tests to determine the direction of effects in linear regression models. *British Journal of Mathematical and Statistical Psychology, 68*, 116–141. https://doi.org/10.1111/bmsp.12037

Wiedermann, W., Artner, R., & von Eye, A. (2017). Heteroscedasticity as a basis of direction dependence in reversible linear regression models. *Multivariate Behavioral Research, 52*(2), 222–241. https://doi.org/10.1080/00273171.2016.1275498

Wiedermann, W., Reinke, W., & Herman, K. (2020). Prosocial skills causally mediate the relation between effective classroom management and academic competence: An application of direction dependence analysis. *Developmental Psychology, 56*(9), 1723–1735.

Wiedermann, W., Kim, D., Sungur, E. A., & von Eye, A. (2021). *Direction dependence in statistical models: Methods of analysis*. Wiley.

Zhang, K., & Hyvärinen, A. (2010). Distinguishing causes from effects using nonlinear acyclic causal models. *JMLR Workshop and Conference Proceedings, 6*, 157–164.

Zhang, K., & Hyvärinen, A. (2016). Nonlinear functional causal models for distinguishing cause from effect. In W. Wiedermann & A. von Eye (Eds.), *Wiley series in probability and statistics* (pp. 185–201). Wiley. https://doi.org/10.1002/9781118947074.ch8

Zhang, R., Sun, X., Huang, Z., Pan, Y., Westbrook, A., Li, S., Bazzano, L., Chen, W., He, J., Kelly, T., & Li, C. (2022). Examination of serum metabolome altered by cigarette smoking identifies novel metabolites mediating smoking-BMI association. *Obesity, 30*(4), 943–952. https://doi.org/10.1002/oby.23386

Chapter 10
Exploration of Dependence Structures in Longitudinal Categorical Data with Ordinal Responses

Li Wang, Shu-Min Liao, and Daeyoung Kim

10.1 Introduction

Longitudinal categorical data with ordinal outcomes is a type of data where a set of ordinal outcome variables of interest are measured repeatedly on subjects over time, along with categorical (nominal/ordinal) covariates. The analysis of such longitudinal data is an important task in various areas of science such as social, medical, biological, health, biomedical, and behavioral science, education, economy, and finance, just to name a few. A variety of statistical methods designed for longitudinal categorical data with ordinal outcomes have been developed for the purpose of formal statistical inference via explanatory and/or predictive modeling (see Bartolucci et al., 2012; Bergsma et al., 2009; Diggle et al., 2002; Hagenaars, 1990; Molenberghs & Verbeke, 2005; Sutradhar, 2014, and references therein).

A primary step to properly analyze longitudinal categorical data is Exploratory Data Analysis (EDA) to collect information on complex dependence structure inherent in the data. To this end, one may employ graphical or non-graphical exploratory methods. Graphical techniques include (but are not limited to) lorelogram (Heagerty & Zeger, 1998), mosaic plot (Friendly, 1994), parallelogram (Kosara et al., 2006), and lasagne plot (Jones et al., 2014). Non-graphical approaches include calculating non-model-based association measures such as ordinal odds ratios (Goodman, 1979,

L. Wang
Applied Scientist II, Microsoft, Redmond, WA, USA

S.-M. Liao
Department of Mathematics and Statistics, Amherst College, Amherst, MA, USA
e-mail: sliao@amherst.edu

D. Kim (✉)
Department of Mathematics and Statistics, University of Massachusetts, Amherst, MA, USA
e-mail: dkim1@umass.edu

M. Stemmler et al. (eds.), *Dependent Data in Social Sciences Research*,
https://doi.org/10.1007/978-3-031-56318-8_10

1983; McCullagh, 1980), generalized Cochran-Mantel-Haenszel methods (Mantel, 1963; Landis et al., 1978), Kendall's tau (Kendall, 1938) and its variants (Kendall, 1945; Stuart, 1953), Goodman and Kruskal's gamma (Goodman & Kruskal, 1954), Somers' D (Somers, 1962), Spearman's rank-based correlation (Agresti, 2010), and Kendall's partial tau and its extension (Kendall, 1948; Hawkes, 1971).

Yet, it is unclear how to utilize the non-model-based measures above in the process of EDA when interest is in exploring and capturing time association between ordinal outcome variables and its relationship with categorical covariates in longitudinal ordinal categorical data. The main reasons for this challenge are twofold: first, these measures mainly focus on bivariate association between two ordinal outcome variables of interest, unconditional/conditional on the other categorical variables; second, some of them assume that two ordinal variables are symmetrically associated with each other (i.e., they influence each other with the same magnitude). It is worth mentioning that Metsämuuronen (2021) recently discussed a directional property of Goodman-Kruskal's gamma, which is usually taken as a symmetric measure, and its relationship with Somers' D.

In this chapter, we propose using a newly developed model-free measure of association, the *Scaled Checkerboard Copula Regression Association Measure (SCCRAM)* (Wei & Kim, 2021), as a data-driven exploratory tool for longitudinal ordinal categorical data with ordinal outcome variables and categorical (nominal or ordinal) covariates. The SCCRAM is based on the checkerboard copula (Schweizer & Sklar, 1974; Nelsen, 2006; Genest et al., 2014), which is known to well describe the dependence structure among multivariate discrete variables. It was designed to quantify the regression dependence between an ordinal response variable and a set of categorical independent variables in a multi-way contingency table. Like other non-model-based measures mentioned above, the SCCRAM does not require any parametric specification of the underlying association structure among the variables. Note that, as proved in Wei and Kim (2021), the SCCRAM can be used when independent variables are not only ordinal but nominal.

The distinctive features of the SCCRAM are as follows: First, the SCCRAM does distinguish which ordinal outcome variable is to be considered as a putative dependent variable and which one is to be considered as a putative independent variable. Thus, one can reveal potential asymmetry between two ordinal outcome variables of interest while taking into account other covariates through the regression framework. Second, the intrinsic ordering of the categories of an ordinal dependent variable is taken into account in the computation of the SCCRAM. Third, the SCCRAM calibrates the degree of the regression association by computing the average proportion of variance in the ordinal dependent variable (with respect to its ordering information and its marginal distribution) explained by the categorical independent variable. Thus, one can evaluate the contribution of newly added independent categorical variables by comparing the difference in the two SCCRAMs, one without and the other with newly added independent variables.

For the purpose of facilitating the proper use of the SCCRAM for longitudinal ordinal categorical data in a model-free and exploratory fashion, we provide a set of SCCRAM-based strategies that account for four aspects: time dependence, data format, potential of asymmetric dependence, and model-free inference. As with time dependence, we consider two types of longitudinal data, *trend data* and *panel data*. While the trend data consists of the outcome variables repeatedly measured over time from different samples from the same population, the panel data has repeated measures of outcomes collected from the same sample of respondents over time. Regarding data format, we consider two formats, the *wide table format* and the *long table format*, whose difference lies in whether or not the repeated measures of an outcome variable is represented as a single variable in the data matrix, and the time can be explicitly specified as an ordinal covariate. For a given type of time dependence and data format, we present different forms of two-way or higher-way ordinal contingency tables and the subsequent asymmetric association structures that can be investigated in the regression framework of the SCCRAM. Lastly, in the spirit of principle of learning from data, plug-in estimation and resampling methods (bootstrap and permutation methods) are used for statistical inference on the SCCRAM such as point/interval estimation and hypothesis testing.

The rest of the article is organized as follows: Sect. 10.2 briefly reviews the SCCRAM (Wei & Kim, 2021) for a multi-way contingency table with an ordinal dependent variable and categorical independent variables and discusses statistical inference for the SCCRAM. Section 10.3 proposes the utility of the SCCRAM for the exploration of dependence between ordinal outcome variables and their relations with categorical covariates in longitudinal ordinal categorical data. Section 10.4 presents two case studies illustrating the application of the SCCRAM, one for trend data and the other for panel data. Finally, we close the chapter with a brief summary and a discussion on the potential extension of the proposed method in terms of panel data.

10.2 Review on the Scaled Checkerboard Copula Regression Association Measure

In this section, we first review the checkerboard copula for multivariate ordinal categorical variables and give an overview on the Scaled Checkerboard Copula Regression Association Measure (SCCRAM) (Wei & Kim, 2021) for a multivariate contingency table with an ordinal dependent variable and a set of categorical independent variables. We then discuss the theoretical properties and the statistical inference for the SCCRAM.

Checkerboard Copula

Denote Y and $\boldsymbol{X}=(X_1, \ldots, X_\ell, \ldots, X_d)$ to be an ordinal variable with J ordered categories $\{y_1 < \cdots < y_j < \cdots < y_J\}$ and a d-dimensional vector of ordinal variables with each X_ℓ having I_ℓ ordered categories $\{x_1^\ell < \cdots < x_{i_\ell}^\ell < \cdots < x_{I_\ell}^\ell\}$, respectively, where $j = 1, \ldots, J$, $\ell = 1, \ldots, d$, $i_\ell = 1, \ldots, I_\ell$, and " $<$ " represent the ordering information of categories in each variable. Let the joint probability mass function (p.m.f.) for $(\boldsymbol{X}^\top, Y)$ be $p_{i_1,\ldots,i_\ell,\ldots,i_d,j} = p_{\boldsymbol{i},j} = P(\boldsymbol{X} = \boldsymbol{x_i}, Y = y_j)$ where $\boldsymbol{i} = (i_1, \ldots, i_\ell, \ldots, i_d)$ and $\boldsymbol{x_i} = (x^1_{i_1}, \ldots, x^\ell_{i_\ell}, \ldots, x^d_{i_d})$. The p.m.f.s for $\boldsymbol{X}$, X_ℓ, and Y and the conditional p.m.f. for Y given $\boldsymbol{X}$ are denoted by $p_{\boldsymbol{i}+}$, $p_{+i_\ell+}$, p_{+j}, and $p_{j|\boldsymbol{i}} = p_{\boldsymbol{i},j}/p_{\boldsymbol{i}+}$, respectively. Furthermore, let the joint cumulative distribution function (c.d.f.) of $(\boldsymbol{X}^\top, Y)$ and the marginal c.d.f.s of X_ℓ and Y be $G_{\boldsymbol{X},Y}$, F_{X_ℓ}, and F_Y, respectively.

Define a $(d+1)$-dimensional vector $(\boldsymbol{U}^\top, V)$ with $\boldsymbol{U} = (U_1, \ldots, U_\ell, \ldots, U_d)$ to be $U_\ell = F_{X_\ell}(X_\ell)$ and $V = F_Y(Y)$. Then $(\boldsymbol{U}^\top, V)$ is the random vector with the domains $\left(\prod_{\ell=1}^d R_{X_\ell}\right) \times R_Y$ where $R_{X_\ell} = \{u^\ell_0, u^\ell_1, \ldots, u^\ell_{i_\ell}, \ldots, u^\ell_{I_\ell}\}$, $R_Y = \{v_0, v_1, \ldots, v_j, \ldots, v_J\}$, $u^\ell_0 = v_0 = 0$, $u^\ell_{I_\ell} = v_J = 1$, $u^\ell_{i_\ell} = F_{X_\ell}(x^\ell_{i_\ell}) = \sum_{h_\ell=1}^{i_\ell} p_{+h_\ell+}$, and $v_j = F_Y(y_j) = \sum_{h=1}^{j} p_{+h}$. By Sklar (1959), there exists a function called *subcopula* C^S associated with $(\boldsymbol{X}^\top, Y)$ on $\left(\prod_{\ell=1}^d R_{X_\ell}\right) \times R_Y$ such that

$$\begin{aligned} G_{\boldsymbol{X},Y}(x^1_{i_1}, \ldots, x^d_{i_d}, y_j) &= C^S(F_{X_1}(x^1_{i_1}), \ldots, F_{X_d}(x^d_{i_d}), F_Y(y_j)) \\ &= \sum_{h_1=1}^{i_1} \cdots \sum_{h_d=1}^{i_d} \sum_{h=1}^{j} p_{h_1,\ldots,h_d,h}. \end{aligned} \tag{10.1}$$

The C^S in Eq. (10.1) is uniquely determined on $\left(\prod_{\ell=1}^d R_{X_\ell}\right) \times R_Y$, which is a strict subset of $[0, 1]^{d+1}$.

If $X_1, \ldots, X_d, Y$ are continuous, each element in $(\boldsymbol{U}^\top, V)$ has a standard uniform distribution $U(0, 1)$, and a subcopula C^S is known as a *copula* on $[0, 1]^{d+1}$. For ordinal random variables, however, each element of $(\boldsymbol{U}^\top, V)$ is not uniformly distributed over $[0, 1]$. The domain of the subcopula C^s in Eq. (10.1) needs to be estimated because it depends on unknown marginal distributions. To resolve this issue, several methods have been proposed to extend a subcopula to a copula on $[0, 1]^{d+1}$ (Schweizer & Sklar, 1974; Carley, 2002; Genest & Nešlehová, 2007; de Amo et al., 2012). A frequently employed approach is to transform the marginal distribution of an ordinal random variable to the standard uniform distribution via the distributional transform of an ordinal random variable (Nešlehová, 2007; Rüschendorf, 2009):

$$\begin{aligned} U^*_\ell &= F_{X_\ell}(X_\ell-) + \{F_{X_\ell}(X_\ell) - F_{X_\ell}(X_\ell-)\}W_\ell, \\ V^* &= F_Y(Y-) + \{F_Y(Y) - F_Y(Y-)\}W_{d+1}, \end{aligned} \tag{10.2}$$

where $\ell = 1, \ldots, d$, $F(x-)$ refers to the left limit of F and $W_1, \ldots, W_{d+1}$ denote independent uniform random variables on $[0, 1]$ that are independent of $(\boldsymbol{X}^\top, Y)$. The continued random variables U^*_ℓ and V^* in Eq. (10.2) are standard uniform random variables, and the $(d+1)$-dimensional random vector $(\boldsymbol{U}^{*\top}, V^*)$ with $\boldsymbol{U}^* = (U_1^*, \ldots, U_\ell^*, \ldots, U_d^*)$ has a unique copula on $[0,1]^d$, known as the *checkerboard copula*, denoted as C^*. It is known that the checkerboard copula C^* represents dependencies of $(\boldsymbol{X}^\top, Y)$ in a contingency table (Schweizer & Sklar, 1974; Nelsen, 2006; Genest et al., 2014, 2017), and its density function c^* can be written as a function of the joint p.m.f. for $(\boldsymbol{X}^\top, Y)$ and their marginal p.m.f.s:

$$c^*(\boldsymbol{u}^*, v^*) = \frac{p_{\boldsymbol{i},j}}{\left(\prod_{\ell=1}^{d} p_{+i_\ell+}\right) p_{+j}} \quad \text{for } u_{i_\ell-1}^\ell < u_\ell^* \le u_{i_\ell}^\ell \text{ and } v_{j-1} < v^* \le v_j, \tag{10.3}$$

where $\ell = 1, \ldots, d$, $\boldsymbol{u}^* = (u_1^*, \ldots, u_\ell^*, \ldots, u_d^*) \in [0,1]^d$, $u_{i_\ell}^\ell \in R_{X_\ell}$, and $v_j \in R_Y$. The marginal density of $\boldsymbol{U}^*$ and the conditional density of V^* given $\boldsymbol{U}^*$ are obtained from the checkerboard copula density in Eq. (10.3): for $u_{i_\ell-1}^\ell < u_\ell^* \le u_{i_\ell}^\ell$ with $\ell = 1, \ldots, d$,

$$c^*(\boldsymbol{u}^*) = \frac{p_{\boldsymbol{i}+}}{\prod_{\ell=1}^{d} p_{+i_\ell+}}, \qquad c^*_{V^*|\boldsymbol{U}^*}(v^* \mid \boldsymbol{u}^*) = \frac{c^*(\boldsymbol{u}^*, v^*)}{c^*(\boldsymbol{u}^*)} = \frac{p_{j|\boldsymbol{i}}}{p_{+j}}.$$

Scaled Checkerboard Copula Regression Association Measure

By employing the checkerboard copula for a multi-dimensional contingency table, Wei and Kim (2021) proposed the *Scaled Checkerboard Copula Regression Association Measure (SCCRAM)*, a model-free measure to identify and quantify the regression association between an ordinal dependent variable and categorical independent variables.

Suppose that Y is a dependent variable and $\boldsymbol{X}$ is a d-dimensional vector of independent variables in a $(d+1)$-dimensional contingency table of the ordinal vector $(\boldsymbol{X}^\top, Y)$ with the joint probability mass function (p.m.f.) $p_{\boldsymbol{i},j} = P(\boldsymbol{X} = \boldsymbol{x_i}, Y = y_j)$, where $\boldsymbol{i} = (i_1, \ldots, i_\ell, \ldots, i_d)$, $i_\ell = 1, \ldots, I_\ell$, and $\boldsymbol{x_i} = (x_{i_1}^1, \ldots, x_{i_\ell}^\ell, \ldots, x_{i_d}^d)$. Recall that the p.m.f. for $\boldsymbol{X}$, the p.m.f.s for X_ℓ and Y, and the conditional p.m.f. for Y given $\boldsymbol{X}$ are denoted by $p_{\boldsymbol{i}+}$, $p_{+i_\ell+}$, p_{+j}, and $p_{j|\boldsymbol{i}} = p_{\boldsymbol{i},j}/p_{\boldsymbol{i}+}$, respectively.

Definition 10.2.1 (Wei & Kim, 2021) The **Checkerboard Copula Regression Association Measure (CCRAM)** of Y on $\boldsymbol{X}$ via their distributional transforms $\boldsymbol{U}^*$ and V^* in Eq. (10.2) is defined to be

$$\rho^2_{(\boldsymbol{X}\to Y)} \equiv \frac{\mathrm{Var}\left\{r_{V^*|\boldsymbol{U}^*}(\boldsymbol{U}^*)\right\}}{\mathrm{Var}(V^*)} = \frac{\mathrm{E}\left[\left\{r_{V^*|\boldsymbol{U}^*}(\boldsymbol{U}^*) - 1/2\right\}^2\right]}{1/12}$$

$$= 12\sum_{i=1}^{I}\left(\sum_{j=1}^{J} s_j p_{j|i} - 1/2\right)^2 p_{i+} \tag{10.4}$$

where $\mathbf{1} = (1, \ldots, 1)$ is of length d, $\boldsymbol{I} = (I_1, \ldots, I_d)$, $\{s_j\}_{j=1}^{J}$ with $s_j = (v_{j-1} + v_j)/2$ and $v_j \in R_Y$ is the checkerboard copula score set for Y, and $r_{V^*|U^*}(\boldsymbol{U}^*)$ is the checkerboard copula regression of Y on $\boldsymbol{X}$ via $(\boldsymbol{U}^{*\top}, V^*)$ in Eq. (10.2) that is defined as follows: for $u^{\ell}_{i_\ell - 1} < u^*_\ell \le u^{\ell}_{i_\ell}$,

$$r_{V^*|U^*}(\boldsymbol{U}^*) \equiv \mathrm{E}^{C^*}_{V^*|U^*}(V^*|\boldsymbol{U}^* = \boldsymbol{u}^*) = \int_0^1 v^* c^*_{V^*|U^*}(v^*|\boldsymbol{u}^*)dv^* = \sum_{j=1}^{J} s_j p_{j|i}.$$

We briefly summarize the properties of the CCRAM in Eq. (10.4), which Wei and Kim (2021) thoroughly investigated. First, the range of CCRAM $\rho^2_{(X\to Y)}$ is between 0 and $12\sigma^2_Y$ where $\sigma^2_Y = \sum_{j=1}^{J} v_{j-1} v_j p_{+j}/4$ is the variance of Y with respect to its checkerboard copula score set $\{s_j\}$ and its marginal p.m.f. $\{p_{+j}\}$. Note that the checkerboard copula score set $\{s_j\}$ of Y has the same ordering information as the categories of Y. Second, $\rho^2_{(X\to Y)}$ represents the lower bound on the average proportion of variance for Y equipped with its checkerboard copula score and its marginal distribution that can be explained by $\boldsymbol{X}$ in the checkerboard copula regression $r_{V^*|U^*}(\boldsymbol{U}^*)$. Note that the checkerboard copula regression is the mean checkerboard copula score of Y with respect to the conditional distribution of Y for any given combination of categories in $\boldsymbol{X}$. Third, the zero value of $\rho^2_{(X\to Y)}$ indicates no explanatory power of $\boldsymbol{X}$ in explaining the variability of Y, and $\rho^2_{(X\to Y)}$ is equal to its upper bound $12\sigma^2_Y$ if and only if Y is a function of $\boldsymbol{X}$ almost surely (i.e., for a given combination of categories in $\boldsymbol{X}$, there is one and only one category of Y whose corresponding joint probability is non-zero, not vice versa). Note that if Y is independent of $\boldsymbol{X}$, $\rho^2_{(X\to Y)}$ equals zero. Fourth, $\rho^2_{(X\to Y)}$ is invariant over the permutation of the categories of any independent variable in $\boldsymbol{X}$, and thus it can be employed when any variables in $\boldsymbol{X}$ are nominal.

It is clear that $\rho^2_{(X\to Y)}$ is dependent on the marginal distribution of Y due to its upper bound, $12\sigma^2_Y$. In order to properly evaluate the magnitude of regression association, the scaled version of $\rho^2_{(X\to Y)}$, named as *Scaled Checkerboard Copula Regression Association Measure* **(SCCRAM)**, can be computed :

$$\rho^{2\,*}_{(X\to Y)} = \frac{\rho^2_{(X\to Y)}}{12\sigma^2_Y}. \tag{10.5}$$

The SCCRAM $\rho^{2\,*}_{(X\to Y)}$ ranges from 0 to 1, and it also satisfies the second, third, and fourth theoretical properties of the CCRAM $\rho^2_{(X\to Y)}$ described above.

In the case studies of Sect. 10.4, we will mainly present the results concerning the SCCRAM in Eq. (10.5), $\rho^{2\,*}_{(X\to Y)}$.

Statistical Inference

To explore regression dependence between Y and $\boldsymbol{X}$ in a model-free manner, to the fullest extent possible, the statistical inference for the CCRAM in Eq. (10.4) and the SCCRAM in Eq. (10.5) can be performed by the plug-in principle and resampling methods (bootstrap method and permutation test) using relative frequencies in a multivariate contingency table.

Let $n_{\boldsymbol{i},j}$ be the frequency of the $(\boldsymbol{i}, j)$-th cell in an observed $(d+1)$-dimensional contingency table of sample size $n = \sum_{\boldsymbol{i}=\boldsymbol{1}}^{\boldsymbol{I}} \sum_{j=1}^{J} n_{\boldsymbol{i},j}$ for $(\boldsymbol{X}^\top, Y)$ where $\boldsymbol{i} = (i_1, \ldots, i_\ell, \ldots, i_d)$, $i_\ell = 1, \ldots, I_\ell$, $\ell = 1, \ldots, d$, $j = 1, \ldots, J$, $\boldsymbol{1} = (1, \ldots, 1)$, and $\boldsymbol{I} = (I_1, \ldots, I_d)$. We can then compute the estimators of the joint p.m.f. for $(\boldsymbol{X}^\top, Y)$, the marginal p.m.f.s for $\boldsymbol{X}$, X_ℓ, Y, the conditional p.m.f. for Y given $\boldsymbol{X}$, and the checkerboard copula score set for Y:

$$\hat{p}_{\boldsymbol{i},j} = \frac{n_{\boldsymbol{i},j}}{n}, \quad \hat{p}_{\boldsymbol{i}+} = \frac{n_{\boldsymbol{i}+}}{n}, \quad \hat{p}_{+i_\ell+} = \frac{n_{+i_\ell+}}{n}, \quad \hat{p}_{+j} = \frac{n_{+j}}{n}, \quad \hat{p}_{j|\boldsymbol{i}} = \frac{\hat{p}_{\boldsymbol{i},j}}{\hat{p}_{\boldsymbol{i}+}},$$

$$\hat{v}_j = \hat{F}_Y(y_j) = \sum_{h=1}^{j} \hat{p}_{+h}, \qquad \hat{s}_j = (\hat{v}_{j-1} + \hat{v}_j)/2,$$

where $n_{\boldsymbol{i}+}$, $n_{+i_\ell+}$, and n_{+j} are the marginal frequencies for $\boldsymbol{X}$, X_ℓ, and Y, respectively.

We finally obtain the plug-in estimators of the CCRAM in Eq. (10.4) and the SCCRAM in Eq. (10.5):

$$\hat{\rho}^2_{(\boldsymbol{X}\to Y)} = 12 \sum_{\boldsymbol{i}=\boldsymbol{1}}^{\boldsymbol{I}} \left(\sum_{j=1}^{J} \hat{s}_j \hat{p}_{j|\boldsymbol{i}} - 1/2 \right)^2 \hat{p}_{\boldsymbol{i}+}, \qquad \hat{\rho}^{2\,*}_{(\boldsymbol{X}\to Y)} = \frac{\hat{\rho}^2_{(\boldsymbol{X}\to Y)}}{12\hat{\sigma}^2_Y}, \tag{10.6}$$

where $\hat{\sigma}^2_Y = \sum_{j=1}^{J} \hat{v}_{j-1}\hat{v}_j \hat{p}_{+j}/4$.

To properly account for the effect of finite sample size on the point estimates in Eq. (10.6), we employ a nonparametric bootstrap method (Efron & Tibshirani, 1993) for the construction of the confidence interval. That is, we first convert the observed contingency table into an individual dataset, perform the nonparametric bootstrap on it (with replacement) to get B bootstrap samples, and convert them back to table-form datasets. Then we compute the bias-corrected and accelerated (BCa) bootstrap interval for the SCCRAM in Eq. (10.5) based on B bootstrapped contingency tables.

When one is interested in statistical inference on the SCCRAM and the difference in the SCCRAMs computed using two sets of independent categorical variables such as $H_0 : \rho^{2\,*}_{(\boldsymbol{X}\to Y)} = 0$ and $H_0 : \rho^{2\,*}_{(\boldsymbol{X}_{sub}\to Y)} = \rho^{2\,*}_{(\boldsymbol{X}\to Y)}$ where $\boldsymbol{X}_{sub}$ is a subset of $\boldsymbol{X}$, the permutation method (Good, 2000; Berry et al., 2018) can be utilized to provide p-values associated with the hypotheses of interest. That is, generate M permutation samples from the original contingency table under the null hypothesis of no regression dependence $H_0 : \rho^{2\,*}_{(X\to Y)} = 0$ or no difference $H_0 : \rho^{2\,*}_{(X_{sub}\to Y)} = \rho^{2\,*}_{(X\to Y)}$ (Doran et al., 2014), obtain the null permutation distributions of the association measures, and compute the p-values by comparing the observed SCCRAMs of interest for the original dataset to the null permutation distributions. Note that $H_0 : \rho^{2\,*}_{(X_{sub}\to Y)} = \rho^{2\,*}_{(X\to Y)}$ may provide evidence about the statistical significance of the contribution of newly added independent variables $\boldsymbol{X} \setminus \boldsymbol{X}_{sub}$ to the overall regression association, $\rho^{2\,*}_{(X\to Y)}$.

10.3 Utility of SCCRAM for Ordinal Longitudinal Data

In this section, we propose the utility of the SCCRAM introduced in Sect. 10.2 as an exploratory tool to learn regression dependence in longitudinal data with ordinal outcome variables and categorical covariates in a model-free fashion.

Aspects of Ordinal Longitudinal Categorical Data

A main goal in the process of EDA for the longitudinal data with ordinal repeated measurements of outcome variables and (time-varying or time-invariant) categorical covariates is to study potential associations of repeated measurements with categorical covariates while taking into account the fact that repeated measures of outcomes may be associated with each other either at the same time point or across time points.

To this end, we consider two basic aspects of longitudinal data, *time dependence in data* and *data format*. As with time dependence, there are two types of commonly encountered longitudinal data, *trend data* and *panel data* (Bergsma et al., 2009). In trend or repeated cross-sectional designs, data are repeatedly collected over time from different (independent) samples of respondents from the same population. This means that the dependence between different outcome variables and their association with covariates occurs at the same time points. On the other hand, panel data consist of repeated measures for the same outcome variables on the same sample of respondents over time. Thus, in addition to the dependence at the same time points, panel data has more complex dependence that has to be explored, dynamic dependence patterns across time points (i.e., past affects future within the same outcome variable or between different outcome variables either at the same or different time points).

Regarding data format, longitudinal data can be recorded in two formats, the *wide table format* and the *long table format* (Liu, 2015). In the wide table format, a respondent has a single row of data, with the measurement taken repeatedly at each time point being assigned to a separate column. That is, the repeated measurements of an outcome variable for each respondent are recorded horizontally as multiple variables in different columns of the same row of the data matrix and the variable name for each column typically has a suffix/subscript indicating the time point. Note that a time-invariant covariate that does not change across time requires only a single column. In this case, longitudinal categorical data with repeated measures of ordinal outcome variables can be summarized into multiple turnover contingency tables for outcome variables where responses are classified by two or more time points, and so transitions between the responses are revealed. On the other hand, in the long table format, each respondent has multiple rows of data with the measurement at each time point in one row. In other words, the repeated measurements for each respondent are represented as a single variable and are recorded in a single column in the data matrix. Note that any time-invariant covariate has the same value in all the rows. A distinctive feature of the long table format is that the time can be explicitly specified as an ordinal covariate in a single column, unlike the wide table format. Thus, the longitudinal categorical data with ordinal repeated measurements of outcome variables and covariates can be presented in a form of two-or-higher-dimensional contingency tables for outcome variables and the covariates including the time covariate.

Application of the SCCRAM to Ordinal Longitudinal Categorical Data

To clearly explain the utility of the SCCRAM taking into account two aspects of longitudinal data described in Sect. 10.3, we consider a simple, yet practically useful longitudinal setting: there are two ordinal outcome variables that are repeatedly measured over L time points, denoted as W_1 and W_2 with the number of categories equal to L_1 and L_2, respectively, and one time-invariant (nominal/ordinal) covariate Z with the number of categories equal to Q.

We first present forms of the contingency tables that can be obtained when the data of the longitudinal categorical setting described above is recorded in each of two formats, long table format and wide table format. In the long table format where the time can be explicitly specified as an ordinal covariate, we can construct a four-dimensional contingency table, denoted as $Z \times T \times W_1 \times W_2$, by classifying all the respondents according to Z, T, W_1, and W_2, where T indicates the time variable with L categories and Z and T are considered as covariates. As to the wide table format, we can obtain two vectors of ordinal outcome variables $\underline{\boldsymbol{W}}_1 = (W_{1,1}, \ldots, W_{1,t}, \ldots, W_{1,L})^\top$ and $\underline{\boldsymbol{W}}_2 = (W_{2,1}, \ldots, W_{2,t}, \ldots, W_{2,L})^\top$, where $t = 1, \ldots, L$ and the number of categories in $W_{1,t}$ and $W_{2,t}$ are L_1 and L_2, respectively, throughout L time points. Then the longitudinal categorical data for $\underline{\boldsymbol{W}}_1$, $\underline{\boldsymbol{W}}_2$, and Z can be represented in two-or-higher-dimensional turnover contingency tables, depending on the data dependence and the goals of the analysis.

In order to succinctly present the dependence structures of the ordinal longitudinal data that will be explored by the SCCRAM, we specify the notation of the regression association,

$$X \rightarrow Y.$$

Here, Y represents a putatively dependent variable of interest selected from either W_1 and W_2 or an element in $\underline{\boldsymbol{W}}_1$ and $\underline{\boldsymbol{W}}_2$, and $\boldsymbol{X}$ stands for a collection of putative independent variables selected from either Z, T, W_1, and W_2 or $\underline{\boldsymbol{W}}_1$, $\underline{\boldsymbol{W}}_2$, and Z that may influence Y. It is worth noting that there are often situations where association between two ordinal outcome variables is not necessarily symmetric in that the two variables affect each other, but the magnitude of influence is not necessarily the same. To investigate the potential asymmetric dependence, we employ the SCCRAM in an asymmetric manner by considering two regression association structures, one treating either W_1 or an element in $\underline{\boldsymbol{W}}_1$ as a putative dependent variable and the other considering either W_2 or an element in $\underline{\boldsymbol{W}}_2$ as a putative dependent variable.

To be specific, when W_1, W_2, and Z (from either trend data or panel data) are recorded in the long table format and are organized into a four-dimensional table $Z \times T \times W_1 \times W_2$, one can explore the following types of regression association in the framework of SCCRAM :

$$(Z, T, W_1) \rightarrow W_2, \quad (Z, W_1) \rightarrow W_2, \quad (T, W_1) \rightarrow W_2, \quad W_1 \rightarrow W_2, \tag{10.7}$$

$$(Z, T, W_2) \rightarrow W_1, \quad (Z, W_2) \rightarrow W_1, \quad (T, W_2) \rightarrow W_1, \quad W_2 \rightarrow W_1. \tag{10.8}$$

They will enable to examine the associations between W_1 and W_2 in an overall and marginal sense (i.e., overall association in the presence of both covariates Z and T and marginal associations in the absence of one or both covariates). Note that for marginal associations in Eq. (10.8) and (10.7), the SCCRAM will be applied to three- and two-dimensional contingency tables, respectively, depending on the number of covariates considered.

For a case of the wide table format, W_1, W_2, and Z are presented as L variables in each of $\underline{\boldsymbol{W}}_1 = (W_{1,1}, \ldots, W_{1,t}, \ldots, W_{1,L})^\top$ and $\underline{\boldsymbol{W}}_2 = (W_{2,1}, \ldots, W_{2,t}, \ldots, W_{2,L})^\top$ and a single variable, respectively. The subsequent subsections will discuss the regression association structures that can be studied via the SCCRAM, depending on whether or not W_1, W_2, and Z are from trend data or panel data.

Trend Data

As the trend data results from independent samples over time, one can examine patterns of association between W_1 and W_2 by applying the SCCRAM to each turnover table between $W_{1,t}$ in $\underline{\boldsymbol{W}}_1$ and $W_{2,t}$ in $\underline{\boldsymbol{W}}_2$ at each same time point t, without/with a covariate Z:

$$W_{1,t} \rightarrow W_{2,t}, \qquad W_{2,t} \rightarrow W_{1,t}, \tag{10.9}$$

$$(Z, W_{1,t}) \rightarrow W_{2,t}, \quad (Z, W_{2,t}) \rightarrow W_{1,t}, \tag{10.10}$$

where $t = 1, \ldots, L$. For the associations of Eq. (10.9) and Eq. (10.10), a two-way turnover table and a three-way turnover table at each time point t will be constructed, respectively. Note that in this case, it makes little sense to consider associations among two outcome variables over different time points because samples from the same population taken at different times are independent.

Panel Data

As with panel data, one can explore dynamic dependence relationships because $W_{1,t}$ in $\underline{\boldsymbol{W}}_1$ and $W_{2,t}$ in $\underline{\boldsymbol{W}}_2$ are repeatedly measured over time from the same sample and we can utilize the usual assumption that the past influences the future. Specifically, one can first examine the pattern of associations over time for each outcome variable (without/with a covariate):

$$W_{m,t_1} \rightarrow W_{m,t_2}, \tag{10.11}$$

$$(Z, W_{m,t_1}) \rightarrow W_{m,t_2}, \tag{10.12}$$

where $m = 1, 2$ and $t_1, t_2 = 1, \ldots, L$ with $t_1 < t_2$. For the associations given in Eq. (10.11) and (10.12), a two-way turnover table and a three-way turnover table with various time periods will be formed, respectively, for each value of (m, t_1, t_2). Second, the time-dependent transitive association between two outcome variables (without/with a covariate) can be studied:

$$W_{m_1,t_1} \rightarrow W_{m_2,t_2}, \qquad W_{m_2,t_1} \rightarrow W_{m_1,t_2}, \tag{10.13}$$

$$(Z, W_{m_1,t_1}) \rightarrow W_{m_2,t_2}, \qquad (Z, W_{m_2,t_1}) \rightarrow W_{m_1,t_2}, \tag{10.14}$$

where $m_1, m_2 = 1, 2$ with $m_1 \neq m_2$ and $t_1, t_2 = 1, \ldots, L$ with $t_1 \leq t_2$. In this case we can assume that different outcome variables ($m_1 \neq m_2$) are associated with each other at the same time point ($t_1 = t_2$). In order to carry out the analysis concerning the regression associations in Eqs. (10.13) and (10.14), two- and three-dimensional contingency tables will be formed for each value of (m_1, m_2, t_1, t_2).

There is one important point to make from the utility of the SCCRAM shown in Eq (10.8)–(10.14). One can quantify and assess the contribution of covariate(s) to the regression association of interest by the difference in the two SCCRAMs, one without the covariates and another with them. For example, $\rho^{2\,*}_{((Z,T,W_2)\rightarrow W_1)} - \rho^{2\,*}_{(W_2\rightarrow W_1)}$ from Eq. (10.8) and $\rho^{2\,*}_{((Z,W_{m_1,t_1})\rightarrow W_{m_2,t_2})} - \rho^{2\,*}_{(W_{m_1,t_1}\rightarrow W_{m_2,t_2})}$ from Eq. (10.14) can measure contributions of (Z, T) to the regression association between W_1 and W_2 and Z to the association between W_{m_2,t_2} and W_{m_1,t_1}, respectively.

10.4 Case Studies

In this section, we will demonstrate the applicability of the SCCRAM using two real-world ordinal longitudinal categorical datasets (Bergsma et al., 2009). The first one consists of trend data on concerns about social security and crime obtained from randomly selected Dutch people at different time points. The second one consists of panel data on marijuana and alcohol use during adolescence obtained by interviewing the same youths repeatedly over time.

In each of two case studies, we will briefly describe the data and variables and provide the main research questions concerning the regression associations to be investigated by the SCCRAM. We will then present the point estimates for the SCCRAMs that enable to explore potential associations of interest relevant to the research questions. In order to provide information on the uncertainties and statistical significances of the estimated regression associations, we will also compute 95% BCa bootstrap intervals for the estimated quantities of interest using 10^4 bootstrap samples and perform a permutation test for each hypothesis of interest based 10^5 random permutations. For hypothesis testing, we declare a result to be statistically significant if the permutation test p-value is less than $\alpha = 0.01$. Note that for multiple hypothesis testing, a p-value adjustment method such as Benjamini and Hochberg (1995) will be employed to control the false discovery rate.

Abuse of Social Security Concern and Crime Concern in the Netherlands

The first dataset, presented in Table 10.1, is the trend data obtained from the Amsterdam Continuous Survey conducted by the University of Amsterdam (Bergsma et al.,

Table 10.1 Concern about crime and social security in the Netherlands. Adapted from Marginal models for dependent, clustered, and longitudinal categorical data, by W.P. Bergsma, M.A. Croon, and J.A.P. Hagenaars, 2009, p. 100

		T									
S	C	1	2	3	4	5	6	7	8	9	10
NB	NB	12	26	28	38	38	25	33	27	45	26
NB	B	26	37	29	39	35	38	35	37	30	52
NB	VB	35	44	32	46	35	22	19	12	19	20
B	NB	6	9	4	16	11	12	13	18	22	18
B	B	61	74	80	80	62	96	98	101	84	127
B	VB	86	82	75	77	72	82	77	71	87	95
VB	NB	11	11	9	15	15	11	8	9	11	11
VB	B	37	39	35	51	55	56	37	62	43	60
VB	VB	249	265	303	222	268	267	247	265	238	186
Total		523	587	595	584	591	609	567	602	579	s595

2009, p.100). Respondents were randomly selected from the Dutch population at each of ten time points from November 1977 to July 1981 (i.e., independent samples at each time point), and they were asked to answer two questions regarding concern about social security and concern about crime (Van der Eijk, 1980). Note that a concern about crime and abuse of social security was considered to indicate the rising conservative ideology in the late 1970s of Dutch society (Hagenaars, 1990, p.289). Table 10.1 is a three-way contingency table with three ordinal categorical variables (obtained from a long table format with a column for the time variable): *Time* (denoted as T) with ten levels coded as 1 (Nov, 1977), 2 (Jan, 1978), 3 (Jun, 1978), 4 (Nov, 1978), 5 (Mar, 1979), 6 (Oct, 1979), 7 (Apr, 1980), 8 (Oct, 1980), 9 (Dec, 1980), and 10 (Jan, 1981) and the concern about social security (denoted as S) and the concern about crime (denoted as C), each with three ordered categories, NB (No Big problem), B (Big problem), and VB (Very Big problem).

The main research questions of interest are twofold. The first one is, "What is the overall association between two outcome variables, concern about crime (C) and concern about social security (S), with and without accounting for the time (T)?" The second one is, "How are patterns of association between C and S changed throughout the study period?" In order to investigate the first question, we will apply the SCCRAM to the three-way contingency table, denoted as $T \times C \times S$ table (taking into account the time factor), and two-way marginal table, denoted as $C \times S$ table (ignoring the time factor). As to the second question, we will analyze ten two-way turnover tables for C and S at each of ten time points, denoted $C_t \times S_t$ where $t = 1, \ldots, 10$, using the SCCRAM and explore changes in the regression association over time.

Tables 10.2 and 10.3 show the results concerning the first question, the global association between C and S under two association structures, $(S, T) \to C$ (treating C as a putative dependent variable and (S, T) as putative independent variables) and $(C, T) \to S$ (where S is considered as a putative dependent variable). First, the point estimates for $\rho^{2\,*}_{((S,T)\to C)}$ and $\rho^{2\,*}_{((C,T)\to S)}$ are 0.2154 and 0.2078, respectively. These values represent that the lower bounds on the average proportion of variance in the dependent variable explained by two independent variables are 21.54% for $(S, T) \to C$ and 20.78% for $(C, T) \to S$. According to the permutation test p-values for $H_0 : \rho^{2\,*}_{((S,T)\to C)} = 0$ and $H_0 : \rho^{2\,*}_{((C,T)\to S)} = 0$ (which are 0.00001 and 0.00001, respectively), the overall regression associations between S and C are statistically significant while taking into account T. As the 95% confidence interval for $\rho^{2\,*}_{((S,T)\to C)}$ contains the point estimate for $\rho^{2\,*}_{((C,T)\to S)}$ and vice versa, the

Table 10.2 Overall regression associations between concern about crime (C) and concern about social security (S) with/without Time (T)

SCCRAM $\rho^{2\,*}_{(X\to Y)}$	Estimate	95% BCa bootstrap interval	SCCRAM $\rho^{2\,*}_{(X\to Y)}$	Estimate	95% BCa bootstrap interval
$(S, T) \to C$	0.2154	(0.1918, 0.2325)	$(C, T) \to S$	0.2078	(0.1834, 0.2241)
$S \to C$	0.2009	(0.1806, 0.2216)	$C \to S$	0.2011	(0.1806, 0.2222)

Table 10.3 Permutation tests for the hypotheses concerning the overall regression associations considered in Table 10.2

Hypotheses	Unadjusted p-value	Adjusted p-value	Hypotheses	Unadjusted p-value	Adjusted p-value
$H_0 : \rho^{2\,*}_{((S,T)\to C)} = 0$	0.00001	–	$H_0 : \rho^{2\,*}_{((C,T)\to S)} = 0$	0.00001	–
$H_0 : \rho^{2\,*}_{(S\to C)} = 0$	0.00001	0.000015	$H_0 : \rho^{2\,*}_{(C\to S)} = 0$	0.00001	0.00003
$H_0 : \rho^{2\,*}_{((S,T)\to C)} = \rho^{2\,*}_{(S\to C)}$	0.00001	0.000015	$H_0 : \rho^{2\,*}_{((C,T)\to S)} = \rho^{2\,*}_{(C\to S)}$	0.01349	0.02023

Table 10.4 Regression association between S_t and C_t at each time point

$T=t$ $\rho^{2\,*}_{(X\to Y)}$	1	2	3	4	5	6	7	8	9	10
$S_t \to C_t$	0.1097	0.1662	0.2343	0.1424	0.1776	0.2071	0.2909	0.2676	0.2676	0.1713
$C_t \to S_t$	0.1110	0.1658	0.2349	0.1432	0.1773	0.2061	0.2902	0.2656	0.2634	0.1706

difference between the estimates of $\rho^{2\,*}_{((S,T)\to C)}$ and $\rho^{2\,*}_{((C,T)\to S)}$ may not be significant. This indicates that the association between C and S is reciprocal in that they influence each other with the same magnitude.

Second, we also observe the similar results when the time factor T was not accounted for under two marginal association structures, $S \to C$ and $C \to S$. That is, the p-value for $H_0 : \rho^{2\,*}_{(S\to C)} = 0$ is 0.00001 along with $\hat{\rho}^{2\,*}_{(S\to C)}$=0.2009, and the p-value for $H_0 : \rho^{2\,*}_{(C\to S)} = 0$ is 0.00001 along with $\hat{\rho}^{2\,*}_{(C\to S)}$=0.2011. Thus, two marginal regression associations without the time factor T are also significant. Moreover, from the 95% confidence interval for $\rho^{2\,*}_{(S\to C)}$ and $\rho^{2\,*}_{(C\to S)}$, the difference between the estimates of $\rho^{2\,*}_{(S\to C)}$ and $\rho^{2\,*}_{(C\to S)}$ may not be significant.

Third, we investigate the contribution of the time effect (T) under each regression association structure by testing the multiple hypotheses, (i) no marginal association between S and C and (ii) the equality of the two regression associations, one with T and the other without T: ($H_0 : \rho^{2\,*}_{(S\to C)} = 0$ and $H_0 : \rho^{2\,*}_{((S,T)\to C)} = \rho^{2\,*}_{(S\to C)}$) for $(S, T) \to C$ and ($H_0 : \rho^{2\,*}_{(C\to S)} = 0$ and $H_0 : \rho^{2\,*}_{((C,T)\to S)} = \rho^{2\,*}_{(C\to S)}$) for $(C, T) \to S$. It turns out that the time effect (T) has significant contribution to the association structure $(S, T) \to C$, whereas its contribution appears to be insignificant for the association structure $(C, T) \to S$ because the adjusted permutation p-value for $H_0 : \rho^{2\,*}_{((S,T)\to C)} = \rho^{2\,*}_{(S\to C)}$ is 0.000015, while the adjusted p-value for $H_0 : \rho^{2\,*}_{((C,T)\to S)} = \rho^{2\,*}_{(C\to S)}$ is 0.02023.

As to the second question concerning the change of the association in the trend data, we present in Table 10.4 and Fig. 10.1 the point estimates of the SCCRAM between S_t and C_t at time t under two regression associations $S_t \to C_t$ and $C_t \to S_t$ along with the corresponding 95% BCa bootstrap intervals. We also performed

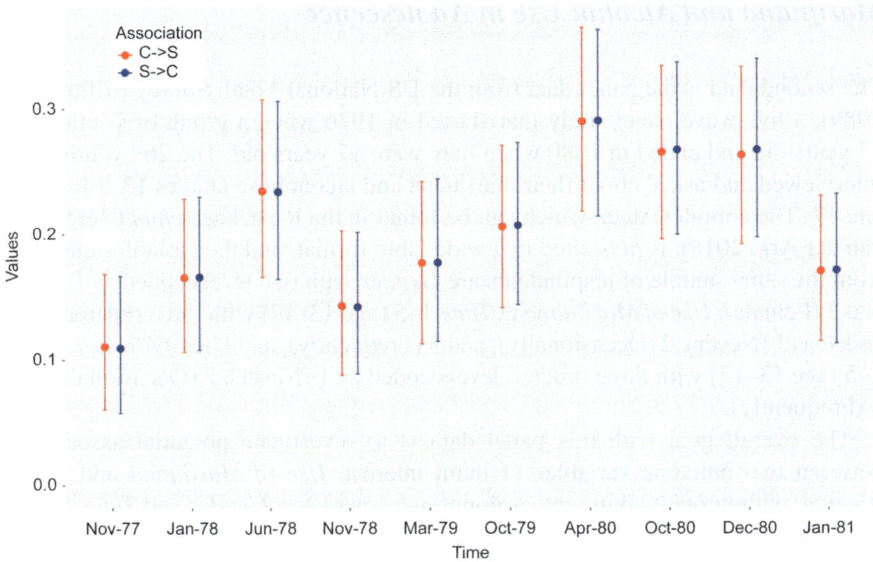

Fig. 10.1 Time trajectory of regression associations between concern about social security (S_t) and concern about crime (C_t) where $t = 1, \ldots, 10$, $\hat{\rho}^{2\,*}_{(S_t \to C_t)}$ (blue) and $\hat{\rho}^{2\,*}_{(C_t \to S_t)}$ (red), along with the corresponding 95% BCa bootstrap intervals

the permutation tests to determine the significance of the regression associations throughout the study period. It turns out that the adjusted p-value for each hypothesis $H_0 : \rho^{2\,*}_{(S_t \to C_t)} = 0$ obtained from multiple hypotheses testing over 10 time points is 0.000029, and we obtained the same adjusted p-value for each of ten hypotheses, $H_0 : \rho^{2\,*}_{(C_t \to S_t)} = 0$.

Based on these results, we make a few interesting observations. First, according to the results of the permutation tests, each of the two regression associations between S_t and C_t across ten time points is statistically significant. Second, the magnitude of the association between S_t and C_t is very similar for both regression structures $S_t \to C_t$ and $C_t \to S_t$ at every time point. Since the 95% confidence intervals for $\rho^{2\,*}_{(S_t \to C_t)}$ and $\rho^{2\,*}_{(C_t \to S_t)}$ include the point estimates for $\rho^{2\,*}_{(C_t \to S_t)}$ and $\rho^{2\,*}_{(S_t \to C_t)}$, respectively, at each t, we may find insignificant difference between $\rho^{2\,*}_{(C_t \to S_t)}$ and $\rho^{2\,*}_{(S_t \to C_t)}$, indicating symmetrical regression association between S_t and C_t. Third, the symmetric regression association between S_t and C_t ranges from about 0.1 to 0.29, which indicates the range of the lower bounds on the average proportion of variance in one outcome variable explained by the other outcome variable. Fourth, although the overall pattern of association between S_t and C_t appears to be increasing, changes in the association between S_t and C_t for both regression structures occur in a non-linear fashion throughout the study period.

Marijuana and Alcohol Use in Adolescence

The second data is the panel data from the US National Youth Survey (Elliot et al., 1989), a five-wave panel study that started in 1976 when a group of youths were 13 years old and ended in 1980 when they were 17 years old. The 269 youths were interviewed and asked about their marijuana and alcohol use at ages 13, 14, 15, 16, and 17. The complete data, which can be found in the R-package *cmm* (Bergsma & van der Ark, 2018), is presented in a wide table format, and the variables measured from the same sample of respondents are *Gender* with two levels coded as 1 (Male) and 2 (Female), *Use of Marijuana* at *Time* 1–5 (age 13–17) with three ordered levels coded as 1 (Never), 2 (Occasionally), and 3 (Frequently), and *Use of Alcohol* at *Time* 1–5 (age 13–17) with three ordered levels coded as 1 (Never), 2 (Occasionally) and 3 (Frequently).

The overall goal with this panel data is to investigate potential associations between two outcome variables of main interest, *Use of Marijuana* and *Use of Alcohol*, without/with taking into account two covariates, *Gender* and *Time*. Since a key feature of this data is that the outcome variables are repeatedly measured from the same youth over 5 years, there are (at least) two types of time dependence that can be considered. One is the transitional change in responses of each outcome variable throughout the study period, and the other is the dynamic association between two outcome variables over time. Thus, specific questions that we will discuss here are threefold. The first one is, "Is there any global regression association between two outcome variables, *Marijuana use* and *Alcohol use*, without/with two covariates, *Time* and *Gender*?" The second question is, "Does the response in each of the two outcome variables show transitional changes over time, without/with factoring in *Gender*?" The third question of interest is, "Is one outcome variable dynamically associated with the other response variable over time, without/with taking account of the effect of *Gender*?"

Note that potential asymmetry in the association between *Marijuana use* and *Alcohol Use* may be worth investigating because it is possible that one may show the same patterns over time when getting used to whatever drug one starts using or one may adopt different patterns for different drugs. Thus, we will consider two regression association structures in the first and third questions: one structure treats *Marijuana use* as a putative dependent variable and *Alcohol Use* as a putative independent variable, while the other is the other way around.

In order to examine each of the three questions above by utilizing the SCCRAMs, we first describe the form of the data that will be used for each question. As to the first question, we will consider the four-way contingency table classified by *Time (T)*, *Gender (G)*, *Use of Alcohol (A)*, and *Use of Marijuana (M)*, denoted as $T \times G \times A \times M$. We then investigate the regression associations between A and M, with/without considering the effects of T and G. With regard to the second question, we will analyze multiple turnover tables for each outcome variable where responses are classified by two time points, earlier and later time points, denoted as (t_1, t_2) where $1 \leq t_1 < t_2 \leq 5$. We will denote three-way turnover tables with *Gender* for *Marijuana Use* and for *Alcohol Use* as $G \times M_{t_1} \times M_{t_2}$ and $G \times A_{t_1} \times A_{t_2}$, respectively, and their respective two-way turnover tables without *Gender* as $M_{t_1} \times M_{t_2}$ and $A_{t_1} \times A_{t_2}$, where M_t and A_t represent *Marijuana Use* and *Alcohol Use* measured at time t. Note that for each outcome variable, we have 20 turnover tables, 10 for three-way tables and 10 for two-way tables. Regarding the third question, we will construct three-way turnover tables with *Gender* and two-way turnover tables without *Gender* cross-classified by both *Marijuana use* and *Alcohol use* at two time points (t_1, t_2) satisfying $1 \leq t_1 \leq t_2 \leq 5$. In this case, the responses of two outcome variables may be associated with each other not only between earlier and later time points but also at the same time points. The three-way tables will be denoted as $G \times A_{t_1} \times M_{t_2}$ and $G \times M_{t_1} \times A_{t_2}$, and their respective two-way tables will denoted as $A_{t_1} \times M_{t_2}$ and $M_{t_1} \times A_{t_2}$.

Tables 10.5 and 10.6 present the results concerning the first issue, "The global regression association between *Marijuana use* and *Alcohol use*," obtained from the application of the SCCRAM to the $T \times G \times A \times M$ table. We first observe that associations between M and A (without/with T and G) are statistically significant under both regression structures according to the permutation test p-values given in Table 10.6 (a-1) and (a-2). The global association between M and A appears to be larger for $(M, T, G) \rightarrow A$ (0.3765) than $(A, T, G) \rightarrow M$ (0.3349), while the marginal association seems to be larger for $A \rightarrow M$ (0.3035) than $M \rightarrow A$ (0.2846). Note that differences in the (global/marginal) associations measured by SCCRAM under two regression structures may not be significant because the 95% confidence interval for the SCCRAM under one regression structure contained the point estimate for the SCCRAM under the other regression structure.

We also find that changes in the magnitude of association between M and A depend on the regression association structure and independent variables considered in the structure. When A is the putative dependent variable, the degree of regression association ranges from 0.2846 to 0.3765, and when T was added to a set of putative independent variables, the regression association was increased by about 30% (from 0.2846 for $M \rightarrow A$ to 0.3684 for $(M, T) \rightarrow A$ and from 0.2873 for $(M, G) \rightarrow A$ to 0.3765 for $(M, T, G) \rightarrow A$), while the increment in the association by the addition of G appears to be negligible (less than 3%): from 0.2846 for $M \rightarrow A$ to 0.2873 for $(M, G) \rightarrow A$ and from 0.3684 for $(M, T) \rightarrow A$ to 0.3765 for $(M, T, G) \rightarrow A$). According to the adjusted permutation p-values for

Table 10.5 Global regression associations between Marijuana use (M) and Alcohol use (A) with/without Time (T) and Gender (G)

SCCRAM $\rho^{2*}_{(X \rightarrow Y)}$	Estimate	95% BCa bootstrap interval	SCCRAM $\rho^{2*}_{(X \rightarrow Y)}$	Estimate	95% BCa bootstrap interval
$(M, T, G) \rightarrow A$	0.3765	(0.3184, 0.4112)	$(A, T, G) \rightarrow M$	0.3349	(0.2573, 0.3730)
$(M, T) \rightarrow A$	0.3684	(0.3154, 0.4097)	$(A, T) \rightarrow M$	0.3249	(0.2603, 0.3731)
$(M, G) \rightarrow A$	0.2873	(0.2424, 0.3307)	$(A, G) \rightarrow M$	0.3101	(0.2517, 0.3651)
$M \rightarrow A$	0.2846	(0.2404, 0.3303)	$A \rightarrow M$	0.3035	(0.2480, 0.3598)

Table 10.6 Permutation tests for the hypotheses concerning the global regression associations considered in Table 10.5

Hypotheses	Unadjusted p-value	Adjusted p-value	Hypotheses	Unadjusted p-value	Adjusted p-value
(a-1)			(a-2)		
$H_0 : \rho^{2\,*}_{((M,T,G)\to A)} = 0$	0.00001	–	$H_0 : \rho^{2\,*}_{((A,T,G)\to M)} = 0$	0.00001	–
$H_0 : \rho^{2\,*}_{(M\to A)} = 0$	0.00001	–	$H_0 : \rho^{2\,*}_{(A\to M)} = 0$	0.00001	–
(b-1)			(b-2)		
$H_0 : \rho^{2\,*}_{((M,T,G)\to A)} = \rho^{2\,*}_{((M,T)\to A)}$	0.30562	0.560298	$H_0 : \rho^{2\,*}_{((A,T,G)\to M)} = \rho^{2\,*}_{((A,T)\to M)}$	0.43894	0.804715
$H_0 : \rho^{2\,*}_{((M,T)\to A)} = \rho^{2\,*}_{(M\to A)}$	0.00001	0.000027	$H_0 : \rho^{2\,*}_{((A,T)\to M)} = \rho^{2\,*}_{(A\to M)}$	0.01343	0.036932
$H_0 : \rho^{2\,*}_{(M\to A)} = 0$	0.00001	0.000027	$H_0 : \rho^{2\,*}_{(A\to M)} = 0$	0.00001	0.000055
(c-1)			(c-2)		
$H_0 : \rho^{2\,*}_{((M,T,G)\to A)} = \rho^{2\,*}_{((M,G)\to A)}$	0.00001	0.000027	$H_0 : \rho^{2\,*}_{((A,T,G)\to M)} = \rho^{2\,*}_{((A,G)\to M)}$	0.09838	0.180362
$H_0 : \rho^{2\,*}_{((M,G)\to A)} = \rho^{2\,*}_{(M\to A)}$	0.15266	0.279892	$H_0 : \rho^{2\,*}_{((A,G)\to M)} = \rho^{2\,*}_{(A\to M)}$	0.05300	0.145749
$H_0 : \rho^{2\,*}_{(M\to A)} = 0$	0.00001	0.000027	$H_0 : \rho^{2\,*}_{(A\to M)} = 0$	0.00001	0.000055

multiple hypotheses shown in Table 10.6 (b-1) and (c-1), the contribution of T to the regression association is statistically significant, unlike the contribution of G, regardless of the order of addition of T and G. On the other hand, we have different observations for the regression structure where M is the putative dependent variable. That is, the range of the regression association is between 0.3035 for $A \to M$ and 0.3349 for $(A, T, G) \to M$ (increase by 10%), and the increments in the association by the addition of G and T are by at most 3% (from 0.3035 for $A \to M$ to 0.3101 for $(A, G) \to M$ and from 0.3249 for $(A, T) \to M$ to 0.3349 for $(A, T, G) \to M$) and at most 8% (from 0.3035 for $A \to M$ to 0.3249 for $(A, T) \to M$ and from 0.3101 for $(A, G) \to M$ to 0.3349 for $(A, T, G) \to M$), respectively. The adjusted permutation p-values for multiple hypotheses in Table 10.6 (b-2) and (c-2) indicate that the contributions of T and G to the regression association are not statistically significant, irrespective of the order of addition of T and G.

As to the second concern, "Transitional changes in responses of each of *Marijuana use* and *Alcohol use*," we present the results in Table 10.7. Note that these results were computed by applying the SCCRAM to the four types of turnover tables with each representing each of the four regression structures: for (t_1, t_2) satisfying $1 \le t_1 < t_2 \le 5$,

Table 10.7 Transition over time in *Marijuana use* and *Alcohol use* without (Table 10.7a) /with (Table 10.7b) accounting for *Gender*

(a) $\rho^{2*}_{(M_{t_1}\to M_{t_2})}$ (lower off-diagonal) and $\rho^{2*}_{(A_{t_1}\to A_{t_2})}$ (upper off-diagonal)

A_{t_2}/M_{t_1}					
A_{t_1}/M_{t_2}	1	2	3	4	5
1	–	0.25876	0.18490	0.08411	0.09645
2	0.14002	–	0.32692	0.21974	0.22048
3	0.10798	0.36546	–	0.40049	0.28854
4	0.08231	0.19588	0.43690	–	0.39118
5	0.06306	0.16509	0.31413	0.44791	–

(b) $\rho^{2*}_{((G,M_{t_1})\to M_{t_2})}$ (lower off-diagonal) and $\rho^{2*}_{((G,A_{t_1})\to A_{t_2})}$ (upper off-diagonal)

A_{t_2}/M_{t_1}					
A_{t_1}/M_{t_2}	1	2	3	4	5
1	–	0.27036	0.19338	0.08649	0.09680
2	0.16817	–	0.36286	0.22432	0.22231
3	0.12481	0.37451	–	0.40595	0.29048
4	0.11327	0.20253	0.43909	–	0.40703
5	0.09066	0.16656	0.31450	0.44984	–

$$G \times M_{t_1} \times M_{t_2} \text{ for } (G, M_{t_1}) \to M_{t_2}, \quad G \times A_{t_1} \times A_{t_2} \text{ for } (G, A_{t_1}) \to A_{t_2}$$
$$M_{t_1} \times M_{t_2} \quad \text{for } M_{t_1} \to M_{t_2}, \quad A_{t_1} \times A_{t_2} \quad \text{for } A_{t_1} \to A_{t_2}$$

The numbers shown in the lower (upper) off-diagonal of Table 10.7 (a) and (b) represent the regression association for $M_{t_1} \to M_{t_2}$ and $(G, M_{t_1}) \to M_{t_2}$ ($A_{t_1} \to A_{t_2}$ and $(G, A_{t_1}) \to A_{t_2}$), respectively.

From the results in Table 10.7, we draw a few interesting observations. First, the patterns of the transitional associations over time between the responses in the *Marijuana use* appear to be similar to those in the *Alcohol use*, whether or not the *Gender* (G) is added in the computation of the SCCRAM. At any fixed time lag $h(= t_2 - t_1)$, the magnitude of the association between the responses at two time points (t_1, t_2) becomes larger for two later time points than for two earlier time points. As the time lag h increases, the degree of the association between two time points tends to decrease, and the amount of decrement varies depending on time points. Notice that under each of the four association structures, the transitional associations at (t_1, t_2) with $t_1 = 1$ tend to be smaller than those measured at (t_1, t_2) with $2 \le t_1 \le 5$, except for $(t_1, t_2) = (1, 2)$ under $A_{t_1} \to A_{t_2}$ and $(G, A_{t_1}) \to A_{t_2}$. Second, the magnitudes of transitional associations between the responses at all pairs of time-points in each outcome variable appear to be non-negligible regardless of the inclusion of G, which are confirmed by the permutation tests. That is, the adjusted permutation p-values for each of the four groups of multiple hypotheses for all (t_1, t_2)s, $H_0 : \rho^{2*}_{(A_{t_1}\to A_{t_2})} = 0$, $H_0 : \rho^{2*}_{(M_{t_1}\to M_{t_2})} = 0$, $H_0 : \rho^{2*}_{((G,A_{t_1})\to A_{t_2})} =$

0, and $H_0 : \rho^{2*}_{((G,,M_{t_1})\rightarrow M_{t_2})} = 0$ are all less than 0.0005, 0.005, 0.005, and 0.005, respectively. Third, it seems that *Gender* has no major contribution to the magnitudes of transitional associations between the responses, and we confirm this observation by permutation tests. In other words, the adjusted permutation p-values for $H_0 : \rho^{2*}_{((G,A_{t_1})\rightarrow A_{t_2})} = \rho^{2*}_{(A_{t_1}\rightarrow A_{t_2})}$ and $H_0 : \rho^{2*}_{((G,M_{t_1})\rightarrow M_{t_2})} = \rho^{2*}_{(M_{t_1}\rightarrow M_{t_2})}$ for all (t_1, t_2)s are all larger than 0.4 (most of them are close to 1) and all larger than 0.9, respectively.

For the third research concern, "Dynamic regression association between *Marijuana use* and *Alcohol use*," the corresponding outputs are given in Table 10.8. Note that the SCCRAM was applied to four types of contingency tables for every pair of (t_1, t_2) where $1 \leq t_1 \leq t_2 \leq 5$, properly reflecting the four regression association structures:

$$G \times A_{t_1} \times M_{t_2} \text{ for } (G, A_{t_1}) \rightarrow M_{t_2}, \quad G \times M_{t_1} \times A_{t_2} \text{ for } (G, M_{t_1}) \rightarrow A_{t_2},$$
$$A_{t_1} \times M_{t_2} \quad \text{for } A_{t_1} \rightarrow M_{t_2}, \quad M_{t_1} \times A_{t_2} \quad \text{for } M_{t_1} \rightarrow A_{t_2}.$$

Based on the results in Table 10.8, we have several key findings. First, when $t_1 = t_2$, the association between *Marijuana use* and *Alcohol use* under each of the four regression structures shows a nonlinear behavior in that it nonlinearly increases until the 4th year and then decreases at the 5th year, down to lower than the association at the 3rd year. The magnitudes of the association under $(G, A_{t_1}) \rightarrow M_{t_2}$ and $A_{t_1} \rightarrow M_{t_2}$ are numerically similar to those under $(G, M_{t_1}) \rightarrow A_{t_2}$ and $M_{t_1} \rightarrow A_{t_2}$, respectively, which may indicate symmetric association. Second, when $t_1 < t_2$, the association under $(G, A_{t_1}) \rightarrow M_{t_2}$ and $A_{t_1} \rightarrow M_{t_2}$ tends to be larger than the association under $(G, M_{t_1}) \rightarrow A_{t_2}$ and $M_{t_1} \rightarrow A_{t_2}$, respectively, at most pairs of time points, except for two pairs of $(t_1, t_2) = (2, 4), (2, 5)$. Third, as the time lag $h(= t_2 - t_1)$ changes from 0 to 4, changes in the association differ by the association structure and the time points. Under the regression structures $M_{t_1} \rightarrow A_{t_2}$ and $(G, M_{t_1}) \rightarrow A_{t_2}$, when the time lag h increases, the association either decreases sharply and stay flat for $t_1 = 1$ or gradually decreases for $t_1 = (3, 4)$, except for $t_1 = 2$ at which the association almost stays constant from the beginning. On the other hand, under the other two structures $A_{t_1} \rightarrow M_{t_2}$ and $(G, A_{t_1}) \rightarrow M_{t_2}$, the association either slightly increases and decreases for $t_1 = (1, 2)$ or moderately decreases for $t_1 = (3, 4)$.

Fourth, in order to determine whether or not the association between *Marijuana use* and *Alcohol use* at each pair of time points under each association structure is statistically different from zero, we carried out the permutation tests for the four groups of multiple hypotheses for all (t_1, t_2)s, $H_0 : \rho^{2*}_{(M_{t_1}\rightarrow A_{t_2})} = 0$, $H_0 : \rho^{2*}_{(A_{t_1}\rightarrow M_{t_2})} = 0$, $H_0 : \rho^{2*}_{((G,M_{t_1})\rightarrow A_{t_2})} = 0$, and $H_0 : \rho^{2*}_{((G,A_{t_1})\rightarrow M_{t_2})} = 0$. The values of the SCCRAMs that are statistically significant (adjusted permutation p-values $< \alpha = 0.01$) are highlighted with red color in Table 10.8a and 10.8b. Note that the adjusted p-values for significant cases in $H_0 : \rho^{2*}_{(M_{t_1}\rightarrow A_{t_2})} = 0$,

Table 10.8 Dynamic (time-dependent) association between *Marijuana use* and *Alcohol use* without (Table 10.8a)/with (Table 10.8b) accounting for *Gender*

(a) $\rho^{2*}_{(M_{t_1}\to A_{t_2})}$ (upper triangular) and $\rho^{2*}_{(A_{t_1}\to M_{t_2})}$ (lower triangular) where $1 \le t_1 < t_2 \le 5$, and $\rho^{2*}_{(M_{t_1}\to A_{t_2})}$ (upper main diagonal) and $\rho^{2*}_{(A_{t_1}\to M_{t_2})}$ (lower main diagonal) where $t_1 = t_2 = 1,\ldots,5$

A_{t_2}/A_{t_1} \ M_{t_1}/M_{t_2}	1	2	3	4	5
1	0.12331 0.12249	0.01970	0.03723	0.040641	0.03619
2	0.14404	0.14685 0.15099	0.13490	0.15417	0.14010
3	0.12105	0.19023	0.28216 0.30772	0.20233	0.11160
4	0.07402	0.07025	0.21219	0.34669 0.38265	0.20098
5	0.04224	0.07117	0.17931	0.26394	0.25321 0.24435

(b) $\rho^{2*}_{((G,M_{t_1})\to A_{t_2})}$ (upper triangular) and $\rho^{2*}_{((G,A_{t_1})\to M_{t_2})}$ (lower triangular) where $1 \le t_1 < t_2 \le 5$, and $\rho^{2*}_{((G,M_{t_1})\to A_{t_2})}$ (upper main diagonal) and $\rho^{2*}_{((G,A_{t_1})\to M_{t_2})}$ (lower main diagonal) where $t_1 = t_2 = 1,\ldots,5$

A_{t_2}/A_{t_1} \ M_{t_1}/M_{t_2}	1	2	3	4	5
1	0.14387 0.12516	0.05248	0.06274	0.04750	0.04065
2	0.15882	0.16436 0.17679	0.14377	0.15531	0.14289
3	0.12989	0.19845	0.28711 0.30892	0.20632	0.13836
4	0.08558	0.07651	0.21370	0.34884 0.39400	0.20112
5	0.04862	0.07432	0.18356	0.26860	0.26348 0.25951

$H_0 : \rho^{2*}_{(A_{t_1}\to M_{t_2})} = 0$, $H_0 : \rho^{2*}_{((G,M_{t_1})\to A_{t_2})} = 0$, and $H_0 : \rho^{2*}_{((G,A_{t_1})\to M_{t_2})} = 0$ are all less than 0.0005, 0.005, 0.0005, and 0.005, respectively. We see that *Marijuana use* and *Alcohol use* are statistically associated with each other at most pairs of (t_1, t_2), except for a few cases where the time point t_1 is earlier ($t_1 = 1$ or $2 < t_2$). Finally, it appears that the contribution of G to the association between *Marijuana use* and *Alcohol use* at any pair of time points (t_1, t_2) is not substantial. This is because the adjusted permutation p-values for $H_0 : \rho^{2*}_{((G,A_{t_1})\to A_{t_2})} = \rho^{2*}_{(A_{t_1}\to A_{t_2})}$ and $H_0 : \rho^{2*}_{((G,A_{t_1})\to M_{t_2})} = \rho^{2*}_{(A_{t_1}\to M_{t_2})}$ for any (t_1, t_2) are all larger than 0.8 and are all close to 1, respectively.

10.5 Conclusion

In this chapter, we have proposed the application of the *Scaled Checkerboard Copula Regression Association Measure (SCCRAM)* developed by Wei and Kim (2021) to facilitate learning of the dependence structure in longitudinal ordinal categorical data. In order to effectively detect and quantify (symmetric/asymmetric) associations between repeatedly measured ordinal outcomes and their relationship with categorical covariates, we have provided a set of SCCRAM-based strategies that take into account types of time dependence (trend data and panel data), data format (the wide table format and the long table format), regression structures, and resampling methods (bootstrap method and permutation tests). We then have demonstrated the utility of the proposed strategies using two real-world longitudinal datasets.

The proposed SCCRAM-based methodology is a model-free and data-driven approach in that parametric assumption on the dependence structures of longitudinal ordinal responses is not needed and statistical inference is based on plug-in estimation and resampling. Thus, the proposed method is able to provide information on dependence structures that may be useful for the purpose of explanatory or predictive inference.

For the analysis of panel data in this chapter, we have focused on turnover tables with one or two outcome variables at two time points t_1 (earlier) and t_2 (later). As a valuable extension of current work, one can consider using higher-dimensional turnover tables with one or more outcome variables at current or several previous time points as long as these tables are not sparse. For example, as an extension of Eq. (10.11), one may explore the regression association between an outcome variable measured at s-th time point and the same variable measured during previous $s-1$ time periods:

$$(W_{m,t_1}, \ldots, W_{m,t_{s-1}}) \rightarrow W_{m,t_s},$$

where $1 \leq t_1 < \cdots < t_{s-1} < t_s \leq T$. One can also extend Eq. (10.13) in a way to investigate transitive dependency relationship of a putative dependent variable at s-th time point with the same variable measured at several previous time points and putative independent variables measured at current and several previous time points:

$$(W_{m_1,t_1}, \ldots, W_{m_1,t_{s-1}}, W_{m_1,t_s}, W_{m_2,t_1}, \ldots, W_{m_2,t_{s-1}}) \rightarrow W_{m_2,t_s}.$$

References

Agresti, A. (2010). *Analysis of ordinal categorical data* (2nd ed.). Wiley.

Bartolucci, F., Farcomeni, A., & Pennoni, F. (2012). *Latent Markov Models for Longitudinal Data*. CRC Press.

Benjamini, Y., & Hochberg, Y. (1995). Controlling the false discovery rate: A practical and powerful approach to multiple testing. *Journal of the Royal Statistical Society, Series B, 57*, 289–300.

Bergsma, W. P., & Andries van der Ark, L. (2018). *CMM: Categorical marginal models*. R package version 0.12.

Bergsma, W. P., Croon, M. A., & Hagenaars, J. A. P. (2009). *Marginal models for dependent, clustered, and longitudinal categorical data*. Springer.

Berry, K. J., Johnston, J. E., & Mielke, J. P. W. (2018). *The measurement of association: A permutation statistical approach*. Springer.

Carley, H. (2002). Maximum and minimum extensions of finite subcopulas. *Communications in Statistics – Theory and Methods, 31*(12), 2151–2166.

de Amo, E., Carrillo, M,. D., & FernáNdez-SáNchez, J. (2012). Characterization of all copulas associated with non-continuous random variables. *Fuzzy Sets and Systems, 191*, 103–112.

Diggle, P. J., Heagerty, P., Liang, K., Zeger, S. (2002). *Analysis of longitudinal data* (2nd ed.). Oxford statistical science series.

Doran, G., Muandet, K., Zhang, K., & Schölkopf, B. (2014). A permutation-based kernel conditional independence test. In *Proceedings of the 30th Conference on Uncertainty in Artificial Intelligence (UAI2014)* (pp. 132–141). AUAI Press.

Efron, B., & Tibshirani, R. (1993). *An introduction to the bootstrap*. Chapman & Hall.

Elliott, D. S., Huizinga, D., & Menard, S. (2012). *Multiple problem youth: Delinquency, substance use, and mental health problems*. Springer.

Friendly, M. (1994). Mosaic displays for multi-way contingency tables. *Journal of the American Statistical Association, 89*, 190–200.

Genest, C., Nešlehová, J. (2007). A primer on copulas for count data. *ASTIN Bulletin: The Journal of the IAA, 37*(2), 475–515.

Genest, C., Nešlehová, J. G., & Rémillard, B. (2014). On the empirical multilinear copula process for count data. *Bernoulli, 20*(3), 1344–1371.

Genest, C., Nešlehová, J. G., & Rémillard, B. (2017). Asymptotic behavior of the empirical multilinear copula process under broad conditions. *Journal of Multivariate Analysis, 159*, 82–110.

Good, P. (2000). *Permutation tests: A practical guide to resampling methods for testing hypotheses*. Springer.

Goodman, L.A. (1979). Simple models for the analysis of association in cross-classifications having ordered categories. *Journal of the American Statistical Association, 74*, 537–552.

Goodman, L.A. (1983). The analysis of dependence in cross-classifications having ordered categories, using log-linear models for frequencies and log-linear models for odds. *Biometrics, 39*, 149–160.

Goodman, L. A., & Kruskal, W. H. (1954). Measures of association for cross classifications. *Journal of the American statistical association, 49*(268), 732–764.

Hagenaars, J. A. P. (1990). *Categorical longitudinal data: Log-linear panel, trend, and cohort analysis*. Sage.

Hawkes, R. K. (1971). The multivariate analysis of ordinal measures. *American Journal of Sociology, 76*, 908–926.

Heagerty, P. J., & Zeger, S. L. (1998). Lorelogram: A regression approach to exploring dependence in longitudinal categorical responses. *Journal of the American Statistical Association, 93*, 150–162.

Jones, M., Hockey, R., Mishra, G. D., & Dobson, A. (2014). Visualising and modelling changes in categorical variables in longitudinal studies. *BMC Medical Research Methodology, 14*, 1–8.

Kendall, M. G. (1938). A new measure of rank correlation. *Biometrika, 30*(1/2), 81–93.

Kendall, M. G. (1945). The treatment of ties in ranking problems. *Biometrika, 33*, 239–251.

Kendall, M. G. (1948). *Rank correlation methods*. C. Griffin.

Kosara, R., Bendix, F., & Hauser, H. (2006). Parallel sets: interactive exploration and visual analysis of categorical data. *IEEE Transactions on Visualization and Computer Graphics, 12*, 558–568.

Landis, J. R., Heyman, E. R., & Koch, G. G. (1978). Average partial association in three-way contingency tables: A review and discussion of alternative tests. *International Statistical Review, 46*, 237–254.

Liu, X. (2015). Introduction to longitudinal data analysis in psychiatric research. *Shanghai Arch Psychiatry, 27*, 256–259.

Mantel, N. (1963). Chi-square tests with one degree of freedom: extensions of the Mantel-Haenszel procedure. *Journal of the American Statistical Association, 58*, 690–700.

McCullagh, P. (1980). Regression models for ordinal data. *Journal of the Royal Statistical Society, Series B, 42*, 109–142.

Metsämuuronen, J. (2021). Directional nature of goodmankruskal gamma and some consequences: Identity of goodmankruskal gamma and somers delta, and their connection to jonckheereterpstra test statistic. *Behaviormetrika, 48*, 283–307.

Molenberghs, G., & Verbeke, G. (2005). *Models for discrete longitudinal data*. Springer.

Nešlehová, J. (2007). On rank correlation measures for non-continuous random variables. *Journal of Multivariate Analysis, 98*(3), 544–567.

Nelsen, R. B. (2006). *An introduction to copulas*. Springer.

Rüschendorf, L. (2009). On the distributional transform, Sklar's theorem, and the empirical copula process. *Journal of Statistical Planning and Inference, 139*, 3921–3927.

Schweizer, B., & Sklar, A. (1974). Operations on distribution functions not derivable from operations on random variables. *Studia Mathematica, 52*(1), 43–52.

Sklar, A. (1959). Fonctions de répartition à*n* dimensions et leurs marges. *Publications de l'Institut de Statistique de L'Université de Paris, 8*, 229–231.

Somers, R. H. (1962). A new asymmetric measure of association for ordinal variables. *American Sociological Review* 799–811.

Stuart, A. (1953). The estimation and comparison of strengths of association in contingency tables. *Biometrika, 40*, 105–110.

Sutradhar, B. C. (2014). *Longitudinal categorical data analysis*. Springer.

Van der Eijk, C. (1980). *Longitudinaal enquete onderzoek: mogelijkheden en problemen*. Universiteit van Amsterdam.

Wei, Z., & Kim, D. (2021). On exploratory analytic method for multi-way contingency tables with an ordinal response variable and categorical explanatory variables. *Journal of Multivariate Analysis, 186*, 104793.

Chapter 11
Bayesian Network for Discovering the Potential Causal Structure in Observational Data

Cody S. Ding

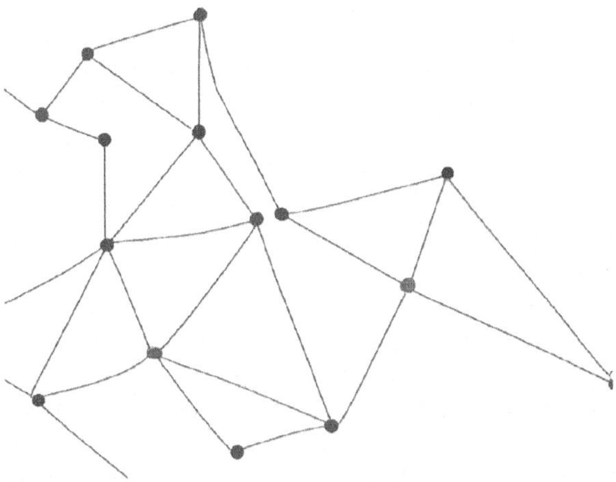

The new computational technologies (e.g., AI and machine learning) and newly found avenues have dramatically expanded analytical possibilities for scientific research, including social sciences (Ahmed et al., 2017). While traditional social scientific analysis approaches, such as latent variable modeling or analysis of variance, are still at the core of data analytical choice, network analysis and predictive modeling are two areas of machine learning affected by new computational technologies, which are known as computational social science (Lazer et al., 2009). King (2011) stated that the data-driven scientific approach gradually increases

C. S. Ding (✉)
University of Missouri-St. Louis, St. Louis, MO, USA
e-mail: dingc@umsl.edu

M. Stemmler et al. (eds.), *Dependent Data in Social Sciences Research*,
https://doi.org/10.1007/978-3-031-56318-8_11

within social science, brought by interdisciplinary research teams. Many new data analytics used by computational social researchers are from computer science or technology, and machine learning is one such type of data-driven technique. Since many machine learning algorithms are available to computational social researchers, the choice of a machine learning method will depend on the research interest. For example, the Bayesian network is one of these machine learning algorithms when the research interest is to identify the potential causal structure in observational data. It is an effective machine learning method used to express uncertainty and probability reasoning of a system in the field of artificial intelligence (Castillo et al., 1997; Pearl, 1988, 2000). Such an explorative approach allows us to build causal psychopathological models (Danks et al., 2010) to conduct bidirectional uncertainty analysis for prediction and diagnosis, for instance.

Human beings know causal effect relations, whose typical form is "if x happens, then y will happen." We could predict some future events from one or more current occurrences. However, identifying causal relations in the observational data is more complicated; for example, many possible causes are in the background, which is not captured in our data. Many theoretical articles and books, past and present, have been dedicated to the discussion of causal relations (e.g., Cox & Wermuth, 2004; Frosini, 2006; Pearl, 2000), and our interest here is not a theoretical or conceptual discussion of causal relations but rather a more data-driven-oriented perspective. Given that much information in social science comes from observational data instead of experiments, our focus is how we could identify potential causal structures, acknowledgingly preliminary, from the observational data using Bayesian network analysis. Borsuk et al. (2004) suggested "A Bayesian network consists of a graphical structure and a probabilistic description of the relationships among variables in a system. The graphical structure explicitly represents cause-and-effect assumptions that allow a complex causal chain linking actions to outcomes to be factored into an articulated series of conditional relationships" (p. 219). The Bayesian network's core is conditional independence or dependence, and Bayesian networks encode these conditional independencies with direct acyclic graphs (DAG) structure (Pe'er, 2005). The DAG structure seems to be the most suitable method to explore possibilities concerning cause (Frosini, 2006), *not proof of causality*, and such a graph was first introduced by Wright (1921) as path analysis.

A DAG structure is something like Fig. 11.1, representing causal relationships between four variables and being defined as a pair $G = (V, E)$, with V being a set of nodes (variables) and E being a set of edges or arrows. A direct edge between variables indicates some direct causal association between the variables involved. For example, *felt helplessness -> depressed* indicates how being depressed happens; in probability terms, $p(\text{depressed} \mid \text{felt helplessness}) \neq p(\text{felt helplessness} \mid \text{depressed})$. In DAG terminology, we say that *felt helplessness* is a parent (ancestor) of *depressed* or that *depressed* is a child (descendent) of *felt helplessness*. Using terminology from path analysis or structural equation modeling, *felt helplessness* is an exogenous variable, and *depressed* is an endogenous variable. Loosely speaking, Fig. 11.1 represents the influence of two exogenous variables on two endogenous variables. The DAG is considered a causal structure in that if x -> y,

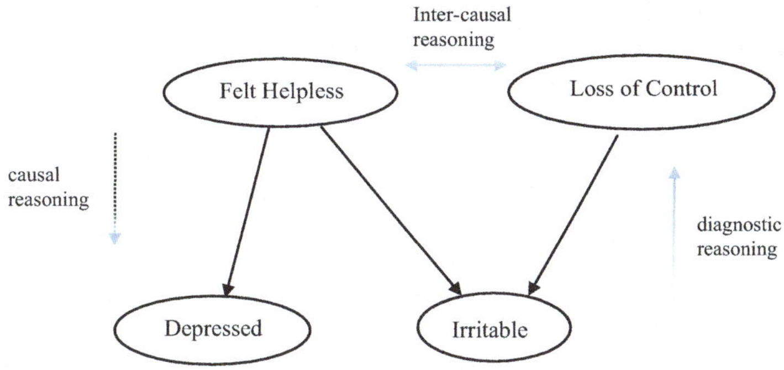

Fig. 11.1 Bayesian networks with deductive (causal), abductive (diagnostic), and inter-causal (explaining away) reasoning

then interventions exist on x that will directly change the distribution of y if all other variables are fixed (Pearl, 2000).

In addition, a DAG like Fig. 11.1 is characterized by the factorization of the conjoint density according to the presence and direction of the edges. For example, the density relating to the DAG in Fig. 11.1 may be written as $p(\text{A, B, C, D}) = p(\text{A})p(\text{B})p(\text{C} \mid \text{A})p(\text{D} \mid \text{A, B})$. In other words, the probability densities of the respective variable are conditional on the values of its parents. The assumptions that relate a probability density to causal relations include (1) causal Markov condition (each variable is independent of its non-descendant conditional on its parents), (2) causal faithfulness (Bayesian network is faithful if and only if the conditional independence relations are exactly those produced by the factorization of joint density regarding G), and (3) casual sufficiency (no unmeasured common causes and no unmeasured selection variables) (Spirtes & Zhang, 2016). The Markov condition can be considered a generalization of an experimental inferential principle: fixing variables that directly influence Y, and we can separate more remote possible causes that can only influence Y via more direct causes (Glymour et al., 2019). The direct edge X –> Y in DAG indicates the experimental claim that if all other variables in the graph are held fixed or controlled for, Y will covary if X is made to vary, but not vice versus, and this condition can be computed from the probability density (Spirtes et al., 2001).

The causal relationships expressed by direct edges in Fig. 11.1 could correspond to the path model or structural model in the structural equation modeling (SEM):

$$\text{C} = f_{\text{C}}(\text{A})$$

$$\text{D} = f_{\text{D}}(\text{A, B})$$

As Spirtes and Zhang (2016) indicated, each path model or SEM is associated with a DAG representing the causal structure and the form of the structural equations. If there are no direct cycles and no double-headed arrows in the path model, the graph is a DAG. DAG is generally a direct graphic causal model (DGCM) type, which can also be represented as an SEM or functional causal model (Glymour et al., 2019). It is worth noting that SEM is not designed to discover causal structures like DAG. However, given the underlying causal structure identified, SEM can estimate the effect strength and assess model fit (Shen et al., 2020). In other words, assuming a causal structure based on prior knowledge or data, SEM can be used as a confirmatory method to test the identified causal structure from search algorithms.

It can be argued that actual casual relations can only be found using randomized experiments. Although this may be true, such a method may be, in many cases, too expensive or impossible. In addition, the threats to internal validity may not be completely ruled out in a field setting. In the lab setting, relevant but extraneous variables are controlled in an experiment, limiting the findings' ecological usefulness, particularly in social science. Therefore, given the theoretical stipulations in the literature cited here, identifying causal information by analyzing observational data using causal discovery algorithms (e.g., constrained optimization techniques) might be a viable alternative, also known as causal discovery (Glymour et al., 2019). One type of causal discovery algorithm is constraint-based and score-based methods, exploiting conditional independence relationships in the data or maximizing the probability of causal structure given the data. Typical constraint-based algorithms include Peter-Clark (PC) and Fast Causal Inference (FCI) (Spirtes et al., 2000). There are other variants of PC algorithms. PC assumes no confounders (all the confounding variables are measured in the dataset), while FCI does not have this assumption. Both are asymptotically correct and can handle many types of data and causal relations based on conditional independence test methods (Glymour et al., 2019). However, they may not differentiate the equivalent classes (causal structures with the same conditional independence). In contrast, algorithms based on linear non-Gaussian models (Shimizu et al., 2006) or nonlinear models (Hoyer et al., 2009) can differentiate different DAGs in the same equivalent class.

Score-based learning algorithms aim to find the causal structure by optimizing the score function in the space of possible structures. In other words, identifying the structure that best fits the data can be defined as an optimization problem where the score is to be optimized. The typical score used to determine the best causal structure includes the Bayesian Information Criterion (BIC) score, the Minimum Description Length (MDL) score, the Akaike Information Criterion (AIC) score, and the Bayesian Dirichlet equivalent uniform (BDeu) score. Among the score-based algorithms, Greedy Equivalent Search (GES) (Chickering, 2002) and its variant are well-known methods that directly search over the space of equivalent classes to maximize a score that reflects how well the structure fits the data. Other algorithms include simulated annealing, genetic algorithm, and tabu search (Chickering, 2002).[1] In contrast, the Hill-Climbing algorithm and its variants (e.g., RSMAX2, max-min Hill-Climbing) are hybrid algorithms that first reduce the space

[1] There are many score-based search algorithms and we did not list all of them here.

of possible DAG using a constraint-based approach and then find the optimal DAG using a score-based approach in the restricted space.

It should be noted that the literature on causal discovery modeling and search algorithms is voluminous.[2] We do not intend to provide a thorough review, and instead, we highlight some key theoretical perspectives on Bayesian network analysis as a causal discovery tool. In addition, some articles provide a comprehensive review of search algorithms (e.g., Russell & Norvig, 2009b), while others review some specific set of search algorithms (e.g., Glymour et al., 2019; Shen et al., 2020). Since many search algorithms exist, deciding which search algorithm to use in research can be challenging. Scutari et al. (2018) found that constraint-based algorithms are more accurate than score-based algorithms in small samples, and both are as accurate as hybrid algorithms. However, score-based algorithms tend to yield the largest network. In comparing FCI (constraint-based) with Fast Greedy Equivalent search (FGES, Score-based), Shen et al. (2020) found that the performance of FGES was higher and more stable than the FCI in their study, which was due to FCI's incorrect independence tests introduced by selection bias or data artifacts. In contrast, score-based algorithms consider the likelihood of the global structure, which localizes these biases or artifacts. The trade-off is that FCI relaxes the assumption of no unmeasured confounders, which is essential if we try to identify unmeasured confounders or unfounded causal structures.

This chapter uses a real-world example to demonstrate the Bayesian network as a causal structure discovery tool in observational data. Three steps were involved. The first step was the causal structure discovery, which involved identifying causal structures using search algorithms. In the second step, we used SEM as a confirmatory tool to assess the model-data fit of the identified causal structures. We used SEM to assess the model fit because the Bayesian network algorithms did not provide model fit indices that were useful to assess model fit directly. In the third step, we made inferences on identified Bayesian network structures based on *conditional probability* (CP) and *maximum* a posteriori (MAP) queries (e.g., Neapolitan, 2004; Yuan et al., 2004). The general idea of the inference is that if we have some *evidence* (i.e., we know the values of some particular variables), we want to look into the probability of some other *events* involved conditional on the evidence (i.e., the value) we have on this or these particular variables. In contrast, the goal of a MAP query is to find the highest probability value of the variable (or a set of variables) based on some evidence we have for a particular variable (or set of variables). Alternatively, a MAP query can also be used for a classic prediction problem if the evidence is used to predict the possible outcomes for individuals.

[2] Readers can also consult references such as Korb and Nicholson (2010); Russell and Norvig (2009a); Friedman et al. (1997); or Neapolitan (2004) to understand the Bayesian network.

11.1 Example: Effects of Social Adversity on Posttraumatic Stress and Mood Symptoms

Most individuals typically encounter adversities or stressors within their daily social interactions. While some can constructively manage stress, others may be psychologically bothered, leading to the most common mental disorders of anxiety and depression (Clark & Beck, 2010; Sartorius et al., 1996), in addition to fear or even suicide. One factor warranting further exploration is the etiological role of social adversities in mental distress. It is particularly fundamental to understand whether these daily feelings of stress can be perceived as traumatic and associated with posttraumatic stress symptoms as well as mood and anxiety symptoms.

Although typical social adversities are less intense, they can be very emotional and personal, particularly when prolonged, affecting one's sense of control and ability to cope with stress. One reason social adversities may cause stress is that our thoughts and behaviors become negatively captured by three essential elements: narrowing of attention, perceived lack of control, and change in emotional states (Kessler, 2015). Although we often experience stress from social interaction that commands our attention, *negative capture* involves more than just typical attention. It alters our emotions by evoking memory, imagination, fear, or anger. For example, one may think that a person calling our name or saying things in a particular tone is not as it appears. Instead, it could indicate a criticism or personal slight, leaving us feeling insecurity or anxiety, which becomes our predominant thought. This preoccupation with thought is an example of negative capture.

There are many examples of such capture situations in our daily lives, as indicated by Kessler (2015), all of which could potentially lead to significant psychological suffering. Previous studies have found that psychological maltreatment is related to posttraumatic stress symptoms (e.g., Caston & Frazier, 2013; Gilman et al., 2013; Hodgdon et al., 2019; McIlveen et al., 2019). Thus, viewing from the lens of negative capture, we expect that social adversity may make us overly focus attention on the negative aspects of an event, which further reinforces our negative schema. This can result in feelings of helplessness, significantly altering their emotional state. In this example, we want to explore how social adversities, posttraumatic stress symptoms, and mood and anxiety symptoms form a complex ecological system using Bayesian network analysis.

Method

Participants

The data used in this example included 478 adults recruited online from several communities in China. Participants were provided a link to Survey Monkey's online software program and asked to complete the survey voluntarily. The age of the participants ranged from 20 to 63 ($M = 30.27$, $SD = 7.46$), and 42% were males. Among them, 19% had an education in high school or below, 61% had a college education, and 20% had a post-college education. Most participants reported having a job (74%), 18% were students, and 8% were unemployed.

Measures

Perceived Chronic Social Adversity Scale-Revised Perceived social adversity events were assessed using the Perceived Chronic Social Adversity Scale-Revised (PCSAS-R), which consists of 30 items that assess social adversity events commonly experienced in daily life (see Appendix A). Four aspects of each social adversity event were assessed: (a) whether an event occurred (0 = no, 1 = yes); (b) whether the event was perceived to occur frequently (i.e., chronicity on a scale of 0 = do not feel it happens frequently and 1 = feel it happens frequently); (c) whether the event felt traumatic (i.e., traumaticity on a scale of 0 = do not feel it to be traumatic and 1 = feel it to be traumatic); and (d) perceived level of distress caused by the event (i.e., event distress on a scale of 1 = not distressed to 5 = very much distressed). Based on the response to each aspect of these event items, we computed four scores: (1) event occurrence, (2) event chronicity, (3) event traumaticity, and (4) event distress. The scores ranged from 0 to 30 for event occurrence, chronicity, and traumaticity; for event distress, the score ranged from 1 to 5. A high score would reflect a higher level of (1) events experienced, (2) event chronicity, (3) event traumaticity, and (4) event distress.

PTSD Checklist for DSM-5 (PCL-5) (Weathers et al., 2013). The PCL-5 is a 20-item self-report measure that assesses the four types of DSM-5 PTSD symptoms: intrusion (e.g., repeated, disturbing, and unwanted memories of the stressful experience); avoidance behavior (e.g., avoiding memories, thoughts, or feelings related to the stressful experience); negative alterations in cognitions and mood (e.g., having strong negative beliefs about yourself, other people, or the world); and alterations in arousal and reactivity (i.e., irritable behavior, angry outbursts, or acting aggressively). Participants rated each item on a 4-point scale, ranging from 0 (not at all) to 4 (extremely). In this study, Cronbach's alpha was 0.86, 0.84, 0.90, and 0.89 for the four subscales. An average score was calculated for each subscale, with a high score indicating higher PTSD symptoms.

Shapiro control inventory (SCI) (Shapiro, 1994). The SCI is a test that provides a multi-dimensional control profile of an individual. It has nine subscales; however, the current study used only General Domain Negative Sense of Control (5 items) to measure individual perceived controllability. Cronbach's alpha for a negative sense of control scale was 0.79 in the current sample.

Short adaptation of the mood and anxiety symptoms questionnaire (MASQ-D30) (Wardenaar et al., 2010). The MASQ-D30 has three subscales: General Distress (10 items), Anxious Arousal (10 items), and Anhedonic Depression (10 items). Items are rated on a 4-point scale, from 1 (not at all) to 4 (always). An average score was calculated for each subscale, with a high score indicating higher mood and anxiety symptoms. For the three subscales, Cronbach's alpha was 0.81, 0.80, and 0.79, respectively.

Analysis Strategies

Causal structure discovery In this example, we used a score-based algorithm, Hill-Climbing (HC) with BIC score, to perform structure learning. Then, SEM path analysis was used as a supplementary analysis to assess the model fit. The model fit indices used included χ^2, CFI (>.90), TLI (>.90), RMSEA (<.06), and SRMR (<.08). As mentioned previously, one reason for doing so is that although search algorithms are based on score functions, the score function cannot be used alone to assess model fit. However, it should be noted that since search algorithms are designed based on different statistical criteria, it makes more sense to refer to an optimal network as "a best" instead of "the best" (Chickering, 1995). Whenever "a best" network is determined, it does not matter which one is selected since the end result will still be a best network (Silander & Myllymaki, 2006).

Since the original data contained the continuous variable, we did not discretize our original variables in structure learning to preserve all the information. However, we discretized all variables before parameter learning after identifying the best causal model. This was necessary because conditional probability may work best with discrete variables due to computational complexity, and it was also easier to make inferences and predictions. Sun and Shenoy (2007) indicated that the discrete variables with binary and three values usually result in the best prediction performance in a Bayesian network. Therefore, we discretized (1) the item *event occurrence* as low (= eight or less than eight events) or high (= more than eight events) and (2) the items of *event frequency, traumaticity, and distress* in a three-value discrete state based on the score distribution, with scores in the bottom 30% qualified as low, the middle 30% to 40% as medium, and the top 30% scores as high.

Parameter learning For a given Bayesian network structure, parameter learning determines the conditional probability table (CPT) at each node of the network model. The learning algorithm used was the maximum likelihood estimate (Furlotte et al., 2014).

Model validation In addition to model fit statistics from SEM, model validation was conducted using the *K*-fold cross-validation method, in which the data are divided into *K* parts, the network structure is trained on $K - 1$ part, and the last K^{th} part is tested. This process is repeated *K* times, with a different part of the data for testing. The extreme of this method is the leave-one-out technique. In addition, we also examine the prediction accuracy using the confusion matrix and the receiver operating characteristic (ROC) curve. Typically, the ROC curve assumes that the outcome events or variables are binary. When the outcome variables are not binary, they can be changed to binary by taking a state of interest and lumping all other states into complementing that state.

Inferences from conditional probability (CP) and maximum a posteriori (MAP) queries We provided examples of inferences based on posterior probability. In addition, we examined the most likely configuration of a set of child nodes (i.e., outcome states) given evidence or observations of a set of parent nodes (i.e., a precedent state) (i.e., MAP). To solve the MAP, we used the annealed MAP algorithm[3] to find it using an approximate optimization procedure of simulated annealing, which works well in practice (Yuan et al., 2004).

We used the *R* packages *bnlearn* (Scutari, 2019) and *lavaan* (Rosseel et al., 2021) for structural learning and model assessment. For parameter learning and inferences, we used *GeNIe* 4.0 (BayesFusion, 2019).

Results

Causal Structure

The result of structure learning using the HC algorithm is shown in panel A of Fig. 11.2, with corresponding edges indicating their dependency. It was interesting to see how event occurrence, perceived traumaticity, and distress influenced PTSD stress and mood symptoms. First, for the nodes related to social adversity, *event occurrences* impacts individuals' perception of *events traumaticity*, which impacts reported *event distress.* All these three nodes impact PTSD intrusion symptoms, which in turn impact other PTSD symptoms. Second, PTSD cognitive alternation and arousal symptoms directly impact mood symptoms. Third, a sense of negative control and general anxiety symptoms influence depression and anxiety arousal symptoms. Fourth, an individual's perception of event frequency is influenced by event traumaticity, PTSD cognitive alternation, and a sense of negative control.

Another interesting finding to note is how PTSD symptoms manifested in the participants. It appeared that PTSD intrusion was a gateway that led to other PTSD symptoms such as avoidance, arousal, and alteration. In addition, a sense of negative control was a gateway that led to other mood symptoms such as anhedonic depression and anxiety arousal. Therefore, a hierarchically causal structure seemed to be social adversities -> PTSD stress symptoms -> mood symptoms. Each clustered structure seemed to have a gateway symptom, such as event occurrence, PTDS intrusion, and a sense of negative control.

Panel B of Fig. 11.2 shows the path model corresponding to the identified causal structure. The coefficients in the path model were identical to those estimated from Bayesian network analysis, and all coefficients were statistically significant at the

[3] Different algorithms can be used for estimating maximum a posteriori probability.

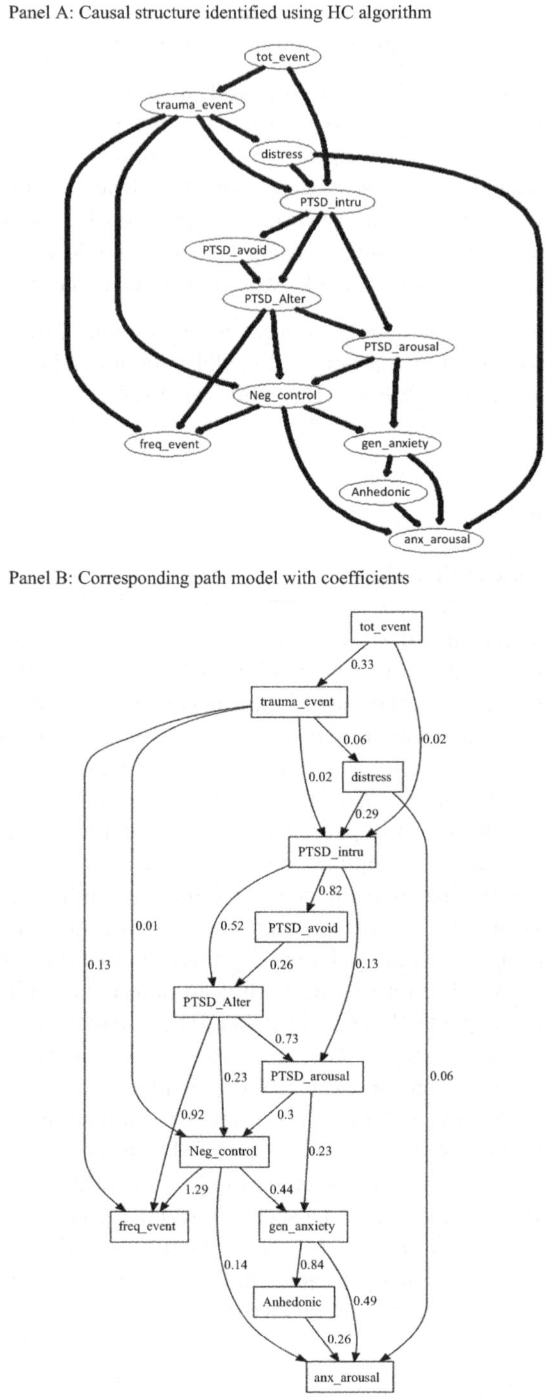

Fig. 11.2 Causal structural model of social adversity and its correlates. Panel A Causal structure identified using HC algorithm. Panel B Corresponding path model with coefficients. *Note*. The coefficients shown in the path model are identical to those estimated from the Bayesian network structure. All the coefficients are standardized and significant at the .001 level. The thick lines indicate the strength of dependencies

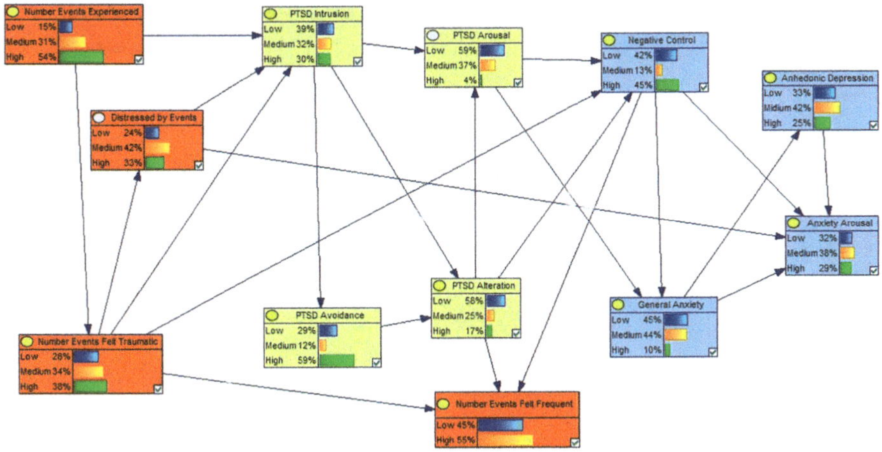

Fig. 11.3 Conditional probability distribution of identified causal structure. *Note*. The different line thicknesses show the strength of influence between the nodes that they connect. It essentially expresses the distance between various conditional probability distributions over the child node conditional on the states of the parent node. The orange-colored nodes indicate social adversity events and their perception, the yellow-colored nodes indicate PTSD symptoms, and the blue-colored nodes indicate mood and anxiety symptoms

.001 level, shown as thick lines in Panel A. The model fit indices indicated that the identified structure model fit the data well, $\chi^2_{(df = 42)} = 70.84$, $p = .004$, CFI = .99, TLI = .98, RMSEA = .038 (CI: .022 ~ .053), and SRMR = .037.

Parameter Learning and Model Validation

Given the causal structure, we performed parameter learning using discretized variables.

Figure 11.3 shows the conditional probability of the causal structure, with the different line thicknesses showing the normalized[4] weighted strength of influence between the nodes they connect (weighted strength weighs the distances by the marginal probability of the parent node). It essentially expresses the distance between various conditional probability distributions over the child node conditional on the states of the parent node.

Model validation We first performed the model validation using the *K-fold* leave-one-out method to examine the predictive accuracy of "anxiety arousal," which is at the end of downstream in the network. The overall prediction accuracy was 0.72, indicating 72% accuracy in predicting the correct state of anxiety arousal. The ROC

[4] For normalized weight, the thickest possible arc is given to that arc that has the highest strength of influence. The thicknesses of all other arcs are calculated proportionally to the thickest arc.

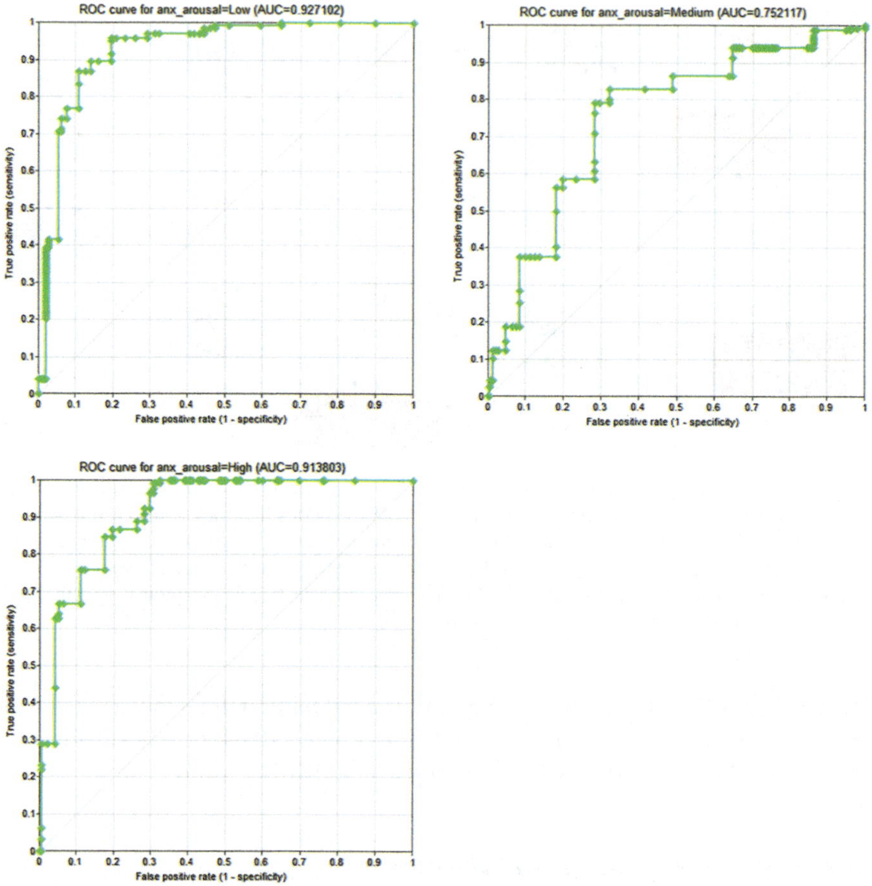

Fig. 11.4 ROC for each of the three states of anxiety arousal

curve shows that the area under the curve (AUC) was .93 for the low state, .75 for the medium state, and .91 for the high state of anxiety arousal (see Fig. 11.4).

Inferences

We mentioned that causal inferences could be made via conditional probability (CP) and maximum a posteriori (MAP) queries based on posterior probability. We must select a target node or a set of target nodes to make inferences via conditional probability. Given a piece of evidence, what would be the posterior probability distribution of the target node? This process can be considered causal probability reasoning, including diagnosis and predicting the posterior probability of state for a given node after specific evidence of a specific source node is provided. Using the posterior probabilities to infer the results from the cause (predicting) and the cause from the result (diagnosis) is often called causal inference. Specifically, we can input the evidence of a particular node (i.e., evidence variable) so that it becomes a problem to solve the posterior probabilities of other nodes with the known status of the evidence variables.

Predicting We could select any nodes to be our target nodes in the Bayesian network, although the space limit did not allow us to show the results related to all possible target nodes.

For example, we can select four nodes of PTSD symptoms as target nodes since they involve Markov blanket connection, linking social adversity to PTSD symptoms. Panel A in Fig. 11.5 shows this Markov blanket connection. Panel B of Fig. 11.5 shows the probability changes of PTSD symptoms and mood symptoms if we have evidence of the high level of social adversity regarding the occurrence, traumaticity, and distress (red-colored nodes). As shown in panel B of Fig. 11.5, the probability of a high level of PTSD stress symptoms increased by 20%, except for PTSD arousal (the probability of a medium level increased by 17%), which in turn caused a probability increase of 20% in the high level of a sense of negative control. Given the evidence, these increases in the posterior probability of high level seemed to suggest that social adversities had predictive power of PTSD stress, which is consistent with the literature on trauma events on psychological functioning. In addition, it is interesting to note that the posterior probability of a high level of nodes "number events felt frequent" also increased from 55% to 74%, suggesting that an individual's perception of event chronicity is influenced by social adversity perceived as traumatic, cognitive alteration, and a sense of negative control.

Diagnosis Another important aspect of posterior probability reasoning is diagnosing a cause given a symptom. The bidirectional reasoning of the Bayesian network can calculate not only the posterior probability of symptoms given a cause but also the posterior probabilities of cause given the evidence of the symptoms. For example, after we specify a high level of "anxiety arousal" as symptom evidence for the faulty condition, Fig. 11.6 shows the diagnostic value of the nodes that ranked from the most to least informative in diagnosing the causes. Unsurprisingly, the most informative nodes directly link to anxiety arousal, but it was interesting to see that PTSD cognitive alteration and intrusion symptoms are also somewhat informative.

The most probable configuration Using the annealed MAP algorithm, we estimated MAP. It explicitly shows the most probable explanation and causal inference about the set of variables that is most likely to occur in terms of the posterior probability. For example, when we input the evidence to be "high" for event occurrence, traumaticity, and distress, Table 11.1 shows the most likely state of mood and anxiety symptoms. The findings indicated that when individuals experienced high levels of social adversity concerning total number, traumaticity, and distress, the likelihood for these individuals to experience a high level of one of the symptoms was relatively high, ranging from 43% to 66%.

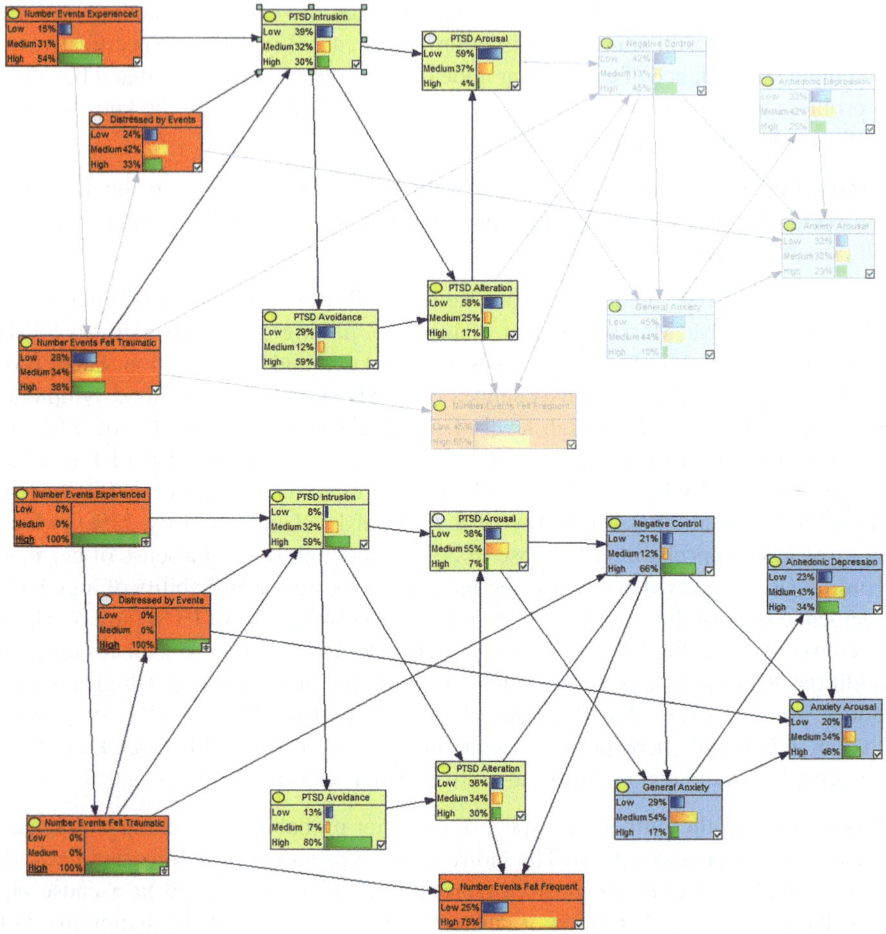

Fig. 11.5 Posterior probability distribution after evidence of high event occurrence, traumaticity, and distress. Panel A Markov blanket connection of PTSD symptoms. Panel B Posterior probability of PTSD and mood symptoms given the evidence

11.2 Discussion

The primary purpose of the current chapter was to illustrate the utility of the Bayesian network as a causal structure discovery tool. We highlighted the fundamental ideas and assumptions using the Bayesian network to identify the possible causal structure in the observational data. Bayesian network analysis is a machine learning, data-driven, and computational social science technique. Admittedly, finding actual causal structures in social science based on observational data is challenging, even with experimental data. However, this should not impede us from exploring ways to approximate reality, *not proof of causality*. For example, if a social

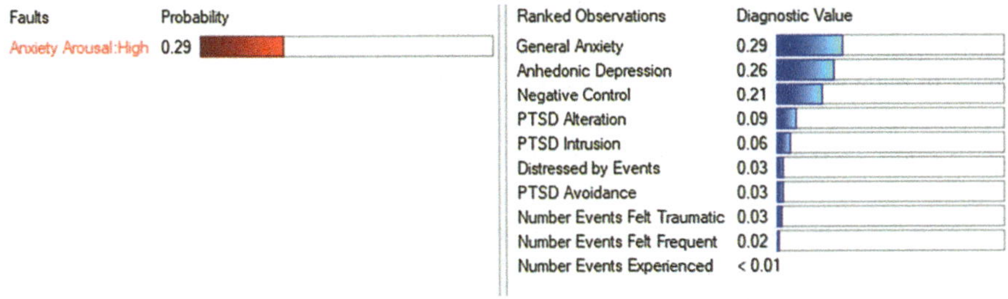

Fig. 11.6 Diagnostic information for the high level of anxiety arousal. *Note*. The left side shows high anxiety arousal as a faulty condition. The right side shows the variables ranked from the most to the least informative in diagnosing a cause

Table 11.1 Most likely configuration of mood symptom states given the state of evidence

Evidence node	Evidence state	
Number total events	High	
Felt distressed by events	High	
A number of events felt traumatic	High	
Symptom nodes	Estimated state	P(map|e)
Anhedonic depression	Medium	.43
Anxiety arousal	High	.46
General anxiety	Medium	.54
Negative control	High	.66

adversity event can strongly predict anxiety arousal, is this social adversity event a cause of anxiety arousal? Of course, other factors, such as resilience, social support, and coping strategies, also play a causal role.

Nevertheless, from a pragmatic perspective, social adversity events are one of many possible causes while holding other factors constant. Thus, we should not avoid using appropriate data analysis techniques for causal discovery, not proof of causality. In fact, we already use SEM modeling for causal discovery, such as modifying model specifications to identify the best-fitting structure model, although it is designed to be a confirmatory tool. As Shen et al. (2020) stated, dedicated causal discovery algorithms outperformed SEM in discovering the possible causal structure. Here, we highlighted the similarity between the Bayesian network and the path model, but the Bayesian network is designed to be a discovery tool, which can be complemented with the path model for network model data fit testing.

In this chapter, we demonstrated this combined approach of discovering causal structure using the Bayesian network and testing network model fit using fit indices from the path model analysis. Based on real data, the results suggested that the causal structure identified by the search algorithm fit the data well. The findings were consistent with those from previous studies indicating social adversities were associated with traumatic stresses, which in turn were associated with mood symptoms. This consistency in findings provides some confidence about the validity of the causal structure. Moreover, the Bayesian network details the possible causal pathways of how different symptoms influence each other. For example, the results show that PTSD intrusion leads to other PTSD stress symptoms. If we design an intervention, it is critical to minimize intrusion behaviors. This strategy makes sense because if no intrusion occurs, the effects of social adversities may not be detrimental.

From the substantive perspective, the findings from the identified possible causal structure seem consistent with the theory of capture (Kessler, 2015). According to this perspective, social adversity events can narrow our attention and affect our feelings and actions. Our findings indicate that as individuals experienced more adversity events, they tended to perceive them as traumatic, frequent, and distressing, which increased the likelihood of them reporting high levels of symptoms related to posttraumatic stress and mood and anxiety. Thus, our findings suggest we should not overlook or neglect the effects of these so-called regular or normal stress experiences on individuals' emotions and behaviors. One key difference between the current study and previous studies on trauma (e.g., Green, 1996; Holmes & Johnson, 2019; McIlveen et al., 2019) is that we focus on adversity events that seem less intense than trauma, such as abuse or violence. However, these everyday adversity events and their corresponding perceptions can predict posttraumatic stress symptoms, whether we consider these events potentially traumatic or not.

Limitations

Using the Bayesian network analysis to identify possible causal structure depends on some assumptions of causal relations, and these assumptions may not be met, casting uncertainty on its ability for causal discovery. It should also be stressed that the Bayesian network's purpose is not to prove causality. In addition, many search algorithms are available, and the network structure identified may differ depending on the particular search algorithm used. Thus, the current Bayesian network can only be viewed as "a best" solution given the estimation algorithm used in the analysis, even if the network model is deemed a good fit for the data. Thus, it is essential to view the validity of the identified causal model concerning theory, previous findings, and interpretation. Since observational data will not disappear, using causal discovery analytical tools may provide insights into human behaviors, with findings viewed in light of these circumstances.

In conclusion, using the Bayesian network analysis for causal structure discovery is legitimate, outperforming SEM in discovering the causal structure. We need to be aware of its limitations and interpret the results in the context of theory and previous findings.

Appendix A: Negative Social Events Scale

Direction

Below is a list of social events you may encounter or experience daily. Please respond to questions for each event. For each event, please be sure that you consider your experiences up to this moment. There are no right or wrong answers; you need to check the response option that you honestly feel describes you.

1. Feel abandoned.

   A. No Yes
   B. Please indicate whether you feel that this experience happens to you frequently or not.

      ① I do not feel it happens frequently. ② I feel it happens frequently.

   C. Please indicate whether you feel this experience is traumatic to you or not.

      ① I do not find it to be traumatic. ② I find it to be traumatic.

   D. How distressed do you feel by this experience?

      ① Not distressed ② A bit distressed ③ Moderately distressed ④ Clearly distressed ⑤ Very much distressed

2. Feel being unfairly treated.

   A. No Yes
   B. Please indicate whether you feel that this experience happens to you frequently or not.

      ① I do not feel it happens frequently. ② I feel it happens frequently.

   C. Please indicate whether you feel this experience is traumatic to you or not.

      ① I do not find it to be traumatic. ② I find it to be traumatic.

   D. How distressed do you feel by this experience?

      ① Not distressed ② A bit distressed ③ Moderately distressed ④ Clearly distressed ⑤ Very much distressed

3. Feel ignored or left out.

   A. No Yes
   B. Please indicate whether you feel that this experience happens to you frequently or not.

      ① I do not feel it happens frequently. ② I feel it happens frequently.

   C. Please indicate whether you feel this experience is traumatic to you or not.

① I do not find it to be traumatic. ② I find it to be traumatic.

D. How distressed do you feel by this experience?

① Not distressed ② A bit distressed ③ Moderately distressed ④ Clearly distressed ⑤ Very much distressed

4. Feel cheated or deceived.

A. No Yes
B. Please indicate whether you feel that this experience happens to you frequently or not.

① I do not feel it happens frequently. ② I feel it happens frequently.

C. Please indicate whether you feel this experience is traumatic to you or not.

① I do not find it to be traumatic. ② I find it to be traumatic.

D. How distressed do you feel by this experience?

① Not distressed ② A bit distressed ③ Moderately distressed ④ Clearly distressed ⑤ Very much distressed

5. Feel excluded or isolated.

A. No Yes
B. Please indicate whether you feel that this experience happens to you frequently or not.

① I do not feel it happens frequently. ② I feel it happens frequently.

C. Please indicate whether you feel this experience is traumatic to you or not.

① I do not find it to be traumatic. ② I find it to be traumatic.

D. How distressed do you feel by this experience?

① Not distressed ② A bit distressed ③ Moderately distressed ④ Clearly distressed ⑤ Very much distressed

6. Feel denied.
   A. No Yes
   B. Please indicate whether you feel that this experience happens to you frequently or not.

      ① I do not feel it happens frequently. ② I feel it happens frequently.

   C. Please indicate whether you feel this experience is traumatic to you or not.

      ① I do not find it to be traumatic. ② I find it to be traumatic.

   D. How distressed do you feel by this experience?

      ① Not distressed ② A bit distressed ③ Moderately distressed ④ Clearly distressed ⑤ Very much distressed

7. Feel being laughed at, teased, or humiliated.
   A. No Yes
   B. Please indicate whether you feel that this experience happens to you frequently or not.

      ① I do not feel it happens frequently. ② I feel it happens frequently.

   C. Please indicate whether you feel this experience is traumatic to you or not.

      ① I do not find it to be traumatic. ② I find it to be traumatic.

   D. How distressed do you feel by this experience?

      ① Not distressed ② A bit distressed ③ Moderately distressed ④ Clearly distressed ⑤ Very much distressed

8. Feel being belittled.
   A. No Yes
   B. Please indicate whether you feel that this experience happens to you frequently or not.

      ① I do not feel it happens frequently. ② I feel it happens frequently.

   C. Please indicate whether you feel this experience is traumatic to you or not.

      ① I do not find it to be traumatic. ② I find it to be traumatic.

   D. How distressed do you feel by this experience?

      ① Not distressed ② A bit distressed ③ Moderately distressed ④ Clearly distressed ⑤ Very much distressed

9. Being slandered.

   A. No Yes
   B. Please indicate whether you feel that this experience happens to you frequently or not.

      ① I do not feel it happens frequently. ② I feel it happens frequently.

   C. Please indicate whether you feel this experience is traumatic to you or not.

      ① I do not find it to be traumatic. ② I find it to be traumatic.

   D. How distressed do you feel by this experience?

      ① Not distressed ② A bit distressed ③ Moderately distressed ④ Clearly distressed ⑤ Very much distressed

10. Feel doubted or distrusted.

   A. No Yes
   B. Please indicate whether you feel that this experience happens to you frequently or not.

      ① I do not feel it happens frequently. ② I feel it happens frequently.

   C. Please indicate whether you feel this experience is traumatic to you or not.

      ① I do not find it to be traumatic. ② I find it to be traumatic.

   D. How distressed do you feel by this experience?

      ① Not distressed ② A bit distressed ③ Moderately distressed ④ Clearly distressed ⑤ Very much distressed

11. My privacy is violated.

   A. No Yes
   B. Please indicate whether you feel that this experience happens to you frequently or not.

      ① I do not feel it happens frequently. ② I feel it happens frequently.

   C. Please indicate whether you feel this experience is traumatic to you or not.

      ① I do not find it to be traumatic. ② I find it to be traumatic.

   D. How distressed do you feel by this experience?

      ① Not distressed ② A bit distressed ③ Moderately distressed ④ Clearly distressed ⑤ Very much distressed

12. Somebody is always trying to change my feelings and thoughts.

   A. No Yes

B. Please indicate whether you feel that this experience happens to you frequently or not.

① I do not feel it happens frequently. ② I feel it happens frequently.

C. Please indicate whether you feel this experience is traumatic to you or not.

① I do not find it to be traumatic. ② I find it to be traumatic.

D. How distressed do you feel by this experience?

① Not distressed ② A bit distressed ③ Moderately distressed ④ Clearly distressed ⑤ Very much distressed

13. My views or behaviors are challenged or questioned, even though when I feel that they are right.

A. No Yes

B. Please indicate whether you feel that this experience happens to you frequently or not.

① I do not feel it happens frequently. ② I feel it happens frequently.

C. Please indicate whether you feel this experience is traumatic to you or not.

① I do not find it to be traumatic. ② I find it to be traumatic.

D. How distressed do you feel by this experience?

① Not distressed ② A bit distressed ③ Moderately distressed ④ Clearly distressed ⑤ Very much distressed

14. My opportunity to express is constrained, such as my talk is limited or interrupted or my talking point is being changed.

A. No Yes

B. Please indicate whether you feel that this experience happens to you frequently or not.

① I do not feel it happens frequently. ② I feel it happens frequently.

C. Please indicate whether you feel this experience is traumatic to you or not.

① I do not find it to be traumatic. ② I find it to be traumatic.

D. How distressed do you feel by this experience?

① Not distressed ② A bit distressed ③ Moderately distressed ④ Clearly distressed ⑤ Very much distressed

15. The opportunity for me to make decisions is constrained.

A. No Yes

B. Please indicate whether you feel that this experience happens to you frequently or not.

① I do not feel it happens frequently. ② I feel it happens frequently.

C. Please indicate whether you feel this experience is traumatic to you or not.

① I do not find it to be traumatic. ② I find it to be traumatic.

D. How distressed do you feel by this experience?

① Not distressed ② A bit distressed ③ Moderately distressed ④ Clearly distressed ⑤ Very much distressed

16. I feel someone can always come up with reasons to turn on me.

A. No Yes
B. Please indicate whether you feel that this experience happens to you frequently or not.

① I do not feel it happens frequently. ② I feel it happens frequently.

C. Please indicate whether you feel this experience is traumatic to you or not.

① I do not find it to be traumatic. ② I find it to be traumatic.

D. How distressed do you feel by this experience?

① Not distressed ② A bit distressed ③ Moderately distressed ④ Clearly distressed ⑤ Very much distressed

17. I feel that my freedom to do social activities is restricted.

A. No Yes
B. Please indicate whether you feel that this experience happens to you frequently or not.

① I do not feel it happens frequently. ② I feel it happens frequently.

C. Please indicate whether you feel this experience is traumatic to you or not.

① I do not find it to be traumatic. ② I find it to be traumatic.

D. How distressed do you feel by this experience?

① Not distressed ② A bit distressed ③ Moderately distressed ④ Clearly distressed ⑤ Very much distressed

18. Being blamed for someone else's problem.

A. No Yes
B. Please indicate whether you feel that this experience happens to you frequently or not.

① I do not feel it happens frequently. ② I feel it happens frequently.

C. Please indicate whether you feel this experience is traumatic to you or not.

① I do not find it to be traumatic. ② I find it to be traumatic.

D. How distressed do you feel by this experience?

① Not distressed ② A bit distressed ③ Moderately distressed ④ Clearly distressed ⑤ Very much distressed

19. My past mistakes are brought up when I argue with others.

A. No Yes
B. Please indicate whether you feel that this experience happens to you frequently or not.

① I do not feel it happens frequently. ② I feel it happens frequently.

C. Please indicate whether you feel this experience is traumatic to you or not.

① I do not find it to be traumatic. ② I find it to be traumatic.

D. How distressed do you feel by this experience?

① Not distressed ② A bit distressed ③ Moderately distressed ④ Clearly distressed ⑤ Very much distressed

20. I feel that others create adverse situations for me to achieve their goals.

A. No Yes
B. Please indicate whether you feel that this experience happens to you frequently or not.

① I do not feel it happens frequently. ② I feel it happens frequently.

C. Please indicate whether you feel this experience is traumatic to you or not.

① I do not find it to be traumatic. ② I find it to be traumatic.

D. How distressed do you feel by this experience?

① Not distressed ② A bit distressed ③ Moderately distressed ④ Clearly distressed ⑤ Very much distressed

21. Others attempt to have me follow their way or ideas by creating negative consequences for me.

A. No Yes
B. Please indicate whether you feel that this experience happens to you frequently or not.

① I do not feel it happens frequently. ② I feel it happens frequently.

C. Please indicate whether you feel this experience is traumatic to you or not.

① I do not find it to be traumatic. ② I find it to be traumatic.

D. How distressed do you feel by this experience?

① Not distressed ② A bit distressed ③ Moderately distressed ④ Clearly distressed ⑤ Very much distressed

22. Others try to deprive me of what I care about if I do not follow them.

A. No Yes
B. Please indicate whether you feel that this experience happens to you frequently or not.

① I do not feel it happens frequently. ② I feel it happens frequently.

C. Please indicate whether you feel this experience is traumatic to you or not.

① I do not find it to be traumatic. ② I find it to be traumatic.

D. How distressed do you feel by this experience?

① Not distressed ② A bit distressed ③ Moderately distressed ④ Clearly distressed ⑤ Very much distressed

23. Blame me for their distress.

A. No Yes
B. Please indicate whether you feel that this experience happens to you frequently or not.

① I do not feel it happens frequently. ② I feel it happens frequently.

C. Please indicate whether you feel this experience is traumatic to you or not.

① I do not find it to be traumatic. ② I find it to be traumatic.

D. How distressed do you feel by this experience?

① Not distressed ② A bit distressed ③ Moderately distressed ④ Clearly distressed ⑤ Very much distressed

24. I feel that I tend to underperform at work or school.

A. No Yes
B. Please indicate whether you feel that this experience happens to you frequently or not.

① I do not feel it happens frequently. ② I feel it happens frequently.

C. Please indicate whether you feel this experience is traumatic to you or not.

① I do not find it to be traumatic. ② I find it to be traumatic.

D. How distressed do you feel by this experience?

① Not distressed ② A bit distressed ③ Moderately distressed ④ Clearly distressed ⑤ Very much distressed

25. I tend to fail in my application to schools or positions at work.

A. No Yes
B. Please indicate whether you feel that this experience happens to you frequently or not.

① I do not feel it happens frequently. ② I feel it happens frequently.

C. Please indicate whether you feel this experience is traumatic to you or not.

① I do not find it to be traumatic. ② I find it to be traumatic.

D. How distressed do you feel by this experience?

① Not distressed ② A bit distressed ③ Moderately distressed ④ Clearly distressed ⑤ Very much distressed

26. I feel to be fired or dismissed from work.

A. No Yes
B. Please indicate whether you feel that this experience happens to you frequently or not.

① I do not feel it happens frequently. ② I feel it happens frequently.

C. Please indicate whether you feel this experience is traumatic to you or not.

① I do not find it to be traumatic. ② I find it to be traumatic.

D. How distressed do you feel by this experience?

① Not distressed ② A bit distressed ③ Moderately distressed ④ Clearly distressed ⑤ Very much distressed

27. I fail in competition or contest.

A. No Yes
B. Please indicate whether you feel that this experience happens to you frequently or not.

① I do not feel it happens frequently. ② I feel it happens frequently.

C. Please indicate whether you feel this experience is traumatic to you or not.

① I do not find it to be traumatic. ② I find it to be traumatic.

D. How distressed do you feel by this experience?

① Not distressed ② A bit distressed ③ Moderately distressed ④ Clearly distressed ⑤ Very much distressed

28. I feel that I am not being respected.

A. No Yes
B. Please indicate whether you feel that this experience happens to you frequently or not.

① I do not feel it happens frequently. ② I feel it happens frequently.

C. Please indicate whether you feel this experience is traumatic to you or not.

① I do not find it to be traumatic. ② I find it to be traumatic.

D. How distressed do you feel by this experience?

① Not distressed ② A bit distressed ③ Moderately distressed ④ Clearly distressed ⑤ Very much distressed

29. Others tend to ignore my presence.

A. No Yes
B. Please indicate whether you feel that this experience happens to you frequently or not.

① I do not feel it happens frequently. ② I feel it happens frequently.

C. Please indicate whether you feel this experience is traumatic to you or not.

① I do not find it to be traumatic. ② I find it to be traumatic.

D. How distressed do you feel by this experience?

① Not distressed ② A bit distressed ③ Moderately distressed ④ Clearly distressed ⑤ Very much distressed

30. My interests or rights are not being protected.

A. No Yes
B. Please indicate whether you feel that this experience happens to you frequently or not.

① I do not feel it happens frequently. ② I feel it happens frequently.

C. Please indicate whether you feel this experience is traumatic to you or not.

① I do not find it to be traumatic. ② I find it to be traumatic.

D. How distressed do you feel by this experience?

① Not distressed ② A bit distressed ③ Moderately distressed ④ Clearly distressed ⑤ Very much distressed

References

Ahmed, I., Proulx, J., & Pilny, A. (2017). Causal inference using Bayesian networks. In A. Pilny & M. S. Poole (Eds.), *Group processes* (pp. 29–49). Springer.

BayesFusion. (2019). *GeNIe Modeler: User manual.*

Borsuk, M. E., Stow, C. A., & Reckhow, K. H. (2004). A Bayesian network of eutrophication models for synthesis, prediction, and uncertainty analysis. *Ecological Modelling, 173*(2), 219–239.

Castillo, E., Gutoerre, J. M., & Hadi, A. S. (1997). *Expert systems and probabilistic network models*. Spinger-Verlag.

Caston, J. S., & Frazier, P. (2013). *Perceived control over traumatic events: a study across events.* Paper presented at the National Conference On Undergraduate Research, April 11–13, University of Wisconsin La Crosse, Wisconsin.

Chickering, D. M. (1995). *A transformational characterization of equivalent Bayesian network structures.* Paper presented at the In Proceedings of the 11th annual conference on uncertainty in artificial intelligence (UAI-95) (p. 87–98), San Francisco, CA.

Chickering, D. M. (2002). Learning equivalence classes of Bayesian-network structures. *Journal of Machine Learning Research, 2*, 445–498.

Clark, D. A., & Beck, A. T. (2010). Cognitive theory and therapy of anxiety and depression: Convergence with neurobiological findings. *Trends in Cognitive Sciences, 14*(9), 418–424. https://doi.org/10.1016/j.tics.2010.06.007

Cox, D. R., & Wermuth, N. (2004). Causality: A statistical view. *International Statistical Review, 72*, 285–305.

Danks, D., Fancsali, S., Glymour, C., & Scheines, R. (2010). Comorbid science? *Behavior & Brain Science, 33*, 153–154.

Friedman, N., Geiger, D., & Goldszmidt, M. (1997). Bayesian network classifiers. *Machine Learning, 29*, 131–163.

Frosini, B. V. (2006). Causality and causal models: A conceptual perspective. *International Statistical Review, 74*(3), 305–334.

Furlotte, N. A., Heckerman, D., & Lippert, C. (2014). Quantifying the uncertainty in heritability. *Journal of Human Genetics, 59*(2), 269–275.

Gilman, R., Carter-Sowell, A., DeWall, C. N., Adams, R. E., & Carboni, I. (2013). Validation of the ostracism experience scale for adolescents. *Psychological Assessment, 25*, 319–330. https://doi.org/10.1037/a0030913

Glymour, C., Zhang, K., & Spirtes, P. (2019). Review of causal discovery methods based on graphical models. *Frontiers in Genetics, 10*, 524. https://doi.org/10.3389/fgene.2019.00524

Green, B. L. (1996). Trauma history questionnaire. In B. H. Stamm (Ed.), *Measurement of stress, trauma, and adaptation* (pp. 366–369). Sidran Press.

Hodgdon, H. B., Suvak, M., Zinoviev, D. Y., Liebman, R. E., Briggs, E. C., & Spinazzola, J. (2019). Network analysis of exposure to trauma and childhood adversities in a clinical sample of youth. *Psychological Assessment, 31*(11), 1294–1306.

Holmes, S. C., & Johnson, N. L. (2019). Understanding the relationship between interpersonal trauma and disordered eating: An extension of the model of psychological adaptation. *Psychological Trauma: Theory, Research, Practice, and Policy*. Advance online publication. https://doi.org/10.1037/tra0000533

Hoyer, P., Janzing, D., Mooji, J., Peters, J., & Schölkopf, B. (2009). Nonlinear causal discovery with additive noise models. In *Advances in neural information processing systems 21.*

Kessler, D. (2015). *Capture: Unraveling the mystery of mental suffering*. HarperCollins Publisher.

King, G. (2011). Ensuring the data-rich future of the social sciences. *Science, 331*(6018), 719–721.

Korb, K., & Nicholson, A. E. (2010). *Bayesian artificial intelligence* (2nd ed.). Chapman & Hall.

Lazer, D., Pentland, A. S., Adamic, L., Aral, S., Barabasi, A. L., Brewer, D., & Jebara, T. (2009). Life in the network: The coming age of computational social science. *Science, 323*(5915), 721.

McIlveen, R., Mitchell, R., Curran, D., Dyer, K., Corry, M., DePrince, A., et al. (2019). Exploring the relationship between alienation appraisals, trauma, posttraumatic stress, and depression. *Psychological Trauma: Theory, Research, Practice, and Policy*. Advance online publication. https://doi.org/10.1037/tra0000523

Neapolitan, R. E. (2004). *Learning Bayesian networks*. Pearson Prentice Hall.

Pe'er, D. (2005). Bayesian network analysis of signaling networks: A primer. *Science's STKE, 281*, 1–12.

Pearl, J. (1988). *Probabilistic reasoning in intelligent systems*. Morgan Kaufmann.

Pearl, J. (2000). *Causality models, reasoning, and inference*. Cambridge University Press.

Rosseel, Y., Jorgensen, T. D., & Rockwood, N. (2021). *Package: lavaan. Latent variable analysis*.

Russell, S., & Norvig, P. (2009a). *Artificial intelligence: A modern approach* (3rd ed.). Prentice Hall.

Russell, S. J., & Norvig, P. (2009b). *Artificial intelligence: A modern approach (3rd edition)*. Prentice Hall.

Sartorius, N., Üstün, T. B., Lecrubier, Y., & Wittchen, H.-U. (1996). Depression comorbid with anxiety: Results from the WHO study on psychological disorders in primary health care. *British Journal of Psychiatry, 168*, 38–43.

Scutari, M. (2019). *Package 'bnlearn': Bayesian network structure learning, parameter learning and inference*.

Scutari, M., Graafland, C. E., & Guti´errez, J. E. M. (2018). Who learns better Bayesian network structures: Constraint-based, score-based or hybrid algorithms? *Proceedings of Machine Learning Research, 72*, 416–427.

Shapiro, D. H. (1994). *Manual for the Shapiro control inventory*. Behaviordata.

Shen, X., Ma, S., Vemuri, P., Simon, G., & the Alzheimer's Disease Neuroimaging Initiative. (2020). Challenges and opportunities with causal discovery algorithms: Application to Alzheimer's pathophysiology. *Scientific Reports, 10*, 2975. https://doi.org/10.1038/s41598-020-59669-x

Shimizu, S., Hoyer, P., Hyvärinen, A., & Kerminen, A. (2006). A linear non-Gaussian acyclic model for causal discovery. *The Journal of Machine Learning Research, 7*, 2003–2030.

Silander, T., & Myllymaki, P. (2006). *A simple approach for finding the globally optimal Bayesian network structure*. Paper presented at the In Proceedings of the Twenty-Second Conference on Uncertainty in Artificial Intelligence (UAI2006), Arlington, Virginia.

Spirtes, P., & Zhang, K. (2016). Causal discovery and inference: Concepts and recent methodological advances. *Applied Informatics, 3*(3), 1–28. https://doi.org/10.1186/s40535-016-0018-x

Spirtes, P., Glymour, C., & Scheines, R. (2000). *Constructing bayesian networks models of gene expression networks from microarray data*. Paper presented at the Proceedings of the Atlantic Symposium on Computational Biology, North Carolina.

Spirtes, P., Glymour, C., & Scheines, R. (2001). *Causation, prediction, and search*. MIT Press.

Sun, L., & Shenoy, P. P. (2007). Using Bayesian networks for bankruptcy prediction: Some methodological issues. *European Journal of Operational Research, 2*(180), 738–753.

Wardenaar, K. J., van Veen, T., Giltay, E. J., de Beurs, E., & Zitman, F. G. (2010). Development and validation of a 30-item short adaptation of the Mood and Anxiety Symptoms Questionnaire (MASQ). *Psychiatric Research, 179*, 101–106. https://doi.org/10.1016/j.psychres.2009.03.005

Weathers, F. W., Litz, B. T., Keane, T. M., Palmieri, P. A., Marx, B. P., & Schnurr, P. P. (2013). *The PTSD Checklist for DSM-5 (PCL-5)*. Available online at: http://www.ptsd.va.gov

Wright, S. (1921). Correlation and causation. *Journal of Agricultural Research, 20*, 557–585.

Yuan, C., Lu, T.-C., & Druzdzel, M. J. (2004). *Annealed MAP*. Paper presented at the Proceedings of the 20th Annual Conference on Uncertainty in Artificial Intelligence (UAI-04), Arlington, Virginia.

Part III
Multilevel Analysis

Chapter 12
Missing Data in the Analysis of Multilevel and Dependent Data

Simon Grund , Oliver Lüdtke , and Alexander Robitzsch

12.1 Missing Data in the Analysis of Multilevel and Dependent Data

Missing data are a common problem in multilevel and other types of dependent data (Enders, 2022). In multilevel data, dealing with missing data can be particularly cumbersome, and recent years have seen the emergence of new and extremely flexible methods for handling missing data that are especially useful for multilevel data (e.g., Enders et al., 2016; Lüdtke et al., 2017; see also Enders, 2022). These methods include model-based methods based on maximum-likelihood and Bayesian estimation techniques and imputation-based methods such as multiple imputation (for a more general overview, see Schafer & Graham, 2002; Little & Rubin, 2019). As a result, researchers today can choose from a growing number of options for handling missing data with regard to both statistical methods and software.

The present chapter pursues three major goals. First, we aim to summarize the current state of the literature on the treatment of missing data in dependent data. In this context, we focus on multilevel data and outline how these issues extend to other types of dependent data. Second, we clarify the features of multilevel data that must be considered when handling missing data and the different options that imputation- and model-based methods offer for doing so. And third, we illustrate the different procedures with a series of examples and provide recommendations for practice. In the following sections, we first provide a short overview of the structure and analysis

S. Grund (✉)
University of Hamburg, Hamburg, Germany
e-mail: simon.grund@uni-hamburg.de

O. Lüdtke · A. Robitzsch
Leibniz Institute for Science and Mathematics Education, Kiel, Germany

Centre for International Student Assessment, Kiel, Germany

M. Stemmler et al. (eds.), *Dependent Data in Social Sciences Research*,
https://doi.org/10.1007/978-3-031-56318-8_12

of multilevel data, before we consider the problem of missing values in multilevel data and the imputation- and model-based methods that can be used to handle them.

12.2 Multilevel Data

Multilevel data are characterized by a hierarchical structure, in which lower-level units are nested within higher-level collectives (Raudenbush & Bryk, 2002). For example, students can be nested in classrooms, employees can be nested in teams or enterprises, and—in longitudinal data—repeated measurements can be nested in persons (see also Hoffman, 2015). Figure 12.1 provides a schematic illustration with ten persons ($i = 1, \ldots, n_j$) nested in three groups ($j = 1, \ldots, J$) and with three variables, two of which were measured at the person level (x and y, Level 1) and one at the group level (w, Level 2). For the remainder of this chapter, we focus on applications with two-level data and assume that the data consist of persons (e.g., students; Level 1) nested within groups (e.g., classrooms; Level 2), although similar considerations hold for longitudinal and other types of dependent data.

Analysis of Multilevel Data

To analyze multilevel data, researchers often use multilevel models (Goldstein, 2011; Raudenbush & Bryk, 2002; see also Hoffman, 2015). These models allow researchers to investigate complex associations between variables both at the person

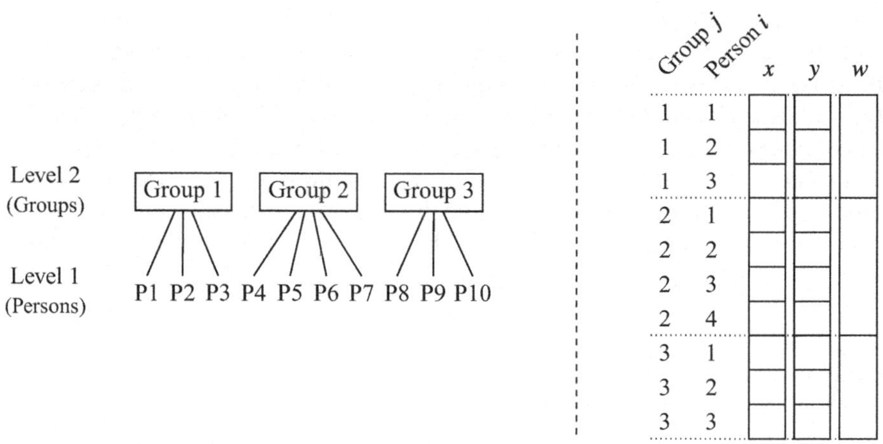

Fig. 12.1 Schematic illustration of the structure and format of multilevel data with persons (Level 1) nested within groups (Level 2) and with two variables at Level 1 (x and y) and one variable at Level 2 (w)

and at the group level and to what extent person- or individual-level effects are homogeneous or heterogeneous across groups. Based on the example above, suppose that a researcher investigates the effects of students' academic achievement (x) on their academic self-concept (y, both at Level 1) while controlling for classroom-level characteristics w (e.g., teachers' use of group-based activities) with the following model:

$$y_{ij} = \beta_0 + \beta_1(x_{ij} - \bar{x}_{\bullet j}) + \beta_2\bar{x}_{\bullet j} + \beta_3 w_j + u_{0j} + e_{ij} \; . \tag{12.1}$$

In this model, the values x_{ij} have been centered at their group means, $\bar{x}_{\bullet j}$, and the regression coefficients β_1 and β_2 represent the within- and between-group effects of x, respectively, whereas β_3 represents the effect of w. The regression coefficients are assumed to be fixed across groups with the exception of the intercept, which is assumed to vary across groups as indicated by the random intercepts, u_{0j}. The random intercepts at Level 2, u_{0j}, and the residuals at Level 1, e_{ij}, are typically assumed to be normally distributed with means of zero and variances denoted by τ^2 and σ^2, respectively.

The random-intercept model above can be extended to allow the within-group effects of the explanatory variable to vary across groups by introducing additional random effects and interaction terms into the model:

$$\begin{aligned} y_{ij} = {} & \beta_0 + \beta_1(x_{ij} - \bar{x}_{\bullet j}) + \beta_2\bar{x}_{\bullet j} + \beta_3 w_j + \beta_4(x_{ij} - \bar{x}_{\bullet j})w_j + \beta_5\bar{x}_{\bullet j}w_j + {} \\ & u_{0j} + u_{1j}(x_{ij} - \bar{x}_{\bullet j}) + e_{ij} \; . \end{aligned} \tag{12.2}$$

In this model, the within-group effect of x varies both randomly across groups, as indicated by the random slopes, u_{1j}, and as a function of w, as indicated by the cross-level interaction (CLI) effect β_4 (see Snijders & Bosker, 2012). In addition, the between-group effect of x also varies as a function of w, as indicated by a group-level interaction (β_5). More generally, multilevel models are extremely flexible and can incorporate a variety of both linear and nonlinear associations as well as heterogeneity in both the associations between variables and the amount of variability within groups (e.g., Gasimova et al., 2014; Hedeker et al., 2008; Leckie et al., 2014).

12.3 Missing Data

Multilevel data are often incomplete, for example, when the persons (e.g., students) who are enrolled in a study fail to answer certain items on a questionnaire or when information at the group level (e.g., from teachers or schools) is unavailable (see also van Buuren, 2011). In the following, we provide a short overview of different mechanisms and patterns of missing data that can occur in multilevel data.

Missing Data Mechanisms

In the missing data literature, several missing data *mechanisms* are often distinguished to describe the relationship between the observable data and the occurrence of missing data. Suppose that each person has a hypothetical complete set of observations $\mathbf{y}_{ij}$ on a number of variables, $\mathbf{y}_{ij} = (y_{ij}1, \ldots, y_{ijP})$, and that there is a set of indicators $\mathbf{r}_{ij}$ that describe which elements in $\mathbf{y}_{ij}$ are observed ($r_{ijp} = 1$ if the pth element in $\mathbf{y}_{ij}$ is observed, and $r_{ijp} = 0$ if it is missing, for $p = 1, \ldots, P$). Consequently, the data can be decomposed into an observed part $\mathbf{y}_{\text{obs}}$ (where $\mathbf{r} = 1$) and an unobserved part $\mathbf{y}_{\text{mis}}$ (where $\mathbf{r} = 0$). Following Little and Rubin (2019), three missing data mechanisms can be distinguished: missing *completely* at random (MCAR), missing at random (MAR), or missing *not* at random (MNAR). When the data are MCAR, missing values occur in a completely random fashion in the sense that missingness depends on neither the observed nor unobserved data, $P(\mathbf{r}|\mathbf{y}_{\text{obs}}, \mathbf{y}_{\text{mis}}) = P(\mathbf{r})$. Under MAR, the missingness depends on the observed but not on the unobserved data, that is, $P(\mathbf{r}|\mathbf{y}_{\text{obs}}, \mathbf{y}_{\text{mis}}) = P(\mathbf{r}|\mathbf{y}_{\text{obs}})$. Finally, when the data are MNAR, missingness also depends on the unobserved data in addition to the observed data.

To illustrate the three mechanisms, consider the example above, where $\mathbf{y}$ would include the students' hypothetical data on their academic achievement (x), self-concept (y), and the characteristics of their learning environment (w), and suppose that only the students' self-concept has missing data (r_y). This scenario is illustrated in Fig. 12.2. In this scenario, an MCAR mechanism would imply that a student's tendency to have missing values in self-concept is unrelated to their self-concept, achievement, and characteristics of their learning environment. A MAR mechanism would imply that this tendency is related to the student's achievement or learning environment but unrelated to their (unobserved) self-concept, once their achievement and environment are taken into account. Finally, an MNAR mechanism would imply that missingness on students' self-concept is related to the unobserved values themselves, even after controlling for differences in academic achievement and learning environment characteristics.

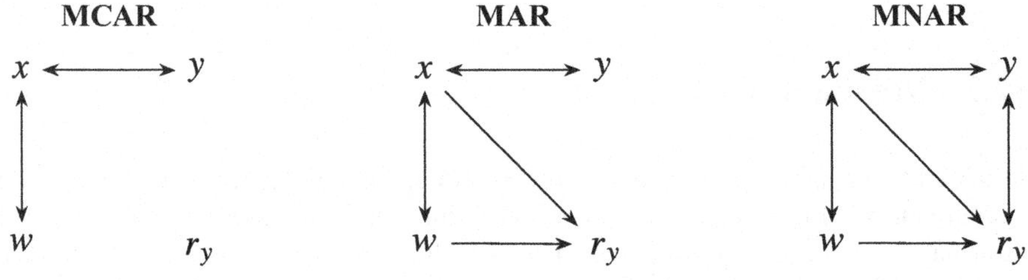

Fig. 12.2 Schematic illustration of missing data mechanisms in a hypothetical example with two variables at Level 1 (x and y), one variable at Level 2 (w), and univariate missing data in y (r_y). MCAR = missing completely at random, MAR = missing at random, MNAR = missing not at random

In distinguishing different missing data mechanisms, it is important to emphasize that these mechanisms represent untestable assumptions about hypothetical relationships between the variables and the occurrence of missing data (Schafer & Graham, 2002). Although the observed data provide some limited utility for investigating which mechanisms might be consistent with the data (e.g., Little's "MCAR test"; Little, 1988), the plausibility of these assumptions can ultimately not be inferred from the observed data alone (see also Carpenter & Kenward, 2013). The statistical methods that we consider in this chapter assume that the data are MAR (Enders, 2022). This is in contrast with traditional methods for handling missing data such as *pairwise* and *listwise deletion*, which remove cases with missing values from the analysis and typically require the stronger assumption that the data are MCAR. Statistical analyses under MNAR are also possible but often require specific assumptions about the missing data mechanisms and should be regarded primarily as sensitivity analyses (e.g., Carpenter & Kenward, 2013; see also Du et al., 2022; Ibrahim et al., 2005; Wu, 2010).

Missing Data Patterns

In addition to these hypothetical mechanisms, missing data can also be classified by their *pattern* of occurrence. For example, Rubin (1987) distinguished between unit and item nonresponse, where item nonresponse refers to cases where only some of the responses from a given person are missing, and unit nonresponse refers to cases where all data from a given person are missing. Similarly, in multilevel data, Resche-Rigon and White (2018) refer to data as *sporadically* missing when only some persons in a group have missing values on a given variable and as *systematically* missing when all values in a group are missing. Systematically missing data are relatively rare in most types of multilevel data and occur more frequently in specific contexts such as in individual-participant-data meta-analyses (see also Jolani et al., 2015). In addition, when higher-level variables are missing for a given group, all persons within this group have incomplete data (see also Cheung, 2007; Grund et al., 2018a). Continuing the example above, Fig. 12.3 provides an illustration of three possible patterns of missing data with (a) sporadic missing values only in students' academic self-concept (y) at Level 1, (b) both sporadic and systematic missing data in the students' academic achievement and self-concept (x and y) at Level 1 and the characteristics of their learning environments (w) at Level 2, and (c) unit nonresponse for all students in the first group.

Several statistical methods have been recommended for handling incomplete multilevel data. To organize the following discussion, we distinguish between

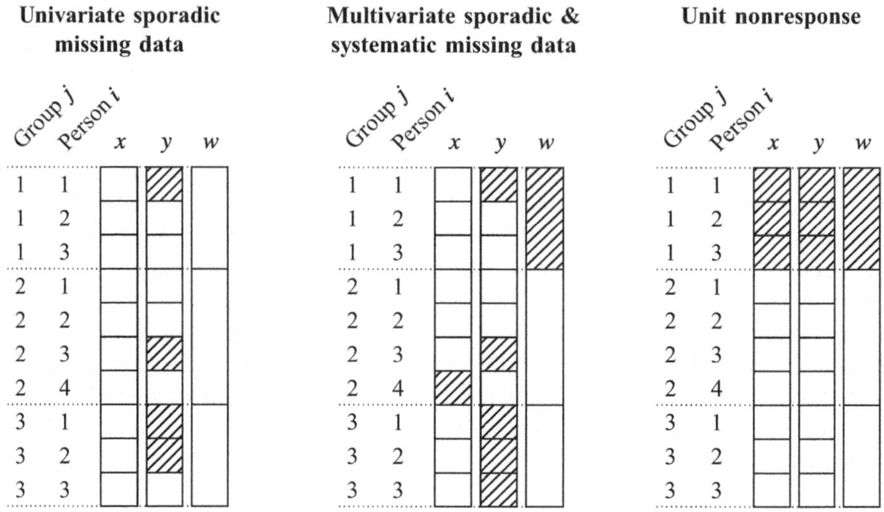

Fig. 12.3 Schematic illustration of different patterns of missing data with univariate sporadic missing, multivariate sporadic/systematic missing data, and unit nonresponse in two variables at Level 1 (x and y) and one variable at Level 2 (w). The hatched areas indicate missing data

imputation-based methods on the one hand, which we consider first, and model-based methods on the other hand, which we consider afterwards.[1]

12.4 Multiple Imputation

The main idea of imputation-based methods is to "fill in" the missing data $\mathbf{y}_{\text{mis}}$ with plausible replacements that are generated on the basis of the observed data and a statistical model (the imputation model; Rubin, 1987). In multiple imputation (MI), this technique is used repeatedly to generate multiple copies $\mathbf{y}_{\text{imp}}^{(1)}, \ldots, \mathbf{y}_{\text{imp}}^{(M)}$ of the data, each with a different set of "filled-in" values for the missing data. To obtain the imputed values, these methods iterate between two steps, by first estimating the parameters of the imputation model, $\boldsymbol{\theta}$, based on the observed data and the currently filled-in values, and then filling in new values by drawing them from the (posterior) predictive distribution of the missing data, given the observed data and

[1] This distinction between imputation- and model-based methods is not the only possible one and is used here primarily for didactic purposes. Specifically, we call methods "imputation-based" when they focus on creating completed versions of the data by "filling in" the missing values, and we call them "model-based" when they focus on parameter estimation (see also Little & Rubin, 2019). Both types of methods ultimately rely on statistical models, but they employ these models in different ways.

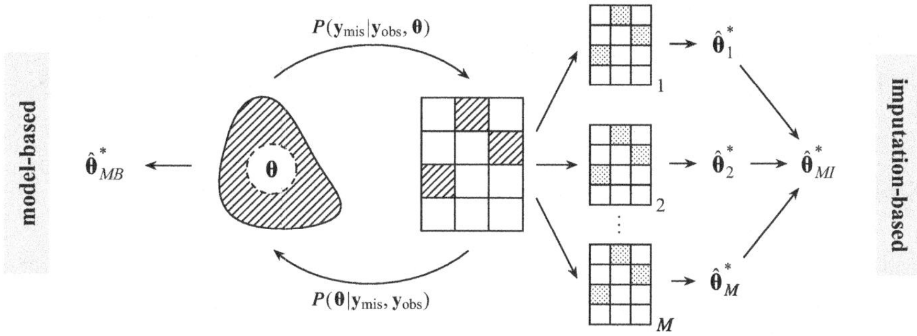

Fig. 12.4 Schematic illustration of the relation between imputation-based and model-based inference (based on Bayesian estimation). The unknown values of the parameters and the missing data are shown as hatched areas. The imputed data are shown as dotted areas. Statistical parameters with an asterisk ($\boldsymbol{\theta}^*$) denote the parameters of the substantive analysis model, which may or may not coincide with the parameters of the joint model ($\boldsymbol{\theta}$)

the estimated parameters, $P(\mathbf{y}_{\text{mis}}|\mathbf{y}_{\text{obs}}, \boldsymbol{\theta})$. Each data set is then analyzed separately, whereby multiple estimates of the parameters for a substantive analysis model are obtained, and the results are pooled to obtain a final set of results. An illustration of MI is shown in the right half of Fig. 12.4, where the iterative steps are shown in the center, and a model-based treatment is shown on the left (for a detailed introduction to MI, see also Schafer & Graham, 2002).

One of the main benefits that are often attributed to imputation-based methods is that the treatment of missing data is separated from the analysis (e.g., Collins et al., 2001; Schafer and Graham, 2002). In other words, the imputation model that the "imputer" uses to generate the filled-in values can differ from the substantive analysis model that the "analyst" uses to obtain their results (see Fig. 12.4). This can be beneficial, because it allows the imputer to specify a rich imputation model that represents the associations among the variables in a very flexible manner and that can include additional variables in the treatment of missing data even if these are not used in subsequent analyses (so-called *auxiliary* variables; Collins et al., 2001). Conversely, it allows the analyst to use the same set of imputed data for different analyses without having to deal with the missing data for each one (for further discussion, see Collins et al., 2001; Schafer, 2003).

However, imputation-based treatments of missing data also come with at least three major challenges. First, the imputation model must be correctly specified in the sense that the relations specified therein match the "true" relations between the variables at least approximately (Robins & Wang, 2000; Rubin, 1996). Second, the imputation model must match the intended analyses in the sense that all relations that are relevant in the analyses are represented in the imputation model (Meng, 1994). And third, the imputation model must include all relevant information from the data that is needed to justify the MAR assumption (e.g., Collins et al., 2001). Meeting these requirements is crucial to ensure that the analysis of the imputed data leads to unbiased results, but doing so can be difficult due to both the structure of multilevel data and the often complex analyses that are conducted with them (see also Du et al., 2022). In the following, we first outline three strategies for specifying the imputation model and explain how these strategies address these challenges.

Specification of the Imputation Model

The key component of MI is the imputation model, by which the imputer aims to capture the relations that exist between the variables in the shape of a multivariate statistical model. Going forward, we refer to the multivariate distribution of the data as $P(\mathbf{y})$ and to the corresponding density (or "model") as $f(\mathbf{y}; \boldsymbol{\theta})$, where $\boldsymbol{\theta}$ represents a set of unknown model parameters. Following Murray (2018), we distinguish three broad strategies for specifying this model.

Joint Modeling

In joint modeling (JM), a multivariate model $f(\mathbf{y}; \boldsymbol{\theta})$ is specified directly, and this model is used to generate replacements for all incomplete variables simultaneously (e.g., Goldstein et al., 2009; Schafer & Yucel, 2002). For example, when all variables are continuous and measured at Level 1, then $f(\mathbf{y}; \boldsymbol{\theta})$ can be specified as a multivariate multilevel model (Schafer & Yucel, 2002). Extensions of this approach have been developed to accommodate level-2 variables (Carpenter & Kenward, 2013; Goldstein et al., 2009), categorical variables (Asparouhov & Muthén, 2010; Carpenter & Kenward, 2013; Goldstein et al., 2009; Enders et al., 2018; Quartagno & Carpenter, 2019; Vidotto, Vermunt, & van Deun, 2018), and cases in which the associations between the variables vary across groups (e.g., random slopes; see Quartagno & Carpenter, 2016; Yucel, 2011). The main benefit of JM is that only a single imputation model is needed to generate imputations for all variables with missing data. However, specifying this model can become increasingly difficult as the number of variables increases.

Sequential Modeling

In sequential modeling (SM), the multivariate imputation model $f(\mathbf{y}; \boldsymbol{\theta})$ is specified indirectly, by factorizing it into a sequence of univariate conditional models (Ibrahim et al., 2005). This sequence still corresponds to a multivariate model but one that has been re-expressed in terms of multiple factors (or "submodels") whose product is the multivariate model for all variables. The sequence can include the variables in any order and has the form:

$$f(\mathbf{y}; \boldsymbol{\theta}) = \prod_{p=1}^{P} f_p(y_p | y_1, \ldots, y_{p-1}; \boldsymbol{\theta}_p) , \tag{12.3}$$

where f_p is the submodel for the pth variable conditional on the variables that were placed earlier in the sequence, and $\boldsymbol{\theta}_p$ is a vector of parameters for this model. Each submodel is specific to only one variable and can be selected to match the respective variable type. For example, for binary variable at Level 1, a multilevel probit model can be used, and other options have been developed for other variable types (see also Enders et al., 2020; Erler et al., 2017; Ibrahim et al., 2001; Lüdtke et al., 2020b). This specificity can greatly simplify the specification of the imputation model for variables of different types, because a potentially complex multivariate model is re-expressed as a sequence of simpler univariate models. However, the factorization also introduces ambiguity, because the order with which the variables are entered into the sequence affects the meaning of the overall model (Ibrahim et al., 2002).

Fully Conditional Specification

Finally, in the fully conditional specification (FCS), the imputation model also consists of a sequence of univariate models, but each model is specified as a *separate* model that conditions on some or all of the other variables. The models are tied together by an iterative procedure that generates imputations for each variable in a step-by-step manner (Raghunathan et al., 2001; van Buuren et al., 2006). Specifically, the models have the form:

$$f_p(y_p | y_1, \ldots, y_{p-1}, y_{p+1}, \ldots, y_P; \boldsymbol{\theta}_p) , \quad (12.4)$$

where f_p is the model for the pth variable with missing data conditional on the other variables, and the $\boldsymbol{\theta}_p$ are model-specific parameters. As in SM, a suitable model can be selected for each variable, and a variety of methods have been developed for variables of different types and at different levels (Audigier et al., 2018; Enders et al., 2018; Grund et al., 2018a).

The emphasis of FCS on separate univariate models can again greatly simplify the specification of the imputation model. However, in contrast with the SM approach, the separate models in FCS do not always correspond to a well-defined multivariate distribution for all variables (Du et al., 2022; Liu et al., 2014). Several authors have argued that this difference in the model specification rarely affects the results and that the FCS approach can be regarded as *approximating* this distribution with an ensemble of separate yet connected models (e.g., Hughes et al., 2014; van Buuren et al., 2006). However, other authors have found that the FCS approach can yield biased results when the relationships between the variables are nonlinear or vary across groups (e.g., Bartlett et al., 2015; Du et al., 2022).

In the following, we now discuss the three major challenges for applications of multilevel MI relating to the role of (1) the multilevel structure, (2) the substantive analyses, and (3) auxiliary variables. For each challenge, we outline how the three imputation strategies—JM, SM, and FCS—for multilevel MI address them, and we discuss their relative strengths and weaknesses.

Challenge 1: Multilevel Structure

The first major challenge in multilevel MI is to ensure that the imputation model is correctly specified so that it matches the "true" model underlying the data as closely as possible. For instance, in the examples above with students nested in classrooms and three incomplete variables (x, y, and w), we assumed that there were separate between- and within-group relations between the variables (Eq. 12.1) or heterogeneous relations across groups (Eq. 12.2). In the following, we discuss how the MI approaches aim to address these aspects of multilevel data.

Between- and Within-Group Effects

Because the relations between the variables can differ in strength and direction depending on the level of analysis (between vs. within groups) and variables can be measured at different levels, preserving the multilevel structure requires an imputation model that is flexible enough to accommodate the variability and the relations between variables at each level (e.g., Enders et al., 2016; Grund et al., 2018b). In JM, this is achieved by specifying a multivariate model for all variables that separates the between- and within-group components of the variables and that takes the relations between them into account. For the three variables x, y, and w, this model can be expressed as

$$f(x, y, w; \boldsymbol{\theta}) : \begin{array}{l} \begin{bmatrix} x_{ij} \\ y_{ij} \end{bmatrix} = \begin{bmatrix} \mu_x \\ \mu_y \end{bmatrix} + \begin{bmatrix} u_{xj} \\ u_{yj} \end{bmatrix} + \begin{bmatrix} e_{xij} \\ e_{yij} \end{bmatrix} \\ [w_j] = [\mu_w] + [u_{wj}] , \end{array} \tag{12.5}$$

where the $\mathbf{u}_j = (u_{xj}, u_{yj}, u_{wj})$ represent the random intercepts (for x and y) and residuals (w) at Level 2 and are assumed to follow a multivariate normal distribution with zero means and a covariance matrix $\mathbf{T}$, and the $\mathbf{e}_{ij} = (e_{xij}, e_{yij})$ represent residuals at Level 1 that are assumed to follow a multivariate normal distribution with zero means and a covariance matrix $\boldsymbol{\Sigma}$. This model effectively decomposes the variables into (latent) between- and within-group components that are represented by the random intercepts $\mathbf{u}_j$ and residuals $\mathbf{e}_{ij}$, respectively. In addition, the model accommodates the between- and within-group relations between the variables by allowing mutual correlations among the $\mathbf{u}_j$ at Level 2 (in $\mathbf{T}$) and the $\mathbf{e}_{ij}$ at Level 1 (in $\boldsymbol{\Sigma}$).

In SM, a sequence of univariate models is specified, and each model conditions on the variables placed earlier in the sequence in a manner that takes the between- and within-group effects into account. For example, the multivariate model $f(x, y, w; \boldsymbol{\theta})$ can be expressed as $f(x, y, w; \boldsymbol{\theta}) = f_w(w; \boldsymbol{\theta}_w) f_x(x|w; \boldsymbol{\theta}_x)$ $f_y(y|x, w; \boldsymbol{\theta}_y)$, and the three submodels can be specified as

$$\begin{aligned} f_w: &\quad w_j = \beta_{w0} + u_{wj} \\ f_x: &\quad x_{ij} = \beta_{x0} + \beta_{x1} w_j + u_{xj} + e_{xij} \\ f_y: &\quad y_{ij} = \beta_{y0} + \beta_{y1}(x_{ij} - \bar{x}_{\bullet j}) + \beta_{y2}\bar{x}_{\bullet j} + \beta_{y3} w_j + u_{yj} + e_{yij} \, , \end{aligned} \tag{12.6}$$

where f_x and f_y are multilevel random intercept models (see Eq. 12.1), and f_w is a group-level regression model. As in JM, the SM approach allows taking the between- and within-group relations between the variables into account. In the specification above, this is achieved by including the group means of the predictor variables at Level 1 (e.g., $\bar{x}_{\bullet j}$) in each submodel and by centering their values at their respective group means. The order with which the variables are entered in the sequence is flexible. Nonetheless, Ibrahim et al. (2005) provided a rule of thumb and recommended that the variables be ordered with regard to their type and percentage of missing data. Specifically, they recommended that variables at Level 2 and categorical variables should be entered first (i.e., as models with fewer predictors) and before continuous variables and variables at Level 1; similarly, for each variable type, those with more missing values should be entered last (i.e., as models with many predictors; see also Enders, 2022; Lüdtke et al., 2020b).

Finally, in FCS, a sequence of separate univariate models is specified that condition on all other variables and the full set of within- and between-group effects. Specifically, to approximate the multivariate model $f(x, y, w; \boldsymbol{\theta})$, the three imputation models $f_x(x|w, y; \boldsymbol{\theta}_x)$, $f_y(y|x, w; \boldsymbol{\theta}_y)$, and $f_w(w|x, y; \boldsymbol{\theta}_w)$ can be specified as

$$\begin{aligned} f_x: &\quad x_{ij} = \beta_{x0} + \beta_{x1}(y_{ij} - \bar{y}_{\bullet j}) + \beta_{x2}\bar{y}_{\bullet j} + \beta_{x3} w_j + u_{xj} + e_{xij} \\ f_y: &\quad y_{ij} = \beta_{y0} + \beta_{y1}(x_{ij} - \bar{x}_{\bullet j}) + \beta_{y2}\bar{x}_{\bullet j} + \beta_{y3} w_j + u_{yj} + e_{yij} \\ f_w: &\quad w_j = \beta_{w0} + \beta_{w1}\bar{x}_{\bullet j} + \beta_{w2}\bar{y}_{\bullet j} + u_{wj} \, . \end{aligned} \tag{12.7}$$

As before, these models include the group means and the centered values of the predictor variables to accommodate the between- and within-group relations between the variables (see also Grund et al., 2018b). The FCS approach imputes the missing values in each variable in an iterative manner, and the most recent imputations for a given variable are used in subsequent iterations of the procedure. Similar to SM, the order of the imputation models in the FCS approach is flexible. However, in contrast with SM, it has been argued that the order typically does not have a meaningful impact on the results (Hughes et al., 2014).

Although JM, SM, and FCS all accommodate between- and within-group effects, they do so in slightly different ways. Specifically, whereas the JM approach is based on a decomposition of the variables into *latent* between- and within-group components, the SM and FCS approaches typically accommodate these effects by including *manifest* group means in the specification of the imputation model (see also Lüdtke et al., 2008). Previous research has shown that these two representations of the between- and within-group effects are not equivalent but that they tend to produce essentially the same results unless in extreme scenarios with strongly imbalanced group sizes (Grund et al., 2018a; Mistler & Enders, 2017; Resche-Rigon & White, 2018). In addition, it is worth noting that SM and FCS are flexible in this regard and can be used with either a latent or a manifest decomposition of the between- and within-group components (Enders et al., 2018; Grund et al., 2018a).

Heterogeneous and Nonlinear Effects

Further complications arise when the associations between variables are heterogeneous across groups. For instance, in Eq. 12.2, the within-group effect of an explanatory variable is assumed to vary across groups. In such a case, the specification of the imputation model must be extended to also preserve the heterogeneity of the associations between variables. As before, the different approaches provide different options for doing so. To accommodate heterogeneous associations in JM, the within-group covariance matrices ($\mathbf{\Sigma}$) can be permitted to vary across groups (Quartagno & Carpenter, 2016; Yucel, 2011). In SM and FCS, random slopes can be incorporated in the specification of the submodels (Enders et al., 2016; Grund et al., 2018b). Similarly, interactions and other types of nonlinear effects can be incorporated in different ways. For example, in JM, nonlinear terms can be calculated beforehand and treated as "just another variable" (JAV; Seaman et al., 2012; von Hippel, 2009). Similarly, in FCS, nonlinear terms can be included with the JAV approach or by updating the nonlinear terms through "passive" imputation steps (Royston, 2005). Finally, in SM, nonlinear effects of the predictors in each submodel can be incorporated by including them directly in the submodel specification. For heterogeneous effects involving categorical variables, some authors have also proposed stratification approaches, whereby the imputation is carried out separately in different subsets of the data (e.g., Kim et al., 2018; von Hippel, 2009).

In contrast to applications with only varying intercepts and otherwise constant effects, these approaches differ fairly strongly in their representation of varying or nonlinear effects and can therefore also provide different results. For example, several studies have shown that some of the approaches for multilevel MI, such as FCS, can still lead to biased results, when missing data occur in explanatory variables with random slopes or nonlinear effects, even if their imputation models take these features into account (e.g., Enders et al., 2018; Enders et al., 2020; Grund et al., 2018b). This is because the imputation models specified in this manner sometimes still do not reflect the actual distribution of the incomplete explanatory variables when the true relations vary across groups (but see Quartagno & Carpenter, 2016). To address this shortcoming of conventional MI approaches, other approaches have been developed that aim to accommodate the substantive analysis model during the imputation and that we now consider in more detail.

Challenge 2: Substantive Analysis Model

The second major challenge in multilevel MI is to ensure that the imputation model "fits" the intended analysis, which can be difficult if the analysis includes random slopes or nonlinear effects (see Eq. 12.2). Several authors have recommended that so-called *substantive-model-compatible* imputation (SMC-MI) approaches be used in this case, which incorporate the analysis directly in the imputation model (Bartlett et al., 2015; Goldstein et al., 2014; Ibrahim et al., 2005). The main idea behind SMC-MI is to factorize the joint distribution $f(\mathbf{y}; \boldsymbol{\theta})$ into two parts, one pertaining to the intended analysis and the other pertaining to the remaining variables. Specifically, if $\mathbf{y}$ consists of an outcome variable y and a number of covariates $\mathbf{x}$, then the joint model can be expressed as

$$f(\mathbf{y}; \boldsymbol{\theta}) = f_y(y|\mathbf{x}; \boldsymbol{\theta}_y) f_{\mathbf{x}}(\mathbf{x}; \boldsymbol{\theta}_{\mathbf{x}}) \ . \tag{12.8}$$

This factorization can easily accommodate random slopes and nonlinear effects by including them in the specification of the outcome model f_y. For example, for the analysis in Eq. 12.2, f_y can be specified as

$$\begin{aligned} f_y : \ y_{ij} = \beta_{y0} + \beta_{y1}(x_{ij} - \bar{x}_{\bullet j}) + \beta_{y2}\bar{x}_{\bullet j} + \beta_{y3}w_j + \beta_{y4}w_j(x_{ij} - \bar{x}_{\bullet j}) \\ + \beta_{y5}w_j\bar{x}_{\bullet j} + u_{y0j} + u_{y1j}(x_{ij} - \bar{x}_{\bullet j}) + e_{yij} \ , \end{aligned} \tag{12.9}$$

which is identical to the substantive analysis (Eq. 12.2). To draw imputations for the missing values, SMC-MI approaches rely on Markov Chain Monte Carlo (MCMC) techniques such as the Metropolis Hastings (MH) algorithm, whereby imputations are accepted or rejected depending on their likelihood under the specified model (see also Goldstein et al., 2014). In doing so, SMC-MI approaches ensure that the imputed values are compatible with the intended analysis.

SMC-JM, SMC-SM, and SMC-FCS

Although SMC-MI approaches all include an outcome model f_y in the specification of the imputation model, the covariate model $f_{\mathbf{x}}$ can be specified in different ways, with either a joint model (SMC-JM), a sequential model (SMC-SM), or the FCS approach (SMC-FCS; for an overview, see Lüdtke et al., 2020b). In SMC-JM, the factorization in Eq. 12.8 is left as is, and the substantive model, f_y, is combined with a multivariate model for the covariates, $f_{\mathbf{x}}$, similar to the one that would be used in a conventional JM approach (Carpenter & Kenward, 2013; Goldstein et al., 2014). In SMC-SM, the factorization is extended to include a univariate submodel for each covariate with missing data:

$$f(\mathbf{y};\boldsymbol{\theta}) = f_y(y|\mathbf{x};\boldsymbol{\theta}_y)\prod_{q=1}^{Q} f_q(x_q|x_1,\ldots,x_{q-1};\boldsymbol{\theta}_q)\ , \tag{12.10}$$

where f_q is the submodel for the qth covariate conditional on all covariates placed earlier in the sequence. In this context, it is worth noting that SMC-SM is really just a special case of SM with a specific order of variables that allows the submodel for y to be specified in accordance with the substantive analysis model. Finally, in SMC-FCS, the models for the covariates are again specified as a sequence of separate conditional models:

$$f_y(y|\mathbf{x})f_q(x_q|x_1,\ldots,x_{q-1},x_{q+1},\ldots,x_Q;\boldsymbol{\theta}_q)\ , \tag{12.11}$$

where f_q is the model for the qth covariate conditional on all other covariates (Bartlett et al., 2015). SMC-JM, SMC-SM, and SMC-FCS differ primarily in their representation of the multivariate distribution of the covariates. For example, when the relations between the covariates are themselves nonlinear, SMC-JM and SMC-FCS can suffer from similar shortcomings as their non-SMC counterparts. By contrast, SMC-SM inherits the flexibility of SM and thus allows random slopes and nonlinear effects of the predictors to be included in any of the submodels, given the order in which the submodels were arranged (Enders et al., 2020; Grund et al., 2021). Consequently, the three SMC-MI approaches—SMC-JM, SMC-SM, and SMC-FCS—again provide different ways to address these aspects of multilevel data.

Several authors have shown that multilevel SMC-MI can provide more accurate results than conventional approaches to MI in multilevel analyses with random slopes or nonlinear effects (e.g., Enders et al., 2020; Erler et al., 2017; Grund et al., 2021; Quartagno and Carpenter, 2022). However, it is important to emphasize that matching the substantive analysis alone does not guarantee unbiased results and that SMC-MI approaches can still produce biased results if the specified model fails to accommodate the "true" relations between the variables or if it fails to include all relevant information (e.g., auxiliary variables). Specifically, previous research has shown that SMC-MI can produce biased results if the imputation model for the covariates does not capture the "true" associations between them or if the "true" outcome model differs from the one that was specified in SMC-MI (Lüdtke et al., 2020b; see also Grund et al., 2021). This requirement is not unique to SMC-MI; however, researchers who use SMC-MI should still be mindful that accommodating the substantive model is only one part of the challenge. On the one hand, SMC-MI offers a powerful set of tools for tailoring the imputation model to a specific analysis, which can be extremely useful for complex multilevel analyses. On the other hand, conventional MI approaches are more limited in their ability to accommodate a *specific* analysis model, but they allow many *different*—though usually simpler—analyses to be conducted with the same data. This can be particularly useful in applications, in which the statistical analyses involve fitting multiple different models that can be difficult to accommodate all at once with SMC-MI.

Challenge 3: Auxiliary Variables

The third major challenge in multilevel MI is the selection of variables to be included in the imputation model. Several authors have argued that researchers should adopt an "inclusive" strategy for MI, whereby the imputation model is extended with additional variables than can be useful for the treatment of missing data even if they are not explicitly used in the analyses (Collins et al., 2001). These so-called *auxiliary* variables are especially useful if they are associated with (ideally) both the occurrence of missing data and the incomplete variables, because they can increase the plausibility of the MAR assumption and help researchers make full use of the available data (however, see also Thoemmes & Rose, 2014). Conversely, omitting relevant auxiliary variables has the potential to bias the results that are obtained in subsequent analyses.

All of the imputation approaches—JM, SM, and FCS—can include auxiliary variables, but they do so in different ways. In JM, FCS, SMC-JM, and SMC-FCS, auxiliary variables are included like any other variable. By contrast, in SM and SMC-SM, the auxiliary variables can be entered at different points in the model sequence, and this affects the order and contents of the submodels and the joint model that they imply (Daniels et al., 2014; Ibrahim et al., 2005). For example, Lüdtke et al. (2020b) outlined two options for including auxiliary variables in SMC-SM (Eq. 12.10). Suppose that there is only a single outcome, covariate, and auxiliary variable (y, x, and a). The first option is to include a as an additional outcome variable and to include the outcome and the covariates as its predictors, that is, $f_a(a|y, x; \boldsymbol{\theta}_a) f_y(y|x; \boldsymbol{\theta}_y) f_x(x; \boldsymbol{\theta}_x)$. The second option is to include a as an additional covariate, that is, $f_y(y|x, a; \boldsymbol{\theta}_y) f_x(x|a; \boldsymbol{\theta}_x) f_a(a; \boldsymbol{\theta}_a)$ or with the order of x and a reversed. In addition, when there are multiple auxiliary variables or covariates, different variations of these basic options are possible, and the usual rules of thumb can be used to determine a sensible order of the variables in SM and SMC-SM (e.g., variables at Level 2 before Level 1, categorical before continuous variables).

12.5 Model-Based Methods

The main idea of model-based methods is to estimate the parameters of an analysis model directly on the basis of the observed data (Anderson, 1957; Little & Rubin, 2019). Model-based methods "handle" missing data by extending the analysis model with additional distributional assumptions about the variables with missing data (for an overview, see Enders, 2022). In doing so, model-based methods ensure that only the observed data contribute to the estimation of the parameters, while the uncertainty that is due to missing data is taken into account. Model-based methods can be used with maximum-likelihood or Bayesian estimation (MLE or BE), and an illustration of BE is shown in the left half of Fig. 12.4, with MI shown on the right. In the following, we discuss the model-based treatment of missing data in multilevel analyses with MLE and BE and their advantages and disadvantages as compared with MI.

Maximum-Likelihood Estimation

In MLE, the model parameters are estimated by finding the values of the unknown parameters that maximize the likelihood of the data, $L(\boldsymbol{\theta}; \mathbf{y}) = P(\mathbf{y}|\boldsymbol{\theta})$. In cases with missing data, this principle is applied to only the observed part of the data, whereas the missing data are "integrated out" of the likelihood function (Little & Rubin, 2019). Using a simplified notation, where $\mathbf{y}_{\text{obs},j}$ and $\mathbf{y}_{\text{mis},j}$ denote the observed and the missing parts of the data in group j, the likelihood of the observed data can be written as

$$L(\boldsymbol{\theta}; \mathbf{y}_{\text{obs}}) = \prod_j P(\mathbf{y}_{\text{obs},j}|\boldsymbol{\theta}) = \prod_j \int P(\mathbf{y}_{\text{obs},j}, \mathbf{y}_{\text{mis},j}|\boldsymbol{\theta})\, d\mathbf{y}_{\text{mis},j} \;. \tag{12.12}$$

Conceptually, the observed-data likelihood "averages" over the possible values of the missing data, which requires that the substantive analysis model is extended to include distributional assumptions for the variables with missing data (see also Little & Rubin, 2019). There is a rich literature on using MLE in multilevel analyses, and previous applications include two-level models with random intercepts (McDonald, 1993; Rabe-Hesketh et al., 2004; Shin & Raudenbush, 2010), random slopes (Ibrahim et al., 2001; Rockwood, 2020; Stubbendick & Ibrahim, 2003), categorical data (Lee & Shi, 2001; Rockwood, 2021), and latent class variables (e.g., Vermunt, 2003).

Bayesian Estimation

In BE, parameters are estimated by sampling values from the posterior distribution of the parameters given the data, $P(\boldsymbol{\theta}|\mathbf{y}) \propto P(\mathbf{y}|\boldsymbol{\theta})P(\boldsymbol{\theta})$, which combines the likelihood of the data with prior information about the parameters (e.g., Gelman et al., 2014). When the data are incomplete, the parameters are estimated from the observed data posterior distribution, which again requires "integrating out" the missing data:

$$P(\boldsymbol{\theta}|\mathbf{y}_{\text{obs}}) \propto \left[\prod_j P(\mathbf{y}_{\text{obs},j}|\boldsymbol{\theta})\right] P(\boldsymbol{\theta}) = \left[\prod_j \int P(\mathbf{y}_{\text{mis},j}, \mathbf{y}_{\text{obs},j}|\boldsymbol{\theta})\, d\mathbf{y}_{\text{mis},j}\right] P(\boldsymbol{\theta}) \;. \tag{12.13}$$

Sampling from the observed-data posterior distribution is often not straightforward and usually requires iterative methods that use "data augmentation" to approximate the posterior distribution of the parameters with sampled (filled-in) values for the missing data in a two-step procedure (Tanner & Wong, 1987). At iteration t, the two steps create random draws for the parameters and the missing values as follows:

$$\begin{aligned} \mathbf{y}_{\text{mis}}^{(t+1)} &\sim P(\mathbf{y}_{\text{mis}} | \boldsymbol{\theta}^{(t)}, \mathbf{y}_{\text{obs}}) \\ \boldsymbol{\theta}^{(t+1)} &\sim P(\boldsymbol{\theta} | \mathbf{y}_{\text{mis}}^{(t+1)}, \mathbf{y}_{\text{obs}}). \end{aligned} \tag{12.14}$$

The resulting draws for $\boldsymbol{\theta}$ can be regarded as draws from the observed-data posterior distribution (Little & Rubin, 2019). The fact that BE requires sampling both the missing values and unknown parameter values makes BE phenomenologically similar to MI, as shown in Fig. 12.4. The main focus of BE as a model-based method is to estimate the model parameters directly from the observed data while treating the missing values as nuisance parameters. In principle, however, BE can be used as an imputation-based method, if the sampled values for the missing data are used to create imputed data sets that are analyzed in a subsequent step (Erler et al., 2016). In the literature on multilevel analysis with missing data, BE has been studied in a variety of contexts with different variable types and random effects configurations (e.g., Erler et al., 2017, 2016; Goldstein et al., 2014, 2009; Grund et al., 2018b).

Model Specification

Regardless of the estimation method used, the main challenge for using model-based methods in multilevel data is that they require the specification of a multivariate model that encompasses the outcome, the explanatory variables, and the auxiliary variables. This can be difficult because this "extended" model has to not only represent the relations between the variables—both substantive and auxiliary—but must do so without changing the meaning of the parameters of interest as well (as compared with the intended analysis). A particularly popular approach to conduct multilevel analyses with a model-based treatment of missing data is in the context of structural equation modeling (SEM, e.g., Arbuckle, 1996). In this framework, the multivariate model, $f(\mathbf{y}; \boldsymbol{\theta})$, is specified directly, and auxiliary variables can be added to this model with the so-called saturated correlates approach (Graham, 2003). As an alternative, Ibrahim et al. (2005) recommended factorizing the multivariate model into a sequence of univariate models, $f(\mathbf{y}; \boldsymbol{\theta}) = \prod_{p=1}^{P} f_p(y_p | y_1, \ldots, y_{p-1}; \boldsymbol{\theta}_p)$ (see also Daniels et al., 2014; Erler et al., 2016). In this approach, a univariate submodel is defined for the outcome, explanatory, and auxiliary variables, and the contents of each submodel can be adjusted to match the presumed relations between the variables (e.g., by including random slopes and nonlinear effects).

12.6 Comparison of Imputation-Based and Model-Based Methods

Imputation- and model-based methods for handling missing data have different advantages and disadvantages. The key advantage of imputation-based methods such as MI is that they separate the treatment of missing data from the analysis (Schafer & Graham, 2002). By specifying a rich imputation model, this allows—at least in principle—to make full use of the available data and to use the same set of imputed data for different analyses. However, the separation of these two tasks also creates challenges. As outlined above, this refers primarily to the fact that the imputation model should match the "true" data-generating model and the substantive analyses as closely as possible. In addition, using MI can also complicate certain aspects of the analysis such as model comparisons or the assessment of model fit due to the pooling phase involved in MI (e.g., Enders & Mansolf, 2018; Grund et al., 2023b).

Model-based methods such as MLE are often regarded as more user-friendly, because they combine the missing data treatment and the analyses in a single procedure (Enders, 2022). Software packages for multilevel SEM, in particular, often provide the ability to handle missing data with MLE or BE right out of the box (e.g., Asparouhov & Muthén, 2021; Muthén & Asparouhov, 2012; Rabe-Hesketh et al., 2004). However, there are challenges associated with model-based methods as well. First, model-based methods tie the treatment of missing data to the analysis, and changing the analysis (e.g., removal of a covariate) can inadvertently change the treatment of missing data. Second, model-based methods can create new issues in multilevel analyses. For example, Grund et al. (2019) showed that multilevel analyses with MLE can be difficult when the model includes manifest group means, because the distributional assumptions needed to "deal" with missing data can alter the meaning of the group-level effects (see also Lüdtke et al., 2008). And finally, model-based methods sometimes provide more limited options for including auxiliary variables (Daniels et al., 2014; Lüdtke et al., 2020b). This is because, in model-based analyses, the auxiliary variables must be included in such a way that the substantive meaning of the parameters of interest remains unchanged (e.g., as additional outcome variables or as saturated correlates; Graham, 2003; Lüdtke et al., 2020a). By comparison, imputation-based methods are more flexible and often provide different options for including auxiliary variables (see also Daniels et al., 2014; Lüdtke et al., 2020b). SMC-MI can be regarded as a hybrid between imputation-based and model-based methods: Despite being imputation-based, SMC-MI procedures tailor the missing data treatment to a specific analysis. This has advantages and disadvantages. On the one hand, complex analyses can be accommodated more effectively by SMC-MI than by conventional MI. On the other hand, imputed data that were created with SMC-MI can be more of a "one-trick pony," as the close fit with a specific analysis can reduce the utility of the imputed data for running other analyses. In comparison with model-based methods, SMC-MI has two advantages, namely that it (a) allows the imputation model for the outcome variable to be more general than the intended analysis (e.g., an imputation model with nonlinear terms for an analysis with only linear effects) and (b) simplifies the inclusion of auxiliary variables (Lüdtke et al., 2020b).

12.7 Example Analyses

To illustrate the imputation- and model-based treatment of missing data, we now provide two examples with simulated data. In Example 12.7, we consider a relatively simple case, in which the relations between the variables are all linear and homogeneous across groups, and the substantive analysis model is a multilevel random-intercept model. In Example 12.7, we then consider a case, in which the relations are heterogeneous across groups, and the substantive analysis model is a multilevel model with random slopes and nonlinear effects. The two examples were chosen because they illustrate the similarities and differences between the methods for handling missing data in situations where the methods allow accommodating the multilevel structure in either similar (Example 12.7) or different ways (Example 12.7) and in situations where the substantive analysis model aligns more (Example 12.7) or less (Example 12.7) well with the "true" model that generated the data. The simulated data and the compute code for the two examples are available on Zenodo (https://zenodo.org/record/8168221; Grund et al., 2023a).

Example 1 (Homogeneous Case)

The data for this example were simulated from a two-level model with two continuous variables (x and y) at Level 1 and one binary variable (w) at Level 2. To reduce the impact of sampling error, we simulated a large sample with 2000 groups, where each group included between 3 and 20 persons (n_j). To generate the three variables, we first generated the values for w for each group, where both values (zero and one) occurred with a probability of 50%. Then, we generated the values for x and y from the following model. For each person $i = 1, \ldots, n_j$ in each group $j = 1, \ldots, J$,

$$\begin{bmatrix} x_{ij} \\ y_{ij} \end{bmatrix} = \begin{bmatrix} 0 \\ 0.20 \end{bmatrix} w_j + \begin{bmatrix} u_{xj} \\ u_{yj} \end{bmatrix} + \begin{bmatrix} e_{xij} \\ e_{yij} \end{bmatrix} , \tag{12.15}$$

where the random intercepts $\mathbf{u}_j = (u_{xj}, u_{yj})$ were normally distributed, $\mathbf{u}_j \sim N(\mathbf{0}, \mathbf{T})$, with a variances of 0.30 and a correlation of 0.20, and the residuals $\mathbf{e}_{ij} = (e_{xij}, e_{yij})$ were normally distributed, $\mathbf{e}_{ij} \sim N(\mathbf{0}, \mathbf{\Sigma})$, with variances of 0.70 and a correlation of 0.50. As a consequence, x and y were related at both Levels 1 and 2, albeit more strongly at Level 1, and y was also related with w. Finally, we induced missing data in all three variables by randomly deleting 10% of the values in each variable in accordance with an MCAR mechanism.

In this example, we assumed that the substantive analysis model is the random-intercept model that is shown in Eq. 12.1 and reproduced here for clarity:

$$y_{ij} = \beta_0 + \beta_1(x_{ij} - \bar{x}_{\bullet j}) + \beta_2\bar{x}_{\bullet j} + \beta_3 w_j + u_{0j} + e_{ij} \,.$$

This model closely matched the way in which we generated the data, with the exception that the between- and within-group relations were represented by regression coefficients (β_1, β_2, and β_3) rather than correlations (Grund et al., 2018a; Resche-Rigon & White, 2018).

Treatment of Missing Data

To handle the missing data, we used three imputation-based methods (JM, FCS, and SMC-SM) and one model-based method (BE). The imputation-based methods were used to generate 100 imputed data sets each, whereas BE was used to estimate the parameters of the analysis model directly from the incomplete data. All methods were implemented in the statistical software R, and we provide a shortened version of the computer code in the Appendix (the full version is available on Zenodo; https://zenodo.org/record/8168221).

JM For JM, we used the R package `jomo` (Quartagno & Carpenter, 2023) and specified a multivariate imputation model that included the variables x and y at Level 1 and w at Level 2 as shown in Eq. 12.5. To include the binary variable w alongside the continuous variables x and y, this method uses the so-called *latent normal* approach, whereby binary variables are regarded as manifestations of an underlying continuous (normal) variable that has been categorized by some threshold (similar to a probit model; see Carpenter & Kenward, 2013; Quartagno & Carpenter, 2019). This specification accommodates the between- and within-group relations between all variables and thus "fits" both the data-generating model and the substantive analysis.

FCS For the FCS approach, we used the R package `mice` (van Buuren & Groothuis-Oudshoorn, 2011) and specified the imputation model as shown in Eq. 12.7. Specifically, the imputation models for x and y were multilevel random-intercept models that included the other variables (and their group means) as predictors. For w, the imputation model likewise included the group means of x and y but used *predictive mean matching* to ensure that the imputed values did not fall outside the observed range (zero and one). This specification, like JM, matched both substantive analysis and the "true" model that generated data.

SMC-SM For SMC-SM, we used the R package `mdmb` (Robitzsch & Lüdtke, 2022) and specified the imputation model as a sequence of univariate models as shown in

Eq. 12.6. Specifically, the model for w was a logistic regression model without any predictors; the models for x and y were then multilevel random-intercept models, where the model for x included only w as a predictor, and the model for y included both x (after centering) and w as well as the group means of x. This order also corresponded to the rules of thumb by Ibrahim et al. (2005).

Bayesian Estimation In addition to the imputation-based methods above, we treated the missing data in a model-based manner with BE, which we also conducted with the R package `mdmb`. To this end, we specified the same model sequence and the same submodels as we did in SMC-SM. However, in contrast with SMC-SM, where we used this model to generate imputations for subsequent analyses, we used BE to fit the model directly to the incomplete data, which allowed us to extract the parameter estimates for the substantive analysis from the respective submodel in the sequence (i.e., the submodel for y).

All of the procedures described above closely matched both the "true" model that generated the data and the substantive analysis model. In addition, because SMC-SM includes the substantive analysis model as one of its submodels, the imputations are guaranteed to be compatible with the substantive analysis. For this reason, we expected all methods to provide similar results.

Results

Table 12.1 summarizes the estimated parameters and standard errors obtained from each of the missing data handling techniques. In addition, the results obtained from the complete data (CD) are included for comparison. All methods produced results that were extremely close to CD, and the results were virtually identical across the missing data handling techniques. The largest differences occurred for the effects of w, which was slightly higher in SMC-SM and BE as compared with the other methods. Overall, the strong similarity across the procedures is not surprising, given that they were all specified in a manner that closely matched the multilevel structure and the substantive analysis.

Example 2 (Heterogeneous Case)

In the second example, we adapted the simulation procedure described above to allow the relations between the variables to vary across groups. Specifically, we again simulated the data from a two-level model with three variables (x, y, and w) and generated the values for w as before. However, the values for x and y were now simulated as follows:

$$\begin{bmatrix} x_{ij} \\ y_{ij} \end{bmatrix} = \begin{bmatrix} 0 \\ 0 \end{bmatrix} w_j + \begin{bmatrix} u_{xj} \\ u_{yj} \end{bmatrix} + \begin{bmatrix} e_{xij} \\ e_{yij} \end{bmatrix} , \tag{12.16}$$

Table 12.1 Results for the different missing data handling techniques in Example 12.7

	CD		JM		FCS		SMC-SM		MB-BE	
Par.	Est.	SE	Est.	SE	Est.	SE	Est.	SE	Est.	SE
Intercept	−0.022	0.019	−0.019	0.019	−0.020	0.019	−0.044	0.019	−0.038	0.024
$(x_{ij} - \bar{x}_{\bullet j})$	0.499	0.006	0.496	0.006	0.496	0.006	0.496	0.006	0.496	0.007
$\bar{x}_{\bullet j}$	0.270	0.021	0.274	0.022	0.273	0.022	0.272	0.022	0.273	0.022
w_j	0.233	0.026	0.225	0.028	0.226	0.028	0.272	0.027	0.262	0.024
Variance components										
Intercept	0.288		0.290		0.290		0.284		0.285	
Residual	0.520		0.522		0.522		0.523		0.523	

Note: CD = complete data, JM = multilevel MI with joint modeling, FCS = multilevel MI with fully conditional specification, SMC-SM = multilevel substantive-model-compatible MI with sequential modeling, MB-BE = model-based treatment with Bayesian estimation where the random intercepts were normally distributed as before, but the covariance matrix of the residuals, $\mathbf{e}_{ij} = (e_{xij}, e_{yij})$, was now assumed to vary across groups, that is, $\mathbf{e}_{ij} \sim N(\mathbf{0}, \boldsymbol{\Sigma}_j)$ with a group-specific covariance matrix $\boldsymbol{\Sigma}_j$. The covariance matrices $\boldsymbol{\Sigma}_j$ were assumed to follow an inverse-Wishart distribution,

$$\boldsymbol{\Sigma}_j \sim W^{-1}(\mathbf{S}_j, \nu) , \tag{12.17}$$

where the scale matrix $\mathbf{S}_j$ was chosen such that the average $\boldsymbol{\Sigma}_j$ comprised variances of 0.70 and correlations that depended on w. Specifically, when $w_j = 0$, this correlation was set to 0.30, and when $w_j = 1$, it was set to 0.70. The degrees of freedom were set to $\nu = 10$. As a consequence, the within-group relations between x and y varied across groups both due to random variation and as a function of w. Finally, in contrast with the previous example, we induced missing data only in x and deleted 25% of the values in accordance with a MAR mechanism, where the probability of missing values depended on y.

The substantive analysis model in this example was the multilevel random coefficients model shown in Eq. 12.2, which we again reproduce here for clarity:

$$\begin{aligned} y_{ij} = \beta_0 + \beta_1(x_{ij} - \bar{x}_{\bullet j}) + \beta_2\bar{x}_{\bullet j} + \beta_3 w_j + \beta_4(x_{ij} - \bar{x}_{\bullet j})w_j + \beta_5\bar{x}_{\bullet j}w_j + \\ u_{0j} + u_{1j}(x_{ij} - \bar{x}_{\bullet j}) + e_{ij} \ . \end{aligned}$$

This model differs from the "true" model that generated the data but still mirrors the key aspects of the multilevel structure. Specifically, this model assumes that the between- and within-group effect of x on y varies across groups both as a result of random variation (u_{1j}) and as a function of w (β_4 and β_5).

Treatment of Missing Data

To handle the missing data, we again used JM, FCS, SMC-SM, and BE. However, in comparison with the previous example, the specification of these methods was more challenging, because the relations between the variables were now not only *heterogeneous* across groups but also the "true" model now *differed* from the substantive analysis model. For this reason, we implemented the missing data handling techniques with different model specifications to illustrate their potential impact on the results. The computer code for this example is available on Zenodo (https://zenodo.org/record/8168221).

For JM, we specified the imputation model in two different ways. The first specification was identical to the previous example and assumed that the relations between the variables were homogeneous across groups. The second specification assumed that the relations were heterogeneous by allowing the residual covariance matrix to vary (Yucel, 2011). For FCS, we also considered two specifications. The first specification was identical to before. In the second specification, the imputation models (here, only for x) were extended to include not only the linear between-

and within-group effects of the predictors but also their interactions and quadratic effects (for a similar approach, see Grund et al., 2018b). Finally, for SMC-SM and BE, we specified the model sequence as in the previous example, but we extended the submodel for y to match the substantive analysis model, whereas the submodels for x and w were specified as before.

These methods and their specifications accommodate the multilevel structure and the substantive analysis to very different extents. For example, the JM and FCS specifications with homogeneous and linear effects do not accommodate the heterogeneous associations between the variables at all. Similarly, the extended JM approach accommodates primarily the random variation in the relations but not the systematic part that is due to w, whereas the extended FCS approach primarily accommodates variation in the relations that is due to nonlinear effects but not the random part. Finally, SMC-SM and BE explicitly ensure compatibility with the substantive analysis model but with the caveat that this model is not fully in line with the "true" multilevel structure. For this reason, we expected that the results differed more strongly across the procedures than they did before.

Results

Table 12.2 summarizes the estimated parameters and standard errors for each method. As expected, the results differed more strongly across the procedures. For JM, the results obtained with both the homogeneous and the heterogeneous specification differed noticeably from those obtained from CD. The same was true for the FCS specification that included only linear effects between the variables. By contrast, the FCS specification with nonlinear effects, SMC-SM, and BE all provided results that were similar to those in CD. The differences between these procedures were usually small and pertained primarily to the regression coefficient of the group-level interaction, where FCS appeared to perform better than SMC-SM and BE, and the slope variance, where SMC-SM and BE appeared to perform better than FCS. In contrast with the previous example, where accommodating the multilevel structure and substantive analysis was fairly straightforward with *all* methods considered, the present example illustrates that this no longer holds in cases with heterogeneous relations across groups, where the procedures differ more strongly and, consequently, can yield different results.

12.8 Software

Researchers today can select from a wide variety of statistical software that provides imputation- or model-based methods for handling missing data. In addition, the options available for multilevel and other types of dependent data have improved dramatically in the past years. In this section, we give a short overview of current

Table 12.2 Results for the different missing data handling techniques in Example 12.7

	CD		JM1		JM2		FCS1		FCS2		SMC-SM		MB-BE	
Par.	Est.	SE	Est.	SE	Est.	SE	Est.	SE	Est.	SE	Est.	SE	Est.	SE
Intercept	0.012	0.019	0.004	0.019	0.021	0.020	0.004	0.019	0.015	0.019	0.010	0.019	0.010	0.025
$(x_{ij} - \bar{x}_{\bullet j})$	0.297	0.013	0.391	0.013	0.228	0.071	0.390	0.013	0.327	0.014	0.340	0.016	0.336	0.019
$\bar{x}_{\bullet j}$	0.202	0.030	0.217	0.030	0.129	0.042	0.218	0.030	0.185	0.031	0.207	0.031	0.204	0.038
w_j	−0.021	0.027	−0.002	0.027	−0.011	0.028	−0.002	0.027	−0.023	0.027	−0.012	0.027	−0.012	0.024
$(x_{ij} - \bar{x}_{\bullet j})w_j$	0.398	0.018	0.254	0.018	0.271	0.028	0.256	0.018	0.336	0.018	0.336	0.021	0.339	0.018
$\bar{x}_{\bullet j} w_j$	0.115	0.044	0.063	0.045	0.033	0.047	0.058	0.045	0.120	0.045	0.083	0.045	0.085	0.039
Variance components														
Intercept	0.322		0.322		0.326		0.322		0.321		0.325		0.327	
$(x_{ij} - \bar{x}_{\bullet j})$	0.081		0.057		0.054		0.057		0.050		0.099		0.102	
Residual	0.433		0.458		0.553		0.458		0.463		0.427		0.427	

Note: CD = complete data, JM = multilevel MI with joint modeling (1 = homogeneous, 2 = heterogeneous) FCS = multilevel MI with fully conditional specification (1 = linear, 2 = nonlinear), SMC-SM = substantive-model-compatible MI with sequential modeling, MB-BE = model-based treatment with Bayesian estimation

software and the types of methods, data structures, and analysis models that they support.

Considering software for multilevel MI, a partial list of available software packages is shown in Table 12.3. These packages differ in terms of both their approach to specifying the imputation model and the types of data structures, variables, and models they can accommodate. In the statistical software R, packages for multilevel MI include `jomo` and `pan` (Schafer & Yucel, 2002) for JM (see also Grund et al., 2016; Quartagno et al., 2019), `mdmb` and `jointAI` (Erler et al., 2021) for SM, and `mice` and its extension packages for FCS. In addition, `jomo` also supports SMC-JM, and `mdmb` and `jointAI` also support SMC-SM. Standalone packages that can be used for multilevel MI include Blimp (Keller & Enders, 2022), which supports all of the methods considered in this chapter, M*plus* (Muthén & Muthén, 2017), which supports JM, SM, SMC-JM, and SMC-SM, and MLwiN (Charlton et al., 2022) and Latent Gold (Vermunt & Magidson, 2013), which support JM. Some packages also provide a more specific set of features beyond standard applications of multilevel MI. For example, while most software packages can handle missing data in variables in both continuous and categorical variables and at both Levels 1 and 2 (only `pan` does not), some packages such as `mice`, `mdmb`, and `jointAI` as well as Blimp and M*plus* can also handle other types of dependent structures such as three-level and cross-classified data. In addition, `jomo` and M*plus* also support specifications with heterogeneous variances and covariances, and limited support for these types of models is also provided by `mice`, `mdmb`, and Blimp. Latent Gold provides imputation approaches for categorical variables that are based on multilevel latent class analyses (Vidotto, Vermunt, & van Deun, 2018).

For a model-based treatment of missing data in multilevel analyses, a large variety of software can be used, so we focus on software packages that can handle arbitrary patterns of missing data in multilevel analyses. For MLE, this includes software for multilevel SEM, such as M*plus*, gllamm (Rabe-Hesketh et al., 2004), or Latent Gold. In R, multilevel analyses and multilevel SEM can be conducted with the packages `lavaan` (Rosseel, 2012) and `OpenMx` (Boker et al., 2011). Most of these packages allow users to extend the model specification with auxiliary variables, but their support for categorical variables or multilevel models with random slopes and nonlinear effects is often more limited (but see Asparouhov & Muthén, 2019; Rockwood, 2020).

Many of the aforementioned packages also provide model-based treatments using BE. In R, this includes the packages `mdmb` and `jointAI`. In addition, M*plus*, Blimp, and MLwiN are standalone packages that support BE. Finally, general-purpose software for Bayesian data analyses such as WinBUGS/OpenBUGS (Lunn et al., 2000), JAGS (Plummer, 2017), Stan (Stan Development Team, 2022), or the R package `nimble` (de Valpine et al., 2017) can also be used for BE. In comparison with MLE, software routines for BE are sometimes more flexible and can accommodate multilevel analyses with random slopes and other types of heterogeneous effects, either directly or by adopting sequential approach to model specification (e.g., see Erler et al., 2021; Lüdtke et al., 2020b).

Table 12.3 Overview of software packages for multilevel MI and supported methods, data structures, and models

		Supported data and models					
Software	Supported methods	CON	CAT	L2	3+	CC	HET
pan (R)	JM	✓					
jomo (R)	JM, SMC-JM	✓	✓	✓			✓
mice (R)	FCS	✓	✓	✓	✓	✓	✓[a]
mdmb (R)	SM, SMC-SM	✓	✓	✓	✓		✓[a]
jointAI (R)	SM, SMC-SM	✓	✓	✓	✓	✓	
Blimp	JM, SM, FCS, SMC-JM, SMC-SM, SMC-FCS	✓	✓	✓	✓		✓[a]
M*plus*	JM, SM, SMC-JM[b], SMC-SM[b]	✓	✓	✓	✓[b]	✓[b]	✓[b]
MLwiN	JM	✓	✓	✓			
Latent Gold	JM		✓	✓			

Note: CON = continuous variables, CAT = categorical variables, L2 = level-2 (or higher-level) variables, 3+ = hierarchical data with 3+ levels, CC = cross-classified data, HET = models with heterogeneous (co-)variance structures

[a] Partially supported (for specific model types or applications)

[b] Supported through Bayesian estimation procedure (H0 imputation)

12.9 Discussion

In this chapter, we considered a variety of methods for handling missing values in multilevel data. Successful applications of the methods require that the relevant features of the data and the intended analyses are taken into account. Throughout the chapter, we distinguished between imputation- and model-based methods that provide two different perspectives on handling missing data, where imputation-based methods such as MI aim to separate the treatment of missing data from the analysis, and model-based methods combine them in a single step. We outlined different strategies for multilevel MI—JM, SM, and FCS—with particular emphasis on how they specify the imputation model and the options they provide for accommodating the multilevel structure and the substantive analyses. In addition, we considered novel methods for SMC-MI, which accommodate the substantive analysis by incorporating it into the imputation model, thus providing a middle ground between the imputation- and model-based perspectives on missing data. We discussed the pros and cons of the methods and illustrated their application in two examples with simulated data.

For most parts of this chapter, we focused on two-level data, which has been the primary focus of research on missing data analysis and arguably represent the most common type of dependent data in cross-sectional and longitudinal research. In principle, the methods considered in this chapter can also be applied to other types of dependent data such as three-level, cross-classified, or multiple-membership data (for an overview, see Goldstein, 2011). For example, Wijesuriya et al. (2020) and Wijesuriya et al. (2022b) considered a variety of MI approaches for three-level data, and Grund et al. (2023c) proposed JM and FCS approaches for the imputation of cross-classified data (see also Wijesuriya et al., 2022a; Yucel et al., 2008). Especially in cross-classified data, the treatment of missing data comes with additional challenges due to the nonhierarchical structure of the data and the more complex pattern of dependency that they imply (Luo & Kwok, 2009; Meyers & Beretvas, 2006; Shi et al., 2010). Similarly, model-based methods such as MLE and BE can also be used in these applications (e.g., Browne et al., 2001). Nonetheless, software options to handle these types of dependent data structures are still relatively scarce, making these applications an important area for future research.

Furthermore, there are a number of challenges associated with multilevel and dependent data that we did not consider in detail. First, multilevel data often include a large number of variables, which can further complicate the treatment of missing data. Some of the methods discussed above can be combined with dimension reduction techniques such as principal components analysis (PCA) or partial least squares (PLS) that could allow a larger number of variables to be included in multilevel MI (e.g., Audigier et al., 2016; Kim et al., 2021). Future research should further explore these options. Second, when working with multilevel data, researchers sometimes use analyses other than multilevel analyses, for example, analyses with cluster-robust standard errors, generalized estimating equations (GEEs), or resampling methods (e.g., Cameron and Miller, 2015). In such a case, the multilevel structure of the data still needs to be taken into account, but it is still an open question how best to accommodate these types of analyses, for example, when using MI. Third, and related to this point, recent research has also demonstrated that MI can be combined with resampling procedures to improve the robustness of MI to misspecifications in the imputation model (Bartlett & Hughes, 2020), and future research should consider these options in more detail for multilevel MI.

Summing up, the statistical methods for handling missing values in multilevel and other types of dependent data have developed significantly over the past years, and researchers can now choose from an ever-growing variety of imputation- and model-based methods handling missing data. These methods often follow different paths to the treatment of missing data and provide different options for accommodating multilevel data. We hope that the present chapter will support the successful application of these methods in research with multilevel and other types of dependent data.

Appendix

Computer Code for Example 12.7

This appendix shows the computer code for the missing data handling techniques in Example 12.7. In each example, we assume that the incomplete data set is named `dat` and that the R packages `mice`, `miceadds`, `jomo`, `mdmb`, and `mitml` are available.

JM

```
# convert w into a factor
dat.jm <- within(dat, w <- factor(w, levels = c(0, 1)))

# specify imputation model
fml.jm <- list(
  x + y ~ 1 + (1|ID),
  w ~ 1
)

# run JM
imp.jm <- mitml::jomoImpute(
  data = dat.jm, formula = fml.jm, n.burn = 2500, n.iter = 50, m = 100
)
```

FCS

```
# specify imputation model
pred <- mice::make.predictorMatrix(dat)
pred[, "ID"] <- -2
pred["x", c("y", "w")] <- c(3, 1)
pred["y", c("x", "w")] <- c(3, 1)
pred["w", c("x", "y")] <- c(1, 1)

meth <- mice::make.method(dat)
meth[c("x", "y")] <- "2l.pan"
meth["w"] <- "2lonly.pmm"

# run FCS
imp.fcs <- mice::mice(
  data = dat, method = meth, predictorMatrix = pred, m = 100, maxit = 10
)
```

SMC-SM

```
# specify model sequence
y.fml <- y ~ 1 + cwc(x, ID) + gm(x, ID) + w + (1|ID)
x.fml <- x ~ 1 + w + (1|ID)
w.fml <- w ~ 1

y.mod <- list(model = "mlreg", formula = y.fml, sampling_level = "ID")
x.mod <- list(model = "mlreg", formula = x.fml, sampling_level = "ID")
w.mod <- list(model = "logistic", formula = w.fml, variable_level = "ID")

# run SMC-SM
imp.sm <- mdmb::frm_fb(
  dat, dep = y.mod, ind = list(x = x.mod, w = w.mod), burnin = 2500, iter = 7500,
  Nimp = 100, aggregation = TRUE
)
```

BE

```
# specify model sequence
y.fml <- y ~ 1 + cwc(x, ID) + gm(x, ID) + w + (1|ID)
x.fml <- x ~ 1 + w + (1|ID)
w.fml <- w ~ 1

y.mod <- list(model = "mlreg", formula = y.fml, sampling_level = "ID")
x.mod <- list(model = "mlreg", formula = x.fml, sampling_level = "ID")
w.mod <- list(model = "logistic", formula = w.fml, variable_level = "ID")

# fit analysis model with BE
fit.be <- mdmb::frm_fb(
  dat, dep = y.mod, ind = list(x = x.mod, w = w.mod), burnin = 2500, iter = 7500,
  aggregation = TRUE
)
```

References

Anderson, T. W. (1957). Maximum likelihood estimates for a multivariate normal distribution when some observations are missing. *Journal of the American Statistical Association*, *52*(278), 200–203. https://doi.org/10.2307/2280845

Arbuckle, J. L. (1996). Full information estimation in the presence of incomplete data. In G. A. Marcoulides & R. E. Schumacker (Eds.), *Advanced structural equation modeling: Issues and techniques*. Erlbaum.

Asparouhov, T., & Muthén, B. (2019). Latent variable centering of predictors and mediators in multilevel and time-series models. *Structural Equation Modeling: A Multidisciplinary Journal*, *26*, 119–142. https://doi.org/10.1080/10705511.2018.1511375

Asparouhov, T., & Muthén, B. (2021). Bayesian estimation of single and multilevel models with latent variable interactions. *Structural Equation Modeling: A Multidisciplinary Journal*, *28*, 314–328. https://doi.org/10.1080/10705511.2020.1761808

Asparouhov, T., & Muthén, B. O. (2010). *Multiple imputation with Mplus* (Technical Appendix). http://statmodel.com/

Audigier, V., Husson, F., & Josse, J. (2016). Multiple imputation for continuous variables using a Bayesian principal component analysis. *Journal of Statistical Computation and Simulation*, *86*, 2140–2156. https://doi.org/10.1080/00949655.2015.1104683

Audigier, V., White, I. R., Jolani, S., Debray, T. P. A., Quartagno, M., Carpenter, J., van Buuren, S., & Resche-Rigon, M. (2018). Multiple imputation for multilevel data with continuous and binary variables. *Statistical Science*, *33*, 160–183. https://doi.org/10.1214/18-STS646

Bartlett, J. W., & Hughes, R. A. (2020). Bootstrap inference for multiple imputation under uncongeniality and misspecification. *Statistical Methods in Medical Research*, *29*, 3533–3546. https://doi.org/10.1177/0962280220932189

Bartlett, J. W., Seaman, S. R., White, I. R., & Carpenter, J. R. (2015). Multiple imputation of covariates by fully conditional specification: Accommodating the substantive model. *Statistical Methods in Medical Research*, *24*, 462–487. https://doi.org/10.1177/0962280214521348

Boker, S., Neale, M., Maes, H., Wilde, M., Spiegel, M., Brick, T., Spies, J., Estabrook, R., Kenny, S., Bates, T., Mehta, P., & Fox, J. (2011). OpenMx: An open source extended structural equation modeling framework. *Psychometrika*, *76*, 306–317. https://doi.org/10.1007/s11336-010-9200-6

Browne, W. J., Goldstein, H., & Rasbash, J. (2001). Multiple membership multiple classification (MMMC) models. *Statistical Modelling*, *1*, 103–124. https://doi.org/10.1177/1471082X0100100202

Cameron, A. C., & Miller, D. L. (2015). A practitioner's guide to cluster-robust inference. *Journal of Human Resources*, *50*, 317–372. https://doi.org/10.3368/jhr.50.2.317

Carpenter, J. R., & Kenward, M. G. (2013). *Multiple imputation and its application*. Wiley.

Charlton, C., Rasbash, J., Browne, W. J., Healy, M., & Cameron, B. (2022). *MLwiN* (Version 3.06). Centre for Multilevel Modelling, University of Bristol, Bristol, UK.

Cheung, M. W.-L. (2007). Comparison of methods of handling missing time-invariant covariates in latent growth models under the assumption of missing completely at random. *Organizational Research Methods*, *10*, 609–634. https://doi.org/10.1177/1094428106295499

Collins, L. M., Schafer, J. L., & Kam, C.-M. (2001). A comparison of inclusive and restrictive strategies in modern missing data procedures. *Psychological Methods*, *6*, 330–351. https://doi.org/10.1037/1082-989X.6.4.330

Daniels, M., Wang, C., & Marcus, B. (2014). Fully Bayesian inference under ignorable missingness in the presence of auxiliary covariates. *Biometrics*, *70*, 62–72. https://doi.org/10.1111/biom.12121

de Valpine, P., Turek, D., Paciorek, C. J., Anderson-Bergman, C., Lang, D. T., & Bodik, R. (2017). Programming with models: Writing statistical algorithms for general model structures with NIMBLE. *Journal of Computational and Graphical Statistics*, *26*, 403–413. https://doi.org/10.1080/10618600.2016.1172487

Du, H., Alacam, E., Mena, S., & Keller, B. T. (2022a). Compatibility in imputation specification. *Behavior Research Methods*. https://doi.org/10.3758/s13428-021-01749-5

Du, H., Enders, C., Keller, B. T., Bradbury, T. N., & Karney, B. R. (2022b). A Bayesian latent variable selection model for nonignorable missingness. *Multivariate Behavioral Research*, *57*, 478–512. https://doi.org/10.1080/00273171.2021.1874259

Enders, C. K. (2022). *Applied missing data analysis* (2nd ed.). Guilford Press.

Enders, C. K., Du, H., & Keller, B. T. (2020). A model-based imputation procedure for multilevel regression models with random coefficients, interaction effects, and nonlinear terms. *Psychological Methods*, *25*, 88–112. https://doi.org/10.1037/met0000228

Enders, C. K., Hayes, T., & Du, H. (2018a). A comparison of multilevel imputation schemes for random coefficient models: Fully conditional specification and joint model imputation with random covariance matrices. *Multivariate Behavioral Research*, *53*, 695–713. https://doi.org/10.1080/00273171.2018.1477040

Enders, C. K., Keller, B. T., & Levy, R. (2018b). A fully conditional specification approach to multilevel imputation of categorical and continuous variables. *Psychological Methods*, *23*, 298–317. https://doi.org/10.1037/met0000148

Enders, C. K., & Mansolf, M. (2018). Assessing the fit of structural equation models with multiply imputed data. *Psychological Methods*, *23*, 76–93. https://doi.org/10.1037/met0000102

Enders, C. K., Mistler, S. A., & Keller, B. T. (2016). Multilevel multiple imputation: A review and evaluation of joint modeling and chained equations imputation. *Psychological Methods*, *21*, 222–240. https://doi.org/10.1037/met0000063

Erler, N. S., Rizopoulos, D., Jaddoe, V. W., Franco, O. H., & Lesaffre, E. M. (2017). Bayesian imputation of time-varying covariates in linear mixed models. *Statistical Methods in Medical Research*, *28*, 555–568. https://doi.org/10.1177/0962280217730851

Erler, N. S., Rizopoulos, D., & Lesaffre, E. M. E. H. (2021). JointAI: Joint analysis and imputation of incomplete data in R. *Journal of Statistical Software*, *100*(20), 1–56. https://doi.org/10.18637/jss.v100.i20

Erler, N. S., Rizopoulos, D., van Rosmalen, J., Jaddoe, V. W. V., Franco, O. H., & Lesaffre, E. M. E. H. (2016). Dealing with missing covariates in epidemiologic studies: A comparison between multiple imputation and a full Bayesian approach. *Statistics in Medicine*, *35*, 2955–2974. https://doi.org/10.1002/sim.6944

Gasimova, F., Robitzsch, A., Wilhelm, O., & Hülür, G. (2014). A hierarchical Bayesian model with correlated residuals for investigating stability and change in intensive longitudinal data settings. *Methodology*, *10*(4), 126–137. https://doi.org/10.1027/1614-2241/a000083

Gelman, A., Carlin, J. B., Stern, H. S., Dunson, D., Vehtari, A., & Rubin, D. B. (2014). *Bayesian data analysis* (3rd). CRC press.

Goldstein, H. (2011). *Multilevel statistical models* (4th ed.). Wiley.

Goldstein, H., Carpenter, J. R., & Browne, W. J. (2014). Fitting multilevel multivariate models with missing data in responses and covariates that may include interactions and non-linear terms. *Journal of the Royal Statistical Society: Series A (Statistics in Society)*, *177*, 553–564. https://doi.org/10.1111/rssa.12022

Goldstein, H., Carpenter, J. R., Kenward, M. G., & Levin, K. A. (2009). Multilevel models with multivariate mixed response types. *Statistical Modelling*, *9*, 173–197. https://doi.org/10.1177/1471082X0800900301

Graham, J. W. (2003). Adding missing-data-relevant variables to FIML-based structural equation models. *Structural Equation Modeling: A Multidisciplinary Journal*, *10*, 80–100. https://doi.org/10.1207/S15328007SEM1001_4

Grund, S., Lüdtke, O., & Robitzsch, A. (2016). Multiple imputation of multilevel missing data: An introduction to the R package pan. *SAGE Open*, *6*(4), 1–17. https://doi.org/10.1177/2158244016668220

Grund, S., Lüdtke, O., & Robitzsch, A. (2018a). Multiple imputation of missing data at level 2: A comparison of fully conditional and joint modeling in multilevel designs. *Journal of Educational and Behavioral Statistics*, *43*, 316–353. https://doi.org/10.3102/1076998617738087

Grund, S., Lüdtke, O., & Robitzsch, A. (2018b). Multiple imputation of missing data for multilevel models: Simulations and recommendations. *Organizational Research Methods*, *21*, 111–149. https://doi.org/10.1177/1094428117703686

Grund, S., Lüdtke, O., & Robitzsch, A. (2019). Missing data in multilevel research. In S. E. Humphrey & J. M. LeBreton (Eds.), *Handbook for multilevel theory, measurement, and analysis* (pp. 365–386). American Psychological Association.

Grund, S., Lüdtke, O., & Robitzsch, A. (2021). Multiple imputation of missing data in multilevel models with the R package mdmb: A flexible sequential modeling approach. *Behavior Research Methods*, *53*, 2631–2649. https://doi.org/10.3758/s13428-020-01530-0

Grund, S., Lüdtke, O., & Robitzsch, A. (2023a, July 20). *Missing data in the analysis of multilevel and dependent data (Examples)*. https://doi.org/10.5281/zenodo.8168221

Grund, S., Lüdtke, O., & Robitzsch, A. (2023b). Pooling methods for likelihood ratio tests in multiply imputed data sets. *Psychological Methods*, *28*, 1207–1221. https://doi.org/10.1037/met0000556

Grund, S., Lüdtke, O., & Robitzsch, A. (2023c). Handling missing data in cross-classified multilevel analyses: An evaluation of different multiple imputation approaches. *Journal of Educational and Behavioral Statistics*, *48*, 454–489. https://doi.org/10.3102/10769986231151224

Hedeker, D., Mermelstein, R. J., & Demirtas, H. (2008). An application of a mixed-effects location scale model for analysis of ecological momentary assessment (EMA) data. *Biometrics*, *64*, 627–634. https://doi.org/10.1111/j.1541-0420.2007.00924.x

Hoffman, L. (2015). *Longitudinal analysis: Modeling within-person fluctuation and change* (1st ed.). Routledge.

Hughes, R. A., White, I. R., Seaman, S. R., Carpenter, J. R., Tilling, K., & Sterne, J. A. (2014). Joint modelling rationale for chained equations. *BMC Medical Research Methodology*, *14*(28), 1–10. https://doi.org/10.1186/1471-2288-14-28

Ibrahim, J. G., Chen, M.-H., & Lipsitz, S. R. (2001). Missing responses in generalised linear mixed models when the missing data mechanism is nonignorable. *Biometrika*, *88*, 551–564. https://doi.org/10.1093/biomet/88.2.551

Ibrahim, J. G., Chen, M.-H., & Lipsitz, S. R. (2002). Bayesian methods for generalized linear models with covariates missing at random. *Canadian Journal of Statistics*, *30*, 55–78. https://doi.org/10.2307/3315865

Ibrahim, J. G., Chen, M.-H., Lipsitz, S. R., & Herring, A. H. (2005). Missing-data methods for generalized linear models: A comparative review. *Journal of the American Statistical Association*, *100*, 332–346. https://doi.org/10.1198/016214504000001844

Jolani, S., Debray, T. P. A., Koffijberg, H., van Buuren, S., & Moons, K. G. M. (2015). Imputation of systematically missing predictors in an individual participant data meta-analysis: A generalized approach using MICE. *Statistics in Medicine*, *34*, 1841–1863. https://doi.org/10.1002/sim.6451

Keller, B. T., & Enders, C. K. (2022). *Blimp user's manual* (Version 3.0). Los Angeles, CA.

Kim, S., Belin, T. R., & Sugar, C. A. (2018). Multiple imputation with non-additively related variables: Joint-modeling and approximations. *Statistical Methods in Medical Research*, *27*, 1683–1694. https://doi.org/10.1177/0962280216667763

Kim, Y., Lee, J. H., & Little, T. (2021). Multiple imputation with principal components for non-normal categorical data. *Multivariate Behavioral Research*, *56*, 165–166. https://doi.org/10.1080/00273171.2020.1869516

Leckie, G., French, R., Charlton, C., & Browne, W. (2014). Modeling heterogeneous variance-covariance components in two-level models. *Journal of Educational and Behavioral Statistics*, *39*(5), 307–332. https://doi.org/10.3102/1076998614546494

Lee, S.-Y., & Shi, J.-Q. (2001). Maximum likelihood estimation of two-level latent variable models with mixed continuous and polytomous data. *Biometrics*, *57*, 787–794. https://doi.org/10.1111/j.0006-341X.2001.00787.x

Little, R. J. A. (1988). A test of missing completely at random for multivariate data with missing values. *Journal of the American Statistical Association*, *83*, 1198–1202. https://doi.org/10.2307/2290157

Little, R. J. A., & Rubin, D. B. (2019). *Statistical analysis with missing data*. Wiley.
Liu, J., Gelman, A., Hill, J., Su, Y.-S., & Kropko, J. (2014). On the stationary distribution of iterative imputations. *Biometrika*, *101*, 155–173. https://doi.org/10.1093/biomet/ast044
Lüdtke, O., Marsh, H. W., Robitzsch, A., Trautwein, U., Asparouhov, T., & Muthén, B. O. (2008). The multilevel latent covariate model: A new, more reliable approach to group-level effects in contextual studies. *Psychological Methods*, *13*, 203–229. https://doi.org/10.1037/a0012869
Lüdtke, O., Robitzsch, A., & Grund, S. (2017). Multiple imputation of missing data in multilevel designs: A comparison of different strategies. *Psychological Methods*, *22*, 141–165. https://doi.org/10.1037/met0000096
Lüdtke, O., Robitzsch, A., & West, S. G. (2020a). Analysis of interactions and nonlinear effects with missing data: A factored regression modeling approach using maximum likelihood estimation. *Multivariate Behavioral Research*, *55*, 361–381. https://doi.org/10.1080/00273171.2019.1640104
Lüdtke, O., Robitzsch, A., & West, S. G. (2020b). Regression models involving nonlinear effects with missing data: A sequential modeling approach using Bayesian estimation. *Psychological Methods*, *25*, 157–181. https://doi.org/10.1037/met0000233
Lunn, D. J., Thomas, A., Best, N., & Spiegelhalter, D. (2000). WinBUGS - A Bayesian modelling framework: Concepts, structure, and extensibility. *Statistics and Computing*, *10*, 325–337. https://doi.org/10.1023/A:1008929526011
Luo, W., & Kwok, O.-M. (2009). The impacts of ignoring a crossed factor in analyzing cross- classified data. *Multivariate Behavioral Research*, *44*, 182–212. https://doi.org/10.1080/00273170902794214
McDonald, R. P. (1993). A general model for two-level data with responses missing at random. *Psychometrika*, *58*, 575–585. http://link.springer.com/article/10.1007/BF02294828
Meng, X.-L. (1994). Multiple-imputation inferences with uncongenial sources of input. *Statistical Science*, *9*(4), 538–558. https://doi.org/10.1214/ss/1177010269
Meyers, J. L., & Beretvas, S. N. (2006). The impact of inappropriate modeling of cross-classified data structures. *Multivariate Behavioral Research*, *41*, 473–497. https://doi.org/10.1207/s15327906mbr4104_3
Mistler, S. A., & Enders, C. K. (2017). A comparison of joint model and fully conditional specification imputation for multilevel missing data. *Journal of Educational and Behavioral Statistics*, *42*, 432–466. https://doi.org/10.3102/1076998617690869
Murray, J. S. (2018). Multiple imputation: A review of practical and theoretical findings. *Statistical Science*, *33*, 142–159. https://doi.org/10.1214/18-STS644
Muthén, B. O., & Asparouhov, T. (2012). Bayesian structural equation modeling: A more flexible representation of substantive theory. *Psychological Methods*, *17*, 313–335. https://doi.org/10.1037/a0026802
Muthén, L. K., & Muthén, B. O. (2017). *Mplus user's guide* (8th ed.). Muthén & Muthén. https://www.statmodel.com
Plummer, M. (2017). *JAGS: A program for analysis of Bayesian graphical models using Gibbs sampling* (Version 4.3.0). http://sourceforge.net/projects/mcmc-jags/
Quartagno, M., & Carpenter, J. R. (2016). Multiple imputation for IPD meta-analysis: Allowing for heterogeneity and studies with missing covariates. *Statistics in Medicine*, *35*, 2938–2954. https://doi.org/10.1002/sim.6837
Quartagno, M., & Carpenter, J. R. (2019). Multiple imputation for discrete data: Evaluation of the joint latent normal model. *Biometrical Journal*, *61*, 1003–1019. https://doi.org/10.1002/bimj.201800222
Quartagno, M., & Carpenter, J. R. (2022). Substantive model compatible multilevel multiple imputation: A joint modeling approach. *Statistics in Medicine*, *41*, 5000–5015. https://doi.org/10.1002/sim.9549
Quartagno, M., & Carpenter, J. R. (2023). *jomo: A package for multilevel joint modelling multiple imputation (Version 2.7-6)*. http://CRAN.R-project.org/package=jomo

Quartagno, M., Grund, S., & Carpenter, J. (2019). jomo: A flexible package for two-level joint modelling multiple imputation. *R Journal, 11*(2), 205–228. https://doi.org/10.32614/RJ-2019-028

Rabe-Hesketh, S., Skrondal, A., & Pickles, A. (2004). Generalized multilevel structural equation modeling. *Psychometrika, 69*, 167–190. https://doi.org/10.1007/BF02295939

Raghunathan, T. E., Lepkowski, J. M., van Hoewyk, J., & Solenberger, P. (2001). A multivariate technique for multiply imputing missing values using a sequence of regression models. *Survey Methodology, 27*(1), 85–96. http://www.statcan.gc.ca/

Raudenbush, S. W., & Bryk, A. S. (2002). *Hierarchical linear models: Applications and data analysis methods* (2nd ed.). Sage.

Resche-Rigon, M., & White, I. R. (2018). Multiple imputation by chained equations for systematically and sporadically missing multilevel data. *Statistical Methods in Medical Research, 27*, 1634–1649. https://doi.org/10.1177/0962280216666564

Robins, J., & Wang, N. (2000). Inference for imputation estimators. *Biometrika, 87*, 113–124. https://doi.org/10.1093/biomet/87.1.113

Robitzsch, A., & Lüdtke, O. (2022). *mdmb: Model based treatment of missing data* (Version 1.6-5). https://CRAN.R-project.org/package=mdmb

Rockwood, N. J. (2020). Maximum likelihood estimation of multilevel structural equation models with random slopes for latent covariates. *Psychometrika, 85*, 275–300. https://doi.org/10.1007/s11336-020-09702-9

Rockwood, N. J. (2021). Efficient likelihood estimation of generalized structural equation models with a mix of normal and nonnormal responses. *Psychometrika, 86*, 642–667. https://doi.org/10.1007/s11336-021-09770-5

Rosseel, Y. (2012). lavaan: An R package for structural equation modeling. *Journal of Statistical Software, 48*(2), 1–36. https://doi.org/10.18637/jss.v048.i02

Royston, P. (2005). Multiple imputation of missing values: Update. *The Stata Journal, 5*, 188–201. http://www.stata-journal.com

Rubin, D. B. (1987). *Multiple imputation for nonresponse in surveys*. Wiley.

Rubin, D. B. (1996). Multiple imputation after 18+ years. *Journal of the American Statistical Association, 91*, 473–489. https://doi.org/10.1080/01621459.1996.10476908

Schafer, J. L. (2003). Multiple imputation in multivariate problems when the imputation and analysis models differ. *Statistica Neerlandica, 57*, 19–35. https://doi.org/10.1111/1467-9574.00218

Schafer, J. L., & Graham, J. W. (2002). Missing data: Our view of the state of the art. *Psychological Methods, 7*, 147–177. https://doi.org/10.1037//1082-989X.7.2.147

Schafer, J. L., & Yucel, R. M. (2002). Computational strategies for multivariate linear mixed-effects models with missing values. *Journal of Computational and Graphical Statistics, 11*, 437–457. https://doi.org/10.1198/106186002760180608

Seaman, S. R., Bartlett, J. W., & White, I. R. (2012). Multiple imputation of missing covariates with non-linear effects and interactions: An evaluation of statistical methods. *BMC medical research methodology, 12*(46), 1–13. https://doi.org/10.1186/1471-2288-12-46

Shi, Y., Leite, W., & Algina, J. (2010). The impact of omitting the interaction between crossed factors in cross-classified random effects modelling. *British Journal of Mathematical and Statistical Psychology, 63*, 1–15. https://doi.org/10.1348/000711008X398968

Shin, Y., & Raudenbush, S. W. (2010). A latent cluster-mean approach to the contextual effects model with missing data. *Journal of Educational and Behavioral Statistics, 35*, 26–53. https://doi.org/10.3102/1076998609345252

Snijders, T. A. B., & Bosker, R. J. (2012). Components of variance. In *Multilevel analysis: An introduction to basic and advanced multilevel modeling* (pp. 114–117). Sage.

Stan Development Team. (2022). *Stan modeling language user's guide and reference manual* (Version 2.31). http://mc-stan.org

Stubbendick, A. L., & Ibrahim, J. G. (2003). Maximum likelihood methods for nonignorable missing responses and covariates in random effects models. *Biometrics, 59*, 1140–1150. https://doi.org/10.1111/j.0006-341X.2003.00131.x

Tanner, M. A., & Wong, W. H. (1987). The calculation of posterior distributions by data augmentation. *Journal of the American Statistical Association*, *82*, 528–540. https://doi.org/10.2307/2289457

Thoemmes, F., & Rose, N. (2014). A cautious note on auxiliary variables that can increase bias in missing data problems. *Multivariate Behavioral Research*, *49*, 443–459. https://doi.org/10.1080/00273171.2014.931799

van Buuren, S. (2011). Multiple imputation of multilevel data. In J. J. Hox (Ed.), *Handbook of advanced multilevel analysis* (pp. 173–196). Routledge.

van Buuren, S., Brand, J. P. L., Groothuis-Oudshoorn, C. G. M., & Rubin, D. B. (2006). Fully conditional specification in multivariate imputation. *Journal of Statistical Computation and Simulation*, *76*, 1049–1064. https://doi.org/10.1080/10629360600810434

van Buuren, S., & Groothuis-Oudshoorn, K. (2011). mice: Multivariate imputation by chained equations in R. *Journal of Statistical Software*, *45*(3), 1–67. https://doi.org/10.18637/jss.v045.i03

Vermunt, J. K. (2003). Multilevel latent class models. *Sociological Methodology*, *33*, 213–239. https://doi.org/10.1111/j.0081-1750.2003.t01-1-00131.x

Vermunt, J. K., & Magidson, J. (2013). *Latent GOLD (Version 5.0)*.

Vidotto, D., Vermunt, J. K., & van Deun, K. (2018). Bayesian multilevel latent class models for the multiple imputation of nested categorical data. *Journal of Educational and Behavioral Statistics*, *43*, 511–539. https://doi.org/10.3102/1076998618769871

von Hippel, P. T. (2009). How to impute interactions, squares, and other transformed variables. *Sociological Methodology*, *39*, 265–291. https://doi.org/10.1111/j.1467-9531.2009.01215.x

Wijesuriya, R., Moreno-Betancur, M., Carlin, J. B., De Silva, A. P., & Lee, K. J. (2022a). Multiple imputation approaches for handling incomplete three-level data with time-varying cluster-memberships. *Statistics in Medicine*, *41*, 4385–4402. https://doi.org/10.1002/sim.9515

Wijesuriya, R., Moreno-Betancur, M., Carlin, J. B., De Silva, A. P., & Lee, K. J. (2022b). Evaluation of approaches for accommodating interactions and non-linear terms in multiple imputation of incomplete three-level data. *Biometrical Journal*, *64*(8), 1404–1425. https://doi.org/10.1002/bimj.202000343

Wijesuriya, R., Moreno-Betancur, M., Carlin, J. B., & Lee, K. J. (2020). Evaluation of approaches for multiple imputation of three-level data. *BMC Medical Research Methodology*, *20*(207), 1–15. https://doi.org/10.1186/s12874-020-01079-8

Wu, L. (2010). *Mixed effects models for complex data*. CRC Press.

Yucel, R. M. (2011). Random covariances and mixed-effects models for imputing multivariate multilevel continuous data. *Statistical Modelling*, *11*, 351–370. https://doi.org/10.1177/1471082X1001100404

Yucel, R. M., Ding, H., Uludag, A. K., & Tomaskovic-Devey, D. (2008). Multiple imputation in multiple classification and multiple-membership structures. *Proceedings of the Section on Bayesian Statistical Science of the American Statistical Association*. http://www.asasrms.org/Proceedings/y2008f.html

Chapter 13
Bootstrap Methods for Robust Multilevel Analysis

Yichi Zhang, Winnie Wing-Yee Tse, and Mark H. C. Lai

13.1 An Overview of Bootstrap Methods

Data used in social and behavioral science research commonly have multilevel structures, where units are nested within clusters. For example, students are nested within classrooms in educational studies, or participants are repeatedly measured at different time points in longitudinal studies. Multilevel models (MLM), also known as hierarchical linear models and mixed effects models, have been widely used to account for clustered data. The most common estimation method for MLM is based on maximum likelihood (ML) estimation, which generates the most probable value of model parameters given the observed data (Raudenbush & Bryk, 2002). The accuracy of the ML estimates and the associated statistical inferences depend on whether the underlying assumptions are satisfied. Specifically, ML assumes that the data have sufficiently large sample sizes and residuals at each level follow independent and identical distributions (i.i.d.; Van der Leeden et al., 2008). When these two assumptions are violated, the desirable properties of ML, namely, consistency, asymptotic efficiency, and asymptotic normality, may not hold, resulting in biased standard error estimates with inaccurate inferences. The bootstrap, a resampling method that approximates the sampling distribution of statistics, can be used to obtain more accurate standard error estimates, confidence intervals, and statistical inferences when assumptions are violated (Busing, 1993).

In this chapter, we explain different types of bootstrapping procedures for MLM and illustrate how and when researchers can use these bootstrap methods in applied research. Section 13.1 is an overview of assumption violations in MLM, followed by a discussion of the general procedure involved in bootstrap methods

Y. Zhang · W. W.-Y. Tse · M. H. C. Lai (✉)
Department of Psychology, University of Southern California, Los Angeles, CA, USA
e-mail: hokchiol@usc.edu

M. Stemmler et al. (eds.), *Dependent Data in Social Sciences Research*,
https://doi.org/10.1007/978-3-031-56318-8_13

for regression models. In Sect. 13.2, we discuss the challenges of bootstrapping in MLM and introduce several bootstrapping procedures. Section 13.3 is a discussion of bootstrap confidence intervals and comparisons of some commonly available options. Section 13.4 demonstrates the use of various bootstrapping procedures on an empirical example, discussed in Snijders and Bosker (2012), which investigated whether school differences exist in students' language proficiency. The last section provides suggestions on available software and resources for MLM bootstrap methods.

Model Equation and Assumptions

This chapter focuses on two-level models, where participants or observations (level-1 units) are nested within clusters (level-2 units). For J clusters of $n_1, n_2, \ldots, n_j$ participants, and $p - 1$ predictors, a two-level model is given by

$$\mathbf{y}_j = \mathbf{X}_j \boldsymbol{\beta} + \mathbf{Z}_j \mathbf{U}_j + \boldsymbol{\epsilon}_j \tag{13.1}$$

where $\mathbf{y}_j$ is an $n_j \times 1$ outcome vector for the jth cluster, $\mathbf{X}_j$ is an $n_j \times p$ design matrix for the unknown fixed parameters (including the intercept and predictors), $\boldsymbol{\beta}$ is an $p \times 1$ vector for the fixed effects, $\mathbf{Z}_j$ is an $n_j \times r$ design matrix for r random effects, $\mathbf{U}_j$ is an $r \times 1$ vector of the unknown random components (including the random intercept and random slopes), and $\boldsymbol{\epsilon}_j$ is an $n_j \times 1$ error vector. $\mathbf{U}_j$ and $\boldsymbol{\epsilon}_j$ are assumed to follow a multivariate normal distribution:

$$\begin{pmatrix} \boldsymbol{\epsilon}_j \\ \mathbf{U}_j \end{pmatrix} \sim \mathcal{N}\left[\begin{pmatrix} \mathbf{0} \\ \mathbf{0} \end{pmatrix}, \begin{pmatrix} \boldsymbol{\Sigma} & \mathbf{0} \\ \mathbf{0} & \mathbf{T} \end{pmatrix}\right] \tag{13.2}$$

where $\mathbf{T}$ is the variance-covariance matrix of the random components, and assuming homoscedasticity, $\boldsymbol{\Sigma} = \sigma^2 \mathbf{I}$ for σ^2 being the error variance.

Model (13.1) implies five main assumptions: linearity, independence of errors, homoscedasticity, normality, and correct model specification. First, the model assumes the relationship between the predictors and outcome is linear. This assumption can be relaxed with a non-linear multilevel model (e.g., Davidian, 2017; Lindstrom & Bates, 1990). Second, the independence of errors assumption denotes that level-2 random components, $\mathbf{U}_j$, are independent between clusters and of level-1 errors, $\boldsymbol{\epsilon}_j$ (Snijders & Bosker, 2012).

Third, the homoscedasticity assumption requires equal variances of errors and random effects across all values of the predictors. Heteroscedasticity, meaning unequal variances, happens when the total variance of $\mathbf{y}_j$, given the values of $\mathbf{X}_j = \mathbf{x}_j$, depends on $\mathbf{x}_j$. For example, in a model with age as a predictor, heteroscedasticity occurs when the error variance is different for participants of different ages. Violation of homoscedasticity can result in biased estimation of the fixed effects and their standard errors and hence erroneous statistical inferences

(Huang et al., 2022; Raudenbush & Bryk, 2002; Snijders & Bosker, 2012; Zhang & Lai, 2024). Fourth, the error and random components are assumed to be multivariate normal, which may be violated with skewed distributions and the presence of outliers or influential observations. Finally, the estimates of model parameters are meaningful only if the model is correctly specified. The correct specification assumption requires that the model includes relevant predictors, cluster means of participant-level predictors, and relevant random slope terms (Snijders & Bosker, 2012).

Basics of Bootstrap Methods

Bootstrap methods are useful for computing standard errors and confidence intervals (CIs) and testing hypotheses for an estimator when some model assumptions are violated (Davison & Hinkley, 1997; Efron & Tibshirani, 1993; Wilcox, 2017). The general procedure of bootstrapping is to repeatedly resample observations and compute the estimator for each resample. The bootstrap process yields an approximate distribution of the estimator, from which the standard errors and CIs can be calculated. Bootstrap methods have been shown to perform well in controlling Type I error rates and maintaining accurate CI coverages such as when normality is violated (e.g., Efron, 1988; Strube, 1988) and homoscedasticity fails (e.g., Krishnamoorthy et al., 2007; Wilcox & Muska, 2001). As we will discuss in the following sections, there are many bootstrap procedures, some better than others in controlling Type I error rates in certain situations. We begin with an introduction of the percentile bootstrap for computing a CI and a standard error for the sample mean.

Consider that we collected data with n observations, $X_1, \ldots, X_n$, and aim to test against the null hypothesis, $H_0 : \mu = \mu_0 = 0$, that the population mean is zero. The student t-statistic is

$$T = \frac{\hat{\mu} - \mu_0}{\hat{se}_N}, \tag{13.3}$$

where $\hat{\mu} = \frac{1}{n}\sum_i^n X_i$ is the sample mean, μ_0 is the null value of the population mean, $\hat{se}_N = s/\sqrt{n}$ is the sample standard error of the mean, and s is the sample standard deviation. The 95% CI for $\hat{\mu}$ can be constructed as

$$[\hat{\mu} - t_{1-\frac{1}{2}\alpha}\hat{se}_N, \hat{\mu} + t_{1-\frac{1}{2}\alpha}\hat{se}_N], \tag{13.4}$$

where $t_{1-\frac{1}{2}\alpha}$ is the $1 - \frac{1}{2}\alpha$ quantile of the central t distribution.

However, suppose our data have a skewed distribution and violate the normality assumption. To account for nonnormality, we can perform bootstrapping using the following procedure (Efron & Tibshirani, 1993; Wilcox, 2017):

1. Randomly draw n observations from the observed data, *with replacement*, to create a bootstrap sample, $X_1^*, \ldots, X_n^*$.
2. For the bootstrap sample, compute the *bootstrap sample mean*, $\hat{\mu}_b^* = \frac{1}{n}\sum_i^n X_i^*$.
3. Repeat steps 1 and 2 B times to obtain B bootstrap sample means, $\hat{\mu}_1^*, \ldots, \hat{\mu}_B^*$.
4. Arrange the bootstrap sample means in ascending order to get $\hat{\mu}_{(1)}^* \leq \hat{\mu}_{(2)}^* \leq \cdots \leq \hat{\mu}_{(B)}^*$. Based on the bootstrap samples, we can estimate μ with the average of the bootstrap sample means, $\hat{\mu}^* = \frac{1}{B}\sum_b^B \hat{\mu}_b^*$, and the standard error of $\hat{\mu}$ with the standard deviation of the bootstrap sample means, $\hat{se}^* = \sqrt{\frac{1}{B}\sum_b^B (\hat{\mu}_b^* - \hat{\mu}^*)^2}$. The bias of the sample mean, $\hat{\mu}$, is estimated by the bias of $\hat{\mu}^*$ (Efron & Tibshirani, 1993; Van der Leeden et al., 2008), given by

$$\hat{\text{Bias}}(\hat{\mu}) = \hat{\text{Bias}}(\hat{\mu}^*) = \hat{\mu}^* - \hat{\mu}, \tag{13.5}$$

and the bias-corrected estimator of $\hat{\mu}$ is

$$\hat{\mu}_{corrected} = 2\hat{\mu} - \hat{\mu}^* = \hat{\mu} - \hat{\text{Bias}}(\hat{\mu}). \tag{13.6}$$

An approximate 95% CI for the sample mean is

$$[\hat{\mu}_{\frac{1}{2}\alpha}^*; \hat{\mu}_{1-\frac{1}{2}\alpha}^*] \tag{13.7}$$

where $\hat{\mu}_{\frac{1}{2}\alpha}^*$ and $\hat{\mu}_{1-\frac{1}{2}\alpha}^*$ are the $\frac{1}{2}\alpha$ and $1 - \frac{1}{2}\alpha$ quantiles of the bootstrap distributions.

13.2 Bootstrapping for Multilevel Models

Because multilevel data involve units at multiple levels, bootstrapping is more complex as it can involve resampling at different levels. For a simple example of N students nested with J schools, there are at least three resampling schemes, (a) resample only schools, (b) resample only students, and (c) resample schools first, and then students. An additional concern, especially for cases bootstrap (also known as the cluster/block/pairs bootstrap), is that students do not constitute independent observations, as they are clustered within schools. As a basic principle, the bootstrap method should honor the nesting structure of the data. Therefore, we should not directly resample students in our example, as that would break the nesting structure such that students from the same school may end up in different schools in the bootstrap sample.

As a result, several bootstrap algorithms have been proposed to account for nesting structures, including parametric, residual, wild, and cases bootstrap methods (Davison & Hinkley, 1997; Goldstein, 2011; Modugno & Giannerini, 2015; Van der Leeden et al., 2008), which will be further discussed below. In general, the parametric, residual, and wild bootstrap preserve the clustered structure by simulating new error terms ($\boldsymbol{\epsilon}_j$) and random effects (the $\mathbf{U}_j$ terms), because they are presumed independent in the statistical model for MLM. These three methods are thus *model-based* because they require specifying a model to obtain quantities that are assumed independent (i.e., the $\boldsymbol{\epsilon}$'s and the $\mathbf{U}$'s). Each new bootstrap sample is formed by generating a new sample of only the response variable (Y), using the deterministic component of the model, e.g., $\mathbf{X}\boldsymbol{\beta}$, the fixed effect component in Eq. (13.1), and the simulated errors and random effects. The parametric and the residual bootstrap differ in terms of whether the errors/random effects are sampled from an assumed parametric distribution (e.g., normal) or the empirical distribution of the model estimates. The wild bootstrap further relaxes the homoscedasticity assumption. On the other hand, the cases bootstrap starts by resampling the top level of observations, which are assumed independent. While one version of the cases bootstrap resamples only the top level, another version further resamples the units with a cluster and repeats for every cluster, as long as the observations are assumed independent and exchangeable within each cluster.

We should also point out that the parametric, residual, and wild bootstrap only simulate new Y, whereas the cases bootstrap gives new samples of all variables in the data, including predictors X. Because of this, the former methods assume fixed X across repetitions, making it more suitable for designs where the predictors are fixed (e.g., randomized experiments with a fixed manipulated variable). On the other hand, the cases bootstrap allows the predictors to vary across the bootstrap samples. In some situations, the inferences could be different when predictors are considered to be either fixed or random (see Gelman & Hill, 2007, Sect. 11.4, for more discussion). While we treat predictors as fixed in this chapter, one can specify random components for X and adapt the parametric and the residual bootstrap to resample the random components of X. Similarly, one can introduce stratifying variables to cases bootstrap to make the bootstrap scheme resemble fixed predictors. A detailed discussion of each of these four bootstrap methods follows below.

Types of Bootstrapping

This section describes the procedures of parametric bootstrap, residual bootstrap, wild bootstrap, and cases bootstrap in detail. A two-level random-effect model is used to illustrate these different types of bootstrapping algorithms.

Parametric Bootstrap

The parametric bootstrap has the strongest assumption among all methods, as it assumes a correct function form and residual distributions (Van der Leeden et al., 2008). Specifically, level-1 residuals are assumed to follow a normal distribution,

$\epsilon_j \sim N(\mathbf{0}, \sigma^2 \mathbf{I}_{n_j})$, while level-2 residuals are assumed to follow a multivariate normal distribution, $\mathbf{U}_j \sim N(\mathbf{0}, \mathbf{T})$. Let $\hat{\boldsymbol{\beta}}$ be the model estimates from the linear mixed-effects model. The procedure for the parametric bootstrap is described below.

1. Fit a two-level model to data and obtain estimates for the fixed effects and random effects.
2. Draw J sets of level-2 residuals $\mathbf{U}_j^*$ from a multivariate normal distribution.
3. Draw J sets of level-1 residuals ϵ_j^* of size n_j from a normal distribution.
4. Generate the bootstrap samples $\mathbf{y}_j^*$, with $\mathbf{y}_j^* = \mathbf{X}_j\hat{\boldsymbol{\beta}} + \mathbf{Z}_j\mathbf{U}_j^* + \epsilon_j^*$.
5. Refit the two-level model to the bootstrap samples and obtain parameter estimates.
6. Repeat steps 2–5 B times to obtain a bootstrap distribution of parameter estimates.

Residual Bootstrap

Unlike the parametric bootstrap, the residual bootstrap does not assume that residuals follow certain distributions. However, the homoscedasticity of residuals assumption still needs to hold (Van der Leeden et al., 2008). The procedure for the residual bootstrap is similar to that of the parametric bootstrap, except the residuals are not drawn from assumed distributions (e.g., normal). One issue is that the empirical residuals, extracted from a fitted model, have smaller variances than the population variances (Goldstein, 2011). Thus, these residuals are shrinkage estimates and will lead to biased results if used directly to form new bootstrapped samples. To overcome this problem, the residual bootstrap first reflates these shrunken residuals so that their model-implied variances match the population variance (Carpenter et al., 2003; Davison & Hinkley, 1997) and then draws residuals from these restored shrunken estimates with replacements (Carpenter et al., 2003). For level-2 residuals $\hat{\mathbf{U}}_j$, the model-implied covariance matrix of the estimated residuals at level-2 is denoted as $\mathbf{T}_{\hat{U}}$. We transform $\hat{\mathbf{U}}_j$ to $\mathbf{L}\mathbf{L}_{\hat{U}}^{-1}\hat{\mathbf{U}}$, where $\mathbf{L}_{\hat{U}}$ is the lower triangular matrix for the Cholesky decomposition of $\mathbf{T}_{\hat{U}}$ such that $\mathbf{L}_{\hat{U}}\mathbf{L}_{\hat{U}}^{\top} = \mathbf{T}_{\hat{U}}$ and $\mathbf{L}$ is the Cholesky decomposition of $\mathbf{T}$. The level-1 residuals can be reflated similarly (Lai, 2021). The procedure for the residual bootstrap is as follows:

1. Fit a two-level model to data and obtain the estimates for fixed effects and random effects.
2. Reflate the centered empirical residuals using the procedure described above.
3. Draw J sets of level-2 residuals $\mathbf{U}_j^*$ from the reflated empirical residuals with replacement.

4. Draw J sets of level-1 residuals $\boldsymbol{\epsilon}_j^*$ of size n_j from the reflated empirical residuals with replacement.
5. Generate the bootstrap samples, $\mathbf{y}_j^*$, with $\mathbf{y}_j^* = \mathbf{X}_j\hat{\boldsymbol{\beta}} + \mathbf{Z}_j\mathbf{U}_j^* + \boldsymbol{\epsilon}_j^*$.
6. Refit the two-level model to the bootstrap data and obtain parameter estimates.
7. Repeat steps 3–6 B times to obtain a bootstrap distribution of parameter estimates.

Note here we focus on one type of reflating process. There are alternative procedures, such as the procedure described in Carpenter et al. (2003) and the transformational residual bootstrap by Van der Leeden et al. (2008).

Wild Bootstrap

The wild bootstrap is a resampling algorithm developed for models with heteroscedastic errors (Modugno & Giannerini, 2015; Wu, 1986). The central idea of the wild bootstrap is to replace the heteroscedastic residuals $\hat{\mathbf{v}}_j = \mathbf{y}_j - \mathbf{X}_j\hat{\boldsymbol{\beta}}$ with the heteroscedasticity consistent (HC) covariance matrix estimator proposed by MacKinnon and White (1985) and multiply the residuals with a random variable m drawn from an auxiliary distribution with mean zero and unit variance. In other words, instead of resampling the residuals directly, the wild bootstrap transforms the residuals and draws bootstrap errors from auxiliary distributions. Then bootstrap responses are generated from this joint distribution, $m\mathbf{v}$, and used to compute the statistic of interest (Liu, 1988). The commonly used covariance matrix estimators are HC_2 and HC_3, which have been shown to outperform other HC covariance matrix estimators in power (Davidson & MacKinnon, 1993; MacKinnon & White, 1985). Assume $\mathbf{H}_j = \mathbf{X}_j(\mathbf{X}^T\mathbf{X})^{-1}\mathbf{X}_j^T$ is the jth diagonal block of the orthogonal project matrix:

$$HC_2 : \tilde{\mathbf{v}}_j = \mathrm{diag}(\mathbf{I}_{n_j} - \mathbf{H}_j)^{-1/2} \circ \hat{\mathbf{v}}_j \tag{13.8}$$

$$HC_3 : \tilde{\mathbf{v}}_j = \mathrm{diag}(\mathbf{I}_{n_j} - \mathbf{H}_j) \circ \hat{\mathbf{v}}_j, \tag{13.9}$$

where $\circ$ is the entry-wise product (Modugno & Giannerini, 2015).[1] The two commonly used auxiliary distributions are F_1, proposed by Mammen (1993) and F_2 (also known as the Rademacher distribution), proposed by Liu (1988), where

[1] At the time of writing, the *lmeresampler* R package is the only R package that implements the wild bootstrap for MLM, which uses the ordinary least square (OLS) estimators but not generalized least square (GLS). Thus, the hat matrix, H_j, does not include the matrix that represents random effects.

$$F_1 : m = \begin{cases} -(\sqrt{5}-1)/2 & \text{with probability} \quad p = (\sqrt{5}+1)/2\sqrt{5}) \\ (\sqrt{5}+1)/2 & \text{with probability} \quad p = 1-p \end{cases} \tag{13.10}$$

and

$$F_2 : m = \begin{cases} 1 & \text{with probability} \quad p = \frac{1}{2} \\ -1 & \text{with probability} \quad p = \frac{1}{2}. \end{cases} \tag{13.11}$$

The specific procedure for multilevel wild bootstrap is listed below (see Modugno & Giannerini, 2015, pp. 4817–4818):

1. Draw a random sample m_j with J values from the auxiliary distribution that has a mean zero and unit variance such as F_1 and F_2.
2. Compute the marginal residuals $\hat{\mathbf{v}}_j = \mathbf{y}_j - \mathbf{X}_j\hat{\boldsymbol{\beta}}$, and transform the residuals to be heteroscedastic consistent $\tilde{\mathbf{v}}_j$.
3. Generate the bootstrap samples $\mathbf{y}_j^* = \mathbf{X}_j\hat{\boldsymbol{\beta}} + \tilde{\mathbf{v}}_j m_j$.
4. Refit the two-level model to the bootstrap data and obtain parameter estimates.
5. Repeat steps 1–4 B times to obtain a bootstrap distribution of bias-corrected parameter estimates.

The wild bootstrap is expected to perform better than the other bootstraps under heteroscedasticity because it uses a two-point distribution to generate errors (Hapuhinna, 2021; MacKinnon & Webb, 2017). Other bootstraps, such as the parametric bootstrap, assume the error follows i.i.d with equal variance. Under these assumptions, the generated responses based on the resampled residuals are close to the true responses. However, when heteroscedasticity exists, the variance depends on the predictor, causing the generated responses to differ from the true responses (Hapuhinna, 2021). In contrast, the wild bootstrap uses a two-point distribution such that the generated residuals preserve the variances and the within-cluster covariances of the errors (MacKinnon & Webb, 2017).

Cases Bootstrap

Similar to the wild bootstrap, the cases bootstrap can overcome the problem of heteroscedasticity. The cases bootstrap has the least restrictive assumption among all bootstrapping procedures because it only assumes a correct specification of hierarchical structure. It resamples the observations (including predictors contained in $\mathbf{X}_j$ and $\mathbf{Z}_j$) with replacements (Davison & Hinkley, 1997). There are multiple versions of cases bootstrap for multilevel data (Van der Leeden et al., 2008). For any version, the cases bootstrap keeps each observed response y_{ij} pairing with the observed predictors $\mathbf{X}_{ij}$ and $\mathbf{Z}_{ij}$. Note that the size of each bootstrap sample could be different from the original sample size when the clusters are unbalanced. One version of cases bootstrap is to only resample level-2 units and keeps level-1 units intact.

1. Draw a sample of size J, $\{j_k^*, k = 1, \ldots, J\}$, with replacement from the identifiers of level-2 units.
2. For each k, take the sample of entire cases, $(y_{ik}^*, \mathbf{X}_{ik}^*, \mathbf{Z}_{ik}^*)$, where $i = 1, \ldots, n_k^*$, from the selected level-2 unit $j = j_k^*$.
3. Fit the two-level model to the bootstrap data and obtain parameter estimates.
4. Repeat steps 1–3 B times to obtain a bootstrap distribution of parameter estimates.

An alternative procedure is to first resample the level-2 units and then resample level-1 units within each cluster.

1. Draw one entire level-2 unit, $(\mathbf{y}_j, \mathbf{X}_j, \mathbf{Z}_j)$, with replacement.
2. Draw a sample of level-1 observations with size n_j from the selected level-2 unit.
3. Repeat steps 1–2 J times.
4. Fit the two-level model to the bootstrap data and obtain parameter estimates.
5. Repeat steps 1–4 B times to obtain a bootstrap distribution of parameter estimates.

Selecting the appropriate cases bootstrap procedure depends on the degree of randomness of the sampling and the sample size for both levels (Van der Leeden et al., 2008). For example, if the repeated measures from individuals are collected, then it is appropriate to resample only individuals (level-2). However, for another example, if individuals' responses from multiple countries are collected, researchers could consider first resampling the countries with replacements and then resampling the individuals within selected countries with replacements. Compared to other bootstrap algorithms, cases bootstrap is less efficient but provides consistent estimators when the normality or homoscedasticity assumptions are violated (Van der Leeden et al., 2008).

13.3 Confidence Intervals

In addition to obtaining bias-corrected standard errors, bootstrapping is widely used for computing confidence intervals (CIs). In this section, we discuss different types of two-sided bootstrap confidence intervals. Let θ be the parameter of interest; $\hat{\theta}$ be a sample estimator of θ; and $\hat{\theta}^*$ be the bootstrap estimate of $\hat{\theta}$. Assume the target confidence level (i.e., probability of a random confidence interval containing the true value of θ) is $1-\alpha$. Let $\Phi(z)$ be the cumulative density function of the standard normal distribution, and then z_α represents the α quantile of the standard normal distribution such that $z_\alpha = \Phi^{-1}(\alpha)$. A confidence interval is defined by limits $\hat{\theta}_{\frac{1}{2}\alpha}$ and $\hat{\theta}_{1-\frac{1}{2}\alpha}$ such that $\Pr(\hat{\theta}_{\frac{1}{2}\alpha} < \theta < \hat{\theta}_{1-\frac{1}{2}\alpha}) = \alpha$.

Bootstrap Normal CI

If normality holds, then the common approximate $(1 - \alpha) \times 100\%$ symmetric CI is

$$[\hat{\theta} - z_{1-\frac{1}{2}\alpha}\hat{se}_N, \hat{\theta} + z_{1-\frac{1}{2}\alpha}\hat{se}_N],$$

where $\hat{\theta}$ is the estimated parameter and $\hat{se}_N$ is the sample standard error of $\hat{\theta}$. For bootstrap normal confidence interval, $\hat{se}_N$ is replaced with $\hat{se}^*$ (Van der Leeden et al., 2008), the bootstrap standard error of $\hat{\theta}$, so the bootstrap normal CI can be constructed as

$$[\hat{\theta} - z_{1-\frac{1}{2}\alpha}\hat{se}^*, \hat{\theta} + z_{1-\frac{1}{2}\alpha}\hat{se}^*].$$

Basic/Hall's Percentile

The basic CI, also called Hall's percentile CI, uses quantiles of the bootstrapped distribution instead of the normal distribution to obtain more accurate results when $\hat{\theta}$ does not follow a normal distribution (Davison & Hinkley, 1997; Hall, 1992; Van der Leeden et al., 2008). For the bootstrap distribution of bias, $\hat{\theta}^* - \hat{\theta}$, $\frac{1}{2}\alpha$ and $1 - \frac{1}{2}\alpha$ quantiles can be computed by $\hat{q}_{\frac{1}{2}\alpha} = \hat{\theta}^*_{\frac{1}{2}\alpha} - \hat{\theta}$ and $\hat{q}_{1-\frac{1}{2}\alpha} = \hat{\theta}^*_{1-\frac{1}{2}\alpha} - \hat{\theta}$. Replacing the quantiles from the bootstrap distribution of bias, the basic CI is

$$[2\hat{\theta} - \hat{\theta}^*_{1-\frac{1}{2}\alpha}, 2\hat{\theta} - \hat{\theta}^*_{\frac{1}{2}\alpha}].$$

Studentized/Bootstrap-t CI

Similar to the basic CI, bootstrap-t CI replaces the quantiles of the normal distribution by quantiles of the bootstrap distribution. The difference between bootstrap-t CI and the basic CI is that bootstrap-t uses the bootstrap distribution of $t^* = (\hat{\theta} - \hat{\theta}^*)/\hat{se}^*$ (Davison & Hinkley, 1997; Van der Leeden et al., 2008). The bootstrap-t confidence interval can be obtained by replacing $z_{1-\frac{1}{2}\alpha}$ by $t^*_{1-\frac{1}{2}\alpha}$, the $1 - \frac{1}{2}\alpha$th quantiles of t^*,

$$[\hat{\theta} - t^*_{1-\frac{1}{2}\alpha}\hat{se}^*, \hat{\theta} + t^*_{1-\frac{1}{2}\alpha}\hat{se}^*],$$

where $\hat{se}^*$ is the bootstrap standard error of $\hat{\theta}$.

Percentile CI

Percentile CI is formed by using quantiles directly from the bootstrapped distribution (Efron, 1982).

$$[\hat{\theta}^*_{\frac{1}{2}\alpha}; \hat{\theta}^*_{1-\frac{1}{2}\alpha}]$$

Although this percentile CI does not assume asymptotic normality of $\hat{\theta}$, it has been found to generate biased results (Efron & Tibshirani, 1993).

Bias-Corrected and Bias-Corrected and Accelerated CI

To correct the bias contained in the percentile CI, Efron and Tibshirani (1993) proposed the bias-corrected and accelerated (BCa) CI. They defined two quantiles, α_L and α_U, as

$$\alpha_L = \Phi\left(w + \frac{w + z_{\alpha/2}}{1 - \alpha(w + z_{\alpha/2})}\right) \tag{13.12}$$

$$\alpha_U = \Phi\left(w + \frac{w + z_{1-\alpha/2}}{1 - \alpha(w + z_{1-\alpha/2})}\right), \tag{13.13}$$

and suggested using these two quantiles to construct the BCa CI. Here w is the bias correction for the median, and α is the acceleration (skewness) correction for the distribution of the estimator. These two correcting factors can be computed using the formulas below:

$$w = \Phi^{-1}\left(\frac{\sum_{r=1}^{R} I_{\{\hat{\theta}^* < \theta\}}}{R + 1}\right) \tag{13.14}$$

$$\alpha = \frac{\sum_{j=1}^{J} l_j^3}{6(\sum_{j=1}^{J} l_j^2)^{\frac{3}{2}}}. \tag{13.15}$$

In the above equations, $I_{\{\hat{\theta}^* < \theta\}}$ is an indicator function, which equals 1 if $\hat{\theta}^* < \theta$ and 0 otherwise. In addition, l_j^3 represents an influence value for cluster j and can be estimated by the grouped jackknife method that accounts for unequal group size (Lai, 2021; Van der Leeden et al., 2008).

Comparing Different Bootstrap Confidence Intervals

Arguably, the most common usage of the bootstrap is to obtain CIs for model parameters (e.g., fixed and random effects) and derived quantities (e.g., effect size, indirect effects, ICC) that are functions of these parameters. Bootstrap CI is most useful for quantities that have nonnormal sampling distributions in finite samples, such as random-effect variances and ICC. For example, Efron (1988) showed that, for correlation coefficients, the percentile CI is a better alternative than the analytic CI based on normal approximation. In behavioral sciences, bootstrap CI has become the standard for studying mediation and making inferences for indirect effects (e.g., Shrout & Bolger, 2002). For multilevel data, there have been several studies on bootstrap CIs for ICC (Braschel et al., 2016; e.g., Ukoumunne et al., 2003).

While there have been few systematic studies for CIs with multilevel data, we expect that recommendations based on single-level data would still hold (e.g., Carpenter & Bithell, 2000; Diciccio & Romano, 1988). The choice should depend on whether the sample estimate is expected to be biased and how much the sampling distribution of the target quantity deviates from normal. As shown by Hall (1992), the studentized and BCa are second-order accurate, compared to percentile, normal, and basic CIs, which are first-order accurate. This means that studentized and BCa CIs, in theory, would have coverage levels converging to the nominal level (e.g., 95%) faster than the other CIs. The drawback for studentized and BCa CIs, however, is that they rely on additional estimates from the sample that may lead to instability: studentized CI requires an estimate of the sampling variance, and BCa CI requires estimating the acceleration constant. This is partly the reason why studentized CI and BCa were not consistently found to outperform other CIs, especially in small samples (e.g., Diciccio & Efron, 1996; Good, 2005, Chapter 3).

The studentized, normal, and basic CIs all assume that the target sampling distribution is symmetric in some fashion and would likely perform better when such an assumption holds. When the target quantity is expected to be skewed, BCa and percentile CIs may have an advantage as they do not assume symmetric distributions. That said studentized, normal, and basic CIs can be improved if there exists a transformation so that the transformed sampling distribution is close to symmetric, such as Fisher's z transformation for correlation and log transformation for ICC. Transformation can also be used to avoid out-of-bound confidence limits for bounded quantities; an example is the use of logit transformation for R^2 so that the back-transformed confidence limits are between 0 and 1.

While a 95% CI is valid when its coverage is 95% or above, when considering CIs with similar coverage properties, the one with a shorter length is usually preferred (Efron, 2003). For example, Efron (1988) gave an example where the percentile CI had a much shorter width than the normal-theory-based CI.

To our knowledge, there has only been limited research evaluating different CIs for multilevel data in finite samples, and earlier studies focused almost exclusively on estimating the ICC. Ukoumunne et al. (2003) found that, among several bootstrap methods, the studentized CI using a variance stabilizing transformation gave the best CI coverage for ICC across conditions, especially when the number of clusters is small (see also Braschel et al., 2016). Reiser et al. (2017) compared studentized, percentile, and BCa CIs for fixed-effect and random-effect variance estimates and found the three methods to be adequate and similar for fixed effects and that BCa CIs did not seem to be better than percentile CIs, especially in small samples. Burch (2012) found that normal CIs with a log transformation gave better coverage than BCa for variance components, and both CIs were better than the percentile CIs. Lai (2021) showed that with non-normal data, basic, and studentized CIs gave the best coverage for standardized mean difference effect size, although studentized CIs tended to be slightly wider than other CIs. More research is needed to compare different bootstrap CIs in the multilevel modeling context.

So based on the very limited literature, we recommend researchers use, when possible, studentized CI for quantities with a relatively symmetric sampling distribution, although all intervals should perform similarly if the sampling distributions are close to normal. For quantities with highly skewed sampling distributions (e.g., variance parameters in small samples) or with bounds (e.g., variance accounted for effect sizes for single- and multilevel analyses, R^2, such as R^2_{GLMM} by Nakagawa and Schielzeth, 2013), one can use studentized CIs on the transformed parameter (e.g., logit transformation for R^2) or use BCa or percentile CIs if studentized CIs cannot be computed and/or an appropriate transformation is not available.

13.4 Illustration

In this section, we illustrate four bootstrap approaches (i.e., parametric, residual, wild, and cases) with the *bootmlm* R package (Lai, 2021) and the wild bootstrap with the *lmeresampler* R package (Loy et al., 2022) on an empirical example. In particular, we will obtain bias-corrected estimates, standard errors, and CIs for the fixed effects, random effects, intraclass correlation (ICC), and R^2 (Nakagawa & Schielzeth, 2013).

The empirical example is an elementary school study by Knuver and Brandsma (1993) and Doolaard (1999), whose data was shared and used in Snijders and Bosker (2012). This study contains data for 3,758 students (after a list-wise deletion of 258 students), from 211 schools with class sizes ranging between 5 and 36. As shown in Snijders & Bosker (2012, Chapters 8 and 10), the intelligence variable in the data appeared to violate both assumptions of normality and homoscedasticity.

We adopt the two-level model in Snijders and Bosker (2012), with students nested within schools, to predict language test scores (lang_{ij}) with students' intelligence (IQ_{ij}), gender (gender_{ij}), and family's socioeconomic status (SES_{ij}). For the continuous level-1 (student-level) predictors, the standard practice is to disaggregate the between- and within-effects by including the level-2 mean (school-level mean) of each predictor in the model and centering the level-1 predictors by the level-2 means. In this example, school_IQ_j is the school mean of intelligence, and school_SES_j is the school mean of socioeconomic status. The equation of the full model is

$$
\begin{aligned}
\text{Level 1: } \text{lang}_{ij} &= \beta_{0j} + \beta_{1j}\text{IQ}_{ij} + \beta_{2j}\text{SES}_{ij} + \beta_{3j}\text{gender}_{ij} + \beta_{4j}\text{IQ}_{ij} \\
&\quad \times \text{SES}_{ij} + \epsilon_{ij}, \quad \epsilon_{ij} \sim N(0, \sigma^2) \\
\text{Level 2: } \beta_{0j} &= \gamma_{00} + \gamma_{01}\text{school\_IQ}_j + \gamma_{02}\text{school\_SES}_j + \gamma_{03}\text{school\_IQ}_j \\
&\quad \times \text{school\_SES}_j + u_{0j} \\
\beta_{1j} &= \gamma_{10} + u_{1j} \\
\beta_{2j} &= \gamma_{20} \\
\beta_{3j} &= \gamma_{30} \\
\beta_{4j} &= \gamma_{40} \\
\begin{pmatrix} u_{0j} \\ u_{1j} \end{pmatrix} &\sim N\left(\begin{bmatrix} 0 \\ 0 \end{bmatrix}, \begin{bmatrix} \tau_{00}^2 & \\ \tau_{01} & \tau_{11}^2 \end{bmatrix}\right)
\end{aligned}
\tag{13.16}
$$

$$
\begin{aligned}
\text{Combined: } \text{lang}_{ij} &= \gamma_{00} + \gamma_{10}\text{IQ}_{ij} + \gamma_{20}\text{SES}_{ij} + \gamma_{30}\text{gender}_{ij} + \gamma_{40}\text{IQ}_{ij} \times \text{SES}_{ij} + \\
&\quad \gamma_{01}\text{school\_IQ}_j + \gamma_{02}\text{school\_SES}_j + \gamma_{03}\text{school\_IQ}_j \times \text{school\_SES}_j + \\
&\quad u_{0j} + u_{1j}\text{IQ}_{ij} + \epsilon_{ij},
\end{aligned}
\tag{13.17}
$$

where γ_{00} is the grand intercept of language scores; γ_{01} and γ_{02} are the school-level fixed effects of intelligence and SES, respectively; γ_{03} is the school-level interaction between intelligence and SES; γ_{10} and γ_{20} are the student-level fixed effects of intelligence and SES, respectively; γ_{30} is the student-level interaction between intelligence and SES; u_{0j} is the random effect from the grand intercept; and u_{1j} is the random effect from the student-level intelligence. For reproducibility, we use the following syntax to download and import the dataset from the website of Snijders and Bosker (2012) to our local R environment.

```
temp <- tempfile()
download.file("https://www.stats.ox.ac.uk/~snijders/mlbook2_r_dat.zip", temp)
sj_dat <- read.table(unz(temp, "mlbook2_r.dat"), header = TRUE)
```

In this illustration, we use five R packages: *lme4* (version 1.1.30; Bates et al., 2015) for multilevel analysis; *performance* (version 0.10.2; Bartoń, 2022) for computing R^2; *bootmlm* (version 0.0.1.1000; Lai, 2021) for parametric, residual, and cases bootstrap; *lmeresampler* (version 0.2.2; Loy et al., 2022) for wild bootstrap; and *boot* (version 1.3.28; Canty & Ripley, 2021; Davison & Hinkley, 1997) for computing bootstrap CIs. We perform two models in *lme4*: (a) the full model in Snijders and Bosker (2012), as shown in Model (13.17), and (b) the null model (the unconditional model) that has no predictors, given by

$$\begin{aligned} \text{lang}_{ij} &= \gamma_{00} + u_{0j} + \epsilon_{ij} \\ u_{0j} &\sim N(0, \tau^2_{00|U}), \quad \epsilon_{ij} \sim N(0, \sigma^2_{|U}), \end{aligned} \tag{13.18}$$

to obtain the ICC: $\rho = \frac{\tau^2_{00|U}}{\tau^2_{00|U}+\sigma^2_{|U}}$.

```
# unconditional model
mod_0 <- lmer(langPOST ~ (1 | schoolnr), data = sj_dat)
# full model
mod_f <- lmer(langPOST ~ IQ_verb*ses + sex + sch_iqv*sch_ses +
                (IQ_verb | schoolnr),
              data = sj_dat)
```

We first define two functions: `eff_rsq()` takes the full model to compute the fixed effects, random effects, and R^2, and `icc()` takes the null model to compute ICC. `eff_rsq()` yields estimates for 13 parameters, including 8 fixed effects ($\gamma_{00}, \gamma_{01}, \gamma_{02}, \gamma_{03}, \gamma_{10}, \gamma_{20}, \gamma_{30}, \gamma_{40}$), 4 random effects ($\tau^2_{00}, \tau^2_{11}, \tau_{01}, \sigma^2$), and R^2. In `icc()`, we compute the ICC and additionally its asymptotic sampling variance using the `vcov_vc()` function from *bootmlm*, which will be needed to construct the studentized/bootstrap-*t* CI.

```
eff_rsq <- function(mod) {
  # Fixed effects
  fix_eff <- fixef(mod)
  # Random effects
  ran_eff <- as.data.frame(VarCorr(mod))$vcov
  # R^2
  # Alternatively, we can get the same R^2 with MuMIn::r.squaredGLMM()
  rsq <- performance::r2_nakagawa(mod)[[1]]
  c(fix_eff = fix_eff, ran_eff = ran_eff, rsq = rsq)
}

icc <- function(mod) {
  # ICC estimate
  vc <- as.data.frame(VarCorr(mod))
  tau00sq <- vc[1, "vcov"]
  sigmasq <- vc[2, "vcov"]
  icc_est <- tau00sq / (tau00sq + sigmasq)
  # ICC variance with delta method
  dG_tau00sq <- sigmasq / (tau00sq + sigmasq)^2
  dG_sigmasq <- - tau00sq / (tau00sq + sigmasq)^2
  grad <- c(dG_tau00sq, dG_sigmasq)
  vcov_ranef <- vcov_vc(mod, sd_cor = FALSE)
  icc_var <- t(grad) %*% vcov_ranef %*% grad
  c(icc_est, icc_var)
}
```

Executing `eff_rsq(mod_f)` and `icc(mod_0)` gives us the following parameter estimates from the two models.

```
##     (Intercept)         IQ_verb             ses             sex
##         40.4257          2.2496          0.1714          2.4063
##         sch_iqv         sch_ses     IQ_verb:ses sch_iqv:sch_ses
##          0.7711         -0.0930         -0.0195         -0.1052
##         tau00sq         tau11sq           tau01         sigmasq
##          8.5035          0.1516         -0.9079         36.0218
##             Rsq             ICC        Var(ICC)
##          0.5474          0.2249          0.0005
```

Parametric Bootstrap

At the time of writing, the `bootstrap_mer()` function in *bootmlm* supports three main types of bootstrapping—parametric, residual, and cases. To perform parametric bootstrap, we denote `type = "parametric"` in the function and specify the fitted model object from *lmer* (i.e., `mod_f` and `mod_0`) along with the defined function for obtaining the parameter estimates (i.e., `eff_rsq()` and `icc()`). The `nsim` option refers to the number of bootstrap samples (B), which is generally recommended to be large for stable estimates. In this illustration, we use $B = 1,999$ because existing literature suggests $.95(B+1)$ should be an integer when computing the 95% interval for an exact test (Davidson & MacKinnon, 2000; Wilcox, 2010).

```
boo_par_eff <- bootstrap_mer(mod_f, eff_rsq, nsim = 1999L, type = "parametric")
boo_par_icc <- bootstrap_mer(mod_0, icc, nsim = 1999L, type = "parametric")
```

After running bootstrap resampling, we can use the `boot.ci()` function in *boot* to compute the CIs for the bootstrap samples. This function provides five ways to construct CIs: normal, basic, percentile, studentized, and BCa. Computing studentized and BCa CIs require additional inputs, which are the bootstrap samples of an estimator's variance and the influence values, respectively.

With parametric bootstrap, we demonstrate how to obtain normal, basic, and percentile CIs for the fixed effects and R^2. In `boot.ci()`, we specify the types of CIs `type = c("norm", "basic", "perc")` to be computed. The index denotes the position of the estimator of interest in the output from `bootstrap_mer()`. As `boo_par_eff` consists of the bootstrap samples for 13 parameters, `lapply()` allows us to compute CIs for all of them simultaneously.

```
boo_par_eff_ci <- lapply(1:13, function(i) {
  boot.ci(boo_par_eff, type = c("norm", "basic", "perc"), index = i)
})
```

For the ICC, we further demonstrate how to construct the studentized CI. Since `icc()` computes both the estimate and asymptotic sampling variance of ICC, `boo_par_icc` contains the bootstrap samples of both estimators, i.e., $\hat{\rho}^*$ and

$\hat{\text{Var}}(\hat{\rho}^*)$. We specify the argument `type = "stud"` in `boot.ci()`, which takes the bootstrap samples of $\hat{\rho}^*$ and $\hat{\text{Var}}(\hat{\rho}^*)$ to construct the studentized CI.

```
boo_par_icc_ci <- boot.ci(boo_par_icc, type = c("norm", "basic", "perc", "stud"))
```

Residual Bootstrap

As discussed in Sect. 13.2, because the residuals in MLM are shrinkage estimates, their sampling variabilities are much smaller than the population random effects (e.g., τ_{00}^2 and σ^2). Residual bootstrap accounts for the small sampling variabilities by rescaling the residuals. To perform residual bootstrap, we specify `type = "residual"` in `bootmlm::bootstrap_mer()`.

```
# Reflated residual
boo_res_eff <- bootstrap_mer(mod_f, eff_rsq, nsim = 1999L, type = "residual")
boo_res_icc <- bootstrap_mer(mod_0, icc, nsim = 1999L, type = "residual")
```

```
inf_val_eff <- lapply(1:13, function(i) {
  empinf_mer(mod_f, eff_rsq, index = i)
})
inf_val_icc <- empinf_mer(mod_0, icc, index = 1)
```

For the fixed effects, random effects, and R^2, we exemplify four ways to construct CIs with `boot::boot.ci()` with the residual bootstrap methods: normal, basic, percentile, and, additionally, BCa. For the ICC, we also illustrate the studentized CI. As discussed in Sect. 13.3, the influence values are needed to compute CIs with BCa and can be obtained using the function `empinf_mer()` in *bootmlm*. This function requires the inputs of the fitted model (`mod_f` and `mod_0`), the defined function that computes the parameter estimate(s) (`eff_rsq()` and `icc()`), and the index of the statistic in the output of the defined function. As `eff_rsq()` yields 13 statistics, we can use `lapply()` to conveniently compute the influence values for all of them at once.

```
# Residual bootstrap
boo_res_eff_ci <- lapply(1:13, function(i) {
  boot::boot.ci(boo_res_eff, type = c("norm", "basic", "perc", "bca"),
                index = i, L = inf_val_eff[[i]])
})
boo_res_icc_ci <- boot::boot.ci(boo_res_icc,
                                type = c("norm", "basic", "perc", "stud", "bca"),
                                L = inf_val_icc)
```

We additionally specify `"bca"` in `type` to compute the CI with BCa.

```
boo_wil_eff <- wild_bootstrap(mod_f, .f = eff_rsq, B = 1999L)
boo_wil_icc <- wild_bootstrap(mod_0, .f = icc, B = 1999L)
```

Wild Bootstrap

While wild bootstrap is unavailable in *bootmlm*, we can use the function `wild_bootstrap()` in the *lmeresampler* R package. Similar to those in `bootstrap_mer()`, the required input arguments in `wild_bootstrap()` are the fitted model object, the defined function that computes the parameter estimates (`.f`), and the number of bootstrap samples (`B`).

```
boo_wil_eff_ci <- confint(boo_wil_eff, type = "all")
boo_wil_icc_ci <- confint(boo_wil_icc, type = "all")
```

At the time of writing, the bootstrap output from *lmeresampler* (version 0.2.4) can be summarized with the `confint()` function, which supports three ways of constructing CIs with wild bootstrap, normal, basic, and percentile, with the argument `type = "all"`.

```
# Fixed effects and Rsq
boo_cas_eff <- bootstrap_mer(mod_f, eff_rsq, nsim = 1999L, type = "case")
boo_cas1_eff <- bootstrap_mer(mod_f, eff_rsq, nsim = 1999L, type = "case",
                              lv1_resample = TRUE)
# ICC
boo_cas_icc <- bootstrap_mer(mod_0, icc, nsim = 1999L, type = "case")
boo_cas1_icc <- bootstrap_mer(mod_0, icc, nsim = 1999L, type = "case",
                              lv1_resample = TRUE)
```

Cases Bootstrap

With cases bootstrap in multilevel modeling, clusters are sampled with replacement using the argument of `type = "case"` in `bootmlm::bootstrap_mer()`. Optionally, we can further resample the cases (observations) within each cluster, which can be done using the `lv1_resample = TRUE` argument. Unlike the other bootstrap methods, *bootmlm* currently supports the cases bootstrap for only two-level models.

```
# Only resampling clusters
boo_cas_eff_ci <- lapply(1:13, function(i) {
  boot::boot.ci(boo_cas_eff, type = c("norm", "basic", "perc", "bca"),
                index = i, L = inf_val_eff[[i]])
})
boo_cas_icc_ci <- boot::boot.ci(boo_cas_icc,
                                type = c("norm", "basic", "perc", "stud", "bca"),
                                index = 1L, L = inf_val_icc)
```

The `boot::boot.ci()` function also supports five ways to compute CI, including normal, basic, percentile, studentized, and BCa. Again, the influence values are needed to compute CIs with BCa, and we use the `lapply()` function to compute the CIs for all 13 statistics defined in `eff_rsq()` in one command.

```
# Resampling both clusters and individuals
boo_cas1_eff_ci <- lapply(1:13, function(i) {
  boot::boot.ci(boo_cas1_eff, type = c("norm", "basic", "perc", "bca"),
                index = i, L = inf_val_eff[[i]])
})
boo_cas1_icc_ci <- boot::boot.ci(boo_cas1_icc,
                                 type = c("norm", "basic", "perc", "stud", "bca"),
                                 index = 1L, L = inf_val_icc)
```

Comparison

Tables 13.1, 13.2, 13.3, and 13.4 summarize the standard errors and CIs for all parameters with the illustrated bootstrap approaches. The tables also include the uncorrected estimates and asymptotic standard errors from the *lme4* model output, as well as the confidence intervals based on profile likelihood using `confint(profile(mod_f, prof.scale = "varcov"))`. Note that with non-bootstrap methods, it is generally difficult to obtain the standard errors and confidence intervals for ICC and R^2. Moreover, as there are no direct ways to obtain standard errors of the variance components for the random effects from *lme4* at the time of writing, we relied on the `vcov_vc()` function from *bootmlm* to compute these standard errors.

Recall that the parametric bootstrap assumes normality; the residual bootstrap relaxes the normality assumption but still relies on the homoscedasticity assumption; the wild bootstrap relaxes the homoscedasticity assumption but relies on the normality assumption; and the cases bootstrap relaxes both assumptions but requires more time to perform with lower computational efficiency. Since the intelligence variable violates both the normality and homoscedasticity assumptions, we expect to see differences in the standard errors and CIs between methods for the fixed effects related to intelligence.

For the school-level fixed effects, the asymptotic standard errors and likelihood-based CIs are similar to the bootstrap standard errors and CIs, except for the school-mean intelligence (γ_{01}). The bootstrap CIs are generally narrower than the likelihood-based CI for γ_{01}. For the student-level fixed effects, the standard errors and CIs are similar across methods, except for the fixed effect of student-level intelligence (γ_{10}) and the interaction between the student-level intelligence and socio-economic status (γ_{40}). Similarly, the cases bootstrap with resampling at both levels gives larger standard errors and wider CIs. For school-mean intelligence, student-level intelligence, and the interaction between student-level intelligence and SES ($\gamma_{01}, \gamma_{10}, \gamma_{40}$), the cases bootstrap with resampling at both levels shows the most drastic differences with larger standard errors and wider CIs.

For the between-variances and the covariance between the random effects (τ_{11}^2, τ_{01}), the bootstrap standard errors and CIs are similar to the asymptotic standard errors and likelihood based CIs, as well as across bootstrap methods. For the within-variances and residual variances (τ_{00}^2, σ^2), however, the bootstrap

Table 13.1 Bias-corrected estimates, standard errors, and confidence intervals for school-level fixed effects

		School-level fixed effects			
	Type	Intercept	IQ	SES	IQ × SES
Est.		40.43	0.77	−0.09	−0.11
Bias-Corrected Est.	Parametric	40.43	0.78	−0.09	−0.11
	Residual	40.45	0.77	−0.09	−0.10
	Wild	40.43	0.76	−0.09	−0.10
	Cases (level-2)	40.41	0.75	−0.09	−0.10
	Cases (both levels)	40.40	0.66	−0.09	−0.10
Asymptotic SE		0.27	0.30	0.04	0.03
Bootstrap SE	Parametric	0.27	0.30	0.04	0.03
	Residual	0.26	0.29	0.04	0.03
	Wild	0.29	0.27	0.04	0.04
	Cases (level-2)	0.29	0.29	0.04	0.04
	Cases (both levels)	0.32	0.34	0.05	0.04
Likelihood-based CI		(39.90, 40.95)	(0.19, 1.36)	(−0.18, −0.01)	(−0.17, −0.04)
Normal	Parametric	(39.89, 40.96)	(0.20, 1.36)	(−0.18, −0.01)	(−0.17, −0.04)
	Residual	(39.93, 40.96)	(0.20, 1.35)	(−0.17, −0.01)	(−0.17, −0.04)
	Wild	(39.86, 40.99)	(0.22, 1.30)	(−0.17, −0.01)	(−0.17, −0.04)
	Cases (level-2)	(39.85, 40.97)	(0.18, 1.32)	(−0.17, −0.01)	(−0.18, −0.03)
	Cases (both levels)	(39.77, 41.03)	(−0.00, 1.31)	(−0.18, 0.00)	(−0.19, −0.01)
Basic	Parametric	(39.90, 40.97)	(0.19, 1.36)	(−0.18, −0.01)	(−0.17, −0.04)
	Residual	(39.94, 40.95)	(0.19, 1.34)	(−0.17, −0.01)	(−0.17, −0.04)
	Wild	(39.85, 40.99)	(0.22, 1.30)	(−0.18, −0.01)	(−0.18, −0.04)
	Cases (level-2)	(39.84, 41.01)	(0.15, 1.31)	(−0.17, −0.01)	(−0.17, −0.02)
	Cases (both levels)	(39.77, 41.03)	(0.00, 1.31)	(−0.18, 0.00)	(−0.18, −0.01)
Percentile	Parametric	(39.88, 40.95)	(0.18, 1.35)	(−0.18, −0.01)	(−0.17, −0.04)
	Residual	(39.90, 40.92)	(0.21, 1.35)	(−0.17, −0.01)	(−0.17, −0.04)
	Wild	(39.86, 41.00)	(0.24, 1.32)	(−0.17, −0.01)	(−0.17, −0.03)
	Cases (level-2)	(39.84, 41.01)	(0.24, 1.40)	(−0.17, −0.02)	(−0.19, −0.04)
	Cases (both levels)	(39.82, 41.08)	(0.24, 1.54)	(−0.19, −0.01)	(−0.20, −0.03)
BCa	Parametric	–	–	–	–
	Residual	(39.94, 40.93)	(0.20, 1.34)	(−0.17, −0.01)	(−0.16, −0.03)
	Wild	–	–	–	–
	Cases (level-2)	(39.80, 40.98)	(0.19, 1.33)	(−0.17, −0.01)	(−0.17, −0.02)
	Cases (both levels)	(39.73, 41.00)	(−0.05, 1.28)	(−0.18, 0.01)	(−0.18, −0.01)

Table 13.2 Bias-corrected estimates, standard errors, and confidence intervals for student-level fixed effects

		Student-level fixed effects			
	Type	IQ	SES	Gender	IQ × SES
Est.		2.25	0.17	−0.02	2.41
Bias−Corrected Est.	Parametric	2.25	0.17	−0.02	2.41
	Residual	2.24	0.17	−0.02	2.41
	Wild	2.25	0.17	−0.02	2.41
	Cases (level-2)	2.25	0.17	−0.02	2.41
	Cases (both levels)	2.20	0.17	−0.02	2.44
Asymptotic SE		0.06	0.01	0.00	0.20
Bootstrap SE	Parametric	0.06	0.01	0.00	0.21
	Residual	0.06	0.01	0.00	0.20
	Wild	0.07	0.01	0.01	0.21
	Cases (level-2)	0.06	0.01	0.01	0.20
	Cases (both levels)	0.09	0.02	0.01	0.29
Likelihood-based CI		(2.13, 2.37)	(0.15, 0.19)	(−0.03, −0.01)	(2.01, 2.80)
Normal	Parametric	(2.13, 2.37)	(0.15, 0.19)	(−0.03, −0.01)	(2.01, 2.81)
	Residual	(2.12, 2.35)	(0.15, 0.19)	(−0.03, −0.01)	(2.02, 2.79)
	Wild	(2.12, 2.38)	(0.15, 0.20)	(−0.03, −0.01)	(2.00, 2.82)
	Cases (level-2)	(2.12, 2.37)	(0.15, 0.19)	(−0.03, −0.01)	(2.02, 2.80)
	Cases (both levels)	(2.03, 2.37)	(0.14, 0.21)	(−0.04, −0.01)	(1.88, 3.00)
Basic	Parametric	(2.13, 2.37)	(0.15, 0.19)	(−0.03, −0.01)	(2.01, 2.82)
	Residual	(2.11, 2.35)	(0.15, 0.19)	(−0.03, −0.01)	(2.02, 2.80)
	Wild	(2.11, 2.37)	(0.15, 0.19)	(−0.03, −0.01)	(2.00, 2.83)
	Cases (level-2)	(2.13, 2.37)	(0.15, 0.19)	(−0.03, −0.01)	(2.01, 2.80)
	Cases (both levels)	(2.03, 2.37)	(0.14, 0.20)	(−0.04, −0.01)	(1.87, 3.00)
Percentile	Parametric	(2.13, 2.37)	(0.15, 0.19)	(−0.03, −0.01)	(2.00, 2.80)
	Residual	(2.15, 2.39)	(0.15, 0.19)	(−0.03, −0.01)	(2.02, 2.79)
	Wild	(2.12, 2.38)	(0.15, 0.20)	(−0.03, −0.01)	(1.98, 2.81)
	Cases (level-2)	(2.12, 2.37)	(0.15, 0.20)	(−0.03, −0.01)	(2.02, 2.81)
	Cases (both levels)	(2.13, 2.47)	(0.14, 0.20)	(−0.03, −0.00)	(1.82, 2.95)
BCa	Parametric	–	–	–	–
	Residual	(2.10, 2.35)	(0.15, 0.19)	(−0.03, −0.01)	(2.03, 2.80)
	Wild	–	–	–	–
	Cases (level-2)	(2.11, 2.36)	(0.15, 0.20)	(−0.03, −0.01)	(2.03, 2.84)
	Cases (both levels)	(2.02, 2.37)	(0.14, 0.21)	(−0.03, −0.01)	(1.89, 3.03)

Table 13.3 Bias-corrected estimates, standard errors, and confidence intervals for random effects

		Random effects			
	Type	τ_{00}^2	τ_{11}^2	τ_{01}	σ^2
Est.		8.50	0.15	−0.91	36.02
Bias-corrected Est.	Parametric	8.48	0.15	−0.91	36.06
	Residual	8.43	0.14	−0.88	35.94
	Wild	8.64	0.15	−0.91	36.00
	Cases (level-2)	8.62	0.15	−0.92	36.09
	Cases (both levels)	6.30	−0.30	−0.72	39.61
Asymptotic SE		1.07	0.07	0.21	0.87
Bootstrap SE	Parametric	1.08	0.06	0.20	0.85
	Residual	1.03	0.06	0.20	0.94
	Wild	1.22	0.06	0.22	2.87
	Cases (level-2)	1.04	0.06	0.21	0.93
	Cases (both levels)	1.34	0.13	0.32	1.18
Likelihood-based CI		(6.52, 10.66)	(0.00, 0.30)	(−1.35, −0.52)	(34.33, 37.76)
Normal	Parametric	(6.37, 10.60)	(0.03, 0.27)	(−1.30, −0.52)	(34.39, 37.72)
	Residual	(6.42, 10.44)	(0.01, 0.26)	(−1.27, −0.49)	(34.11, 37.77)
	Wild	(6.25, 11.03)	(0.03, 0.27)	(−1.34, −0.49)	(30.38, 41.62)
	Cases (level-2)	(6.58, 10.67)	(0.03, 0.27)	(−1.34, −0.50)	(34.26, 37.92)
	Cases (both levels)	(3.68, 8.92)	(−0.55, −0.04)	(−1.35, −0.10)	(37.30, 41.92)
Basic	Parametric	(6.40, 10.56)	(0.01, 0.25)	(−1.30, −0.51)	(34.30, 37.67)
	Residual	(6.32, 10.46)	(0.00, 0.24)	(−1.26, −0.47)	(34.08, 37.80)
	Wild	(6.22, 10.93)	(0.03, 0.25)	(−1.30, −0.46)	(30.05, 41.54)
	Cases (level-2)	(6.52, 10.67)	(0.02, 0.26)	(−1.34, −0.49)	(34.27, 37.91)
	Cases (both levels)	(3.59, 8.74)	(−0.57, −0.06)	(−1.32, −0.08)	(37.27, 41.95)
Percentile	Parametric	(6.44, 10.61)	(0.05, 0.29)	(−1.30, −0.52)	(34.37, 37.75)
	Residual	(6.55, 10.69)	(0.06, 0.30)	(−1.34, −0.55)	(34.24, 37.96)
	Wild	(6.08, 10.79)	(0.05, 0.28)	(−1.36, −0.51)	(30.51, 41.99)
	Cases (level-2)	(6.34, 10.49)	(0.05, 0.28)	(−1.32, −0.48)	(34.13, 37.77)
	Cases (both levels)	(8.27, 13.42)	(0.36, 0.87)	(−1.74, −0.50)	(30.09, 34.77)
BCa	Parametric	–	–	–	–
	Residual	(6.60, 10.72)	(0.06, 0.30)	(−1.31, −0.53)	(34.26, 37.98)
	Wild	–	–	–	–
	Cases (level-2)	(6.74, 10.91)	(0.06, 0.32)	(−1.41, −0.55)	(34.39, 38.17)
	Cases (both levels)	(6.81, 8.76)	–	(−1.37, −0.16)	–

Table 13.4 Bias-corrected estimates, standard errors, and confidence intervals for R^2 and ICC

	Type	R^2	R^2 (transformed)	ICC
Est.		0.43	–	0.22
Bias-Corrected Est.	Parametric	0.43	–	0.23
	Residual	0.43	–	0.22
	Wild	0.43	–	0.23
	Cases (level-2)	0.43	–	0.23
	Cases (both levels)	0.43	–	0.17
Bootstrap SE	Parametric	0.01	–	0.02
	Residual	0.01	–	0.02
	Wild	0.02	–	0.03
	Cases (level-2)	0.02	–	0.03
	Cases (both levels)	0.02	–	0.03
Likelihood-based CI		–	–	–
Normal	Parametric	(0.41, 0.46)	(0.41, 0.46)	(0.18, 0.27)
	Residual	(0.40, 0.46)	(0.40, 0.46)	(0.17, 0.27)
	Wild	(0.39, 0.48)	–	(0.18, 0.28)
	Cases (level-2)	(0.40, 0.46)	(0.40, 0.47)	(0.18, 0.28)
	Cases (both levels)	(0.38, 0.47)	(0.38, 0.47)	(0.12, 0.23)
Basic	Parametric	(0.40, 0.46)	(0.41, 0.46)	(0.18, 0.27)
	Residual	(0.40, 0.46)	(0.41, 0.46)	(0.17, 0.27)
	Wild	(0.38, 0.48)	–	(0.17, 0.27)
	Cases (level-2)	(0.40, 0.46)	(0.40, 0.46)	(0.17, 0.28)
	Cases (both levels)	(0.38, 0.47)	(0.38, 0.47)	(0.11, 0.23)
Percentile	Parametric	(0.41, 0.46)	(0.41, 0.46)	(0.18, 0.27)
	Residual	(0.41, 0.46)	(0.41, 0.46)	(0.18, 0.28)
	Wild	(0.39, 0.48)	–	(0.18, 0.28)
	Cases (level-2)	(0.40, 0.47)	(0.40, 0.47)	(0.17, 0.28)
	Cases (both levels)	(0.40, 0.48)	(0.40, 0.48)	(0.22, 0.34)
Studentized	Parametric	–	–	(0.18, 0.27)
	Residual	–	–	(0.17, 0.27)
	Wild	–	–	–
	Cases (level-2)	–	–	(0.18, 0.28)
	Cases (both levels)	–	–	(0.13, 0.23)
BCa	Parametric	–	–	–
	Residual	(0.41, 0.46)	(0.41, 0.46)	(0.18, 0.28)
	Wild	–	–	–
	Cases (level-2)	(0.40, 0.47)	(0.40, 0.47)	(0.19, 0.30)
	Cases (both levels)	(0.39, 0.47)	(0.39, 0.47)	(0.19, 0.23)

CIs are generally narrower than the likelihood-based CIs, and the standard errors and bootstrap CIs slightly differ across methods. The residual bootstrap yields the smallest SE and narrowest CIs for τ_{00}^2 but a larger SE and wider CIs for σ^2. The wild bootstrap has a larger SE and wider CIs than all methods, except cases bootstrap with resampling at both levels, for τ_{00}^2 and σ^2. The cases bootstrap produces unusually small bias-corrected estimates, large SEs, and wide CIs, compared to other methods. As the cases bootstrap has also been found to perform poorly for variance component estimation (Van der Leeden et al., 1997, 2008), we do not recommend cases bootstrap at both levels for the random effect variance components. More studies are needed to evaluate the performance of the cases bootstrap.

For the R^2 and ICC, although the asymptotic standard errors and likelihood-based CIs are generally unobtainable, we can use bootstrap methods to estimate the standard errors and CIs. All bootstrap methods yield similar standard errors and CIs, but for ICC in particular, the cases bootstrap at both levels tend to give narrowest intervals with the normal and basic CIs.

Bootstrap CI with Transformation

For parameters with highly skewed sampling distributions or with bounds (e.g., R^2 and ICC), obtaining the CI from the transformed parameters may give better coverage at the 95% intervals (Ukoumunne et al., 2003). For example, we can transform R^2 to an unbounded scale with logit transformation, obtain the CIs on the transformed scale, and back-transform the CIs to the bounded scale between 0 and 1. One can also use the same procedure to transform ICC with the stabilizing transformation suggested by Ukoumunne et al. (2003).

Here we demonstrate how to obtain CIs for R^2 with the logit transformation ($f[x] = x/[1 - x]$). In `boot::boot.ci()`, we additionally provide three inputs: (a) the logit function, `qlogis`, (b) the first derivative of the logit function function `function(x) 1 / (x - x^2)`, and (c) the inverse of the logit function, `plogis`. As R^2 is the 13th parameter in our defined function `eff_rsq()` for bootstrapping, we specify `index = 13` to compute the CI for R^2. Below is the syntax for applying logit transformation for the bootstrap samples of R^2 with parametric, residual, and cases bootstrap.

```
# Parametric
boo_par_rsq_ci <- boot.ci(boo_par_eff, h = qlogis,
                          hdot = function(x) 1 / (x - x^2),
                          hinv = plogis,
                          index = 13, type = c("norm", "basic", "perc"))
# Residual
boo_res_rsq_ci <- boot.ci(boo_res_eff, L = inf_val_fr[[13]], h = qlogis,
                          hdot = function(x) 1 / (x - x^2),
                          hinv = plogis,
                          index = 13, type = c("norm", "basic", "perc", "bca"))
# Cases (resampling only clusters)
boo_cas_rsq_ci <- boot.ci(boo_cas_eff, L = inf_val_fr[[13]], h = qlogis,
                          hdot = function(x) 1 / (x - x^2),
                          hinv = plogis,
                          index = 13, type = c("norm", "basic", "perc", "bca"))
# Cases (resampling clusters and individuals)
boo_cas1_rsq_ci <- boot.ci(boo_cas1_eff, L = inf_val_fr[[13]], h = qlogis,
                           hdot = function(x) 1 / (x - x^2),
                           hinv = plogis,
                           index = 13, type = c("norm", "basic", "perc", "bca"))
```

In this example, the CIs for R^2 with transformation are similar to the CIs without transformation, as shown in Table 13.3.

Software for Multilevel Bootstrap

Several statistical software programs support multilevel bootstrapping, including R, SPSS, and SAS, as summarized in Table 13.5. We discuss these software programs in this section and refer readers to Van der Leeden et al. (2008) for details about two other programs, *MLwiN* and *MLA*.

The *lme4* (Bates et al., 2015), *bootmlm* (Lai, 2021), and *lmeresampler* (Loy et al., 2022) are packages in R that implement the bootstrap methods for *lme4* multilevel model outputs. The *lme4* package provides parametric and residual bootstrap, referred to as semiparametric bootstrap. The residual bootstrap procedure in this package is currently at an experimental stage and does not support resampling from reflated empirical residuals. The *bootmlm* package, as demonstrated in Sect. 13.4, allows parametric, residual, and cases bootstrap at level-2 or both levels. For the bootstrap samples from *lme4* and *bootmlm*, one can use the *boot* package to obtain normal, basic, percentile, studentized, and BCa CIs. The *lmeresampler* package additionally implements the wild bootstrap, whereas its output can currently be summarized using the *stats* R package, which supports normal, basic, and percentile CIs.

SPSS and SAS are two commercial and proprietary statistical software programs that offer multilevel bootstrap functionalities. SPSS has the options of parametric and cases bootstrap at level-2 and provides percentile and BCa CIs (IBM Corp, 2021). In SAS, one can perform parametric, residual (known as semiparametric),

Table 13.5 Summary of software for multilevel bootstrapping

Software	Supported bootstrap types	Supported confidence interval methods
R (lme4)	Parametric, residual	Normal, basic, percentile, studentized, BCa (via the boot R package)
R (bootmlm)	Parametric, residual, cases	Normal, basic, percentile, studentized, BCa (via the boot R package)
R (lmeresampler)	Parametric, residual, cases, wild	Normal, basic, percentile (via the stats R package)
SPSS	Parametric, cases	Percentile, BCa
SAS	Parametric, residual, cases	Percentile

and cases (known as nonparametric) bootstraps and obtain the percentile CIs (Appiah, 2018).

13.5 Summary

In this chapter, we introduced different multilevel bootstrap procedures and described how and when to use them. Specifically, we reviewed parametric, residual, wild, and cases bootstraps. We discussed five ways of constructing confidence intervals (CIs) after obtaining a bootstrap sampling distribution: bootstrap normal CI, basic CI, studentized CI, percentile CI, and bias-corrected and accelerated (BCa) CI. In an empirical example, we illustrated how to obtain these bootstrap CIs with R packages for the fixed effects, random effects, intraclass correlation, and R^2. Lastly, we detailed statistical software programs that support bootstrap procedures for multilevel data. In general, our suggestion is to use the residual bootstrap with studentized CI or BCa CI for (transformed) parameters if sample size is more than 70 clusters or has more than 25 observations for the average cluster size (see Lai, 2021 for a detailed discussion). For smaller sample sizes, we recommend using the residual bootstrap with the basic CI. However, each type of bootstrapping procedure has different assumptions, so researchers should first check if distributional assumptions are tenable and then select the appropriate procedure. In terms of bootstrap CIs, we recommend that researchers compare different CIs and report BCa or studentized CIs if they are obtainable.

References

Appiah, A. K. (2018). *Bootstrap linear mixed-effects models using SAS®procedures*, Vol. 17. SAS Institute.

Bartoń, K. (2022). *MuMIn: Multi-model inference*. Retrieved from https://CRAN.R-project.org/package=MuMIn

Bates, D., Mächler, M., Bolker, B., & Walker, S. (2015). Fitting linear mixed-effects models using lme4. *Journal of Statistical Software, 67*(1). https://doi.org/10.18637/jss.v067.i01

Braschel, M. C., Svec, I., Darlington, G. A., & Donner, A. (2016). A comparison of confidence interval methods for the intraclass correlation coefficient in community-based cluster randomization trials with a binary outcome. *Clinical Trials, 13*(2), 180–187. https://doi.org/10.1177/1740774515606377

Burch, B. D. (2012). Nonparametric bootstrap confidence intervals for variance components applied to interlaboratory comparisons. *Journal of Agricultural, Biological, and Environmental Statistics, 17*(2), 228–245. https://doi.org/10.1007/s13253-012-0087-9

Busing, F. (1993). *Distribution characteristics of variance estimates in two-level models: A Monte Carlo study* (PhD thesis). https://doi.org/10.13140/RG.2.2.18116.94088

Canty, A., & Ripley, B. D. (2021). *Boot: Bootstrap r (s-plus) functions.*

Carpenter, J., & Bithell, J. (2000). Bootstrap confidence intervals: When, which, what? A practical guide for medical statisticians. *Statistics in Medicine, 19*(9), 1141–1164. https://doi.org/10.1002/(sici)1097-0258(20000515)19:9%3C1141::aid-sim479%3E3.0.co;2-f

Carpenter, J., Goldstein, H., & Rasbash, J. (2003). A novel bootstrap procedure for assessing the relationship between class size and achievement. *Journal of the Royal Statistical Society: Series C (Applied Statistics), 52*(4), 431–443. https://doi.org/10.1111/1467-9876.00415

Davidian, M. (2017). *Nonlinear models for repeated measurement data.* Routledge.

Davidson, R., & MacKinnon, J. G. (1993). *Estimation and inference in econometrics* (1st ed.). Oxford University Press.

Davidson, R., & MacKinnon, J. G. (2000). Bootstrap tests: How many bootstraps? *Econometric Reviews, 19*(1), 55–68. https://doi.org/10.1080/07474930008800459

Davison, A. C., & Hinkley, D. V. (1997). *Bootstrap methods and their application.* Cambridge University Press.

Diciccio, T. J., & Efron, B. (1996). Bootstrap confidence intervals. *Statistical Science, 11*(3). https://doi.org/10.1214/ss/1032280214

Diciccio, T. J., & Romano, J. P. (1988). A review of bootstrap confidence intervals. *Journal of the Royal Statistical Society: Series B (Methodological), 50*(3), 338–354. https://doi.org/10.1111/j.2517-6161.1988.tb01732.x

Doolaard, S. (1999). *Schools in change or schools in chains? The development of educational effectiveness in a longitudinal perspective* (PhD thesis). University of Twente, Netherlands.

Efron, B. (1982). *The jackknife, the bootstrap and other resampling plans. Philadelphia: Society for Industrial and Applied Mathematics.* http://dx.doi.org/10.1137/1.9781611970319

Efron, B. (1988). Bootstrap confidence intervals: Good or bad? *Psychological Bulletin, 104*, 293–296. https://doi.org/10.1037/0033-2909.104.2.293

Efron, B. (2003). Second thoughts on the bootstrap. *Statistical Science, 18*(2), 135–140. https://doi.org/10.1214/ss/1063994968

Efron, B., & Tibshirani, R. J. (1993). *An Introduction to the Bootstrap.* Springer US. https://doi.org/10.1007/978-1-4899-4541-9

Gelman, A., & Hill, J. (2007). *Data analysis using regression and multilevel/hierarchical models.* Cambridge University Press.

Goldstein, H. (2011). Bootstrapping in multilevel models. In *European Association for Methodology Series. Handbook for advanced multilevel analysis* (pp. 163–171). Routledge/Taylor & Francis Group.

Good, P. I. (2005). *Permutation, parametric and bootstrap tests of hypotheses* (3rd ed.). Springer.

Hall, P. (1992). On bootstrap confidence intervals in nonparametric regression. *The Annals of Statistics, 20*(2), 695–711. https://doi.org/10.1214/aos/1176348652

Hapuhinna, N. S. S. M. (2021). Bootstrap methods for estimation in linear mixed models with heteroscedasticity. *Mathematics Ph.D. Dissertations., 82*, 138.

Huang, F. L., Wiedermann, W., & Zhang, B. (2022). Accounting for heteroskedasticity resulting from between-group differences in multilevel models. *Multivariate Behavioral Research, 58*(3), 637–657. https://doi.org/10.1080/00273171.2022.2077290

IBM Corp. (2021). *IBM SPSS statistics for macintosh, version 28.0.* IBM Corp.

Knuver, A. W., & Brandsma, H. P. (1993). Cognitive and affective outcomes in school effectiveness research. *School Effectiveness and School Improvement*, *4*(3), 189–204.

Krishnamoorthy, K., Lu, F., & Mathew, T. (2007). A parametric bootstrap approach for ANOVA with unequal variances: Fixed and random models. *Computational Statistics & Data Analysis*, *51*(12), 5731–5742. https://doi.org/10.1016/j.csda.2006.09.039

Lai, M. H. C. (2021). Bootstrap confidence intervals for multilevel standardized effect size. *Multivariate Behavioral Research*, *56*(4), 558–578. https://doi.org/10.1080/00273171.2020.1746902

Lindstrom, M. J., & Bates, D. M. (1990). Nonlinear mixed effects models for repeated measures data. *Biometrics*, 673–687.

Liu, R. Y. (1988). Bootstrap procedures under some non-i.i.d. models. *The Annals of Statistics*, *16*(4), 1696–1708. https://doi.org/10.1214/aos/1176351062

Loy, A., Steele, S., & Korobova, J. (2022). *Lmeresampler: Bootstrap methods for nested linear mixed-effects models*. https://doi.org/10.48550/arXiv.2106.06568

MacKinnon, J. G., & Webb, M. D. (2017). Wild bootstrap inference for wildly different cluster sizes. *Journal of Applied Econometrics*, *32*(2), 233–254. https://doi.org/10.1002/jae.2508

MacKinnon, J. G., & White, Halbert. (1985). Some heteroskedasticity-consistent covariance matrix estimators with improved finite sample properties. *Journal of Econometrics*, *29*(3), 305–325. https://doi.org/10.1016/0304-4076(85)90158-7

Mammen, E. (1993). Bootstrap and wild bootstrap for high dimensional linear models. *The Annals of Statistics*, *21*(1), 255–285. https://doi.org/10.1007/978-1-4612-2950-6_9

Modugno, L., & Giannerini, S. (2015). The wild bootstrap for multilevel models. *Communications in Statistics - Theory and Methods*, *44*(22), 4812–4825. https://doi.org/10.1080/03610926.2013.802807

Nakagawa, S., & Schielzeth, H. (2013). A general and simple method for obtaining R^2 from generalized linear mixed-effects models. *Methods in Ecology and Evolution*, *4*(2), 133–142. https://doi.org/10.1111/j.2041-210x.2012.00261.x

Raudenbush, S. W., & Bryk, A. S. (2002). *Hierarchical linear models: Applications and data analysis methods* (2nd ed.). Thousand Oaks: SAGE Publications, Inc.

Reiser, M., Yao, L., Wang, X., Wilcox, J., & Gray, S. (2017). A comparison of bootstrap confidence intervals for multi-level longitudinal data using Monte-Carlo simulation. In *ICSA Book Series in Statistics. Monte-Carlo Simulation-Based Statistical Modeling* (pp. 367–403). Springer. https://doi.org/10.1007/978-981-10-3307-0_17

Shrout, P. E., & Bolger, N. (2002). Mediation in experimental and nonexperimental studies: New procedures and recommendations. *Psychological Methods*, *7*(4), 422–445. https://doi.org/10.1037/1082-989X.7.4.422

Snijders, T. A. B., & Bosker, R. J. (2012). *Multilevel analysis: An introduction to basic and advanced multilevel modeling*. SAGE.

Strube, M. J. (1988). Bootstrap Type I error rates for the correlation coefficient: An examination of alternate procedures. *Psychological Bulletin*, *104*, 290–292. https://doi.org/10.1037/0033-2909.104.2.290

Ukoumunne, O. C., Davison, A. C., Gulliford, M. C., & Chinn, S. (2003). Non-parametric bootstrap confidence intervals for the intraclass correlation coefficient. *Statistics in Medicine*, *22*(24), 3805–3821. https://doi.org/10.1002/sim.1643

Van der Leeden, R., Meijer, E., & Busing, F. M. T. A. (1997). *Bootstrap methods for two-level models* (Technical Report No. PRM 97-04). Leiden: Leiden University, Department of Psychology.

Van der Leeden, R., Meijer, E., & Busing, F. M. T. A. (2008). Resampling Multilevel Models. In J. de Leeuw & E. Meijer (Eds.), *Handbook of multilevel analysis* (pp. 401–433). Springer. https://doi.org/10.1007/978-0-387-73186-5_11

Wilcox, R. (2010). *Fundamentals of modern statistical methods: Substantially improving power and accuracy*. Springer.

Wilcox, R. (2017). *Modern statistics for the social and behavioral sciences: A practical introduction* (2nd ed.). Chapman; Hall/CRC. https://doi.org/10.1201/9781315154480

Wilcox, R., & Muska, J. (2001). Inferences about correlations when there is heteroscedasticity. *British Journal of Mathematical and Statistical Psychology*, *54*(1), 39–47. https://doi.org/10.1348/000711001159410

Wu, C. F. J. (1986). Jackknife, bootstrap and other resampling methods in regression analysis. *The Annals of Statistics, 14*(4), 1261–1295. https://doi.org/10.1214/aos/1176350142

Zhang, Y., & Lai, M. H. C. (2024). Evaluating two small-sample corrections for fixed-effects standard errors and inferences in multilevel models with heteroscedastic, unbalanced, clustered data. *Behavior Research Methods*. Advance online publication. https://doi.org/10.3758/s13428-023-02325-9

Chapter 14
Investigating the Use of Robust Standard Errors to Account for Two-Way Clustering in Cross-Classified Data Structures

Bixi Zhang and Francis L. Huang

Multilevel models (MLMs) have often been used to account for dependent data structures in the educational, psychological, and social sciences. In particular, data in these areas often have nested data structures which violate the assumption of observation independence. In these instances, MLMs can capture the effects of lower- and higher-level units with random effects associated with the clusters (Raudenbush & Bryk, 2002). Thus, MLMs can make the estimation and inference more reliable by partitioning the variance appropriately at the different levels.

Traditional MLMs have been used when lower-level units are completely nested within higher-level units (e.g., students nested within classrooms nested within schools). However, more complex, "messy" multilevel data exist in natural settings. For instance, observations can be nested within two dimensions of clusters simultaneously, but the two dimensions do not necessarily need to have a pure nested, hierarchical structure (i.e., one is not nested within the other). For example, Raudenbush (1993) discussed the effects of neighborhoods and schools on student educational attainment. Students in the same neighborhood could attend different schools, and a school could have students from multiple neighborhoods. In this case, students are nested in both schools and neighborhoods, but the two dimensions of clustering do not have a pure nested structure. Another example comes from Gregory and Huang (2013) who investigated college-going expectations of high school students who were nested within their English and mathematics teachers. The English and mathematics teachers do not have a purely nested relationship. These data examples are known to be cross-classified by two dimensions of clusters.

B. Zhang (✉)
The Graduate Center, City University of New York, New York, NY, USA
e-mail: bzhang2@gc.cuny.edu

F. L. Huang
Educational, School and Counseling Psychology, University of Missouri, Columbia, MO, USA

M. Stemmler et al. (eds.), *Dependent Data in Social Sciences Research*,
https://doi.org/10.1007/978-3-031-56318-8_14

For data with observations that simultaneously belong to multiple settings, ignoring one dimension of clusters might influence the relationship between the outcome and predictors, possibly leading to biased regression coefficients (with unbalanced data; Luo & Kwok, 2009) and/or affecting the precision of the effects (Doedens et al., 2022). Previous evidence shows that misspecification of the cross-classification often biases the standard errors for the fixed effects of both the observation- and cluster-level predictors (Kim et al., 2021). Specifically, ignoring one of the two dimensions of clustering underestimates the standard errors of the cluster-level predictors, narrows the confidence intervals, and inflates the variance component of the clusters that was properly accounted for (Doedens et al., 2022; Dunn et al., 2015; Meyers & Beretvas, 2006).

To avoid biased estimates, cross-classified random effects models (CCREMs), also referred to as cross-classified multilevel models (CCMMs), have been developed to capture the described situations (Goldstein, 1987; Rasbash & Goldstein, 1994; Raudenbush, 1993). Specifically, CCREMs can account for the random effects from multiple dimensions of clusters and estimate the unique variance from each dimension appropriately. CCREMs can be treated as a generalization of MLMs but incorporate a non-hierarchical nested data structure.

Although CCREMs work well, as model complexity increases, so does the likelihood of model nonconvergence (Lee & Pustejovsky, 2023; Matuschek et al., 2017). However, an alternative to CCREMs would be to use a linear regression model (i.e., ordinary least squares; OLS) with cluster-robust standard errors (CRSEs) (Liang & Zeger, 1986). Unlike MLMs, OLS with CRSEs obtains asymptotically consistent standard errors, thus getting unbiased estimations of the standard errors and correct statistical inference with the appropriate degrees of freedom. Cameron et al. (2011) introduced robust inference using adjusted standard errors accounting for multiway clustering, at times also referred to as double clustering (Thompson, 2011). Although the notation may differ, the two-way clustering case is equivalent to the cross-classified data structure. Lee and Pustejovsky (2023) show that two-way OLS regression with CRSEs (referred to as two-way OLS-CRSE) may perform better than CCREMs when the assumption of homoscedasticity is violated for a level-1 predictor. However, the better performance of two-way OLS-CRSE applies only when there are a large number of clusters (i.e., >70). Compared with CCREMs, few studies in education and psychology use the simpler two-way OLS-CRSE to account for the potential variance underestimation,[1] and CCREMs have often been used to account for the cross-classified nature of the data, with only a few studies (<5%) bothering to use CRSEs instead (Lee & Pustejovsky, 2023).

However, it is well known that CRSEs may still misestimate the standard errors when a small number of clusters are present (e.g., <40) (Bell & McCaffrey, 2002; Cameron & Miller, 2015). This could very well be an issue with cross-classified

[1] In contrast, in disciplines such as finance, two-way clustering using CRSE is more common, and in a review of the literature, Petersen (2009) noted that less than 3% of studies used random effects models using generalized least-square estimates (i.e., a multilevel model).

data. Although one dimension of clustering may have an adequate number of clusters, the other clustering dimension may have a more limited number of clusters. For example, the classic example of Rasbash and Goldstein (1994) used to discuss the importance of CCREMs analyzed a dataset where students were nested within 148 primary schools but only 19 secondary schools. Several other examples can be found: Carroll-Scott et al. (2015) tested associations of school and neighborhood effects on fifth and sixth grade students' body mass index using data from 12 schools and 25 neighborhoods. Claus et al. (2020) highlighted the importance of CCREMs in industrial/organizational psychology and, as an example, analyzed scores using nested data with 127 participants and 12 raters. In a CCREM simulation design, 50 assessors and 20 proposals were assumed to estimate an assessment of ratings (Luo & Kwok, 2009). As cross-classified data may at times have a limited number of clusters, small sample corrections might be considered when using alternative approaches to CCREMs (e.g., two-way OLS-CRSE) because a small number of clusters may lead to underestimating the standard errors (Cameron & Miller, 2015), potentially leading to higher Type I error rates.

Fortunately, few cluster adjustments have been around for two decades though they have seen limited use in the applied literature and may not be readily available in some commercial statistical software. Bell and McCaffrey (2002) proposed the biased-reduced linearization (BRL) estimator to correct the estimation bias that result from analyzing data with a small number of clusters. The correction approach is referred to as CV2VE (Cameron & Miller, 2015), CR2 estimator (Pustejovsky & Tipton, 2018), or LZ2 (Imbens & Kolesár, 2016). Recently, the CR2 adjustment has been successfully used in linear and generalized linear models with only a few clusters (i.e., 10) with applied examples in education and social sciences (Huang et al., 2022; Huang & Li, 2022). However, it is still unclear whether the CR2 adjustment may work well with cross-classified data structures with a small number of clusters (Cameron et al., 2011). Additionally, previous studies focused more on testing the performance for level-1 predictors in cross-classified data structures (Barr et al., 2013; Cameron et al., 2011; DeBruine & Barr, 2021; Lee & Pustejovsky, 2023), but a researcher's interest may lie in the higher-level predictors (e.g., Claus et al., 2020). Therefore, this current study compares the performance between CCREM and two-way OLS with the CRSE-CR2 adjustment (two-way OLS-CR2) for cross-classified data with a small number of clusters and a focus on cluster-level predictors.

14.1 Background on Cross-Classified Random Effects Models

CCREMs are an extension of MLMs which allows the lower-level observations to be cross-classified by higher-level clusters. CCREMs have been covered in many multilevel modeling books (e.g., Raudenbush & Bryk, 2002; Snijders & Bosker, 1999), and several articles in different fields on why CCREMs should be used have been published (Claus et al., 2020; Meyers & Beretvas, 2006; Rasbash & Goldstein, 1994). Recent reviews of CCREMs usage have shown an increase in its use in the past decade (Barker et al., 2020; Claus et al., 2020) though some may indicate that CCREMs are still underutilized in applied research (Huang & Jeon, 2022).

In a two-way case of a CCREM, the unconditional model (i.e., a model without any predictors) can be specified into two levels (Raudenbush & Bryk, 2002):

$$\text{Level 1 model}: Y_{ijk} = \pi_{0jk} + e_{ijk},$$

$$\text{Level 2 model}: \pi_{0jk} = \theta_0 + b_{00j} + c_{00k},$$

where Y_{ijk} is the outcome for observation i nested in cluster j ($j = 1, \ldots, J$) and cluster k ($k = 1, \ldots, K$), π_{0jk} is the intercept of the level 1 model, and the level 1 error term e_{ijk} follows a normal distribution with a mean of 0 and a variance of σ^2, $e_{ijk} \sim N(0, \sigma^2)$. Of note is that cluster j and k are not nested within each other as is the case in a purely hierarchical data structure. The intercept is modeled by two higher-level dimensions of clusters in the level 2 model. More specifically, two random effects terms are introduced in the model (b_{00j} and c_{00k}). Each random effect term is associated with each dimension of clustering and follows a normal distribution, where $b_{00j} \sim N\left(0, \tau_{b00}^2\right)$ and $c_{00k} \sim N\left(0, \tau_{c00}^2\right)$. Independence of the two random effects terms and independence of error terms are assumed. Thus, the unconditional CCREM can be written in a single equation as:

$$Y_{ijk} = \theta_0 + b_{00j} + c_{00k} + e_{ijk}. \tag{14.1}$$

The three random effect terms capture the variations from within-cluster and two cross-classified clusters. The variance partitioning allows for the computation of the intra-unit correlation coefficient (IUCC),[2] which represents the amount of variance associated with cluster j or k:

$$IUCC_j = \frac{\tau_{b00}^2}{\sigma^2 + \tau_{b00}^2 + \tau_{c00}^2} \text{ and } IUCC_k = \frac{\tau_{c00}^2}{\sigma^2 + \tau_{b00}^2 + \tau_{c00}^2}.$$

Note that the models can easily be extended to accommodate more than two dimensions of clustering (see examples in Barker et al., 2020 for a review of CCREM studies in health research). CCREMs are flexible and can also incorporate additional levels of nesting (e.g., students nested within two teachers nested within schools).

Furthermore, a random intercept model can be fit by adding predictors at level 1 and level 2. For example, if interested in three predictors, one at level 1 and the other two at level 2 from two different dimensions of clustering, the model can be written as:

[2] At times, this is referred to simply as the intraclass correlation coefficient (ICC).

$$\text{Level 1 model}: Y_{ijk} = \pi_{0jk} + \pi_{1jk}X_{ijk} + e_{ijk},$$

$$\text{Level 2 model}: \pi_{0jk} = \theta_0 + \beta_1 W_j + \gamma_1 W_k + b_{00j} + c_{00k},$$

$$\pi_{1jk} = \theta_1,$$

where X_{ijk} is the level 1 predictor, W_j is the level 2 predictor associated with cluster j, and W_k is the other level 2 predictor associated with cluster k. Three random effect terms capture the variations from different levels and dimensions of clustering. A combined model can be written as:

$$Y_{ijk} = \theta_0 + \theta_1 X_{ijk} + \beta_1 W_j + \gamma_1 W_k + b_{00j} + c_{00k} + e_{ijk}. \tag{14.2}$$

If researchers are interested in heterogenous effects of the level 1 predictors, a random slope can be added. For example, a random slope term (i.e., b_{10j}) can be introduced in the level 2 model:

$$\pi_{0jk} = \theta_0 + \beta_1 W_j + \gamma_1 W_k + b_{00j} + c_{00k},$$

$$\pi_{1jk} = \theta_1 + b_{10j}.$$

In this case, the effect of X_{ijk} can randomly vary by cluster j. The two random terms from cluster j follow a multivariate normal distribution, i.e.,

$$\begin{pmatrix} b_{00j} \\ b_{10j} \end{pmatrix} \sim N\left(\begin{pmatrix} 0 \\ 0 \end{pmatrix}, \begin{pmatrix} \tau_{b00}^2 & \tau_{b00,b10} \\ \tau_{b00,b10} & \tau_{b10}^2 \end{pmatrix}\right).$$

CCREMs can provide less biased inference of the fixed effects because it captures the heterogeneities from both clustering dimensions. CCREMs can be estimated using maximum likelihood (ML) or restricted ML (REML), though prior studies have shown that REML performs better when only a few clusters are present (Huang, 2016).

However, successful usage and consistent inferences of CCREMs are based on model assumptions. The four assumptions are as follows: (1) the error variance at level 1 should be constant, which indicates the assumption of homoscedasticity; (2) the random effects terms from different dimensions of clusters are assumed to be independent (i.e., $b_{00j} \perp c_{00k}$); (3) the random effects should follow normal distributions; (4) the predictors in the model should be independent of the random effect terms. Violation of these assumptions can lead to biased statistical inference. Thus, a model with weaker assumptions may be suitable for cross-classified data and can be useful as a more parsimonious alternative if researchers are not interested in the random effects. The current study investigates the performance of an OLS model with CRSEs that accounts for two-way clustering as an alternative to CCREMs with a focus on higher-level predictors.

Cross-Classified Data Structures

Cross-classified data structures vary by research area and are often categorized into two types: complete and partial cross-classification (Luo & Kwok, 2009). With complete cross-classification, the probability of a lower-level unit being nested in a cell (which represents an intersection between two dimensions of clustering) is the same across all units. This type of structure is frequently seen in experimental psychology when investigating responses/scores nested with subject and item (Barr et al., 2013; Singmann & Kellen, 2019). For example, participants (i.e., subjects) are shown a set of stimulus materials (i.e., items), and responses are nested within the subject-item intersection (a crossed random effect or a crossed factor). In addition to the random intercept, these models may also include a random slope. DeBruine and Barr (2021) provide several examples of experimental designs that sample both subjects and stimuli often used in experimental research in social cognition and psycholinguistics.

An example of complete classification may be seen when grant reviewers all rate (or score) a set of grant proposals. In such an instance, the review scores are nested within reviewer and proposal. Researchers may be interested in the characteristics of the reviewer and/or the proposal that may be related to the outcome (e.g., Is reviewer seniority related to the rating? Does the prestige of an institution where the proposal originated from matter?).

In contrast, partially cross-classified designs are often seen in different types of studies such as school-neighborhood research (e.g., Raudenbush, 1993). For this example, children in a geographic area can only attend particular public schools, and students attending certain schools can only come from a set of *feeder* neighborhoods. Units in one cluster can only be associated with some (not all) clusters in the other factor. However, since not all the clusters are completely crossed (as in the complete cross-classification designs), there are several empty cells as students in certain schools will not be able to come from (be fed from) specific neighborhoods. Due to the empty cells, this results in a condition of data sparseness. The number of empty cells was not found to bias results (Meyers & Beretvas, 2006) though Luo and Kwok (2009) indicated otherwise.

14.2 Ordinary Least Squares Regression with Cluster Robust Standard Errors (OLS-CRSE)

OLS regression is widely used in social and behavioral sciences and, as a single-level model, has fewer assumptions compared to MLMs. An OLS model with one level 1 predictor (X_{ijk}) and two level 2 predictors (W_j and W_k) can be written as:

$$Y_{ijk} = \beta_0 + \beta_1 W_j + \beta_2 W_k + \beta_3 X_{ijk} + e_{ijk}, where\ e_{ijk} \sim N\left(0, \sigma^2\right).$$

However, simply computing the parameters from OLS results in biased estimates of the standard errors due to the violation of observation independence. Thus, standard error corrections are needed for accurate statistical inference. Cameron et al. (2011) proposed a robust inference approach for cross-classified data in OLS regression models (i.e., two-way clustering) as an extension of CRSEs (Liang & Zeger, 1986).

The default OLS variance is $\hat{\sigma}^2(X'X)^{-1}$ where $\hat{\sigma}^2$ is the estimated error variance. However, the OLS estimated variance is inconsistent if the observations are not independent. To adjust the heterogeneity from higher-level clusters, a commonly used cluster-robust variance matrix (Liang & Zeger, 1986) is shown as:

$$\hat{V}\left(\hat{\beta}\right) = (X'X)^{-1}\hat{B}(X'X)^{-1},$$

where $\hat{B} = \sum_{j=1}^{J} X'_j\hat{\varepsilon}_j\hat{\varepsilon}'_j X_j$ and j indicates jth cluster from the higher-level units. Using another format of notation, $\hat{B} = X'\left(\hat{\varepsilon}\hat{\varepsilon}' \cdot S^J\right)X$, where S^J is an N by N indicator matrix and $\cdot$ represents element-wise multiplication (Cameron et al., 2011). Specifically, the stth element of S^J is equal to one if the sth and tth observations are in the same cluster j; otherwise the stth elements are equal to zero. The variance is consistently estimated in this way for OLS models with clustered data structures. Further, the estimated standard errors of the fixed effects can be obtained from the square root of elements on the diagonal of the variance matrix $\hat{V}\left(\hat{\beta}\right)$.

However, CRSEs work well for traditionally nested data without cross-classified clusters. By using the method described by Cameron et al. (2011), it is possible to estimate variance consistently from the data with two non-hierarchical structured dimensions of clustering (i.e., two-way OLS-CRSE). Suppose the cross-classified data have two dimensions of clusters, J and K, $\hat{B}$ can be specified as $X'\left(\hat{\varepsilon}\hat{\varepsilon}' \cdot S^{JK}\right)X$ where S^{JK} is a matrix indicating the observation's membership in two higher-level dimensions and $\cdot$ represents element-wise multiplication. In particular, if the dataset includes N observations, S^{JK} is an N by N matrix. The stth element is equal to one if the sth and tth observations share any cluster; otherwise it is equal to zero. S^{JK} can be also computed by three parts of information, $S^{JK} = S^J + S^K - S^{J\cap K}$. This formulation for multiway clustering has been used by others as well (e.g., Petersen, 2009; Thompson, 2011).

In the expression, the stth element of S^J is equal to one if the sth and tth observations are in the same cluster g, and the stth element of S^K is equal to one if the sth and tth observations are in the same cluster h. And finally, the stth element of $S^{J\cap K}$ is equal to one if the sth and tth observations are in both cluster j and cluster k. Otherwise, other elements from the three matrices are equal to zero. Therefore, the estimated variance of two-way clustering (i.e., two-way OLS-CRVE) becomes the combination of three single cluster robust variances:

$$\hat{V}\left(\hat{\beta}\right)_{TW-CR0} = \left(X'X\right)^{-1}X'\left(\hat{\varepsilon}\hat{\varepsilon}'\cdot S^{J}\right)X\left(X'X\right)^{-1} + \left(X'X\right)^{-1}X'\left(\hat{\varepsilon}\hat{\varepsilon}'\cdot S^{K}\right)X\left(X'X\right)^{-1} - \left(X'X\right)^{-1}X'\left(\hat{\varepsilon}\hat{\varepsilon}'\cdot S^{J\cap K}\right)X\left(X'X\right)^{-1}.$$

14.3 CR2 Correction for a Small Number of Cluster

The type of CRSEs without any small sample adjustment is referred to as CRSEs with a CR0 correction in OLS regression (OLS-CR0). The OLS-CR0 still results in biased standard errors for the higher-level predictors when the data have a few clusters (Cameron & Miller, 2015). A few small sample corrections for CRSEs were developed to address this problem. For instance, the CR2 correction was developed by Bell and McCaffrey (2002) and shows good performance in recent methodological work (e.g., F. L. Huang & Li, 2022; Imbens & Kolesár, 2016). With a CR2 adjustment in a two-level model, $\hat{B}_{CR2}$ becomes $\sum_{j=1}^{J} X'_j\left[I_{N_j} - H_j\right]^{-\frac{1}{2}}\hat{\varepsilon}_j\hat{\varepsilon}'_j\left[I_{N_j} - H_j\right]^{-\frac{1}{2}}X_j$, where $\left[I_{N_j} - H_j\right]^{-\frac{1}{2}}$ is the inverse of the square root of the matrix $I_{N_j} - H_j$. The component I_{N_j} is an identity matrix for cluster j with N_j observations and H_j is the hat matrix of cluster j (which can be computed using $X_j\left(X'_jX_j\right)^{-1}X'_j$). In a two-way clustering case, similar to the two-way OLS-CR0, the computation includes three parts of information. We first consider the cluster robust variance matrix for each dimension of clustering:

$$\hat{V}\left(\hat{\beta}\right)_J = \left(X'X\right)^{-1}X'\left(\sum_{j=1}^{J} X'_j\left[I_{N_j} - H_j\right]^{-\frac{1}{2}}\hat{\varepsilon}_j\hat{\varepsilon}'_j\left[I_{N_j} - H_j\right]^{-\frac{1}{2}}X_j\right)X\left(X'X\right)^{-1},$$

$$\hat{V}\left(\hat{\beta}\right)_K = \left(X'X\right)^{-1}X'\left(\sum_{k=1}^{K} X'_k\left[I_{N_k} - H_k\right]^{-\frac{1}{2}}\hat{\varepsilon}_k\hat{\varepsilon}'_k\left[I_{N_k} - H_k\right]^{-\frac{1}{2}}X_k\right)X\left(X'X\right)^{-1}.$$

The $\hat{V}\left(\hat{\beta}\right)_J$ only contains the cluster information for one dimension J, and the $\hat{V}\left(\hat{\beta}\right)_K$ only contains the cluster information for the other dimension K. The third part indicates the cluster-robust variance matrix information for the observations sharing the same jth and kth cluster:

$$\hat{V}\left(\hat{\beta}\right)_{J\cap K} = \left(X'X\right)^{-1}X'\left(\sum_{jk=1}^{J*K} X'_{jk}\left[I_{N_{jk}} - H_{jk}\right]^{-\frac{1}{2}}\hat{\varepsilon}_{jk}\hat{\varepsilon}'_{jk}\left[I_{N_{jk}} - H_{jk}\right]^{-\frac{1}{2}}X_{jk}\right) X\left(X'X\right)^{-1}.$$

Thus, the two-way OLS-CR2 can be constructed as:

$$\hat{V}\left(\hat{\beta}\right)_{TW-CR2} = \hat{V}\left(\hat{\beta}\right)_{J} + \hat{V}\left(\hat{\beta}\right)_{K} - \hat{V}\left(\hat{\beta}\right)_{J\cap K}.$$

The two-way OLS-CR2 has been implemented in the *sandwich* package (Zeileis et al., 2020) for R (referred to as the HC2 adjustment) (R Core Team, 2023).

14.4 The Current Study

The current study evaluates the performance of the two-way OLS-CR2 in comparison to the standard CCREM in evaluating cross-classified data with a small number of clusters in one or two dimensions. Specifically, we conducted a Monte Carlo simulation study comparing the performance of CCREMs, two-way OLS-CR0, and two-way OLS-CR2 for cluster-level predictors using two types of cross-classified data structures and varying the number of clusters. We include an analysis of an applied dataset as well.

14.5 Methods

Data Generating Process (DGP)

To compare the model performance for cross-classified data with a small number of clusters in one or two dimensions, we conducted a Monte Carlo simulation. The data generating process (DGP) simulated a continuous outcome (Y_{ijk}) for observation i simultaneously nested in level 2 clusters j and k. Three independent variables were included in the DGP—one level 1 continuous predictor (X_{ijk}) and two level 2 continuous predictors (W_j and W_k). For the level 2 predictors, one (i.e., W_j) is associated with the jth cluster in one dimension of clustering, and the other one (i.e., W_k) is associated with the kth cluster in the other dimension. Thus, for a random intercept model, the level 1 model was specified as:

$$Y_{ijk} = \pi_{0jk} + \pi_{1jk}X_{ijk} + e_{ijk}, \text{ where } e_{ijk} \sim N\left(0, \sigma^2\right),$$

and the level 2 models were constructed to:

$$\pi_{0jk} = \theta_0 + \beta_1 W_j + \gamma_1 W_k + b_{00j} + c_{00k},$$

$$\pi_{1jk} = \theta_1,$$

$$where \begin{pmatrix} b_{00j} \\ c_{00k} \end{pmatrix} \sim N\left(\begin{pmatrix} 0 \\ 0 \end{pmatrix}, \begin{pmatrix} \tau_{b00}^2 & 0 \\ 0 & \tau_{c00}^2 \end{pmatrix}\right).$$

We combined level 1 and level 2 models into a combined model including both random intercepts in each dimension of clustering:

$$Y_{ijk} = \theta_0 + \theta_1 X_{ijk} + \beta_1 W_j + \gamma_1 W_k + b_{00j} + c_{00k} + e_{ijk}.$$

From the above equation, the outcome was constructed based on four fixed terms and three random terms.

In the simulation design, we set the value of the intercept (θ_0) to 1 and the value of all fixed parameters to 0.3 (i.e., $\theta_1 = \beta_1 = \gamma_1 = 0.3$). All continuous predictors and the level 1 error term (i.e., e_{ijk}) were generated from a normal distribution with a mean of 0 and a variance of 1. The level 2 error terms (b_{00j} and c_{00k}) followed a normal distribution with a mean of 0 and a variance of 0.10 (i.e., $\tau_{b00}^2 = \tau_{c00}^2 = 0.10$).

Additionally, the present simulation considered two cross-classified data structures. In the first scenario, the simulation aimed to mimic the completely cross-classified data structure such as those found in experimental psychology where scores are nested in items and subjects. The data structure is more likely to have a dense matrix, and the number of observations per cell is usually small. We assumed the number of clusters from two higher-level dimensions (J and K) was 10, 20, 30, 40, or 50. To avoid redundant conditions, in each combined cluster condition, K was assumed to be smaller or equal to J. For example, the condition $J = 20$ and $K = 30$ equals $J = 30$ and $K = 20$. Overall, there were 15 unique level 2 sample size conditions. The number of observations per cell was set to 3. To induce some form of imbalance, approximately 10% of observations were randomly dropped. Figure 14.1 shows an example of the complete cross-classification data structure

		D2 1	2	3	4	5	6	7	8	9	10
D1	1	2	3	3	3	2	3	3	3	2	3
	2	3	3	2	2	3	3	2	3	3	3
	3	3	3	3	3	3	2	2	3	2	3
	4	2	3	3	3	3	2	3	3	3	3
	5	3	3	3	3	3	3	3	2	3	3
	6	3	2	3	2	3	3	3	3	2	3
	7	3	3	3	3	1	3	3	3	3	3
	8	3	3	3	3	3	2	2	2	3	2
	9	3	3	3	3	3	3	2	2	2	3
	10	3	3	1	3	3	2	3	3	2	1

Fig. 14.1 Simulated complete cross-classified data structure in scenario 1. D1 and D2 represent the dimensions of clustering, and numbers indicate observations within cells

	D2	1	2	3	4	5	6	7	8	9	10
D1	1	0	12	18	0	0	0	0	0	0	0
	2	0	0	12	18	0	0	0	0	0	0
	3	0	0	0	13	17	0	0	0	0	0
	4	0	0	0	0	14	16	0	0	0	0
	5	0	0	0	0	0	11	19	0	0	0
	6	0	0	0	0	0	0	17	13	0	0
	7	0	0	0	0	0	0	0	14	16	0
	8	0	0	0	0	0	0	0	0	15	15
	9	12	0	0	0	0	0	0	0	0	18
	10	11	19	0	0	0	0	0	0	0	0

Fig. 14.2 Simulated partial cross-classified data structure example in scenario 2. D1 and D2 represent the dimensions of clustering, and numbers indicate observations within cells

with ten clusters in each dimension of clustering with at most three observations per observed cell in the first scenario.

In the second scenario, the simulation mimicked a partially cross-classified data structure where students are nested in schools and neighborhoods or students are nested in primary and secondary schools. In this scenario, the data structure tends to be sparse where there are many cells with no observations. The number of clusters from two higher-level dimensions (J and K; or schools and neighborhoods) was set to 10, 20, 30, 40, or 50. In each combined condition with clusters, K was assumed to be smaller or equal to J. The number of observations in each feeder cluster (i.e., in each school k) was 30 or 100. The sparse ratio, which indicates the unbalanced status (i.e., the numbers of observations per feeder cluster are unbalanced), was set to 0.2 to make a more realistic setting (in our case, the number of schools in each neighborhood become $0.2 \times J$, thus with a range from 2 to 10). Therefore, there were 30 conditions (15 conditions of combining the number of clusters from each dimension $\times$ 2 numbers of observations per feeder cluster) in total. Figure 14.2 shows an example of this partial cross-classification data structure with 10 clusters in each dimension of clustering with 30 observations for each feeder cluster (e.g., D1 here) in the second scenario.

Analytic Strategy

Two thousand datasets were generated in each condition under each scenario. For each simulated cross-classified dataset, CCREM was applied to predict the outcome using level 1 and level 2 predictors in the DGP. Restricted maximum likelihood (REML) using the *lme4* (Bates et al., 2015) and *lmerTest* (Kuznetsova et al., 2017) packages in R was used to estimate the models.

Second, OLS regression was also used. Because unadjusted single-level linear regression violates the assumption of observation independence, the standard errors of the fixed terms are biased. To test our main goals of the study, two-way clustering was applied to the model with both the CR0 and CR2 corrections using the *sandwich* package in R (Zeileis et al., 2020). For each standard error, statistical significance was evaluated using a t statistic with empirical degrees of freedom using Satterthwaite approximation from the *lmerTest* package. Overall, two OLS related models were used in the comparison (i.e., two-way OLS-CR0 and two-way OLS-CR2).

To test the standard error performance, bias in standard errors was computed from the simulation. The bias of standard errors was shown as the deviation between the empirical standard errors θ and the estimated standard errors $\hat{\theta}$ divided by the empirical standard errors (i.e., $\frac{\theta-\hat{\theta}}{\theta}$). Specifically, the empirical standard errors were calculated using the standard deviations of the estimated regression coefficients. Typically, the absolute value of bias is smaller than 10% indicating an acceptable bias (Muthén & Muthén, 2002).

Coverage rates were also computed. It combines the use of the standard errors and the degrees of freedom (df). Prior studies have shown that the choice of df can impact Type I error rates (Huang & Li, 2022), and in the study, the df choice (i.e., Satterthwaite df) was kept constant. The coverage rate is referred to as the proportion of samples where the confidence interval contains the population parameter (Carsey & Harden, 2013). A coverage rate close to .95 level is desirable which also indicates a nominal Type I error rate of 5% (1–coverage rate) (Harwell, 2020). Threshold boundaries of .925 to .975 were deemed acceptable using Bradley's (1978) liberal criteria.

In additional to the performance indices, convergence rates were evaluated because CCREMs had potential non-convergence issue. The small sample size also carries a risk of obtaining negative variance estimates from the computation of two-way clustering for cross-classified data (Cameron et al., 2011). The rate of negative variance occurrences was also evaluated to indicate the performance of the two-way clustering adjustments (i.e., two-way OLS-CR0 and two-way OLS-CR2).

14.6 Results

Scenario 1: Complete Cross-Classification

The standard error bias and coverage rates are shown in Tables 14.1 and 14.2. The uncorrected OLS model did not work well for estimating standard errors of all predictors (results not shown) because of the failure to account for the cluster-level random effects (as expected). Standard error bias indicated that the CCREM standard errors were generally unbiased for both level 1 and level 2 predictors, even when the number of clusters in one dimension was small (i.e., 10). It also aligns with the assumption because the DGP was based on a CCREM. For the level 1 predictor, the two-way OLS-CR0 and two-way OLS-CR2 showed low bias for the standard

Table 14.1 Standard error bias in the complete cross classification scenario

		CCREM			TW-CR0			TW-CR2		
J	K	X	W_j	W_k	X	W_j	W_k	X	W_j	W_k
10	10	**−0.011**	**−0.074**	**−0.079**	**−0.031**	−0.303	−0.297	**−0.019**	−0.176	−0.181
	20	**−0.050**	**−0.007**	**−0.044**	**−0.092**	−0.219	−0.190	**−0.082**	**−0.082**	−0.121
	30	**−0.014**	**−0.060**	**−0.024**	**−0.051**	−0.254	−0.154	**−0.043**	−0.126	−0.109
	40	**0.012**	**−0.064**	**−0.004**	**−0.028**	−0.250	−0.118	**−0.020**	−0.126	**−0.081**
	50	**0.042**	−0.109	**0.001**	**0.001**	−0.281	−0.110	**0.009**	−0.160	**−0.081**
20	20	**0.001**	**−0.034**	**−0.024**	**−0.018**	−0.162	−0.154	**−0.011**	**−0.095**	**−0.086**
	30	**−0.004**	**0.017**	**−0.021**	**−0.040**	−0.110	−0.120	**−0.034**	**−0.039**	**−0.073**
	40	**−0.033**	**−0.032**	**−0.026**	**−0.062**	−0.148	−0.111	**−0.056**	**−0.080**	**−0.075**
	50	**0.022**	**−0.015**	**0.015**	**−0.001**	−0.122	**−0.060**	**0.004**	**−0.054**	**−0.030**
30	30	**0.013**	**−0.041**	**0.009**	**−0.009**	−0.128	**−0.083**	**−0.004**	**−0.083**	**−0.035**
	40	**0.004**	**−0.025**	**−0.017**	**−0.031**	−0.107	**−0.095**	**−0.027**	**−0.061**	**−0.059**
	50	**−0.012**	**−0.045**	**−0.014**	**−0.024**	−0.124	**−0.085**	**−0.020**	**−0.078**	**−0.056**
40	40	**0.023**	**−0.015**	**−0.038**	**−0.008**	**−0.084**	−0.105	**−0.004**	**−0.049**	**−0.070**
	50	**−0.007**	**−0.008**	**0.002**	**−0.036**	**−0.073**	**−0.063**	**−0.032**	**−0.037**	**−0.034**
50	50	**−0.016**	**−0.004**	**−0.008**	**−0.012**	**−0.064**	**−0.062**	**−0.009**	**−0.035**	**−0.033**

Note. Bold numbers indicate bias within [−0.1, 0.1]. TW-CR0 = two-way OLS-CR0; TW-CR2 = two-way OLS-CR2. X = Level-1 predictor; W_j, W_k = Level-2 predictors associated with dimension J/K

Table 14.2 Coverage rate (%) in the complete cross classification scenario

		CCREM			TW-CR0			TW-CR2		
J	K	X	W_j	W_k	X	W_j	W_k	X	W_j	W_k
10	10	**95.2**	**95.2**	**94.9**	89.8	84.7	84.2	90.2	88.6	88.6
	20	**93.6**	**95.2**	**95.2**	89.2	87.9	88.3	89.6	91.8	90.6
	30	**94.7**	**94.7**	**94.6**	90.4	86.4	89.1	90.8	90.1	90.0
	40	**94.8**	**95.5**	**95.6**	90.7	87.7	91.3	90.9	90.9	91.9
	50	**95.9**	**94.4**	**95.1**	91.7	86.1	91.6	91.8	89.9	92.4
20	20	**95.9**	**95.5**	**94.9**	92.0	88.6	91.1	92.4	90.7	92.4
	30	**95.0**	**95.7**	**94.5**	92.3	90.9	90.5	**92.6**	**92.8**	91.8
	40	**93.5**	**94.7**	**95.2**	91.0	89.3	92.1	91.3	91.7	**93.1**
	50	**95.0**	**94.7**	**95.8**	92.4	90.3	92.0	**92.7**	92.2	**93.1**
30	30	**95.7**	**94.7**	**95.8**	**93.7**	90.4	91.8	**93.9**	91.5	**92.7**
	40	**94.6**	**95.4**	**95.2**	**93.0**	92.1	92.3	**93.3**	**93.2**	**93.4**
	50	**95.2**	**95.1**	**94.9**	**92.8**	91.0	92.1	**92.8**	92.1	**93.2**
40	40	**95.6**	**94.5**	**94.3**	**93.3**	92.2	91.6	**93.3**	**93.1**	**92.7**
	50	**94.4**	**95.3**	**95.1**	**92.8**	**92.6**	**93.0**	**92.8**	**93.4**	**93.7**
50	50	**94.5**	**95.8**	**95.0**	**93.5**	**93.3**	**93.2**	**93.6**	**93.9**	**93.7**

Note. Bold numbers indicate coverage rates within [92.5%, 97.5%]. TW-CR0 = two-way OLS-CR0; TW-CR2 = two-way OLS-CR2. X = Level-1 predictor; W_j, W_k = Level-2 predictors associated with dimension J/K

errors. However, the bias of level 2 predictors was acceptable for two-way OLS-CR0 only in the conditions with a relatively large number of clusters (e.g., 50) in both dimensions of clustering. Compared with two-way OLS-CR0, two-way OLS-CR2 showed less bias for all conditions. When the number of clusters was greater than 20, the small sample adjustment showed standard error biases less than 10%.

The results of the coverage rate showed that CCREMs performed well in all conditions with a small number of clusters. However, two-way OLS-CR2 showed an acceptable coverage rate (i.e., between 92.5% and 97.5%) only when the number of clusters in one dimension was greater than 30. Two-way OLS-CR2 did not work well in the case with an extremely small number of clusters for both level 1 and level 2 predictors.

Moreover, the non-convergence rate in all conditions was rare at less than 0.1%. The issue of having negative variance for the CRSEs may happen in the cases with ten clusters in one dimension of clustering. However, with the most extreme case—where both dimensions only had ten clusters each—approximately 3% of replications had negative variance when computing two-way OLS-CR0 and two-way OLS-CR2. Those models did not have an issue of negative variances when increasing the number of clusters. Therefore, the results combining standard error bias and coverage rate indicated that two-way OLS-CR2 could work for level 2 predictors with at least 30 clusters in one dimension of clustering. CCREMs with REML estimation work well with the random intercept models even when the number of clusters is as low as 10.

Scenario 2: Partial Cross-Classification

The performance of standard error bias and coverage rate is shown in Tables 14.3 and 14.4 for the partial cross-classification data structure. Compared with the first scenario, this scenario has more cells with no observations (i.e., a sparse data structure). Similar to the completely cross-classification structure, the CCREM results had low standard error bias for both level 1 and level 2 predictors even with a small numbers of clusters (e.g., 10). Two-way OLS-CR2 showed a small standard error bias for the level 1 predictor for all conditions. Nevertheless, for level 2 predictors, two-way OLS-CR2 showed minimal standard error bias when there were 20 clusters in one dimension and more than 20 clusters in the other dimension. The performance was similar to the results in scenario 1.

However, the two-way OLS-CR0 did not work well for level 2 predictors in all conditions. The coverage rate results in Table 14.4 indicated that CCREMs showed good coverage for all predictors. The two-way OLS-CR2 had acceptable coverage rates when the number of clusters was larger than 30 in each dimension of clustering.

Overall, the performance results were similar to those in the first scenario. However, the sparse data structure condition faced more non-convergence cases. The more zero cells in the data, the more non-convergence cases were shown in the simulation. Moreover, the number of observations per feeder cluster influenced the

Table 14.3 Standard error bias in the partial cross classification scenario

			CCREM			TW-CR0			TW-CR2		
GS per K	J	K	X	W_j	W_k	X	W_j	W_k	X	W_j	W_k
30	10	10	**0.088**	−0.120	−0.108	−0.491	−0.706	−0.723	**−0.054**	**−0.092**	−0.133
	20	10	0.136	−0.134	**−0.043**	−0.474	−0.695	−0.597	**−0.016**	−0.114	**−0.094**
		20	**0.064**	**−0.047**	**−0.021**	−0.279	−0.510	−0.523	**−0.035**	**−0.047**	**−0.078**
	30	10	**0.085**	−0.122	**−0.032**	−0.495	−0.69	−0.526	**−0.051**	−0.129	**−0.057**
		20	**0.053**	**−0.068**	**−0.024**	−0.281	−0.502	−0.429	**−0.034**	**−0.062**	**−0.052**
		30	**0.068**	**−0.035**	**−0.028**	**−0.085**	−0.363	−0.401	**0.004**	**−0.034**	**−0.060**
	40	10	**0.094**	−0.140	**−0.034**	−0.476	−0.687	−0.489	**−0.018**	−0.117	**−0.059**
		20	**0.092**	**−0.07**	**−0.027**	−0.261	−0.505	−0.373	**−0.003**	**−0.077**	**−0.054**
		30	**0.022**	**−0.066**	**−0.040**	−0.137	−0.383	−0.333	**−0.047**	**−0.078**	**−0.067**
		40	**0.036**	**−0.046**	**−0.031**	**0.014**	−0.274	−0.289	**−0.020**	**−0.058**	**−0.048**
	50	10	**0.085**	−0.115	**0.000**	−0.495	−0.669	−0.452	**−0.043**	**−0.085**	**−0.043**
		20	**0.078**	**−0.063**	**−0.023**	−0.260	−0.490	−0.334	**−0.007**	**−0.041**	**−0.052**
		30	**0.088**	**−0.048**	**−0.012**	**−0.086**	−0.363	−0.257	**0.017**	**−0.038**	**−0.032**
		40	**0.05**	**−0.041**	**−0.004**	**0.030**	−0.253	−0.210	**−0.009**	**−0.029**	**−0.027**
		50	**0.034**	**−0.026**	**−0.003**	0.115	−0.164	−0.180	**−0.039**	**−0.025**	**−0.033**
100	10	10	**0.099**	−0.164	**−0.095**	**−0.062**	−0.688	−0.706	**−0.055**	−0.125	−0.139
	20	10	0.118	−0.131	**−0.046**	**−0.044**	−0.669	−0.535	**−0.022**	−0.114	−0.103
		20	**0.081**	**−0.087**	**−0.014**	0.363	−0.476	−0.470	**−0.002**	**−0.088**	**−0.050**
	30	10	**0.081**	−0.108	**−0.009**	**−0.073**	−0.640	−0.418	**−0.057**	**−0.080**	**−0.060**
		20	**0.068**	**−0.086**	**−0.032**	0.332	−0.473	−0.361	**−0.016**	**−0.089**	**−0.060**
		30	**0.061**	**−0.040**	**−0.054**	0.596	−0.315	−0.345	**−0.035**	**−0.046**	**−0.072**
	40	10	0.105	−0.133	**0.001**	**−0.058**	−0.649	−0.349	**−0.035**	−0.115	**−0.062**
		20	**0.064**	**−0.063**	**0.002**	0.331	−0.448	−0.258	**−0.014**	**−0.057**	**−0.036**
		30	**0.083**	**−0.054**	**−0.024**	0.667	−0.310	−0.230	**0.010**	**−0.059**	**−0.057**
		40	**0.045**	**−0.055**	**−0.016**	0.855	−0.203	−0.213	**−0.028**	**−0.052**	**−0.051**
	50	10	0.117	−0.145	**−0.029**	**−0.043**	−0.647	−0.304	**−0.014**	−0.114	**−0.081**
		20	**0.059**	**−0.057**	**−0.036**	0.317	−0.446	−0.211	**−0.029**	**−0.052**	**−0.079**
		30	**0.069**	**−0.050**	**−0.014**	0.633	−0.299	−0.129	**−0.012**	**−0.030**	**−0.033**
		40	**0.030**	**−0.030**	**−0.024**	0.841	−0.180	−0.126	**−0.034**	**−0.035**	**−0.051**
		50	**0.024**	**−0.045**	**−0.029**	1.074	**−0.095**	−0.120	**−0.021**	**−0.052**	**−0.049**

Note. Bold numbers indicate bias between [−0.1, 0.1]. TW-CR0 = two-way OLS-CR0; TW-CR2 = two-way OLS-CR2. X = Level-1 predictor; W_j, W_k = Level-2 predictors associated with dimension J/K

non-convergence rate. For example, in the case with 50 clusters in each dimension and 30 observations per feeder cluster, the non-convergence rate was approximately 2%. However, when the number of observations per feeder cluster increases to 100, the non-convergence rate is approximately 10%. Negative variance estimates for the CRSE occurred only in the conditions with an extremely small number of clusters in each dimension. However, the rate was low (about 2%).

Table 14.4 Coverage rate (%) in the partial cross classification scenario

			CCREM			TW-CR0			TW-CR2		
GS per K	J	K	X	W_j	W_k	X	W_j	W_k	X	W_j	W_k
30	10	10	**97.0**	**93.5**	**93.6**	71.6	55.5	53.3	90.9	91.1	89.9
	20	10	97.6	**92.9**	**94.5**	72.7	56.1	61.3	92.2	91.9	90.9
		20	**96.4**	**94.2**	**94.8**	85.6	70.8	70.9	92.1	**92.5**	92.1
	30	10	**97.1**	**92.9**	**95.4**	70.5	57.3	68.2	91.6	90.5	92.2
		20	**97.0**	**94.3**	**95.3**	85.2	72.0	76.8	92.0	**92.8**	92.4
		30	**96.4**	**94.4**	**94.9**	**93.1**	80.9	78.1	**93.7**	**93.3**	91.9
	40	10	**97.2**	**94.1**	**94.2**	73.2	56.4	71.4	**93.0**	90.8	91.5
		20	**97.1**	**94.2**	**94.8**	86.7	71.1	80.7	**93.4**	91.4	92.4
		30	**95.3**	**93.7**	**95.1**	91.2	79.2	82.3	**92.6**	91.8	**93.2**
		40	**95.8**	**93.9**	**95.1**	**95.4**	85.7	85.5	**93.5**	**92.7**	**92.9**
	50	10	**97.3**	**94.0**	**95.2**	72.5	56.9	73.0	91.9	91.5	91.6
		20	**96.5**	**94.1**	**94.7**	86.6	73.1	82.4	**93.5**	**92.7**	**93.7**
		30	**96.4**	**93.9**	**94.6**	**93.2**	82.2	87.4	**94.8**	**93.3**	**94.2**
		40	**96.3**	**94.8**	**95.6**	**95.5**	87.3	89.5	**94.0**	**93.9**	**93.7**
		50	**95.7**	**94.5**	**94.5**	**97.2**	90.9	89.5	**93.6**	**94.1**	92.1
100	10	10	**97.5**	**93.1**	**94.7**	**95.3**	56.4	53.9	91.0	90.9	90.3
	20	10	**97.5**	**93.4**	**94.4**	**95.3**	59.1	68.8	92.0	91.0	90.6
		20	**96.2**	**93.3**	**95.2**	**99.5**	74.1	76.0	92.4	91.1	92.1
	30	10	**97.2**	**93.3**	**95.5**	**94.3**	61.6	76.9	91.0	91.7	92.0
		20	**97.0**	**93.6**	**94.6**	**99.2**	74.4	81.4	**93.0**	92.1	92.2
		30	**95.9**	**94.4**	**94.3**	**99.8**	84.0	82.1	**93.2**	**93.5**	92.0
	40	10	**97.4**	**94.1**	**93.9**	**94.5**	61.4	81.7	91.0	91.2	91.4
		20	**96.4**	**93.4**	**95.7**	**99.3**	76.2	87.6	**92.8**	92.1	**93.6**
		30	**96.8**	**94.0**	**95.1**	**99.9**	84.7	88.2	**93.8**	**92.9**	**92.8**
		40	**96.0**	**93.0**	**94.6**	**100.0**	89.4	89.1	**92.5**	92.2	**93.2**
	50	10	**97.2**	**92.8**	**95.2**	**95.4**	61.3	84.2	**92.7**	90.5	91.9
		20	**96.4**	**94.6**	**93.8**	**99.1**	76.8	89.1	**93.1**	**92.9**	91.6
		30	**96.8**	**93.8**	**94.7**	**99.9**	84.6	91.6	**94.0**	**93.1**	**94.1**
		40	**96.1**	**94.6**	**94.5**	**100.0**	90.4	**92.7**	**93.5**	**93.9**	**93.8**
		50	**95.6**	**94.7**	**95.1**	**100.0**	**93.4**	92.4	**93.5**	**93.8**	**92.5**

Note. The bold number indicates the coverage rate is in [92.5%, 97.5%]. TW-CR0 = two-way OLS-CR0; TW-CR2 = two-way OLS-CR2. X = Level-1 predictor; W_j, W_k = Level-2 predictors associated with dimension J/K

Applied Example

For an applied example, we use the dataset of Hox (2002), which was used in his book on multilevel modeling to introduce the concept of CCREMs.[3] The simulated

[3] The dataset and tutorials on using CCREM are available at https://stats.oarc.ucla.edu/r/examples/mlm-ma-hox/r-mlm-chp-7/

dataset consisted of 1000 students who attended 50 different primary schools and later went on to attend 30 secondary schools. The outcome variable is achievement at the secondary school level (range = 3.90 to 9.90, M = 6.34, SD = 0.87). Similar to the case of students nested within schools and neighborhoods, students are nested within primary and secondary schools. Predictors at level 1 included gender (female = 1, male = 0) and student socioeconomic status (SES; on a six-point scale). At the second level, of interest was if the primary and secondary schools attended were public (denom = 0) or denominational (denom = 1). Based on a null model, the ICC was .23 and .09 for primary and secondary schools, respectively.

In Hox's (2002) book chapter, CCREMs are shown for a random intercept (RI) model and if SES was allowed to vary by secondary school (a random slope [RS] model). Our results (shown below) differ slightly from what Hox presented originally as we use restricted maximum likelihood (instead of ML) and we use a modified χ^2 difference test (LaHuis & Ferguson, 2009) to evaluate if including the random slope resulted in improved model fit (as we are testing whether the variability was greater than zero as variance cannot be less than zero). A 50:50 χ^2 mixture distribution is used instead.

In Table 14.5, we show results using a CCREM (RI), CCREM (RS), OLS, and two-way OLS with a CR2 (two-way OLS-CR2) adjustment. Based on the modified χ^2 difference test, the inclusion of the random slope was warranted; $\chi^2 = 5.69$, $p = .04$. Comparing model results shows that the point estimates are similar to each other. A difference is that with the OLS model—which ignores any form of clustering—both the level 2 predictors are statistically significant at the conventional .05 level as standard errors are underestimated. The two-way OLS-CR2 results are comparable to the CCREMs as well.

Table 14.5 Results comparing methods for analyzing cross-classified data (n = 1000)

	CCREM (RI)	CCREM (RS)	OLS	TW-CR2
(Intercept)	5.519***	5.524***	5.520***	5.520***
	(0.143)	(0.138)	(0.100)	(0.164)
Female	0.263***	0.260***	0.271***	0.271***
	(0.046)	(0.045)	(0.053)	(0.044)
SES	0.114***	0.113***	0.115***	0.115***
	(0.016)	(0.020)	(0.019)	(0.027)
Denomination (P)	0.204	0.210+	0.221***	0.221
	(0.126)	(0.126)	(0.054)	(0.135)
Denomination (S)	0.176+	0.165+	0.146**	0.146
	(0.096)	(0.091)	(0.056)	(0.109)

Notes. SES = socioeconomic status. P = primary school (n = 50). S = secondary school (n = 30). CCREM = cross classified random effects model. RI = random intercept. RS = random slope. OLS = ordinary least squares. TW-CR2 = OLS with the two-way CR2 adjustment. + $p < .10$, * $p < .05$, ** $p < .01$, *** $p < .001$

14.7 Discussion

The current study tested alternative ways to model partially and completely cross-classified data with both level 1 and level 2 predictors in the conditions with a small number of clusters in one or both clustering dimensions. The traditional way, often used in education or psychology, to deal with cross-classified data structures is using a CCREM. However, multilevel models can be more complex compared to single-level models that use standard error adjustments. Therefore, the current study compared the performance between CCREMs and OLS with CRSEs in conditions with a few clusters in each dimension of clustering.

Simulation results indicated good performance of the standard error corrections and statistical inference of CCREMs with a small number of clusters in one dimension (i.e., 20 clusters). Alternatively, two-way OLS-CR2 worked for level 1 and level 2 predictors when the number of clusters in each dimension was at least 30 or greater. Lee and Pustejovsky (2023) suggested that two-way OLS-CR0 could be an alternative for analyzing cross-classified data when the number of clusters was larger than 70, and we extend their findings by investigating cluster-level predictors.

Our findings show that the CR2 small sample correction can work for cross-classified data when the number of clusters is at least 30 in each dimension of clustering and if the focus is on the cluster-level predictors. Although the CR2 was found to work with at least ten clusters with traditionally nested data (Huang & Li, 2022), this did not work quite as well when two dimensions of clustering were of consideration.

Furthermore, the current study investigated two different data structures—complete and partial cross-classification. The two different scenarios depend on the research field. For example, in education, students are nested in schools and neighborhoods exhibiting partial cross-classification. Using CCREM to model partial cross-classified data with many zero cells potentially has a greater risk of experiencing non-convergence issue. In this case, if the number of clusters in each dimension is more than 30, two-way OLS-CR2 could be a viable alternative.

In the simulation, two-way OLS-CR0 and two-way OLS-CR2 might have an issue with negative variances due to an extremely small number of clusters. Cameron et al. (2011) suggested that in cases with negative variance estimates, due to subtracting the $\hat{V}\left(\hat{\beta}\right)_{J \cap K}$ matrix, this could be remedied by replacing negative values with zero, and this is an option when using the *sandwich* package (Zeileis et al., 2020). However, negative variance occurrences were rare when the number of clusters was at least 30.

There are several potential limitations and future directions based on the current study. First, only random intercept models were tested in the simulation. Thus, future studies could focus on the performance of the random slope models in the cross-classified data when the number of clusters is small. The additional heterogeneity might influence the performance of CCREMs and two-way OLS-CR2. Second, although CCREMs work well even with a small number of clusters, the extreme cases (ten clusters in one clustering dimension) may have limited power to detect effects. DeBruine and Barr (2021) discuss how to compute power using simulations using cross-classified data. Third, the simulation study considered the cases with at most 50 clusters in each dimension. Future study could test how two-way OLS-CR2 works for the cases when one dimension has a small number of clusters and another dimension has a relatively large number of clusters.

In conclusion, the current study (based on the simulation and applied example) shows that the two-way OLS-CR2 may work well for a small number of clusters, but the minimum number of clusters should be at least 30 per dimension. However, if using a CCREM is possible given the data, we suggest using CCREMs to model the cross-classified data even if one dimension only has a few clusters. Two-way OLS-CR2 might work as a sensitivity check (e.g., whether the results are consistent with the CCREM results) to get robust inferences in this case.

References

Barker, K. M., Dunn, E. C., Richmond, T. K., Ahmed, S., Hawrilenko, M., & Evans, C. R. (2020). Cross-classified multilevel models (CCMM) in health research: A systematic review of published empirical studies and recommendations for best practices. *SSM – Population Health, 12*, 100661. https://doi.org/10.1016/j.ssmph.2020.100661

Barr, D. J., Levy, R., Scheepers, C., & Tily, H. J. (2013). Random effects structure for confirmatory hypothesis testing: Keep it maximal. *Journal of Memory and Language, 68*(3), 255–278. https://doi.org/10.1016/j.jml.2012.11.001

Bates, D., Mächler, M., Bolker, B., & Walker, S. (2015). Fitting linear mixed-effects models using lme4. *Journal of Statistical Software, 67*(1), 1–48. https://doi.org/10.18637/jss.v067.i01

Bell, R., & McCaffrey, D. (2002). Bias reduction in standard errors for linear regression with multi-stage samples. *Survey Methodology, 28*(2), 169–181.

Bradley, J. V. (1978). Robustness? *British Journal of Mathematical and Statistical Psychology, 31*(2), 144–152. https://doi.org/10.1111/j.2044-8317.1978.tb00581.x

Cameron, A. C., & Miller, D. L. (2015). A practitioner's guide to cluster-robust inference. *Journal of Human Resources, 50*(2), 317–372. https://doi.org/10.3368/jhr.50.2.317

Cameron, A. C., Gelbach, J. B., & Miller, D. L. (2011). Robust inference with multiway clustering. *Journal of Business & Economic Statistics, 29*(2), 238–249. https://doi.org/10.1198/jbes.2010.07136

Carroll-Scott, A., Gilstad-Hayden, K., Rosenthal, L., Eldahan, A., McCaslin, C., Peters, S. M., & Ickovics, J. R. (2015). Associations of neighborhood and school socioeconomic and social contexts with body mass index among urban preadolescent students. *American Journal of Public Health, 105*(12), 2496–2502. https://doi.org/10.2105/AJPH.2015.302882

Carsey, T. M., & Harden, J. J. (2013). *Monte Carlo simulation and resampling methods for social science*. SAGE Publications.

Claus, A. M., Arend, M. G., Burk, C. L., Kiefer, C., & Wiese, B. S. (2020). Cross-classified models in I/O psychology. *Journal of Vocational Behavior, 120*, 103447. https://doi.org/10.1016/j.jvb.2020.103447

DeBruine, L. M., & Barr, D. J. (2021). Understanding mixed-effects models through data simulation. *Advances in Methods and Practices in Psychological Science, 4*(1), 2515245920965119. https://doi.org/10.1177/2515245920965119

Doedens, P., ter Riet, G., Boyette, L.-L., Latour, C., de Haan, L., & Twisk, J. (2022). Cross-classified multilevel models improved standard error estimates of covariates in clinical outcomes – A simulation study. *Journal of Clinical Epidemiology, 145*, 39–46. https://doi.org/10.1016/j.jclinepi.2022.01.005

Dunn, E. C., Richmond, T. K., Milliren, C. E., & Subramanian, S. V. (2015). Using cross-classified multilevel models to disentangle school and neighborhood effects: An example focusing on smoking behaviors among adolescents in the United States. *Health & Place, 31*, 224–232. https://doi.org/10.1016/j.healthplace.2014.12.001

Goldstein, H. (1987). Multilevel covariance component models. *Biometrika, 74*(2), 430–431. https://doi.org/10.2307/2336159

Gregory, A., & Huang, F. (2013). It takes a village: The effects of 10th grade college-going expectations of students, parents, and teachers four years later. *American Journal of Community Psychology, 52*(1–2), 41–55. https://doi.org/10.1007/s10464-013-9575-5

Harwell, M. (2020). The importance of type I error rates when studying bias in Monte Carlo studies in statistics. *Journal of Modern Applied Statistical Methods, 18*(1). https://doi.org/10.22237/jmasm/1556670360

Hox, J. J. (2002). *Multilevel analysis: Techniques and applications*. Lawrence Erlbaum.

Huang, F. L. (2016). Alternatives to multilevel modeling for the analysis of clustered data. *The Journal of Experimental Education, 84*(1), 175–196. https://doi.org/10.1080/00220973.2014.952397

Huang, S., & Jeon, M. (2022). Modern applications of cross-classified random effects models in social and behavioral research: Illustration with R package PLmixed. *Frontiers in Psychology, 13*, 976964. https://doi.org/10.3389/fpsyg.2022.976964

Huang, F. L., & Li, X. (2022). Using cluster-robust standard errors when analyzing group-randomized trials with few clusters. *Behavior Research Methods, 54*, 1181–1199. https://doi.org/10.3758/s13428-021-01627-0

Huang, F. L., Zhang, B., & Li, X. (2022). Using robust standard errors for the analysis of binary outcomes with a small number of clusters. *Journal of Research on Educational Effectiveness*, 1–33. https://doi.org/10.1080/19345747.2022.2100301

Imbens, G. W., & Kolesár, M. (2016). Robust standard errors in small samples: Some practical advice. *Review of Economics and Statistics, 98*(4), 701–712. https://doi.org/10.1162/REST_a_00552

Kim, S., Jeong, Y., & Hong, S. (2021). The impact of ignoring a crossed factor in cross-classified multilevel modeling. *Frontiers in Psychology, 12*, 637645. https://doi.org/10.3389/fpsyg.2021.637645

Kuznetsova, A., Brockhoff, P. B., & Christensen, R. H. B. (2017). lmerTest Package: Tests in linear mixed effects models. *Journal of Statistical Software, 82*(13). https://doi.org/10.18637/jss.v082.i13

LaHuis, D. M., & Ferguson, M. W. (2009). The accuracy of significance tests for slope variance components in multilevel random coefficient models. *Organizational Research Methods, 12*(3), 418–435.

Lee, Y. R., & Pustejovsky, J. E. (2023). Comparing random effects models, ordinary least squares, or fixed effects with cluster robust standard errors for cross-classified data. *Psychological Methods*. Advance online publication. https://doi.org/10.1037/met0000538

Liang, K.-Y., & Zeger, S. L. (1986). Longitudinal data analysis using generalized linear models. *Biometrika, 73*(1), 13–22.

Luo, W., & Kwok, O. (2009). The impacts of ignoring a crossed factor in analyzing cross-classified data. *Multivariate Behavioral Research, 44*(2), 182–212. https://doi.org/10.1080/00273170902794214

Matuschek, H., Kliegl, R., Vasishth, S., Baayen, H., & Bates, D. (2017). Balancing Type I error and power in linear mixed models. *Journal of Memory and Language, 94*, 305–315. https://doi.org/10.1016/j.jml.2017.01.001

Meyers, J. L., & Beretvas, S. N. (2006). The impact of inappropriate modeling of cross-classified data structures. *Multivariate Behavioral Research, 41*(4), 473–497. https://doi.org/10.1207/s15327906mbr4104_3

Muthén, L. K., & Muthén, B. O. (2002). How to use a Monte Carlo study to decide on sample size and determine power. *Structural Equation Modeling: A Multidisciplinary Journal, 9*(4), 599–620. https://doi.org/10.1207/S15328007SEM0904_8

Petersen, M. A. (2009). Estimating standard errors in finance panel data sets: Comparing approaches. *Review of Financial Studies, 22*(1), 435–480. https://doi.org/10.1093/rfs/hhn053

Pustejovsky, J. E., & Tipton, E. (2018). Small-sample methods for cluster-robust variance estimation and hypothesis testing in fixed effects models. *Journal of Business & Economic Statistics, 36*(4), 672–683. https://doi.org/10.1080/07350015.2016.1247004

R Core Team. (2023). *R: A language and environment for statistical computing*. R Foundation for Statistical Computing. https://www.r-project.org/

Rasbash, J., & Goldstein, H. (1994). Efficient analysis of mixed hierarchical and cross-classified random structures using a multilevel model. *Journal of Educational and Behavioral Statistics, 15*(4), 337–350.

Raudenbush, S. W. (1993). A crossed random effects model for unbalanced data with applications in cross-sectional and longitudinal research. *Journal of Educational Statistics, 18*(4), 321–349. https://doi.org/10.2307/1165158

Raudenbush, S. W., & Bryk, A. S. (2002). *Hierarchical linear models: Applications and data analysis methods*. SAGE.

Singmann, H., & Kellen, D. (2019). An introduction to mixed models for experimental psychology. In D. Spieler & E. Schumacher (Eds.), *New methods in cognitive psychology* (1st ed., pp. 4–31). Routledge. https://doi.org/10.4324/9780429318405-2

Snijders, T., & Bosker, R. (1999). *Multilevel analysis: An introduction to basic and advanced multilevel modeling*. SAGE.

Thompson, S. B. (2011). Simple formulas for standard errors that cluster by both firm and time. *Journal of Financial Economics, 99*(1), 1–10. https://doi.org/10.1016/j.jfineco.2010.08.016

Zeileis, A., Köll, S., & Graham, N. (2020). Various versatile variances: An object-oriented implementation of clustered covariances in *R*. *Journal of Statistical Software, 95*(1). https://doi.org/10.18637/jss.v095.i01

Chapter 15
Self-Normalized, Score-Based Tests of Parameter Heterogeneity in Mixed Models

Ting Wang and Edgar C. Merkle

Score-based tests have been used to study parameter heterogeneity in a variety of fields including economics (e.g., Andrews, 1993; Hjort & Koning, 2002), policy analysis (e.g., Zeileis & Hornik, 2007), and psychometrics (e.g., Debelak & Strobl, 2019; Merkle et al., 2014; Merkle & Zeileis, 2013; Schneider et al., 2022; Wang et al., 2018). The essential idea is to detect whether one or more model parameters vary with an unmodeled, auxiliary variable. This approach has the advantage of simplifying the statistical model of interest (auxiliary variables are not included in the model) and providing information about values of the auxiliary variable for which model parameters differ. For example, in the context of factor analysis, we can use score-based tests to study measurement non-invariance. The tests help us study whether or not factor loadings vary with respect to a continuous variable, such as respondent age. If the loadings do vary, then the tests also provide information about age groups for which loadings differ (Merkle & Zeileis, 2013). Additionally, if there exist multiple auxiliary variables (more than just "age"), researchers can consider tree approaches that use score-based tests to select important auxiliary variables (e.g., Arnold et al. 2021; Brandmaier et al. 2013; Fokkema et al. 2018; Henninger et al. 2022; Strobl et al. 2015; Zeileis et al. 2008).

The above applications almost always involve the assumption that the scores entering into the test are independent. Such an assumption is satisfied for regression

This paper was partially supported by NSF grant 1460719. The authors thank Professor Xianyang Zhang for comments and code, as well as Rudolf Debelak for comments on the paper. All remaining errors are the authors' responsibility.

T. Wang (✉)
American Board of Family Medicine, Lexington, KY, USA

E. C. Merkle
Department of Psychological Sciences, University of Missouri, Columbia, MO, USA

M. Stemmler et al. (eds.), *Dependent Data in Social Sciences Research*,
https://doi.org/10.1007/978-3-031-56318-8_15

and generalized linear models, where cases are assumed to be independent. But the assumption can be violated in mixed models, where cases within the same group (cluster) are correlated. If we ignore this dependence, then the score-based tests will exhibit suboptimal statistical properties (low power, incorrect Type I error rates; Wang et al., 2018).

A small number of methods have been proposed to handle score dependence in mixed models. For models with nested random effects, we can aggregate the scores to the highest level so that they are again independent (e.g., Wang et al., 2021). This aggregation to the highest level implicitly happens in many psychometric applications, where the highest level is the person. But it is restrictive because we can only study parameter heterogeneity with respect to the highest level of the data. To address this issue, Fokkema et al. (2018) propose a two-step procedure where a generalized linear mixed model is first estimated, and then the model is "downgraded" to a generalized linear model by conditioning on predictions of the random effects. The scores are independent in this downgraded model, so that the score-based tests can be carried out as usual. While this solution is sufficient for many applications, it also ignores uncertainty in the random effect predictions and cannot be used to study heterogeneity in random effect (co)variance parameters. It seems worthwhile to find a solution that can be applied to all mixed models and that does not require us to condition on random effects.

In this chapter, we apply a self-normalization procedure (Shao, 2010; Shao & Zhang, 2010; Zhang et al., 2011) to handle case dependence in score-based tests for linear mixed models. The term *self-normalization* generally means that we use some aspect of the observed data to standardize and decorrelate a focal quantity. Shao (2015) discusses the fact that the Student t-test is the earliest example of self-normalization, whereby a mean difference is standardized by the sample standard error. The situation becomes more complicated in time series analysis, where we must consider autocorrelation in addition to variability. But the papers cited at the start of the paragraph describe methods for detecting change points in time series, where the self-normalization handles unmodeled dependence between cases. That is, instead of assuming a particular autoregressive or moving average structure, the self-normalization procedure allows the data to "speak for themselves" when accounting for dependence and studying change points. We use a similar strategy here, where we do not attempt to use any analytic results pertaining to case dependence in mixed models. Instead, we use self-normalization to characterize the empirical dependence between cases, which allows us to decorrelate the scores.

In the current chapter, we first provide specific background on score-based tests and on the self-normalization procedure. We then conduct simulation studies to compare traditional score-based tests with self-normalized, score-based tests. Next, we apply the self-normalized tests to empirical data. Finally, we discuss the potential broad application of the self-normalized, score-based tests in linear mixed models and other models where cases exhibit dependence.

15.1 Score-Based Tests

This section includes a brief introduction to score-based tests; for a more detailed introduction, see Wang et al. (2014).

In the usual maximum likelihood estimation framework, the goal is to find the parameter vector $\hat{\boldsymbol{\theta}}$ (say, of length q) that maximizes the model's log likelihood ℓ:

$$\hat{\boldsymbol{\theta}} = \underset{\boldsymbol{\theta}}{\operatorname{argmax}}\, \ell(\boldsymbol{\theta}; \boldsymbol{y}_1, \ldots, \boldsymbol{y}_n), \tag{15.1}$$

where $\boldsymbol{y}_1, \ldots, \boldsymbol{y}_n$ represents the observed data. Maximization of the likelihood function is equivalent to finding $\hat{\boldsymbol{\theta}}$ such that the *score* vectors satisfy the following condition:

$$\sum_{i=1}^{n} s(\hat{\boldsymbol{\theta}}; \boldsymbol{y}_i) = \mathbf{0}, \tag{15.2}$$

where a *score* is the partial first derivative of the model's log likelihood for case i:

$$s(\boldsymbol{\theta}; \boldsymbol{y}_i) = \left(\frac{\partial \ell(\boldsymbol{\theta}; \boldsymbol{y}_i)}{\partial \theta_1}, \ldots, \frac{\partial \ell(\boldsymbol{\theta}; \boldsymbol{y}_i)}{\partial \theta_j} \right)^T. \tag{15.3}$$

Score-based tests are used to study how model parameters fluctuate with unmodeled, auxiliary variables. To carry out a test, we organize the score vectors into a scaled, cumulative score process defined as

$$\boldsymbol{B}(t; \hat{\boldsymbol{\theta}}) = n^{-1/2} \hat{\boldsymbol{I}}^{-1/2} \sum_{i=1}^{\lfloor n \cdot t \rfloor} s(\hat{\boldsymbol{\theta}}; \boldsymbol{y}_{(i)}), \tag{15.4}$$

where t is in [0, 1]; $\boldsymbol{y}_{(i)}$ represents the ith-smallest observation (ordered with respect to the unmodeled auxiliary variable); and $\hat{\boldsymbol{I}}$ represents the information matrix, which has an analytical expression for linear mixed models (see Wang & Merkle, 2018, Section 3.3). The purpose of $\hat{\boldsymbol{I}}$ is to decorrelate the cumulative score matrix's columns. In other words, a case's score with respect to one parameter is typically correlated with its score with respect to other parameters. Premultiplication by $\hat{\boldsymbol{I}}^{-1/2}$ decorrelates these scores, which facilitates development of score-based tests.

Using this scaled cumulative score process, multiple test statistics can be computed. We will focus on one, the *Cramér von Mises* (CvM) statistic, because it is similar to the self-normalized statistic that we describe later. The CvM statistic averages the sum of squared values of the cumulative scores, across all points of the cumulative score process. To facilitate comparison to the proposed statistic, the CvM statistic can be written as

$$CvM = n^{-1}\sum_{k=1}^{n} \boldsymbol{B}(k/n;\hat{\boldsymbol{\theta}})^T \boldsymbol{B}(k/n;\hat{\boldsymbol{\theta}}) \tag{15.5}$$

$$= n^{-1}\sum_{k=1}^{n}\left[\left\{n^{-1/2}\hat{\boldsymbol{I}}^{-1/2}\sum_{i=1}^{k}\boldsymbol{s}(\hat{\boldsymbol{\theta}};y_{(i)})\right\}^T\left\{n^{-1/2}\hat{\boldsymbol{I}}^{-1/2}\sum_{i=1}^{k}\boldsymbol{s}(\hat{\boldsymbol{\theta}};y_{(i)})\right\}\right] \tag{15.6}$$

$$= n^{-2}\sum_{k=1}^{n}\left[\left\{\sum_{i=1}^{k}\boldsymbol{s}(\hat{\boldsymbol{\theta}};y_{(i)})\right\}^T\hat{\boldsymbol{I}}^{-1}\left\{\sum_{i=1}^{k}\boldsymbol{s}(\hat{\boldsymbol{\theta}};y_{(i)})\right\}\right]. \tag{15.7}$$

We further define partial cumulative scores as

$$\boldsymbol{B}^{\star}_{a,b} = \sum_{i=a}^{b}\boldsymbol{s}(\hat{\boldsymbol{\theta}};y_{(i)}) \quad a \leq b \tag{15.8}$$

$$\boldsymbol{B}^{\star}_{a,b} = \sum_{i=b}^{a}\boldsymbol{s}(\hat{\boldsymbol{\theta}};y_{(a+b-i)}) \quad a > b. \tag{15.9}$$

The first line, where $a \leq b$, is the usual cumulative sum starting at the score associated with $y_{(a)}$ and ending at the score associated with $y_{(b)}$. The second line, where $a > b$, is a backward cumulative sum starting with the score associated with $y_{(a)}$ and ending at the score associated with $y_{(b)}$. Using this notation, we can rewrite Eq. (15.7) as

$$CvM = n^{-2}\sum_{k=1}^{n}\boldsymbol{B}^{\star T}_{1,k}\hat{\boldsymbol{I}}^{-1}\boldsymbol{B}^{\star}_{1,k}. \tag{15.10}$$

This form of the CvM statistic will be helpful to keep in mind, as we consider the self-normalization statistic below.

15.2 Self-Normalization

While the information matrix decorrelates column-wise dependence among scores (i.e., dependence between parameters), it does not address row-wise dependence (between cases). As mentioned earlier, this is a problem for mixed modeling, where cases within the same cluster are correlated. If we applied the traditional, score-based test statistics to situations where row-wise dependence exists, the resulting test statistic will be incorrect (Perron and et al., 2006). Andrews (1991) proposed substitutes for $\hat{\boldsymbol{I}}$ in this situation, but these substitutes introduce an extra tuning parameter, which is difficult to implement in practice.

The method that we study here comes from a series of papers by Shao and Zhang (Shao & Zhang, 2010; Zhang et al., 2011), who proposed a self-normalization approach for time series data that is related to the previous work by Lobato (2001) and others (Kiefer & Vogelsang, 2005; Kiefer et al., 2000). A key idea is to further decorrelate dependence of the scores as a function of k, where $k \in 1, \ldots, n-1$ represents each possible changing point. Formally, the statistic is defined as

$$SN = \sup_{k=1,\ldots,n-1} \boldsymbol{T}_n(k)^T \boldsymbol{V}_n^{-1}(k) \boldsymbol{T}_n(k), \tag{15.11}$$

which resembles the CvM statistic from Eq. (15.10). The $\boldsymbol{T}_n(k)$ and $\boldsymbol{V}_n(k)$ are specifically defined below.

The $\boldsymbol{T}_n(k)$ vector is the difference between the cumulative score process at point k and its expected value under the hypothesis of no changing points. This can be written as

$$\boldsymbol{T}_n(k) = n^{-1/2}\left(\boldsymbol{B}^{\star}_{1,k} - \frac{k}{n}\boldsymbol{B}^{\star}_{1,n}\right). \tag{15.12}$$

Because the sum of scores across all cases is by definition zero (e.g., $\boldsymbol{B}^{\star}_{1,n} = \boldsymbol{0}$), the second term in parentheses disappears in most score-based applications. The only exception here may be situations where the scores involve an integral approximation, as in GLMMs (Wang et al., 2022). In these situations, the scores may not sum to exactly zero due to the approximation.

Next, $\boldsymbol{V}_n(k)$ involves the (co)variance in the cumulative scores, considered separately before and after point k. Before point k, we compute covariances between the usual cumulative score, whereas after point k, we compute covariances between the backward cumulative score that starts at point n and ends at point $k+1$.

As an intermediate step toward defining $\boldsymbol{V}_n(k)$, let $\boldsymbol{C}_{a,b}$ be a matrix containing deviations of the cumulative sum $\boldsymbol{B}^{\star}$ from its expected value between points a and b. Specifically, for $a \leq b$, we have a $q \times (b-a+1)$ matrix $\boldsymbol{C}_{a,b}$ whose jth column is

$$\boldsymbol{B}^{\star}_{a,(a+j-1)} - \frac{j}{(b-a+1)}\boldsymbol{B}^{\star}_{a,b}$$

for $j = 1, \ldots, (b-a+1)$. Conversely, for the backward cumulative sum with $a > b$, $\boldsymbol{C}_{a,b}$ is a $q \times (a-b+1)$ matrix whose jth column is

$$\boldsymbol{B}^{\star}_{a,(a-j+1)} - \frac{j}{(a-b+1)}\boldsymbol{B}^{\star}_{a,b} \quad a > b$$

for $j = 1, \ldots, (a-b+1)$. Using these intermediate matrices, we write $\boldsymbol{V}_n(k)$ as

$$\boldsymbol{V}_n(k) = n^{-2}\left[\boldsymbol{C}_{1,k}\boldsymbol{C}^T_{1,k} + \boldsymbol{C}_{n,k+1}\boldsymbol{C}^T_{n,k+1}\right], \tag{15.13}$$

which includes the covariance of entries of the cumulative scores up to point k, along with the covariance of entries of the backward cumulative scores down to point $k+1$. This form of the matrix was developed for vector time series applications where the autocovariance function is unknown. Multiplication by n^{-2} facilitates computation of the resulting statistic's null distribution. The $\boldsymbol{V}_n(k)$ matrix basically is chosen so that, when we take its inverse, it cancels out the covariance that exists in the cumulative scores $\boldsymbol{T}_n(k)$. See Shao (2015) for further discussion and alternate expressions for $\boldsymbol{V}_n(k)$.

Comparing the CvM from Eq. (15.10) to the self-normalization statistic from Eq. (15.11), we can observe three main differences. First, the self-normalization statistic involves the dynamic matrix $\boldsymbol{V}_n(k)$ that changes with k and that can decorrelate ("self-normalize") both row and column dependencies in the cumulative score matrix. Second, the summing operation from the CvM statistic has changed into a maximum (sup). It seems possible that we could also use a summing operator in the self-normalization procedure, but we do not consider it here. We return to this issue in the General Discussion. Third, the multiplication by n differs across the two statistics. This is not a crucial difference because critical values of the SN statistic are obtained via simulation, as described next.

The SN statistic converges to a function of a Brownian bridge (for details, see Shao & Zhang, 2010; Zhang et al., 2011). We realize that some readers may be unfamiliar with a Brownian bridge, and we refer those readers to Wang et al. (2014) for further discussion and additional references. For our purposes here, it is sufficient to know that the Brownian bridge is a well-studied statistical process that is used in other score-based tests and that allows us to derive critical values and p-values under the null hypothesis. For the SN statistic defined above, we can compute critical values by repeatedly simulating the Brownian bridge and computing the SN statistic. This is similar to how some other score-based test statistics are obtained (e.g., Merkle et al., 2014).

In the following section, we will demonstrate problems with applying traditional score-based tests to correlated cases, and we will study the performance of the self-normalized test that we just described.

15.3 Simulation

In two-level, linear mixed models, observations in the same cluster are correlated with one another. This causes problems for traditional score-based tests but not necessarily for self-normalized tests. We now illustrate these issues via simulation, with the code being available at http://semtools.r-forge.r-project.org/.

Method

The simulation setup is similar to Wang et al. (2021), with the only difference being that the level 1 scores are being used in place of aggregated level 2 scores (where the level 2 scores are independent of each other). Our model is similar to that used for the *sleepstudy* data (Belenky et al., 2003) that are included with *lme4* (Bates et al., 2015). These data come from a longitudinal study of the association between sleep deprivation and reaction time, with each subject contributing ten observations. The dependent variable is the reaction time, and the predictor is *Days* of sleep deprivation. The fixed effects are therefore an intercept, β_0, and a slope for *Days*, β_1. We also have covarying intercept and slope random effects, leading to two random effect variances (σ_0^2, σ_1^2) and one covariance (σ_{01}). The variance not captured by the random effects is modeled by the residual variance σ_r^2. The model can be expressed as

$$\text{Reaction Time}_j \mid b_{0j}, b_{1j} \sim N(\beta_0 + b_{0j} + (\beta_1 + b_{1j})Days, \boldsymbol{R}_j) \quad (15.14)$$

$$\begin{pmatrix} b_{0j} \\ b_{1j} \end{pmatrix} \sim N\left(\mathbf{0}, \begin{bmatrix} \sigma_0^2 & \sigma_{01} \\ \sigma_{01} & \sigma_1^2 \end{bmatrix}\right) \quad (15.15)$$

$$\boldsymbol{R}_j = \sigma_r^2 \boldsymbol{I}_{10}, \quad (15.16)$$

where j indicates subject j, *Days* is a vector containing the values 0 to 9, and $\boldsymbol{I}_{10}$ denotes identity matrix with dimension of 10.

For simplicity, we focus on instability in fixed effect parameters with respect to an unmodeled continuous variable, which we loosely call *cognitive ability*. In one condition, the fixed intercept differs for individuals below the median level of cognitive ability, as compared to individuals above the median level of cognitive ability. In a second condition, the fixed slope differs instead of the fixed intercept. The magnitude of differences between parameters is based on a d parameter. When d is 0, it represents no change in the corresponding parameter, which serves as the baseline; when d is greater than 0, it represents differences in parameter values, with larger d indicating more severe parameter change. In this simulation, $d \in 0, 1, 2, 3, 4$, which roughly indicates the number of standard errors by which parameters differ across groups.

The data generating model's true parameter values were set to be the same as the estimates from the *sleepstudy* data. Model estimation proceeded via marginal maximum likelihood, where the form of the data generating model matched that of the fitted model (except for parameter differences between groups). Parameter instability was tested individually in the fixed intercept parameter and in the fixed slope parameter.

For both tests, we examine power and Type I error across three sample sizes (24, 48, or 96 individuals, contributing a total of $n = 240, 480$, or 960 observations), five magnitudes of parameter change, and two parameter conditions (change in either the fixed intercept or fixed slope). For each combination of conditions, we generate

1,000 datasets. For traditional score-based tests, we examine three continuous statistics. Specifically, in addition to the CvM statistic mentioned above, we also compute the double maximum statistic (DM) and maximum Lagrange multiplier test (maxLM). These two statistics differ from CvM on different aggregation strategies for the cumulative score matrix $\boldsymbol{B}(t; \boldsymbol{\theta})$; for more details, see Merkle & Zeileis (2013). The main focus of this simulation is to compare all of these statistics to the self-normalized statistic proposed in this chapter.

Results

The full simulation results for β_0 and β_1 are demonstrated in Figs. 15.1, 15.2, 15.3, and 15.4. These figures are arranged similarly, with the first two showing results for the traditional statistics (which assume independence of scores) and the second two showing results for the self-normalized statistic. In all four figures, the panel titles indicate the sample size and the tested parameter, and the y-axis indicates power (using $\alpha = 0.05$). Figures 15.1 and 15.3 show results when β_0 exhibits instability, and Figs. 15.2 and 15.4 show results when β_1 exhibits instability.

Looking at the first two figures, it is clear that the traditional statistics exhibit virtually no power in this scenario. The scores' lack of independence lead to this problem. On the contrary, the self-normalized results in Figs. 15.3 and 15.4 exhibit good power for the parameter that is truly changing. The power is monotonic with d and increases with sample size. When $d \geq 3$, even the smallest sample size of 240 demonstrates high power (power greater than 95%). Additionally, we can see the power is slightly higher for parameter β_0 compared to β_1. This phenomenon was also observed in Wang & Merkle (2018).

Tables 15.1 and 15.2 show the exact numbers underlying Figs. 15.3 and 15.4, which is especially useful for examining Type I error rates (at $d = 0$). Specifically, Type I error is equal to power when $d = 0$ (when parameters do not change) or when tests are applied to parameters that do not change. We can see that Type I error rates are generally below or approaching 5%. These results provide some evidence that self-normalized, score-based tests can be applied to models whose scores exhibit dependence.

15.4 Application

Wang et al. (2021) provided an application of score-based tests to linear mixed models, studying heterogeneity in the relationship between socioeconomic status and math test scores. They showed that the tests could detect heterogeneity in both fixed effects and random effect variances, when the heterogeneity occurred with respect to a level-2 auxiliary variable. Those tests were restricted to level-2 auxiliary variables, though, due to the dependence issue described throughout this chapter. In

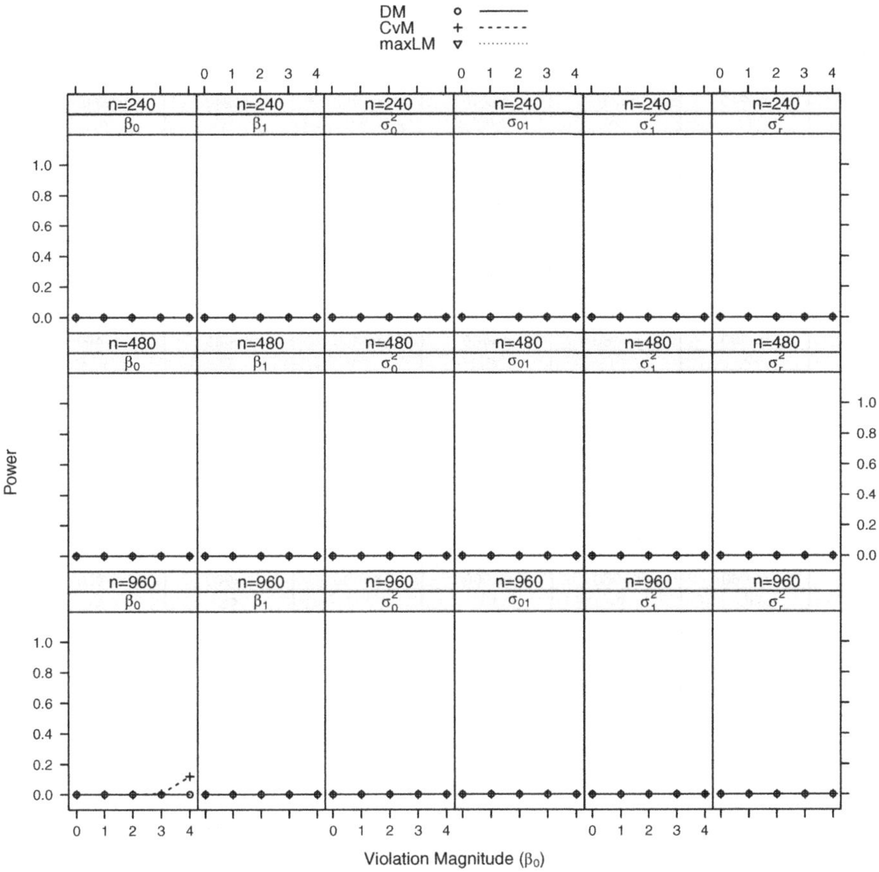

Fig. 15.1 Simulated power curves for CvM, DM, maxLM across parameter change of 0–4 asymptotic standard error. The changing parameter is β_0. Panel labels denote the parameter being tested along with sample size

this section, we use self-normalization to conduct similar tests with respect to a level-1 auxiliary variable.

Method

We use the *bdf* dataset (Snijders & Bosker, 2011) from R package *mlmRev* (Bates et al., 2020). The dataset contains language test scores of 2287 students from 131 schools. Variables in the dataset include students' verbal IQs, along with language test scores at two timepoints ("pre" and "post").

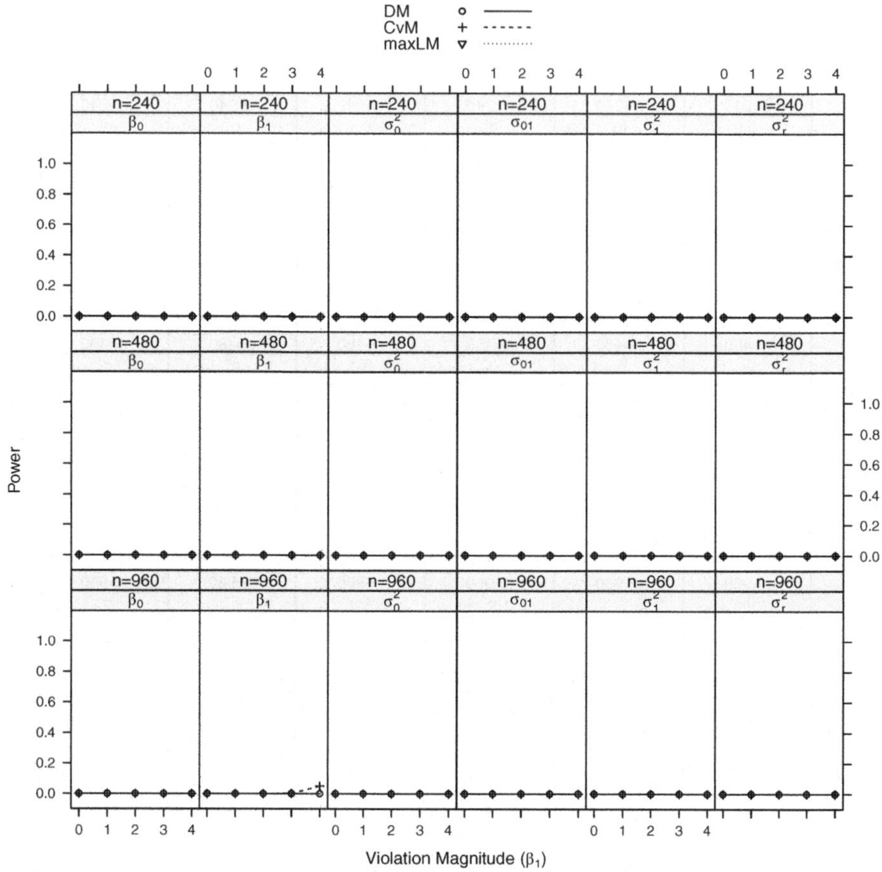

Fig. 15.2 Simulated power curves for CvM, DM, maxLM across parameter change of 0–4 asymptotic standard error. The changing parameter is β_1. Panel labels denote the parameter being tested along with sample size

The aim of the current analysis is to determine how students' language scores at time 2 (denoted as *langPOST* in the dataset) are associated with their language scores at time 1, along with their verbal IQs. It is plausible that the relationship between time 1 and 2 language scores differs for students with different verbal IQs. As illustrated in the previous research (Wang et al., 2021), such an interaction can be insignificant due to heterogeneity in random effect variances. Furthermore, because both covariates are at level 1, we have dependence in the scores. We use the self-normalized score test to study heterogeneity in mixed model parameters with respect to verbal IQ. The model specifications are shown via code in Figs. 15.5 and 15.6.

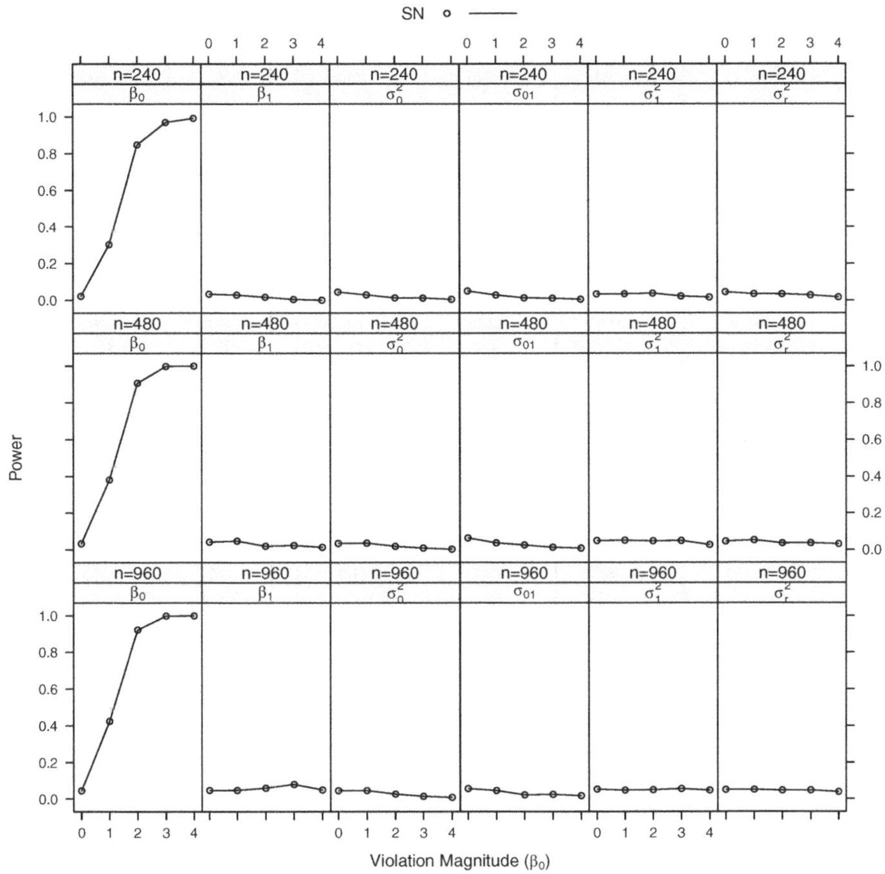

Fig. 15.3 Simulated power curves for SN, across parameter change of 0–4 standard errors. The changing parameter is β_0. Panel labels denote the parameter being tested along with sample size

Results

The most common approach to testing the interaction between verbal IQ and time 1 scores involves including the interaction directly in the mixed model, as shown in Fig. 15.5. In estimating this model, we observe that the coefficient for the interaction term is not significant ($t(2264) = -1.182$, $p = 0.24$). But the significance test for the interaction might be impacted by variance/covariance heterogeneity in random effects (Wang et al., 2021). Thus, we use the self-normalized score-based tests to distinguish between the level-1 interaction and variance heterogeneity.

To conduct the score-based tests, we first fit a model with time 1 scores as the only covariate, as shown in Fig. 15.6. The score test is then carried out with verbal IQ as the auxiliary variable. Because verbal IQ is a level 1 covariate, the traditional score-based tests are not appropriate due to dependence of the scores.

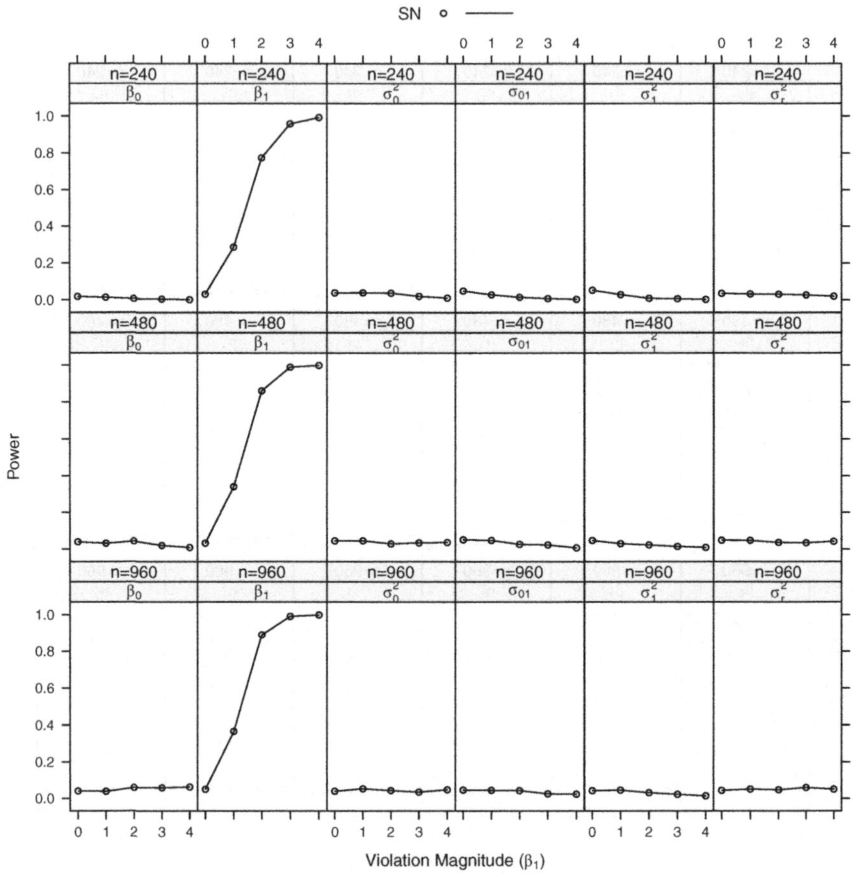

Fig. 15.4 Simulated power curves for SN, across parameter change of 0–4 standard errors. The changing parameter is β_1. Panel labels denote the parameter being tested along with sample size

We compute a self-normalized score test statistic separately for the fixed effect of time 1 test scores and for the residual variance. The corresponding code is shown in Fig. 15.7. The resulting fluctuation processes are demonstrated in Fig. 15.8, where the left panel is for the fixed effect and the right panel is for the residual variance. The black line shows the value of the SN statistic at different values of verbal IQ, with the maximum (the highest point) being the official value of the test statistic. The red line indicates the critical value, which is obtained from Table 1 (Shao & Zhang, 2010) based on the number of testing parameters and alpha level. Because the black line crosses the red line in both cases, we conclude that both the fixed effect of time 1 scores and the residual variance fluctuate with verbal IQ.

Because both black lines of Fig. 15.8 peak around verbal IQs of 13, we can conclude that students with verbal IQs below 13 have different parameter values than students with verbal IQs above 13. This leads us to estimate two separate

Table 15.1 Simulated power for self-normalized score-based test across three sample sizes n, five magnitudes of parameter change from zero to four times of asymptotic standard error of true changing parameter β_0, and six subsets of tested parameters. See Fig. 15.3 for a visualization

Observation	Tested parameter	Statistic	0	1	2	3	4
$n = 240$	β_0	SN	2.2	30.3	84.8	96.9	99.2
	β_1	SN	3.3	2.9	1.6	0.4	0.0
	σ_0^2	SN	4.4	2.9	1.2	1.1	0.4
	σ_{01}	SN	5.0	2.8	1.2	1.0	0.4
	σ_1^2	SN	3.3	3.4	3.7	2.2	1.6
	σ_r^2	SN	4.4	3.5	3.4	2.8	1.7
$n = 480$	β_0	SN	3.3	38.1	90.7	99.8	100.0
	β_1	SN	4.1	4.7	1.9	2.3	1.3
	σ_0^2	SN	3.4	3.5	1.8	0.8	0.2
	σ_{01}	SN	6.3	3.6	2.5	1.2	0.7
	σ_1^2	SN	4.8	5.0	4.7	4.8	2.6
	σ_r^2	SN	4.5	5.2	3.5	3.6	3.1
$n = 960$	β_0	SN	4.4	42.4	92.4	99.9	100.0
	β_1	SN	4.7	4.7	5.8	7.9	4.8
	σ_0^2	SN	4.4	4.5	2.5	1.3	0.7
	σ_{01}	SN	5.5	4.5	2.1	2.4	1.7
	σ_1^2	SN	5.2	4.7	4.9	5.5	4.6
	σ_r^2	SN	5.0	5.0	4.6	4.7	3.8

models, one for students with IQs less than or equal to 13 and one for students with IQs greater than 13. In estimating these separate models, we find that the fixed effect of time 1 scores is estimated at 0.89 and 0.69 for students with verbal IQs lower than 13 and higher than 13, respectively. The residual variance for students with verbal IQs below 13 is estimated at 33, while the residual variance for students with verbal IQs above 13 is estimated at 17.

These numerical results indicate that, while the association between time 1 and time 2 scores is somewhat lower for students with high verbal IQ, those students have lower residual variance and generally obtain high test scores. On the other hand, while time 1 scores are more helpful for predicting time 2 scores of students with low verbal IQ, those students also exhibit more residual variability in their scores. These types of results might help practitioners adopt different strategies for different types of students.

Table 15.2 Simulated power for self-normalized score-based test across three sample sizes n, five magnitudes of parameter change from 0 to 4 times of asymptotic standard error of true changing parameter β_1, and six subsets of tested parameters. See Fig. 15.4 for a visualization

Observation	Tested parameter	Statistic	0	1	2	3	4
$n = 240$	β_0	SN	1.9	1.4	0.7	0.5	0.2
	β_1	SN	3.2	28.7	77.4	95.9	99.3
	σ_0^2	SN	3.9	4.0	3.8	2.0	1.1
	σ_{01}	SN	5.0	2.9	1.5	0.8	0.4
	σ_1^2	SN	5.4	3.0	0.9	0.7	0.3
	σ_r^2	SN	3.6	3.3	3.2	2.7	2.0
$n = 480$	β_0	SN	4.0	3.2	4.4	2.0	0.9
	β_1	SN	3.3	34.1	86.1	99.0	99.9
	σ_0^2	SN	4.5	4.6	2.9	3.4	3.7
	σ_{01}	SN	5.1	4.8	2.6	2.3	0.7
	σ_1^2	SN	4.7	3.0	2.3	1.5	0.9
	σ_r^2	SN	4.8	4.8	3.6	3.4	4.2
$n = 960$	β_0	SN	4.0	3.9	5.9	5.8	6.2
	β_1	SN	5.0	36.5	89.0	99.2	99.9
	σ_0^2	SN	3.9	5.3	4.3	3.5	4.7
	σ_{01}	SN	4.6	4.4	4.3	2.5	2.3
	σ_1^2	SN	4.2	4.5	3.1	2.2	1.4
	σ_r^2	SN	4.3	5.0	4.7	5.9	4.9

```
> library("mlmRev")
> library("lmerTest")
> data("bdf")
> m1 <- lmer(langPOST ~ IQ.verb * langPRET + (1 | schoolNR), data = bdf,
+            REML = FALSE)
```

Fig. 15.5 Code for including an interaction directly in the linear mixed model

```
> m2 <- lmer(langPOST ~ langPRET + (1 | schoolNR), data = bdf,
+            REML = FALSE)
```

Fig. 15.6 Code for fitting a model with only one fixed effect

```
> library("strucchange")
> library("merDeriv")
> source("mzall.R")
> psi1 <- estfun.lmerMod(m2, level = 1)
> mz_gefp1  <- as.matrix(psi1[,2])
> mz_gefp2  <- as.matrix(psi1[,4])
> statsval1 <- calc.self.stats.cont(mz_gefp1, bdf$IQ.verb)
> statsval2 <-  calc.self.stats.cont(mz_gefp2, bdf$IQ.verb)
```

Fig. 15.7 Code to generate self-normalized statistics

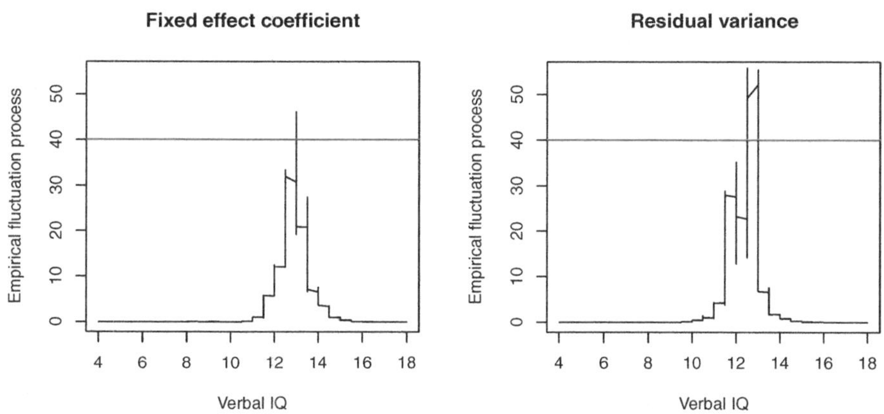

Fig. 15.8 Empirical statistics process based on the self-normalization approach

15.5 General Discussion

In this chapter, we demonstrated that the self-normalization approach can be used to carry out score-based tests in situations where scores exhibit dependence. This allows researchers to apply score-based tests to mixed models and other models where observations are not independent. It extends related work on score-based tests for mixed models (Fokkema et al., 2018; Wang et al., 2022, 2021), allowing us to test for heterogeneity with respect to level-1 auxiliary variables. While this chapter provides evidence that self-normalization is promising for score-based tests, it leaves open many issues for future work. We briefly describe some issues in the sections below.

Extensions of Self-Normalization

In this chapter, we used self-normalization to test one model parameter at a time. We also used the statistic that was previously proposed by Zhang et al. (2011). In future work, it would be worthwhile to test multiple parameters at a time and to use self-normalization in other statistics.

If we test for multiple parameters at a time, it is unclear whether we should first decorrelate the cumulative scores via $\hat{\boldsymbol{I}}$, prior to self-normalization. This is because the self-normalization procedure can potentially address both row-wise (cases) and column-wise (parameters) score dependence. However, because the information matrix is often obtained analytically (as opposed to approximated numerically), we may see improved power from a self-normalization test that first decorrelates via $\hat{\boldsymbol{I}}$.

Additionally, while we focused on the SN statistic from Eq. (15.11), it appears that $\boldsymbol{V}_n(k)$ could also replace $\hat{\boldsymbol{I}}$ in the CvM statistic of Eq. (15.10). The $\boldsymbol{V}_n(k)$ matrix may similarly replace $\hat{\boldsymbol{I}}$ in other score-based statistics, leading to a family of self-normalized, score-based tests that is similar to the traditional family of score-based tests. The family of self-normalized tests could be applied to a wider variety of statistical models.

Weighted Statistics

Shao and Zhang (2010) and Shao (2015) discuss how we can include weights in the SN statistic, which would be useful for carrying out score-based tests with respect to ordinal variables (Merkle et al., 2014). Specifically, let $w(t), t \in [0, 1]$ be the weight function, then the general form of the weighted SN statistic is

$$SN_w = \sup_{1 \le k \le n-1} w(k/n)\, \boldsymbol{T}_n(k)^T \boldsymbol{V}_n^{-1}(k) \boldsymbol{T}_n(k). \tag{15.17}$$

Merkle et al. (2014) described weights that allow for score-based tests with respect to ordinal auxiliary variables, which could be inserted into the above equation as

$$SN_{ord} = \sup_{k \in k_1, \ldots, k_{m-1}} \left\{ \frac{k}{n} \left(1 - \frac{k}{n} \right) \right\}^{-1} \boldsymbol{T}_n(k)^T \boldsymbol{V}_n^{-1}(k) \boldsymbol{T}_n(k), \tag{15.18}$$

where m represents the number of ordinal levels in the auxiliary variable. This means that the self-normalized statistics could flexibly handle many aspects of traditional score-based statistics.

Computation

Assuming that the ideas in the previous two subsections work as expected, we imagine an implementation similar to the functionality of the *strucchange* package (Zeileis et al., 2002). This would provide users with score-based functionality for a wider variety of mixed modeling situations. With this in mind, our code is currently far from optimal and could be improved in a variety of manners. Here, we consider recursive computations of the self-normalized test statistic.

Recursive computations of the $\boldsymbol{V}_n(k)$ matrix from Eq. (15.13) may be helpful, because that matrix must be computed and inverted for each value of $k = 1, \ldots, (n-1)$. The inversion is not a major problem if we test one parameter at a time (as we did in this chapter), but it could become computationally slow if we test many parameters at a time. To speed this up, it appears possible to write $\boldsymbol{V}_n(k)$ as a function of $\boldsymbol{V}_n(k-1)$ using identities from Welford (1962, also see the helpful Wikipedia entry on Welford's online algorithm). That result might then be plugged in to the Sherman-Morrison formula, allowing us to compute $\boldsymbol{V}_n^{-1}(k)$ from $\boldsymbol{V}_n^{-1}(k-1)$. This could enable us to invert the $q \times q$ matrix for a single value of k and then obtain subsequent inverses through simple matrix multiplications. Of course, the full derivations remain to be worked out.

Summary

Given the flexibility of the self-normalization approach and the existing applications of score-based tests, the combination of these two methods appears fruitful. We expect that further development of the ideas presented here can provide enhanced score-based tools to researchers in the social sciences and beyond.

15.6 Computational Details

All results were obtained using the R system for statistical computing (R Core Team, 2022), version 4.3.1, especially including packages *lme4* 1.1–33 (Bates et al., 2015) for model estimation, *merDeriv* 0.2–4 (Wang & Merkle, 2018) for score computations, and *strucchange* 1.5–3 (Zeileis et al., 2002) for score-based tests. Code for reproducing our results is available at http://semtools.r-forge.r-project.org/.

R and the aforementioned packages are freely available under the General Public License 2 from the Comprehensive R Archive Network at https://CRAN.R-project.org/.

References

Andrews, D. W. (1991). Heteroskedasticity and autocorrelation consistent covariance matrix estimation. *Econometrica: Journal of the Econometric Society*, 817–858.

Andrews, D. W. (1993). Tests for parameter instability and structural change with unknown change point. *Econometrica: Journal of the Econometric Society*, 821–856.

Arnold, M., Voelkle, M. C., & Brandmaier, A. M. (2021). Score-guided structural equation model trees. *Frontiers in Psychology, 11*, 564403. https://doi.org/10.3389/fpsyg.2020.564403

Bates, D., Mächler, M., Bolker, B., & Walker, S. (2015). Fitting linear mixed-effects models using lme4. *Journal of Statistical Software, 67*(1), 1–48. https://doi.org/10.18637/jss.v067.i01

Bates, D., Maechler, M., & Bolker, B. (2020). *mlmRev:* Examples from multilevel modelling software review [Computer software manual]. Retrieved from https://CRAN.R-project.org/package=mlmRev (R package version 1.0-8)

Belenky, G., Wesensten, N. J., Thorne, D. R., Thomas, M. L., Sing, H. C., Redmond, D. P., ...Balkin, T. J. (2003). Patterns of performance degradation and restoration during sleep restriction and subsequent recovery: A sleep dose-response study. *Journal of Sleep Research, 12*(1), 1–12.

Brandmaier, A. M., von Oertzen, T., McArdle, J. J., & Lindenberger, U. (2013). Structural equation model trees. *Psychological Methods, 18*, 71–86.

Debelak, R., & Strobl, C. (2019). Investigating measurement invariance by means of parameter instability tests for 2PL and 3PL models. *Educational and Psychological Measurement, 79*(2), 385–398. https://doi.org/10.1177/0013164418777784

Fokkema, M., Smits, N., Zeileis, A., Hothorn, T., & Kelderman, H. (2018). Detecting treatment-subgroup interactions in clustered data with generalized linear mixed-effects model trees. *Behavior Research Methods, 50*(5), 2016–2034.

Henninger, M., Debelak, R., & Strobl, C. (2022). A new stopping criterion for Rasch trees based on the Mantel-Haenszel effect size measure for differential item functioning. *Educational and Psychological Measurement*. https://doi.org/10.1177/00131644221077135

Hjort, N. L., & Koning, A. (2002). Tests for constancy of model parameters over time. *Journal of Nonparametric Statistics, 14*(1–2), 113–132.

Kiefer, N. M., & Vogelsang, T. J. (2005). A new asymptotic theory for heteroskedasticity-autocorrelation robust tests. *Econometric Theory, 21*(6), 1130–1164.

Kiefer, N. M., Vogelsang, T. J., & Bunzel, H. (2000). Simple robust testing of regression hypotheses. *Econometrica, 68*(3), 695–714.

Lobato, I. N. (2001). Testing that a dependent process is uncorrelated. *Journal of the American Statistical Association, 96*, 1066–1076.

Merkle, E. C., Fan, J., & Zeileis, A. (2014). Testing for measurement invariance with respect to an ordinal variable. *Psychometrika, 79*, 569–584.

Merkle, E. C., & Zeileis, A. (2013). Tests of measurement invariance without subgroups: A generalization of classical methods. *Psychometrika, 78*(1), 59–82.

Perron, P., et al. (2006). Dealing with structural breaks. *Palgrave Handbook of Econometrics, 1*(2), 278–352.

R Core Team. (2022). R: A language and environment for statistical computing [Computer software manual]. Vienna, Austria. Retrieved from https://www.R-project.org/

Schneider, L., Strobl, C., Zeileis, A., & Debelak, R. (2022). An R toolbox for score-based measurement invariance tests in IRT models. *Behavior Research Methods*, 2101–2113. https://doi.org/10.3758/s13428-021-01689-0

Shao, X. (2010). A self-normalized approach to confidence interval construction in time series. *Journal of the Royal Statistical Society. Series B (Statistical Methodology), 72*(3), 343–366.

Shao, X. (2015). Self-normalization for time series: A review of recent developments. *Journal of the American Statistical Association, 110*(512), 1797–1817.

Shao, X., & Zhang, X. (2010). Testing for change points in time series. *Journal of the American Statistical Association, 105*(491), 1228–1240.

Snijders, T. A., & Bosker, R. J. (2011). *Multilevel analysis: An introduction to basic and advanced multilevel modeling*. Sage.

Strobl, C., Kopf, J., & Zeileis, A. (2015). Rasch trees: A new method for detecting differential item functioning in the Rasch model. *Psychometrika, 80*(2), 289–316.

Wang, T., Graves, B., Rosseel, Y., & Merkle, E. C. (2022). Computation and application of generalized linear mixed model derivatives using lme4. *Psychometrika, 87*, 1173–1193.

Wang, T., & Merkle, E. C. (2018). *merDeriv:* Derivative computations for linear mixed effects models with application to robust standard errors. *Journal of Statistical Software, 87*, 1–16.

Wang, T., Merkle, E. C., Anguera, J. A., & Turner, B. M. (2021). Score-based tests for detecting heterogeneity in linear mixed models. *Behavior Research Methods, 53*(1), 216–231.

Wang, T., Merkle, E. C., & Zeileis, A. (2014). Score-based tests of measurement invariance: Use in practice. *Frontiers in Psychology, 5*(438), 1–11. https://doi.org/10.3389/fpsyg.2014.00438

Wang, T., Strobl, C., Zeileis, A., & Merkle, E. C. (2018). Score-based tests of differential item functioning via pairwise maximum likelihood estimation. *Psychometrika, 83*(1), 132–155.

Welford, B. P. (1962). Note on a method for calculating corrected sums of squares and products. *Technometrics, 4*(3), 419–420. https://doi.org/10.1080/00401706.1962.10490022

Zeileis, A., & Hornik, K. (2007). Generalized M-fluctuation tests for parameter instability. *Statistica Neerlandica, 61*(4), 488–508.

Zeileis, A., Hothorn, T., & Hornik, K. (2008). Model-based recursive partitioning. *Journal of Computational and Graphical Statistics, 17*(2), 492–514.

Zeileis, A., Leisch, F., Hornik, K., & Kleiber, C. (2002). strucchange: An R package for testing for structural change in linear regression models. *Journal of Statistical Software, 7*(2), 1–38. https://doi.org/10.18637/jss.v007.i02

Zhang, X., Shao, X., Hayhoe, K., & Wuebbles, D. J. (2011). Testing the structural stability of temporally dependent functional observations and application to climate projections. *Electronic Journal of Statistics, 5*, 1765–1796.

Chapter 16
Statistical Power in Cross-Sectional Multilevel Experiments in Education

Spyros Konstantopoulos, Wei Li, and Bixi Zhang

One important aspect of experimental design is about ensuring the design is sufficiently sensitive to detect an anticipated treatment effect of interest. During the planning stage of an experiment, researchers should make certain that the statistical test used to determine the statistical significance of the treatment effect is adequately powered. The degree to which the sensitivity of an experimental design is high enough to detect an expected treatment effect translates into assuring that the level of statistical power of the statistical test of the treatment effect is acceptable.

Statistical power is defined as the probability of correctly rejecting the null hypothesis when it is false. By definition, statistical power is the difference between the total probability under the curve of the alternative distribution and the Type II error probability. Because it is a probability, statistical power ranges between zero and one. However, intuitively, researchers wish statistical power to be closer to one, which subsequently means that the Type II error probability will be closer to zero. Commonly, a probability of .80 has been designated as the lower bound of statistical power of the statistical test of a treatment effect (Cohen, 1988). This indicates that for a treatment effect of a specific size, the study will have at least an 80% chance of detecting that effect. Statistical power is calculated specifically for a theorized effect of a certain size that is derived from prior and current empirical work. That is, researchers need to identify the minimum value of the treatment effect that is substantively meaningful within a particular context of interest. Overall, as

---

S. Konstantopoulos (✉)
Michigan State University, East Lansing, MI, USA
e-mail: spyros@msu.edu

W. Li
University of Florida, Gainesville, FL, USA

B. Zhang
The Graduate Center, City University of New York, New York, NY, USA
e-mail: bzhang2@gc.cuny.edu

M. Stemmler et al. (eds.), *Dependent Data in Social Sciences Research*,
https://doi.org/10.1007/978-3-031-56318-8_16

the size of the treatment effect changes (e.g., increases), power changes as well (e.g., increases), other things being equal.

Power analysis is valuable during the planning phase of an experiment, because it helps researchers calculate the sample size required to achieve power greater than .8. This type of power analysis is frequently called prospective or a priori power analysis and should always be integrated in the designing stage of an experiment. Prior to carrying out an experiment that requires a considerable investment in money, resources, time, effort, and coordination, it is judicious to conduct power analysis to ensure the sample size of the study is large enough to detect a presumed non-zero treatment effect of a certain magnitude. Arguably, researchers would be unwilling to carry out an experiment had they known they would have a low chance of detecting the treatment effect of interest. Statistical power analysis provides guidance to researchers to determine the adequate sample size for their study (i.e., the number of units required to detect an anticipated treatment effect with probability greater than .80). For example, an education researcher might ask how many students should I include in my sample to be able to detect an intervention effect on an important outcome such as student achievement? Power analysis addresses this question satisfactorily.

It is noteworthy that the magnitude of the treatment effect plays a critical role in attaining adequate power. In single-level designs, for example, when the magnitude of the treatment effect (e.g., a standardized mean difference) is large (e.g., .50 standard deviations), a two-tailed t-test that examines the difference between two means has adequate statistical power of .80 with a small sample size of 65 observations (see Cohen, 1988). However, when the magnitude of the treatment effect is reduced considerably (e.g., .20 standard deviations), the power of a two-tailed t-test that examines the difference between two means reaches adequate statistical power of .80 when the sample size is more than six times as large (i.e., 400 observations) (Cohen, 1988). When the magnitude of a standardized treatment effect is further decreased (e.g., .10 standard deviations), the statistical power of the statistical test will probably be low, and therefore statistical significance may not be attained, unless the sample size is fairly large (e.g., $\geq$1000 observations) (Cohen, 1988). When statistical significance is not reached, it is challenging for researchers to interpret properly the non-significant results because failure to reach statistical significance would simply be attributed to a low power probability due to a small effect and a small sample size. The results consequently are inconclusive and non-informative. However, when a study has prospectively identified the necessary sample size to detect a hypothesized treatment effect of a certain magnitude, the results of the study should reach statistical significance.

Much of the literature in the early stages of statistical power analysis involved power computations of studies that used a single-level simple random sampling of units (e.g., students) and did not take into account potential nesting or clustering in the power computations (e.g., Cohen, 1988; Lipsey, 1990; Murphy & Myors, 2004). Software packages for computing statistical power in single-level designs are broadly available (e.g., Borenstein et al., 2001; Mayr et al., 2007). In single-level designs, the statistical power of the statistical test of a treatment effect is a function of the Type I error probability (i.e., the significance level), the magnitude of a standardized measure of the treatment effect or effect size (e.g., a standardized mean difference), and the sample size (i.e., the number of units in the sample). Keeping the significance level constant (e.g., .05 level) and for a presumed size of the treatment effect, as the study sample size increases, statistical power also increases.

Nonetheless, in the last two decades, many randomized experiments in education, social, and health sciences have been large-scale studies that often times follow a multilevel structure that involves nesting of lower-level units (e.g., students) within higher-level units (e.g., schools) (Donner & Klar, 2000; Hedges & Rhoads, 2010; Konstantopoulos, 2008a; Murray, 1998; Murray et al., 2004; Raudenbush, 1997). Statistical power in multilevel experiments entails challenges not encountered in single-level designs. First, there are sample sizes in more than one level that might affect power differently. Second, there are variances in more than one level and also might affect power differently. The effect size, the sample sizes, and the variances at higher levels of the hierarchy (i.e., the clustering) are the design parameters that impact the statistical power of the statistical test of the treatment effect in multilevel designs.

The purpose of this chapter is to discuss a priori statistical power analysis for large-scale multilevel experiments in education. The designs are cross-sectional, involve a dichotomous treatment variable (i.e., one treatment and one control condition), and are balanced (i.e., the same number of units is included in the treatment and the control conditions). However, we also briefly discuss unbalanced designs where clusters have the same sample size and the proportion of units assigned randomly to a treatment or a control condition is the same for all clusters. We discuss power computations for two- and three-level experimental designs. Specifically, both cluster randomized designs, where clusters are randomly assigned to a treatment and a control condition, and block randomized designs, also known as multi-site trials, where units within clusters are randomly assigned to a treatment and a control group, are described. For simplicity, the treatment or intervention variable is assumed to have two arms, an experimental and a control condition.

This chapter is structured as follows. First, multilevel randomized designs are described. Second, statistical power in two- and three-level cluster randomized designs is discussed. Third, statistical power in two- and three-level block randomized designs is delineated. Fourth, specific cases of unbalanced designs are outlined. Fifth, the optimal sampling of units of both types of designs is outlined. Sixth, the impact of the sample sizes in different levels of the hierarchy is discussed. Lastly, a discussion with concluding remarks is offered.

16.1 Multilevel Randomized Designs

To illustrate nesting or clustering, we draw examples from the field of education. In particular, education students at the first level are nested within classrooms at the second level which are in turn nested within schools at the third level and so forth. Classrooms and schools in this instance are called clusters, which are broadly defined as higher-level, larger units that comprise of multiple lower-level, smaller units. Lower-level units within a higher-level unit have outcome values that are more similar compared to outcome values of lower-level units in different higher-level units. That is, the units within a specific cluster are more alike than the units in different clusters. This grouping of lower-level units within higher-level units results in data dependencies that need to be taken into account both in power analysis and statistical analysis. For example, the outcome values of students grouped in a specific classroom may be correlated, and this data dependency is eventually translated into correlated residuals (i.e., residuals are not independently distributed).

When nesting or clustering is incorporated in experimental design, power computations will be impacted by variance components introduced at different levels of the hierarchy. Specifically, the clusters (e.g., classrooms or schools) will be modeled as random effects which are normally distributed with zero means and constant variances. When an effect is random across classrooms, its variance captures the between-classroom variability or differences between classrooms in the outcome. Similarly, when an effect is random across schools, its variance captures the between-school variability or differences between schools in the outcome. The effects of these variances capture nesting or clustering that can be denoted statistically by a standardized index called the intraclass correlation coefficient (ICC). ICC is a variance ratio, ranging from zero to one, that shows the degree to which clusters are similar or dissimilar to each other with respect to the outcome variable (see detailed, formal definition of the ICC in the following section). Intuitively, when clusters are similar, the ICC is close to zero, and when clusters are dissimilar, the ICC deviates from zero. Other things being equal, as ICC values increase, statistical power decreases.

The multilevel structure of the design also introduces sample sizes at different levels of the hierarchy that may affect power differently. Assuming a three-level structure in education, at the first level, the sample size often comprises of the number of students in a classroom. At the second level, the sample size is often the number of classrooms in a school. At the third level, the sample size is typically the number of schools. Thus, in multilevel designs education researchers ought to estimate the number of students, classes, and schools necessary to detect the anticipated treatment effect with a high probability (>.80). Also, power may be influenced differently by sample sizes at different levels of the hierarchy.

The Sampling Design

Sampling design refers to the procedure employed to select units in a sample from a population of interest. Typically, simple random sampling is supposed where a subset of units in the population are selected randomly (have the same chance of being selected randomly) into a sample. In multilevel designs however, the sampling is multi-layered and involves sampling of units of different sizes at different stages or levels. Large-scale survey research provides a useful framework that illustrates the complexity of such sampling designs. Suppose the objective of a study is to select randomly kindergartners in a particular US state. One effective way to accomplish this task is to use random sampling in multiple stages. A two-stage sampling scheme, for example, would involve simple random sampling of schools within each geographic region in the particular state at the first stage and then simple random sampling of kindergarten students within each selected school at the second stage. A three-stage sampling scheme would involve simple random sampling of schools within each geographic region in the state at the first stage, simple random sampling of kindergarten classrooms within each selected school at the second stage, and simple random sampling of students within each selected classroom in the third stage. In a large-scale survey research, these complicated designs are known as multi-stage sampling designs. In the example above, the geographic region is used for stratification purposes, which suggests that sampling systems that follow multi-stage sampling may include stratification to ensure representation of units across a state, a country, etc.

Cluster sampling is typically involved in multi-stage sampling and sometimes in more than one stage. In a two-stage sampling design, cluster sampling suggests that clusters such as schools are randomly selected from a population of clusters at the first stage. In three- or four-stage sampling schemes, random sampling of clusters of different sizes may take place at various stages (e.g., sampling of classrooms, schools, school districts). For example, in a four-stage sampling scheme, cluster sampling may be involved in the first three stages, namely, school districts are sampled at the first stage, schools are sampled within each selected school district at the second stage, and classrooms are selected within each selected school at the third stage.

Multi-stage cluster sampling of units introduces variability across units at different levels. Sampling schools introduces between-school variability for instance. In turn, sampling classrooms within these schools introduces between-classroom within-school variability. The corresponding linear statistical model would treat schools and classrooms as random effects, and the variances of these random effects represent the differences between schools and classrooms, respectively, for an outcome of interest. These variance components at different levels capture what we call clustering. In the last 20 years, the development of multilevel modeling that account for clustering has allowed researchers to appropriately analyze data with nested structures (e.g., Goldstein, 2011; Raudenbush & Bryk, 2002). Methods for computing power in two-level and three-level experimental designs have also been documented the last two decades (e.g., Donner & Klar, 2000; Hedges & Rhoads, 2010; Konstantopoulos, 2008a, b; Murray, 1998; Murray et al., 2004; Raudenbush, 1997; Raudenbush & Liu, 2000), and statistical software packages about a priori power analysis have been widely available (e.g., Borenstein et al., 2001; Dong & Maynard, 2013; Spybrook et al., 2011).

Clustering

Clustering is frequently quantified by the ICC. Consider first a two-level population structure where students at the first level are nested within schools (the clusters) at the second level. Suppose the population level 2 variance is σ_2^2, the population level 1 variance is σ_1^2, and the subscripts 1 and 2 indicate first and second levels, respectively. The total variance is the sum of the two variance components, namely, $\sigma_T^2 = \sigma_1^2 + \sigma_2^2$, and the ICC is defined as

$$\rho = \frac{\sigma_2^2}{\sigma_T^2}. \tag{16.1}$$

ICC values close to zero suggest similar clusters and much larger variability within clusters than between clusters, while ICC values farther from zero indicate non-trivial between-cluster variability (i.e., dissimilar clusters) and thus larger clustering.

Consider second a three-level population where students at the first level are nested within classrooms (sub-clusters) at the second level and classrooms are in turn nested within schools (clusters) at the third level. Suppose the level 1 population variance is σ_1^2, the level 2 population variance is σ_2^2, the level 3 population variance is σ_3^2, and subscripts 2 and 3 indicate second and third levels, respectively. The total variance is the sum of all three variance components, namely, $\sigma_T^2 = \sigma_1^2 + \sigma_2^2 + \sigma_3^2$, and two ICCs can be defined. At the third level, the ICC is defined as

$$\rho_3 = \frac{\sigma_3^2}{\sigma_T^2} \tag{16.2}$$

and at the second level the ICC is defined as

$$\rho_2 = \frac{\sigma_2^2}{\sigma_T^2}, \tag{16.3}$$

where the subscripts 2 and 3 indicate the corresponding level of the ICC. The ICCs defined in Eqs. (16.1) to (16.3) range between 0 and 1 with values close to zero indicating low levels of clustering and values close to one high levels of clustering.

Types of Experimental Designs

A completely randomized design is a single-level design where units (e.g., individuals) are assigned to a treatment or a control group via some random allocation mechanism (e.g., table of random numbers). Clustering is not taken into account in this design, and any potential between-cluster variability is assumed to be zero (i.e., simple random sampling of units is assumed at a single-level only).

This chapter focuses on multilevel randomized designs that incorporate cluster sampling. The first multilevel design we discuss is known as the cluster, group, or hierarchical design (Donner & Klar, 2000; Hedges & Rhoads, 2010; Kirk, 2012; Konstantopoulos, 2008a; Murray, 1998; Raudenbush, 1997). In this design, clusters at the higher level of the hierarchy are randomly assigned to a treatment or a control condition. For example, some schools are randomly assigned to a group that receives a novel whole school assessment program (the intervention condition), and other schools are randomly assigned to a group that does not receive the specific intervention (the control condition). Two types of cluster randomized designs are covered. First, a two-level cluster randomized design (e.g., students nested within schools) is described where cluster sampling takes place at the second level and second-level units are assigned to a treatment or a control arm. Second, a three-level cluster randomized design (e.g., students nested within classrooms and classrooms nested within schools) is presented where cluster sampling takes place at the second and the third levels and the treatment is assigned to units at the third level. Regardless of whether cluster sampling takes place at one or two levels, random assignment always takes place at the top level in cluster randomized designs.

The second multilevel design we describe is known as the block randomized or multisite design where the top-level unit is the block or site (see Hedges & Rhoads, 2010; Kirk, 2012; Konstantopoulos, 2008b; Raudenbush & Liu, 2000). In this design cluster sampling is assumed at the top level (e.g., sampling of schools), that is, the blocks at the top level are random effects. Within each block (e.g., the school), lower-level units (e.g., students) are randomly assigned to a treatment or a control condition. Naturally, in this design, the intervention is randomly assigned to units a level below the top level. For example, students within a school are randomly assigned to smaller (treatment condition) or regular size (control condition) classrooms. Three types of block randomized designs are discussed. First, a two-level block randomized design (e.g., students nested within schools) is outlined where cluster sampling takes place at the second level (e.g., school level) and level 1 units within blocks are assigned randomly to treatment or control conditions. Second, a three-level block randomized design (e.g., students nested within classrooms and classrooms nested within schools) is defined where cluster sampling takes place at the top and middle levels (e.g., schools and classrooms) and second level sub-clusters such as classrooms within schools (the blocks) are randomly assigned to treatment or control conditions. Third, a slightly different three-level block randomized design (e.g., students nested within classrooms and classrooms nested within schools) is described where cluster sampling takes place at the middle and top levels (e.g., classrooms and schools) and first-level units such as students within classrooms and schools are randomly assigned to treatment or control conditions. In this design both second- and third-level units (e.g., classrooms and schools) are blocks. Regardless of whether cluster sampling takes place at one or two levels, random assignment takes place always below the top level in block randomized designs. For example, schools in this design are blocks or sites within which the experiment takes place. Thus, conceptually this design is a series of experiments that take place within sites, and the average treatment effect can be viewed as a combined average of site-specific treatment effects (Konstantopoulos, 2011a).

16.2 Cluster Randomized Designs

Two Levels

In this design level 2 units are randomly assigned to a treatment or a control group (i.e., the treatment variable is dichotomous). Consider a two-level design where m level 2 units (e.g., schools) are nested within the two treatment conditions and n level 1 units (e.g., students) are nested within the jth level 2 unit and one of the two treatment conditions (Kirk, 2012). The total sample size at the second level is $2m$, and the total sample size across both levels is $2mn$. The level 2 units (the clusters) are random effects (i.e., cluster sampling is assumed for level 2 units). The linear model for outcome y_{jl} in ANCOVA notation is

$$y_{jl} = \mu + \alpha_A + \boldsymbol{X}_{jl}^T \boldsymbol{B}_1 + \boldsymbol{Z}_j^T \boldsymbol{B}_2 + \upsilon_{Aj} + \varepsilon_{Ajl}, \tag{16.4}$$

where l represents level 1 units ($l = 1, \ldots, n$), j represents level 2 units ($j = 1, \ldots, m$), μ is the grand mean, α_A is the covariate adjusted effect, $\boldsymbol{B}_1$ is a column vector of r level 1 covariate parameters, $\boldsymbol{B}_2$ is a column vector of w level 2 covariate parameters, $\boldsymbol{X}_{jl}^T$ is a row vector of r level 1 covariates (e.g., student characteristics), $\boldsymbol{Z}_j^T$ is a row vector of w level 2 covariates (e.g., school characteristics), υ_{Aj} is a level 2 random effect with a mean 0 and a variance σ_{r2}^2, and ε_{Ajl} is the level 1 residual term with a mean 0 and variance σ_{r1}^2. The superscript T indicates the transpose of a vector (or matrix), the subscript A indicates adjustment because of the covariates, and the subscript r indicates residual variances.[1] The superscript T and the subscripts A and r are used in the same way in the four subsequent ANCOVA models in cluster and block randomized designs. Level 1 covariates are assumed to be centered around their corresponding cluster means, and thus, they explain variance only at the first level. Level 2 covariates explain variance only at the second level.

When the alternative hypothesis is true, the test statistic that examines the significance of a treatment effect follows a non-central t-distribution with $2(m - 1) - w$ degrees of freedom (*df*) and a non-centrality parameter (*NCP*) λ_{1A} defined as

$$\lambda_{1A} = \sqrt{\frac{mn}{2}} \sqrt{\frac{1}{[1 + (n-1)\rho] - [\pi_1 + (n\pi_2 - \pi_1)\rho]}}\delta, \tag{16.5}$$

[1] The subscript A is used in all ANCOVA models in all cluster and block randomized designs presented in this chapter to indicate adjustment of coefficients, error terms, and *NCP*s due to covariates. In ANOVA models that do not include covariates, the subscript A is omitted.

where $\delta = (\alpha_1 - \alpha_2)/\sigma_T$ is a standardized mean difference or effect size; π_1 and π_2 are proportions of variance explained by covariates at the first and second levels, respectively; and all other terms have been defined previously (see Hedges & Rhoads, 2010; Konstantopoulos, 2008a; Murray, 1998). The component $[\pi_1 + (n\pi_2 - \pi_1)\rho]$ represents the impact of covariates in reducing variances at the first and second levels. The ICC, ρ, is defined in Eq. (16.1). To illustrate the notion of the proportions, when the level 1 covariates explain 50% of the variance in the first level the proportion $\pi_1 = .50$, and when the level 2 covariates explain 60% of the variance in the second level, the proportion $\pi_2 = .6$. When the model does not include covariates $\pi_1 = \pi_2 = 0$ and the *NCP* is simplified as

$$\lambda_1 = \sqrt{\frac{mn}{2}}\sqrt{\frac{1}{[1 + (n-1)\rho]}}\delta. \tag{16.6}$$

The statistical power of a two-tailed *t*-test at significance level α is

$$\begin{aligned} p_1 = 1 - \mathrm{H}\ &[c(\alpha/2, 2(m-1) - w), (2(m-1) - w), \lambda_{1A}] \\ + \mathrm{H}\ &[-c(\alpha/2, 2(m-1) - w), (2(m-1) - w), \lambda_{1A}], \end{aligned} \tag{16.7}$$

and the statistical power of a one-tailed *t*-test is

$$p_1 = 1 - \mathrm{H}\ [c(\alpha, 2(m-1) - w), (2(m-1) - w), \lambda_{1A}], \tag{16.8}$$

where $c(\alpha, \nu)$ is the one-tailed (upper tail) critical value of the *t*-distribution with ν *df* at significance level α [e.g., $c(.05, 60) = 1.67$] and $\mathrm{H}(x, \nu, \lambda)$ is the cumulative distribution function of the non-central *t*-distribution with ν *df* and *NCP* λ_{IA}. When the model does not include covariates, $w = 0$ and λ_I is used to compute statistical power.

Illustrative Example

Suppose the outcome is reading achievement, the magnitude of a meaningful intervention effect size (a standardized mean difference) is one-quarter of a standard deviation ($\delta = .25$), there are $m = 25$ clusters (e.g., schools) in each intervention arm (i.e., 50 schools altogether), $n = 30$ students per school, the ICC is $\rho = .15$, and no covariates are included in the model. The *NCP* is calculated as

$$\lambda_1 = \sqrt{\frac{25 * 30}{2}} * \sqrt{\frac{1}{[1 + (30-1) * .15]}} * .25 = 2.09,$$

where the symbol * indicates multiplication. The statistical power of the one-tailed *t*-test is

$$p_1 = 1 - \text{H}\ [\mathit{1.68, 48, 2.09}] = 1 - .34 = .66,$$

and the statistical power of the two-tailed t-test is

$$p_1 = 1 - \text{H}\,[\mathit{2.01, 48, 2.09}] + \text{H}\,[\mathit{-2.01, 48, 2.09}] = 1 - .46 + 0 = .54.$$

The power of the one-tailed test is greater than that of the two-tailed test because the critical value of the one-tailed test is smaller than that of the two-tailed test, and, as a result, the value of the cumulative distribution function of the non-central *t*-distribution is smaller in the one-tailed test than in the two-tailed test. In our example, the value of the cumulative distribution function of the non-central *t*-distribution in the one-tailed test is .34, while in the two-tailed test is .46. Nonetheless, both power estimates are smaller than the threshold level of power, namely, .8.

To improve the power probability, a researcher would naturally include level 1 and level 2 covariates in the model and assume that they explain a considerable proportion of the variance at each level. Suppose, for example, that one level 1 covariate (e.g., pretest scores) explains 50% of the variance at the first level and one level 2 covariate ($w = 1$) such as school SES explains 50% of the variance at the second level, that is, $\pi_1 = \pi_2 = .5$. The *NCP* is calculated as

$$\lambda_{1A} = \sqrt{\frac{25 * 30}{2}} * \sqrt{\frac{1}{\left(\begin{array}{l}[1 + (30 - 1) * .15] \\ \quad - [.50 + (30 * .50 - .50) * .15]\end{array}\right)}} * .25 = 2.96,$$

where the symbol * indicates multiplication. The statistical power of the one-tailed *t*-test is

$$p_1 = 1 - \text{H}\ [\mathit{1.68, 47, 2.96}] = 1 - .10 = .90,$$

and the statistical power of the two-tailed t-test is

$$p_1 = 1 - \text{H}\,[\mathit{2.01, 47, 2.96}] + \text{H}\,[\mathit{-2.01, 47, 2.96}] = 1 - .17 + 0 = .83.$$

The proportions of variances explained by the covariates at both levels resulted in an increase of the value of the *NCP* and eventually of statistical power. That is, when variances are reduced, the *NCP* increases, and as a result, power increases other things being constant. Now, both power values are above the lower bound of power which indicates that the researcher has a considerable chance of detecting a treatment effect of one-quarter of a standard deviation given the values of all other design parameters. Notice also that the $df = 47$ are slightly reduced compared to the model that did not include covariates ($df = 48$), because the *df* are a function of the number of the top (second) level covariates included in the model and in this case $w = 1$. However, reducing the *df* slightly by including one level 2 covariate in the model does not have much impact on power. In our example, even when 10 level 2 covariates are included at the second level (i.e., $w = 10$ and $df = 38$) and the proportion of variance explained at the second level is still .50, namely $\pi_2 = .50$, the power is essentially the same (.82 for the two-tailed test). This result indicates that covariates that explain a considerable amount of variance at the first and second levels increase power considerably other things being equal, while the corresponding reduction of df due to including top-level covariates in the model has a relatively small impact on power.

However, level 2 covariates have a stronger impact on power than level 1 covariates other things being constant (see also Bulus & Sahin, 2019). To demonstrate, suppose the values of the design parameters in the example above remain the same (i.e., $\delta = .25$, $m = 25$, $n = 30$, and $\rho = .15$). First, suppose that only one level 1 covariate is included at the first level and explains 50% of the first-level variance and there are no covariates at the second level (i.e., $\pi_1 = .50$, $\pi_2 = 0$). In this case, the power of a two-tailed test is .57, and the power of a one-tailed test is .69. Second, suppose that only one level 2 covariate is included at the second level and explains 50% of the second-level variance and there are no covariates at the first level (i.e., $\pi_1 = 0$, $\pi_2 = .50$). In this instance, the power of a two-tailed test is .77, and the power of a one-tailed test is .86, which suggests an increase in power by more than 20%.

Three Levels

In this design level 3 units are randomly assigned to a treatment or a control group (i.e., the treatment variable is dichotomous). Consider a three-level design where m level 3 units (e.g., schools) are nested within the two treatment conditions of the dichotomous treatment variable, p level 2 units (e.g., classrooms) are nested within the jth level 3 unit, and n level 1 units (e.g., students) are nested within the kth level 2 unit, the jth level 3 unit, and the two treatment conditions (Kirk, 2012). The total sample size at the top level is $2m$, and the total sample size across all levels is $2mpn$. Cluster sampling of units at the second and third levels is assumed, which indicates the level 2 and level 3 units (e.g., classrooms, schools) are random effects. The linear model for outcome y_{jkl} in ANCOVA notation is

$$y_{jkl} = \mu + \alpha_A + \boldsymbol{X}_{jkl}^T \boldsymbol{B}_1 + \boldsymbol{W}_{jk}^T \boldsymbol{B}_2 + \boldsymbol{Z}_j^T \boldsymbol{B}_3 + \upsilon_{Aj} + \xi_{Ajk} + \varepsilon_{Ajkl}, \qquad (16.9)$$

where l represents level 1 units within a level 2 unit ($l = 1, \ldots, n$), k represents level 2 units within a level 3 unit ($k = 1, \ldots, p$), j represents level 3 units ($j = 1, \ldots, m$) within a treatment arm, μ is the grand mean, α_A is the covariate adjusted effect, $\boldsymbol{B}_1$ is a column vector of r level 1 covariate parameters, $\boldsymbol{B}_2$ is a column vector of q level 2 covariate parameters, $\boldsymbol{B}_3$ is a column vector of w level 3 covariate parameters, $\boldsymbol{X}_{jkl}^T$ is a row vector of r level 1 covariates (e.g., student characteristics), $\boldsymbol{W}_{jk}^T$ is a row vector of q level 2 covariates (e.g., classroom characteristics), $\mathbf{Z}_j^T$ is a row vector of w level 3 covariates (e.g., school characteristics), υ_{Aj} is a level 3 random effect, ξ_{Ajk} is a level 2 random effect, and ε_{Ajkl} is the level 1 residual term. The corresponding variances of these random effects are σ_{r3}^2, σ_{r2}^2, and σ_{r1}^2, respectively. Subscript r indicates residual variance because of covariates. Level 1 covariates are assumed to be centered around their corresponding level 2 cluster means, and level 2 covariates are assumed to be centered around their corresponding level 3 cluster means, and thus, level 1 and level 2 covariates explain variance only at the respective corresponding levels (i.e., level 1 covariates explain variance only at the first level, and level 2 covariates explain variance only at the second level).

When the alternative hypothesis is true, the test statistic that examines the significance of the treatment effect follows a non-central t-distribution with $2(m-1)-w$ df and a NCP λ_{2A} defined as

$$\lambda_{2A} = \sqrt{\frac{mpn}{2}}\sqrt{\frac{1}{\begin{pmatrix}[1+(n-1)\,\rho_2+(pn-1)\,\rho_3] \\ -[\pi_1+(n\pi_2-\pi_1)\,\rho_2+(pn\pi_3-\pi_1)\,\rho_3]\end{pmatrix}}}\delta, \tag{16.10}$$

where $\delta = (\alpha_1 - \alpha_2)/\sigma_T$, π_1, π_2, and π_3 are proportions of variance explained by covariates at the first, second, and third level, respectively; ICCs, ρ_2, ρ_3, are defined in Eqs. (16.2) and (16.3); and all other terms have been defined previously (see also Konstantopoulos, 2008a). The component $\pi_1 + (n\pi_2 - \pi_1)\rho_2 + (pn\pi_3 - \pi_1)\rho_3$ represents the impact of covariates in reducing variance at the first, second, and third levels. To exemplify, when the level 1 covariates explain 40% of the variance in the first level, the proportion $\pi_1 = .40$; when the level 2 covariates explain 70% of the variance in the second level, $\pi_2 = .70$; and when the level 3 covariates explain 60% of the variance in the third level, $\pi_3 = .60$. When the model does not include covariates $\pi_1 = \pi_2 = \pi_3 = 0$ and the NCP becomes

$$\lambda_2 = \sqrt{\frac{mpn}{2}}\sqrt{\frac{1}{[1+(n-1)\,\rho_2+(pn-1)\,\rho_3]}}\delta. \tag{16.11}$$

To compute statistical power of a two-tailed t-test at significance level α, Eq. (16.7) can be used. To compute statistical power of a one-tailed t-test at significance level α, Eq. (16.8) can be used. When the model does not include covariates, $w = 0$ and λ_2 is used to compute power.

Illustrative Example

Once more, suppose the outcome is reading achievement, the magnitude of a meaningful intervention effect (a standardized mean difference) is one-quarter of a standard deviation ($\delta = .25$), there are $m = 25$ level 3 clusters (e.g., schools) in each intervention arm (i.e., 50 schools altogether), $p = 3$ level 2 clusters (e.g., classrooms) within each level 3 cluster, $n = 10$ level 1 units (e.g., students) within each level 2 and level 3 units, the ICC's are $\rho_2 = .10$ and $\rho_3 = .15$, and no covariates are included. The *NCP* is calculated as

$$\lambda_1 = \sqrt{\frac{25 * 3 * 10}{2}} * \sqrt{\frac{1}{[1 + (10 - 1) * .10 + (3 * 10 - 1) * .15]}} * .25 = 1.94,$$

where the symbol * indicates multiplication. The statistical power of the one-tailed *t*-test is

$$p_1 = 1 - \mathrm{H}\ [1.68, 48, 1.94] = 1 - .40 = .60,$$

and the statistical power of the two-tailed t-test is

$$p_1 = 1 - \mathrm{H}[2.01, 48, 1.94] + \mathrm{H}[-2.01, 48, 1.94] = 1 - .52 + 0 = .48.$$

The power values in the three-level model are smaller than those in the two-level model. The total number of students is the same in both cases (1500), the magnitude of the effect size is the same, and the ICC at the top level is the same. The ICC in the middle level is assumed to be two-thirds of the ICC in the third level. This computational example suggests that clustering in two levels of the hierarchy instead of clustering in one level only will result in lower power estimates other things being equal.

Both power estimates are smaller than .80, and thus to increase the power probability, researchers would naturally include level 1, level 2, and level 3 covariates in the model and assume they explain a considerable proportion of the variance at each level. Suppose one covariate is used at each level and each covariate explains 50% of the variance at the corresponding level, that is, $\pi_1 = \pi_2 = \pi_3 = .5$. The *df* of the test are slightly modified ($df = 47$) because $w = 1$. The *NCP* is calculated as

$$\lambda_{2A} = \sqrt{\frac{750}{2}} * \sqrt{\frac{1}{6.25 - 3.13}} * .25 = 2.74,$$

where the symbol * indicates multiplication. The statistical power of the one-tailed *t*-test is

$$p_1 = 1 - \mathrm{H}\ [1.68, 47, 2.74] = 1 - .15 = .85,$$

and the statistical power of the two-tailed t-test is

$$p_1 = 1 - \mathrm{H}[2.01, 47, 2.74] + \mathrm{H}[-2.01, 47, 2.74] = 1 - .23 + 0 = .77.$$

These power estimates are closer to the lower bound of power of .8. When level-specific covariates explain 50% of the variance at each level, the values of the *NCP* and power increase. The *df* are slightly reduced compared to the model that does not include covariates (a reduction from 48 to 47 *df*), but power is hardly impacted. In this example, even when 10 level 3 covariates are included at the third level (i.e., $w = 10$ and $df = 38$) and the proportion of variance explained at the second level was still .50, namely, $\pi_3 = .50$, the power is essentially the same (.76. for the two-tailed test). As in the two-level case, this result indicates that covariates that explain a considerable amount of variance at the first, second, and third levels increase power substantially other things being equal and that the corresponding reduction of *df* has a relatively small impact on power.

However, level 3 covariates have a stronger impact on power than level 1 or level 2 covariates other things being fixed (see also Bulus & Sahin, 2019). To demonstrate, suppose the values of the design parameters in the example above remain the same (i.e., $\delta = .25$, $m = 25$, $p = 3$, $n = 10$, $\rho_2 = .10$, and $\rho_3 = .15$). First, suppose that only one level 1 covariate is included at the first level and explains 50% of the first-level variance and there are no covariates at the second and third levels (i.e., $\pi_1 = .50$, $\pi_2 = \pi_3 = 0$). In this case, the power of a two-tailed test is .50 and the power of a one-tailed test is .63. Second, suppose that only one level 2 covariate is included at the second level and explains 50% of the second-level variance and there are no covariates at the first and third levels (i.e., $\pi_1 = \pi_3 = 0$, $\pi_2 = .50$). In this instance, the power of a two-tailed test is .51 and the power of a one-tailed test is .64. Third, suppose that only one level 3 covariate is included at the third level and explains 50% of the third-level variance and there are no covariates at the first and second levels (i.e., $\pi_1 = \pi_3 = 0$, $\pi_2 = .50$). Now, the power of a two-tailed test is .66 and the power of a one-tailed test is .77. This result holds for various values of ICCs at the second and third levels.

16.3 Block Randomized Designs

Two Levels

In this design level 1 units are nested within two treatment conditions which in turn are nested within level 2 units. Cluster sampling is assumed at the second level. For example, students are nested within a treatment or a control condition and schools, and the schools are random effects. Random assignment takes place at the first level within each level 2 block (e.g., school). The clusters in this design are called blocks or sites (Hedges & Rhoads, 2010; Kirk, 2012; Konstantopoulos, 2008b; Raudenbush & Liu, 2000). The linear model for outcome y_{il} in ANCOVA notation is

$$y_{il} = \mu + \alpha_A + \boldsymbol{X}_{il}^T \boldsymbol{\Delta}_1 + \boldsymbol{Z}_i^T \boldsymbol{\Delta}_2 + \alpha_A \boldsymbol{Z}_i^T \boldsymbol{\Delta}_3 + \upsilon_{Ai} + \alpha_A \upsilon_{Ai} + \varepsilon_{Ail}, \quad (16.12)$$

where l represents level 1 units within one of the two treatment conditions ($l = 1, \ldots, n$), i represents level 2 units ($i = 1, \ldots, m$), μ is the grand mean, α_A is the covariate adjusted effect, $\boldsymbol{X}_{il}^T$ is a row vector of r level 1 covariates (e.g., pretest scores, socioeconomic status) centered around their respective level 2 cluster means, $\boldsymbol{Z}_i^T$ is a row vector of w level 2 covariates (e.g., school composition), $\alpha_A \boldsymbol{Z}_i^T$ indicates fixed effects interaction terms between the treatment and level 2 covariates, $\boldsymbol{\Delta}_1$ is a column vector of r level 1 covariate parameters, $\boldsymbol{\Delta}_2$ is a column vector of w level 2 covariate parameters, $\boldsymbol{\Delta}_3$ includes interaction parameters between level 2 covariates and the treatment variable, and the last three terms, υ_{Ai}, $\alpha_A \upsilon_{Ai}$, ε_{Ail}, are level 2, treatment by level 2 unit interaction, and level 1 random effects, respectively. The treatment by level 2 unit interaction random effect is a cross-level interaction random effect that is intrinsic in the two-level block randomized design. The cross-level interaction random effect $\alpha_A \upsilon_{Ai}$ is normally distributed with a mean of zero and a variance σ_{ra}^2 which captures the variance of the treatment effect across level 2 units. Notice that variance σ_{ra}^2 is adjusted because of the covariates in $\boldsymbol{Z}_i^T$, and therefore it is a residual variance. Similarly, the variances of υ_{Ai} and ε_{Ail}, namely, σ_{r2}^2 and σ_{r1}^2, are also residual variances because of the covariates included at the corresponding levels.

Suppose there are m clusters at the second level and n level 1 units in each of the two treatment arms in each level 2 cluster, which means the total number of level 1 units in each cluster is $2n$ and the total number of level 1 units across m clusters is $2mn$. When the alternative hypothesis is true, the test statistic that examines the significance of the treatment effect follows a non-central t-distribution with $m - 1 - w$ *df* and a *NCP* λ_{3A} defined as

$$\lambda_{3A} = \sqrt{\frac{mn}{2}} \sqrt{\frac{1}{\begin{pmatrix} [1 + (n\vartheta_1 - 1)\,\rho] \\ -\,[\pi_1 + (n\vartheta_1\pi_a - \pi_1)\,\rho] \end{pmatrix}}}\,\delta, \tag{16.13}$$

where $\delta = (\alpha_1 - \alpha_2)/\sigma_T$, $\vartheta_1 = \frac{1}{2}\psi_1$, $\psi_1 = \sigma_{ra}^2/\sigma_{r2}^2$ is the variance ratio of the treatment effect variance across level 2 clusters to the total level 2 variance, π_a is the proportion of the variance of the treatment effect across level 2 clusters explained by the level 2 covariates, and all other terms have been defined previously. The component $[\pi_1 + (n\vartheta_1\pi_a - \pi_1)\rho]$ captures the impact of covariates with respect to reducing the level 1 variance and the level 2 treatment effect variance. When the model does not include covariates $\pi_1 = \pi_a = 0$ and the *NCP* becomes

$$\lambda_3 = \sqrt{\frac{mn}{2}} \sqrt{\frac{1}{[1 + (n\vartheta_1 - 1)\,\rho]}}\,\delta. \tag{16.14}$$

The statistical power of a two-tailed t-test at significance level α is

$$p_3 = 1 - \mathrm{H}\left[c(\alpha/2, m-1-w), (m-1-w), \lambda_{3A}\right] + \mathrm{H}\left[-c(\alpha/2, m-1-w), (m-1-w), \lambda_{3A}\right]. \quad (16.15)$$

and the statistical power of a one-tailed *t*-test is

$$p_3 = 1 - \mathrm{H}\left[c(\alpha, m-1-w), (m-1-w), \lambda_{3A}\right]. \quad (16.16)$$

When the model does not include covariates, $w = 0$ and λ_3 is used to compute power.

Illustrative Example

Suppose the outcome is again reading achievement; the magnitude of a meaningful intervention effect size (the standardized mean difference) is one-quarter of a standard deviation ($\delta = .25$); there are $m = 50$ clusters (e.g., schools), 30 students per school, and $n = 15$ students per treatment arm within each school; the ICC is $\rho = .15$, $\vartheta_1 = .30$; and covariates are not included in any level. The *NCP* is calculated as

$$\lambda_{3A} = \sqrt{\frac{50 * 15}{2}} * \sqrt{\frac{1}{[1 + (15 * .30 - 1) * .15]}} * .25 = 3.92,$$

where the symbol * indicates multiplication. The statistical power of the one-tailed *t*-test is

$$p_3 = 1 - \mathrm{H}\ [1.68, 49, 3.92] = 1 - .01 = .99,$$

and the statistical power of the two-tailed t-test is

$$p_3 = 1 - \mathrm{H}[2.01, 49, 3.92] + \mathrm{H}[-2.01, 49, 3.92] = 1 - .03 + 0 = .97.$$

Both power estimates are very close to the probability of one which indicates an experiment would be greatly powered. The power in this block randomized design is higher than that in the two-level cluster randomized design. The main reason is that ϑ_1 is much smaller than the total between-cluster variance at the second level, namely, σ_2^2. In addition, the *df* in block randomized designs are always slightly larger than the *df* in cluster randomized designs when the number of clusters and the number of covariates at the top level remain constant for both designs.

Three Levels

Treatment at the Second Level

Suppose that treatment variable is dichotomous (e.g., a treatment and a control condition) and the design is balanced. Cluster sampling is assumed in the second and third levels (i.e., second- and third-level units are random effects). Level 2 sub-clusters are nested within the two treatments and level 3 clusters, and level 1 units are nested within level 2 sub-clusters (Kirk, 2012). Specifically, within each level 3 cluster, level 2 sub-clusters are randomly assigned to a treatment or a control group (e.g., classrooms receive a new math curriculum or not). The linear model for the outcome y_{ikl} in ANCOVA notation is

$$\begin{aligned} y_{ikl} &= \mu + \alpha_A + \boldsymbol{X}_{ikl}^T \boldsymbol{\Delta}_1 + \boldsymbol{W}_{ik}^T \boldsymbol{\Delta}_2 + \boldsymbol{Z}_i^T \boldsymbol{\Delta}_3 + \alpha_A \boldsymbol{Z}_i^T \boldsymbol{\Delta}_4 \\ &\quad + \upsilon_{Ai} + \alpha_A \upsilon_{Ai} + \xi_{Aik} + \varepsilon_{Aikl}, \end{aligned} \tag{16.17}$$

where l represents level 1 units within a level 2 unit ($l = 1, \ldots, n$), k represents level 2 units within one of the two treatment conditions ($k = 1, \ldots, p$),, i represents level 3 units ($i = 1, \ldots, m$), μ is the grand mean, and α_A is the covariate adjusted effect. There are r level 1 covariates represented by $\boldsymbol{X}_{ikl}^T$, q level 2 covariates represented by $\boldsymbol{W}_{ik}^T$, and w level 3 covariates are represented by $\boldsymbol{Z}_i^T$. The covariate parameters at the first, second, and third levels are represented by $\boldsymbol{\Delta}_1$, $\boldsymbol{\Delta}_2$, and $\boldsymbol{\Delta}_3$, respectively, the term $\alpha_A \boldsymbol{Z}_i^T$ indicates fixed effects interaction terms between the treatment and level 3 covariates, and $\boldsymbol{\Delta}_4$ entails the interaction parameters. The last four terms, $\upsilon_{Ai}, \alpha_A \upsilon_{Ai}, \xi_{Aik}, \varepsilon_{Aikl}$, are level 3, treatment by level 3 unit interaction, level 2, and level 1 random effects, respectively. The corresponding variances of these random effects are $\sigma_{r3}^2, \sigma_{ra3}^2, \sigma_{r2}^2$, and σ_{r1}^2, and subscript r indicates residual variances. The variance σ_{ra3}^2 in particular is the variance of the treatment effect across the level 3 clusters that is adjusted because of the level 3 covariates in $\boldsymbol{Z}_i^T$.

Suppose there are m clusters at the third level, two treatment arms at the second level, p level 2 units in each treatment arm at the second level, and n level 1 units in each level 2 unit, which means the total number of level 1 units across all clusters and sub-clusters is *2mpn*. When the alternative hypothesis is true, the test statistic that examines the significance of the treatment effect follows a non-central t-distribution with $m - 1 - w$ *df* and a *NCP* λ_{4A} defined as

$$\lambda_{4A} = \sqrt{\frac{mpn}{2}} \sqrt{\frac{1}{\left(\begin{array}{l} [1 + (n-1)\rho_2 + (pn\vartheta_3 - 1)\rho_3] \\ \quad - [\pi_1 + (n\pi_2 - \pi_1)\rho_2 + (pn\vartheta_3 \pi_{a3} - \pi_1)\rho_3] \end{array} \right)}} \delta, \tag{16.18}$$

where $\delta = (\alpha_1 - \alpha_2)/\sigma_T$, $\vartheta_3 = \frac{1}{2}\psi_3$, and $\psi_3 = \sigma_{ra3}^2/\sigma_{r3}^2$ is the variance ratio of the treatment effect variance across level 3 clusters to the total level 3 variance,

π_{a3} is the proportion of the variance of the treatment effect across level 3 clusters explained by the level 3 covariates, and all other terms have been defined previously. The component $[\pi_1 + (n\pi_2 - \pi_1)\rho_2 + (pn\vartheta_3\pi_{a3} - \pi_1)\rho_3]$ captures the effects of covariates with respect to reducing variances at the first and second levels and the treatment effect across level 3 unit variance at the third level. When the model does not include covariates $\pi_1 = \pi_2 = \pi_{a3} = 0$ and the *NCP* becomes

$$\lambda_4 = \sqrt{\frac{mpn}{2}}\sqrt{\frac{1}{[1 + (n-1)\,\rho_2 + (pn\vartheta_3 - 1)\,\rho_3]}}\delta. \tag{16.19}$$

Equations (16.15) and (16.16) can be used to compute power for one-tailed and two-tailed *t*-tests. When the model does not include covariates, $w = 0$ and λ_4 is used to compute power.

Illustrative Example

Suppose the outcome is reading achievement; the magnitude of a meaningful intervention effect size (a standardized mean difference) is one-quarter of a standard deviation ($\delta = .25$); there are $m = 50$ clusters (e.g., schools); four classrooms in each school, two classrooms per treatment arm ($p = 2$), and $n = 10$ students per classroom; the two ICCs are $\rho_2 = .10$ and $\rho_3 = .15$; $\vartheta_3 = .30$; and covariates are not included in any level. The *NCP* is calculated as

$$\lambda_4 = \sqrt{\frac{50 * 2 * 10}{2}} * \sqrt{\frac{1}{\left(\begin{array}{l}[1 + (10-1) * .10 \\ \quad + (2 * 10 * .30 - 1) * .15]\end{array}\right)}} * .25 = 3.43,$$

where the symbol * indicates multiplication. The statistical power of the one-tailed *t*-test is

$$p_4 = 1 - \mathrm{H}\ [\mathit{1.68, 49, 3.43}] = 1 - .04 = .96,$$

and the statistical power of the two-tailed t-test is

$$p_4 = 1 - \mathrm{H}[\mathit{2.01, 49, 3.43}] + \mathrm{H}[\mathit{-2.01, 49, 3.43}] = 1 - .08 + 0 = .92.$$

The power estimates are much greater than .80; however, they are slightly smaller than the power estimates in the two-level block randomized design, which suggests that adding levels and therefore variances to the hierarchy decreases power other things being equal. When covariates are included in the model, $\pi_1, \pi_2, \pi_{a3} = .50$ and $w = 1$, the power estimates for a two- and a one-tailed test are essentially equal to one.

Treatment at the First Level

Suppose again the treatment variable is dichotomous, the design is balanced, and cluster sampling is assumed at the second and third levels (i.e., second- and third-level units are random effects). Level 2 sub-clusters are nested within level 3 clusters, and level 1 units are nested within a treatment or a control group and level 2 and level 3 units (Kirk, 2012). Specifically, within each level 2 sub-cluster, level 1 units are randomly assigned to a treatment or a control group (e.g., within a classroom, students receive a reading intervention or not). The linear model for outcome y_{ijl} in ANCOVA notation is

$$\begin{aligned} y_{ijl} &= \mu + \alpha_A + \boldsymbol{X}_{ijl}^T \boldsymbol{\Delta}_1 + \boldsymbol{W}_{ij}^T \boldsymbol{\Delta}_2 + \boldsymbol{Z}_i^T \boldsymbol{\Delta}_3 + \alpha_A \boldsymbol{W}_{ij}^T \boldsymbol{\Delta}_4 + \alpha_A \boldsymbol{Z}_i^T \boldsymbol{\Delta}_5 \\ &\quad + \upsilon_{Ai} + \alpha_A \upsilon_{Ai} + \xi_{Aij} + \alpha_A \xi_{Aij} + \varepsilon_{Aijl}, \end{aligned} \tag{16.20}$$

where l represents level 1 units within one of the two treatment conditions ($l = 1, \ldots, n$), j represents the level 2 units within a level 3 unit ($j = 1, \ldots, p$), i represents level 3 units ($i = 1, \ldots, m$), μ is the grand mean, and α_A is the covariate adjusted effect. There are r level 1 covariates represented by $\boldsymbol{X}_{ijl}^T$, q level 2 covariates are represented by $\boldsymbol{W}_{ij}^T$, and w level 3 covariates are represented by $\boldsymbol{Z}_i^T$. The covariate parameters at the first, second, and third levels are represented by $\boldsymbol{\Delta}_1$, $\boldsymbol{\Delta}_2$, and $\boldsymbol{\Delta}_3$, respectively, the term $\alpha_A \boldsymbol{Z}_i^T$ indicates fixed effects interaction terms between the treatment and level 3 covariates, and $\boldsymbol{\Delta}_5$ entails the corresponding interaction parameters. The term $\alpha_A \boldsymbol{W}_{ij}^T$ indicates fixed effects interaction terms between the treatment and level 2 covariates, and $\boldsymbol{\Delta}_4$ includes the corresponding interaction parameters. The last five terms, υ_{Ai}, $\alpha_A \upsilon_{Ai}$, ξ_{Aij}, $\alpha_A \xi_{Aij}$, ε_{Aijl}, are level 3, treatment by level 3 unit interaction, level 2, treatment by level 2 unit interaction, and level 1 random effects, respectively. The corresponding variances of these random effects are σ_{r3}^2, σ_{ra3}^2, σ_{r2}^2, σ_{ra2}^2, and σ_{r1}^2, and subscript r indicates residual variances. The variance σ_{ra2}^2 in particular is the variance of the treatment effect across the level 2 clusters that is adjusted for level 2 covariates in $\boldsymbol{W}_{ij}^T$, and it is therefore a residual variance. In this model because the treatment is that first level it can vary across level 2 and level 3 units.

Suppose there are m clusters at the third level, p sub-clusters within each third level cluster at the second level, two treatment arms at the first level, and n level 1 units within each treatment arm within level 2 and level 3 units, which means the total number of level 1 units across all clusters and sub-clusters is *2mpn*. When the alternative hypothesis is true, the test statistic that examines the significance of the treatment effect follows a non-central t-distribution with $m - 1 - w$ *df* and a *NCP* λ_{5A} defined as

$$\lambda_{5A} = \sqrt{\frac{mpn}{2}} \sqrt{\frac{1}{\begin{pmatrix} [1 + (n\vartheta_2 - 1)\,\rho_2 + (pn\vartheta_3 - 1)\,\rho_3] \\ -[\pi_1 + (n\vartheta_2\pi_{a2} - \pi_1)\,\rho_2 + (pn\vartheta_3\pi_{a3} - \pi_1)\,\rho_3] \end{pmatrix}}}\delta, \tag{16.21}$$

where $\delta = (\alpha_1 - \alpha_2)/\sigma_T$, $\vartheta_2 = \frac{1}{2}\psi_2$, and $\psi_2 = \sigma_{ra2}^2/\sigma_{r2}^2$ is the variance ratio of the treatment effect variance across level 2 sub-clusters to the total level 2 variance, π_{a2} is the proportion of the variance of the treatment effect across level 2 sub-clusters explained by level 2 covariates, and all other terms have been defined previously. The component $[\pi_1 + (n\vartheta_2\pi_{a2} - \pi_1)\rho_2 + (pn\vartheta_3\pi_{a3} - \pi_1)\rho_3]$ captures the effects of covariates with respect to reducing the first-level variance and the treatment effect variances at the second and third levels. When the model does not include covariates $\pi_1 = \pi_{a2} = \pi_{a3} = 0$ and the *NCP* becomes

$$\lambda_5 = \sqrt{\frac{mpn}{2}} \sqrt{\frac{1}{[1 + (n\vartheta_2 - 1)\,\rho_2 + (pn\vartheta_3 - 1)\,\rho_3]}}\delta, \tag{16.22}$$

Equations (16.15) and (16.16) can be used to compute power for one-tailed and two-tailed t-tests. When the model does not include covariates, $w = 0$ and λ_5 is used to compute power.

Illustrative Example

Suppose the outcome is reading achievement; the magnitude of a meaningful intervention effect size (a standardized mean difference) is one-quarter of a standard deviation ($\delta = .25$); there are $m = 50$ clusters (e.g., schools), four classrooms in each school ($p = 4$), and $n = 5$ students in each treatment arm within each classroom; the ICCs are $\rho_2 = .10$ and $\rho_3 = .15$; $\vartheta_2 = \vartheta_3 = .30$, and covariates are not included in any level. The *NCP* is calculated as

$$\lambda_5 = \sqrt{\frac{50 * 4 * 5}{2}} * \sqrt{\frac{1}{\begin{pmatrix} [1 + (5 * .30 - 1) * .10 \\ + (4 * 5 * .30 - 1) * .15] \end{pmatrix}}} * .25 = 4.17,$$

where the symbol * indicates multiplication. The statistical power of the one-tailed t-test is

$$p_4 = 1 - \mathrm{H}\ [\mathit{1.68, 49, 4.17}] = 1 - .01 = .99,$$

and the statistical power of the two-tailed t-test is

$$p_4 = 1 - \mathrm{H}[\mathit{2.01, 49, 4.17}] + \mathrm{H}[\mathit{-2.01, 49, 4.17}] = 1 - .02 + 0 = .98.$$

Both power estimates are very close to one which suggests that the experiment will be greatly powered. These power estimates are slightly larger than the power estimates in the three-level block randomized design where assignment to treatment took place at the middle level and are similar to the power estimates obtained from the two-level block randomized design. This result suggests that power estimates are larger in three-level block randomized designs where treatment assignment takes place at the first level (see also Bulus & Dong, 2022). The main reason is that the treatment effect variance across level 2 units is much smaller than the total level 2 variance. In general, so long as the treatment effect variances across level 2 and level 3 units are much smaller than the total variances at the second and third levels, the three-level block randomized designs will produce larger power estimates than in cluster randomized designs. Also, the *df* in cluster randomized designs are slightly smaller than the *df* in block randomized designs when the number of clusters and covariates at the top level are fixed in both designs.

Key facets of all five designs we discussed are summarized in Tables 16.1 and 16.2. Specifically, Table 16.1 reports the designs, multilevel models, *NCP* parameters, and *df* when covariates are included in all levels. Table 16.2 summarizes the designs, multilevel models, *NCP* parameters, and *df* when covariates are not included in the models.

16.4 Unbalanced Designs

Sometimes because of practical constraints, the design might not be balanced. In cluster randomized designs, for instance, the numbers of the clusters assigned to a treatment or a control group might not be the same (e.g., 40% of schools are assigned to the treatment group, 60% to the control group), although the sample sizes of the clusters remain the same in the treatment and control conditions (e.g., each school has 30 students). In a cluster randomized design, suppose the total number of clusters in both conditions is M and the proportion of clusters assigned to the treatment group is P_C where subscript C indicates a cluster design. Then, the total number of level 1 units in the treatment group is $P_C Mn$ for two-level cluster randomized designs, where n represents the number of level 1 units within each level 2 unit. In three-level cluster randomized designs, the total number of level 1 units in the treatment group is $P_C Mpn$, where n represents the number of level 1 units within each level 2 unit and p represents the number of level 2 units within each level 3 unit. Similarly, in block randomized designs, the number of lower-level units within a block assigned to a treatment or a control condition might not be the same (e.g., 40% of students within a school are assigned to the treatment group, 60% to the control group), although the sample sizes of the blocks remain the same (e.g., each school has 30 students). In a two-level block randomized design, suppose the total number of level 1 units within each block is N, the total number of blocks (level 2 units) is m, and the proportion of level 1 units assigned to the treatment group within each block is P_{B2} (i.e., the proportion of assignment to treatment within blocks remains

Table 16.1 Summary of balanced designs, models, non-centrality parameters, and degrees of freedom: covariates in all levels

Design	Multilevel model	Non-centrality parameter (λ_A)	Degrees of freedom (df)
Two-level cluster design	$y_{jl}=\mu+\alpha_A+\boldsymbol{X}_{jl}^T\boldsymbol{B}_1+\boldsymbol{Z}_j^T\boldsymbol{B}_2+\upsilon_{Aj}+\varepsilon_{Ajl}$	$\lambda_{1A}=\sqrt{\frac{mn}{2}}\sqrt{\frac{1}{[1+(n-1)\rho]-[\pi_1+(n\pi_2-\pi_1)\rho]}}\delta$	$2(m-1)-w$
Three-level cluster design	$y_{jkl}=\mu+\alpha_A+\boldsymbol{X}_{jkl}^T\boldsymbol{B}_1+\boldsymbol{W}_{jk}^T\boldsymbol{B}_2$ $+\boldsymbol{Z}_j^T\boldsymbol{B}_3+\upsilon_{Aj}+\xi_{Ajk}+\varepsilon_{Ajkl}$	$\lambda_{2A}=\sqrt{\frac{mpn}{2}}\sqrt{\frac{1}{[1+(n-1)\rho_2+(pn-1)\rho_3]-[\pi_1+(n\pi_2-\pi_1)\rho_2+(pn\pi_3-\pi_1)\rho_3]}}\delta$	$2(m-1)-w$
Two-level block design	$y_{il}=\mu+\alpha_A+\boldsymbol{X}_{il}^T\boldsymbol{\Delta}_1+\boldsymbol{Z}_i^T\boldsymbol{\Delta}_2$ $+\alpha_A\boldsymbol{Z}_i^T\boldsymbol{\Delta}_3+\upsilon_{Ai}+\alpha_A\upsilon_{Ai}+\varepsilon_{Ail}$	$\lambda_{3A}=\sqrt{\frac{mn}{2}}\sqrt{\frac{1}{[1+(n\vartheta_1-1)\rho]-[\pi_1+(n\vartheta_1\pi_a-\pi_1)\rho]}}\delta$	$m-1-w$
Three-level block design: treatment at second level	$y_{ikl}=\mu+\alpha_A+\boldsymbol{X}_{ikl}^T\boldsymbol{\Delta}_1+\boldsymbol{W}_{ik}^T\boldsymbol{\Delta}_2$ $+\boldsymbol{Z}_i^T\boldsymbol{\Delta}_3+\alpha_A\boldsymbol{Z}_i^T\boldsymbol{\Delta}_4+\upsilon_{Ai}+\alpha_A\upsilon_{Ai}$ $+\xi_{Aik}+\varepsilon_{Aikl}$	$\lambda_{4A}=$ $\sqrt{\frac{mpn}{2}}\sqrt{\frac{1}{\left(\begin{array}{l}[1+(n-1)\rho_2+(pn\vartheta_3-1)\rho_3]\\-[\pi_1+(n\pi_2-\pi_1)\rho_2+(pn\vartheta_3\pi_{a3}-\pi_1)\rho_3]\end{array}\right)}}\delta$	$m-1-w$
Three-level block design: treatment at first level	$y_{ijl}=\mu+\alpha_A+\boldsymbol{X}_{ijl}^T\boldsymbol{\Delta}_1+\boldsymbol{W}_{ij}^T\boldsymbol{\Delta}_2$ $+\boldsymbol{Z}_i^T\boldsymbol{\Delta}_3+\alpha_A\boldsymbol{W}_{ij}^T\boldsymbol{\Delta}_4+\alpha_A\boldsymbol{Z}_i^T\boldsymbol{\Delta}_5+\upsilon_{Ai}$ $+\alpha_A\upsilon_{Ai}+\xi_{Aij}+\alpha_A\xi_{Aij}+\varepsilon_{Aijl}$	$\lambda_{5A}=$ $\sqrt{\frac{mpn}{2}}\sqrt{\frac{1}{\left(\begin{array}{l}[1+(n\vartheta_2-1)\rho_2+(pn\vartheta_3-1)\rho_3]\\-[\pi_1+(n\vartheta_2\pi_{a2}-\pi_1)\rho_2+(pn\vartheta_3\pi_{a3}-\pi_1)\rho_3]\end{array}\right)}}\delta$	$m-1-w$

Table 16.2 Summary of balanced designs, models, non-centrality parameters, and degrees of freedom: no covariates

Design	Multilevel model	Non-centrality parameter (λ)	Degrees of freedom (*df*)
Two-level cluster design	$y_{jl} = \mu + \alpha + \upsilon_j + \varepsilon_{jl}$	$\lambda_1 = \sqrt{\frac{mn}{2}}\sqrt{\frac{1}{1+(n-1)\rho}}\delta$	$2(m-1)$
Three-level cluster design	$y_{jkl} = \mu + \alpha + \upsilon_j + \xi_{jk} + \varepsilon_{jkl}$	$\lambda_2 = \sqrt{\frac{mpn}{2}}\sqrt{\frac{1}{1+(n-1)\rho_2+(pn-1)\rho_3}}\delta$	$2(m-1)$
Two-level block design	$y_{il} = \mu + \alpha + \upsilon_i + \alpha\upsilon_i + \varepsilon_{il}$	$\lambda_3 = \sqrt{\frac{mn}{2}}\sqrt{\frac{1}{1+(n\vartheta_1-1)\rho}}\delta$	$m-1$
Three-level block design: treatment at second level	$y_{ikl} = \mu + \alpha + \upsilon_i + \alpha\upsilon_i + \xi_{ik} + \varepsilon_{ikl}$	$\lambda_4 = \sqrt{\frac{mpn}{2}}\sqrt{\frac{1}{1+(n-1)\rho_2+(pn\vartheta_3-1)\rho_3}}\delta$	$m-1$
Three-level block design: treatment at first level	$y_{ijl} = \mu + \alpha + \upsilon_i + \alpha\upsilon_i + \xi_{ij} + \alpha\xi_{ij} + \varepsilon_{ijl}$	$\lambda_5 = \sqrt{\frac{mpn}{2}}\sqrt{\frac{1}{1+(n\vartheta_2-1)\rho_2+(pn\vartheta_3-1)\rho_3}}\delta$	$m-1$

the same for all blocks). Then, the total number of level 1 units in the treatment group is $P_{B2}mN$. Now, in a three-level block randomized design where the treatment is at the first level, suppose the total number of level 1 units within a level 2 unit is N, the proportion of level 1 units assigned to the treatment group within each block (level 2 unit) is P_{B31}, the total number of level 2 units within a level 3 unit is P, and there are a total of m blocks at the third level. Then, the total number of level 1 units in the treatment group is $P_{B31}mPN$. In three-level designs where the treatment is at the second level, suppose the total number of level 2 units for each level 3 unit is P, the total number of level 3 units is m, N is the total number of students within each level 2 unit, and P_{B32} is the proportion of level 2 units assigned to a treatment or a control condition within each level 3 unit. Then, the total number of level 1 units in the treatment group is $P_{B32}mPN$. Table 16.3 summarizes the *NCP*s for these cases of unbalanced designs with covariates, and Table 16.4 summarizes the *NCP*s for these cases without any covariates (see also Bloom, 2005). To compute power estimates for each of the five unbalanced designs, we use the same power formulas, namely, Eqs. (16.7), (16.8), (16.15), and (16.16).

Illustrative Example

We offer an example for a two-level cluster randomized design to demonstrate the computation of the *NCP* and power in the case of imbalance. Suppose the outcome is reading achievement; the magnitude of a meaningful intervention effect size (a standardized mean difference) is one-quarter of a standard deviation ($\delta = .25$); there are $M = 50$ clusters (e.g., schools) in total and $n = 30$ students per school; the *ICC* is $\rho = .15$; and no covariates are included in the model. Suppose that the proportion of clusters assigned to the treatment group is $P_C = .40$. Then, the total sample size in the treatment group is $P_C Mn = .40*50*30 = 600$ (symbol * indicates multiplication), which means that the total sample size in the control group is 1500–600 = 900. Based on the formula provided in Table 16.4 for two-level cluster randomized unbalanced designs, the *NCP* is calculated as

$$\lambda_{1u} = \sqrt{\frac{.40 * (1 - .40) * 50 * 30}{[1 + (30 - 1) * .15]}} * .25 = 2.05,$$

where the symbol * indicates multiplication and the subscript u indicates unbalanced design. The statistical power of the one-tailed t-test is

$$p_{1u} = 1 - \mathrm{H}\ [1.68, 48, 2.05] = 1 - .35 = .65,$$

and the statistical power of the two-tailed t-test is

$$p_{1u} = 1 - \mathrm{H}[2.01, 48, 2.05] + \mathrm{H}[-2.01, 48, 2.05] = 1 - .48 + 0 = .52.$$

Table 16.3 Summary of non-centrality parameters in certain unbalanced designs: covariates in all levels

Design	Non-centrality parameter (λ_A)	Degrees of freedom (*df*)
Two-level cluster design	$\lambda_{1uA} = \sqrt{P_C(1-P_C)Mn}\sqrt{\frac{1}{[1+(n-1)\rho]-[\pi_1+(n\pi_2-\pi_1)\rho]}}\delta$	$M-2-w$
Three-level cluster design	$\lambda_{2uA} = \sqrt{P_C(1-P_C)Mpn}\sqrt{\frac{1}{[1+(n-1)\rho_2+(pn-1)\rho_3]-[\pi_1+(n\pi_2-\pi_1)\rho_2+(pn\pi_3-\pi_1)\rho_3]}}\delta$	$M-2-w$
Two-level block design	$\lambda_{3uA} = \sqrt{P_{B2}(1-P_{B2})mN}\sqrt{\frac{1}{[1+(P_{B2}(1-P_{B2})N\psi_1-1)\rho]-[\pi_1+(P_{B2}(1-P_{B2})N\psi_1\pi_a-\pi_1)\rho]}}\delta$	$m-1-w$
Three-level block design: treatment at the middle level	$\lambda_{4uA} = \sqrt{P_{B32}\mathrm{Pr}(1-P_{B32})mPN}\sqrt{\frac{1}{\left(\begin{array}{l}1+(N-1)\rho_2+[P_{B32}(1-P_{B32})PN\psi_3-1]\\ \rho_3-[\pi_1+(N\pi_2-\pi_1)\\ \rho_2+(P_{B32}(1-P_{B32})PN\psi_3\pi_{a3}-\pi_1)\rho_3]\end{array}\right)}}\delta$	$m-1-w$
Three-level block design: treatment at the first level	$\lambda_{5uA} = \sqrt{P_{B31}(1-P_{B31})mPN}\sqrt{\frac{1}{\left(\begin{array}{l}1+[P_{B31}(1-P_{B31})N\psi_2-1]\\ \rho_2+[P_{B31}(1-P_{B31})PN\psi_3-1]\\ \rho_3-[\pi_1+(P_{B31}(1-P_{B31})N\psi_2\pi_{a2}-\pi_1)\\ \rho_2+(P_{B31}(1-P_{B31})PN\psi_3\pi_{a3}-\pi_1)\rho_3]\end{array}\right)}}\delta$	$m-1-w$

Note: (1) For cluster designs, the total number of clusters is *M*. (2) For the two-level block design and the three-level block design with treatment at the first level, the number of level 1 units per level 2 unit is *N*. (3) For the two three-level block design, the number of level 2 units per level 3 unit is *P*. (4) For all designs *w* represents the number of covariates at the top level. (5) The subscript *u* indicates unbalanced designs

Table 16.4 Summary of non-centrality parameters in certain unbalanced designs: no covariates

Design	Non-centrality parameter (λ)	Degrees of freedom (df)
Two-level cluster design	$\lambda_{1u} = \sqrt{P_C\,(1 - P_C)\,Mn}\sqrt{\frac{1}{1+(n-1)\rho}}\delta$	$M - 2$
Three-level cluster design	$\lambda_{2u} = \sqrt{P_C\,(1 - P_C)\,Mpn}\sqrt{\frac{1}{\left[1+(n-1)\rho_2+(pn-1)\rho_3\right.}}\delta$	$M - 2$
Two-level block design	$\lambda_{3u} = \sqrt{P_{B2}\,(1 - P_{B2})\,mN}\sqrt{\frac{1}{1+[P_{B2}(1-P_{B2})N\psi_1-1]\rho}}\delta$	$m - 1$
Three-level block design: treatment at the middle level	$\lambda_{4u} = \sqrt{P_{B32}\,(1 - P_{B32})\,mPN}\sqrt{\frac{1}{1+(N-1)\rho_2+[P_{B32}(1-P_{B32})PN\psi_3-1]\rho_3}}\delta$	$m - 1$
Three-level block design: treatment at the first level	$\lambda_{5u} = \sqrt{P_{B31}\,(1 - P_{B31})\,mPN}\sqrt{\frac{1}{1+[P_{B31}(1-P_{B31})N\psi_2-1]\rho_2+[P_{B31}(1-P_{B31})PN\psi_3-1]\rho_3}}\delta$	$m - 1$

Note: (1) For cluster designs, the total number of clusters is M. (2) For the two-level block design and the three-level block design with treatment at the first level, the number of level 1 units per level 2 unit is N. (3) For the two three-level block designs, the number of level 2 units per level 3 unit is P. (4) The subscript u indicates unbalanced designs

Notice that both power estimates in this unbalanced design are slightly smaller than those in the balanced two-level cluster randomized design. In fact, as imbalance becomes more extreme (e.g., $P_C = .20$), other things being equal, the power estimates become even smaller. In our example, when $P_C = .20$, the power estimates are decreased to .38 for a two-tailed test and .50 for a one-tailed test. In general, unbalanced designs produced smaller power estimates than balanced designs across all five designs covered in this chapter. In addition, the larger the imbalance (i.e., the smaller the proportion of random allocation of units to a treatment group), the smaller the power estimates.

16.5 Optimal Sampling of Units in Balanced Designs

A key component of designing multilevel experiments is to determine the sample sizes necessary at various levels of the hierarchy: the number of level 1 units within level 2 units, the number of level 2 units within level 3 units, the number of level 3 units, and so forth. The choice of sampling more units in one level than another depends on the cost of the level-specific unit because there are always budget constraints when carrying out an experiment. In survey research, there is a long tradition of optimal sampling in two-stage cluster designs (Cochran, 1977; Lohr, 2022). Prior research has discussed optimal allocation and power analysis in generalizability studies and measurement designs with budget constraints (e.g., Marcoulides, 1993, 1997). In psychology researchers have discussed optimal allocation methods for study facets such as the number of individuals in different treatment levels and the number of repeated measurements within individuals (Allison et al., 1997; McClelland, 1997). In education optimal allocation methods have been provided in two-, three-, and four-level cluster and block randomized designs (Bulus & Dong, 2021; Hedges & Borenstein, 2014; Konstantopoulos, 2009, 2011b, 2013; Liu, 2003; Raudenbush, 1997; Raudenbush & Liu, 2000). Below we discuss optimal sampling of units in two- and three-level cluster and block randomized designs.

Cluster Randomized Designs

Following previous work (e.g., Hedges & Borenstein, 2014; Konstantopoulos, 2009; Raudenbush, 1997), suppose the linear cost function for the total cost of three-level cluster randomized design is

$$C_T = MpnC_1 + MpC_2 + MC_3 \tag{16.23}$$

where C_T is the total cost for all units in all levels, $M = 2m$ is the total number of level 3 units (m units per treatment condition), C_1 is the cost of each level 1 unit, C_2 is the cost of each level 2 unit, and C_3 is the cost of each level 3 unit. Solving Eq.

(16.23) for $m = M/2$ yields

$$m = \frac{C_T}{2\,(pnC_1 + pC_2 + C_3)}. \tag{16.24}$$

Suppose for simplicity the total cost as well as the cost of each unit at each level is fixed and that the cost of the units in the treatment and control groups is the same. The objective is to determine the optimal number of level 1, level 2, and level 3 units that maximizes power. In order to accomplish the objective, one needs to maximize the *NCP* with respect to level 1 and level 2 units. Suppose the *NCP* in a three-level cluster randomized design with covariates in all three levels is

$$\lambda_{3Acrd} = \sqrt{\frac{mpn}{2}}\sqrt{\frac{1}{\eta_1 + (n\eta_2 - \eta_1)\,\rho_2 + (pn\eta_3 - \eta_1)\,\rho_3]}}\delta, \tag{16.25}$$

where $\eta_1 = 1 - \pi_1$, $\eta_2 = 1 - \pi_2$, $\eta_3 = 1 - \pi_3$, and ηs represent proportions of residual variances and are used here instead of πs, which are proportions of variance explained. The subscript indicates a three-level cluster randomized design with covariates. Notice that Eq. (16.25) is equivalent to Eq. (16.10) when the η's are substituted by π's. Using Eq. (16.24) to substitute m changes Eq. (16.25) to

$$\lambda_{3Acrd} = \sqrt{\frac{C_T\,pn}{4\,(pnC_1 + pC_2 + C_3)}}\sqrt{\frac{1}{\eta_1 + (n\eta_2 - \eta_1)\,\rho_2 + (pn\eta_3 - \eta_1)\,\rho_3}}\delta. \tag{16.26}$$

Maximizing Eq. (16.26) with respect to n and p produces an optimal n of

$$n_{3crd} = \sqrt{\frac{C_2}{C_1}}\sqrt{\frac{\eta_1\,(1 - \rho_2 - \rho_3)}{\eta_2\rho_2}}, \tag{16.27}$$

and an optimal p of

$$p_{3crd} = \sqrt{\frac{C_3}{C_2}}\sqrt{\frac{\eta_2\rho_2}{\eta_3\rho_3}}. \tag{16.28}$$

Substituting Eqs. (16.27) and (16.28) into Eq. (16.24) produces

$$m_{3crd} = \frac{C_T}{2\Big(p_{3crd}n_{3crd}C_1 + p_{3crd}C_2 + C_3\Big)}. \tag{16.29}$$

In the two-level cluster randomized design, the optimal number of level 1 units within each level 2 unit is

$$n_{2crd} = \sqrt{\frac{C_2}{C_1}}\sqrt{\frac{\eta_1(1-\rho)}{\eta_2\rho}}, \tag{16.30}$$

and the number of level 2 units in the treatment arm is computed as

$$m_{2crd} = \frac{C_T}{2(n_{2crd}C_1 + C_2)}, \tag{16.31}$$

where the subscript *2crd* indicates that the terms correspond to a two-level cluster randomized design. The *NCP* in the two-level case is

$$\lambda_{2crd} = \sqrt{\frac{mn}{2}}\sqrt{\frac{1}{\eta_1 + (n\eta_2 - \eta_1)\rho_2}}\delta \tag{16.32}$$

which is equivalent to Eq. (16.5) when the η's are substituted by π's. When covariates are not included in any level, Eqs. (16.26) to (16.32) are slightly simplified because $\eta_1 = \eta_2 = \eta_3 = 1$.

Illustrative Example

Consider a simple three-level example where the total cost is $C_T = 1000$, the cost of level 1 units is $C_1 = 1$, the cost of level 2 units is $C_2 = 5C_1$, and the cost of level 3 units is $C_3 = 5C_2$. Suppose the magnitude of a meaningful intervention effect size is one-third of a standard deviation ($\delta = .33$), the ICCs are $\rho_2 = .10$ and $\rho_3 = .15$, and the covariates explain 50% of the variance at each level, that is, $\eta_1 = \eta_2 = \eta_3 = .50$. One covariate is presumed at the top level (i.e., $w = 1$). The optimal level 1 and level 2 units are computed as

$$n_{3crd} = \sqrt{\frac{5*1}{1}} * \sqrt{\frac{.50*(1-.10-.15)}{.50*.10}} = 6.12 \approx 6,$$

$$p_{3crd} = \sqrt{\frac{5*5*1}{5*1}} * \sqrt{\frac{.50*.10}{.50*.15}} = 1.83 \approx 2,$$

and using Eq. (16.29)

$$m_{3crd} = \frac{1000}{2*(6*2*1 + 2*5*1 + 5*5*1)} \approx 11.$$

Using Eq. (16.26) the *NCP* is

$$\lambda_{3Acrd} = \sqrt{\frac{11 * 2 * 6}{2}} * \sqrt{\frac{1}{\begin{pmatrix} .50 + (6 * .50 - .50) * .10 \\ + (2 * 6 * .50 - .50) * .15 \end{pmatrix}}} * .33 = 2.14$$

and the power of a one-tailed t-test using Eq. (16.8) is

$$pwr_{3crd} = 1 - \mathrm{H}\ [\mathit{1.73, 19, 2.14}] = 1 - .34 = .66.$$

Now consider a two-level example where the total cost is $C_T = 1000$, the cost of level 1 units is $C_1 = 1$, and the cost of level 2 units is $C_2 = 15C_1$. Suppose the magnitude of the treatment effect is one-third of a standard deviation ($\delta = .33$), and the ICC is $\rho = .20$, and the covariates explain 50% of the variance at each level, that is, $\eta_1 = \eta_2 = .50$. One covariate is presumed at the top level (i.e., $w = 1$). The optimal level 1 units are computed as

$$n_{2crd} = \sqrt{\frac{15 * 1}{1}} * \sqrt{\frac{.50 * (1 - .20)}{.50 * .20}} = 7.75 \approx 8,$$

and using Eq. (16.31)

$$m_{2crd} = \frac{1000}{2 * (8 * 1 + 15 * 1)} = 21.74 \approx 22.$$

Using Eq. (16.32) the *NCP* is equal to

$$\lambda_{2crd} = \sqrt{\frac{22 * 8}{2}} \sqrt{\frac{1}{.50 + (8 * .50 - .50) * .20}} * .33 = 2.83$$

and the power of a one-tailed t-test is

$$pwr_{2crd} = 1 - \mathrm{H}\ [2.\mathit{13}, \mathit{41}, 2.\mathit{83}] = 1 - .24 = .76.$$

Block Randomized Designs

Treatment at the Second Level

Following previous work (Hedges & Borenstein, 2014; Konstantopoulos, 2013; Raudenbush & Liu, 2000), consider a linear cost function for the total cost of the study

$$C_T = 2mpnC_1 + 2mpC_2 + mC_3 \tag{16.33}$$

where C_T is the total cost for all units in all levels, m is now the total number of level 3 units, p is the number of level 2 units within a treatment condition, n is the number of level 1 units within a level 2 unit, C_1 is the cost of each level 1 unit, C_2 is the cost of each level 2 unit, and C_3 is the cost of each level 3 unit.

Suppose once more that the total cost as well as the cost for each unit at each level is fixed and that the cost for units in the treatment and the control group is the same. To achieve the objective to determine the optimal number of level 1, level 2, and level 3 units that maximizes power, one needs to maximize the *NCP* (see Eq. 16.18) with respect to n and p (see Hedges & Borenstein, 2014; Konstantopoulos, 2013; Raudenbush & Liu, 2000). Solving Eq. (16.33) for m yields $m = C_T/(2pnC_1 + 2pC_2 + C_3)$. Substituting m in the *NCP* formula (Eq. 16.18) when ηs are used instead of πs gives

$$\lambda_{3Abrd1} = \sqrt{\frac{C_T pn}{2\left(2pnC_1 + 2pC_2 + C_3\right)}}\sqrt{\frac{1}{\eta_1 + \left(n\eta_2 - \eta_1\right)\rho_2 + \left(pn\vartheta_3\eta_{a3} - \eta_1\right)\rho_3}}\,\delta, \tag{16.34}$$

where $\eta_{a3} = 1 - \pi_{a3}$ is the proportion of the residual variance of the treatment effect across level 3 units and all other terms have been previously defined. Taking partial derivatives with respect to n and p provides the solutions

$$n_{3brd1} = \sqrt{\frac{C_2}{C_1}}\sqrt{\frac{\eta_1\left(1 - \rho_2 - \rho_3\right)}{\eta_2\rho_2}},\quad p_{3brd1} = \sqrt{\frac{C_3}{2C_2}}\sqrt{\frac{\eta_2\rho_2}{\eta_{a3}\vartheta_3\rho_3}}. \tag{16.35}$$

The total number of level-3 units is determined as

$$m_{3brd1} = \frac{C_T}{2p_{3brd1}n_{3brd1}C_1 + 2p_{3brd1}C_2 + C_3}. \tag{16.36}$$

When covariates are not included in any level, Eqs. (16.34) to (16.36) are slightly simplified because $\eta_1 = \eta_2 = \eta_{a3} = 1$.

Illustrative Example

Consider a simple three-level example where the total cost is $C_T = 1000$, the cost of level 1 units is $C_1 = 1$, the cost of level 2 units is $C_2 = 5C_1$, and the cost of level 3 units is $C_3 = 5C_2$. Suppose the magnitude of a meaningful intervention effect size is one-third of a standard deviation ($\delta = .33$), the ICC's are $\rho_2 = .10$ and $\rho_3 = .15$, covariates explain 50% of the variance at each level, that is, $\eta_1 = \eta_2 = \eta_{a3} = .50$ and $\vartheta_3 = .25$. One covariate is presumed at the top level (i.e., $w = 1$). The optimal level 1 and level 2 units are computed as

$$n_{3brd1} = \sqrt{\frac{5*1}{1}} * \sqrt{\frac{.50\,(1-.10-.15)}{.50*.10}} = 6.12 \approx 6,$$

$$p_{3brd1} = \sqrt{\frac{5*5*1}{2*5*1}} * \sqrt{\frac{.50*.10}{.50*.25*.15}} = 2.36 \approx 2.$$

and the total number of level 3 units is determined using Eq. (16.36) as

$$m = \frac{1000}{2*2*6*1+2*2*5*1+5*5*1} = 14.49 \approx 14 \text{ (rounding down to 14)}.$$

Using Eq. (16.34) the *NCP* is

$$\lambda_{3Abrd1} = \sqrt{\frac{14*2*6}{2}} * \sqrt{\frac{1}{\begin{pmatrix} .50+(6*.50-.50)\,.10 \\ +(2*6*.25*.50-.50)*.15 \end{pmatrix}}} * .33 = 3.19$$

and the power of a one-tailed t-test using Eq. (16.16) is

$$pwr_{3brd1} = 1 - \mathrm{H}\ [\mathit{1.78}, \mathit{12}, \mathit{3.19}] = 1 - .09 = .91.$$

Treatment at the First Level

Following prior research (Hedges & Borenstein, 2014; Konstantopoulos, 2013; Raudenbush & Liu, 2000), consider a linear cost function for the total cost of the study

$$C_T = 2mpnC_1 + mpC_2 + mC_3 \tag{16.37}$$

where C_T is the total cost for all units in all levels, m is the total number of level 3 units, p is the number of level 2 units within a level 3 unit, n is the number of level 1 units within a treatment condition within a level 2 unit and within a level 3 unit, C_1 is the cost of each level 1 unit, C_2 is the cost of each level 2 unit, and C_3 is the cost of each level 3 unit. Solving Eq. 16.35 for m gives $m = C_T/(2pnC_1 + pC_2 + C_3)$. Substituting m in the *NCP* formula (Eq. 16.21) gives

$$\lambda_{3Abrd2} = \sqrt{\frac{C_T pn}{2\,(2pnC_1 + pC_2 + C_3)}} \sqrt{\frac{1}{\eta_1 + (n\vartheta_2\eta_{a2} - \eta_1)\,\rho_2 + (pn\vartheta_3\eta_{a3} - \eta_1)\,\rho_3}}\,\delta, \tag{16.38}$$

where $\eta_{a2} = 1 - \pi_{a2}$ is the proportion of the residual variance of the treatment effect across level 2 units and all other terms have been previously defined. Taking partial derivatives with respect to n and p provides the solutions

$$n_{3brd2} = \sqrt{\frac{C_2}{2C_1}}\sqrt{\frac{\eta_1 (1 - \rho_2 - \rho_3)}{\eta_{a2}\vartheta_2\rho_2}} \text{ and } p_{3brd2} = \sqrt{\frac{C_3}{C_2}}\sqrt{\frac{\eta_{a2}\vartheta_2\rho_2}{\eta_{a3}\vartheta_3\rho_3}}, \tag{16.39}$$

and the total number of level 3 units is determined as

$$m_{3brd2} = \frac{C_T}{2p_{3brd2}n_{3brd2}C_1 + p_{3bdr2}C_2 + C_3}. \tag{16.40}$$

The optimal n and p formulas in Eq. (16.39) are used in Eq. 16.38 to compute the *NCP*.

In the two-level case (i.e., when the third level is omitted), the optimal n formula is

$$n_{2brd} = \sqrt{\frac{C_2}{2C_1}}\sqrt{\frac{\eta_1 (1 - \rho)}{\eta_a\vartheta_1\rho}}, \tag{16.41}$$

where $\eta_1 = 1 - \pi_1$, $\eta_\alpha = 1 - \pi_a$, and all other terms have been defined previously in the two-level block randomized design section. The second level units are computed as

$$m_{2brd} = \frac{C_T}{2n_{2brd}C_1 + C_2}. \tag{16.42}$$

The *NCP* in a two-level block randomized design can be expressed as

$$\lambda_{2brd} = \sqrt{\frac{mn}{2}}\sqrt{\frac{1}{[\eta_1 + (n\vartheta_1\eta_\alpha - \eta_1)\rho]}}\delta, \tag{16.43}$$

which is equivalent to Eq. (16.13) when $\eta_1 = 1 - \pi_1$ and $\eta_\alpha = 1 - \pi_a$.

Using Eq. (16.42) the *NCP* in the two-level case becomes

$$\lambda_{2brd} = \sqrt{\frac{C_T n_{2brd}}{2(2n_{2brd}C_1 + C_2)}}\sqrt{\frac{1}{[\eta_1 + (n\vartheta_1\eta_\alpha - \eta_1)\rho]}}\delta, \tag{16.44}$$

When covariates are not included in any level, Eqs. (16.38) through (16.44) are slightly simplified because $\eta_1 = \eta_{a2} = \eta_{a3} = \eta_a = 1$.

Illustrative Example

Consider a simple three-level example where the total cost is $C_T = 1000$, the cost of level 1 units is $C_I = 1$, the cost of level 2 units is $C_2 = 5C_I$. and the cost of level 3 units is $C_3 = 5C_2$. Suppose the magnitude of a meaningful intervention effect size is one-third of a standard deviation ($\delta = .33$); the ICCs are $\rho_2 = .10$ and $\rho_3 = .15$; covariates explain 50% of the variance at each level, that is, $\eta_1, = \eta_{a2} = \eta_{a3} = .50$, and $\vartheta_2 = \vartheta_3 = .30$. One covariate is presumed at the top level (i.e., $w = 1$). The optimal level 1 and level 2 units are computed as

$$n_{3brd2} = \sqrt{\frac{5*1}{2*1}} * \sqrt{\frac{.50\,(1-.10-.15)}{.50*.30*.10}} = 7.91 \approx 8,$$

$$p_{3brd2} = \sqrt{\frac{5*5*1}{5*1}} * \sqrt{\frac{.50*.30*.10}{.50*.30*.15}} = 1.83 \approx 2.$$

and the total number of level 3 units is calculated using Eq. (16.40) as

$$m = \frac{1000}{(2**2*8*1+2*5*1+5*5*1)} = 14.93 \approx 15.$$

Using Eq. (16.38) the *NCP* is

$$\lambda_{3Abrd2} = \sqrt{\frac{15*2*8}{2}} * \sqrt{\frac{1}{\begin{pmatrix} .50 + (8*.30*.50-.50)\,.10 \\ +\,(2*8*.30*.50-.50)*.15 \end{pmatrix}}} * .33 = 3.91$$

and the power of a one-tailed t-test using Eq. (16.16) is

$$pwr_{3brd1} = 1 - \mathrm{H}\ [1.77\ 13, 3.91] = 1 - .02 = .98.$$

Now consider a two-level example where the total cost is $C_T = 1000$, the cost of level 1 units is $C_I = 1$, and the cost of level 2 units is $C_2 = 15C_I$. Suppose the magnitude of the treatment effect is one-third of a standard deviation ($\delta = .33$), and the ICC is $\rho = .20$, and the covariates explain 50% of the variance at each level, that is, $\eta_1 = \eta_\alpha = .50$ and $\vartheta_1 = .30$. One covariate is presumed at the top level (i.e., $w = 1$). The optimal level 1 units are computed as

$$n_{2brd} = \sqrt{\frac{15*1}{2*1}} * \sqrt{\frac{.50*(1-.20)}{.50*.30*.20}} = 10,$$

and using Eq. (16.42)

$$m_{2brd} = \frac{1000}{2 * 10 * 1 + 15 * 1} = 28.57 \approx 29 \text{ (rounding up)}.$$

Using Eq. (16.43) the *NCP* is equal to

$$\lambda_{2brd} = \sqrt{\frac{29 * 10}{2}} \sqrt{\frac{1}{.50 + (10 * .30 * .50 - .50) * .20}} * .33 = 4.75$$

and the power of a one-tailed t-test is

$$pwr_{2brd} = 1 - \text{H}\ [1.70, 27, 4.75] = 1 - .001 = .999.$$

Table 16.5 provides a summary of all five balanced designs discussed including cost functions and optimal sample sizes at various levels. Specifically, the third column of Table 16.5 presents optimal sample sizes when covariates are included in all levels and the fourth column displays optimal sample sizes when covariates are not included in the models.

16.6 The Impact of Sample Sizes on Power in Balanced Designs

Sample sizes are essential design parameters in prospective power analysis. Multilevel designs comprise of multiple levels, and consequently there are multiple sample sizes that may impact power differently. For example, in education, researchers would want to know how many schools, classrooms, and students should be part of their sample to detect the anticipated treatment effect of interest with a probability larger than .80. If the number of top-level clusters such as schools, for instance, has a greater impact on power than classrooms and students, other things being equal, researchers should recruit more schools in their sample when budget and resources are sufficient.

One way to demonstrate theoretically the impact of level-specific sample sizes on statistical power is through the estimation of the *NCP*. Specifically, one can estimate the maximum *NCP* for each design when the number of level 1 units, level 2 units, or level 3 units becomes infinitely large. The objective of this analysis is to offer some intuition about how the number of units at different levels could influence power other things being fixed, using a simple mechanism that capitalizes on maximizing the *NCP*. Table 16.6 summarizes the formulas of the maximum *NCPs* for each of the five balanced designs discussed in this chapter when covariates are included in the models in all levels. For fixed values of the number of units at lower levels of the hierarchy, regardless of the design, Table 16.6 provides consistent evidence that when the number of units (clusters) at the top level becomes infinitely large, the *NCPs* also become infinitely large, and eventually power becomes equal to 1.

Table 16.5 Summary of optimal sampling of units in balanced designs

Design	Cost function	Optimal sample sizes: covariates in all levels	Optimal sample sizes: no covariates
Two-level cluster design	$C_T = 2mnC_1 + 2mC_2$	$n_{2crd} = \sqrt{\frac{C_2}{C_1}}\sqrt{\frac{\eta_1(1-\rho)}{\eta_2\rho}}$ $m_{2crd} = \frac{C_T}{2(n_{2crd}C_1+C_2)}$	$n_{2crd} = \sqrt{\frac{C_2}{C_1}}\sqrt{\frac{(1-\rho)}{\rho}}$ $m_{2crd} = \frac{C_T}{2(n_{2crd}C_1+C_2)}$
Three-level cluster design	$C_T = 2mpnC_1 + 2mpC_2 + 2mC_3$	$n_{3crd} = \sqrt{\frac{C_2}{C_1}}\sqrt{\frac{\eta_1(1-\rho_2-\rho_3)}{\eta_2\rho_2}}$ $p_{3crd} = \sqrt{\frac{C_3}{C_2}}\sqrt{\frac{\eta_2\rho_2}{\eta_3\rho_3}}$ $m_{3crd} = \frac{C_T}{2(p_{3crd}n_{3crd}C_1+p_{3crd}C_2+C_3)}$	$n_{3crd} = \sqrt{\frac{C_2}{C_1}}\sqrt{\frac{(1-\rho_2-\rho_3)}{\rho_2}}$ $p_{3crd} = \sqrt{\frac{C_3}{C_2}}\sqrt{\frac{\rho_2}{\rho_3}}$ $m_{3crd} = \frac{C_T}{2(p_{3crd}n_{3crd}C_1+p_{3crd}C_2+C_3)}$
Two-level block design	$C_T = 2mnC_1 + mC_2$	$n_{2brd} = \sqrt{\frac{C_2}{2C_1}}\sqrt{\frac{\eta_1(1-\rho)}{\eta_a\vartheta_1\rho}}$ $m_{2brd} = \frac{C_T}{2n_{2brd}C_1+C_2}$	$n_{2brd} = \sqrt{\frac{C_2}{2C_1}}\sqrt{\frac{(1-\rho)}{\vartheta_1\rho}}$ $m_{2brd} = \frac{C_T}{2n_{2brd}C_1+C_2}$
Three-level block design: treatment at second level	$C_T = 2mpnC_1 + 2mpC_2 + mC_3$	$n_{3brd1} = \sqrt{\frac{C_2}{C_1}}\sqrt{\frac{\eta_1(1-\rho_2-\rho_3)}{\eta_2\rho_2}}$ $p_{3brd1} = \sqrt{\frac{C_3}{2C_2}}\sqrt{\frac{\eta_2\rho_2}{\eta_{a3}\vartheta_3\rho_3}}$ $m_{3brd1} = \frac{C_T}{2p_{3brd1}n_{3brd1}C_1+2p_{3brd1}C_2+C_3}$	$n_{3brd1} = \sqrt{\frac{C_2}{C_1}}\sqrt{\frac{(1-\rho_2-\rho_3)}{\rho_2}}$ $p_{3brd1} = \sqrt{\frac{C_3}{2C_2}}\sqrt{\frac{\rho_2}{\vartheta_3\rho_3}}$ $m_{3brd1} =$ $\frac{C_T}{2p_{3brd1}n_{3brd1}C_1+2p_{3brd1}C_2+C_3}$
Three-level block design: treatment at first level	$C_T = 2mpnC_1 + mpC_2 + mC_3$	$n_{3brd2} = \sqrt{\frac{C_2}{2C_1}}\sqrt{\frac{\eta_1(1-\rho_2-\rho_3)}{\eta_{a2}\vartheta_2\rho_2}}$ $p_{3brd2} = \sqrt{\frac{C_3}{C_2}}\sqrt{\frac{\eta_{a2}\vartheta_2\rho_2}{\eta_{a3}\vartheta_3\rho_3}}$ $m_{3brd2} = \frac{C_T}{2p_{3brd2}n_{3brd2}C_1+p_{3bdr2}C_2+C_3}$	$n_{3brd2} = \sqrt{\frac{C_2}{2C_1}}\sqrt{\frac{(1-\rho_2-\rho_3)}{\vartheta_2\rho_2}}$ $p_{3brd2} = \sqrt{\frac{C_3}{C_2}}\sqrt{\frac{\vartheta_2\rho_2}{\vartheta_3\rho_3}}$ $m_{3brd2} = \frac{C_T}{2p_{3brd2}n_{3brd2}C_1+p_{3bdr2}C_2+C_3}$

For a preset number of units at the top level, when the number of units at lower levels becomes infinitely large, the *NCPs* reach a specific upper limit as the formulas indicate, but do not become infinitely large. Thus, whether the design involves two or three levels or whether it is a cluster or a block randomized design, the sample size at the top level impacts power more than the sample sizes in lower levels other things being equal. For example, education researchers who want to plan a field experiment would preferably want to sample more schools than classrooms or students, so long as the budget and the resources allow for that.

Table 16.7 summarizes the formulas of the maximum *NCPs* for each of the five balanced designs outlined when covariates are not included in the models. The patterns are the same with those in Table 16.6. Again, for fixed numbers of lower-level units, when the number of top-level units becomes infinitely large, the *NCPs* become infinitely larger, and power tends to one. For a fixed number of third-level units however, when lower-level units become infinitely large, the *NCPs* reach specific maximum limits, but power does not necessarily tend to one.

It is noteworthy that the sample size at the top level also influences power through the *df*. In fact, power is a direct function of the *df*, and thus as the *df* increase, power increases. In particular, in all five designs delineated, the *df* are a function of the number of units at the top level, and the larger the sample size at the top level, the larger the *df* and as a result the larger the power of the statistical test, other things being equal.

From a practical perspective, level 1 and level 2 units can't be infinitely large because they are bounded by organizational structure (e.g., there is a limit on the number of students in a classroom or the number of classrooms in a school). Similarly, units at the third level (e.g., schools) can't be infinitely large either in practice. Thus, Tables 16.6 and 16.7 provide a conceptual perspective about how the number of units at each level could impact statistical power. From a pragmatical viewpoint however, although the number of top-level units has a stronger impact on power than the number of lower-level units other things being equal, researchers have to take into account budget and resource constraints when making decisions about sample sizes in their projects. For example, because of budget constraints and limited resources, researchers may have to sample fewer schools and more classes or students per school.

16.7 Discussion

A priori statistical power analysis is a crucial part of the planning stage of a study. In fact, power analysis is informative to researchers who want to conduct experiments that entail clustering and wish to determine the sample sizes necessary to achieve a high probability (e.g., $> .80$) of detecting an anticipated treatment effect. A priori power analysis is essentially about ascertaining the sample sizes needed in an experiment in order to detect a treatment effect of a certain magnitude.

Table 16.6 Impact of sample sizes on non-centrality parameters in balanced designs: covariates in all levels

Design	n → ∞	p → ∞	m → ∞
Two-level cluster design	$\lambda_{1Amax} = \sqrt{\frac{m}{2}}\sqrt{\frac{1}{\rho - \pi_2\rho}}\delta$	NA	$\lambda_{1max} \to \infty$
Three-level cluster design	$\lambda_{2Amax} = \sqrt{\frac{mn}{2}}\sqrt{\frac{1}{[\rho_2 + p\rho_3] - [\pi_2\rho_2 + p\pi_3\rho_3]}}\delta$	$\lambda_{2Amax} = \sqrt{\frac{mn}{2}}\sqrt{\frac{1}{n\rho_3 - n\pi_3\rho_3}}\delta$	$\lambda_{2max} \to \infty$
Two-level block design	$\lambda_{3Amax} = \sqrt{\frac{m}{2}}\sqrt{\frac{1}{\vartheta_1\rho - \vartheta_1\pi_a\rho}}\delta$	NA	$\lambda_{3max} \to \infty$
Three-level block design: treatment at second level	$\lambda_{4Amax} = \sqrt{\frac{mn}{2}}\sqrt{\frac{1}{[\rho_2 + p\vartheta_3\rho_3] - [\pi_2\rho_2 + p\vartheta_3\pi_{a3}\rho_3]}}\delta$	$\lambda_{4Amax} = \sqrt{\frac{mn}{2}}\sqrt{\frac{1}{n\vartheta_3\rho_3 - n\vartheta_3\pi_{a3}\rho_3}}\delta$	$\lambda_{4max} \to \infty$
Three-level block design: treatment at first level	$\lambda_{5Amax} = \sqrt{\frac{mn}{2}}\sqrt{\frac{1}{[\vartheta_2\rho_2 + p\vartheta_3\rho_3] - [\vartheta_2\pi_{a2}\rho_2 + p\vartheta_3\pi_{a3}\rho_3]}}\delta$	$\lambda_{5Amax} = \sqrt{\frac{mn}{2}}\sqrt{\frac{1}{n\vartheta_3\rho_3 - n\vartheta_3\pi_{a3}\rho_3}}\delta$	$\lambda_{5max} \to \infty$

Table 16.7 Impact of sample sizes on non-centrality parameters in balanced designs: no covariates

Design	$n \to \infty$	$p \to \infty$	$m \to \infty$
Two-level cluster design	$\lambda_{1\max} = \sqrt{\frac{m}{2}}\sqrt{\frac{1}{\rho}}\delta$	NA	$\lambda_{1max} \to \infty$
Three-level cluster design	$\lambda_{2\max} = \sqrt{\frac{mp}{2}}\sqrt{\frac{1}{\rho_2+p\rho_3}}\delta$	$\lambda_{2\max} = \sqrt{\frac{mn}{2}}\sqrt{\frac{1}{n\rho_3}}\delta$	$\lambda_{2max} \to \infty$
Two-level block design	$\lambda_{3\max} = \sqrt{\frac{m}{2}}\sqrt{\frac{1}{\vartheta_1\rho}}\delta$	NA	$\lambda_{3max} \to \infty$
Three-level block design: treatment at second level	$\lambda_{4\max} = \sqrt{\frac{mp}{2}}\sqrt{\frac{1}{\rho_2+p\vartheta_3\rho_3}}\delta$	$\lambda_{4\max} = \sqrt{\frac{mn}{2}}\sqrt{\frac{1}{n\vartheta_3\rho_3}}\delta$	$\lambda_{4max} \to \infty$
Three-level block design: treatment at first level	$\lambda_{5\max} = \sqrt{\frac{mp}{2}}\sqrt{\frac{1}{\vartheta_2\rho_2+p\vartheta_3\rho_3}}\delta$	$\lambda_{5\max} = \sqrt{\frac{mn}{2}}\sqrt{\frac{1}{n\vartheta_3\rho_3}}\delta$	$\lambda_{5max} \to \infty$

Several results emerge from our discussion. First, clustering as manifested via ICCs constitutes an important design parameter of prospective power analysis and is inversely related to the *NCP* and ultimately power. Table 16.8 displays design effects for all five balanced designs we outlined. In survey research, the design effect is a variance correction factor that captures clustering due to cluster sampling and adjusts the variance of the average effect (Kish, 1965; Lohr, 2022). In prospective statistical power analysis, the design effect is a direct function of the ICC, and so long as the ICC $\neq 0$ and the sample sizes at the lower levels of the hierarchy are larger than 1, the design effect is a positive number larger than 1 (see column 2 of Table 16.8). Other things being equal, as the clustering (i.e., the ICCs) becomes larger, the design effect becomes larger, and subsequently the *NCP* and ultimately statistical power become smaller.

Second, the importance of covariates in explaining variances at various levels of the hierarchy is demonstrated in the third column of Table 16.8. Specifically, these design effect formulas show an adjustment because of the covariates' effects. For fixed values of design parameters, the values of the design effects calculated using the formulas in column 2 will be larger than those of the design effects using the formulas in column 3. This reduction of the design effects, due to variance explained because of covariates, corresponds to an increase of the *NCPs* and finally power. Therefore, including covariates that explain considerable proportion of the variances at different levels of the hierarchy is critical for reducing clustering, design effects, and concurrently increasing the *NCPs* and power. Under the assumption that lower-level covariates are centered at their cluster means, our results indicate that in two- and three-level cluster designs, top-level covariates influence power more than lower-level covariates. Previous work has provided a compendium of design parameter values for ICCs and covariate effects in education (e.g., Hedges & Hedberg, 2007). The design effects of the unbalanced designs we briefly outlined are the denominators of the *NCPs* reported in Tables 16.3 and 16.4.

Third, the effect size (e.g., a standardized mean difference between treatment and control groups) is also an important design parameter that influences statistical power. In particular, as the magnitude of the effect size, δ, increases, the *NCP* increases, and thus power increases, other things being constant. It is prudent that a priori power analysis incorporates informed values of the magnitude of the effect size. Results from prior and current relevant research, primary studies and systematic meta-analytic reviews, could provide a realistic collection of plausible values of treatment effects.

Fourth, the Type I error probability known as the significance level can also influence power. Specifically, as the Type I error probability increases, power increases other things being equal. In contrast, as the significance level becomes smaller, power decreases. It is recommended that the significance level is set at a specific probability (e.g., .05 which is considered the standard) when conducting a priori power analysis. A related point is about whether the test of the treatment effect is a one- or a two-tailed test. This is about identifying the percentile at which the critical value of the *t*-test is calculated. Other things remaining constant, larger critical values of *t*-tests correspond to lower values of power (see Eqs. 16.7, 16.8,

Table 16.8 Design effects in balanced designs

Design	Design effect: no covariates	Design effect: covariates in all levels
Two-level cluster design	$DEFF_1 = 1 + (n-1)\rho$	$DEFF_{1A} = [1 + (n-1)\rho] - [\pi_1 + (n\pi_2 - \pi_1)\rho]$
Three-level cluster design	$DEFF_2 = 1 + (n-1)\rho_2 + (pn-1)\rho_3$	$DEFF_{2A} = [1 + (n-1)\rho_2 + (pn-1)\rho_3]$ $- [\pi_1 + (n\pi_2 - \pi_1)\rho_2 + (pn\pi_3 - \pi_1)\rho_3]$
Two-level block design	$DEFF_3 = 1 + (n\vartheta_1 - 1)\rho$	$DEFF_{3A} = [1 + (n\vartheta_1 - 1)\rho] - [\pi_1 + (n\vartheta_1\pi_a - \pi_1)\rho]$
Three-level block design: treatment at second level	$DEFF_4 = 1 + (n-1)\rho_2 + (pn\vartheta_3 - 1)\rho_3$	$DEFF_{4A} = [1 + (n-1)\rho_2 + (pn\vartheta_3 - 1)\rho_3]$ $- [\pi_1 + (n\pi_2 - \pi_1)\rho_2 + (pn\vartheta_3\pi_{a3} - \pi_1)\rho_3]$
Three-level block design: treatment at first level	$DEFF_5 = 1 + (n\vartheta_2 - 1)\rho_2 + (pn\vartheta_3 - 1)\rho_3$	$DEFF_{5A} = [1 + (n\vartheta_2 - 1)\rho_2 + (pn\vartheta_3 - 1)\rho_3]$ $- [\pi_1 + (n\vartheta_2\pi_{a2} - \pi_1)\rho_2 + (pn\vartheta_3\pi_{a3} - \pi_1)\rho_3]$

16.15, and 16.16). For the purposes of being conservative, it is recommended to assume a two-tailed test when conducting power computations.

Fifth, as the design effect formulas suggest (Table 16.8), *t*-tests in two-level designs (whether cluster or block designs) should in principle have higher statistical power than *t*-tests in three-level designs, when other design parameters are fixed. The underlying assumption is that clustering in three-level designs is typically larger than clustering in two-level designs. This assumption holds when the middle level (and its variance) in a three-level design is omitted. For instance, prior research in education has shown that the sum of the variances at the second and third levels in three-level designs is nearly 50% larger than the second-level variance in two-level designs, where the middle level is omitted (e.g., Nye et al., 2000). However, when the top level (and its variance) is omitted in a three-level design, the second level variance in the ensuing two-level design is equal to the sum of the variances at the middle and top levels in the three-level design. In this case, the power in three level designs will not be necessarily smaller than that in two-level designs.

Sixth, the power in block randomized designs is typically larger than the power in cluster randomized designs both for two- and three-level designs. The main reason for this result is theta, ϑ, which is defined as one-half of the proportion of the treatment effect variance to the total variance at a specific level. Specifically, at any level, ϑ is arguably smaller than .50 under the assumption that the ratio of the treatment effect variance to the total variance at a specific level is less than 1, which is a plausible postulation. Because ϑ is a smaller number, closer to zero, the design effects in block designs are reduced further compared to cluster designs, other things being constant. Consequently, reduced design effects correspond to increased values of *NCPs* and ultimately power values. Another secondary reason the block designs have higher power than cluster designs is attributed to the *df* (both in two- and three-level designs). Specifically, under the supposition that the number of top-level units and the number of top-level covariates are the same in both designs, the difference in *df* between block designs and cluster designs will be equal to one (favoring block designs). That is, other things being equal, the *t*-test in the block designs will have larger *df* than in cluster designs. However, the small difference in *df* between the two designs is not as impactful on power as the effect ϑ has on power. Nonetheless, power values should not be the driving force when making decisions about adopting a block versus a cluster design in an experiment. Features such as the suitable level of the treatment (e.g., the school level if it is a whole school intervention) and eliminating contamination between treatment and control groups should be the critical reasons for selecting a design.

Seventh, the power estimates in the unbalanced designs we briefly discussed are smaller than those obtained from balanced designs. The difference in power estimates is more pronounced when the imbalance is extreme (e.g., 20% of units are assigned to the treatment group). Thus, in cases where some imbalance seems inevitable, we recommend that researchers try to contain imbalance to low levels (e.g., the percentage of assigning units to the treatment group should be between 40% and 50%) to avoid large decreases in power estimates.

To conclude, power analysis requires some information about design parameters such as the magnitude of the effect size, clustering (ICCs), and the proportion of variance explained by covariates at each level. In block randomized designs, some knowledge about ϑ is also needed. Such information could be extracted from relevant empirical work that provides estimates about these design parameters. The sample size at each level is key in power analysis of multilevel experimental designs. For both cluster and block designs with either two or three levels, the sample size at the top level affects power more than the sample sizes at lower levels. Thus, to increase power, expanding the sample size of top level clusters is a good strategy other things being equal. Of course, researchers need to take into account constraints such as the budget and available resources to ensure that adding top-level clusters in their sample is affordable and manageable. Lastly, because the values of the parameters in power analysis are not exact, but rather uncertain to some degree, it is recommended to employ values that are plausible and conservative. Perhaps computing a range of plausible power values and identifying a lower bound of power would be a more pragmatic and enlightening approach. It is also recommended to try to achieve power levels that are larger than .80 and preferably closer to one. Power computations that target a power estimate that is exactly equal to .80 may result in smaller sample sizes and ultimately in underpowered experiments because of the inherent uncertainty in the values of the design parameters.

References

Allison, D. B., Allison, R. L., Faith, M. S., Paultre, F., & Pi-Sunyer, F. X. (1997). Power and money: Designing statistically powerful studies while minimizing financial costs. *Psychological Methods, 2*, 20–33.

Bloom, H. S. (2005). *Learning more from social experiments*. Russell Sage.

Borenstein, M., Rothstein, H., & Cohen, J. (2001). *Power and precision*. Biostat, Inc.

Bulus, M., & Dong, N. (2021). Bound constrained optimization of sample sizes subject to monetary restrictions in planning multilevel randomized trials and regression discontinuity studies. *The Journal of Experimental Education, 89*(2), 379–401.

Bulus, M., & Dong, N. (2022). Consequences of ignoring a level of nesting on design and analysis of blocked three-level regression discontinuity designs: Power and type I error rates. *Adıyaman University Journal of Educational Sciences, 12*(1), 42–55.

Bulus, M., & Sahin, S. G. (2019). Estimation and standardization of variance parameters for planning cluster-randomized trials: A short guide for researchers. *Journal of Measurement and Evaluation in Education and Psychology, 10*(2), 179–201.

Cochran, W. G. (1977). *Sampling techniques*. Wiley.

Cohen, J. (1988). *Statistical power analysis for the behavioral sciences* (2nd ed.). Academic Press.

Dong, N., & Maynard, R. (2013). PowerUp!: A tool for calculating minimum detectable effect sizes and minimum required sample sizes for experimental and quasi-experimental design studies. *Journal of Research on Educational Effectiveness, 6*(1), 24–67.

Donner, A., & Klar, N. (2000). *Design and analysis of cluster randomization trials in health research*. Arnold.

Goldstein, H. (2011). *Multilevel Statistical Models* (4th ed.). Wiley.

Hedges, L. V., & Borenstein, M. (2014). Constrained optimal design in three and four level experiments. *Journal of Educational and Behavioral Statistics, 39*, 257–281.

Hedges, L. V., & Hedberg, E. C. (2007). Intraclass correlation values for planning group randomized trials in education. *Educational Evaluation and Policy Analysis, 29*, 60–87.
Hedges, L., & Rhoads, C. (2010). *Statistical power analysis in education research (NCSER 2010–3006)*. National Center for Special Education Research, Institute of Education Sciences, U.S. Department of Education.
Kirk, R. E. (2012). *Experimental design: Procedures for the behavioral sciences* (4th ed.). SAGE.
Kish, L. (1965). *Survey sampling*. John Wiley.
Konstantopoulos, S. (2008a). The power of the test for treatment effects in three-level cluster randomized designs. *Journal of Research on Educational Effectiveness, 1*, 66–88.
Konstantopoulos, S. (2008b). The power of the test for treatment effects in three-level block randomized designs. *Journal of Research on Educational Effectiveness, 1*, 265–288.
Konstantopoulos, S. (2009). Incorporating cost in power analysis for three-level cluster randomized designs. *Evaluation Review, 33*, 335–357.
Konstantopoulos, S. (2011a). How consistent are class size effects? *Evaluation Review, 35*, 71–92.
Konstantopoulos, S. (2011b). Optimal sampling of units in three-level cluster randomized designs: An ANCOVA framework. *Educational and Psychological Measurement, 71*, 798–813.
Konstantopoulos, S. (2013). Optimal design in three-level block randomized designs with two levels of nesting: An ANOVA framework with random effects. *Educational and Psychological Measurement, 73*(5), 784–802.
Lipsey, M. W. (1990). *Design sensitivity: Statistical power analysis for experimental research*. Sage Publications.
Liu, X. (2003). Statistical power and optimum sample allocation ratio for treatment and control having unequal costs per unit of randomization. *Journal of Educational and Behavioral Statistics, 28*, 231–248.
Lohr, S. L. (2022). *Sampling: Design and analysis* (3rd ed.). Chapman and Hall/CRC.
Marcoulides, G. A. (1993). Maximizing power in generalizability studies under budget constraints. *Journal of Educational Statistics, 18*, 197–206.
Marcoulides, G. A. (1997). Optimizing measurement designs with budget constraints: The variable cost case. *Educational and Psychological Measurement, 57*, 800–812.
Mayr, S., Erdfelder, E., Buchner, A., & Faul, F. (2007). A short tutorial of GPower. *Tutorials in Quantitative Methods for Psychology, 3*(2), 51–59.
McClelland, G. H. (1997). Optimal design in psychological research. *Psychological Methods, 2*, 3–19.
Murphy, K. R., & Myors, B. (2004). *Statistical power analysis: A simple and general model for traditional and modern hypothesis tests* (2nd ed.). Lawrence Erlbaum.
Murray, D. M. (1998). *Design and analysis of group-randomized trials*. Oxford University Press.
Murray, D. M., Varnell, S. P., & Blitstein, J. L. (2004). Design and analysis of group-randomized trials: A review of recent methodological developments. *American Journal of Public Health, 94*, 423–432.
Nye, B., Hedges, L. V., & Konstantopoulos, S. (2000). Effects of small classes on academic achievement: The results of the Tennessee class size experiment. *American Educational Research Journal, 37*, 123–151.
Raudenbush, S. W. (1997). Statistical analysis and optimal design for cluster randomized trails. *Psychological Methods, 2*, 173–185.
Raudenbush, S. W., & Bryk, A. S. (2002). *Hierarchical linear models* (2nd ed.). Sage Publications.
Raudenbush, S. W., & Liu, X. (2000). Statistical power and optimal design for multisite randomized trails. *Psychological Methods, 5*, 199–213.
Spybrook, J., Raudenbush, S. W., Liu, X. F., Congdon, R., & Martınez, A. (2011). Optimal design for longitudinal and multilevel research: Documentation for the Optimal Design software version 3.

Part IV
Longitudinal and Cross-Sectional Dependent Categorical Data Analysis and Discrete Sequence Analysis

Chapter 17
Exploring Temporal Pattern of Intergenerational Educational Mobility in Germany: An Application of Configural Frequency Analysis Using Weighted Prediction

Jörg-Henrik Heine, Florian G. Hartmann, and Christian Tarnai

17.1 Introduction

This contribution deals with intergenerational educational mobility and the concept of educational expansion in Germany. It aims at examining configurations of formal educational qualifications from family triads (child, mother, and father) in Germany. In our contribution, this overarching objective is substantiated on the one hand in a methodical research question and on the other hand in content-motivated research questions.

At the level of the content-related research question, we are investigating the extent to which typical and atypical patterns of formal school-leaving qualifications within family triads (child, mother, father) in Germany have developed in the longitudinal course over birth cohorts ranging from the beginning to the end of the previous century. Based on joint and separate analyses of the East and West parts of reunited Germany, we will trace the similarities and differences in the educational biographies.

These authors "Florian G. Hartmann and Christian Tarnai" contributed equally to this work.

J.-H. Heine (✉)
Ludwig-Maximilians-Universität & Gesellschaft für Weiterbildung und sozialwissenschaftliche Forschung (GWSF) e.V., München, Bavaria, Germany
e-mail: joerg.heine@gwsf-muenchen.de

F. G. Hartmann
Department of Psychology, Paris Lodron University of Salzburg, Salzburg, Austria
e-mail: florian.hartmann@plus.ac.at

C. Tarnai
Department of Human Sciences, Universität der Bundeswehr, München, Bavaria, Germany
e-mail: christian.tarnai@unibw.de

M. Stemmler et al. (eds.), *Dependent Data in Social Sciences Research*,
https://doi.org/10.1007/978-3-031-56318-8_17

For our analysis, we use data from a general, representative population survey (see sections "The German General Social Survey and Sample, Data, and Weighting"). During the periodically recurring collection of this data, adjustments are made regarding the population ratios for the respective sample collected via sample weights, which are provided in the data. Although the use of such survey weights in classic variable-centered, in contrast to person-centered, evaluation methods (see e.g. Heine & Stemmler, 2021a) can be regarded as largely established, there is still no widely accepted gold standard for dealing with survey weights in the person-centered analysis principle used in our analyses. For our methodically oriented research question, we take up this existing gap and first show how, in analogy to the variable-centered data analysis perspective, the use of survey weights can also be implemented in person-centered data analysis within the framework of log-linear modeling. On this basis, we perform our analyses comparatively with and without survey weights in order to investigate their effects on statistical inference in person-centered analyses on the population based on representative samples.

To this end, the remainder of this chapter is organized as follows. First, in the following two sections, we provide a substantive theoretical background on the concepts of educational mobility and expansion in Germany. Within these substance-scientific theoretical sections, we also introduce the similarities and differences between the educational practice in West and East Germany. We then present the basic concept of the ALLBUS survey (e.g. GESIS - Leibniz-Institut für Sozialwissenschaften, 2021; Baumann & Thiesen, 2021; Bandilla et al., 1992; Braun et al., 1993), a data source that is particularly suitable for dealing with the present research questions. In the method part of this work, we present the basic principles of configural frequency analysis (CFA) as a person-centered method of data analysis. One focus here is a specific CFA model that is suitable for processing our longitudinal research question. In addition, we present the principle of implementing survey weights for the CFA in the *R* package `confreq` (cf. Heine et al., 2022). Finally, we discuss our findings in the subsequent results section from the perspectives of the content and methodological issues.

Educational Mobility

Institutional education systems in modern societies are typically expected to ensure that access to higher education does not depend on social background or specifically on the educational level of parents. The term "educational mobility" subsumes the concept of a sociographic measure of the permeability of an educational system, assuming a social structure stratified by educational levels (see, e.g., the social theory by Bourdieu, 1986). For Bourdieu (1986), the family is an essential factor in the acquisition of cultural capital, for which he describes the duration of educational acquisition as the least inaccurate measure. Moreover, Blau et al. (1956) outlined a schematic of the career choice and selection process that identifies the historical, sociological, (socio)psychological, and (socio)economic determinants, as well as the individual and environmental conditions, that are important for the educational and vocational development of individuals (cf. Tarnai & Hartmann, 2017). The influence of multiple factors manifests itself in the family as an instance of socialization. Through the parents of a generation, the factors that play a role in the choice of school career of their children become effective. The analysis of educational mobility based on the schooling and occupational position of parents and their children focuses on an important part of the influencing factors. The scheme of occupational choice according to Blau et al. (1956) can also be seen as a stage in the operationalization of theories that seek to explain, within a larger framework, the effects of social conditions on the individual (e.g., Bourdieu's social theory; see Bourdieu, 1986). The family, as the focal point of predominantly social influences, also plays a central role here.

Even though in modern western societies equal opportunities in institutional education systems such as the school system are typically viewed as a desirable goal, it has been shown for Germany that children's educational opportunities depend crucially on their family background, especially the educational level of their parents (e.g., Hillmert, 2012). Against the background of such family structures as a framing context, as outlined above, the extent of social mobility within a society is therefore seen as a difference in social achievement depending on social origin (e.g., Boudon, 1974). The observation of family educational configurations, for example, the proportion of persons with a higher formal school-leaving qualification in a population according to the different educational levels of their parents, can serve as an indicator of the permeability of an education system. Typically, an education system is considered to be equal opportunity when there is a high degree of permeability or mobility (cf. Boudon, 1974; Hillmert, 2012).

Hence, the duration of educational acquisition or the educational attainment of persons of one generation compared to their parents is an appropriate measure for the determination of intergenerational educational mobility. The more independent the educational acquisition is from the parents' education, the higher is the educational mobility. Intragenerational educational mobility can be determined by repeatedly recording the level of education attained and the occupational position. Occupational position, which in the German education system depends to a large extent on the level of education, is important for individuals and their parents. In the case of parents, occupational position plays a facilitating or inhibiting role in the educational attainment of their children, for whom the same applies. By taking into account occupational position, the notion of educational mobility is broadened to the extent that educational attainment can be modified by experience, usually occupational.

For families and their children, the society, the educational system, and its development in particular, in which the acquisition of education and professional position has taken place or is taking place, is of special importance. For Germany, the division into the Federal Republic and the German Democratic Republic (GDR) must be taken into account. The most important differences and developments of the two educational systems are briefly characterized below.

Educational Expansion in Germany

The term "educational expansion" refers to the increasing expansion of the education system in recent decades since around the 1950s. In Germany, this trend is characterized by the expansion of secondary schools for intermediate (e.g. "Realschulen" and "Gesamtschulen") and especially higher secondary education ("Gymnasien"). As a result, more and more young people are attending higher secondary education institutions and acquiring intermediate or higher educational qualifications, thus staying in the education system for an increasingly extended time.

The educational systems of the divided Germany in West and East, initially based on their common origin, have been characterized by different developments and changes since 1945 until the reunification. What both education systems have in common is that they each allow for an increase in the general level of education in the sense of an increase in intermediate and higher educational qualifications. While there is a tripartite school system in various state-specific forms in the West and in the present reunified Germany, a de facto two-tier school system prevailed in the East during the time of division, in which an intermediate level of education was the rule.

Geißler (2014) describes the development of changing formal school-leaving qualifications during the educational expansion in West and East Germany based on official government sources such as Kohler and Schreier (1990), the Statistical Yearbook of the GDR (Staaliche Zentralverwaltung für Statistik, 1989) and the Federal Statistical Office (Statistisches Bundesamt, 2012). According to this overview (see Geißler, 2014, p. 337), the proportion of adolescents in West Germany who left the school system without a secondary school-leaving certificate ("Hauptschulabschluss") decreases monotonically from 17% in 1960 to 8% in 2000, while the share of adolescents with upper secondary education ("Fachhochschulreife" or "Allgemeine Hochschulreife") increases monotonically, in sum from 6% in 1960 to 38% in the year 2000. The figures in Geißler (2014, p.335, and following) reflect that there was an expansive development in the Western part of Germany. In the East part of Germany, the effects of the restrictive policies of the 1970s can be observed in the higher school-leaving qualifications. Following reunification, the trend then converges with West German conditions. The share of higher school graduations rises from 8% in 1960 to 13% in 1970, then drops to 11% in 1980, and finally rises to 34% in 2000. The share of the same age group without a lower secondary school-leaving certificate or without a POS certificate (Polytechnische Oberschule) drops in the East from 18% in 1970 to 8% in 1992, but is back at 12% in 2000.

By comparing the distribution of types of school-leaving qualifications between males and females who have completed school, it is possible to determine whether the gender-specific disadvantage of females has decreased. Geißler (2014) concludes in light of the body of research (Geißler, 1990, 2013; Rodax & Hurrelmann, 1986; Hadjar, 2011; Hille et al., 1990; Statistisches Bundesamt, 2012) that girls initially caught up with boys over time on the basis of their historically better performance in school and subsequently slightly overtook them. With reference to access to the "Erweiterte Oberschule" (EOS) in the GDR, this reversal of the gender distribution of educational opportunities began in accordance with the politically intended goal of equal educational opportunities as early as the 1960s and with respect to the "Gymnasium" in West Germany somewhat later, at the beginning of the 1980s (Helbig, 2012).

The German General Social Survey

The German General Social Survey (ALLBUS) is a trend study that has been collecting data on opinions and attitudes in the Federal Republic of Germany every 2 years since 1980. The ALLBUS pursues the goal of mapping social changes in German society and covers a wide range of topics. Central themes such as political attitudes and political participation (1988, 1998, 2008, 2018), religion and worldview (1982, 1992, 2002, 2012), and social inequality (1984, 1994, 2004, 2014) are repeatedly covered about every 10 years (Siegers et al., 2019). Some variables, such as education, educational background, and indices and aspects of current occupation, are recorded as standard in all survey waves, which makes the ALLBUS data set a well-suited data basis for the present work. Even more, the ALLBUS data is easily publicly accessible and available in German and English at https://www.gesis.org/allbus/allbus. In addition to the data sets of the single survey times, the Leibniz Institute for the Social Sciences (GESIS) also releases cumulative data sets that provide the data from several survey times in a homogenized form (ALLBUS Cumulation 1980-2018, GESIS - Leibniz-Institut für Sozialwissenschaften, 2021).

Up to and including 1990, the population from which the ALLBUS samples were drawn included all persons who lived in private households in West Germany (including West Berlin) and who were entitled to vote. In 1991, in the course of the reunification of Germany, an extra survey was carried out (e.g. Bandilla et al., 1992). Since then, the samples have been drawn from the adult resident population, i.e., all Germans and foreigners residing in private households in the territory of the Federal Republic of Germany at the time of sampling who have reached the age of 18.

By 1990, 3000 people had been interviewed; in 1991, there were 1500 in West Germany and 1500 in East Germany (Bandilla et al., 1992). Since 1992, the goal has been to survey 2400 people in West Germany and about 1100 people in East Germany (cf. Braun et al., 1993). In turn, respondents in East Germany have been included in the sample in a larger proportion than would correspond to their share of the population[1] (oversampling). This oversampling is intended to allow statistically justifiable analyses even for smaller population groups in East Germany (cf. Baumann & Thiesen, 2021). In addition, in the sampling procedure used in

[1] In 2018, for example, 15.8% of the German population lived in East Germany (Statistisches Bundesamt, 2021).

ALLBUS, the sampling probabilities for each individual from the population of Germany are not equal, as it would be the case with a simple random sample. The three-stage sampling design (ADM-Design, Koch, 1997) used in ALLBUS aims at a so-called self-weighted sample only at the household level, in which all private households should have equal chances of being sampled (Bandilla et al., 1992; Braun et al., 1993). This in turn implies that the proportions of respondents for individual characteristic variables (person level) in the sample for Germany as a whole can therefore not be regarded as an estimate of the proportions in the population. Rather, the sampling probability of an individual person depends on the size of the household in which they live (Braun et al., 1993; Bandilla et al., 1992). The overrepresentation of East Germany and the different probability of drawing an individual respondent must be taken into account in the analyses of personal characteristics through appropriate weighting.

The so-called transformation weighting (e.g Sand & Kunz, 2020) is introduced in the ALLBUS for the sampling probability of persons in households. The transformation weighting proportionalizes the sample and transforms it from a household sample to a person sample (Baumann & Thiesen, 2021).

For East-West comparative analyses for all of Germany, a weighting that compensates for the disproportionality of the two sub-samples must be used (see Braun et al., 1993; Gabler, 1994). The purpose of this weighting is to undo the oversampling of respondents from the eastern part of Germany so that the cross-tabulations can be interpreted and compared directly for Germany as a whole (Baumann & Thiesen, 2021).

Overall, the German General Social Survey (ALLBUS) has been a central data source for examining social change in Germany since 1980. The ALLBUS has set methodological standards for conducting representative surveys in Germany (Siegers et al., 2019). Through the regular repetition of thematic focus modules such as education, educational background, and indices and aspects of current occupation, changes in the attitudes and education-related behavior of the German population can be ideally mapped and examined. This content-related orientation and the methodological aspects of data weighting make the ALLBUS data, in addition to their easy availability, ideal as a data basis for processing our research questions, which are methodically motivated on the one hand and content-related on the other hand.

17.2 Methodological Considerations

Sample, Data, and Weighting

As already introduced in the previous section, the basis for our analyses is the cumulative data set (ALLBUS cumulation 1980-2018, GESIS - Leibniz-Institut für Sozialwissenschaften, 2021) of the German General Social Survey.

In order to be able to perform comparative analysis for the East and West part of Germany, we restrict our analysis sample to the period of assessment since the German unification. This survey period includes data from a total of 15 survey waves, which were collected in a 2-year cycle from 1991 to 2018. In each survey wave, between $n = 2820$ (2002) and $n = 3804$ (2000), persons were interviewed in households. The total (cumulative) database selected for our analyses thus comprises $n = 50{,}013$ persons, of whom $n = 24{,}548$ are male and $n = 25{,}465$ female. This sample includes birth cohorts ranging from 1898 (one interviewee) to 1999 (46 interviewees), with an average of 494 interviewees in 1 year of birth and the maximum being 1123 people (born in 1960 and 1962) per year of birth.

In the ALLBUS data, specific weighting variables exist to compensate for the overrepresentation of the East German sub-sample on the one hand and to compensate for varying sampling probabilities at the person level on the other hand (see Table 1 in Baumann & Thiesen, 2021, p. iv, for a comprehensive summary). Even though both weights are named differently in the ALLBUS terminology (disproportionality weight and transformation weight), there are no fundamental functional differences. Both weights are design weights (Dorofeev & Grant, 2006), which are statistically substantiated as the inverse of the inclusion probability (e.g., Gabler et al., 2015) and compensate for the specificities of the ALLBUS sample design at different levels of sampling (see also Faulbaum, 2019, p. 417). The transformation weights are a (special) part of the design weighting (Sand & Kunz, 2020), which transforms the sample drawn at household levels to the person level (Baumann & Thiesen, 2021).

From a practical perspective, both weightings are taken into account in the compound weighting variable `wghtptew` (see Table 1 in Baumann & Thiesen, 2021, p. iv) available in the data. In our analyses, we examine the effects of weighting by taking into account both weightings summarized in the variable `wghtptew` on the one hand and disregarding the design weighting of the sample on the other hand.

It should be noted that the design weights used in ALLBUS at both levels, such as East-West and household, are normalized to the effective sample size, of the respondents (cf. Gabler et al., 2015). Specifically, for the calculation of the (household) transformation weight, the respective household size (number of persons entitled to the sample living in the household) is divided by the corresponding mean value, in each case separately for East and West, so that after weighting the number of cases (sum of weights) in the weighted and unweighted *sample* remains the same (Baumann and Thiesen, 2021, page viii). Similarly, the weights for the East-West proportionality weighting are normalized so that the size (sum of the weights) of the German total sample does not change in its size.

The ALLBUS weighting variables therefore typically have a narrow numerical range of values; specifically, the combined weighting variable `wghtptew`, which combines the household transformation weight and the East-West proportionality weight, ranges numerically between 0 and 5 with a mean of $M_{wghtptew} = 1$ and a standard deviation of $SD_{wghtptew} = 0.38$. This is due to the fact that these weights in ALLBUS do not contain a so-called expansion factor (cf. Dorofeev & Grant, 2006, p. 46), which would transform the weights in such a way that their sum (across the sampled units) reflects the population size. Such *population weights* combining design weighting and an expansion factors merged into one weighting variable are used, in contrast to the ALLBUS, in other large-scale assessments, as, for example, in PISA (e.g., OECD et al., 2002; OECD, 2009). As a result, using or ignoring the ALLBUS design weighting variables possibly might lead to only minor numerical differences between the weighted and unweighted pattern frequencies as compared to survey data sets with such combined population-weighting variables.

Variables for Pattern of Educational Attainment and Covariates

The basis for the analysis of family patterns of formal educational attainment is three variables on general school completion in the ALLBUS data set. These refer to the respondent himself (`educ`) and his two parents—the mother (`meduc`) and the father (`feduc`). Across all survey waves, formal school completion is captured in these variables with seven levels. These include five, ordinal ascending forms of school-leaving qualification—*without school-leaving qualification* ["ohne Schulabschluss"] (1); *elementary or lower secondary school* ["Volksschule oder Hauptschule"] (2); *lower secondary, with access to upper secondary* ["Mittlere Reife"] (3); *upper secondary education entrance qualification to a German University of applied sciences* ["Fachhochschulreife"] (4); *upper secondary academic education* ["Allgemeine Hochschulreife"] (5); and the two categories *other school-leaving qualification* (6) and *still student* (7).

For our analyses of family patterns of formal school-leaving qualifications, the seven levels of variables were, in a first step, each transformed into a three-level category system. In the first step, both categories 6 and 7 were interpreted as missing information on one of the five other (ordinal ascending) school-leaving qualifications and coded accordingly. Levels 1 and 2 were merged into the category "low" educational attainment, level 3 into the category "medium" educational attainment, and finally levels 4 and 5 into the category "high" educational attainment. To reduce the number of possible configurations for the family triads without losing any meaningful information, we combine the three levels of education from the mother and father into a single categorical variable (labeled with `MxF`). This variable combines ordinal information on the school-leaving qualifications of both parents in five nominal categories. Two categories code whether the mother's or father's formal school-leaving qualification is higher than that of the other parent. In three further categories, if the parents have the same level of formal schooling, it is coded whether both school-leaving qualifications can be assigned to the category "high," "medium," or "low."

In our analyses, alongside the three variables representing the family triads of formal school-leaving qualifications, we also utilize additional variables from the ALLBUS dataset. First, we use the variable `yborn`, which is the birth year of the respondent. To (again) reduce the number of possible patterns for our analyses, we group the individual birth cohorts in the variable `yborn` into a total of nine ordinal ascending categories. The classification is based on theoretical considerations about historically salient events for the first three categories. The first category (1898–1913) comprises the birth cohorts before World War I. The second category (1914–1938) comprises the birth cohorts before World War II, and the third category (1939–1944) comprises the birth cohorts which were born during World War II up to its end. The remaining categories were formed in such a way that, on the one hand, the entire decades are represented as far as possible and, on the other hand, largely equal numbers of cases are retained in each category. The latter was achieved for all categories except for the category containing the birth cohorts of the baby boomer generation (1961–1970). The resulting derived categorical variable `YBORN` is used in the prediction CFA as a criterion variable in our analyses to examine how the pattern of family triads of formal school-leaving qualifications changes over the birth cohorts as an expression of educational expansion and the presumed change in educational mobility.

Furthermore, we draw on two groups of variables that on the one hand encode the occupational prestige of the currently held occupations of the respondent's parents and on the other hand encode an index of socioeconomic status for the mother and father of the respondent (see Ganzeboom et al., 1992; Ganzeboom & Treiman, 1996). The first group of variables comprises `msiops88` (mother) and `fsiops88` (father), and relate to the Standard International Occupational Prestige Scale (SIOPS, Ganzeboom & Treiman, 1996) for the parents' respective occupations, which is based on the ISCO-88 occupational classification (see International Labour Office, 1990). The second group of variables (`misei88`, mother and `fisei88`, father) relate to the International Socio-Economic Index (ISEI, Ganzeboom et al., 1992).

In our analyses, we use these four variables, as well as the year of assessment (`year`), as additional covariates in the CFA prediction model presented in the following section.

Configuration Frequency Analysis and the Prediction Model

Configuration frequency analysis (CFA) is a method for multivariate analysis of categorical data, which are typically conceived as multidimensional contingency tables (e.g., Stemmler, 2020; Agresti, 2007, 2015). The CFA was originally developed by Lienert (Lienert, 1971; Krauth & Lienert, 1973) for the analysis of clinical patient data to identify (new) significant syndromes defined as over-random frequent or rare configurations of single symptoms. Thus, the frequencies of such categorical feature combinations of specific groups of people in the available data are analyzed. Lienert himself named such configurations with significantly high frequencies ($f_{obseved} > f_{expected}$) as types and configurations with significantly low frequencies ($f_{obseved} < f_{expected}$) as antitypes (Lienert, 1971). The CFA is also referred to as a person-centered method (e.g., Stemmler, 2020; Heine & Stemmler, 2021a), which, in contrast to the widespread variable-centered analysis methods, does not restrict to the analysis of linear relationships between a selection of single variables (Heine & Stemmler, 2021a). Since its development by Lienert, numerous extensions and variants of the CFA have been developed and proposed (e.g., von Eye & Wiedermann, 2021a), which nevertheless share a common basic principle. This overarching principle of CFA is that its search for types and antitypes represents significant local deviations from a specific null hypothesis. From the perspective of fitting a log-linear model to formulate any CFA model, the presence of types and antitypes implies a (locally) non-fitting model. Thus, CFA is a kind of residual analysis that sheds light on the details of a non-fitting model, which can also help explain nonlinear associations between the variables considered. Within the framework of the log-linear modeling to estimate the expected frequencies for a specific CFA model, different null hypotheses and thus models can be implemented. A basic CFA model is the so-called main effect model or independence model. In this model, the independence of each of the variables involved is assumed, and only the main effects are specified within the framework of its log-linear formulation. In the main effect model, the expected values of the single patterns from combinations of characteristics correspond to the product of the marginal terms of the multidimensional contingency table standardized on the sample size. Another alternative model of CFA is referred to as prediction CFA (P-CFA, cf. von Eye, 2002; von Eye & Wiedermann, 2021a; Stemmler et al., 2008), in which the predictor variables are assumed to be independent of the criterion variable (cf. Stemmler, 2020). In the log-linear formulation, the particular model that represents the underlying null hypothesis of the independence of the criterion variables on the total set of predictor variables is saturated in the set of predictor variables. In the following, we will present the log-linear specification of the two CFA models based on our analysis variables.

In the framework of the generalized linear model (GLM, Nelder & Wedderburn, 1972), a CFA model can be written as a log-linear model assuming a Poisson distribution $O \sim \mathcal{P}(E)$ as $ln(E) = \boldsymbol{\lambda} X = \lambda_0 + \lambda_1^{X_1} + \cdots + \lambda_k^{X_k}$ (cf. e.g., Kawasaki & Zimmermann, 1964; Nelder, 1974), where O and E are the observed and the expected cell counts of the contingency table, respectively, X is the model design matrix, $\boldsymbol{\lambda}$ is the vector of their model parameters, and λ_i are the coefficients. Considering the three variables, gender of the respondent (`SEX`), his or her formal school-leaving qualification (`EDUC`), and the composite variable on the formal school-leaving qualifications of the parents (`MxF`), the CFA base model of independence would look like given in Eq. 17.1:

$$ln(E^{S,E,MxF}) = \lambda_0 + \lambda_1^S + \lambda_2^E + \lambda_3^{MxF}, \tag{17.1}$$

where $ln(E^{S,E,MxF})$ is the natural logarithm of the expected frequencies for the respective configurations built from the three variables, λ_0 is the intercept, and $\lambda_{1,\dots,k}$ are the coefficients of the variables `SEX`, `EDUC`, and `MxF`, respectively. Note that here the superscripts (*S*, *E*, *MxF*) denote the variables involved in an effect and are not exponents. The general assumption implied in such a main effect CFA model is the independency of all variables included in the model.

For the prediction CFA model, we now want to introduce the criteria "year of birth" (YBORN) to be predicted by the configuring variables of formal school-leaving qualifications. For these predicting variables we, therefore, have to consider—next to their main effects—all second- and third-order interaction effects, such that the model is saturated within the set of predictor variables as given in Eq. 17.2:

$$ln(E^{E,M,F}) = \lambda_0 + \lambda_1^Y + \lambda_2^S + \lambda_3^E + \lambda_4^{MxF} + \lambda_{2,3}^{S,E} + \lambda_{3,4}^{E,MxF} + \lambda_{2,4}^{S,MxF} + \lambda_{2,3,4}^{S,E,MxF}, \quad (17.2)$$

Conceptually, one could say that the saturated part of the model equation combines all predictor variables into one block, so that in a sense only this predictor block and the criterion (YBORN) remain as two components of a (higher-level) independence model. The null hypothesis then consists in the assumption of independence of these two model components, whereby any types or antitypes in local model testing would reject the hypothesis of independence and thus indicate any dependence (nonlinear or linear) between the variables involved.

The two CFA models presented so far only consider categorical variables that form the basis for the configurations, in our case the family triads for formal school qualifications. Another variant of a CFA model is the CFA covariate model (e.g., Glück & von Eye, 2000; von Eye & Wiedermann, 2021b). In such a CFA model, appropriate continuous covariates, which are assumed to be possibly related to the categorical variables defining the pattern, can contribute to clarifying the "cause" of the (initially) found as types and antitypes (see Heine & Stemmler, 2021a). The expected frequencies of the single configurations are predicted more precisely by substantial covariates, so that the difference between the observed and expected frequency is reduced and types or antitypes no longer occur (cf. Glück & von Eye, 2000; Heine & Stemmler, 2021b, for practical applications).

In order to introduce covariates into the CFA model, a term for the covariates in the model equation is added as part of the formulation as a log-linear model. Based on the design matrix X, which is represented by the right part of the model equation 17.2 of the prediction CFA model, this model including covariates can be written as given in Eq. 17.3:

$$ln(E^{E,M,F\|C}) = \boldsymbol{\lambda} X + \boldsymbol{\lambda}_C C, \quad (17.3)$$

where X is the design matrix with the vectors for the effect-coded contrasts of the main and higher-order effects of the prediction model, $\boldsymbol{\lambda}$ the vector of their coefficients, and C the matrix of the covariates (`msiops88`, `fsiops88`, `fisei88`, `misei88`, and `year`) and $\boldsymbol{\lambda}_c$ is the vector of the coefficients of those covariates.

The four variables derived from parents' occupation are available only for the ALLBUS waves from the year 2000 onward, with a missing gap for the 2002 wave for the variables related to the mother of the respondent. We thus coded the respective entries in the C matrix of the covariates with zero, which corresponds to a zero effect of the covariate for the corresponding feature combinations (pattern) from the respective survey waves.

Applying the presented CFA models to data that account for subgroup disproportionalities or varying drawing probabilities for the sampling units via survey weights raises questions about their proper implementation in the model. In the following section, we will therefore present the basic principle of implementing the survey weights in the log-linear formulation of an CFA model.

Introducing Survey Weighting into CFA

The use of survey weights in variable-centered analyses can be considered as a widely established practice. The arrangement or form of the underlying data matrices is typically referred to as "wide-form" data (e.g., Heine and Stemmler, 2021a). The term "wide-form" here means that there is a row for each case and a column for each variable (at a specific time of measurement) in the data matrix (see also Young, 1984; Jacoby, 1991; Carroll & Arabie, 1980, for the classification of data structures).

To realize survey weighting in any wide-form data set basically means to introduce a column vector of case weights $\boldsymbol{\omega}_i$ as an "additional variable" to indicate the varying contribution of a respondent to the total sample (see Dorofeev & Grant, 2006, p. 45). In variable-centered analyses, the weights would be used as a multiplying factor for each case when calculating any summary statistics or estimate parameters for any statistical population model based on the weighted sample.

For person-centered methods such as CFA, the underlying data matrices must have a data format which is different from the "wide-form" data used in variable-centered analysis. This is true at least for the internal processing in corresponding software for the calculation of the respective models. Since CFA analyzes multi-dimensional contingency tables, data that are available in the typical "wide format" must first be converted into a tabulated format. The basic principle is to aggregate the wide-format data by all theoretically possible pattern that can arise from the combinations of categories of the categorical variables selected for analysis. In simple terms, we count the number of times each pattern occurs in the data. It is important to note that combinations of characteristics (pattern) that do not occur in the empirical data are nevertheless included in the tabulation result with a frequency of $f_{pat.} = 0$. In turn, the form or the dimensions of the tabulated data matrix have a fixed number of rows and correspond to the design matrix for a CFA model in the context of its log-linear modeling (see, e.g., von Eye, 1988). This number of rows corresponds to the number of possible patterns (configurations) and equals the product of the number of categories of the variables involved in the analysis (see, e.g., Heine & Stemmler, 2021b). For the main effect model comprising three variables for the characteristics gender and formal school qualification of the respondents as well as the combined school qualification of both parents (`SEX`, `EDUC`, and `MxF`) with two, three, and five categories, respectively, there are $2 \times 3 \times 5 = 30$

pattern. For the prediction model, the number of possible pattern increases, since the variable YBORN comprising nine categories is added ($2 \times 3 \times 5 \times 9 = 270$ pattern).

As mentioned above, in variable-centered analysis, a column vector of survey weights ($\boldsymbol{\omega}_i$) is treated as an additional multiplicative supplement to any measure (x_i) when calculating any descriptive or inferential statistics. Given the specific form of the tabulated data to be analyzed in the CFA, the survey weights, which are given in the wide-form data as continuous variables for each individual case, must be aggregated in a meaningful way. Note that the same is true for any other continuous variables to be introduced into any (extended) CFA model, such as the CFA model with covariates. The case weights aggregated from the wide-form data then represent weights ($\boldsymbol{\omega}_p$) for all possible patterns (configurations) from the categorical variables. This vector of configuration weights $\boldsymbol{\omega}_p$ can then be inserted as an additional multiplicative component into the CFA model equation formulated as a log-linear model.

For example, since Eq. 17.2, as introduced in the previous section for the predictive CFA model, has a direct relationship to the tabulated data in the contingency table being analyzed, the equation for estimating the predictive CFA model given the survey weights is as follows:

$$
\begin{aligned}
ln(\omega E^{S,E,MxF}) &= \lambda_0\omega + \lambda_1\omega_p^Y + \lambda_2^S\omega_p + \lambda_3^E\omega_p + \lambda_4^{MxF}\omega_p \\
&\quad + \lambda_{2,3}^{S,E}\omega_p + \lambda_{3,4}^{E,MxF}\omega_p + \lambda_{2,4}^{S,MxF}\omega_p + \lambda_{2,3,4}^{S,E,MxF}\omega_p.
\end{aligned} \tag{17.4}
$$

Note that the vector of weights ($\boldsymbol{\omega}_p$) inserted in Eq. 17.4 is on a structural and functional level not identical with the vector of weights ($\boldsymbol{\omega}_i$) as used in variable-centered approaches. While the length of $\boldsymbol{\omega}_i$ corresponds to the sample size n, thus providing a case weight for each case, the length of $\boldsymbol{\omega}_p$, as inserted in Eq. 17.4, must correspond to the number of rows in the design matrix, which equals to the number of possible patterns from the categorical variables, which is also identical with the number of rows in the matrix of tabulated data.

Considering a CFA model that includes P possible pattern where $\boldsymbol{\psi}_p$ with $p = 1, 2, \ldots, P$ indicates the vector of pattern $\boldsymbol{\psi}$, we need accordingly a vector of weights $\boldsymbol{\omega}_p$ with $p = 1, 2, \ldots, P$. Thus, we changed the lower index from i to p to account for the fact that we are now talking about a new vector of summarized weights $\boldsymbol{\omega}_p$ comprising weights for *groups* of persons who share the same pattern of trait combinations from the categorical variables, aggregated from sections S_p of the vector $\boldsymbol{\omega}_i$ with *individual* case weights. As Eq. 17.5 shows, such aggregation is undertaken by calculating the mean of the weights from the respective sections S_p out of the vector ω_i comprising weights for every person:

$$
\boldsymbol{\omega}_p = \vec{\boldsymbol{\omega}}^{\psi_p}_{p=1;i\in s_p}, \vec{\boldsymbol{\omega}}^{\psi_p}_{p=2;i\in s_p}, \ldots, \vec{\boldsymbol{\omega}}^{\psi_p}_{p=P;i\in s_p} \tag{17.5}
$$

The resulting vector of weights ω_p serves then as a multiplying factor for initially unweighted pattern frequencies. The vector of weighted observed pattern frequencies can accordingly be written as given in Eq. 17.6:

$$\boldsymbol{f}_{\psi_p weighted} = \sum_{p=1;i\in s_p}^{\psi_p} \omega_i^{\psi_p}, \sum_{p=2;i\in s_p}^{\psi_p} \omega_i^{\psi_p}, \ldots, \sum_{p=P;i\in s_p}^{\psi_p} \omega_i^{\psi_p}, \tag{17.6}$$

The new vector $\boldsymbol{f}_{\psi_p weighted}$ of the weighted pattern frequencies results from a sequence of summed sections of individual weights from the vector $\boldsymbol{\omega}_i$ taken from the wide-form data. These single sections S_p correspond to the respective respondents who share the same pattern ψ_p with $p = 1, \ldots, P$ on the categorical variables, respectively. This summation of the weights is justified from the (simple) counting principle in the determination of pattern frequencies in the unweighted case, according to which the absolute pattern frequency is simply the sum of all individuals with one and the same pattern in each case. Formally, for example, for the first pattern ψ_1, the frequency f_{ψ_1} of all respondents i_{ψ_1} sharing the same pattern is given as 17.7

$$f_{\psi_1} = \sum^{\psi_1} \boldsymbol{i}^{\psi_1} \times \mathbf{1}, \tag{17.7}$$

$$f_{\psi_1 weighted} = \sum_{i\in S_p}^{\psi_1} \boldsymbol{i}^{\psi_1} \times \boldsymbol{\omega}_i^{\psi_1} = \sum_{i\in S_p}^{\psi_1} \boldsymbol{\omega}_i^{\psi_1} \tag{17.8}$$

While in the unweighted case each respondent counts the same (cf. the expression "×1" in 17.7), in the weighted case, each responder is instead summed to the proportion corresponding to his weight (see 17.8).

17.3 Summary of Research Questions and Methodological Procedures

The current contribution pursues two main objectives. The first objective deals with the question which significant pattern of formal school qualifications within family triads (child, mother, father) can be observed in Germany in the longitudinal course. This research objective includes questions about how the occurrence of educational family pattern depends on the respondents' gender and residential area (East or West Germany) and to what extent the pattern is related to covariates such as the year of the interview, parents' occupational prestige, or socioeconomic status (see introductory sections "Educational Mobility" and "Educational Expansion in Germany"). To address these research questions, we relate to a sub-sample of data (see section "Sample, Data, and Weighting") from the German General Social

Survey (ALLBUS) as introduced in section "The German General Social Survey". The three different CFA models presented in section "Configuration Frequency Analysis and the Prediction Model" are used in our analyses of the ALLBUS data. Specifically, these models are (1) the CFA main effect model including three categorical variables (see section "Variables for Pattern of Educational Attainment and Covariates"), (2) the CFA prediction model with the (additional) criterion variable birth cohort, and finally (3) the CFA prediction model with covariates, which mainly reflects the occupational prestige of the currently held occupations as well as socioeconomic status of the respondent's parents. The analysis of ALLBUS data potentially requires the consideration of two weights. A transformation (design) weight for analyses on person level (instead of household level), as well as a design weight that takes into account the overrepresentation of East Germans in the sample (see section "Sample, Data, and Weighting"). Since there has been no gold standard software for considering weights in person-centered analyses with CFA, the second objective of the present study is thus methodologically justified and dedicated to investigating the implementation and effects of sample weights in person-centered evaluation procedures. To do so, the analyses are performed comparatively with and without the use of the combined design weight provided in the data (see section "Sample, Data, and Weighting"), respectively. The analyses are performed with the package confreq (Heine et al., 2022) for the free statistical environment R (R Core Team, 2022). The complete R-syntax file and the analyzed data are available at the OSF repository (see https://osf.io/7ykzv; Heine et al., 2023).

17.4 Results

Overall, the consideration of weights in the data leads to different cell frequencies of the multidimensional contingency table as compared to unweighted data. Table 17.1 summarizes these effects on the cell populations based on the total sample including data for East and West Germany for the two types of CFA models (see also Tables 17.2 and 17.3) where the number of cells in the contingency table is different due to a different number of pattern-defining variables. As also defined in Eqs. 17.1 and 17.2, we have to consider for the main effect model in our analyses $2 \times 3 \times 5 = 30$ pattern and for the prediction model $2 \times 3 \times 5 \times 9 = 270$ pattern (see the second column in Table 17.1).

Columns three to eight in Table 17.1 give the maximum, the minimum, and the mean cell (pattern) frequencies for weighted and unweighted data across all patterns for both CFA models, respectively. The next column (number nine) to the right in Table 17.1 provides the mean across all cells of the absolute difference between the unweighted (f_p) and weighted ($\sum \omega_p$) cell frequencies. The latter is simply the respective sum of weights for each pattern occurring in the CFA model, respectively. The last double column of Table 17.1 refers to the difference in cell frequencies between weighted and unweighted data. The left side of this double column shows the mean absolute difference between weighted and unweighted data over all the

Table 17.1 Summary of cell (pattern) frequencies for two types of CFA models for weighted and unweighted data

		Unweighted			Weighted			$\|f_p - \sum \omega_p\|$	
CFA model	P	f^{e}_{max}	f^{e}_{min}	$M f \psi_p$	$max \sum \omega_p$	$min \sum \omega_p$	$M \sum \omega_p$	$M_{f\text{dif.}}$	$M_{\%\text{dif.}}$
Main effects	30	7973[a]	28[b]	1447.97	8211.14[a]	29.29[b]	1454.77	82.53	6.61
Prediction CFA	270	3291[c]	0[d]	160.70	3015.60[c]	0.00[d]	161.43	14.81	11.22

[a] Pattern (unweighted and weighted): "Female low low"
[b] Pattern (unweighted), "Male low high"; pattern (weighted), Female low high
[c] Pattern (unweighted and weighted): "1914–1938 female low low"
[d] 11 Pattern for birth cohort "1898–1913" and one pattern: "1945–1952 male low high" for unweighted and weighted data
[e] Highest frequency from the vector of pattern frequencies ($\max_{p=1,\ldots,P}\{f_{\psi_p}\}$), lowest frequency from the vector of pattern frequencies ($\min_{p=1,\ldots,P}\{f_{\psi_p}\}$)

patterns of the respective model. In the last column, this difference is given as a percentage deviation of the weighted frequencies from the unweighted frequencies.

What we basically can see from Table 17.1 is that the mean pattern (cell) frequencies for weighted ($M_{\sum \omega_p}$) and unweighted data (M_{f_p}) seem to differ not so much at first glance (compare columns five and eight in Table 17.1). However, the mean absolute difference between weighted and unweighted frequency data (see the ninth column) shows noteworthy deviations. Moreover, looking at the percentage for the difference of frequencies between the weighted and unweighted data, we can see that this percentage nearly doubles for the prediction CFA model frequency data, as compared to the frequency data underlying the main effect model.

Table 17.2 shows the results for the main effect model of the CFA on the patterns from the three categorical variables (`SEX`, `EDUC`, and `MxF`) based on the total sample including East and West Germany. The results refer to analyses on the one hand without and on the other hand with consideration of the survey weights. The two forms of analysis yield consistent results except for a single pattern or rather configuration of family triads for formal school qualifications (cf. tenth line in Table 17.2).

The main effect model tests the hypothesis of the independence of the three variables and the resulting types (+) and antitypes (−), thus indicating local deviations from this hypothesis. In this sense, it is shown that female and male respondents with high school-leaving qualifications from homes with high, middle, and different formal school-leaving qualifications of the parents can be observed significantly more often (types) than would be assumed under the null hypothesis. On the other hand, female and male interviewees with a high level of formal school-leaving qualifications from families with a low level of school-leaving qualifications are observed significantly less frequently (antitypes) than expected (cf. first and fourth group of rows in Table 17.2). Exactly the opposite inverse picture emerges, for both sexes equally, for respondents with low formal school qualifications (cf. third and last set of rows in Table 17.2). There are differences between the two sexes only for the group of respondents with an intermediate formal school-leaving

Table 17.2 Main effect CFA (East and West): Types and antitypes for pattern of educational mobility within family triads

		Unweighted		Weighted							
Self	Mother-father	Observed	Expected	Observed	Expected	Type	χ^2	*df*	p_{χ^2}	Z_{χ^2}	pz_{χ^2}
Female high	High	1012	360.630	1048.166	365.629	+	1274.127	1	0.000	35.695	0.000
Female high	M<F	1485	828.049	1563.826	870.773	+	551.604	1	0.000	23.486	0.000
Female high	M>F	656	394.110	660.092	400.120	+	168.912	1	0.000	12.997	0.000
Female high	Medium	723	484.448	697.230	436.796	+	155.280	1	0.000	12.461	0.000
Female high	Low	1889	4201.433	1932.098	4322.754	–	1322.129	1	0.000	−36.361	0.000
Female medium	High	181	413.278	164.913	376.460	–	118.877	1	0.000	−10.903	0.000
Female medium	M<F	1100	948.935	1057.318	896.568	+	28.821	1	0.000	5.369	0.000
Female medium	M>F	566	451.645	544.523	411.973	+	42.647	1	0.000	6.530	0.000
Female medium	Medium	801	555.171	678.763	449.735	+	116.632	1	0.000	10.800	0.000
Female medium	Low	4979	4814.793	4693.510	4450.808	+ .[a]	13.235	1	0.000	3.638	0.000
Female low	High	32	493.110	29.291	511.250	–	454.346	1	0.000	−21.315	0.000
Female low	M<F	351	1132.239	367.201	1217.581	–	593.920	1	0.000	−24.370	0.000
Female low	M>F	165	538.889	174.255	559.478	–	265.242	1	0.000	−16.286	0.000
Female low	Medium	111	662.413	102.757	610.761	–	422.535	1	0.000	−20.556	0.000
Female low	Low	7973	5744.858	8211.143	6044.398	+	776.716	1	0.000	27.870	0.000
Male high	High	1018	350.658	1022.065	362.177	+	1202.320	1	0.000	34.674	0.000
Male high	M<F	1637	805.152	1758.786	862.552	+	931.230	1	0.000	30.516	0.000
Male high	M>F	627	383.212	649.102	396.343	+	161.192	1	0.000	12.696	0.000

(continued)

Table 17.2 (continued)

Self	Mother-father	Unweighted		Weighted		Type	χ^2	*df*	p_{χ^2}	Z_{χ^2}	pz_{χ^2}
		Observed	Expected	Observed	Expected						
Male high	Medium	699	471.052	679.843	432.672	+	141.200	1	0.000	11.883	0.000
Male high	Low	2618	4085.256	2720.551	4281.942	–	569.354	1	0.000	−23.861	0.000
Male medium	High	228	401.850	200.205	372.906	–	79.981	1	0.000	−8.943	0.000
Male medium	M<F	809	922.695	792.842	888.104	–	10.218	1	0.001	−3.197	0.001
Male medium	M>F	518	439.157	485.462	408.083	+	14.672	1	0.000	3.830	0.000
Male medium	Medium	848	539.820	666.728	445.489	+	109.871	1	0.000	10.482	0.000
Male medium	Low	4139	4681.656	3824.650	4408.787	–	77.394	1	0.000	−8.797	0.000
Male low	High	28	479.475	30.204	506.423	–	447.817	1	0.000	−21.162	0.000
Male low	M<F	356	1100.930	401.691	1206.085	–	536.488	1	0.000	−23.162	0.000
Male low	M>F	199	523.988	216.760	554.196	–	205.456	1	0.000	−14.334	0.000
Male low	Medium	175	644.096	155.128	604.995	–	334.515	1	0.000	−18.290	0.000
Male low	Low	7516	5586.003	8114.069	5987.332	+	755.430	1	0.000	27.485	0.000

+ type, – antitype, . not significant at all

Inference based on the χ^2-approximation to the z-test, results for weighted data

Bonferroni adjusted alpha: $\alpha = 0.001666667$

[a] Not significant at all with unweighted data

certificate (see second and fifth group of rows in Table 17.2). While among female respondents a middle school-leaving certificate is associated significantly more frequently than expected (+) with a lower qualification of the mother compared to that of the father, this configuration is observed significantly less often (−) than expected for the male respondents. Furthermore, based on the weighted analysis, it can be seen that females with a medium level of education are significantly more likely than expected to be climbers from families with a low level of formal schooling, and males with the same medium level of education come from families with a low level of education significantly less than expected.

In Table 17.3, we show the results of the prediction CFA model to investigate the question of a temporal dependency of typical patterns of family educational triads on the birth year (birth cohort) of the respondent. The table compares the analysis results based on weighted and unweighted data in nine double columns for birth cohorts combined into nine decades. As an overarching finding, the results in Table 17.3 indicate that for both sexes, the pattern with high formal school qualifications in the family educational triads increase over time and can be observed significantly more frequently than expected.

In particular, among birth cohorts between 1945 and 1970, there is a significant increase in female and male educational ascenders with middle and high formal

Table 17.3 Prediction CFA: Types and antitypes for temporal pattern of educational mobility within family triads across nine decades of birth cohorts (East and West)

Pattern		1898–1913		1914–1938		1939–1944		1945–1952		1953–1960		1961–1970		1971–1980		1981–1990		1991–1999	
Self	Mother–Father	s[a]	w[b]	s[a]	w[b]	s[a]	w[b]	s[a]	w[b]	s[a]	w[b]	s[a]	w[b]	s[a]	w[b]	s[a]	w[b]	s[a]	w[b]
Female high	high	.	.	–	–	–	–	–	–	–	–	.	.	+	+	+	+	+	+
Female high	M<F	.	.	–	–	–	–	.	.	.	.	+	+	+	+	+	+	+	+
Female high	M>F	.	.	–	–	–	–	–	.	.	.	.	.	+	+	+	+	+	+
Female high	medium	.	.	–	–	–	–	–	–	–	–	.	.	+	+	+	+	+	+
Female high	low	.	.	–	–	–	–	.	.	+	+	+	+	.	+	.	.	.	.
Female medium	high	.	.	.	.	.	.	.	.	.	.	+	.	.	.	.	.	.	.
Female medium	M<F	.	.	–	–	.	.	.	.	.	.	.	.	.	.	.	.	.	.
Female medium	M>F	.	.	–	–	–	.	.	.	.	.	.	.	+	+	+	+	.	.
Female medium	medium	.	.	–	–	–	.	–	–	–	–	.	.	+	+	+	+	.	.
Female medium	low	–	.	–	–	–	–	+	+	+	+	+	+	.	.	–	–	–	–
Female low	high	.	.	.	.	.	.	.	.	.	.	.	.	.	.	.	.	.	.
Female low	M<F	.	.	+	+	+	+	.	.	.	.	.	.	.	.	.	.	.	.
Female low	M>F	.	.	.	.	.	.	.	.	.	.	.	.	.	.	.	.	.	.
Female low	medium	.	.	.	.	.	.	.	.	.	.	.	.	.	.	+	+	.	.
Female low	low	+	+	+	+	+	+	.	+	–	–	–	–	–	–	–	–	–	–
Male high	high	.	.	–	–	–	–	–	–	.	–	.	.	+	+	+	+	+	+
Male high	M<F	.	.	–	–	.	.	.	.	.	.	.	.	.	.	.	.	.	.
Male high	M>F	.	.	–	–	–	–	–	–	.	.	.	.	+	+	+	+	+	+
Male high	medium	.	.	–	–	–	–	.	.	–	.	.	.	+	+	+	+	+	+
Male high	low	–	.	–	–	.	.	+	+	+	+	.	+	–	.	–	–	–	–
Male medium	high	.	.	–	–	.	.	–	–	.	.	.	.	+	+	+	+	+	+
Male medium	M<F	.	.	–	–	.	.	.	.	.	.	+	+	+	+	.	.	.	.
Male medium	M>F	.	.	–	–	.	.	.	.	.	–	.	.	+	+	+	+	+	+
Male medium	medium	.	.	–	–	–	–	–	–	–	–	.	.	+	+	+	+	+	+
Male medium	low	–	–	–	–	–	–	+	+	+	+	+	+	–	.	–	–	–	–
Male low	high	.	.	.	.	.	.	.	.	.	.	.	.	+	+	.	.	.	.
Male low	M<F	.	.	.	.	.	.	.	.	.	.	.	.	.	.	.	.	.	.
Male low	M>F	.	.	.	.	.	.	.	.	.	.	.	.	.	.	+	.	.	.
Male low	medium	.	.	–	–	.	.	.	.	.	.	.	.	+	.	+	+	+	+
Male low	low	.	.	+	+	+	+	.	+	–	–	–	–	–	–	–	–	–	–

+ type, – antitype, . not significant at all
ALLBUS (1991–2018, East + West, $n = 50{,}013$)
Inference based on the χ^2-approximation to the z-test
Bonferroni adjusted alpha: $\alpha = 0.0001851852$
[a] Inference based on **unweighted** data [b] Inference based on **weighted** data

school degrees from families with little parental formal schooling (cf. the bottom rows in the first and second as well as the fourth and fifth row groups in Table 17.3), while such a pattern is observed much less frequently than expected for individuals with middle formal school degrees from families with low levels of education in the birth cohorts before and, interestingly, also after, for example, for the birth cohorts from the 1980s onward. Although these overall trends are largely parallel for both sexes, there are also differences between the sexes for certain configurations. For example, for the last three groups of birth cohorts, but especially for the years between 1991 and 1999, we find that males with an intermediate formal school-leaving certificate are significantly more likely than expected to come from either an intermediate or high family background or to have a mother who has a higher school degree than the father. The equivalent patterns for females are not shown to be significant types for the same period.

Given the findings from the analysis so far, based on the total sample (East and West together), that the influence of the survey weights is small for the main effect CFA (see Table 17.2) but quite substantial for the prediction CFA (see Table 17.3 and see also Table 17.5), we report the comparative prediction CFA for East and West separately in double columns (see Table 17.4) based on the weighted data only.

Table 17.4 Prediction CFA: Types and antitypes for temporal pattern of educational mobility within family triads across nine decades of birth cohorts (East vs. West)

Pattern		1898–1913		1914–1938		1939–1944		1945–1952		1953–1960		1961–1970		1971–1980		1981–1990		1991–1999	
Self	Mother–Father	East[a]	West[b]	East[a]	West[b]	East[a]	West[b]	East[a]	West[b]	East[a]	West[b]	East[a]	West[b]	East[a]	West[b]	East[a]	West[b]	East[a]	West[b]
Female high	high	.	.	–	–	.	–	.	–	.	.	.	.	+	+	+	+	+	+
Female high	M<F	.	.	.	–	.	–	.	.	.	.	.	+	.	+	.	+	.	+
Female high	M>F	.	.	–	–	.	–	.	.	.	.	.	.	+	+	+	+	+	+
Female high	medium	.	.	–	–	.	–	.	.	.	–	.	.	+	+	+	+	+	+
Female high	low	.	.	.	–	.	–	+	–	+	+	.	+	–	+	–	.	.	.
Female medium	high	.	.	.	.	.	.	.	.	.	.	+	.	.	.	.	.	.	.
Female medium	M<F	.	.	–	.	.	.	.	.	.	.	+	.	.	.	.	.	.	.
Female medium	M>F	.	.	–	–	.	.	.	.	.	.	+	.	+	.	.	+	.	.
Female medium	medium	.	.	–	.	–	.	–	.	.	–	.	.	+	+	+	+	.	.
Female medium	low	.	.	–	–	.	.	+	.	+	+	+	+	–	.	–	–	.	–
Female low	high	.	.	.	.	.	.	.	.	.	.	.	.	.	.	.	.	.	.
Female low	M<F	.	.	.	.	.	.	.	.	.	.	.	.	.	.	.	.	.	.
Female low	M>F	.	.	.	.	.	.	.	.	.	.	.	.	.	.	.	.	.	.
Female low	medium	.	.	.	.	.	.	.	.	.	.	.	.	.	.	+	.	.	.
Female low	low	+	+	+	+	+	+	–	+	–	.	–	–	–	–	–	–	.	–
Male high	high	.	.	–	–	.	.	.	.	.	.	.	.	+	+	+	+	+	+
Male high	M<F	.	.	.	–	.	.	.	.	.	.	.	.	.	.	.	.	.	.
Male high	M>F	.	.	.	–	.	–	.	–	.	.	.	.	+	+	+	+	+	+
Male high	medium	.	.	–	–	.	.	.	.	.	.	.	+	+	+	+	+	+	+
Male high	low	.	.	.	–	+	.	+	.	.	+	.	+	–	.	–	–	.	–
Male medium	high	.	.	–	.	.	.	.	.	.	.	.	.	+	.	+	+	.	+
Male medium	M<F	.	.	–	–	.	.	.	.	.	.	+	.	.	+	.	.	.	.
Male medium	M>F	.	.	–	–	.	.	.	.	.	–	+	.	+	+	+	+	.	+
Male medium	medium	.	.	–	–	–	.	–	.	.	.	.	.	+	+	+	+	+	+
Male medium	low	.	.	–	–	.	.	+	.	+	+	+	+	–	.	.	.	.	.
Male low	high	.	.	.	.	.	.	.	.	.	.	.	.	.	+	.	.	.	.
Male low	M<F	.	.	.	.	.	.	.	.	.	.	.	.	.	.	.	.	.	.
Male low	M>F	.	.	.	.	.	.	.	.	.	.	.	.	.	.	.	.	.	.
Male low	medium	.	.	.	–	.	.	.	.	.	.	.	.	.	.	+	+	+	.
Male low	low	.	.	+	+	+	+	.	+	–	.	–	–	–	–	–	–	.	–

+ type, – antitype, . not significant at all

ALLBUS (1991–2018, East: $n = 16{,}663$, West: $n = 33{,}350$)

Inference based on the χ^2-approximation to the z-test

Bonferroni adjusted alpha: $\alpha = 0.0001851852$

[a] Inference based on East German sub-sample

[b] Inference based on West German sub-sample

Overall using the prediction model separately for the East and West sub-sample, we find differences between eastern and western educational biographies for 63 configurations of family triads from formal school-leaving qualification. With three of these differences, there is a complete reversal of the expected and observed frequencies to the effect that, for example, a type results for the respective configuration in one sub-sample, whereas the same configuration is identified as an antitype in the other sub-sample. For example, for the direct postwar birth cohort (1945–1952), there is a significant type for females with high formal school qualifications from parents with low qualifications for the East German sub-sample. In contrast, the same configuration is identified as a significant antitype for the West German sub-sample (cf. Table 17.4, fifth line, third double column).

The analyses whose results are reported in Table 17.5 were conducted to examine the extent to which the significant types and antitypes found in the prediction model from Table 17.3 can be explained by five covariates, namely, the year of assessment, the occupational prestige according to the occupational prestige (SIOPS), and the socioeconomic status (ISEI), both for the father and the mother, respectively. For

Table 17.5 Prediction CFA with covariates: Types and antitypes for the temporal pattern of educational mobility within family triads across nine decades of birth cohorts (East and West)

Pattern		1898–1913		1914–1938		1939–1944		1945–1952		1953–1960		1961–1970		1971–1980		1981–1990		1991–1999	
Self	Mother–Father	s[a]	w[b]	s[a]	w[b]	s[a]	w[b]	s[a]	w[b]	s[a]	w[b]	s[a]	w[b]	s[a]	w[b]	s[a]	w[b]	s[a]	w[b]
Female high	high	.	.	–	–	.	.	–	–	.	.	.	.	+	+	+	+	+	+
Female high	M<F	–	–	+	+	–	–	.	.	.	.	.	.	+	+	.	.	.	.
Female high	M>F	.	.	–	–	.	.	.	.	.	.	–	–	.	.	+	+	+	+
Female high	medium	.	.	.	.	.	.	.	.	–	–	.	.	.	.	+	+	.	.
Female high	low	+	.	.	–	.	–	–	–	.	.	.	.	.	.	–	–	–	–
Female medium	high	.	.	.	.	.	.	.	.	–	–	+	.	–	.	.	.	.	.
Female medium	M<F	+	+	.	.	+	+	.	.	+	+	.	.	.	.	–	–	.	.
Female medium	M>F	.	.	.	.	+	+	.	.	.	.	.	.	+	.	+	+	.	.
Female medium	medium	.	.	–	.	–	.	–	–	+	.	.	.	+	+	+	+	.	.
Female medium	low	+	+	–	–	–	–	+	.	+	+	+	+	.	.	–	–	–	–
Female low	high	–	.	.	.	+	+	.	.	+	+	.	.	.	.	+	+	.	.
Female low	M<F	+	+	+	+	+	+	.	.	.	.	.	.	.	.	–	–	.	.
Female low	M>F	.	.	–	–	+	+	.	.	.	.	.	.	.	.	.	.	.	.
Female low	medium	.	.	.	.	.	.	+	+	+	+	.	.	–	–	.	.	.	.
Female low	low	.	.	+	+	+	+	+	+	–	.	–	–	–	–	–	–	–	–
Male high	high	.	.	–	–	–	–	.	.	.	.	.	.	+	+	+	+	+	+
Male high	M<F	–	–	.	.	–	.	.	.	.	.	+	+	+	+	.	.	.	.
Male high	M>F	.	.	.	.	.	.	.	.	.	.	.	.	.	.	+	+	.	.
Male high	medium	.	.	–	–	–	.	.	.	.	.	+	+	+	+	.	.	+	.
Male high	low	.	.	.	.	.	.	.	.	.	+	+	+	.	+	–	.	–	–
Male medium	high	.	.	.	.	+	+	.	.	.	.	.	.	+	+	+	+	.	.
Male medium	M<F	.	.	.	.	–	.	.	.	.	–	+	.	.	.	.	.	–	–
Male medium	M>F	.	.	–	–	+	+	.	.	.	.	–	–	+	+	+	+	+	+
Male medium	medium	.	.	–	–	–	–	–	.	–	–	+	.	+	+	+	+	+	+
Male medium	low	.	.	–	.	.	.	–	–	+	.	–	–	–	–	–	–	.	.
Male low	high	.	.	.	.	+	.	.	.	+	+	.	.	.	.	+	+	.	.
Male low	M<F	.	.	.	.	+	+	.	.	.	.	.	.	–	–	.	.	.	.
Male low	M>F	.	.	+	+	.	.	.	.	.	.	.	.	.	.	+	+	.	.
Male low	medium	.	.	.	.	.	.	.	.	.	.	.	.	.	.	+	+	+	.
Male low	low	+	+	+	+	+	+	.	.	–	–	–	–	–	–	–	–	–	–

+ type, – antitype, . not significant at all

Inference based on the χ^2-approximation to the z-test

Bonferroni adjusted alpha: $\alpha = 0.0001851852$ Covariates introduced into the model: Year of assessment "year," Standard International Occupational Prestige Scale (SIOPS), mother: "msiops08," and father: 'fsiops08';

International Socio-Economic Index (ISEI), mother: "misei88," father: "fisei88"

[a] Inference based on **unweighted** data

[b] Inference based on **weighted** data

these analyses, we again report results based on both weighted and unweighted data comparatively in double columns (see Table 17.5).

For both prediction models, with and without covariates (cf. Tables 17.5 and 17.3), it can be seen that, compared to the main effect model (cf. Table 17.2), the influence of weighting increases considerably. While the main effect model shows differences in the statistical inference between weighted and unweighted data for only one configuration, the prediction models show differences in the significance of types or antitypes for 17 (without covariates, Table 17.3) and 23 (with covariates, Table 17.5) configurations, respectively.

Overall, looking at weighted results, the introduction of the five covariates (year of assessment, SIOPS, and ISEI for father and mother) causes a change in the significance statement for 98 patterns of familial educational triads, while there is no direct change from type to antitype and generally a slight reduction of the number of types and antitypes due to the covariates. Specifically, the introduction of covariates into the prediction model causes the number of types to decrease from 62 to 60 and the number of antitypes to decrease from 61 to 48. Table 17.6 gives a detailed overview to the varying counts of types and antitypes across birth cohorts comparing the outcome of the prediction models without and with covariates.

17.5 Discussion

The current study investigated two main objectives, a content-related objective dealing with educational expansion and educational mobility in Germany and a methodologically founded objective. The latter deals with the application of sample weights in person-centered analyses using different CFA models and their influences on the results in comparison with unweighted analyses. Regarding the content-related objective, we focus on the comparative prediction CFA for East and West Germany (cf. Table 17.4), since these analyses provide the most differentiated results, which can well be related to the results of the previous research. The discussion of the findings from our analyses is therefore also divided into separate sections in accordance with the research questions motivated by content on the one hand and methodology on the other hand.

Effects of Weighted and Unweighted Analysis with Different CFA Models

For investigating the methodological objective, comparative analyses were carried out with and without weights. In addition, analyses with and without covariates were compared for the prediction CFA model, which turned out to show only small overall effects, with greater impact for later birth cohorts. The results show

Table 17.6 Comparison of counts of types and antitypes from pattern of educational mobility within family triads for weighted prediction CFA model with and without covariates across nine decades of birth cohorts

Counts		1898–1913	1914–1938	1939–1944	1945–1952	1953–1960	1961–1970	1971–1980	1981–1990	1991–1999
Without	Types	1	3	3	5	4	6	15	14	11
covariate	Antitypes	1	20	11	7	8	2	2	5	5
With	Types	4	5	10	2	6	4	10	14	5
covariate	Antitypes	2	9	5	4	5	5	5	7	6

that using weights in configural frequency analysis (CFA) is feasible and that it is *necessary* to consider weights in person-centered investigations, since the comparative analyses, especially in the case of the prediction CFA, yielded different results in terms of significant types and antitypes.

The findings from our analyses on the effects of considering sample weights in CFA analyses on the one hand, and ignoring them on the other hand, need to be evaluated and discussed in light of the specific nature of the weights used. The central point here is that the weights provided in the ALLBUS data are pure design weights containing no expansion factor (cf. Dorofeev & Grant, 2006, p. 46), which would ensure that the sum of the weights (thus expanded) corresponds to the size of the population. Instead the ALLBUS weights are normalized to the effective sample size (cf. Gabler et al., 2015) in a way that the sum of weights of the weighted sample and the number of persons being sampled remains the same. In turn, these weights show only a small (numerical) deviations from their expected value $\hat{\omega} = 1$. This special feature of these normalized weights has decisive repercussions on the comparison of the weighted and unweighted frequencies depending on the cell size of the multidimensional contingency table on which the CFA analyses are based. Conferring to the results depicted in Table 17.2, we can see from the comparison of the two columns with the observed pattern counts for the weighted an unweighted case that there are only small differences in the respective frequencies. On the level of the substantive interpretation of the analysis results, this is reflected in the fact that when testing for types and antitypes in the main effect model, a deviating inference would have to be drawn between the weighted and unweighted results only in the case of one configuration (cf. Table 17.2). In summary, these only small deviations between weighted and unweighted frequencies are also reflected in the last column in the first row of Table 17.1, where the percentage deviation of the mean difference between the weighted and unweighted frequencies from the mean pattern frequencies across all patterns is quite small for the main effect model at slightly more than 6%. In comparison, the percentage deviation for the prediction CFA model is almost twice as high at 11.2%. Comparative inspection between weighted and unweighted maximum, minimum, and mean pattern frequencies across the 270 cells of the prediction model (compared to 30 cells in the main effect model) shows that this is accompanied by the lower mean cell frequency in the prediction model. Although the mean absolute deviation between weighted and unweighted frequencies (cf. penultimate column in Table 17.1) is even more than five times smaller in the prediction model, this nevertheless leads to a larger effect between weighted and unweighted analysis results. This is evident regarding the substantive interpretation of the analysis results in Tables 17.3 and 17.5 where, in comparison to the main effect model, more deviating inferences are to be drawn when testing for types and antitypes between the weighted and unweighted results.

The findings from the weighted and unweighted analyses with different CFA models in this paper suggest that the magnitude of the effect of applying sample weights is mainly influenced by two aspects. First, the size of the mean absolute pattern frequencies in the single cells of the analyzed multidimensional contingency table seems to influence the potential effect of weighting in the sense that the expected effect of weighting tends to be larger for smaller pattern frequencies. Second, based on the findings, it can be reasonably assumed that the types of weights, pure design weights on the one hand and population weights including any expansion factor on the other hand, influence the size of the effect of a weighting in CFA models. The difference in the range of values of design weights compared to population weights is likely to lead to the fact that the impact of weights on the analyses is magnified when population weights are used as opposed to pure design weights. In contrast to pure design weights, which only serve to balance the disproportionality of specific subgroups in the sample compared to the population, population weights aim (in addition to balancing such disproportionality) to also correctly depict the underlying size of the population. In this sense, population weights would influence the analyzed frequencies of configurations (cell frequencies) from categorical data to a much more than design weights.

The introduction of covariates into the CFA prediction model initially leads only to a small reduction in the number of types and antitypes. Although the difference between the two models and thus the relevance of the covariates must initially be seen as rather small, a detailed comparison by birth cohorts reveals more distinct differences. It is striking that the reduction of the number of types by the covariates is more evident for later birth cohorts (1971–1980, 1981–1990, and 1991–1999). For the earlier birth cohorts, the change in the number of types and antitypes is distributed rather unsystematically. This finding can easily be attributed to the fact that data for the covariates are only available for the survey waves starting around the year 2000. Any systematic effects of the covariates are thus rarely observable for the earlier birth cohorts.

Educational Mobility in Germany

For the analysis of educational mobility against the backdrop of educational expansion in Germany, Tarnai and Hartmann (2019) also refer to the data of the German General Social Survey (ALLBUS). Tarnai and Hartmann (2019) investigated quite similar research questions but used different methods, i.e., descriptive analyses and logistic regression models. Referring to previous investigations (e.g., Apel, 1992; Blossfeld et al., 2015; Buchheister, 2002; Hadjar & Berger, 2010; Riphahn & Trubswetter, 2013; Ziefle, 2017), they conclude that there was an educational expansion in terms of higher proportions of intermediate and higher school-leaving qualifications in both East and West Germany. The educational mobility, however, was not only limited, but the attainment of higher educational qualifications was also heavily dependent on the education of the parents.

The results of the current study confirm the limitation of the educational mobility. For example, the configuration of a high education of the respondent and a high parental education is a significant type for the 1971–1999 cohorts in both East and West Germany. Also, for females in West and East Germany, the configuration according to which the respondent has a low education and parents with high education was not significant throughout all cohorts. This is similar for males. The type indicating an educational descent from high parental education to low education of the respondent was only significant in West Germany and here only for the 1971–1980 cohort. This indicates the tendency in society to strive to at least maintain the educational status of the parents. At the same time, the types including an intermediate or high school-leaving qualification of the respondent become more frequent from the older to the younger birth cohorts. In a completely permeable society, there would not only be educational advancements but also educational descents to a similar extent. The low level of educational mobility in Germany is thus primarily reflected in educational advancement, which is dependent on parental education, at least for many cohorts in the East and West, especially for the older and younger cohorts. Here, the combination of a low education of the respondent and a low education of the parents is a type (1914–1944) or the configuration of a high education and a low education of the parents is not significant or an antitype (1981–1999) (see also Buchheister, 2002). However, and contrary to the results of Apel (1992), the current study indicates that for some cohorts, there were not just single-stage ascents from one level to the next higher. In West Germany, the combination of a high school-leaving qualification and a low parental education (ascending two stages) appears to be a significant type for females of the cohorts 1953–1980 (no significance in the further course) and for males of the cohorts 1953–1970 (no significance or a significant antitype in the further course). In East Germany, this configuration is only a significant type for women born between 1945 and 1960, and there are no significant types or antitypes in the further course. For men in East Germany, there is a significant type for the cohorts 1939–1952 and without significant types or antitypes in the further course (see also Ziefle, 2017).

Gender Differences in Educational Mobility

Hadjar and Berger (2010) analyzing cohorts from 1925 to 1974 state that in West Germany, the chance to get a high school-leaving qualification increased for females from cohort to cohort but still remained lower than the chance for males. This is partly confirmed by the current study indicating that for females of the cohorts in the period from 1939 to 1952, the configuration of a high education and a low parental education appeared to be an antitype which was not the case for males (c.f. Blossfeld et al., 2015; Tarnai & Hartmann, 2019).

Hadjar and Berger (2010) also show that in East Germany, the chances to get a high school-leaving qualification is similar for males and females of the 1945–1954 cohort and are higher for females of the younger cohorts (1955–1964 and 1965–1974). This is partly confirmed by the results of the CFA according to which for both males and females of the 1945–1952 cohort the combination of a high education and a low parental education is a significant type, whereas this is only the case for females in the following cohort (1953–1960). The present study also shows for the younger cohorts (1971–1980, 1981–1990, 1991–1999) that educational decline, particularly from high to medium, is more of a significant type for males than for females.

Overall, in both East and West Germany, there is no clear gender difference in any of the cohorts, according to which there is a significant type for females and a significant antitype for males or vice versa.

Differences Between East and West Germany

In the current study, the most concise differences appear when there is a significant type for East and a significant antitype for West or vice versa. This can be observed for the 1945–1952 cohort. Here, the configuration of females with a high school-leaving qualification and parents with a low education is a type in East but an antitype in West. For the same cohort, the configuration of a low school-leaving qualification and parents with a low education is an antitype in East but a type in West. This is similar for males, albeit less clear. Here, the configuration of a high school-leaving qualification and parents with a low education is a type in East Germany but not in West Germany, and the configuration of a low school-leaving qualification and parents with a low education is a type in West Germany but not in East Germany. This indicates that the politically motivated promotion of children from lower educated families in East Germany had some success in the early postwar years. Afterward, the only concise difference between East and West Germany appears for females of the 1971–1980 cohort, where the combination of a high school-leaving qualification and parents with a low education is an antitype in East Germany and a type in West Germany. Again, the same tendency can be observed for males, where the first mentioned type (high qualification and low educated parents) is an antitype in East Germany but not in West Germany. A related finding is that the educational advancement from a low qualification of the parents to a middle qualification of the respondent is an antitype for males and females of the cohorts 1971–1980 and 1981–1990 in East Germany. These findings can be traced back to the school system in East Germany that caused the intermediate school-leaving qualification to be the rule in society (so that an advancement from a low qualification is hardly possible; c.f. Tarnai & Hartmann, 2019). Another interesting difference between East and West appears for females in the cohorts 1961–1970, 1971–1980, 1981–1990, and 1991–1999. Here, respondents with a high education and a mother who has a lower education than the father are a type only in West Germany. This may be due to another systematic difference between East and West Germany according to which females were more promoted in East Germany, especially in the early postwar years so that there are less parents with this kind of dissimilar education. As with the study of Tarnai and Hartmann (2019), there is no clear indication that a higher education of mothers is especially effective for their daughters, but, in the younger cohorts (1971–1980, 1981–1990, 1991–1999), is related with a higher education for both males and females.

The present analytical results are consistent with the analyses of Tarnai and Hartmann (2019) to the extent that intergenerational educational mobility in West and East Germany is observed only in the time periods in which the development of educational systems leads to educational expansion, from which women in particular benefit. In the younger cohorts that go through their education shortly before and after reunification, educational mobility decreases and the associated link between education and that of the parents increases.

References

Agresti, A. (2007). *An introduction to categorical data analysis* (2nd ed.). Wiley-Interscience.

Agresti, A. (2015). *Foundations of linear and generalized linear models*. Wiley.

Apel, H. (1992). Intergenerative Bildungsmobilität in den alten und neuen Bundesländern. In J. Zinnecker (Ed.), *Jugend '92. Bd. 2* (pp. 353–370). Leske + Budrich. https://doi.org/10.1007/978-3-322-95886-0_20

Bandilla, W., Gabler, S., Wiedenbeck, M. (1992). *ALLBUS Baseline-Studie 1991: Methodenbericht zum DFG-Projekt* (ZUMA-Arbeitsbericht No. 92/04). Zentrum für Umfragen, Methoden und Analysen (ZUMA).

Baumann, H., & Thiesen, S. (Eds.). (2021). *ALLBUS-Cumulation 1980-2018 - Variable Report* (GESIS-Variable Reports No. 2021|3). GESIS – Leibniz-Institut für Sozialwissenschaften.

Blau, P. M., Gustad, J. W., Jessor, R., Parnes, H. S., & Wilcock, R. C. (1956). Occupational choice: A conceptual framework. *ILR Review, 9*(4), 531–543.

Blossfeld, P. N., Blossfeld, G. J., & Blossfeld, H.-P. (2015). Educational expansion and inequalities in educational opportunity: Long-term changes for East and West Germany. *European Sociological Review, 31*(2), 144–160. https://doi.org/10.1093/esr/jcv017

Boudon, R. (1974). *Education, opportunity, and social inequality: Changing prospects in western society*. Wiley-Interscience.

Bourdieu, P. (1986). The forms of capital. In J. G. Richardson (Ed.), *Handbook of theory and research for the sociology of education* (pp. 229–242). Greenwood Press. (First published in Kapital, Ö., Kapital, K., Kapital, S. (1983). In *Soziale Ungleichheiten* (Soziale Welt, Sonderheft 2), R. Kreckel (Ed.). Otto Schartz & Co).

Braun, M., Eilinghoff, C., & Gabler, S. (1993). *Methodenbericht zur Allgemeinen Bevölkerungsumfrage der Sozialwissenschaften (ALLBUS) 1992* (ZUMA-Arbeitsbericht No. 93/01). Zentrum für Umfragen, Methoden und Analysen (ZUMA).

Buchheister, C. (2002). *Intergenerationale Bildungs- und Erwerbsmobilität in Ost- und Westdeutschland. Eine empirische Analyse*. Universität Potsdam.

Carroll, J. D., & Arabie, P. (1980). Multidimensional scaling. *Annual Review of Psychology, 31*(1), 607–649.

Dorofeev, S., & Grant, P. (2006). *Statistics for real-life sample surveys: Non-simple-random samples and weighted data*. Cambridge University Press.

Faulbaum, F. (2019). *Methodische Grundlagen der Umfrageforschung*. Springer Fachmedien. https://doi.org/10.1007/978-3-531-93278-1

Gabler, S. (1994). ALLBUS-baseline-studie 1991 und ALLBUS 1992: Ost- West-Gewichtung der Daten. *ZUMA Nachrichten, 35*(18), 77–81.

Gabler, S., Kolb, J.-P., Sand, M., & Zins, S. (2015). *Gewichtung* (GESIS - Survey Guidelines). GESIS - Leibniz-Institut für Sozialwissenschaften. https://doi.org/10.15465/gesis-sg_007

Ganzeboom, H. B. G., De Graf, P. M., Treiman, D. J., & de Leeuw, J. (1992). A standard international socio-economic index of occupational status. *Socil Science Review, 21,* 1–56.

Ganzeboom, H. B. G., & Treiman, D. J. (1996). Internationally comparable measures of occupational status for the 1988 international standard classification of occupations. *Social Science Research, 25*(3), 201–239. https://doi.org/10.1006/ssre.1996.0010

Geißler, R. (1990). Soziale Ungleichheiten zwischen Frauen und Männern: Erfolge und Hindernisse auf dem Weg zur Gleichstellung in den beiden deutschen Gesellschaften. *Sozialwissenschaftliche Informationen, 19,* 181–196.

Geißler, R. (2013). Die Metamorphose der Arbeitertochter zum Migrantensohn. Zum Wandel der Chancenstruktur im Bildungssystem nach Schicht, Geschlecht, Ethnie und deren Verknüpfungen. In P. A. Berger & H. Kahlert (Eds.), *Institutionalisierte Ungleichheiten: wie das Bildungswesen Chancen blockiert* (3rd ed., pp. 71–100). Beltz Juventa.

Geißler, R. (2014). *Die Sozialstruktur Deutschlands.* Springer Fachmedien. https://doi.org/10.1007/978-3-531-19151-5

GESIS - Leibniz-Institut für Sozialwissenschaften (2021). *German General Social Survey ALLBUS - Cumulation 1980-2018.* GESIS Datenarchiv, Köln. ZA5276 Datenfile Version 1.1.0. https://doi.org/10.4232/1.13774

Glück, J., & von Eye, A. (2000). Including covariates in configural frequency analysis. *Psychologische Beiträge, 42*(3), 405–417.

Hadjar, A. (2011). *Geschlechtsspezifische Bildungsungleichheiten.* VS Verlag für Sozialwissenschaften. Retrieved 2022-10-20, from http://link.springer.com/book/10.1007/978-3-531-92779-4

Hadjar, A., & Berger, J. (2010). Dauerhafte Bildungsungleichheiten in Westdeutschland, Ostdeutschland und der Schweiz: Eine Kohortenbetrachtung der Ungleichheitsdimensionen soziale Herkunft und Geschlecht. *Zeitschrift für Soziologie, 39*(3), 182–201. https://doi.org/10.1515/zfsoz-2010-0302

Heine, J.-H., Hartmann, F. G., & Tarnai, C. (2023). *Exploring temporal pattern - educational mobility. Accompanying R syntax.* Retrieved from https://osf.io/7ykzv

Heine, J.-H., & Stemmler, M. (2021a). Analysis of Categorical data with the R package confreq. *Psych, 3*(3), 522–541. https://doi.org/10.3390/psych3030034

Heine, J.-H., & Stemmler, M. (2021b). Die (Nicht-)Bedeutsamkeit des "Migrationshintergrundes" für die PISA-Leistung - Eine Analyse mittels KFA und LCA. In J. Reinecke & C. Tarnai (Eds.), *Klassifikationsanalysen in den Sozialwissenschaften* (pp. 75–99). Waxmann Verlag.

Heine, J.-H., Stemmler, M., & Alexandrowicz, R. W. (2022). *Confreq: Configural Frequencies Analysis Using Log-Linear Modeling.* Retrieved 2022-11-9, from https://CRAN.R-project.org/package=confreq. (R-package version 1.6.1-1).

Helbig, M. (2012). Die Umkehrung – Geschlechterungleichgheiten beim Erwerb des Abiturs im Wandel. In R. Becker & H. Solga (Eds.), *Soziologische Bildungsforschung* (pp. 374–392). Springer Fachmedien. https://doi.org/10.1007/978-3-658-00120-9_16

Hille, B. (1990). Mädchen- und Frauenbildung. In O. Anweiler, W. Mitter, H. Peisert, H.-P. Schaefer, & W. Stratenwerth (Eds.), *Vergleich von Bildung und Erziehung in der Bundesrepublik Deutschland und in der Deutschen Demokratischen Republik.* (pp. 581–596). Verlag Wissenschaft und Politik.

Hillmert, S. (2012). Familienstrukturen und soziale Bildungsreproduktion. In Becker & H. Solga (Eds.), *Soziologische Bildungsforschung* (pp. 325–345). Springer Fachmedien. https://doi.org/10.1007/978-3-658-00120-9_14

International Labour Office (1990). *International standard classification of occupations: ISCO-88* ([Rev. ed. 1988] ed.). International Labour Office.

Jacoby, W. G. (1991). *Data theory and dimensional analysis.* SAGE Publications Inc.

Kawasaki, S., & Zimmermann, K. F. (1964). Measuring relationships in the log-linear probability model by some compact measures of association. *Statistische Hefte, 22*(2), 82–109. https://doi.org/10.1007/BF02933546

Koch, A. (1997). ADM-design und einwohnermelderegister-stichprobe. In Gabler & J.H.P. Hoffmeyer-Zlotnik (Eds.), *Stichproben in der Umfragepraxis* (pp. 99–116). VS Verlag für Sozialwissenschaften. https://doi.org/10.1007/978-3-322-86533-5_8

Krauth, J., & Lienert, G. A. (1973). *Die Konfigurationsfrequenzanalyse (KFA) und ihre Anwendung in Psychologie und Medizin: ein multivariates nichtparametrisches Verfahren zur Aufdeckung von Typen und Syndromen; mit 70 Tabellen.* Alber Karl.

Köhler, H., & Schreier, G. (1990). Statistische Grunddaten zum Bildungswesen. In O. Anweiler, W. Mitter, H. Peisert, H.-P. Schaefer, & W. Stratenwerth (Eds.), *Vergleich von Bildung und Erziehung in der Bundesrepublik Deutschland und in der Deutschen Demokratischen Republik.* (pp. 112–155). Verlag Wissenschaft und Politik.

Lienert, G. A. (1971). Die Konfigurationsfrequenzanalyse: I. Ein neuer Weg zu Typen und Syndromen. *Zeitschrift für Klinische Psychologie und Psychotherapie, 19* (2), 99–115.

Nelder, J. A. (1974). Log linear models for contingency tables: A generalization of classical least squares. *Journal of the Royal Statistical Society. Series C (Applied Statistics), 23* (3), 323–329. https://doi.org/10.2307/2347125

Nelder, J. A., & Wedderburn, R. W. M. (1972). Generalized linear models. *Journal of the Royal Statistical Society. Series A (General), 135* (3), 370–384. https://doi.org/10.2307/2344614

OECD, Adams, R. J., & Wu, M. (Eds.). (2002). *PISA 2000 Technical report*. OECD Publishing.

OECD (Ed.). (2009). *PISA data analysis manual. SPSS* (2nd ed.) OECD Publishing.

R Core Team (2022). *R: A language and environment for statistical computing [Computer software manual]*. Vienna, Austria. Retrieved from https://www.R-project.org/

Riphahn, R. T., & Trübswetter, P. (2013). The intergenerational transmission of education and equality of educational opportunity in East and West Germany. *Applied Economics, 45*(22), 3183–3196. https://doi.org/10.1080/00036846.2012.703314

Rodax, K., & Hurrelmann, K. (1986). Die Bildungsbeteiligung der Maedchen und Frauen - ein Indikator fuer wachsende Chancengleichheit? *Zeitschrift für Sozialisationsforschung und Erziehungssoziologie, 6*(1), 138–146.

Sand, M., & Kunz, T. (2020). *Gewichtung in der Praxis* (GESIS - Survey Guidelines). GESIS - Leibniz-Institut für Sozialwissenschaften. https://doi.org/10.15465/gesis-sg_030

Siegers, P., Schulz, S., & Hochman, O. (2019). Einleitung: ALLBUS, IEDI und die Wiederaufnahme der Reihe “Blickpunkt Gesellschaft”. In P. Siegers, S. Schulz, & O. Hochman (Eds.), *Einstellungen und Verhalten der deutschen Bevölkerung: Analysen mit dem ALLBUS* (pp. 1–11). Springer Fachmedien. https://doi.org/10.1007/978-3-658-21999-4_1

Staaliche Zentralverwaltung für Statistik (Ed.). (1989). *Statistisches Jahrbuch der Deutschen Demokratischen Republik* (Vol. 1989). Staatsverlag der Deutschen Demokratischen Republik. Retrieved 2022-10-20, from https://www.statistischebibliothek.de/mir/receive/DEHeft_mods_00143472

Statistisches Bundesamt. (2012). *Allgemeinbildende Schulen - Fachserie 11 Reihe 1 - Schuljahr 2011/2012*. Destatis.

Statistisches Bundesamt (2021). *Bevölkerungsfortschreibung auf Grundlage des Zensus 2011 - Fachserie 1, Reihe 1.3 - 2020*. Destatis. Retrieved 2022-10-27, from https://www.destatis.de/DE/Themen/Gesellschaft-Umwelt/Bevoelkerung/Bevoelkerungsstand/inhalt.html#_vokc7tw51

Stemmler, M. (2020). *Person-centered methods: Configural frequency analysis (CFA) and other methods for the analysis of contingency tables* (2nd ed.). Springer International Publishing. https://doi.org/10.1007/978-3-030-49421-6

Stemmler, M., Lösel, F., Beelmann, A., & Jaursch, S. (2008). A configural perspective on the stability of externalizing problem behavior in children: Results from tbe Erlangen-Nuremberg development and prevention study. In M. Stemmler, E. Lautsch, & D. Martinke (Eds.), *Configural frequency analysis (CFA) and other non-parametrical statistical methods: Gustav A. Lienert memorial issue* (pp. 70–83). Pabst Science Publishers.

Tarnai, C., & Hartmann, F. G. (2017). Intergenerationale Bildungsmobilität und Bildungshomogamie. In T. Eckert & B. Gniewosz (Eds.), *Bildungsgerechtigkeit* (pp. 93–118). Springer Fachmedien. https://doi.org/10.1007/978-3-658-15003-7_7

Tarnai, C., & Hartmann, F. G. (2019). Intergenerationale Bildungsmobilität im Übergang vom geteilten zum wiedervereinigten Deutschland. In P. Siegers, S. Schulz, & O. Hochman (Eds.), *Einstellungen und Verhalten der deutschen Bevölkerung: Analysen mit dem ALLBUS* (pp. 229–267). Springer Fachmedien. https://doi.org/10.1007/978-3-658-21999-4_8

von Eye, A. (2002). *Configural frequency analysis: Methods, models, and applications*. Lawrence Erlbaum Associates.

von Eye, A., & Wiedermann, W. (2021a). *Configural frequency analysis*, A. von Eye & W. Wiedermann (Eds.). Springer. https://doi.org/10.1007/978-3-662-64008-1_8
von Eye, A., & Wiedermann, W. (2021b). Special variables in CFA. In A. von Eye & W. Wiedermann (Eds.), *Configural frequency analysis: Foundations, models, and applications* (pp. 221–268). Springer. https://doi.org/10.1007/978-3-662-64008-1_6
von Eye, A. (1988). The general linear model as a framework for models in configural frequency analysis. *Biometrical Journal, 30*(1), 59–67. https://doi.org/10.1002/bimj.4710300110
Young, F. W. (1984). Scaling. *Annual Review of Psychology, 35*(1), 55–81. https://doi.org/10.1146/annurev.ps.35.020184.000415
Ziefle, A. (2017). Der lange Arm der Bildungsexpansion: Die Bedeutung zunehmender elterlicher Bildungsressourcen für die Bildungsbeteiligung von Frauen in Deutschland. *Kölner Zeitschrift für Soziologie und Sozialpsychologie, 69*(1), 51–77. https://doi.org/10.1007/s11577-016-0402-5

Chapter 18
Configural Frequency Analysis Under Multinormality

Alexander von Eye and Wolfgang Wiedermann

By conducting Configural Frequency Analysis (CFA; Lienert, 1968; Lienert & Krauth, 1975; von Eye & Gutiérrez Peña, 2004; von Eye & Wiedermann, 2022), researchers attempt to identify those sectors in the data space that contradict model estimation. CFA begins with the a priori specification of a data generation process (DGP; von Eye et al., 2021). From the DGP, the researchers derive specific probability models, they are called *base models*, for the estimation of expected frequencies in cross-classifications. Deviations are interpreted with respect to these base models. When, in a particular cell, more cases are observed than expected, this cell is said to constitute a *CFA type*. When fewer cases are observed than expected, the cell is said to constitute a *CFA antitype*.

In a recent chapter (von Eye, et al., 2021), the relation of DGPs and CFA base models was elaborated. Extant base models were grouped into four classes of DGPs. These processes result in base models with which researchers test hypotheses concerning variable relations, distributions of observations, global functions that represent series of observations, and sector-specific functions. In the present chapter, we extend this taxonomy and propose DGPs that result in base models that incorporate information concerning probability distributions of observations in tandem with specific variable relations.

This chapter is structured as follows. In the following section, we review the four steps to be taken when conducting CFA (more detail can be found, e.g., in von Eye & Wiedermann, 2022, or von Eye et al., 2021). This is followed by a section in which we propose new base models. Specifically, we propose models in which

A. von Eye (✉)
Michigan State University, East Lansing, MI, USA

Montpellier, France
e-mail: voneye@msu.edu

W. Wiedermann
Educational, School and Counseling Psychology, University of Missouri, Columbia, MO, USA

M. Stemmler et al. (eds.), *Dependent Data in Social Sciences Research*,
https://doi.org/10.1007/978-3-031-56318-8_18

distributions of observations and variable relations are jointly examined. Third, we present data examples and, to conclude, a discussion.

18.1 The Four Steps of CFA

In the *first step of CFA*, researchers specify a DGP. This specification is based on knowledge concerning the nature of the observations to be analyzed, theory, and prior knowledge. For example, vaccination is known to reduce the incidence rate of a pandemic because the number of individuals left that possibly can be infected diminishes. This reduction decelerates as the vaccination rates increase. One corresponding DGP could incorporate a curve of exponential decay.

Based on this DGP, a base model can be derived. In the base model, all effects are included that are not of relevance for the DGP. When the base model is rejected, effects of relevance exist, statistically. When, for example, the deceleration posited above is examined, a function can be specified that includes a negative regression slope but no deceleration parameter. When this function plus other effects of interest—except the deceleration—fail to describe the change in incidence rates after the onset of vaccination, deceleration does take place, statistically.

In brief, a CFA base model that is derived from a posited DGP has the following properties (see von Eye, 2004; von Eye et al., 2021):

1. *Unique interpretation*: the model must be specified such that there exists only one option for the interpretation of types and antitypes. This option is given by the effects not included in the base model. For example, when a base model is specified that focuses on predictor-outcome relations, it contains all effects within the predictor variables as well as within the outcome variables, but none that describes predictor-outcome relations. When this model fails, predictor-outcome relations are bound to exist.
2. *Completeness*: the base model must contain each of the effects that are not of interest to the researchers (see, however, Point 3, below). When this model fails, only the effects of interest can possibly explain the observations.
3. *Parsimony*: when theoretical knowledge or prior results allow one to disregard effects, they can be omitted from a base model (see, Schuster & von Eye, 2000). This applies, for example, when higher-order effects make no contribution to the explanation of observations (cf. parsimony principle; Wu & Hamada, 2000).
4. *Consideration of sampling scheme*: sampling schemes determine the procedures of data collection. For example, multinomial sampling does not place any constraints on marginal distributions. In contrast, product-multinomial sampling implies that marginal probabilities be reproduced when data are collected (for example, in the context of a balanced randomized controlled trial, the marginal cell probabilities of a binary treatment indicator can be expected to be 0.50). In CFA, the specification of base models takes the sampling scheme into consideration and, therefore, restricts the selection of base models.

In the *second step of CFA*, researchers select a significance test. CFA significance tests are applied to individual cells or groups of cells (smaller than the cross-classification that is being analyzed). The null hypothesis for these tests is H_0 : $E[m_i] = \hat{m}_i$, where E is the expectancy, m_i is the observed frequency of Cell i (for $i = 1, \ldots, t$; where t denotes the number of cells in a cross-classification), and $\hat{m}_i$ is the expected frequency in Cell i. $\hat{m}_i$ is estimated under the CFA base model (see von Eye & Gutiérrez Peña, 2004; von Eye & Wiedermann, 2022).

Cell probability $p_i(m_i - 1)$ indicates the probability that Cell i contains m_i – 1 cases. Cell i is said to constitute a *CFA type* when the null hypothesis is rejected because $p_i(m_i - 1) \geq 1 - \alpha$. In contrast, when the null hypothesis is rejected because $p_i(m_i) \leq \alpha$, Cell i is said to constitute a *CFA antitype*. When the null hypothesis prevails, Cell i constitutes neither a type nor an antitype.

A large number of tests has been proposed for CFA (see also Lautsch & von Weber, 1995; von Eye et al., 2010). These tests differ in many characteristics. Some are exact, others are approximative; some require product-multinomial sampling, others can be applied under multinomial sampling; some suggest conservative decisions concerning the null hypothesis, others tend to suggest non-conservative decisions. Here, we present, as an example of a CFA test, the binomial test. This test is exact. It can be applied under any sampling scheme and results, in the cross-classification that is analyzed, in the frequency distribution

$$P\left(m = m_1, \ldots, m_t = m_r | N, \pi_1, \ldots, \pi_t\right) = \frac{N!}{m_i!, \ldots, m_t!} \sum_{r=1}^{t} \pi_r^{m_r},$$

where t is the number of cells, $\sum_r p_r = 1$, and $\sum_r m_r = N$, where r indexes the cells of the cross-classification, with $r = 1, \ldots, t$. Thus, cell frequency m_r has a binomial distribution, with

$$P\left(m = m_r | N, \pi_r\right) = \frac{N!}{m_r!\,(N - m_r)!} \pi_r^{m_r} (1 - \pi_r)^{N - m_r}.$$

Based on this, the exact binomial test can be given by

$$B_{N,p}(m) = \sum_{j=0}^{m} \frac{N!}{j!\,(N - J)!} p^j (1 - p)^{N-j},$$

with $0 \leq m \leq N$. Usually, p is estimated from the sample.

In the *third step of CFA*, a procedure is selected for the protection of the nominal significance level, α. Protection is needed because cell-wise CFA tests are known to be dependent and because the risk of committing an α error is taken with each test, that is, maximally t times. Here, we present two examples of such procedures. Both are applicable under any sampling scheme.

Best known is the Bonferroni (1936) procedure. This single-step procedure specifies the following:

- The sum of all α_i-values of the t CFA tests cannot be greater than the nominal α, that is, $\sum_i \alpha_i \leq \alpha$, for $i = 1, \ldots, t$.
- The protected α^* is the same for each test.

The value of α^* that fulfills both of these specifications is $\alpha^* = \alpha/t$. Krauth and Lienert (1973) note that this value results in type/antitype decisions that are only slightly more conservative than the nominal α, when all t tests are mutually independent.

$$\alpha_i^* = 1 - (1 - \alpha)^{\frac{1}{r-i+1}}.$$

Less conservative than Bonferroni adjustment is the procedure that was proposed by Holland and Di Ponzio Copenhaver (1987). This is a step-down procedure for which the test statistics must be arranged in descending order. Testing begins with the greatest test statistic. For the ith test that is performed, the protected significance threshold is when $i = 1$, the procedures by Bonferroni and Holland—Di Ponzio Copenhaver operate under the same α^*. Beginning with the second test, Holland and Di Ponzio Copenhaver's procedure is less extreme.

In the *fourth step of CFA*, cell frequencies are estimated under the specified base model, significance tests are performed, and types and antitypes are interpreted with respect to the information that was used to specify the base model.

18.2 Base Models of CFA

In this section, we focus on log-linear base models of the form log $\hat{m} = X\lambda$, where $\hat{m}$ is the vector of expected cell frequencies, X is the design matrix, and λ is the vector of model parameters. The models used in CFA can be hierarchical, non-hierarchical, and non-standard (Mair & von Eye, 2007). Following the hierarchy of CFA base models (see von Eye, 2002; von Eye & Wiedermann, 2022), the simplest model is the no-effect model, or log $\hat{m} = 1\lambda_1$, where λ_1 is a parameter vector with just one entry. It represents the model constant. The next higher model in the hierarchy is the one that considers all main effects, or log $\hat{m} = [1 \mid X_m]\begin{bmatrix}\lambda_1 \\ \lambda_{X_m}\end{bmatrix}$, where X_m indicates the columns of X that represent the main effects, and λ_{X_m} are the parameter vector entries that correspond to the columns for the main effects in X_m. This model is that of *first-order CFA*. This can be extended by also including two-way interactions, that is, by specifying a *second-order CFA* base model, or

$$\log \hat{m} = [1 \mid X_m \mid X_k]\begin{bmatrix}\lambda_1 \\ \lambda_{X_m} \\ \lambda_{X_k}\end{bmatrix},$$

where X_k indicates the columns in X that represent the two-way interactions and λ_{X_k} are the corresponding parameters (see also Victor 1983; Kieser & Victor, 1991).

All of these base models are hierarchical in nature. Based on a method for the detection of local deviations from multinormality (cf. von Eye & Gardiner, 2004; von Eye et al., 2006), we now extend the class of CFA base models by also including a DGP that posits that one or multiple variables of the cross-classification are normally distributed. This renders the base model non-standard (Mair & von Eye, 2007). In the simplest case, there is just one normally distributed variable, and the base model is a null model, that is, $\log\ \hat{m} = [1 \mid X_n]\begin{bmatrix}\lambda_1 \\ \lambda_{X_n}\end{bmatrix}$, where X_n is the vector that contains the cell-wise probabilities that represent the normal distribution and λ_{X_n} is the corresponding parameter. This model can be viewed as parallel to the one used by von Eye and Gardiner (2004). Also taking main effects into account, one obtains

$$\log\ \hat{m} = [1 \mid X_m \mid X_n]\begin{bmatrix}\lambda_1 \\ \lambda_{X_m} \\ \lambda_{X_n}\end{bmatrix}.$$

Adding, in addition, two-way interactions results in

$$\log\ \hat{m} = [1 \mid X_m \mid X_k \mid X_n]\begin{bmatrix}\lambda_1 \\ \lambda_{X_m} \\ \lambda_{X_k} \\ \lambda_{X_n}\end{bmatrix}.$$

This can be extended by including the hypothesis that more than one variable is normally distributed.

In these models, just one variable or multiple individual variables are included that are hypothesized to be normally distributed. Positing that two variables exhibit a bivariate normal distribution results in

$$\log\ \hat{m} = [1 \mid X_m \mid X_k \mid X_{2n}]\begin{bmatrix}\lambda_1 \\ \lambda_{X_m} \\ \lambda_{X_k} \\ \lambda_{X_{2n}}\end{bmatrix},$$

where X_{2n} is the vector that represents the probabilities of the bivariate normal distribution of two variables. Accordingly, multiple variables can be hypothesized to follow a multivariate normal distribution.

Estimation In a way similar to the procedure proposed by von Eye and Gardiner (2004), estimation of the probabilities of a d-variate normal distribution can be performed by taking the following steps:

1. *Binning*: each of the variables is subjected to binning; the number of segments (bins) for variable j is $c_j > 1$, with $j = 1, \ldots, d$.
2. *Crossing*: the thus created categorical variables are completed crossed to result in a d-dimensional cross-classification with $\Pi_{j=1}^{d} c_j$ sectors; these sectors are the configurations of CFA.
3. *Probability estimation*: for each of the sectors, the probability is estimated under the assumption that the distribution is d-variate normal; specifically, the bins of the jth variable have, as their lower limit, $z_{a,j}$, and their upper limit is $z_{e,j}$; the probability of a particular sector is $p\left(z_{a,1} - z_{e,1}, \ldots, z_{a,d} - z_{e,d}\right) = \int_{z_{a,1}}^{z_{e,1}} \ldots \int_{z_{a,d}}^{z_{e,d}} \Psi\left(z_{a,1} - z_{e,1}, \ldots, z_{a,d} - z_{e,d}\right)$; to estimate these probabilities, the numerical solution proposed by Somerville (1998) can be used (for an alternative solution, see Genz, 1992).
4. *Estimation of expected cell frequencies*: for the estimation of the expected cell frequency of Sector $s_{i,j,\ \ldots,d}$ in the space of the d crossed segmented variables, the base model is used, that is, for example, the model $\log \hat{m} =$

$$[1 \mid X_m \mid X_k \mid X_{2n}] \begin{bmatrix} \lambda_1 \\ \lambda_{X_m} \\ \lambda_{X_k} \\ \lambda_{X_{2n}} \end{bmatrix}.$$

This fourth step allows one to extend the base models of CFA under normal distribution assumptions that have been proposed thus far (von Eye & Wiedermann, 2022). These base models took into account either variable relations or the hypothesis of a multivariate normal distribution (von Eye & Gardiner, 2004). The approach proposed here allows researchers to specify base models that include the following:

1. Variable relations as in standard CFA base models (von Eye & Gutiérrez Peña, 2004)
2. Hypotheses concerning uni-, bi-, or multivariate normality of the variables that span the cross-classification that is being analyzed
3. Elements of both 1. and 2.
4. Covariates (Glück & von Eye, 2000)

18.3 Identification

The base models proposed in this chapter will, in most applications, be identified. However, there are exceptions. First, the design matrix of the base model must not be singular. This can be the case when there are linear dependencies among the vectors of the design matrix. In real-world data applications, this, however, is rarely the case. Even high correlations usually pose no problems. The problems that correlations pose concern the interpretation of model parameters. This is of importance in modeling. In CFA, this usually is not an issue because the focus is on individual expected cell frequencies and their differences to observed cell frequencies, not on parameter interpretation. When, however, researchers wish to interpret model parameters in CFA nevertheless, and the columns of the design matrix are correlated, the Schuster transformation (von Eye et al., 1998) can be applied. This transformation has the effects that (1) parameters can be interpreted as specified in the design matrix—even when the vectors in the design matrix are correlated—and (2) the estimates of the expected cell frequencies remain unchanged.

The second exception is far more likely to occur. This exception concerns the nature of the variables that are crossed and the degrees of freedom that are available for the base model. The size of the cross-classification that is being analyzed determines the number of effects that can be taken into consideration. To give an example, consider a $2 \times 2 \times 2$ table. When the model of first-order CFA is estimated for this table, that is, the base model of variable independence, the design matrix of the base model contains four columns (one column for the model constant and three columns for the main effects), and three degrees of freedom are available before the model becomes saturated. The model of multivariate normality cannot be meaningfully applied in this case because it would, above and beyond the main effects, posit that the categories of the variables that were crossed have probabilities that are determined by where on the scales the variables were dichotomized. When they are dichotomized at the median, the categories would be hypothesized to be equiprobable. Therefore, in this case, although there are enough degrees of freedom, the assumption of multinormality might contradict the main effect hypothesis, in particular, when sampling is multinomial.

To give a second example, consider a $2 \times 2 \times 3$ table, for which the base model of second-order CFA is estimated. This model includes, in addition to the constant, four columns for the main effects and five columns for the two-way interactions. Thus, there is one degree of freedom available before the model becomes saturated. In this case, the hypothesis that the three variables exhibit each a univariate normal distribution cannot be tested, for lack of degrees of freedom. In contrast, the model of a trivariate normal distribution—it requires just one column in the design matrix—can be tested when no additional covariate is to be included.

In sum, there clearly are cases, in particular, when tables are small, in which certain base models cannot be used. In the following sections, we present a data example of the extended approach to CFA.

18.4 Data Example

For the following example, we reanalyze data from a prospective longitudinal study on the effects on female partners of violence that was exerted by male partners (Levendosky et al., 2000; Huth-Bocks et al., 2004). 201 women participated in this study. When they were pregnant, they were administered a structured interview to assess their subjective experiences with their unborn children. Here, we analyze

Table 18.1 Univariate descriptive statistics of Fear, Joy, and Anxiety

	Fear	Joy	Anxiety
N of Cases	201	201	201
Arithmetic Mean	2.741	2.796	2.960
Variance	0.453	1.033	0.788
Skewness	0.062	0.275	0.381
Standard Error of Skewness	0.172	0.172	0.172
Kurtosis	0.238	−0.585	−0.386
Standard Error of Kurtosis	0.341	0.341	0.341

three affective features of maternal representations: Fear (F), Joy (J), and Anxiety (A).

The questions we ask in the following analysis are (1) where in the $F \times J \times A$ cross-classification are associations among the three variables most strongly visible (if any), that is, whether there exist local associations in a CFA sense and (2) whether the results of this analysis change when information is taken into account that concerns the trivariate normal distribution of F, J, and A.

The analysis proceeds in three stages. First, we ask whether the $F \times J \times A$ cross-classification exhibits any CFA types and antitypes. Second, we ask whether F, J, and A jointly follow a trivariate normal distribution. Third, we ask whether these types and antitypes can be explained when information about the trivariate distribution of the three variables is taken into account in the CFA. The three variables were binned into three categories each.

Before performing the analyses in these three stages, we describe the three variables and ask whether they follow univariate normal distributions. Table 18.1 shows that they do. For none of the variables, the Mardia (1970, 1980) tests of skewness and kurtosis indicate any violation.

Stage 1: First-order CFA of the F × J × A Cross-Classification

In Stage 1 of the analyses, we ask whether the three variables Fear, Joy, and Anxiety are related such that individual configurations (cells) stand out as deviating from the expectation that was estimated based on the assumption of variable independence. We take the four steps of CFA.

Step 1: Specification of base model The base model used for this stage is $\log \hat{m} = [1 \mid X_m]\begin{bmatrix} \lambda_1 \\ \lambda_{X_m} \end{bmatrix}$, where X_m indicates the columns of X that represent the main effects of F, J, and A, and λ_{X_m} are the parameters that correspond to these columns. This is the base model of first-order CFA.

Steps 2 and 3: Selection of significance test and of procedure to protect α For CFA, we use the binomial test. This test is exact and can be applied when cell frequencies are small (see Table 18.2, below). We protect α using the Bonferroni procedure. The same methods are applied in the subsequent stages of analysis.

Table 18.2 First-order CFA of the cross-classification of Fear, Joy, and Anxiety

Configuration	m	$\hat{m}$	p
111	13.00	10.0697	.21025688
112	11.00	12.3241	.42040742
113	2.00	7.8153	.01450975
121	10.00	6.9801	.16386312
122	8.00	8.5428	.51561301
123	4.00	5.4174	.36744564
131	10.00	5.9502	.07746780
132	11.00	7.2824	.11576976
133	.00	4.6181	.00935371
211	11.00	16.3449	.10092632
212	21.00	20.0043	.44122461
213	18.00	12.6856	.08612840
221	10.00	11.3300	.41666001
222	17.00	13.8666	.22605062
223	6.00	8.7934	.22017210
231	11.00	9.6584	.37321881
232	11.00	11.8207	.47935793
233	7.00	7.4961	.52383966
311	1.00	2.9187	.20941162
312	2.00	3.5722	.30517565
313	9.00	2.2653	.00052081 Type
321	1.00	2.0232	.39837345
322	1.00	2.4762	.29031852
323	4.00	1.5703	.07401399
331	.00	1.7247	.17690331
332	.00	2.1108	.11979172
333	2.00	1.3386	.38717968

From the global correlation tests, one would expect that sectors deviate, because the three variables are weakly to moderately correlated: $r_{FJ} = -0.106$, $r_{FA} = 0.441$, and $r_{JA} = -0.114$. Table 18.2 displays the results of first-order CFA.

The log-linear main effect model that was used to analyze Table 18.2 is, overall, rejected (LR-$X^2 = 54.178$; $df = 20$; $p < 0.001$). We, therefore, expect types or antitypes to emerge. One type did emerge. It is constituted by Configuration 3 1 3. This profile describes women who indicate high levels of Joy and Anxiety but low levels of Fear. Nine women (4.5%) did show this profile, but only 2.265 (i.e., 1.1%) had been expected to show it.

Stage 2: Test of Local Deviations from a Trivariate Normal Distribution

It is well known that variables that follow a multivariate normal distribution are also normally distributed at the univariate level, but not necessarily the other way around. Therefore, we now ask whether Fear, Joy, and Anxiety follow a trivariate normal distribution. The random vector $x(d \times 1)$ of d variables follows a multinormal distribution if its probability density function can be described by

$$f(x) = (2\pi)^{-d/2} \left| \sum \right|^{-0.5} e^{-0.5(x-\mu)' \sum^{-1} (x-\mu)}$$

with positive definite covariance matrix X of rank d.

Specifically, we are interested in finding local deviations, if they exist. To answer this question, we employ the method proposed by von Eye and Gardiner (2004). The model specified for this purpose is, as was indicated above, $\log \hat{m} = [1 \mid X_n] \begin{bmatrix} \lambda_1 \\ \lambda_{X_n} \end{bmatrix}$, where X_n is the vector that contains the cell-wise probabilities of the cross-classification of F, J, and A that represent the trivariate normal distribution and λ_{X_n} is the corresponding parameter. Specifically,

$$X_n = [.024064, .035292, .005822, .050851, .065745, .009539, .023699, .026999,$$
$$.003440, .030234, .109065, .046727, .056700, .179288, .067337,$$
$$.023443, .064984, .021364, .001197, .012238, .013456, .001955,$$
$$.017592, .017118, .000702, .005562, .004784].$$

The binomial test is used to test the hypothesis that a cell does not deviate from normality, and we protect the nominal α using the Bonferroni procedure again. Table 18.3 summarizes the results from Stage 2 of the analyses.

The model that was used to analyze Table 18.3 is, overall, rejected as well (LR-$X^2 = 110.89$; $df = 25$; $p < 0.001$). In addition, the multivariate skewness test suggests deviations from trivariate normality (skewness $= 0.6304$; $p = 0.020$), but the multivariate kurtosis test does not suggest any deviation (kurtosis $= 15.005$; $p = 0.497$). The cell-wise tests of multinormality suggest that the null hypothesis can be rejected in three sectors. The first is sector 1 1 1. This indicates that more women reported low levels of Fear, Joy, as well as Anxiety than is compatible with the hypothesis of trivariate normality. 4.837 women (2.4%) were expected to show this pattern, but 13 (6.5%) did show it. Sector 2 2 2 was shown by 17 women (8.5%), but 36.037 (17.9%) had been expected. These are the women who reported average levels of Fear, Joy, and Anxiety. Sector 3 1 3 contains 9 women (4.5%), but only 2.705 (1.3%) had been expected. These are the women who reported high levels of Fear and Anxiety and low levels of Joy.

It should be emphasized that the questions asked in Stages 1 and 2 of the analyses differ fundamentally. In Stage 1, we asked whether the three variables are related to each other. Overall, there is a significant correlation between Joy and Anxiety. At the level of individual configurations, Configuration 3 1 1 stood out. It contains more women than expected under the assumption of variable independence. In Stage 2, we asked whether Fear, Joy, and Anxiety follow a trivariate normal distribution. The answer was that they, overall, deviate significantly because their distribution is more skewed than compatible with a trivariate normal distribution. This characteristic is

Table 18.3 Cell-wise testing the null hypothesis of trivariate normality of the $F \times J \times A$ cross-classification

Cell FJA	m	$\hat{m}$	p
111	13	4.837	.00129490 Type
112	11	7.094	.10144865
113	2	1.170	.32676828
121	10	10.221	.55470838
122	8	13.215	.08303763
123	4	1.917	.12736658
131	10	4.764	.02255166
132	11	5.427	.02157095
133	0	.691	.50030347
211	11	6.077	.04328915
212	21	21.922	.47370254
213	18	9.392	.00663222
221	10	11.397	.40879100
222	17	36.037	.00011143 Antitype
223	6	13.535	.01606187
231	11	4.712	.00836520
232	11	13.062	.33992133
233	7	4.294	.14130721
311	1	.241	.21396832
312	2	2.460	.55342804
313	9	2.705	.00176963 Type
321	1	.393	.32517256
322	1	3.536	.12983095
323	4	3.441	.45132959
331	0	.141	.86832708
332	0	1.118	.32594512
333	2	.962	.25010252

most prominent in 3 of the 27 cells of the cross-classification. It should be noticed that these deviations were not visible when the univariate normal distribution hypothesis was tested for the three variables individually.

Stage 3: CFA under Consideration of Multivariate Normality
In Stage 3, we ask a different question again. We ask whether adding the information used in the second stage to the base model that was used in the first stage results in a model that explains the $F \times J \times A$ cross-classification and makes the type from first-order CFA disappear. We answer this question first at the global level, by way of fitting the model, and, then, at the individual level, by way of testing individual configurations.

At this stage, we combine the information used in the first two stages. The resulting model is $\log \hat{m} = [1 \mid X_m \mid X_n] \begin{bmatrix} \lambda_1 \\ \lambda_{X_m} \\ \lambda_{X_n} \end{bmatrix}$, where X_m contains the vectors

Table 18.4 First-order CFA of the cross-classification of Fear, Joy, and Anxiety under consideration of the hypothesis of a trivariate normal distribution

Configuration					
F	J	A	m	$\hat{m}$	p
1	1	1	13	11.7689	0.39678087
1	1	2	11	10.3839	0.46612952
1	1	3	2	8.0784	0.01172364
1	2	1	10	7.6559	0.23858600
1	2	2	8	6.9583	0.39495250
1	2	3	4	4.3602	0.44193708
1	3	1	10	7.9089	0.26940533
1	3	2	11	6.5445	0.06609495
1	3	3	0	5.3409	0.00445754
2	1	1	11	14.1566	0.23765091
2	1	2	21	21.5856	0.49580786
2	1	3	18	12.8712	0.09538815
2	2	1	10	9.1852	0.43753428
2	2	2	17	19.9560	0.28846594
2	2	3	6	7.9647	0.31279685
2	3	1	11	9.0313	0.29503306
2	3	2	11	10.1838	0.44051452
2	3	3	7	7.0655	0.41121118
3	1	1	1	3.3177	0.15414674
3	1	2	2	2.9228	0.43914363
3	1	3	9	2.9149	0.00290699
3	2	1	1	1.7483	0.47739254
3	2	2	1	1.5986	0.47539253
3	2	3	4	1.5726	0.07432330
3	3	1	0	2.2272	0.10649830
3	3	2	0	1.8664	0.15333707
3	3	3	2	1.8315	0.45241808

for the main effects of F, J, and A and X_n contains the vector that represents the cell-wise probabilities under trivariate normality. The λs are the corresponding model parameters.

To make results comparable, we protect α again using the Bonferroni procedure and perform binomial tests. Table 18.4 summarizes the results.

The joint model used to analyze Table 18.4 fails to describe the data well (LR-$X^2 = 47.535$; $df = 19$; $p = 0.0002$). However, not a single configuration stands out and deviates from expectation.

We conclude that taking into account the distributional characteristics of the three variables helps explain the deviations from variable independence. In other words, the local deviations from variable independence can be explained from distributional characteristics of the three variables.

18.5 Discussion

In this chapter, we presented an approach to combining Configural Frequency Analysis with a cell-wise test of multinormality. Base models of CFA, thus far, represented either specific variable relations (von Eye, 2002) or processes that underlie the development of observations in longitudinal studies (von Eye et al., 2021). The present approach extends the applicability of CFA by allowing the researcher to include information that does not target variable relations.

Whenever additional information is included in CFA base models, interpretability of types and antitypes needs to be considered. Considering the order of the three stages, we first asked whether there exist types or antitypes that indicate where in the space of the trivariate cross-classification there are significant deviations from independence. Second, we asked whether the three variables follow a trivariate normal distribution. Third, we asked whether also including the information concerning the joint distribution of the variables alters the results from the first step. This was the case. The information concerning the joint distribution of the three variables is, this way, used as a covariate in the base model of first-order CFA (cf. Glück & von Eye, 2000).

The model thus created is parsimonious because just one degree of freedom is needed to remove the type that had emerged from first-order CFA. Not unexpectedly, there are alternatives that have the same effect. For example, using the base model of second-order CFA results in a nonsignificant LR-X^2 as well (LR-$X^2 = 10.083$; $df = 8$; $p = 0.259$), and neither types not antitypes emerge.

When it comes to selecting between the two models that make the type disappear, one can make a decision based on two arguments. First, the model of second-order CFA exclusively uses variable relations to explain the variation in the cross-classification. In contrast, the base model of first-order CFA plus trivariate normal distribution information uses information that reflect distributional characteristics. In brief, depending on whether distributional assumptions can meaningfully be applied, one of these two approaches can be preferred. When variables are nominal level, normal distribution information cannot be applied, but there are sampling distributions that might be applied in this case, e.g., the binomial or the negative binomial distribution.

The second argument is, as was noted above, that of parsimony. The first-order CFA plus normality information is more parsimonious than the model of second-order CFA. Here, the difference is 11 degrees of freedom. The base model of second-order CFA might just be unnecessarily complex and that without taking distributional characteristics of variables into account. Therefore, whenever a small number of covariates can meaningfully be added to a base model, this may be the preferred first option.

As was indicated above, the number of bins needs to be carefully considered as well. When only two bins are created, specifically, when variables are dichotomized at the median, the multinormality test is sensitive only to deviations from symmetry.

The base model of first-order CFA, that is, the main effect model, should be able to capture these variations. Therefore, we recommend creating at least three bins for at least one variable. When the number of bins increases, the number of observations must increase as well for the table not to become overly sparsely populated.

The joint model proposed in this chapter can be extended in many ways. For instance, distributional assumptions might differ over variables. Some variables might be normally distributed, others binomially. All this can be straightforwardly incorporated in a CFA base model. Another extension would be to apply the joint model in moderator CFA. This approach to CFA considers two or more comparison groups (see Krauth & Lienert, 1978; von Eye, 2002). Here, distributional hypotheses can be specified separately for the comparison groups, and it can be asked whether information about the distribution of variables helps explain deviations from the base model.

To conclude, the approach proposed here provides, by way of adding distributional information to a CFA base model, a parsimonious option to explain types and antitypes from CFA.

References

Bonferroni, C. E. (1936). Teoria statistica delle classi e calcolo delle probabilità. *Pubblicazioni del R Istituto Superiore di Scienze Economiche e Commerciali di Firenze, 8*, 3–62.

Genz, A. (1992). Numerical computation of multivariate normal probabilities. *Journal of Computational and Graphical Statistics, 1*, 141–149.

Glück, J., & von Eye, A. (2000). Including covariates in configural frequency analysis. *Psychologische Beiträge, 42*, 405–417.

Holland, B. S., & Di Ponzio Copenhaver, M. (1987). An improved sequentially rejective Bonferroni test procedure. *Biometrics, 43*, 417–423.

Huth-Bocks, A. C., Levendosky, A. A., Theran, S. A., & Bogat, G. A. (2004). The impact of domestic violence on mothers' prenatal representations of their infants. *Infant Mental Health Journal, 25*, 79–98.

Kieser, M., & Victor, N. (1991). A test procedure for an alternative approach to configural frequency analysis. *Methodika, 5*, 87–97.

Krauth, J., & Lienert, G. A. (1973). KFA. In *Die Konfigurationsfrequenzanalyse und ihre Anwendung in Psychologie und Medizin*. Alber.

Krauth, J., & Lienert, G. A. (1978). Nonparametric two-sample comparison of learning curves based on orthogonal polynomials. *Psychological Research, 40*, 159–171.

Lautsch, E., & von Weber, S. (1995). *Methoden und Anwendungen der Konfigurationsfrequenzanalyse (KFA)*. Psychologie Verlags Union.

Levendosky, A. A., Bogat, G. A., Davidson, W. S., & von Eye, A. (2000). *Risk and protective factors for domestic violence*. Centers for Disease Control, Injury Prevention Center, RO1/CCR518519–01.

Lienert, G. A. (1968). *Die "Konfigurationsfrequenzanalyse" als Klassifikationsmethode in der klinischen Psychologie*. Vortrag auf dem 26. Kongress der Deutschen Gesellschaft für Psychologie in Tübingen 1968.

Lienert, G. A., & Krauth, J. (1975). Configural frequency analysis as a statistical tool for defining types. *Educational and Psychological Measurement, 35*, 231–238.

Mair, P., & von Eye, A. (2007). Application scenarios for nonstandard log-linear models. *Psychological Methods, 12*, 139–156.

Mardia, K. V. (1970). Measures of multivariate skewness and kurtosis with applications. *Biometrika, 57*, 519–530.
Mardia, K. V. (1980). Tests of univariate and multivariate normality. In P. R. Krishnaiah (Ed.), *Handbook of statistics* (Vol. 1, pp. 279–320). North-Holland.
Schuster, C., & von Eye, A. (2000). Using log-linear modeling to increase power in two-sample configural frequency analysis. *Psychologische Beiträge, 42*, 273–284.
Somerville, P. N. (1998). Numerical computation of multivariate normal and multivariate t-probabilities over convex regions. *Journal of Computational and Graphical Statistics, 7*, 529–544.
Victor, N. (1983). An alternative approach to configural frequency analysis. *Methodika, 3*, 61–73.
von Eye, A. (2002). *Configural frequency analysis - methods, models, and applications*. Lawrence Erlbaum.
von Eye, A. (2004). Base models for configural frequency analysis. *Psychology Science, 46*, 150–170.
von Eye, A., & Gardiner, J. C. (2004). Locating deviations from multivariate normality. *Understanding Statistics, 3*, 313–331.
von Eye, A., & Gutiérrez Peña, E. (2004). Configural frequency analysis - the search for extreme cells. *Journal of Applied Statistics, 31*, 981–997.
von Eye, A., & Wiedermann, W. (2022). *Configural frequency analysis*. Springer. (in press).
von Eye, A., Schuster, C., & Rogers, W. M. (1998). Modeling synergy using manifest categorical variables. *International Journal of Behavioral Development, 22*, 537–557.
von Eye, A., von Eye, M. J. E., & Bogat, G. A. (2006). Multinormality and symmetry: A comparison of two statistical tests. *Psychology Science, 48*, 419–435.
von Eye, A., Mair, P., & Mun, E.-Y. (2010). *Advances in configural frequency analysis*. Guilford Press.
von Eye, A., Wiedermann, W., Herman, K., & Reinke, W. (2021). Local effects of intervention: A configural analysis. *Prevention Science*. https://doi.org/10.1007/s11121-021-01241-8
Wu, C. F. J., & Hamada, M. (2000). *Experiments: Planning, analysis and parameter design optimization*. Wiley.

Chapter 19
Configural Frequency Analysis Under Multinormality in Incomplete Tables

Alexander von Eye and Wolfgang Wiedermann

Cross-classifications of categorical variables are *incomplete* when they contain cells that cannot contain any count (Bishop et al., 1975). For example, in the cross-classification of species (with categories Human–non-Human) and extremities (with categories "Arms and Legs" and Wings), the cell "human with wings" contains no count. These zeros are called *structural zeros*. There are, however, cells with zero counts for which, under different conditions, cases could have been found. These zeros are called *random zeros*, *observed zeros*, or *empirical zeros* (Andersen, 1997; von Eye & Mun, 2013).

Unless there are so many random zeros in a table that estimation becomes impossible, they usually pose no problems. Frequencies and parameters can be estimated, observed zeros and expected cell frequencies can be compared, and model fit can be evaluated. Of note, for random zeros, expected frequencies are estimated as well. When, however, there are structural zeros, frequencies must be estimated such that expectation for structural zeros is zero. The probability mass that would have been assigned to the cells, had they not represented structural zeros, is then distributed over those cells that contain empirical zero and nonzero counts.

This is important in log-linear modeling (Bishop et al., 1975) because investing estimated frequencies in cells with structural zeros not only would be incorrect, but

A. von Eye (✉)
Michigan State University, East Lansing, MI, USA

Montpellier, France
e-mail: voneye@msu.edu

W. Wiedermann
Educational, School and Counseling Psychology, University of Missouri, Columbia, MO, USA

M. Stemmler et al. (eds.), *Dependent Data in Social Sciences Research*,
https://doi.org/10.1007/978-3-031-56318-8_19

it can also dramatically increase model-data discrepancies and, therefore, require unnecessarily complex models for acceptable fit. In Configural Frequency Analysis (CFA; Lienert, 1968; Lienert & Krauth, 1975; von Eye & Gutiérrez Peña, 2004; von Eye & Wiedermann, 2022), this is equally important because decisions in CFA are based on cell-wise comparisons of observed with expected frequencies.

Specifically, in CFA, the null hypothesis $H_0 : m_i = \hat{m}_i$ is tested, where m_i is the observed frequency of Cell i (for $i = 1, \ldots, t$; where t denotes the number of cells in a cross-classification) and $\hat{m}_i$ is the expected frequency in Cell i. $\hat{m}_i$ is estimated under the CFA base model (see von Eye, 2004; von Eye & Gutiérrez Peña, 2004; von Eye & Wiedermann, 2022), that is, a suitable probability model.

$p_i(m_i - 1)$ indicates the probability that Cell i contains $m_i - 1$ cases. This cell is said to constitute a *CFA type* when the null hypothesis is rejected because $p_i(m_i - 1) \geq 1 - \alpha$. When, however, the null hypothesis is rejected because $p_i(m_i) \leq \alpha$, Cell i is said to constitute a *CFA antitype*. When the null hypothesis prevails, Cell i constitutes neither a type nor an antitype. When applying CFA, researchers aim at identifying types and antitypes.

Types and antitypes are interpreted with respect to CFA base models. These are specified according to a number of criteria (see von Eye, 2004; von Eye & Wiedermann, 2022), the most important of which is that there be only one reason for types and antitypes to emerge. For example, the base model of first-order CFA—it corresponds to a log-linear main effect model—can result in CFA types and antitypes only when variable relations locally exist. Similarly, the base model of prediction CFA can result in types and antitypes only when predictor-criterion relations exist because all effects on the predictor side and all effects on the criterion side are taken into account, and the model posits that predictors and criterion variables are independent.

CFA base models that have been discussed in the literature belong to one of four groups (von Eye et al., 2021). In the first group, variable relations are specified (as in the two examples in the last paragraph). In the second group, variable distributions are taken into account such as the multivariate normal distribution (von Eye et al., 2021). The third group is used to investigate global functions, for instance, developmental functions. The fourth group investigates functions that are specified for particular sectors of series of observations.

In the companion chapter by von Eye and Wiedermann (2023), it has been proposed considering the multivariate normal distribution in tandem with variable relations. This approach will be extended in the present chapter. Specifically, we revisit CFA under a multivariate normal distribution for the case in which tables are incomplete.

The remainder of this chapter is structured as follows. In the next section, we describe Somerville's (1998a, b, 2001) method to estimate probabilities under the assumption of multinormality (for an alternative solution, see Genz, 1992). This is followed by the incorporation of this method into CFA (von Eye et al., 2021; see also von Eye, 2010) of incomplete tables and a real-world data example.

19.1 Estimating Probabilities Under the Assumption of Multinormality

Decisions about the existence of clusters of cases in a multivariate data space are often based on comparisons of the observed density of cases in a particular sector with the observed density in some other, typically neighboring sector. There exist just a few approaches to comparing the observed numbers of cases in clusters with expected numbers. For example, Bock (1996) or Kopp (1976) proposed models under the assumption that observations follow a uniform distribution. von Eye (2010) suggested using the assumption of a multinormal distribution and compared results obtained under the assumptions of a Poisson, a unimodal model, and a multinormal distribution. The existence of clusters would contradict expectations based on these models.

In contrast to cluster analysis, CFA does not only look at agglomerations but also at sparsely populated sectors of the data space. Standard clusters would, therefore, correspond to CFA types, and sparsely populated sectors would correspond to CFA antitypes. Apart from the rare cases in which CFA is applied under the assumption of a uniform distribution (zero order CFA), CFA derives base models from data generation processes that can be very flexible probability models. The model that we use in the present context is that of multinormality.

Also in contrast to cluster analysis, we work, in the present context, with categorical variables. There are two aspects to this difference. First, the variables that can be analyzed with the method proposed here are either ordinal or binned, that is, categorized, interval, or ratio level. This is required for the multinormality hypothesis to be applicable. Second, in Somerville (1998a; cf. Wang & Kennedy, 1992), the sectors to be analyzed are convex and bounded on all sides. Accordingly, von Eye (2010; cf. von Eye, 1978; von Eye & Wirsing, 1978) encircled clusters with convex hulls, that is, ellipsoids or d-dimensional rectangles, and, thus, arrived at the same situation as Somerville (1998a, b). Cross-classified categorized variable bins are convex, by definition.

Now, to describe Somerville's (1998a, b) method, consider the d-dimensional bin that is created by segments on d independent multinormal variables $x' = (x_1, x_2, \ldots, x_d)$. That is, $f(x) = \mathrm{MVN}(\mu, \Sigma\sigma^2)$, where Σ is a known positive definite matrix and σ^2 is a constant representing the global variance term of the multinormal distribution. Let A be the known bounded sector in the d-space, the probability of which,

$$P = \int_A f(x)dx,$$

is to be determined.

When the Poisson model of uniformity is used, this probability simply is given by the relative size of the sector, A. When, however, a multinormal distribution is hypothesized, multivariate integrals need to be evaluated. To perform this in the concrete case, Somerville (1998a, b) proposes bootstrapping procedures. Somerville's method first performs a coordinate transformation (i.e., the Cholesky decomposition). The resulting new coordinate system is spanned by d uncorrelated normal random variables with unit variances. The subregion A to be evaluated is the bin. For each of these, a fixed number of points are placed in random directions. The probability is estimated for each of these points to be placed inside the bin. This process is repeated a fixed number of times for each bin (for the data examples in this chapter, we selected this number to be 10,000). The integral is obtained as the average of the probability estimates (for more technical detail, see Somerville, 1998a, b). In other words, this procedure "shoots" multinormally distributed data points in the direction of A and calculates the probability that a point lands inside of A. The result of this Monte Carlo approach is the estimate of the probability of Sector A.

19.2 CFA Base Models that Consider Multinormality

In the remainder of this chapter, we consider CFA base models of the form $\log \hat{m} = X\lambda$, where $\hat{m}$ is the vector of expected cell frequencies, X is the design matrix, and λ is the vector of model parameters. As was noted in von Eye and Wiedermann (2023), the models used in CFA can be hierarchical, non-hierarchical, and non-standard (Mair & von Eye, 2007). von Eye and Wiedermann (2023) considered non-hierarchical base models that take the probability of each configuration (cell) into account that was estimated using the Somerville (1998a, b) algorithm. These are models of the form $\log \hat{m} = [X_0 \mid X_n] \begin{bmatrix} \lambda_0 \\ \lambda_{X_n} \end{bmatrix}$, where X_0 is the standard base model and X_n the vector(s) that contain the multinormal probabilities. λ_0 contains the parameters for the standard base models and λ_n contains the parameters for the multinormal probability vector(s). When the hypothesis is considered that the variables that span the cross-classification follow a multinormal distribution, λ_n contains just one parameter. When different distributions are tested for different variables, λ_n can contain multiple parameters.

In the present chapter, we extend this model to be able to accommodate incomplete tables. These are tables that contain cells that are empty because they contain structural zeros. For each of these cells, one vector is included in the design matrix. The base model thus becomes

$$\log \hat{m} = [X_0 \mid X_n \mid X_z] \begin{bmatrix} \lambda_0 \\ \lambda_{X_n} \\ \lambda_{X_z} \end{bmatrix},$$

where X_0, X_n, λ_0, and λ_n are defined as before, X_z contains the vectors that indicate the location of the structural zeros, and λ_z contains the parameters. Please note that this model can be applied as well when there are fixed cells, that is, cells that do contain observed frequencies but are excluded from the estimation and model fitting process.

The extension of the base model obviously does not change the form of the model. It comes, however, with a complication that requires calculation of the probabilities in vector X_n under different circumstances. In earlier applications of Somerville's approach (e.g., von Eye, 2010; von Eye & Wiedermann, 2023; see also Genz, 1992; Wang & Kennedy, 1992), the probabilities of each bin were estimated. That is, in the present context, the probability of each configuration is estimated under the assumption of multinormality. The algorithm does not allow one to consider structural cells. Here, structural cells are taken into account.

19.3 Data Example

In the following example, we reanalyze data that were collected in an experiment conducted by Hussy (1991; for an earlier analysis of these data, under a different base model, see von Eye & Wiedermann, 2022). In this experiment, 173 participants played a problem-solving game in which color, and position of pegs on a playing board must be identified in the smallest number of guesses possible. The time needed for each guess was coded as 1 = 0 to 10 seconds (included), 3 = between 10 and 20 seconds (included), and 3 = more than 20 seconds. A fourth category, 4, was used to indicate that a participant had solved the problem. For the following example, we use the problem-solving times in the final two of eight trials. These times are labeled $T7$ and $T8$.

The $T7 \times T8$ cross-classification contains three cells with structural zeros. These are Cells 4 3, 4 2, and 4 1. These profiles describe participants who had solved the problem before trial eight (and, thus, completed the game), but would still have recorded a problem-solving time in trial eight, which is impossible. Participants with Profile 4 4 also solved the problem before the last trial. However, for these participants, no problem-solving time was recorded for trial eight. Therefore, Cell 4 4 does not contain a structural zero. Instead, it contains an observed frequency.

In the following paragraphs, we present results from three approaches to analyzing these data. In the first approach, we ignore that there are three structural zeros and analyze the 4 × 4 cross-classification as if the zeros in Cells 4 1, 4 2, and 4 3 were empirical, that is, random zeros. We use the base model of first-order CFA, that is, $\log \hat{m} = [X_0][\lambda_0]$, where, as before, X_0 is the design matrix for standard first-order CFA—it corresponds to a log-linear main effect model—and λ_0 represents the corresponding parameters. This model is given in the parts before the first vertical separation line in the equation for the model for the third analytic approach, below.

In the present example, X_0 contains seven columns. The first is for the model constant, the following three represent the main effects of trial seven, and the last three represent the main effects of trial eight. λ_0 contains the corresponding seven parameters.

Table 19.1 First-order CFA of the problem-solving times in Trials 7 and 8

Configuration *T7 T8*	m	$\hat{m}$	p	
1 1	22.00	7.8613	.00001424	Type
1 2	9.00	6.4740	.20186777	
1 3	7.00	8.5549	.37345234	
1 4	2.00	17.1098	.00000299	Antitype
2 1	4.00	6.4855	.21996563	
2 2	9.00	5.3410	.08935676	
2 3	15.00	7.0578	.00511261	
2 4	5.00	14.1156	.00391360	
3 1	8.00	7.8613	.53036019	
3 2	10.00	6.4740	.11628355	
3 3	15.00	8.5549	.02542666	
3 4	7.00	17.1098	.00362213	
4 1	.00	11.7919	.00000497	Antitype
4 2	.00	9.7110	.00004566	Antitype
4 3	.00	12.8324	.00000162	Antitype
4 4	60.00	25.6647	.00000000	Type

For each of the three configural analyses that are presented in the following paragraphs, we use the binomial test. This test is exact and can be applied under any base model. We protect the nominal α using the Bonferroni procedure which results in the constant, protected $\alpha^* = 0.003125$. Table 19.1 displays the results from first-order CFA, that is, the one that neglects the status of Cells 4 1, 4 2, and 4 3 as containing structural zeros. Distributional characteristics are not taken into account either.

The base model of first-order CFA fails to describe the data well. Specifically, the likelihood ratio (LR) chi-square goodness-of-fit test suggests that very strong effects exist in the cross-classification of the two variables (LR-$X^2 = 172.80$; $df = 9$; $p < 0.001$). Two types and four antitypes emerge. This indicates that the problem-solving times in the two last trials are associated.

Instead of interpreting the types and antitypes in Table 19.1 in detail, we point at the antitypes that are listed for Configurations 4 1, 4 2, and 4 3. These are the cells that contain structural zeros. Cells with structural zeros cannot constitute types or antitypes. The zeros in these cells are not random, and therefore, there are no expected frequencies from which they could deviate.

Therefore, in the following step of analysis, we take the status of Configurations 4 1, 4 2, and 4 3 as containing structural zeros into account. The base model for this CFA is $\log \hat{m} = [X_0 \mid X_z] \begin{bmatrix} \lambda_0 \\ \lambda_{X_z} \end{bmatrix}$, where X_z contains the vectors that represent the structural zeros and λ_z contains the corresponding parameters. For each structural zero, there is one column in the design matrix. The design matrix for this base model

Table 19.2 First-order CFA of the cross-classification of *T7* with *T8* (structural zeros considered)

Configuration *T7 T8*	*m*	$\hat{m}$	*p*
1 1	22.000	12.035	0.004599
1 2	9.000	9.912	0.465645
1 3	7.000	13.097	0.045154
1 4	2.000	4.957	0.124742
2 1	4.000	9.929	0.027303
2 2	9.000	8.177	0.432720
2 3	15.000	10.805	0.125034
2 4	5.000	4.089	0.388645
3 1	8.000	12.035	0.143402
3 2	10.000	9.912	0.465645
3 3	15.000	13.097	0.331062
3 4	7.000	4.957	0.229151
4 1	0.000	0.000 *	-
4 2	0.000	0.000 *	-
4 3	0.000	0.000 *	-
4 4	60.000	59.998	-

thus contains a total of ten columns. This matrix is also given in the equation for the final model, below. The three vectors for the structural zeros can be found between the two separation lines. Table 19.2 displays the results for this analysis.

In comparison to the first run, the base model describes the data significantly better (ΔLR-$X^2 = 151.54$; $p < 0.001$), but it still fails to describe the frequencies in the *T7* × *T8* cross-classification well (LR-$X^2 = 21.26$; $df = 6$; $p = 0.0017$). In addition, not a single type results. Cell 1 1 comes closest to constituting a type. This profile describes those 22 participants (12.72% of the sample) for which short problem-solving times were recorded in both of the final two trials (12.04 had been expected; 6.96%). The probability for this cell is estimated to be $p = 0.005$, a value that is larger than the threshold of 0.003125. This cell fails to constitute a type even when the adjusted significance threshold is $\alpha^* = 0.05/12 = 0.004166667$—this value results when the actual number of cell-wise tests, $t = 12$, is used to determine the adjusted threshold. Here, $t = 12$ can be used because testing hypotheses for structural zeros is not done.

In the third stage of analysis, we add a second component to the CFA base model. This component consists of the cell probabilities under the assumption of binormality. The base model thus becomes $\log \hat{m} = [X_0 \mid X_z \mid X_n] \begin{bmatrix} \lambda_0 \\ \lambda_{X_z} \\ \lambda_n \end{bmatrix}$, where X_n is the column that represents the probabilities and λ_n represents the corresponding parameter. Explicitly, the design matrix now is, in effect coding, for the third stage of the present example,

$$X = \begin{bmatrix}
1 & 1 & 0 & 0 & 1 & 0 & 0 & | & 0 & 0 & 0 & | & .055802 \\
1 & 0 & 1 & 0 & 1 & 0 & 0 & | & 0 & 0 & 0 & | & .068013 \\
1 & 0 & 0 & 1 & 1 & 0 & 0 & | & 0 & 0 & 0 & | & .046651 \\
1 & -1 & -1 & -1 & 1 & 0 & 0 & | & 0 & 0 & 0 & | & .017994 \\
1 & 1 & 0 & 0 & 0 & 1 & 0 & | & 0 & 0 & 0 & | & .061123 \\
1 & 0 & 1 & 0 & 0 & 1 & 0 & | & 0 & 0 & 0 & | & .091554 \\
1 & 0 & 0 & 1 & 0 & 1 & 0 & | & 0 & 0 & 0 & | & .077181 \\
1 & -1 & -1 & -1 & 0 & 1 & 0 & | & 0 & 0 & 0 & | & .036607 \\
1 & 1 & 0 & 0 & 0 & 0 & 1 & | & 0 & 0 & 0 & | & .034505 \\
1 & 0 & 1 & 0 & 0 & 0 & 1 & | & 0 & 0 & 0 & | & .063532 \\
1 & 0 & 0 & 1 & 0 & 0 & 1 & | & 0 & 0 & 0 & | & .065819 \\
1 & -1 & -1 & -1 & 0 & 0 & 1 & | & 0 & 0 & 0 & | & .038371 \\
1 & 1 & 0 & 0 & -1 & -1 & -1 & | & 1 & 0 & 0 & | & 0 \\
1 & 0 & 1 & 0 & -1 & -1 & -1 & | & 0 & 1 & 0 & | & 0 \\
1 & 0 & 0 & 1 & -1 & -1 & -1 & | & 0 & 0 & 1 & | & 0 \\
1 & -1 & -1 & -1 & -1 & -1 & -1 & | & 0 & 0 & 0 & | & .34326
\end{bmatrix}$$

The first column in this design matrix represents the model constant. The following three columns represent the main effects of *T*7, in effect coding. They are followed by the three columns for the main effects of *T*8. After the separator column—it was inserted for visual clarity and has no mathematical or statistical implication—one finds the three columns that indicate the cells with structural zeros. The last column contains the probabilities that were estimated using Somerville's (2001) algorithm.

This design matrix represents the design matrix for the third configural analysis of the problem-solving times in trials 7 and 8. Table 19.3 displays the results.

The base model of independence of the problem-solving times of *T*7 and *T*8 that takes into account that three cells contain structural zeros and estimate the expected cell frequencies under the assumption of binormality describes the frequencies well (LR-$X^2 = 8.524$; $df = 5$; $p = 0.1296$). Accordingly, there is not a single cell for which the observed and the expected cell frequencies deviate from each other such that it would constitute a CFA type or antitype.

It should be noticed that no CFA test was performed for Cell 4 4, although it contains an empirical frequency. The reason for this is that, when all but one cells for one variable category—here, these are Cells 4 1, 4 2, and 4 3—are fixed, there is no room left for deviations in the last cell—here, this is Cell 4 4. The observed frequency in this cell will always be perfectly reproduced.

From these results, an interesting conclusion can be derived: The problem-solving times in *T*7 and *T*8 are independent. The empirical correlation between the

Table 19.3 First-order CFA of the cross-classification of *T7* with *T8* under binormality (structural zeros considered)

Configuration				
T7	*T8*	*m*	$\hat{m}$	*p*
1	1	22.000	18.498364	0.22556
1	2	9.000	9.990	0.45554
1	3	7.000	8.111	0.43394
1	4	2.000	3.402	0.33667
2	1	4.000	8.606	0.06498
2	2	9.000	10.270	0.41994
2	3	15.000	11.302	0.16154
2	4	5.000	2.822	0.15430
3	1	7.000	6.896	0.38543
3	2	10.000	7.741	0.24829
3	3	15.000	17.588	0.30841
3	4	7.000	7.776	0.48214
4	1	0.000	0.000 *	-
4	2	0.000	0.000 *	-
4	3	0.000	0.000 *	-
4	4	60.000	60.000	-

problem-solving times in these two trials is $r = 0.35$. However, this correlation can be explained from the distributional characteristics of the two variables. These, in tandem with the structural zeros, result in estimated expected cell frequencies, none of which deviates significantly from its corresponding observed frequency.

19.4 Discussion

In this chapter, we extended the arsenal of methods for configural analysis by showing how the assumption of multinormality can be taken into account when tables are incomplete. The results of a data example suggest that CFA (and log-linear models) can produce heavily biased results when only the structural zeros but not the assumption of multinormality are taken into account, above and beyond the base model. In earlier discussions of methods of analysis of cross-classifications, it has been shown already that neglecting information can bias results. For example, von Eye and Wiedermann (2022) illustrated (using the same data as this chapter) that antitypes are bound to emerge when frequencies are estimated for cells with structural zeros. In a similar fashion, von Eye and Wiedermann (2018) demonstrated that ignoring the information that comes with scale levels can lead to misinterpretation of cross-classifications (see also Bergman, 2018).

In this chapter, methods are presented and applied that allow researchers to consider information from two sources. These are (1) the completeness of a table and (2) distributional characteristics of the variables that span a table. Extensions of this approach can include the following. First, when categorized continuous and nominal-level variables are crossed, these variables can follow different distributions. For example, categorized interval-level variables can follow a multinormal distribution. Nominal-level variables, in the same table, might follow a binomial distribution. Information concerning these distributions can be included in the base model as well. Second, scale information can also be incorporated.

Taking into account information as discussed in this chapter serves two purposes. First, in log-linear modeling, this information can have the effect that models that describe data are simpler than without taking into account this information. Second, in configural analysis, types and antitypes might disappear (or freshly emerge) when this information is included in the base model. In both cases, results are bound to be simpler and easier to interpret, and conclusions are avoided that are based on variable relations that, empirically, do not exist.

References

Andersen, E. B. (1997). *Introduction to the statistical analysis of categorical data*. Springer. https://doi.org/10.1007/978-3-642-59123-5_5

Bergman, L. R. (2018). Response to von Eye and Wiedermann's commentary "strengthening arguments based on scale levels?". *Journal for Person-Oriented Research, 6*, 48. https://doi.org/10.17505/jpor.2018.05

Bishop, Y. M. M., Fienberg, S. E., & Holland, P. W. (1975). *Discrete multivariate analysis*. MIT Press.

Bock, H. H. (1996). Probability models in partitional cluster analysis. *Computational Statistics and Data Analysis, 23*, 5–28.

Genz, A. (1992). Numerical computation of multivariate normal probabilities. *Journal of Computational and Graphical Statistics, 1*, 141–149.

Hussy, W. (1991). Eine experimentelle Studie zum Intelligenzkonzept "Verarbeitungskapazität". *Diagnostica, 37*, 314–333.

Kopp, B. (1976). Über ein Verfahren zur Gruppenbildung durch Dichtefunktionen. *Biometrische Zeitschrift, 18*, 291–296.

Lienert, G. A. (1968). *Die "Konfigurationsfrequenzanalyse" als Klassifikationsmethode in der klinischen Psychologie.* Vortrag auf dem 26. Kongress der Deutschen Gesellschaft für Psychologie in Tübingen 1968.

Lienert, G. A., & Krauth, J. (1975). Configural Frequency Analysis as a statistical tool for defining types. *Educational and Psychological Measurement, 35*, 231–238.

Mair, P., & von Eye, A. (2007). Application scenarios for nonstandard log-linear models. *Psychological Methods, 12*, 139–156.

Somerville, P. N. (1998a). Numerical computation of multivariate normal and multivariate t-probabilities over convex regions. *Journal of Computational and Graphical Statistics, 7*, 529–544.

Somerville, P. N. (1998b). A Fortran 90 program to evaluate multivariate normal and multivariate-t integrals over convex regions. *Journal of Statistical Software*. http://www.stat.ucla.edu/journals/jss/v03/i04/. Downloaded in December, 2004.

Somerville, P. N. (2001). Numerical computation of multivariate normal and mutivariate-t probabilities over ellipsoidal regions. *Journal of Statistical Software*. http://www.stat.ucla.edu/www.jstatsoft.org/v06/i08/. Downloaded in December, 2004.

von Eye, A. (1978). Clusteranalyse durch Umhüllung lokaler Dichtezentren. In L. H. Eckensberger (Ed.), *Bericht über den 31. Kongreß der Deutschen Gesellschaft für Psychologie in Mannheim* (pp. 506–508). Hogrefe.

von Eye, A. (2004). Base models for Configural Frequency Analysis. *Psychology Science, 46*, 150–170.

von Eye, A. (2010). Examining distributional characteristics of clusters. *Bulletin de la Société des Sciences Médicales du Grand-Duché de Luxembourg, 1*, 14–39.

von Eye, A., & Gutiérrez Peña, E. (2004). Configural Frequency Analysis – The search for extreme cells. *Journal of Applied Statistics, 31*, 981–997.

von Eye, A., & Mun, E.-Y. (2013). *Log-linear modeling – Concepts, interpretation and applications*. Wiley.

von Eye, A., & Wiedermann, W. (2018). Strengthening arguments based on scale levels? *Journal of Person-Oriented Research, 4*, 45–46. https://doi.org/10.17505/jpor.2018.04

von Eye, A., & Wiedermann, W. (2022). *Configural frequency analysis*. Springer.

von Eye, A., & Wiedermann, W. (2023). Configural Frequency Analysis under multinormality. In M. Stemmler, W. Wiedermann, & F. Huang (Eds.), *Dependent data in social sciences research: Forms, issues, and methods of analysis* (2nd ed.). Springer.

von Eye, A., Wiedermann, W., & von Weber, S. (2021). Configural analysis of oscillating progression. *Journal of Person-Oriented Research, 7*, 14–21.

von Eye, A., & Wirsing, M. (1978). An attempt for a mathematical foundation and evaluation of MACS, a method for multidimensional automatical cluster detection. *Biometrical Journal, 20*, 655–666.

Wang, M., & Kennedy, W. J. (1992). A numerical method for accurately approximating multivariate normal probabilities. *Computational Statistics & Data Analysis, 13*, 197–210.

Chapter 20
Higher-Order Configural Frequency Analysis of Groups of Variables: Dependencies in Test Data

Alexander von Eye and Wolfgang Wiedermann

Configural Frequency Analysis (CFA; Lienert & Krauth, 1975; von Eye & Gutiérrez Peña, 2004; von Eye & Wiedermann, 2021a) allows researchers who take a person-oriented perspective to identify those sectors in the data space that deviate significantly from expectation. These sectors are said to constitute *CFA types* when they contain more cases than expected. They are said to constitute *CFA antitypes* when they contain fewer cases than expected. The interpretation of types and antitypes rests on probability models, termed *CFA base models* (see von Eye, 2004; von Eye & Wiedermann, 2021a; Wiedermann et al., 2021), that are specified to reflect a priori hypotheses. More specifically, base models have as two of their main characteristics that the relations of interest are negated to exist and that there is just one way to deviate from the model. If deviations occur in the form of types and antitypes, the relations of interest statistically exist, locally. In other words, there is just one form of variable relations that can result in types and antitypes, and this form supports the a priori hypotheses.

Consider two examples of base models. The first is that of classical CFA (Lienert, 1968). In this approach, chi-square components are estimated. When these are significant, the assumption of variable independence that underlies Pearson's chi-square test is violated. When types or antitypes emerge, variables must be statistically related, at least locally. These are variable relations of any kind, as long as the chi-square test is sensitive to them. The second example is that of Prediction CFA (P-CFA; see von Eye et al., 2005). The base model of P-CFA includes all possible main effects and interactions among the predictor variables

A. von Eye (✉)
Michigan State University, East Lansing, MI, USA

Montpellier, France
e-mail: voneye@msu.edu

W. Wiedermann
Educational, School and Counseling Psychology, University of Missouri, Columbia, MO, USA

M. Stemmler et al. (eds.), *Dependent Data in Social Sciences Research*,
https://doi.org/10.1007/978-3-031-56318-8_20

and all possible main effects and interactions among the criterion variables, but it proposes independence between these two groups of variables. When, under this base model, types and antitypes emerge, relations among predictors and criterion variables statistically exist, at least locally.

The classical CFA base model has experienced development in the sense that higher-order models have been developed. In these models, first-, second-, or higher-order interactions are taken into account, so that types and antitypes can only emerge when interactions of an even higher order exist (see Lienert et al., 1987; von Eye & Lienert, 1984; von Eye & Wiedermann, 2021a). In contrast, P-CFA and 2-group CFA have not been further developed in this direction. Thus far, relations of any order, that is, first-order, second-order, or higher-order interactions, can result in P-CFA types and antitypes. In the present chapter, we propose higher-order CFA of two groups of variables.

This chapter is structured as follows. First, we present base models for second- and higher-order CFA of two groups of variables. We discuss the models and their interpretation. This is followed by a data example and the discussion.

20.1 Base Models for Two Groups of Variables

Let $Z_1, \ldots, Z_i$ be the I variables in the first group and $Y_1, \ldots, Y_j$ the J variables in the second group. The goal of analysis is the identification of local relations between these two groups of variables that exist regardless of the variable relations within these two groups. In analogy to P-CFA, the simplest base model for these two groups of variables is, in log-linear notation,

$$\log \hat{m} = \Lambda_z \left[Z_1, \ldots, Z_i\right] + \Lambda_Y \left[Y_1, \ldots, Y_j\right],$$

where Λ_Z is a vector that contains all possible main effects and interactions of the Z variables and Λ_Y is a vector that contains all possible main effects and interactions of the Y variables. In this model, there is no provision for Z—Y interactions. Therefore, types and antitypes indicate local relations (interactions) among the Z and the Y variables. In terms of design matrices, this model can equivalently be expressed as

$$\log \hat{m} = \left[X_Z \mid X_Y\right] \begin{bmatrix} \Lambda_Z \\ \Lambda_Y \end{bmatrix},$$

where X_Z is the part of the design matrix that specifies all effects of the Z variables and X_Y is the part of the design matrix that specifies all effects of the Y variables. Here as well, there is no provision for interactions that link the Z and the Y variables.

In the present chapter, we propose taking such interactions into account. The base model then becomes

$$\log \hat{m} = \Lambda_Z [Z_1, \ldots, Z_i] + \Lambda_Y [Y_1, \ldots, Y_j] + \Lambda_{ZY} [Z_1, \ldots, Z_i, Y_1, \ldots, Y_j],$$

where Λ_{ZY} is the vector of all Z—Y interactions that are taken into account. Equivalently, this model can also be expressed as

$$\log \hat{m} = [X_Z \mid X_Y \mid X_{ZY}] \begin{bmatrix} \Lambda_Z \\ \Lambda_Y \\ \Lambda_{ZY} \end{bmatrix},$$

where Λ_{ZY} is the vector of all Z—Y interactions that are taken into account and X_{ZY} is that part of the design matrix of this model that specifies these interactions.

With these models in mind, CFA can proceed as usual. Specifically, the following four steps are taken (see von Eye & Wiedermann, 2021a; Wiedermann et al., 2021):

1. *Specification of base model and estimation of expected cell frequencies.* The base models of interest in the context of higher-order CFA of two groups of variables follow the form given above. The expected cell frequencies are estimated with log-linear models, or, equivalently, generalized linear models (GLMs) with Poisson error structure and a log link. Please note that there exist CFA base models for which the expected cell frequencies are not estimated with log-linear models, e.g., symmetry models, but these are of lesser interest in the present context.
2. *Selection of significance test.* CFA significance test is not goodness-of-fit tests. Instead, they help determine whether individual cells (or groups of cells) constitute types or antitypes. The selection of a test depends on sampling scheme and base model. In the present case, any test can be selected that is admissible under multinomial sampling. These include, for instance, the z-test, the chi-square test, or the binomial test.
3. *Performing CFA under protection of the nominal significance level* α. A number of procedures has been proposed for the protection of α. Among these are the Bonferroni (1936), the Holm (1979), and the Holland and Di Ponzio Copenhaver (1987) procedures. Each of these determines the protected level α in dependence of the number of tests to be performed. When the number of tests is very large, the protected level α can become extreme. Therefore, more liberal procedures can be considered (see below, in the data example).
4. *Interpretation of types and antitypes.* The sampling scheme and the base model are used to interpret CFA types and antitypes. In this chapter, new base models are proposed. Therefore, we place, in the following sections, an emphasis on the interpretation that is based on these new base models. Sampling is assumed to be multinomial.

20.2 Interpretation of Types and Antitypes in Higher-Order CFA of Groups of Variables

As was indicated above, CFA base models are specified such that they take into account every possible effect except the effects of interest. When such a base model is rejected and types and antitypes emerge, the effects of interest statistically exist. In the present context, we ask, what are the effects of interest. The answer to this question is clear. These are the higher-order interactions of variables from one group of variables (above, these were the Z variables) with the other group of variables (above, these were the Y variables).

More specifically, the following effects link the Z variables with the Y variables:

1. First-order interactions between two variables, e.g., an interaction of the form $Z_2 \times Y_3$ (with subscripts indicating the position of the variables in the corresponding design matrices). These interactions are not of particular interest in higher-order CFA of variable groups because they can cause types and antitypes to emerge in standard CFA designs such as P-CFA. In addition, they are taken into account in the base model of higher-order CFA.
2. Second-order interactions among three variables from different groups, e.g., the $Z_2 \times Y_3 \times Y_4$ interaction. Within each of the groups, Z and Y, interactions of any order are taken into account. When variables from different groups interact, this interaction can be the cause for types and antitypes to emerge in higher-order CFA of variable groups.
3. Third- and higher-order interactions among four or more variables from different groups, e.g., the $Z_1 \times Z_2 \times Y_3 \times Y_4$ interaction. These can exist even when second-order interactions are taken into account and are, therefore, of interest in higher-order CFA of variable groups.

The question now is how to interpret types and antitypes when they are caused by second- or higher-order interactions. When there are three binary variables under study, interpretation is straightforward, because there can be only one three-way interaction. Types and antitypes can be caused only by this effect. When, however, the three variables are non-binary or when more than three variables are under study, the data analyst needs to search for the effects that result in types or antitypes. This search can be performed with the tools of exploratory *functional CFA* (*F-CFA*; von Eye & Mair, 2008a, b, c).

F-CFA can be applied with two strategies. The *ascending strategy* involves adding terms (effects) to the base model until types and antitypes disappear. The terms that make types and antitypes disappear are termed *type- (antitype-) generating*. When added terms have no effect on the presence of types and antitypes, they are removed from the model. The *descending strategy* begins with the saturated model. Step-by-step, nonsignificant terms are removed until types and antitypes resurface. The terms that have the effect that types and antitypes reappear are, again, termed type- (antitype-) generating.

With rare exceptions, the two strategies point at the same type-(antitype-) generating terms. One advantage of the ascending strategy is that covariates can be included in the model at any step (cf. Glück & von Eye, 2000). This way, hypotheses can be examined concerning effects of variables that are not used to create the cross-classification under study.

The terms themselves can be interpreted as usual. We give two examples. The first is the three-way interaction of three binary variables. The design matrix of the saturated model for these three variables is, in effect coding

$$X = \begin{bmatrix} 1 & 1 & 1 & 1 & 1 & 1 & 1 & 1 \\ 1 & 1 & 1 & -1 & 1 & -1 & -1 & -1 \\ 1 & 1 & -1 & 1 & -1 & 1 & -1 & -1 \\ 1 & 1 & -1 & -1 & -1 & -1 & 1 & 1 \\ 1 & -1 & 1 & 1 & -1 & -1 & 1 & -1 \\ 1 & -1 & 1 & -1 & -1 & 1 & -1 & 1 \\ 1 & -1 & -1 & 1 & 1 & -1 & -1 & 1 \\ 1 & -1 & -1 & -1 & 1 & 1 & 1 & -1 \end{bmatrix}.$$

The last column of this design matrix contains the vector for the three-way interaction. This vector suggests that the effect that is observed for the first category of the first variable, that is, in the first four cells (this is the effect 1, −1, −1, 1), is reversed for the second category of the first variable (we find −1, 1, 1, −1). This can be described in analogous fashion for the first and the second categories of the other two variables.

Now, when this interaction exists, CFA types and antitypes can emerge for any one (or more than one) of the cells of the cross-classification. Specifically, one would expect that types emerge in cells with a +1 and antitypes in cells with a −1. In studies in which a priori hypotheses exist about which cells are concerned, the observed cell frequencies are directly compared with the expected frequencies in these cells. Otherwise, standard exploratory CFA is performed.

The CFA model for the analysis of Meehl's paradox can be represented using the design matrix just given (cf. Lienert, 1968; von Eye & Wiedermann, 2021a). This paradox describes data situations in which variables are completely independent—in pairs. None of the two-way interactions is significant. And yet, the cross-classification can be perfectly described, even with a less-than-saturated model. In the paradox, just the second-order interaction among all three variables (i.e., the three-way interaction) is significant. It is represented by the last column of the design matrix. When the two-way interactions are nonsignificant, they can be removed without loss of precision of explanation. Removing them results in a nonhierarchical model (see Mair & von Eye, 2007) with three degrees of freedom that fits the data.

In the second example, more than three variables are analyzed, and some of them may have more than two categories. By implication, the number of effects increases. Even for individual variables, the number of effects in an interaction increases. The important element in this example is that, for each interaction with variables that have more than two categories, more than one effect can be part of the model. In F-CFA, these effects can be inserted (in ascending F-CFA) or removed (in descending F-CFA) individually. This allows for a very detailed interpretation of types and antitypes.

20.3 Data Example

For the following example, we reanalyze data from a study on the covariation of scales of Amthauer's intelligence structure test (IST; Amthauer, 1973; Krueger et al., 1979). The data set contains the responses of 997 individuals that took the IST. The responses were dichotomized at the median. The resulting scores indicate submedian performance (score = 1) and at or above median performance (score = 2). The IST is composed of the nine subtests: A = sentence completion, B = word selection, C = analogies, D = similarities, E = calculating, F = number series, G = figures, H = cubes, and I = memory. Crossed, these nine subtests span a table with $2^9 = 512$ cells.

In earlier analyses, Krueger et al. (1979) had performed zero-order CFA (the base model for zero-order CFA is the null model, $\log \hat{m} = 1\lambda$) and identified density centers that point at particularly frequent (or infrequent) response patterns. von Eye and Wiedermann (2021a) performed first-order CFA (the base model for first order CFA is a main effect model, $\log \hat{m} = \lambda + \lambda^A + \lambda^B + \lambda^C + \lambda^D + \lambda^E + \lambda^F + \lambda^G + \lambda^H + \lambda^I$). The Pearson chi-square test of global association shows that strong interactions exist in the cross-classification of the nine IST subtests ($X^2 = 11{,}650.67$; $df = 502$; $p < 0.001$). Nine types emerged and no antitype. Each of these types contradicts the assumption of scale independence.

Here, we ask whether there exist relations among the two groups of the verbal scales (variables, A, B, C, and D) and the nonverbal scales (variables E, F, G, H, and I). We perform two CFAs. In the first, we ask whether types and antitypes remain when first-order interactions among scales from the two groups are included in the base model. If types and antitypes result, interactions higher than first order may exist. This would be interesting considering that first-order relations are relatively strong, in particular, among the verbal subscales (see Table 3.2 in von Eye & Wiedermann, 2021a). In the second analysis, we also include second-order interactions. We take the four steps of CFA.

Step 1: Specification of CFA base model In this first application of higher-order CFA of two groups of variables, we include in the base model the following:

- All possible main effects and interactions in the group of verbal subtests
- All possible main effects and interactions in the group of non-verbal subtests
- All possible two-way interactions among subtests from the two groups.

None of the higher than first-order interactions among subtests from the two groups are included in the base model. When, therefore, types and antitypes emerge, they point at the existence of such interactions. The base model is, thus,

$$\begin{aligned}
\log \hat{m} &= \lambda + \lambda^{A} + \lambda^{B} + \lambda^{C} + \lambda^{D} + \lambda^{E} + \lambda^{F} + \lambda^{G} + \lambda^{H} + \lambda^{I} \\
&+ \lambda^{AB} + \lambda^{AC} + \lambda^{AD} + \lambda^{BC} + \lambda^{BD} + \lambda^{CD} \\
&+ \lambda^{ABC} + \lambda^{ABD} + \lambda^{ACD} + \lambda^{BCD} + \lambda^{ABCD} \\
&+ \lambda^{EF} + \lambda^{EG} + \lambda^{EH} + \lambda^{EI} + \lambda^{FG} + \lambda^{FH} + \lambda^{FI} + \lambda^{GH} + \lambda^{GI} + \lambda^{HI} \\
&+ \lambda^{EFG} + \lambda^{EFH} + \lambda^{EFI} + \lambda^{EGH} + \lambda^{EGI} + \lambda^{EHI} + \lambda^{FGH} \\
&+ \lambda^{FGI} + \lambda^{FHI} + \lambda^{GHI} \\
&+ \lambda^{EFGH} + \lambda^{EFGI} + \lambda^{EFHI} + \lambda^{FGHI} + \lambda^{EFGHI} \\
&+ \lambda^{AE} + \lambda^{AF} + \lambda^{AG} + \lambda^{AH} + \lambda^{AI} + \lambda^{BE} + \lambda^{BF} + \lambda^{BG} + \lambda^{BH} \\
&+ \lambda^{BI} + \lambda^{CE} + \lambda^{CF} + \lambda^{CG} + \lambda^{CH} + \lambda^{CI} + \lambda^{DE} + \lambda^{DF} + \lambda^{DG} + \lambda^{DH} + \lambda^{DI}.
\end{aligned}$$

The first line of this equation of the base model for the analysis of two variable groups contains the model constant and all nine main effects. The second line contains all first-order interactions among the four verbal subtests (A, B, C, and D). The third line contains all three-way and four-way interactions among the verbal subtests. The following three lines contain all first- and higher-order interactions among the five nonverbal subtests (E, F, G, H, and I). In the last two lines, we find all of the 20 two-way interactions between variables from the two different groups. When this model is rejected, types and antitypes point at the existence of three- or higher-way interactions among variables from the two substests.

Step 2: Significance testing For the overall goodness-of-fit tests of the following models, we use the likelihood ratio chi-square test. For the CFA tests of individual cells, we use the z test. This can be justified by the relatively large sample size (please note that, for reasons of comparison, we also used the binomial test; results were largely the same). To protect the nominal significance threshold α, we set a z-score of 3.5. This corresponds to a protected tail probability of 0.00047. This new threshold is applied throughout, thus corresponding to the well-known Bonferroni procedure (many other procedures for the protection of α use thresholds that vary with the number of tests performed; see von Eye & Wiedermann, 2021a).

Step 3: Performing the CFA In the following sections, we present results from two CFA runs. The first is a standard CFA of two groups of variables. The second is a higher-order CFA. We begin with the approach of standard CFA.

The standard base model of CFA of two groups of variables contains just the first six lines of the above equation. The overall goodness-of-fit LR chi-square of 1103.75 ($df = 465$; $p < 0.001$) represents a large and significant improvement over the main effect model (Δ chi-square = 1695.24; $\Delta df = 37$; $p < 0.001$). It is, however, still significant by itself, and one can expect types and antitypes that point at the existence of interactions. Figure 20.1 displays the results of this CFA. The bars indicate the z-scores of the 512 configurations beginning, from left to right,

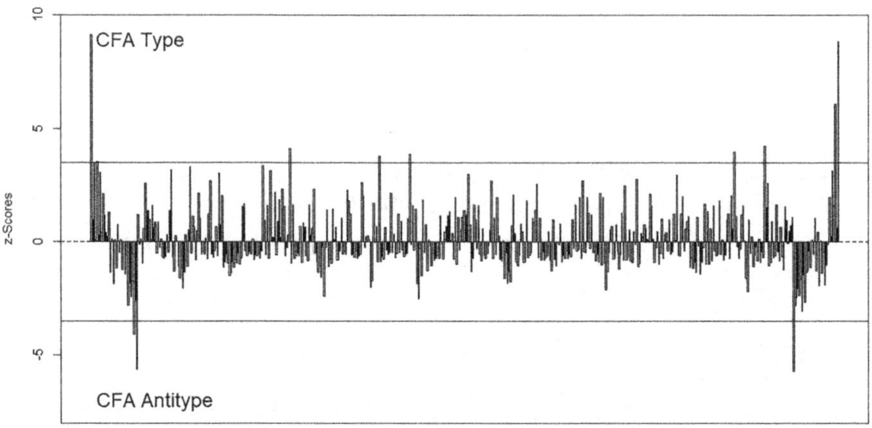

Fig. 20.1 Standard CFA of two groups of variables

with Configuration 1 1 1 1 1 1 1 1 1 and ending with Configuration 2 2 2 2 2 2 2 2 2 (the same results can be requested in tabular form from the authors).

Standard CFA of two groups of variables results in nine types and three antitypes (see Fig. 20.1). Instead of interpreting each of these, we just look at the most extreme ones. The most extreme type is constituted by Profile 1 1 1 1 1 1 1 1 1. These are the 88 participants that perform below the median in all nine subtests (34.93 had been expected). The second most extreme type is constituted by Profile 2 2 2 2 2 2 2 2 2. These are the 99 participants that perform above the median in all nine subtests (41.82 had been expected). Evidently, consistent performance occurs more often than expected from the assumption of subtest independence across the two groups of verbal and nonverbal tasks.

The most extreme antitype is constituted by Profile 2 2 2 2 1 1 1 1 1. These are the two participants that perform below the median in all verbal tasks, but above the median in all nonverbal tasks (36.42 had been expected). Here, we note that extreme inconsistent performance occurs much less frequently than expected from the model assumptions.

Now moving to higher-order CFA of two groups of variables, we see that the complete above model is much improved. Including the 20 two-way interactions between the variables from the two groups reduces the overall goodness-of-fit LR chi-square to 520.09 ($df = 445$; $p = 0.008$). This improvement is significant (Δ chi-square = 583.66; $\Delta df = 20$; $p < 0.001$), but the model-data discrepancies are still significant. Figure 20.2 shows that five types emerged, but no antitype (again, these results can be requested, in tabular form, from the authors). We again interpret only the most extreme two types.

The most extreme type is constituted by the two participants who performed above the median in two verbal and two nonverbal tasks and below the median in all other tasks. This is Profile 1 2 2 1 1 1 2 1 2. Only 0.203 participants had

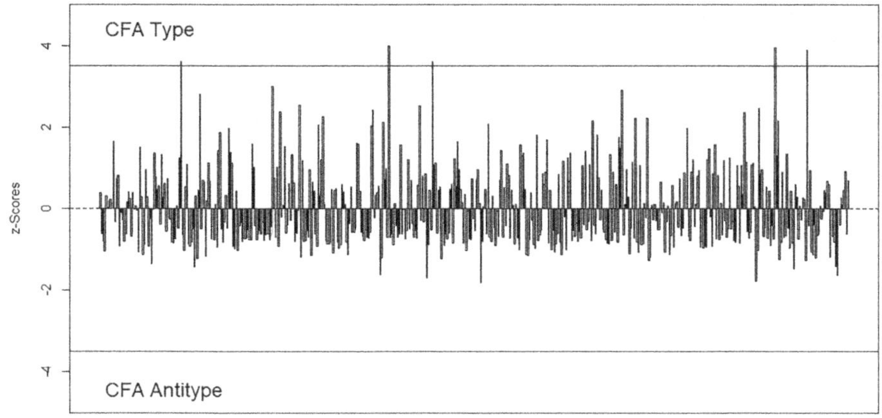

Fig. 20.2 Higher-order CFA of two groups of variables

been expected to show this profile. The second most extreme type is constituted by the three participants who performed above the median in three verbal and two nonverbal tasks and below the median in all other tasks. This is Profile 1 2 2 2 1 1 1 2 2. Only 0.424 participants had been expected to show this profile.

F-CFA for the analysis of the types in Fig. 20.2. In standard CFA, the identification of the types in Fig. 20.2 would conclude the analysis. Here, however, we ask which interactions of second or higher order have the effect that these types and antitypes emerge. To answer this question, we perform F-CFA to find the interactions that have that effect. We proceed in ascending order (von Eye & Mair, 2008a, b, c). That is, we insert (and remove, when an interaction remains nonsignificant) interactions, beginning with the lowest-order interactions, until all types have disappeared. The interactions that have this effect are called *type-generating* (or *antitype-generating*).

In the present data example of higher-order CFA of two groups of variables, the lowest-order interactions that can be included are three-way interactions. We begin with the $A \times B \times E$, that is, the first interaction that includes verbal as well as nonverbal subtests. When evaluating the individual interactions that are added to the base model, it is important to realize that it is not required that the interactions be significant. However, the interactions must eliminate types (and antitypes, when they had emerged). After removing all types (and antitypes), the model usually fits. The order in which the interactions are inserted usually plays no role.

In the example presented in Fig. 20.2, five three-way interactions are needed to remove all types. Table 20.1 shows these interactions along with the corresponding significance statistics.

Overall, the model fits the cross-classification well (LR chi-square = 489.647; df = 440; p = 0.051). Of the five type-generating interactions, only one is significant, but all five are needed to remove all of the types in Fig. 20.2.

Table 20.1 Type-generating interactions for the data example in Fig. 20.2

Interaction	Wald statistic	*df*	*p*
ABF	2.74	1	0.098
AEG	0.08	1	0.772
AFG	15.85	1	<0.001
BEH	3.15	1	0.076
CEF	3.14	1	0.076

Table 20.2 Effect coding of the three-way interaction among F, A, and B

		B (word selection)	
F (number series)	A (sentence completion)	1	2
1	1	1	−1
	2	−1	1
2		−1	1
		1	−1

To interpret the types in even more detail, the design matrix given above can be used. In the last column, the pattern is 1 −1 −1 1 −1 1 1 −1. Inspecting the first four numbers of this effect code pattern, we find the standard interaction in a 2 × 2 table. Inspecting the last four numbers, we see that this pattern is reversed in signs. This corresponds to a three-way interaction in log-linear models and analysis of variances (ANOVA).

To illustrate this, we take the first interaction that was added to the base model of higher-order CFA of two groups of variables, ABF (i.e., sentence completion-word selection-number series). This interaction is shown in Table 20.2.

From Fig. 20.2, we learn that the first type exhibits the ABF configuration 1 1 1 (cf. the most extreme type in Fig. 20.1). From Table 20.2, we see that this corresponds to the italicized effect code in the body of the table. We can, therefore, conclude that the three-way interaction has this pattern. This applies accordingly to the other four added interactions. All five of the three-way interactions are type-generating and needed to explain the local relations visible in the cross-classification of the nine IST subtest results.

20.4 Discussion

Extending the earlier approach to taking into account first-order interactions in standard CFA (Lienert et al., 1987; von Eye & Lienert, 1984), this chapter presents an approach in which interactions of second and higher order can be made part of the CFA base model. In the hierarchy of global CFA models, interactions of various order have already been incorporated (see von Eye & Wiedermann, 2021a). Here, regional CFA models are considered, that is, models in which variables are grouped.

The special case of higher-order CFA for the comparison of two variable groups is discussed.

In the models discussed in this chapter, the variable groups are of equal status. That is, there is no distinction between causes and effects, predictors and outcomes, mediators, moderators, or covariates. In fact, P-CFA still suffers from the issue that the effects that are considered are symmetric. By implication, the results table of P-CFA will be exactly the same when the predicting trajectory is reversed in direction. Krauth (1993) states that the distinction between predictors and outcome variables is purely interpretational or conceptual. This applies as well to the models used by von Eye and Brandtstädter (1997). In the models proposed here, this distinction is not even made. Whether the causal CFA models proposed by von Eye and Wiedermann (2016, 2018, 2021b; Wiedermann & von Eye, 2020) can be extended so that first- and higher-order interactions can be incorporated remains to be elaborated in future work.

When, as was the case in the above data example, groups of items are analyzed, the relations among items usually do not reflect directed effects and dependencies. When items or groups of items are created using factor analytic methods, only first-order relations are taken into account (covariances or correlations). Higher-order relations and dependencies can, therefore, still exist. This was illustrated in the data example in this chapter.

An important issue concerning the interpretation of types and antitypes that are caused by first- or higher-order interaction results from possible mutual dependencies among types and antitypes. These dependencies can result in patterns of cells that constitute types and antitypes. It is important, however, to realize that when Configuration 1 1 1 constitutes a type, Configuration 1 1 2 will not necessarily constitute an antitype. The same applies to the three remaining configurations with a -1 code in Table 20.2. In $2 \times 2 \times 2$ tables, the number of type-antitype patterns is 985 (for more detail on the number of type-antitype patterns, see Krauth, 2003). Only in 2×2 tables that are analyzed under the base model of first-order CFA, that is, the main effect model, a type or antitype in any cell predetermines types and antitypes in all other cells completely (von Weber et al., 2003). In cross-classifications of the size analyzed in the present data example, it is unpredictable which other cell is extreme (if at all) when one cell was identified as type or antitype.

From a methodological perspective, it is important to observe that second- and higher-order interactions can play a role in the interpretation of the structure of scales that were created using standard factor analysis. The standard model of factor analysis is known to be (in LISREL notation; see Jöreskog & Sörbom, 2001)

$$x = \Lambda\xi + \delta,$$

where x is the p-dimensional vector of observed scores, Λ is the matrix of factor loadings, ξ are the latent variables (factors), and δ are the error variables (for introductions to and applications of factor analysis, see Arminger et al., 1995; Bartholomew & Knott, 1999; von Eye et al., 2023). The important part of the factor model is that the unit of analysis is the covariation of variables. Covariances correspond to first-order interactions. Therefore, a rendering of the structure of scales that explains first-order relations leaves second- and higher-order relations untouched. They still can exist. In the present chapter, we showed how they can be analyzed from a person-oriented perspective.

The model presented here can be extended in various ways, three of which are discussed here. A first way concerns the number of variable groups. Multiple groups can be considered. A second way concerns the application of F-CFA. In the present example, type- (and antitype-) generating effects were searched with the goal of eliminating, that is, here, explaining every type (and antitype). The result suggested that five three-way interactions are needed for this goal to be achieved. It could also be asked, however, whether individual types or antitypes can be explained from specific effects or specific sets of effects. When this question is asked, an effect can be considered type- or antitype-generating when the type or antitype disappears that is being examined, even when the resulting model still does not fit. In either case, F-CFA concludes when the model fits and all types and antitypes are explained.

A third way concerns the characteristics of the scales that are analyzed. In many contexts, researchers assume that scales are normally distributed and that even though it is well known this assumption can be confirmed only rarely (see, e.g., Micceri, 1989). In the present volume, approaches are presented in which this assumption can be integrated in the CFA base model (von Eye & Wiedermann, 2024a), even when cross-classifications are incomplete (von Eye & Wiedermann, 2024b) or when higher-order relations are examined, as in the present chapter.

References

Amthauer, R. (1973). *I-S-T 70. Intelligenz-Struktur-Test Handanweisung*. Hogrefe.

Arminger, G., Clogg, C. C., & Sobel, M. (Eds.). (1995). *Handbook of statistical modeling for the social and behavioral sciences*. Plenum Press. https://doi.org/10.1007/978-1-4899-1292-3

Bartholomew, D. J., & Knott, M. (1999). *Latent variable models and factor analysis* (2nd ed.). Arnold.

Bonferroni, C. E. (1936). Teoria statistica delle classi e calcolo delle probabilità. *Pubblicazioni del R Istituto Superiore di Scienze Economiche e Commerciali di Firenze, 8*, 3–62.

Glück, J., & von Eye, A. (2000). Including covariates in Configural Frequency Analysis. *Psychologische Beiträge, 42*, 405–417.

Holland, B. S., & Di Ponzio Copenhaver, M. (1987). An improved sequentially rejective Bonferroni test procedure. *Biometrics, 43*, 417–423.

Holm, S. (1979). A simple sequentially rejective multiple test procedure. *Scandinavian Journal of Statistics, 6*, 65–70.

Jöreskog, K., & Sörbom, D. (2001). *Lisrel 8: User's reference guide*. Scientific Software International.

Krauth, J. (1993). *Einführung in die Konfigurationsfrequenzanalyse*. Beltz Psychologie.

Krauth, J. (2003). Type structures in CFA. *Psychology Science, 45*, 330–338.

Krueger, H.-P., Lienert, G. A., Gebert, A., & von Eye, A. (1979). Eine inferentielle Clusteranalyse für Alternativdaten. *Psychologische Beiträge, 21*, 540–553.

Lienert, G. A. (1968). *Die "Konfigurationsfrequenzanalyse" als Klassifikationsmethode in der klinischen Psychologie.* Vortrag auf dem 26. Kongress der Deutschen Gesellschaft für Psychologie in Tübingen 1968.
Lienert, G. A., & Krauth, J. (1975). Configural Frequency Analysis as a statistical tool for defining types. *Educational and Psychological Measurement, 35*, 231–238.
Lienert, G. A., Netter, P., & von Eye, A. (1987). Die Konfigurationsfrequenzanalyse XXV. Typen und Syndrome höherer Ordnung. *Zeitschrift für Klinische Psychologie, Psychopathologie, und Psychotherapie, 35*, 344–352.
Mair, P., & von Eye, A. (2007). Application scenarios for nonstandard log-linear models. *Psychological Methods, 12*, 139–156.
Micceri, T. (1989). The unicorn, the normal curve, and other improbable creatures. *Psychological Bulletin, 105*(1), 156–166. https://doi.org/10.1037/0033-2909.105.1.156
von Eye, A. (2004). Base models for Configural Frequency Analysis. *Psychology Science, 46*, 150–170.
von Eye, A., & Brandtstädter, J. (1997). Configural Frequency Analysis as a searching device for possible causal relationships. *Methods of Psychological Research – Online, 2*, 1–23.
von Eye, A., & Gutiérrez Peña, E. (2004). Configural Frequency Analysis – The search for extreme cells. *Journal of Applied Statistics, 31*, 981–997.
von Eye, A., & Lienert, G. A. (1984). Die Konfigurationsfrequenzanalyse XX. Typen und Syndrome zweiter Ordnung (Komplextypen). *Zeitschrift für Klinische Psychologie und Psychotherapie, 32*, 345–355.
von Eye, A., & Mair, P. (2008a). A functional approach to Configural Frequency Analysis. *Austrian Journal of Statistics, 37*, 161–173.
von Eye, A., & Mair, P. (2008b). A functional approach to Configural Frequency Analysis – Computational issues. In M. Stemmler, E. Lautsch, & D. Martinke (Eds.), *Configural Frequency Analysis (CFA) and other nonparametric statistical methods: Gustav A. Lienert Memorial Issue* (pp. 84–94). Pabst.
von Eye, A., & Mair, P. (2008c). Functional Configural Frequency Analysis: Explaining types and antitypes. *Bulletin de la Société des Sciences Médicales, Luxembourg, 144*, 35–52.
von Eye, A., & Wiedermann, W. (2016). Direction of effects in categorical variables – A structural perspective. In W. Wiedermann & A. von Eye (Eds.), *Statistics and causality: Methods for applied empirical research* (pp. 107–130). Wiley.
von Eye, A., & Wiedermann, W. (2018). Locating event-based causal effects: A configural perspective. *Integrative Psychological and Behavioral Science, 52*, 307–330. https://doi.org/10.1007/s12124-018-9423-0
von Eye, A., & Wiedermann, W. (2021a). *Configural Frequency Analysis. Foundations, models, and applications*. Springer.
von Eye, A., & Wiedermann, W. (2021b). Locating direction dependence using log-linear modeling, configural frequency analysis, and prediction analysis. In W. Wiedermann, D. Kim, E. A. Sungur, & A. von Eye (Eds.), *Direction dependence in statistical modeling. Methods of analysis* (pp. 185–218). Wiley.
von Eye, A., Wiedermann, W., & von Weber, S. (2023). Latent variables analysis in person-oriented research. *Journal of Person-oriented Research, 9*, 17–28. https://doi.org/10.17505/jpor.2023.25258
von Eye, A., & Wiedermann, W. (2024a). Configural Frequency Analysis under multinormality. In M. Stemmler, W. Wiedermann, & F. Huang (Eds.), *Dependent data in social sciences research: Forms, issues, and methods of analysis* (2nd ed.). Springer.
von Eye, A., & Wiedermann, W. (2024b). Configural Frequency Analysis under multinormality in incomplete tables. In M. Stemmler, W. Wiedermann, & F. Huang (Eds.), *Dependent data in social sciences research: Forms, issues, and methods of analysis* (2nd ed.). Springer.
von Eye, A., Mair, P., & Bogat, G. A. (2005). Prediction models for Configural Frequency Analysis. *Psychology Science, 47*, 342–355.
von Weber, S., Lautsch, E., & von Eye, A. (2003). On the limits of Configural Frequency Analysis: Analyzing small tables. *Psychology Science, 45*, 339–354.

Wiedermann, W., & von Eye, A. (2020). Log-linear models to evaluate direction of effect in binary variables. *Statistical Papers, 61*, 317–346. https://doi.org/10.1007/s00362-017-0936-2
Wiedermann, W., Herman, K., Reinke, W., & von Eye, A. (2021). Configural frequency trees. *Development and Psychopathology, 34*, 1585–1603. https://doi.org/10.1017/S0954579421000018

Chapter 21
Visualization of Dependence in Multidimensional Contingency Tables with an Ordinal Dependent Variable via Copula Regression

Shu-Min Liao, Li Wang, and Daeyoung Kim

21.1 Introduction

Exploratory data analysis (EDA), introduced by John Tukey (1977), is one of the essential activities that need to be done with the data at hand before formal statistical modeling and inference. Exploration and visualization of data in areas of statistics and data science, which require substantial time and effort, enable to not only check basic features of the data but also discover unexpected or unaware ones, and thus it can provide important insights on the dependence structure of the data (Buja et al., 2009; Shmueli, 2010; Gelman and Vehtar, 2021).

In particular, statistical visualization is an important tool in the process of implementing EDA because it can facilitate understanding the dependence structures between variables captured by statistical methods and thus allows us to construct complicated models efficiently and effectively (Tufte, 1983; Donoho, 2017; Gelman and Vehtar, 2021). It could also serve as a complement to (but not a replacement for) quantitative methods designed to quantify the degree of dependence in the data.

Multivariate categorical data, typically summarized into a multidimensional contingency table, is one of common data types occurring in a variety of sciences, including social, economic, political, behavioral, clinical, and biomedical sciences. Many graphical/visual methods for categorical data have been developed to serve

S.-M. Liao
Department of Mathematics and Statistics, Amherst College, Amherst, MA, USA
e-mail: sliao@amherst.edu

L. Wang
Applied Scientist II, Microsoft, Redmond, WA, USA

D. Kim (✉)
Department of Mathematics and Statistics, University of Massachusetts, Amherst, MA, USA
e-mail: dkim1@umass.edu

M. Stemmler et al. (eds.), *Dependent Data in Social Sciences Research*,
https://doi.org/10.1007/978-3-031-56318-8_21

the purpose of EDA. While some methods were adapted from visualization methods designed for continuous data, other methods (such as fourfold plot, sieve diagram, and mosaic plot/matrix) were developed using different and novel graphical schemes/format because of the discrete nature of categorical data that make it distinguishable from continuous data. For detailed information on fundamental principles and ideas of statistical graphics and recent developments in graphical methods for categorical data, refer to Tufte (1983), Cleveland (1985, 1993), Wilkinson (2005), Gower and Hand (1996), Blasius and Greenacre (1998), Friendly (2000), and Friendly and Meyer (2015).

In this paper, as an effort to facilitate visualization in the workflow of EDA for multivariate categorical data, we propose the graphical approach designed to learn and visualize the regression dependence structure of an ordinal dependent variable with a set of independent categorical (ordinal or nominal) variables in a multi-way contingency table. The proposed approach consists of two steps. First, we utilize two copula regressions for ordinal dependent variables, *subcopula regression* (Wei and Kim, 2017; Wang, 2021) and *checkerboard copula regression* (Wei and Kim, 2021), and their prediction of the category of the ordinal dependent variable for any given combination of categories of independent variables and visualize the prediction results. Note that the copula regressions are applicable not only for ordinal independent variables but also for nominal independent variables. Second, we then incorporate resampling techniques, bootstrap method and permutation method, into the copula regression-based prediction to take into account the uncertainty involved in the prediction and calibration of the discovered regression associations in an exploratory and visual manner. These two resampling techniques can be considered as attempts to add some inference to the process of visual evaluation of the regression association identified by the copula regression-based prediction and its significance relative to no regression association. As a whole, the proposed method will be a new tool of EDA for multi-way contingency tables by providing information to aid in setting up one or more initial statistical models for an ordinal dependent variable.

The rest of the article is organized as follows. Section 21.2 briefly reviews subcopula and checkerboard copula for a multidimensional contingency table with an ordinal dependent variable. We then introduce two types of copula regression, *subcopula regression* (Wei and Kim, 2017; Wang, 2021) and *checkerboard copula regression* (Wei and Kim, 2021), and discuss their theoretical properties and the procedure of the copula regression-based prediction for an ordinal dependent variable. Section 21.3 discusses the estimation of the copula regressions and their prediction. We then propose employing two resampling techniques to visually evaluate the regression association identified by the copula regression-based prediction in terms of the prediction uncertainty and significance of the detected regression patterns. In Sect. 21.4, we demonstrate the utility of the copula regression-based prediction and resampling techniques using three real-world data sets for learning and visualizing the regression association structures between an ordinal dependent variable and categorical independent variables in an exploratory and model-free manner. Note that we present the analysis results of the proposed method in different visualization formats: cross-tabulation, bubble plot, and doubledecker plot (Hofmann, 2001; Meyer et al., 2006). Section 21.5 ends the paper with a few concluding remarks and future work.

21.2 Review on the Copula Regressions for Multidimensional Contingency Tables

The copulas developed by Sklar (1959) have been broadly used for modeling statistical dependence for multivariate continuous data in various fields of science (Nelsen, 2006; Joe, 2014; Durante and Sempi, 2015). Recently, several researchers have utilized copulas to explore and model the association structure of discrete variables (Denuit and Lambert, 2005; Nešlehová, 2007; Genest et al., 2014, 2017; Wei and Kim, 2017; Rachasingho and Tasena, 2020; Wang, 2021; Tasena, 2021a,b; Wei and Kim, 2021). In this section, we will briefly review two types of copula for ordinal variables, *subcopula* and *checkerboard copula.* We then introduce the corresponding regressions proposed by Wei and Kim (2017), Wang (2021) and Wei and Kim (2021), *subcopula regression*, and *checkerboard copula regression*, to examine regression association structures for a multidimensional contingency table with an ordinal dependent variable and a set of categorical independent variables.

Subcopula and Checkerboard Copula for Ordinal Contingency Tables

Let Y be an ordinal dependent variable with J ordered categories $\{y_1 < \cdots < y_j < \cdots < y_J\}$ and $\boldsymbol{X}=(X_1, \ldots, X_\ell, \ldots, X_d)$ be a d-dimensional vector of ordinal independent variables with each X_ℓ having I_ℓ ordered categories $\{x_1^\ell < \cdots < x_{i_\ell}^\ell < \cdots < x_{I_\ell}^\ell\}$ where $j = 1, \ldots, J$, $\ell = 1, \ldots, d$, $i_\ell = 1, \ldots, I_\ell$ and " $<$" reflects the ordering of categories in each variable.

Consider a $(d+1)$-dimensional contingency table of the ordinal vector $(\boldsymbol{X}^\top, Y)$ with the joint probability mass function (p.m.f.) $p_{\boldsymbol{i},j} = P(\boldsymbol{X} = \boldsymbol{x_i}, Y = y_j)$ where $\boldsymbol{i} = (i_1, \ldots, i_\ell, \ldots, i_d)$ and $\boldsymbol{x_i} = (x_{i_1}^1, \ldots, x_{i_\ell}^\ell, \ldots, x_{i_d}^d)$. The p.m.f. for $\boldsymbol{X}$, the p.m.f.s for X_ℓ and Y, and the conditional p.m.f. for Y given $\boldsymbol{X}$ are denoted by $p_{\boldsymbol{i}+}$, $p_{+i_\ell+}$, p_{+j}, and $p_{j|\boldsymbol{i}} = p_{\boldsymbol{i},j}/p_{\boldsymbol{i}+}$, respectively.

Let the joint cumulative distribution function (c.d.f.) of $(\boldsymbol{X}^\top, Y)$ and the marginal c.d.fs of X_ℓ and Y be $G_{\boldsymbol{X},Y}$, F_{X_ℓ}, and F_Y, respectively. Define a $(d+1)$-dimensional vector $(\boldsymbol{U}^\top, V)$ with $\boldsymbol{U} = (U_1, \ldots, U_\ell, \ldots, U_d)$ to be

$$U_\ell = F_{X_\ell}(X_\ell), \quad V = F_Y(Y). \tag{21.1}$$

Then $(\boldsymbol{U}^\top, V)$ is the random vector with the domains

$$R_{X_\ell} = \{u_0^\ell, u_1^\ell, \ldots, u_{i_\ell}^\ell, \ldots, u_{I_\ell}^\ell\}, \quad R_Y = \{v_0, v_1, \ldots, v_j, \ldots, v_J\}, \tag{21.2}$$

where $u_0^\ell = v_0 = 0$, $u_{I_\ell}^\ell = v_J = 1$, $u_{i_\ell}^\ell = F_{X_\ell}(x_{i_\ell}^\ell) = \sum_{h_\ell=1}^{i_\ell} p_{+h_\ell+}$, and $v_j = F_Y(y_j) = \sum_{h=1}^{j} p_{+h}$. Note that R_{X_ℓ} and R_Y are finite subsets of [0, 1], and each component in $(\boldsymbol{U}^\top, V)$ does not have a continuous standard uniform distribution.

By Sklar (1959), there exists a function called *subcopula* C^S associated with $(\boldsymbol{X}^\top, Y)$ on $\left(\prod_{\ell=1}^d R_{X_\ell}\right) \times R_Y$ such that

$$\begin{aligned} G_{\boldsymbol{X},Y}(\boldsymbol{x}_i, y_j) &= G_{\boldsymbol{X},Y}(x_{i_1}^1, \ldots, x_{i_d}^d, y_j) = C^S(F_{X_1}(x_{i_1}^1), \ldots, F_{X_d}(x_{i_d}^d), F_Y(y_j)) \\ &= C^S(u_{i_1}^1, \ldots, u_{i_d}^d, v_j) = \sum_{h_1=1}^{i_1} \cdots \sum_{h_d=1}^{i_d} \sum_{h=1}^{j} c^s(u_{h_1}^1, \ldots, u_{h_d}^d, v_h) \end{aligned} \tag{21.3}$$

where $c^s(u_{i_1}^1, \ldots, u_{i_d}^d, v_j) = p_{i_1,\ldots,i_d,j} = p_{\boldsymbol{i},j}$ is the p.m.f. of $C^S(u_{i_1}^1, \ldots, u_{i_d}^d, v_j)$. The subcopula C^S in Eq. (21.3) is uniquely determined on $\left(\prod_{\ell=1}^d R_{X_\ell}\right) \times R_Y$ which is a strict subset of $[0, 1]^{d+1}$ and contains the information on the dependence structure of ordinal variables $(\boldsymbol{X}^\top, Y)$.

If $X_1, \ldots, X_d, Y$ are continuous, each of $(\boldsymbol{U}^\top, V)$ has a standard uniform distribution $U(0, 1)$ and a subcopula C^S is known as a *copula* on $[0, 1]^{d+1}$. In this case, the copula is "margin-free" in that it is invariant under increasing transformations of the marginal distributions and thus copula-based dependence measures such as Kendall's tau or Spearman's rho are margin-free.

For ordinal random variables, however, each element of $(\boldsymbol{U}^\top, V)$ in Eq. (21.1) is not uniformly distributed over [0, 1] and the domain of the subcopula C^s in Eq. (21.3) depends on marginal distributions (i.e., "margin-dependent"). It also needs to be estimated because it is unknown. To resolve this issue, several methods have been proposed to extend a subcopula to a copula on $[0, 1]^{d+1}$ (Schweizer and Sklar, 1974; Carley, 2002; Genest and Nešlehová, 2007; de Amo et al., 2012). One of the commonly used methods is to transform the marginals of the ordinal random variables to uniforms by employing the distributional transform of ordinal random variables (Nešlehová, 2007; Rüschendorf, 2009):

$$\begin{aligned} U_\ell^* &= F_{X_\ell}(X_\ell-) + \{F_{X_\ell}(X_\ell) - F_{X_\ell}(X_\ell-)\}W_\ell, \\ V^* &= F_Y(Y-) + \{F_Y(Y) - F_Y(Y-)\}W_{d+1}, \end{aligned} \tag{21.4}$$

where $\ell = 1, \ldots, d$, $F(x-)$ refers to the left limit of F and $W_1, \ldots, W_{d+1}$ denote independent uniform random variables on [0, 1] that are independent of $(\boldsymbol{X}^\top, Y)$. The continued random variables U_ℓ^* and V^* in Eq. (21.4) are standard uniform random variables, and the $(d+1)$-dimensional random vector $(\boldsymbol{U}^{*\top}, V^*)$

with $\boldsymbol{U}^* = (U_1^*, \ldots, U_\ell^*, \ldots, U_d^*)$ has a unique copula on $[0, 1]^d$, known as the *checkerboard copula*, denoted as C^*. As a smooth version of the subcopula C^S for $(\boldsymbol{X}^\top, Y)$ in a contingency table, the checkerboard copula C^* preserves dependence structures for $(\boldsymbol{X}^\top, Y)$ (Schweizer and Sklar, 1974; Nelsen, 2006; Genest et al., 2014, 2017), and it has the density c^* expressed as a function of the joint p.m.f. for $(\boldsymbol{X}^\top, Y)$ and their marginal p.m.f.s:

$$c^*(\boldsymbol{u}^*, v^*) = \frac{p_{\boldsymbol{i},j}}{\left(\prod_{\ell=1}^{d} p_{+i_\ell+}\right) p_{+j}} \quad \text{for} \quad u_{i_\ell-1}^\ell < u_\ell^* \le u_{i_\ell}^\ell \quad \text{and} \quad v_{j-1} < v^* \le v_j, \tag{21.5}$$

where $\boldsymbol{u}^* = (u_1^*, \ldots, u_\ell^*, \ldots, u_d^*) \in [0, 1]^d$, $u_{i_\ell}^\ell \in R_{X_\ell}$, and $v_j \in R_Y$.

Subcopula Regression and Checkerboard Copula Regression

Using the subcopula and the checkerboard copula for $(\boldsymbol{X}^\top, Y)$ in a multidimensional contingency table, Wei and Kim (2017), Wang (2021) and Wei and Kim (2021) proposed the *subcopula regression* and *checkerboard copula regression* to explore the regression association between Y and $\boldsymbol{X}$.

Definition 21.2.1 (Wei and Kim (2017) and Wang (2021)) The **subcopula regression** of Y on $\boldsymbol{X}$ via $(\boldsymbol{U}^\top, V)$ in Eq. (21.1) is defined as follows:

$$r_{Y|\boldsymbol{X}}^{C^S} \equiv \mathrm{E}_{V|\boldsymbol{U}}^{C^S}(V|\boldsymbol{U} = \boldsymbol{u}) = \sum_{v_j \in R_Y} v_j c_{V|\boldsymbol{U}}^S(v_j \mid \boldsymbol{u_i}) = \sum_{j=1}^{J} v_j p_{j|\boldsymbol{i}}, \tag{21.6}$$

where $\boldsymbol{u_i} = (u_{i_1}, \ldots, u_{i_\ell}, \ldots, u_{i_d})$, $c_{V|\boldsymbol{U}}^S$ is the conditional p.m.f. of V given $\boldsymbol{U}$ obtained from the p.m.f of the subcopula C^S in Eq. (21.3) for $(\boldsymbol{X}^\top, Y)$, and $\{v_j\}_{j=1}^J$ with $v_j \in R_Y$ is the **subcopula score** set for Y.

Definition 21.2.2 (Wei and Kim (2021)) The **checkerboard copula regression** of Y on $\boldsymbol{X}$ via $(\boldsymbol{U}^{*\top}, V^*)$ in Eq. (21.4) is defined as follows:

For $u_{i_\ell-1}^\ell < u_\ell^* \le u_{i_\ell}^\ell$ with $u_{i_\ell}^\ell \in R_{X_\ell}$ and $\ell = 1, \ldots, d$,

$$r_{Y|\boldsymbol{X}}^{C^*} \equiv \mathrm{E}_{V^*|\boldsymbol{U}^*}^{C^*}(V^*|\boldsymbol{U}^* = \boldsymbol{u}^*) = \int_0^1 v^* c_{V^*|\boldsymbol{U}^*}^*(v^*|\boldsymbol{u}^*)dv^* = \sum_{j=1}^{J} s_j p_{j|\boldsymbol{i}}, \tag{21.7}$$

where $\boldsymbol{u}^* = (u_1^*, \ldots, u_\ell^*, \ldots, u_d^*)$, $c_{V^*|\boldsymbol{U}^*}^*$ is the conditional density of V^* given $\boldsymbol{U}^*$ obtained from the checkerboard copula density in Eq. (21.5), and $\{s_j\}_{j=1}^J$ with $s_j = (v_{j-1} + v_j)/2$ and $v_j \in R_Y$ is the **checkerboard copula score** set for Y.

The subcopula regression in Definition 21.2.1 and the checkerboard copula regression in Definition 21.2.2 represent the mean score of Y with respect to its conditional distribution (or equivalently the mean of Y equipped with its scores and its conditional distribution) for any given combination of categories in $\boldsymbol{X}$, the former for the subcopula score $\{v_j\}_{j=1}^J$ and the latter for the checkerboard score $\{s_j\}_{j=1}^J$. The rationales for incorporating these score sets of Y into the copula regression are three folds. First, the use of information on the ordering of categories of Y would be beneficial in understanding the association between Y and $\boldsymbol{X}$ (Agresti, 2010). Second, the distances between the ordered categories in Y are unknown, and even when the categories of Y are numeric, they do not necessarily represent true magnitudes. Note that both the subcopula score set $\{v_j\}_{j=1}^J$ and the checkerboard score set $\{s_j\}_{j=1}^J$ can be estimated from the data. Third, both score sets preserve the natural ordering of the category in Y: $0 < v_1 < \ldots < v_j < \ldots < v_J$ for the subcopula score set and $0 < s_1 < \ldots < s_j < \ldots < s_J$ for the checkerboard score set.

The properties of the scores for Y and the copula regressions in Definitions 21.2.1 and 21.2.2 are given below.

Proposition 21.2.1

(a) *The subcopula score and the checkerboard score are the conditional means of V in Eq.* (21.1) *and V^* in Eq.* (21.4) *given $Y = y_j$, respectively: $E(V \mid Y = y_j) = v_j$ and $E(V^* \mid Y = y_j) = v_{j-1} + (v_j - v_{j-1})/2 = (v_{j-1} + v_j)/2 = s_j$.*

(b) *Each of the copula regressions in Eqs.* (21.6) *and* (21.7) *is invariant under permutation of categories of independent variables. This means that the copula regressions in Eqs.* (21.6) *and* (21.7) *can be applied when independent variables are ordinal as well as nominal (Wang, 2021; Wei and Kim, 2021).*

(c) *When Y is jointly independent with $\boldsymbol{X}$ (i.e., $p_{\boldsymbol{i},j} = p_{\boldsymbol{i}+}p_{+j}$ for all $(\boldsymbol{i}, j)$), the subcopula regression and the checkerboard copula regression become constant over each combination of categories in $\boldsymbol{X}$:*

$$r_{Y|\boldsymbol{X}}^{C^S} = \sum_{j=1}^{J} v_j p_{j|\boldsymbol{i}} = \sum_{j=1}^{J} v_j p_{+j} = 0.5 + 0.5 \sum_{j=1}^{J} p_{+j}^2,$$

$$r_{Y|\boldsymbol{X}}^{C^*} = \sum_{j=1}^{J} s_j p_{j|\boldsymbol{i}} = \sum_{j=1}^{J} s_j p_{+j} = 0.5.$$

Using the copula regressions in Definitions 21.2.1 and 21.2.2, we can also predict the category of Y for a given combination of categories of $\boldsymbol{X}$. The specific steps for the prediction are as follows:

[Step 1] For a given combination of categories of the independent variables in $\boldsymbol{X}$, find the corresponding $\tilde{\boldsymbol{u}}$ from $\prod_{\ell=1}^{d} R_{X_\ell}$ in Eq. (21.2) (the Cartesian product of the ranges of the marginal distributions of $\boldsymbol{X}$),

[Step 2] Obtain the predicted value $\tilde{v}$ of the copula regression at $\tilde{\boldsymbol{u}}$: $\tilde{v} = r^{C^S}_{Y|X}(\tilde{\boldsymbol{u}})$ for the subcopula regression in Eq. (21.6), and $\tilde{v} = r^{C^*}_{Y|X}(\tilde{\boldsymbol{u}})$ for the checkerboard copula regression in Eq. (21.7),

[Step 3] From R_Y in Eq. (21.2) (the range of the marginal distribution of Y), we get the index j^* satisfying $v_{j^*-1} < \tilde{v} \leq v_{j^*}$ and obtain the corresponding predicted category of Y, y_{j^*}.

Note that the prediction using both copula regressions can also be applied when independent variables are nominal because according to Proposition 21.2.1(b), the prediction result obtained from each of both copula regressions in Eqs. (21.6) and (21.7) is invariant under permutation of categories of independent variables.

Using the prediction results, one can reveal the potential association structures between Y and $\boldsymbol{X}$ in a model-free and exploratory manner. That is, one may visualize the predicted category of Y for every combination of the categories of $\boldsymbol{X}$ in either a table or plot format, study the patterns of change in the predicted category of Y across the combinations of categories of $\boldsymbol{X}$, and compare them with the prediction results obtained under the assumption of the joint independence between Y and $\boldsymbol{X}$. Note that Proposition 21.2.1(c) indicates that there would be no difference in the predicted category of Y across combination of categories of $\boldsymbol{X}$ if Y is independent of $\boldsymbol{X}$.

21.3 Statistical Inference

In this section, we discuss the estimation of the copula regressions reviewed in Sect. 21.2 and propose using the resampling techniques, bootstrapping, and permutation methods, to facilitate the understanding of the regression association revealed from the predictions using the estimated copula regressions.

Estimation

The subcopula regression and the checkerboard copula regression are functions of the joint p.m.f. $p_{i,j}$ for $(\boldsymbol{X}^\top, Y)$ in a multidimensional contingency table and the corresponding marginal p.m.f.s for $\boldsymbol{X}$ and Y. For the point estimation of the copula regressions, two approaches can be considered:

- Non-model-based approach: estimate $p_{i,j}$ using the relative frequencies in the observed contingency table.
- Model-based approach: estimate $p_{i,j}$ using the parametric models well fitted to the observed contingency table. Note that when the parametric models are employed to estimate $p_{j|i}$, one also needs to (parametrically/nonparametrically) estimate the marginal p.m.f. for $\boldsymbol{X}$, p_{i+} to estimate $p_{i,j} (= p_{j|i} \times p_{i+})$.

To serve the purpose of "dependence learning in the process of EDA," in this paper, we use a non-model-based approach to estimate $p_{\boldsymbol{i},j}$. For an observed $(d+1)$-dimensional contingency table of $(\boldsymbol{X}^\top, Y)$, let $n_{\boldsymbol{i},j}$ be the frequency in the $(\boldsymbol{i}, j)$ cell where $\boldsymbol{i} = (i_1, \ldots, i_\ell, \ldots, i_d)$, $\ell = 1, \ldots, d$, $i_\ell = 1, \ldots, I_\ell$, $j = 1, \ldots, J$ and $n = \sum_{\boldsymbol{i}=\boldsymbol{1}}^{\boldsymbol{I}} \sum_{j=1}^{J} n_{\boldsymbol{i},j}$ with $\boldsymbol{1} = (1, \ldots, 1)$ and $\boldsymbol{I} = (I_1, \ldots, I_d)$. Then the estimators of the joint p.m.f. $p_{\boldsymbol{i},j}$ for $(\boldsymbol{X}^\top, Y)$, the corresponding marginal p.m.f.s $\boldsymbol{X}$, X_ℓ and Y, and the two copula regressions in Eqs. (21.6) and (21.7) are given below:

$$\hat{p}_{\boldsymbol{i},j} = \frac{n_{\boldsymbol{i},j}}{n}, \quad \hat{p}_{\boldsymbol{i}+} = \frac{n_{\boldsymbol{i}+}}{n}, \quad \hat{p}_{+i_\ell+} = \frac{n_{+i_\ell+}}{n}, \quad \hat{p}_{+j} = \frac{n_{+j}}{n}, \quad \hat{p}_{j|\boldsymbol{i}} = \frac{\hat{p}_{\boldsymbol{i},j}}{\hat{p}_{\boldsymbol{i}+}},$$

$$\hat{R}_{X_\ell} = \{\hat{u}_0^\ell, \hat{u}_1^\ell, \ldots, \hat{u}_{i_\ell}^\ell, \ldots, \hat{u}_{I_\ell}^\ell\}, \quad \hat{u}_{i_\ell}^\ell = \hat{F}_{X_\ell}(x_{i_\ell}^\ell) = \sum_{h_\ell=1}^{i_\ell} \hat{p}_{+h_\ell+},$$

$$\hat{R}_Y = \{\hat{v}_0, \hat{v}_1, \ldots, \hat{v}_j, \ldots, \hat{v}_J\}, \quad \hat{v}_j = \hat{F}_Y(y_j) = \sum_{h=1}^{j} \hat{p}_{+h},$$

$$\widehat{r_{Y|\boldsymbol{X}}^{C^S}} = \sum_{j=1}^{J} \hat{v}_j \hat{p}_{j|\boldsymbol{i}}, \quad \widehat{r_{Y|\boldsymbol{X}}^{C^*}} = \sum_{j=1}^{J} \hat{s}_j \hat{p}_{j|\boldsymbol{i}},$$

where $n_{\boldsymbol{i}+}$, $n_{+i_\ell+}$, and n_{+j} are the marginal frequencies for $\boldsymbol{X}$, X_ℓ, and Y, respectively.

Evaluation of the Prediction by the Estimated Copula Regression

Using the copula regression-based prediction method given in section "Subcopula Regression and Checkerboard Copula Regression" and the estimators given above, we can estimate the predicted category of Y for a given combination of categories of $\boldsymbol{X}$ and then present the estimated prediction results in a table or plot. Note that, according to Proposition 21.2.1(c), we can also estimate the predicted category of Y under the assumption of joint independence by empirically estimating the marginal distribution of Y.

In order to hone the understanding of the regression patterns (visually) discovered from the copula regression-based prediction, we consider two resampling approaches, bootstrap method for characterizing the uncertainty in the prediction and permutation method for visually evaluating the significance of the discovered regression patterns.

To assess uncertainty in the predicted category of Y, one may use the asymptotic distributions for the estimators of the copula regressions (e.g., see Theorem 1 in

Wei and Kim, 2021 for the checkerboard copula regression) and then compute the asymptotic interval for the predicted category of Y. However, in order to properly quantify and visualize uncertainty in the prediction of Y due to the effects of finite sample size, we employ nonparametric bootstrapping. That is, simulate B bootstrap samples of size n from the observed multidimensional contingency table, compute the proportion of each category of Y predicted by the estimated copula regression, and present the results in the same table or plot along with the estimate of the predicted category of Y obtained from the observed contingency table.

To visually calibrate the significance of the regression association structure discovered by the estimated copula regression and the bootstrap method, we use the permutation method under the assumption of the joint independence between Y and $\boldsymbol{X}$. That is, generate B permuted contingency tables of size n from the observed multidimensional contingency table under the assumption of the joint independence (Doran et al., 2014), run the copula regression-based prediction method on the simulated permutation samples, and compute the proportion of each category of Y predicted by the copula regression. The reason for using this strategy is as follows. The permutation samples generated from the joint independence may be able to provide a list of situations where there is no regression associations between Y and $\boldsymbol{X}$. Thus, the copula regression-based prediction results obtained from the permutation samples can be used as a reference to visually assess the regression structure detected from the observed data at hand through the application of the copula regression.

In each case study of Sect. 21.4, we will present two prediction tables/plots using the copula regression-based prediction, one constructed from the estimates of the predicted categories of Y based on the observed contingency table and the associated bootstrap samples, called the *discovery prediction table/plot*, and the other obtained from the estimates of the prediction of Y under the joint independence between Y and $\boldsymbol{X}$ and the associated permutation samples, called the *null reference prediction table/plot*. We will then compare them to see if the regression association detected from the observed data is visually distinguishable from the absence of association structure (expected under the assumption of the joint independence).

21.4 Case Studies

In this section, using three real-world data examples, we illustrate the utility of the copula regression-based prediction reviewed in Sect. 21.2 and the resampling approaches proposed in section "Evaluation of the Prediction by the Estimated Copula Regression" to unravel the regression association structures for contingency tables with an ordinal dependent variable in an exploratory fashion. For the second and third example, we consider the subcopula regression and checkerboard copula regression, respectively, while we use both copula regressions in the first data example. As a tool to visualize the association structures captured by the copula regressions, we employ "cross-tabulation" for the first example, "bubble plot" for the second example, and "doubledecker plot" (Hofmann, 2001; Meyer et al., 2006) for the third example.

Ice Cream Study

The first data, shown in Table 21.1 adapted from Wang (2021), is concerned with the ice cream study where the goal was to find an optimal amount of fat for tasty ice creams (PennState-Statistics, 2022). In this study, two ordinal variables were measured from 493 participants who tasted ice creams: the fat level (F) with 8 categories starting from 0.00 to 0.28 and the rating (R) with 9 categories from 1 (didn't like it all) to 9 (yum yum!). According to the goal of this study, it is reasonable to consider the rating (R) as a dependent variable and the fat level (F) as an independent variable. We observe from Table 21.1 that the regression association appears to show a quadratic pattern in that rating tends to increase initially and then decreases as the fat level increases, and when the fat levels are between 0.04 and 0.16, many participants chose ratings 7 or 8.

We apply two copula regressions to this data and predict the category of rating R at each category of fat F using the prediction method and the resampling techniques. The "cross-tabulation" in Table 21.2 are the "discovery prediction tables" showing the predicted category of R for a given level of F using the copula regressions and prediction uncertainty estimation using the bootstrap method (Table 21.2(a), the subcopula regression; Table 21.2(b), the checkerboard copula regression). The category of R in the column with red-colored bold font number is the estimate of the predicted category of R by the copula regression at each category of F, and the number for each category combination of F and R is the percentage of times a category of R is predicted for a given level of F over 1000 bootstrap samples. It appears that the quadratic pattern between rating and fat level is revealed, and the fat level resulting in the higher average rating appears to be between 0.08 and

Table 21.1 Ice cream study data with two ordinal variables, fat level and rating. Adapted from the dissertation of Li Wang, "Model-Free Descriptive Modeling for Multivariate Categorical Data with An Ordinal Dependent Variable" (PhD dissertation, University of Massachusetts Amherst, 2021)

	Rating (R)								
Fat level (F)	1	2	3	4	5	6	7	8	9
0.00	4	17	8	16	5	6	4	2	1
0.04	1	1	5	6	7	9	21	12	0
0.08	0	2	2	2	4	13	16	21	3
0.12	1	1	1	3	4	11	15	23	4
0.16	0	3	2	6	3	7	17	17	5
0.20	0	1	3	8	4	13	14	11	8
0.24	1	5	4	14	2	13	13	7	2
0.28	4	6	9	11	5	9	7	8	3

0.16. Both copula regressions show the same results for the predicted category of R, except when the level of F is 0.04 (predicted category of R was 7 for the subcopula regression and 6 for the checkerboard copula regression). A main difference lies in the pattern of the uncertainty estimation in prediction. When the level of F is 0.04, 0.20, and 0.28, the checkerboard copula regression showed more uncertainty around the corresponding predicted category than the subcopula regression. On the other hand, when the fat level is between 0.08 and 0.16, the subcopula copula regression selected "8" as a predicted category of R more frequently than the checkerboard copula regression.

Table 21.3 gives the "null reference prediction tables" for the predicted category of R for a given level of F using the copula regressions and the permutation method under the independence between R and F (Table 21.3(a), the subcopula regression; Table 21.3(b), the checkerboard copula regression). Note that the number for each category combination of F and R is the percentage of times a category of R is predicted for a given level of F over 10^4 permutation samples generated under the independence. First, the estimate of the predicted category of R by the copula regression under the independence is 7 for the subcopula regression and 6 for the checkerboard copula regression (represented by the category of R in the column with blue-colored bold font), regardless of a category of F. Second, the pattern of the predicted category of R over the permutation samples is uniform around the estimate of the predicted category of R by each copula regression. That is, the proportions of the predicted category of R equal to 7 for the subcopula regression and 6 for the checkerboard copula regression are around 90% and 80%, respectively, irrespective of the category of F.

From the results of Tables 21.2 and 21.3, we can see that the quadratic regression association identified from the observed table is visually clearly distinguishable from that under the independence.

Back Pain Data

The second data, summarized in a four-dimensional contingency table in Table 21.4 (adapted from Wei and Kim, 2021), is collected from a study of 101 back pain patients (Anderson, 1984). After patients received a treatment for back pain, their progress on pain levels (dependent variable, denoted as Y) was evaluated on six response categories: "Worse" (W), "Same" (S), "Slight Improvement" (SI), "Moderate Improvement" (MODI), "Marked Improvement" (MARI), and "Complete Relief" (CR). At the beginning of the treatment, the three prognostic variables (independent variables) were also measured from the patients: length of previous attack (X_1) with two levels, "Short" (S) and "Long" (L); pain change (X_2) with three levels, "getting Better" (B), "Same" (S), and "Worse" (W); and lordosis (X_3) with two levels, "Absent/Decreasing" (AD) and "Present/Increasing" (PI).

To explore and visualize the regression association structures for this data, we present the "discovery prediction bubble plot" in Fig. 21.1 summarizing the

Table 21.2 Discovery prediction tables for ice cream data—the predicted category of the rating for each fat level and its uncertainty estimation using the copula regressions and bootstrap method. The category of R in the column with red-colored bold font number is the category of R predicted by each of two copula regressions

Subcopula regression

Fat level (F) \ Rating (R)	1	2	3	4	5	6	7	8	9
0.00	0	0	0	**51.7**	45.6	2.6	0	0	0
0.04	0	0	0	0	0	3.8	**96.2**	0	0
0.08	0	0	0	0	0	0	**87.2**	12.8	0
0.12	0	0	0	0	0	0	**72.9**	27.1	0
0.16	0	0	0	0	0	0	**93.7**	6.3	0
0.20	0	0	0	0	0	0.1	**99.4**	0.5	0
0.24	0	0	0	0	0	**66.4**	33.6	0	0
0.28	0	0	0	0	2.9	**93.2**	3.9	0	0

(a)

Checkerboard copula regression

Fat level (F) \ Rating (R)	1	2	3	4	5	6	7	8	9
0.00	0	0	0.3	**94.3**	5.4	0	0	0	0
0.04	0	0	0	0	0	**64.1**	35.9	0	0
0.08	0	0	0	0	0	0.5	**99.5**	0	0
0.12	0	0	0	0	0	0.2	**99.7**	0.1	0
0.16	0	0	0	0	0	3.6	**96.4**	0	0
0.20	0	0	0	0	0	13	**87**	0	0
0.24	0	0	0	0	2.9	**96.4**	0.7	0	0
0.28	0	0	0	1.4	30.1	**68.4**	0.1	0	0

(b)

Table 21.3 Null reference prediction tables for ice cream data—the predicted category of the rating for each fat level and the proportion of each category of the rating for each fat level using the copula regression and the permutation method under the assumption of the independence between fat and rating. The category of R in the column with blue-colored bold font number is the predicted category of R estimated by each of two copula regressions under the independence

Subcopula regression

Fat level (F) \ Rating (R)	1	2	3	4	5	6	7	8	9
0.00	0	0	0	0	0	9.6	**90.4**	0	0
0.04	0	0	0	0	0	9.9	**90.1**	0	0
0.08	0	0	0	0	0	10.1	**89.9**	0	0
0.12	0	0	0	0	0	9.5	**90.5**	0	0
0.16	0	0	0	0	0	10.2	**89.8**	0	0
0.20	0	0	0	0	0	9.5	**90.5**	0	0
0.24	0	0	0	0	0	9.8	**90.2**	0	0
0.28	0	0	0	0	0	9.7	**90.3**	0	0

(a)

Checkerboard copula regression

Fat level (F) \ Rating (R)	1	2	3	4	5	6	7	8	9
0.00	0	0	0	0	0	**80.2**	19.8	0	0
0.04	0	0	0	0	0	**79.1**	20.9	0	0
0.08	0	0	0	0	0	**80.6**	19.4	0	0
0.12	0	0	0	0	0	**79.8**	20.2	0	0
0.16	0	0	0	0	0	**79.7**	20.3	0	0
0.20	0	0	0	0	0	**79.4**	20.3	0	0
0.24	0	0	0	0	0	**79.1**	20.9	0	0
0.28	0	0	0	0	0	**79.9**	20.1	0	0

(b)

Table 21.4 Back pain prognosis data with pain levels (Y) and three prognostic variables (X_1, X_2, X_3). Adapted from "On exploratory analytic method for multi-way contingency tables with an ordinal response variable and categorical explanatory variables" by Z. Wei and D. Kim, 2021, *Journal of Multivariate Analysis*, 186: 104793, p. 10

Length of previous attack (X_1)	Pain change (X_2)	Lordosis (X_3)	Back pain (Y) W	S	SI	MODI	MARI	CR
Short	Better	Absent/Decreasing	0	1	0	0	2	4
Short	Better	Present/Increasing	0	0	0	1	3	0
Short	Same	Absent/Decreasing	0	2	3	0	6	4
Short	Same	Present/Increasing	0	1	0	2	0	1
Short	Worse	Absent/Decreasing	0	0	0	0	2	2
Short	Worse	Present/Increasing	0	0	1	1	3	0
Long	Better	Absent/Decreasing	0	0	3	0	1	2
Long	Better	Present/Increasing	0	1	0	0	3	0
Long	Same	Absent/Decreasing	0	3	4	5	6	2
Long	Same	Present/Increasing	1	4	4	3	0	1
Long	Worse	Absent/Decreasing	2	2	1	5	2	0
Long	Worse	Present/Increasing	2	0	2	3	0	0

prediction results of the pain level (Y) using the checkerboard copula regression-based prediction. Note that the dark dots represent the predicted category of Y at each combination of categories of (X_1, X_2, X_3) and the size of the bubbles (circles) indicates the proportion of each category of Y computed from 10,000 bootstrap resampling which is larger than 5%.

From Fig. 21.1, there appears to be potential three-way interaction effect of three independent variables on the dependent variable. That is, depending on the length of previous attack (X_1), the pain level (Y) is differently affected by the statuses of pain change (X_2) and lordosis (X_3). When the length of the previous attack (X_1) is "Short" (S), the patient's pain level tends to be "Marked Improvement" (MARI) irrespective of pain change (X_2) and lordosis (X_3), except when the pain change

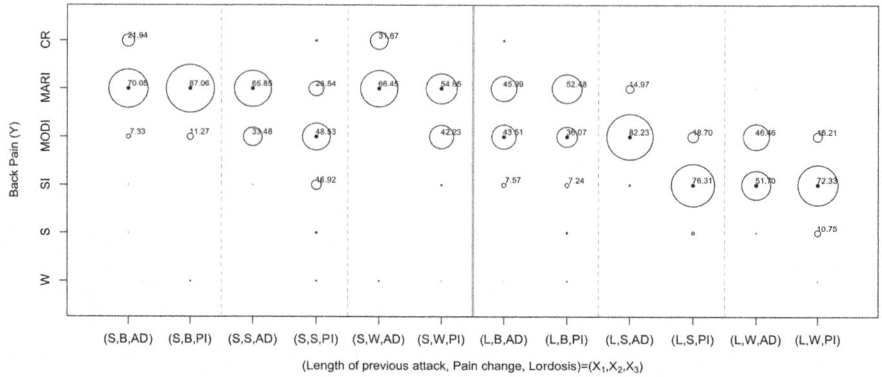

Fig. 21.1 Discovery prediction bubble plot for back pain data—prediction of the category of the pain level Y (represented by dark dots) using the checkerboard copula regression at each combination of categories of (X_1, X_2, X_3) and the proportion of each category of Y predicted by the checkerboard copula regression over 10,000 bootstrap resampling (represented by the bubbles whose sizes are proportional to the numbers)

is "Same" (S) and lordosis is "Present/Increasing" (PI). However, when the length of the previous attack (X_1) is "Long" (L), the pattern of pain level changes from "Marked Improvement" (MARI)/ "Moderate Improvement" (MODI) to "Slight Improvement" (SI) as the status of pain change (X_2) switches from "getting Better" (B) to "Worse" (W). Note that when the pain change is "Same" (S), the back pain is improved from "Slight" (SI) to "Moderate" (MODI) as the status of lordosis (X_3) changes from "Present/Increasing" (PI) to "Absent/Decreasing" (AD).

To visually assess if the regression structures discovered from Fig. 21.1 are different from the reference scenario assuming absence of the regression association, we construct the "null reference prediction bubble plot" in Fig. 21.2 showing the prediction results of the pain level (Y) using the checkerboard copula regression and the permutation method under the joint independence between Y and (X_1, X_2, X_3). Note that the dark dots represent the estimates of predicted categories of Y and the size of the bubbles (circles) indicates the proportion of each category of Y computed from 10,000 permutation samples which is larger than 5%. We see that the prediction results of Y do not show any regression pattern over all combinations of categories of (X_1, X_2, X_3) which is what would be expected under the assumption of joint independence. That is, regardless of the categories of (X_1, X_2, X_3), the predicted categories of Y over the permutation samples are consistently equal to MODI, an estimate of the predicted category of Y obtained by the checkerboard copula regression under the joint independence.

In sum, the two bubble plots in Figs. 21.1 and 21.2 provide visual evidence that the regression dependence in Table 21.4 detected by the checkerboard copula regression is clearly discernible and it certainly is distinctive compared to the pattern expected under the joint independence.

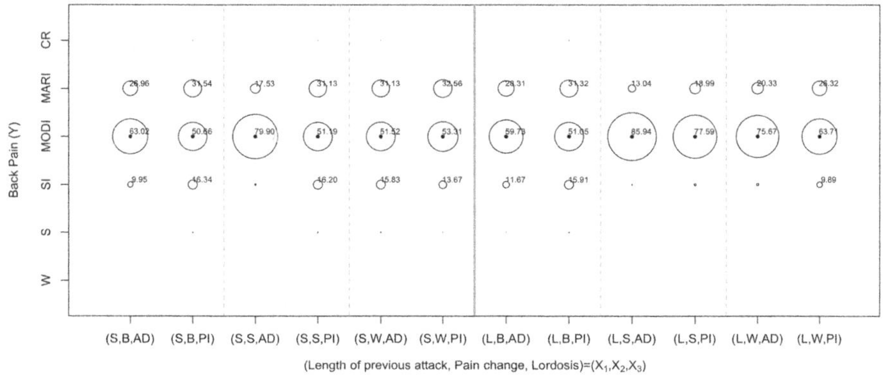

Fig. 21.2 Null reference prediction bubble plot for back pain data—prediction of the category of the pain level Y (represented by dark dots) using the checkerboard copula regression at each combination of categories of (X_1, X_2, X_3) under the joint independence between Y and (X_1, X_2, X_3) and the proportion of each category of Y predicted by the checkerboard copula regression over 10,000 permutation samples generated from the joint independence (represented by the bubbles whose sizes are proportional to the numbers)

Three Mile Island (TMI) Accident Data

The third data set, presented in Table 21.5, is the five-way contingency table data collected from a longitudinal interview study on the stress levels of mothers of young children living within ten miles of the Three Mile Island nuclear power plant after the Spring 1979 accident (Fienberg et al., 1985). The four waves of interviews were performed between Winter 1970 and Autumn 1982. At each interview, the stress levels of mothers, denoted as *WAVE1* (*W1*), *WAVE2* (*W2*), *WAVE3* (*W3*), and *WAVE4* (*W4*), were recorded as one of the three ordered levels: "L" for Low, "M" for Medium, and "H" for High. Furthermore, the information on distance from the nuclear plant, denoted as *DISTANCE* (*D*), was recorded as one of the two levels: "$<$ 5 miles" and "$>$ 5 miles."

The main goal of this study was to understand the patterns of the change in the stress level over time, and thus we consider the regression association structure where *W4* is the dependent variable and (*W3*, *W2*, *W1*, *D*) are the independent variables. In order to effectively present the results of the proposed method, we decide to use the doubledecker plot (Hofmann, 2001; Meyer et al., 2006), a mosaic plot with vertical splits for (*W3*, *W2*, *W1*, *D*) and with horizontal splits for *W4*. This is because it helps to make clear distinctions between the categories of *W4* conditional on combinations of the categories of (*W3*, *W2*, *W1*, *D*) (Friendly and Meyer, 2015).

We first applied the subcopula regression-based prediction and the associated bootstrap method to this data and then construct the "discovery prediction doubledecker plot" in Fig. 21.3 showing the predicted category of *W4* at each combination of categories of (*W3*, *W2*, *W1*, *D*) and its uncertainty estimation.

Table 21.5 A longitudinal interview study on the stress levels of mothers of young children living within ten miles of the Three Mile Island (TMI) nuclear power plant

			<5 miles			>5 miles						<5 miles			>5 miles		
Wave			Wave 4						Wave			Wave 4					
1	2	3	L	*M*	H	L	*M*	H	1	2	3	L	M	*H*	L	M	H
L	L	L	2	0	0	1	2	0	M	M	H	0	2	3	0	5	1
L	L	M	2	3	0	2	0	0	M	H	L	0	0	0	0	0	0
L	L	H	0	0	0	0	0	0	M	H	M	0	2	0	0	1	1
L	M	L	0	1	0	1	0	0	M	H	H	0	1	1	0	3	1
L	M	M	2	4	0	0	3	0	H	L	L	0	0	0	0	0	1
L	M	H	0	0	0	0	0	0	H	L	M	0	0	0	0	0	0
L	H	L	0	0	0	0	0	0	H	L	H	0	0	0	0	0	0
L	H	M	0	0	0	0	0	0	H	M	L	0	0	0	0	0	0
L	H	H	0	0	0	0	0	0	H	M	M	0	4	3	1	13	0
M	L	L	5	1	0	4	4	0	H	M	H	0	1	4	0	0	0
M	L	M	1	4	0	5	15	1	H	H	L	0	0	0	0	0	0
M	L	H	0	0	0	0	0	0	H	H	M	1	2	0	0	7	2
M	M	L	3	2	0	2	2	0	H	H	H	0	5	12	0	2	7
M	M	M	2	38	4	6	53	6									

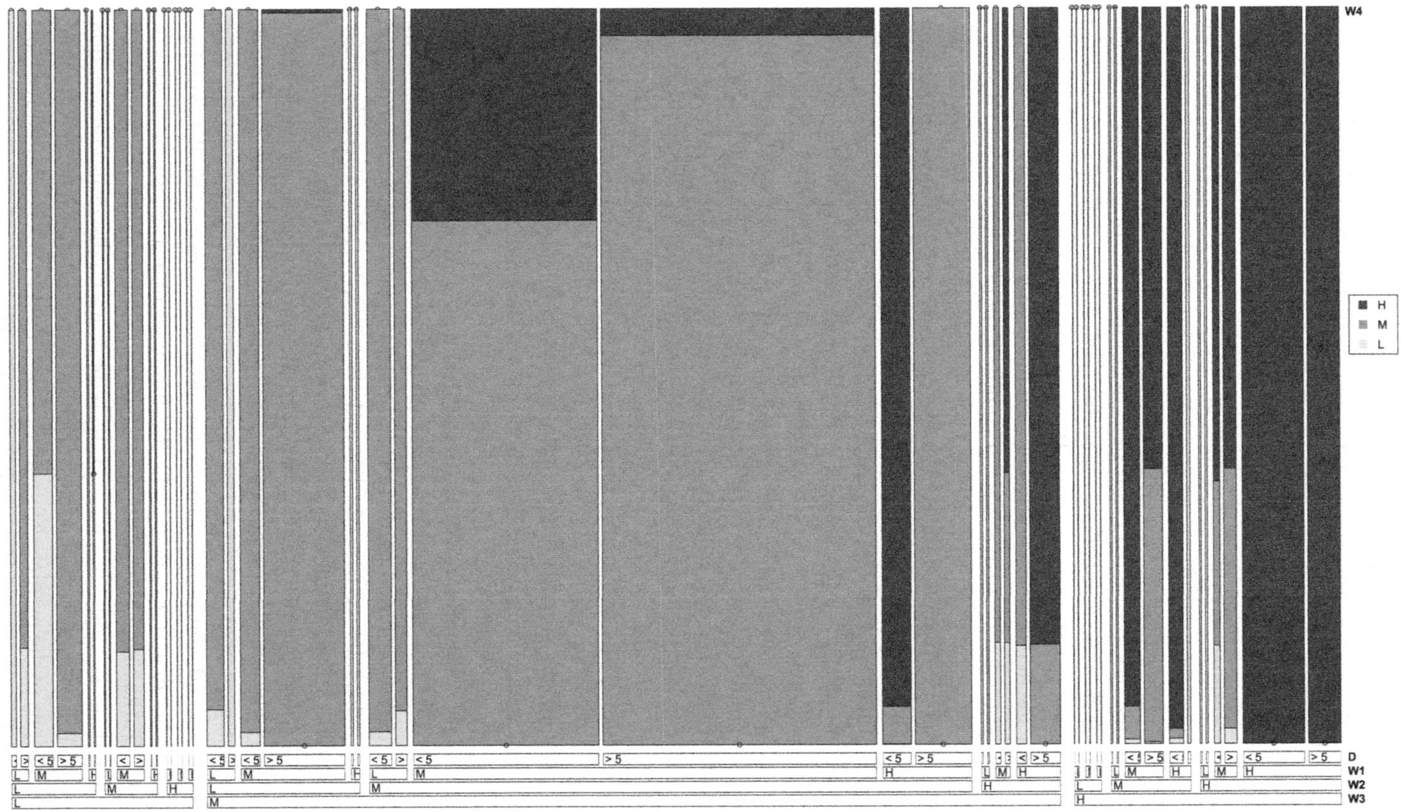

Fig. 21.3 Discovery prediction double decker plot for the TMI data—prediction of the category of $W4$ for each combination of the categories of $(W3, W2, W1, D)$ using the subcopula regression and the bootstrap method. Note that this figure is adapted from the dissertation of Li Wang, "Model-Free Descriptive Modeling for Multivariate Categorical Data with An Ordinal Dependent Variable" (PhD dissertation, University of Massachusetts Amherst, 2021)

For the sake of easy interpretation, we provide a few details on Fig. 21.3. First, the layers of labels along the x-axis show the order of the four independent variables considered in the construction of the doubledecker plot (i.e., *W3/W2/W1/D* for the first/second/third/last variable), which takes into account the time-dependency structure between the waves. Second, the width of each bar for each combination of categories of (*W3*, *W2*, *W1*, *D*) is proportional to the observed frequency of the corresponding combination of the categories of (*W3*, *W2*, *W1*, *D*). Note that when the observed frequencies for all categories of *W4* at any combination of categories of (*W3*, *W2*, *W1*, *D*) are all zero, the corresponding bar is represented by a light gray thin line. Third, the blocks of three levels (dark/medium/light) of gray color in each bar are proportional to the numbers of times that each category of $W4$ is predicted at each combination of categories of (*W3*, *W2*, *W1*, *D*) over 10^4 bootstrap samples. Thus, the longer the height of each gray-colored block in a bar, the less the relative uncertainty in the prediction of *W4* given a combination of categories of (*W3*, *W2*, *W1*, *D*).

When the observed frequency for at least one category of *W4* at any combination of categories of (*W3*, *W2*, *W1*, *D*) is not zero, we make a few interesting observations from Fig. 21.3. First, we find that the category of $W4$ whose height of the gray-colored block in each bar is the longest is equal to the estimate of the predicted category of $W4$ at each combination of categories of (*W3*, *W2*, *W1*, *D*). Second, it appears that *W4* is very associated with *W3*. That is, when the stress level of *W3* is "H" ("M"), the predicted category of *W4* tends to be "H" ("M"). Third, when the stress level of *W3* is "H," *W4* is likely to be closely related to the stress level of *W2* and *W1* because the predicted category of *W4* seems to be "H" when the stress levels of (*W2*, *W1*) are "M" or "H," regardless of *D*. Fourth, when the level of *W3* is "M," the level of *W4* is dominantly predicted to be "M," except for three cases: the predicted category of *W4* is "H" for $(W2, W1, D) =$ ("M," "H," "<5 miles"), ("H," "M," ">5 miles"), and ("H," "H," ">5 miles"). Fifth, when *W3* is "L" or "M," there appears to be more bars where the proportions of the predicted category of *W4* equal to "L" is not negligible.

As a part of the calibration process on the discovered regression dependence, we construct in Fig. 21.4 the "null reference prediction doubledecker plot" using the subcopula regression-based prediction and the associated permutation method under the joint independence between *W4* and (*W3*, *W2*, *W1*, *D*). There are three principal differences that distinguish Fig. 21.4 from Fig. 21.3. First, regardless of combination of categories of (*W3*, *W2*, *W1*, *D*), the longest block in every bar is the medium gray-colored block representing "M" which is same as the estimate of the predicted category of $W4$ obtained by the subcopula regression under the joint independence. Second, the proportions of the predicted category of *W4* equal to "M" is much higher than those of "L" and "H" for all combinations of categories of (*W3*, *W2*, *W1*, *D*) (its range is between 67 and 100%). Third, unlike the "discovery prediction doubledecker plot" in Fig. 21.3, there appears no discernible regression dependence of *W4* with the independent variables, especially (*W3*, *W2*, *W1*). Note that the proportions of the predicted category of *W4* equal to "L" is negligibly small for most combinations of categories of (*W3*, *W2*, *W1*, *D*).

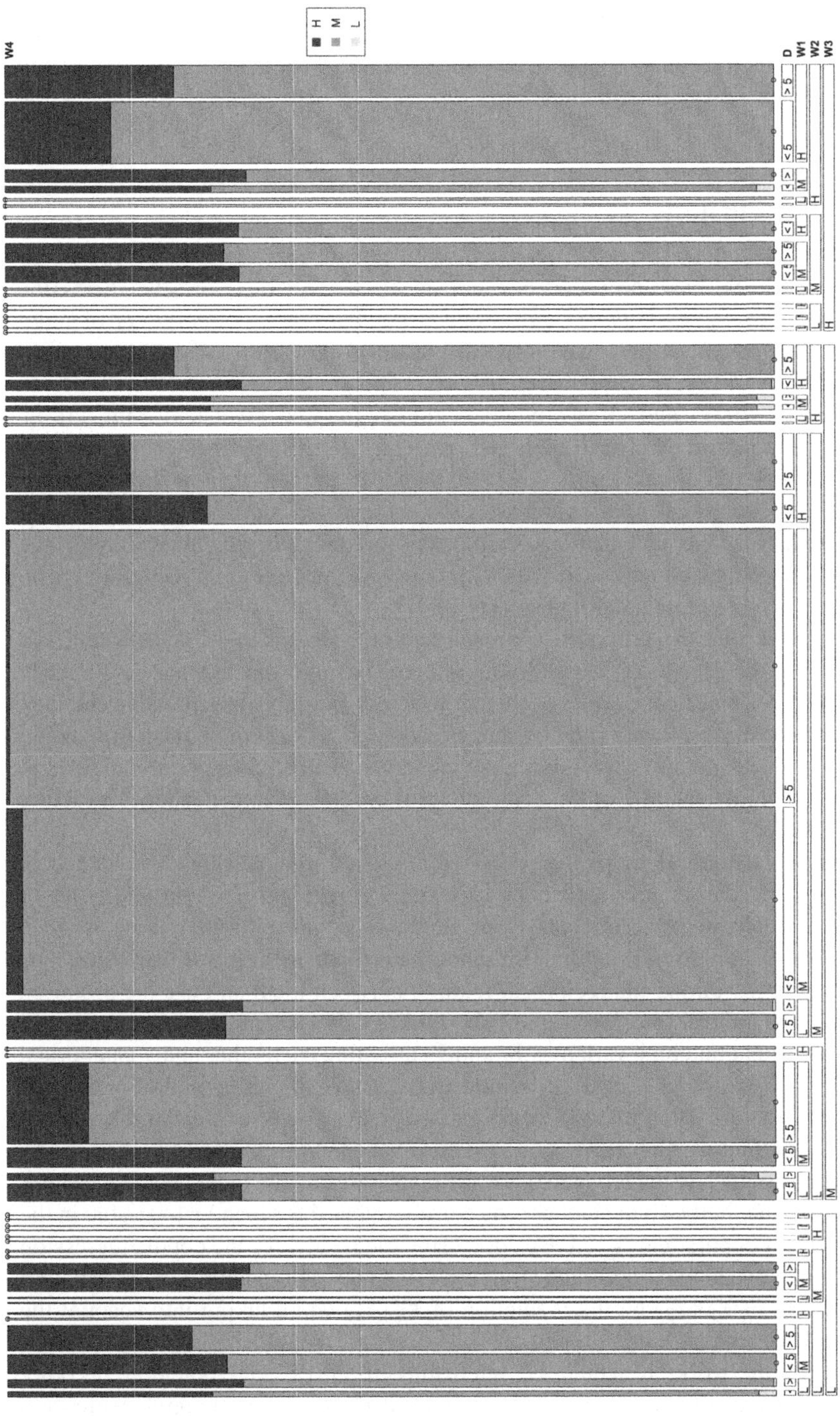

Fig. 21.4 Null reference prediction double decker plot for the TMI data—prediction of the category of $W4$ for each combination of the categories of $(W3, W2, W1, D)$ using the subcopula regression and the permutation method under the joint independence between $W4$ and $(W3, W2, W1, D)$

To sum up, we conclude that the dependence structures identified from the TMI data are the features that are distinctively different from those expected from the assumption of the joint independence and it would beneficial to take them into account for formal statistical modeling.

21.5 Conclusion

In this paper, we have proposed a graphical approach to explore the regression dependence structures between an ordinal dependent variable and a set of independent categorical variable in a multidimensional contingency table. The proposed approach is based on two copula regressions, subcopula regression (Wei and Kim, 2017; Wang, 2021) and checkerboard copula regression (Wei and Kim, 2021), and their prediction. We have also incorporated two resampling techniques into the framework of copula regressions to characterize the uncertainty and assess the visual significance of the regression associations revealed by the copula regression-based prediction. We have demonstrated, using three real-world data examples, the utility of the proposed approach with different visualization tools including cross-tabulation, bubble plot, and doubledecker plot.

We believe that the proposed approach can be a useful tool for exploring and representing the regression dependence inherent in multidimensional contingency tables with ordinal dependent variable in a model-free and interpretable manner. Furthermore, the proposed method can be adopted as a complement to existing quantitative measures of regression association for ordinal categorical variables to enhance the understanding of the complex form of regression dependence structure in the data.

The copula regressions of the proposed approach assumed that knowledge of the underlying causal mechanism of an ordinal dependent variable and a set of categorical independent variables is available in advance. When there is no a priori justification for the underlying causal mechanism of some variables, one may consider assuming the reciprocal relation between them and use the proposed method, with careful interpretation of the findings. For example, suppose that we have two ordinal outcome variables Y_1 and Y_2 affecting each other and d categorical covariates $\boldsymbol{X} = (X_1, \ldots, X_d)$ influencing Y_1 and Y_2 in a multidimensional contingency table. Then one can consider two regression structures and apply the proposed method to each regression structure: $(\boldsymbol{X}, Y_2) \rightarrow Y_1$ and $(\boldsymbol{X}, Y_1) \rightarrow Y_2$ where the variables in the left side of $\rightarrow$ are considered as the independent variables and the variable in the right side of $\rightarrow$ is considered as the dependent variable in the framework of copula regression. By comparing the results for the copula regression prediction between two regression structures, one may obtain empirical information on the effect of Y_1 (Y_2) on the prediction of Y_2 (Y_1) while taking into account the categorial covariates, $\boldsymbol{X}$.

As a future work, it would be interesting to investigate potential of other graphical tools to effectively and efficiently visualize the results of copula regression prediction, depending on the dimension (the number of variables and the number of categories in an ordinal dependent variable and independent categorical variables) and context of the data.

References

Agresti, A. (2010). *Analysis of ordinal categorical data* (2nd ed.). Wiley.

Anderson, J. A. (1984). Regression and ordered categorical variables. *Journal of the Royal Statistical Society: Series B (Methodological), 46*(1), 1–22.

Blasius, J., & Greenacre, M. (1998). *Visualization of categorical data*. Academic.

Buja, A., Cook, D., Hofmann, H., Lawrence, M., Lee, E.-K., Swayne, D. F., & Wickham, H. (2009). Statistical inference for exploratory data analysis and model diagnostics. *Philosophical Transactions of the Royal Society of London, 367,* 4361–4383.

Carley, H. (2002). Maximum and minimum extensions of finite subcopulas. *Communications in Statistics - Theory and Methods, 31*(12), 2151–2166.

Cleveland, W. S. (1985). *The elements of graphing data*. Wads-worth.

Cleveland, W. S. (1993). *Visualizing data*. Hobart Press.

de Amo, E., Díaz Carrillo, M., & FernáNdez-SáNchez, J. (2012). Characterization of all copulas associated with non-continuous random variables. *Fuzzy Sets and Systems, 191*, 103–112.

Denuit, M., & Lambert, P. (2005). Constraints on concordance measures in bivariate discrete data. *Journal of Multivariate Analysis, 93*(1), 40–57.

Donoho, D. L. (2017). 50 years of data science. *Journal of Computational and Graphical Statistics, 26,* 745–766.

Doran, G., Muandet, K., Zhang, K., & Schölkopf, B. (2014). A permutation-based kernel conditional independence test. In *Proceedings of the 30th Conference on Uncertainty in Artificial Intelligence (UAI2014)* (pp. 132–141). AUAI Press.

Durante, F., & Sempi, C. (2015). *Principles of copula theory*. Taylor & Francis.

Fienberg, S. E., Bromet, E. J., Follmann, D., Lambert, D., & May, S. M. (1985). Longitudinal analysis of categorical epidemiological data: A study of three mile island. *Environmental Health Perspectives, 63,* 241–248.

Friendly, M. (2000). *Visualizing categorical data*. SAS Institute.

Friendly, M., & Meyer, D. (2015). *Discrete data analysis with R: Visualization and modeling techniques for categorical and count data*. CRC Press.

Gelman, A., & Vehtar, A. (2021). What are the most important statistical ideas of the past 50 years? *Journal of The American Statistical Association, 116,* 2087–2097.

Genest, C., & Nešlehová, J. (2007). A primer on copulas for count data. *ASTIN Bulletin: The Journal of the IAA, 37*(2), 475–515.

Genest, C., Nešlehová, J. G., & Rémillard, B. (2014). On the empirical multilinear copula process for count data. *Bernoulli, 20*(3), 1344–1371.

Genest, C., Nešlehová, J. G., & Rémillard, B. (2017). Asymptotic behavior of the empirical multilinear copula process under broad conditions. *Journal of Multivariate Analysis, 159*, 82–110.

Gower, J. C., & Hand, D. J. (1996). *Biplots*. Chapman & Hall.

Hofmann, H. (2001). Generalized odds ratios for visual modeling. *Journal of Computational and Graphical Statistics, 10,* 628–640.

Joe, H. (2014). *Dependence modeling with copulas*. Chapman and Hall/CRC.

Meyer, D., Zeileis, A., & Hornik, K. (2006). The strucplot framework: Visualizing multi-way contingency tables with vcd. *Journal of Statistical Software, 17,* 1–48.

Nelsen, R. B. (2006). *An introduction to copulas*. Springer.

Nešlehová, J. (2007). On rank correlation measures for non-continuous random variables. *Journal of Multivariate Analysis, 98*(3), 544–567.
PennState-Statistics (2022). Case study: The ice cream study at penn state. In *STAT 504 Analysis of Discrete Data*. PennState: Statistics Online Courses. https://online.stat.psu.edu/stat504/lesson/case-study-ice-cream-study-penn-state
Rachasingho, J., & Tasena, S. (2020). A metric space of subcopulas - an approach via hausdorff distance. *Fuzzy Sets and Systems, 378*(1), 144–156.
Rüschendorf, L. (2009). On the distributional transform, sklar's theorem, and the empirical copula process. *Journal of Statistical Planning and Inference, 139,* 3921–3927.
Schweizer, B., & Sklar, A. (1974). Operations on distribution functions not derivable from operations on random variables. *Studia Mathematica, 52*(1), 43–52.
Shmueli, G. (2010). To explain or to predict? *Statistical Science, 25*(3), 289–310.
Sklar, A. (1959). Fonctions de répartition à *n* dimensions et leurs marges. *Publications de l'Institut de Statistique de L'Université de Paris, 8,* 229–231.
Tasena, S. (2021a). On a distribution form of subcopulas. *International Journal of Approximate Reasoning, 128*, 1–19.
Tasena, S. (2021b). On metric spaces of subcopulas. *Fuzzy Sets and Systems, 415*(15), 76–88.
Tufte, E. R. (1983). *The visual display of quantitative information*. Graphics Press.
Tukey, J. W. (1977). *Exploratory data analysis*. Addison-Wesley.
Wang, L. (2021). *Model-Free Descriptive Modeling for Multivariate Categorical Data with An Ordinal Dependent Variable*. PhD Thesis, University of Massachusetts Amherst, 7 2021. https://doi.org/10.7275/22254792.0
Wei, Z., & Kim, D. (2017). Subcopula-based measure of asymmetric association for contingency tables. *Statistics in Medicine, 36*(24), 3875–3894.
Wei, Z., & Kim, D. (2021). On exploratory analytic method for multi-way contingency tables with an ordinal response variable and categorical explanatory variables. *Journal of Multivariate Analysis, 186,* 104793.
Wilkinson, L. (2005). *The grammar of graphics* (2nd ed.). Springer.

Chapter 22
Mental Health Symptom Profiles Over Time: A Three-Step Latent Transition Cognitive Diagnosis Modeling Analysis with Covariates

Qianru Liang, Jimmy de la Torre, Mary E. Larimer, and Eun-Young Mun

22.1 Introduction

Cognitive diagnosis models (CDMs) have primarily been utilized to understand students' knowledge, thinking processes, and problem-solving strategies in education. Given a set of fine-grained attributes in a specific domain, CDMs can provide diagnostic information on whether or not a given student has mastered the attributes (de la Torre & Minchen, 2014), facilitating teaching and learning. Examples of CDM applications in educational subjects include Ma et al. (2020) in proportional reasoning attributes (Tjoe & de la Torre, 2014), Chen et al. (2022) in second language reading skills, and Liang et al. (2021) in digital literacy skills. However, the utility of CDMs is not limited to educational research. CDMs can also be used to better characterize attribute profiles of mental health symptoms (Tan et al., 2023) or diagnose mental disorders (e.g., de la Torre et al., 2018; Templin & Henson, 2006).

The project described was supported by grants R01 AA019511 and K02 AA028630 from the National Institute on Alcohol Abuse and Alcoholism (NIAAA). The content is solely the responsibility of the authors and does not necessarily represent the official views of the NIAAA or the National Institutes of Health.

Q. Liang
Guangdong Institute of Smart Education, Jinan University, Guangzhou, China

J. de la Torre
Faculty of Education, The University of Hong Kong, Hong Kong, China

M. E. Larimer
Department of Psychiatry and Behavioral Sciences, University of Washington, Seattle, WA, USA

E.-Y. Mun (✉)
College of Public Health, University of North Texas Health Science Center, Fort Worth, TX, USA
e-mail: eun-young.mun@unthsc.edu

M. Stemmler et al. (eds.), *Dependent Data in Social Sciences Research*,
https://doi.org/10.1007/978-3-031-56318-8_22

The current study aims to provide an introduction to latent transition CDM as a novel tool for classifying individual attribute profiles of mental health symptoms and their transition over time to illustrate its utility in an actual clinical data application. The current study extended the CDM classification work on mental health symptoms assessed in a single time point (Tan et al., 2023) to examine whether a brief alcohol intervention affects the transition in mental health symptom profiles over time in a longitudinal analysis. In addition, whether the intervention effect on transition is moderated by gender was examined. We used a three-step latent transition CDM with covariates, as shown in Liang et al. (2023b). In the following sections, we introduce CDM, including model formulation of commonly used CDMs and a three-step estimation approach to latent transition CDM with covariates. We then demonstrate an application of the three-step approach to mental health symptoms data and its results in the third and fourth sections, respectively. Finally, this chapter concludes by discussing findings and future directions.

22.2 Cognitive Diagnosis Modeling

Model Formulation

CDMs refer to a class of restricted latent class models (LCMs) that can classify or diagnose whether a respondent possesses a given set of *attributes*. In educational research, attributes may represent specific skills or cognitive processes that examinees must master to answer an item correctly. In mental health research, attributes can be the mental health problems or disorders measured by clinical assessment tools (e.g., de la Torre et al., 2018; Tan et al., 2023). In this mental health research setting, the presence or absence of an attribute can be interpreted as having or not having a symptom, problem, or disorder.

Various CDMs exist in the literature (von Davier & Lee, 2019). One of the most often cited CDMs is the generalized deterministic inputs, noisy, and gate (G-DINA; de la Torre, 2011) model. The G-DINA model is a general CDM, from which several other general CDMs and commonly used reduced CDMs can be obtained by using different link functions and setting proper constraints, respectively. For instance, the G-DINA model under the logit link is equivalent to the log-linear CDM (Henson et al., 2009). Examples of the reduced CDMs that the G-DINA model can subsume include (1) the DINA model (Junker & Sijtsma, 2001), which assumes that only the individuals with all the attributes (or disorders) measured by the item are expected to endorse the item; (2) the deterministic inputs, noisy, or gate (DINO; Templin & Henson, 2006) model, which assumes that the individuals with at least one attribute measured by the item are expected to endorse the item; and (3) the additive-CDM (A-CDM; de la Torre, 2011), which assumes that the probability of an item endorsement is related to additive main effects of the attributes that an individual possesses. Moreover, the additive models with the log and logit links are equivalent to the reduced reparameterized unified model (R-RUM; Hartz, 2002) and the linear logistic model (LLM; Maris, 1999), respectively. Within the G-DINA model framework, the reduced CDMs can be compared against the saturated G-DINA model for each item to determine whether a more parsimonious model can be selected (Ma et al., 2016).

As with item response theory models, CDM implementation requires response data stored in an $N \times J$ matrix, where N is the number of respondents (i.e., sample size). For dichotomous CDMs, the response data contain each respondent's responses to J items, where 1s represent item endorsement and 0s otherwise. In addition to response data, most CDMs require a Q-matrix (Tatsuoka, 1983, 2009), which delineates the relationships between the items and the attributes. A Q-matrix is an essential component and provides input for a CDM analysis. In most practical applications, it is usually constructed by domain experts (e.g., Tjoe & de la Torre, 2014). A Q-matrix is typically a $J \times K$ matrix of 1s and 0s, where J is the number of items and K is the number of attributes. The entry of the jth row and kth column in a Q-matrix is 1 if attribute k is measured by item j and 0 otherwise.

Each respondent's attribute profile is estimated by fitting CDMs to response data with a Q-matrix linking items to attributes. A respondent is assigned to a latent class by a binary attribute vector or profile, where 1 and 0 represent the presence and absence of attribute k, respectively. Typically, K attributes produce 2^K latent classes. For example, when $K = 3$, there are $2^3 = 8$ latent classes, namely, (0, 0, 0), (1, 0, 0), (0, 1, 0), (0, 0, 1), (1, 1, 0), (1, 0, 1), (0, 1, 1), and (1, 1, 1), where an attribute profile (0, 0, 0) indicates the respondent does not have any of the three attributes. In contrast, an attribute profile (1, 0, 1) indicates that the respondent has the first and the third attributes but not the second.

When the G-DINA model is used, the probability of respondent i with attribute profile $\boldsymbol{\alpha}_i$ endorsing item j can be written as

$$P\left(Y_{ij} = 1 \middle| \boldsymbol{\alpha}_i\right) = \delta_{j0} + \sum_{k=1}^{K_j} \delta_{jk}\alpha_{ik} + \sum_{k'=k+1}^{K_j} \sum_{k=1}^{K_j-1} \delta_{jkk'}\alpha_{ik}\alpha_{ik'} + \cdots + \delta_{j12\ldots K_j} \prod_{k=1}^{K_j} \alpha_{ik}, \tag{22.1}$$

where K_j is the number of attribute(s) measured by item j; δ_{j0} represents the baseline probability of item j; δ_{jk} is the main effect of attribute k; $\delta_{jkk'}$ is the two-way interaction effect between attribute k and k'; and $\delta_{j12\ldots K_j}$ is the highest-order interaction effect of all measured attributes. In short, Eq. 22.1 expresses a fully saturated model of all possible effects, including terms for main effects, lower-order interaction effects, and the highest-order interaction effect.

A Q-matrix is validated and revised by empirically fitting data with CDMs, discussing results with domain experts, and revising the Q-matrix accordingly (de la Torre & Chiu, 2016; Tjoe & de la Torre, 2014). With the revised Q-matrix, we can compare the reduced CDMs (e.g., DINA, DINO, A-CDM, LLM, and R-RUM) against the saturated G-DINA model. Items potentially measuring multiple attributes can be examined using the Wald test or likelihood ratio test (Ma et al., 2016) to determine whether one of the interpretable reduced models can be used instead of the saturated model. Lastly, the selected CDMs fit response data to obtain item parameter estimates and respondents' attribute profiles. The following fit statistics can evaluate the absolute fit of CDMs: the residual between observed and predicted transformed correlation (r) and the log-odds ratios of item pairs (l) (Chen et al., 2013), the limited information root mean square error approximation ($RMSEA_2$; Maydeu-Olivares & Joe, 2014), and the M_2 statistic (Hansen et al., 2016; Liu et al., 2016; Ma, 2020). Classification accuracy can be estimated to ascertain the utility of applying CDMs to data (Wang et al., 2015).

Three-Step Latent Transition CDM with Covariates

Latent transition analysis (LTA), also referred to as a latent or hidden Markov model (Baum & Petrie, 1966), is a longitudinal analog of the latent class model (LCM). It has been developed to model the transitions of latent class membership over time, where LCM serves as the measurement model (Collins & Lanza, 2010). To distinguish LTA from LCM, we refer to the respondents' temporal state at each time point as a latent "state" rather than a latent class in the current study.

LTA or LCM can further incorporate informative covariates to investigate the association of covariates with latent transition or class membership using a one-step or three-step approach via latent logistic regression (e.g., Humphreys & Janson, 2000; Iaconangelo & de la Torre, 2016; Lanza et al., 2010; Reboussin et al., 1998; Vermunt, 2010; Vermunt et al., 1999). The one-step approach (e.g., Clogg, 1981; Dayton & Macready, 1988; Lanza et al., 2013) simultaneously estimates both the measurement and the logistic regression (structural) models. The one-step approach provides precise estimates of the measurement and structural components. However, it has one notable disadvantage: any modifications to either model require refitting the entire model, which is computationally inconvenient and potentially tricky, especially with high dimensional data. In contrast, a three-step approach (e.g., Bolck et al., 2004; Di Mari et al., 2016; Iaconangelo & de la Torre, 2016; Lu & Thomas, 2008; Vermunt, 2010) is more flexible because the measurement and structural models are estimated in separate sequential steps. They are CDM estimation (step 1), latent state assignment (step 2), and logistic regression analysis (step 3). However, the stepwise three-step estimation approach can yield biased estimates of covariate effects (Bolck et al., 2004; Vermunt, 2010) when classification errors are unaccounted for by directly regressing the estimated attribute status or profile on covariates. To avoid this bias in stepwise estimation, correction weights can be added to latent state assignments in subsequent logistic regression analyses (Bolck et al., 2004; Iaconangelo & de la Torre, 2016; Vermunt, 2010). Recent studies have extended this approach to LTA (Di Mari et al., 2016) and latent transition CDMs (Liang et al., 2023b). In the current study, the *three-step approach* implies that classification error probabilities (CEP) were taken into account. Methodological research on incorporating covariates into longitudinal CDMs is quite limited in the literature. Liang et al. (2023b) recently proposed a three-step approach for latent transition CDMs with covariates under a general CDM framework (i.e., the G-DINA model framework) to strengthen the three-step estimation of latent transition CDM with covariates. In the present study, we adopted this three-step approach shown in Liang et al. (2023b). We examined the transition of respondents' mental health symptoms before and after intervention and explored how informative covariates, such as intervention membership, affect latent states and their transitions over time.

Technical Details

The three-step approach proceeds as follows. In the first step, the repeated response data from the same respondents are treated as independent data sets at each time point and estimated separately using the G-DINA model. In this step, Q-matrix validation and modification, model selection, and item parameter estimation are completed. Note that item parameters are constrained to be equal across time points to ensure longitudinal measurement invariance. Imposing constraints can help avoid classification problems and allow interpretable results over time. In the second step, respondents are classified into latent (discrete) states given their responses at each time point using the expected a posteriori (EAP; Huebner & Wang, 2011) method. The associated CEP at each time point is also computed in this step. Lastly, in the third step, the latent transition CDM corrected with the CEP obtained from step 2 is estimated to derive regression coefficients.

Formally, details of the three-step latent transition CDM with covariates can be presented as follows. Let T denote the number of time points, Y_{it} the respondent i's response at time t, $t = 1, \ldots, T$, $\boldsymbol{Y}_t$ the response data at time t, and $\boldsymbol{Y}$ the full response data at T time points. Denote the attribute profile and attribute status of α_k at time t by $\boldsymbol{\alpha}^{(t)}$ and $\alpha_k^{(t)}$, respectively. The set of covariates related to the classification of latent state membership at time t is denoted by $\boldsymbol{Z}_t$, and the full set of covariates at all T time points is denoted by $\boldsymbol{Z}$. The latent transition CDM specifying the item response probabilities at pretest and posttest, given the time-specific covariates, is formulated as

$$P\left(\boldsymbol{Y}\middle|\boldsymbol{Z}\right) = \sum_{\alpha_k^{(t=1)}=0}^{1} \sum_{\alpha_k^{(t=2)}=0}^{1} P\left(\alpha_k^{(t=1)}\middle|\boldsymbol{Z}_1\right) P\left(\alpha_k^{(t=2)}\middle|\alpha_k^{(t=1)}, \boldsymbol{Z}_2\right) \prod_{t=1}^{T=2} P\left(\boldsymbol{Y}_t\middle|\boldsymbol{\alpha}^{(t)}\right), \tag{22.2}$$

where $P(\boldsymbol{Y}_t|\,\boldsymbol{\alpha}^{(t)})$ is the G-DINA measurement model estimating the latent state membership at each time point from the response data, $P\left(\alpha_k^{(t=1)}\middle|\boldsymbol{Z}_1\right)$ is the initial state probability, and $P\left(\alpha_k^{(t=2)}\middle|\alpha_k^{(t=1)}, \boldsymbol{Z}_2\right)$ is the transition probability between pretest and posttest, given covariates at each time point. Note that covariates can be time-invariant or time-varying depending on the research questions. In the current study, the covariates were time-invariant (e.g., intervention vs. control group; see Fig. 22.1). The initial state probabilities and transition probabilities are associated with the covariates through logistic regression, which can be written as

$$P\left(\alpha_k^{(t=1)} = 1 \middle| \mathbf{Z}_1\right) = \frac{\exp\left(\beta_{0k}^{(t=1)} + \boldsymbol{\beta}_k^{\prime(t=1)} \mathbf{Z}_1\right)}{1+\exp\left(\beta_{0k}^{(t=1)} + \boldsymbol{\beta}_k^{\prime(t=1)} \mathbf{Z}_1\right)} \tag{22.3}$$

and

$$P\left(\alpha_k^{(t=2)} \middle| \alpha_k^{(t=1)}, \mathbf{Z}_2\right) = \frac{\exp\left(\beta_{0k}^{(t=2)} + \boldsymbol{\beta}_k^{\prime(t=2)} \mathbf{Z}_2\right)}{1+\exp\left(\beta_{0k}^{(t=2)} + \boldsymbol{\beta}_k^{\prime(t=2)} \mathbf{Z}_2\right)}, \tag{22.4}$$

respectively, where β_{0k} is the intercept and $\boldsymbol{\beta}_k$ is a vector of remaining regression coefficients. For attribute-level transitions, the transition probability matrix of α_k at two time points for binary states is a 2×2 matrix (see Table 22.1).

To obtain correction weights, a CEP matrix needs to be computed. A CEP matrix estimates the degree of misclassification in the measurement model conditional on the true or latent attribute state status. Denote the estimated attribute status by α_q and

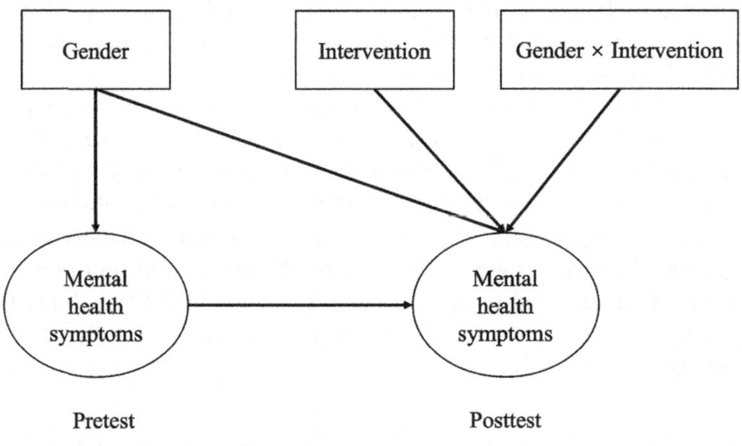

Fig. 22.1 Covariates at pretest and posttest

Table 22.1 Transition probability matrix of α_k

	Attribute state	Posttest	
		0	1
Pretest	0	$1 - P\left(\alpha_k^{(t=2)} = 1 \middle| \alpha_k^{(t=1)} = 0, \mathbf{Z}_2\right)$	$P\left(\alpha_k^{(t=2)} = 1 \middle| \alpha_k^{(t=1)} = 0, \mathbf{Z}_2\right)$
	1	$P\left(\alpha_k^{(t=2)} = 0 \middle| \alpha_k^{(t=1)} = 1, \mathbf{Z}_2\right)$	$1 - P\left(\alpha_k^{(t=2)} = 0 \middle| \alpha_k^{(t=1)} = 1, \mathbf{Z}_2\right)$

the true attribute presence status by α_k. The attribute-level CEP at each time point is a 2×2 matrix representing the probabilities of the true attribute presence status α_k (rows) being classified into the observed attribute status α_q (columns), which is calculated as

$$P\left(\alpha_q^{(t)} \middle| \alpha_k^{(t)}, \boldsymbol{Y}_t\right) = \frac{\frac{1}{NT}\sum_{i=1}^{N}\sum_{t=1}^{T} P\left(\alpha_{ik}^{(t)} \middle| \boldsymbol{Y}_{it}\right) I\left[\alpha_{iq}^{(t)} = \alpha_{ik}^{(t)}\right]}{P\left(\alpha_k^{(t)} \middle| \boldsymbol{Y}_t\right)}, \tag{22.5}$$

where $P\left(\alpha_k^{(t)} \middle| \boldsymbol{Y}_t\right)$ is the posterior proportion of α_k estimated by the G-DINA model at time t and $I[\bullet]$ is an indicator function equal to 1 if α_{iq} is equal to α_{ik} at time t and 0 otherwise.

For example, consider a 2 x 2 CEP matrix for attribute k at time point t:

$$P\left(\alpha_q^{(t)} \middle| \alpha_k^{(t)}, \boldsymbol{Y}_t\right) = \begin{bmatrix} 0.88 & 0.12 \\ 0.24 & 0.76 \end{bmatrix}.$$

In this example, at time point t, 88% of the respondents who do not possess attribute k (first row) will be correctly classified as not having the attribute (first column), whereas 12% will be incorrectly classified as possessing the attribute. Similarly, 76% of the respondents who possess attribute k (second row) will be correctly classified as having the attribute (second column), whereas 24% will be incorrectly classified as not possessing the attribute. Note that the row elements of the CEP matrix sum up to 1. In addition, overall classification accuracy is influenced by the marginals. For didactic purposes, let us assume that the true number of respondents who possess the attribute is $N_+ = \frac{1}{4}N$ and the true number of respondents who do not possess the attribute is $N_- = \frac{3}{4}N$. In such a scenario, only $(88\ \% \times N_-)/(88\ \% \times N_- + 24\ \% \times N_+) = 92\%$ of respondents who were classified as not possessing attribute k do not, in fact, possess the attribute. Similarly, only 68% of respondents who were classified as possessing attribute k do, in fact, possess the attribute. Under ideal situations, the diagonal elements of CEP approximate 1.0, and the off-diagonal elements approximate 0.0, indicating that the respondents are near-perfect classification. In practice, the misclassification rates need to be considered in subsequent analyses.

To do so, the correction weight of respondent i is the element of the CEP matrix at the α_{kt}-th row and α_{iqt}-th column and is given by

$$w_{\alpha_{iq}^{(t)}}^{\alpha_k^{(t)}} = P\left(\alpha_{iq}^{(t)} \middle| \alpha_k^{(t)}, \boldsymbol{Y}_t\right). \tag{22.6}$$

Using the above example, for a respondent classified as possessing α_k attribute at time point t, the correction weights for, in fact, not having and having this attribute are $w_{\alpha_{iq}^{(t)}=1}^{\alpha_k^{(t)}=0} = 0.12$ and $w_{\alpha_{iq}^{(t)}=1}^{\alpha_k^{(t)}=1} = 0.76$, respectively (the second column of the above CEP matrix). These correction weights are used in the latent logistic regression in the final third step of estimation.

Finally, in the third step, we estimate the parameters of the structural model (i.e., $\boldsymbol{\beta}$ in Eqs. 22.3 and 22.4) by maximizing the below objective function:

$$L_{step3} = \sum_{i=1}^{N} \log \sum_{\alpha_k^{(t=1)}=0}^{1} \sum_{\alpha_k^{(t=2)}=0}^{1} P\left(\alpha_k^{(t=1)} \middle| \mathbf{Z}_1\right) \prod_{t=2}^{T} P\left(\alpha_k^{(t=2)} \middle| \alpha_k^{(t=1)}, \mathbf{Z}_2\right) \prod_{t=1}^{T=2} w_{\alpha_{iq}^{(t)}}^{\alpha_k^{(t)}}. \tag{22.7}$$

Note that it is possible to further improve this three-step latent transition CDM by updating classification results using Bayes' theorem on the classification information obtained from the first step of CDM estimation and information from the third step, logistic regression (see Sun & de la Torre, 2020). The updated posterior probability of respondent i having α_k at time t is given by

$$P\left(\alpha_{ik}^{(t)} \middle| \mathbf{Y}_{it}, \mathbf{Z}_{it}\right) = \frac{L\left(\mathbf{Z}_{it} \middle| \alpha_{ik}^{(t)}\right) P\left(\alpha_{ik}^{(t)} \middle| \mathbf{Y}_{it}\right)}{\sum_{\alpha_{ik}^{(t)}=0}^{1} L\left(\mathbf{Z}_{it} \middle| \alpha_{ik}^{(t)}\right) P\left(\alpha_{ik}^{(t)} \middle| \mathbf{Y}_{it}\right)}, \tag{22.8}$$

where $P\left(\alpha_{ik}^{(t)} \middle| \mathbf{Y}_{it}\right)$ is the marginal posterior probability obtained using the G-DINA model. We followed this and updated the results as an ad hoc step.

22.3 Methods

Sample

Data were collected from 4241 respondents (38.0% men) from three university campuses in the USA at two time points 12 months apart (Larimer et al., 2007; Mun et al., 2015). All three universities utilized the same study design and intervention. Tan et al. (2023) reported baseline data before intervention at one of the three universities, a subset of the current sample. In the current study, we analyzed complete response data from 2005 respondents (34.5% men) after excluding missing item responses at both time points. In this final sample, 48.8% of the respondents received a brief alcohol intervention at baseline (pretest) and were followed up after 12 months (posttest). The sample size by gender and intervention status is shown in Table 22.2. Gender was used as the covariate at pretest, and gender and intervention status and their interaction were used as covariates at posttest in this analysis (see Fig. 22.1).

Table 22.2 Sample size by gender and intervention status

Group	Female	Male
Control	661	366
Intervention	652	326
Total	1313	692

Measures

Alcohol-Related Problems

Participants' alcohol-related problems were assessed by the Rutgers Alcohol Problem Index (RAPI; White & Labouvie, 1989), a widely-used questionnaire for adolescent and young adult problem drinking. The RAPI is a 23-item self-reported instrument measuring the frequency of consequences experienced while drinking alcohol or as a result of alcohol use using a five-point scale ranging from 0 (= never) to 4 (= more than 10 times). Cronbach's α and ω for the RAPI with the current sample were 0.90 and 0.91, respectively, at both pretest and posttest.

Psychological Symptoms

Anxiety (6 items), hostility (5 items), and depression (6 items) were measured by the Brief Symptom Inventory (BSI; Derogatis, 1975), a widely used self-reported assessment to identify clinically relevant psychological symptoms in adolescents and adults. Respondents were asked, "During the past 7 days, how much were you distressed by," using a five-point scale ranging from 0 (= not at all) to 4 (= extremely). Based on the current sample, Cronbach's α and ω for the anxiety items were 0.84 and 0.87, respectively, at pretest and 0.85 and 0.88, respectively, at posttest; Cronbach's α and ω for the Hostility items were 0.81 and 0.86, respectively, at pretest and 0.82 and 0.87, respectively, at posttest; and Cronbach's α and ω for the Depression items were 0.90 and 0.92, respectively, at both pretest and posttest.

Data Analysis

The original five-point scaled responses (0–4) were dichotomized into nonzero (= 1) and zero (= 0) responses in the current study. It is reasonable to dichotomize responses since over 82% of the responses were zero, with limited variation in nonzero responses. We analyzed an inventory of 40 items that measured four attributes, namely, alcohol-related problems (AP), anxiety (AN), hostility (HO), and depression (DE). We used the Q-matrix (Table 22.3), which was empirically validated by Tan et al. (2023). As per Tan et al. (2023), we also used the selected CDMs (i.e., A-CDM for item 29; R-RUM for items 21 and 28; LLM for items 26 and 27; and the saturated G-DINA model for the remaining items). The item parameters were obtained for illustration purposes in this longitudinal study. We imposed a monotonicity constraint when fitting the CDMs.

The detailed procedures for analyzing respondents' mental health symptoms over time are demonstrated as follows. First, we fitted selected CDMs to the pretest and posttest data separately, with item parameters constrained to be equal across time to ensure longitudinal measurement invariance. Second, based on the measurement model estimation, we assigned respondents to latent states (i.e., mental health symptom profiles) at each time point and computed the associated CEP and correction weights. Third, we estimated the latent transition CDM with the correction weights to obtain regression coefficients. A quasi-Newton optimization method, namely, Broyden-Fletcher-Goldfarb-Shanno (BFGS; Nash, 1990) with box constraints (i.e., L-BFGS-B; Byrd et al., 1995), was used to maximize the objective function given in Eq. 22.7. Lastly, the classification results were updated based on Eq. 22.8.

Given the updated classification results, the transition probabilities of mental health symptoms can be computed by the following logistic regression function:

$$\text{logit}(P) = \beta_0 + \beta_1 \times \text{Gender} + \beta_2 \times \text{Intervention} + \beta_3 \times \text{Gender} \times \text{Intervention}, \tag{22.9}$$

where P is the transition probability in Eq. 22.4. We included binary covariates in the logistic regression so that Gender $= 1$ indicates males ($0 =$ female) and Intervention $= 1$ ($0 =$ control) indicates intervention group. The logit of transition probabilities by gender and intervention status can be shown in Table 22.4. The significance of the intervention effect on transition probabilities among females can be revealed by β_2.

Note that males in the control and intervention groups can be directly compared by switching the coding of the gender variable (i.e., in this case, Gender $= 1$ for females and 0 for males). In doing so, we can also test whether the transition probabilities between intervention and control groups among males differ by testing the new β_2. We can easily obtain the logit of transition probability by swapping the positions of "Male" and "Female" in the "Gender" column in Table 22.4. Annotated computer code in R and data utilized in the current paper are available at Mendeley Data (https://doi.org/10.17632/kpjp3gnwbt.1; Liang et al., 2023a).

Table 22.3 Q-matrix of the 40-item measure

Item and a brief description	AP	AN	HO	DE
1. Not able to do homework or study	1	0	0	0
2. Got into fights or did mean things	1	0	0	0
3. Missed out on other things	1	0	0	0
4. Went to work or school drunk	1	0	0	0
5. Caused shame to someone	1	0	0	0
6. Neglected responsibilities	1	0	0	0
7. Relative avoided you	1	0	0	0
8. Needed more alcohol than before	1	0	0	0
9. Tried to control drinking	1	0	0	0
10. Had withdrawal symptoms	1	0	0	0
11. Noticed a change in personality	1	0	0	0
12. Felt you had an alcohol problem	1	0	0	0
13. Missed school or work	1	0	0	0
14. Tried to cut down or quit drinking	1	0	0	0
15. Found yourself in a place you don't remember	1	0	0	0
16. Passed out suddenly	1	0	0	0
17. Had a fight or argument with a friend	1	0	0	0
18. Had a fight or argument with a family member	1	0	0	0
19. Kept drinking when meant not to	1	0	0	0
20. Felt going crazy	1	0	0	0
21. Had a bad time	1	0	1	0
22. Felt physically or psychologically dependent	1	0	0	0
23. Told to stop or cut down on drinking	1	0	0	0
24. Nervousness inside	0	1	0	0
25. Feeling scaled for no reason	0	1	0	0
26. Feeling fearful	0	1	0	1
27. Feeling tense	0	1	0	1
28. Spells of terror or panic	0	1	0	1
29. Feeling restless	0	1	1	0
30. Feeling annoyed or irritated	0	0	1	0
31. Temper outburst	0	0	1	0
32. Having urges to harm someone	0	0	1	0
33. Having urges to break things	0	0	1	0
34. Getting into arguments	0	0	1	0
35. Thoughts of ending life	0	0	0	1
36. Feeling lonely	0	0	0	1
37. Feeling blue	0	0	0	1
38. Feeling no interest in things	0	0	0	1
39. Feeling hopeless	0	0	0	1
40. Feeling worthlessness	0	0	0	1

Note. 1 = an item measures the attribute; 0 = an item does not measure the attribute. *AP* alcohol-related problems, *AN* anxiety, *HO* hostility, *DE* depression

Table 22.4 Logit of transition probability by gender and intervention status (Gender = 1 for males)

Gender	Intervention status	Logit(P)
Male	Intervention	$\beta_0 + \beta_1 + \beta_2 + \beta_3$
	Control	$\beta_0 + \beta_1$
Female	Intervention	$\beta_0 + \beta_2$
	Control	β_0

Note. Logit(P) = $\beta_0 + \beta_1 \times$ Gender + $\beta_2 \times$ Intervention + $\beta_3 \times$ Gender $\times$ Intervention

Table 22.5 The CEP matrices of the attribute of alcohol-related problems at pretest and posttest

	Estimated	
True	0	1
Pretest		
0	0.97	0.03
1	0.06	0.93
Posttest		
0	0.97	0.03
1	0.07	0.94

Note. Row totals may not add up to 1.00 due to rounding

22.4 Results

Model Fit, Classification Without Correction, and CEP Matrix (Steps 1 and 2)

In Step 1, the measurement model (i.e., the selected CDMs) showed adequate absolute fit with an $RMSEA_2 = 0.04$ (<0.05; Maydeu-Olivares & Joe, 2014) at both pretest and posttest. In addition, classification accuracy was estimated to ascertain the utility of using the selected CDMs for the data. The classification accuracy indices at the attribute level ranged from 0.91 to 0.96 at both pretest and posttest. The probability of 0.9 or above in accurately classifying respondents' four mental health symptom attributes would be considered high. The hostility attribute had the lowest (0.91), and alcohol-related problems (0.96) had the highest classification accuracy for both time points. These classification accuracy indices indicate the measurement models at both time points were appropriate for analyzing the data.

In Step 2, respondents were classified into latent states based on the measurement model estimated at each time point. In addition, the CEP matrices for each attribute at both time points were also computed. Table 22.5 displays the CEP matrices of the attribute of alcohol-related problems at pretest and posttest. At pretest, 97% of the participants without alcohol-related problems were correctly classified as not having the attribute; similarly, 93% of the participants with alcohol-related problems were classified as having the attribute. Similar results were obtained at posttest. Table 22.8 shows the CEP matrices of the remaining attributes.

Latent Logistic Regression (Step 3)

In the third step, the model shown in Fig. 22.1 was estimated. Table 22.6 shows the regression coefficients of the latent logistic regression for the initial state and transition probabilities of the four attributes of mental health symptoms from pretest and posttest. At the initial state (pretest), gender was negatively related to anxiety (AN) and positively associated with hostility (HO). Specifically, the odds of males being classified as having the anxiety attribute were odds ratio (*OR*) = 0.56 (= $e^{-0.58}$; $p < 0.001$, $d = -0.32$) times of females, indicating that females were 79% (= 1/0.56 = 1.79) more likely to have the anxiety attribute at pretest. Males were more likely to be classified as having the hostility attribute than females, in particular, $OR = 1.23$ (= $e^{0.20}$; $p < 0.05$, $d = 0.11$) times females.

Multiple logistic regressions were conducted at posttest to examine how transition probabilities were associated with covariates. We investigated participants' transition for each attribute from 0 to 1 (i.e., from not possessing to possessing the attribute) and from 1 to 0 (i.e., from possessing to not possessing the attribute). For the transition from 0 to 1 (see Table 22.6; the middle row block), gender was significantly related to the transition probabilities for the alcohol-related problems and depression attributes. A significant interaction term between gender and intervention status was also observed for alcohol-related problems and depression. Specifically, in terms of alcohol-related problems, among males, their odds of transitioning from 0 to 1 in the intervention group were 0.45 (= $e^{\beta_2+\beta_3} = e^{0.09 - 0.88}$) times of those in the control group; and among females, their odds of transitioning from 0 to 1 in the intervention group were 1.09 (= $e^{\beta_2} = e^{0.09}$) times those in the control group. With regard to depression, the odds of males transitioning from 0 to 1 in the intervention group were 1.14 (= $e^{\beta_2+\beta_3} = e^{-0.28 + 0.41}$) compared to those in the control group, and the odds of females transitioning from 0 to 1 in the intervention group were 0.75 (= $e^{\beta_2} = e^{-0.28}$) compared to those in the control group. Participants who received the intervention were less likely to transition from 0 to 1 for the attribute of hostility.

Concerning the transition from 1 to 0 (see Table 22.6; bottom row block), men were more likely to transition from 1 to 0 compared to women for the anxiety attribute. Participants in the intervention (vs. control) group were more likely to transition from having anxiety and depression attributes to not having them.

Because gender and intervention status variables were considered clinically informative covariates in this study, the transition probabilities were calculated using the regression coefficients by gender and intervention status based on Table 22.6 (see Fig. 22.2). The significance of the intervention effects among females or males can be examined using the coefficients of intervention (β_2). Table 22.6 shows the regression coefficients obtained when Gender = 1 for males (see Table 22.9 for the results when Gender = 1 for females). Figure 22.2 shows intervention effects for each attribute and each transition separately for males and females. Overall, the intervention effect on the transition from 1 to 0 was more likely to occur than the transition from 0 to 1. Of the transition from 1 to 0, the intervention benefited

Table 22.6 Coefficients of the latent logistic regression obtained in Step 3

	AP			AN			HO			DE		
	β	*OR*	*d*	β	*OR*	*d*	β	*OR*	*d*	β	*OR*	*d*
Initial state												
Intercept (β_0)	−0.81	0.44	−0.45	−0.83	0.43	−0.46	−0.81	0.45	−0.44	−0.61	0.55	−0.33
Gender (β_1)	0.00	1.00	0.00	−0.58**	0.56	−0.32	0.20*	1.23	0.11	0.03	1.03	0.02
Transition from absence (0) to presence (1)												
Intercept (β_0)	−2.12	0.12	−1.17	−1.36	0.26	−0.75	−1.18	0.31	−0.65	−1.10	0.33	−0.61
Gender (β_1)	0.60**	1.82	0.33	−0.04	0.96	−0.02	0.01	1.01	0.01	−0.28*	0.75	−0.16
Intervention (β_2)	0.09	1.09	0.05	−0.06	0.94	−0.03	−0.30**	0.74	−0.17	−0.01	0.99	−0.01
Gender × intervention (β_3)	−0.88**	0.41	−0.49	−0.25	0.78	−0.14	0.30	1.34	0.16	0.41*	1.51	0.23
Transition from presence (1) to absence (0)												
Intercept (β_0)	−0.78	0.46	−0.43	−0.26	0.77	−0.14	−0.24	0.79	−0.13	−0.49	0.62	−0.27
Gender (β_1)	−0.23	0.80	−0.13	0.54**	1.71	0.30	0.20	1.22	0.11	0.15	1.16	0.08
Intervention (β_2)	−0.19	0.83	−0.10	0.53**	1.69	0.29	0.39	1.47	0.21	0.47*	1.60	0.26
Gender × intervention (β_3)	0.29	1.33	0.16	−0.33	0.72	−0.18	−0.56	0.57	−0.31	0.18	1.20	0.10

Note. Gender = 0 for females and =1 for males; Intervention = 0 for control group and =1 for intervention group; *AP* alcohol-related problems, *AN* anxiety, *HO* hostility, *DE* depression; d = Cohen's $d = \beta \times \frac{\sqrt{3}}{\pi}$ (Sánchez-Meca et al., 2003); *OR* = odds ratio = exp (β); $^{*}p < 0.05$, $^{**}p < 0.01$

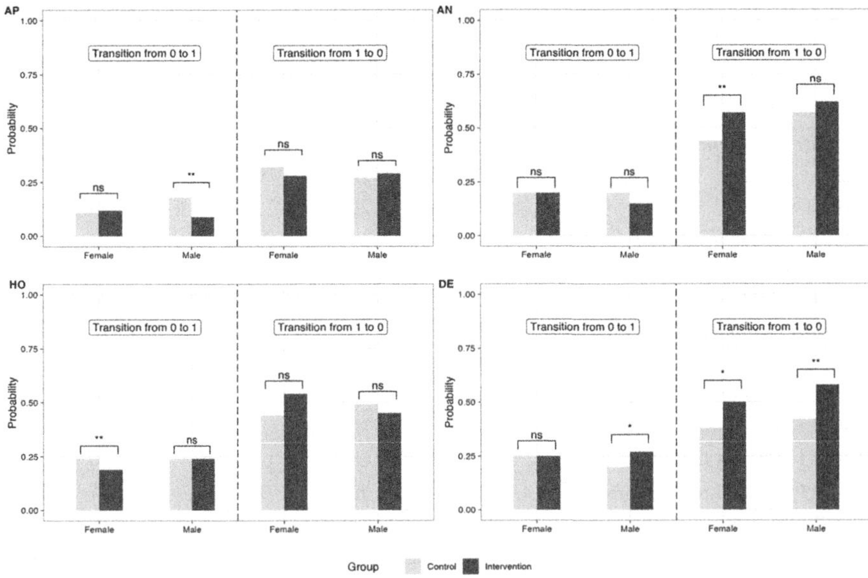

Fig. 22.2 Transition probabilities of mental health symptoms by gender and intervention status. (*Note*. *AP* alcohol-related problems, *AN* anxiety, *HO* hostility, *DE* depression; ns $p > 0.05$, $*p < 0.05$, $**p < 0.01$)

both males and females who had depression at the initial time point; however, the intervention effect on anxiety was found only among females. In contrast, the transition from 0 to 1 was more common for those in the control group than those who received the intervention (i.e., alcohol-related problems among males and hostility among females). However, depression among men showed the opposite pattern—men in the intervention group were more likely to transition from 0 to 1 than those in the control group.

The updated latent state membership was computed using Eq. 22.8, from which the attribute prevalences and the transition probabilities of the mental health symptoms in Table 22.7 were computed. Not surprisingly, the probabilities of transitioning from 1 to 0 were substantially greater than the corresponding probabilities of transitioning from 0 to 1. Regarding the transition probabilities between the control and intervention groups, except for depression, participants in the intervention group were less likely to transition from 0 to 1 compared to those in the control group. Except for alcohol-related problems, participants in the intervention group were more likely to transition from 1 to 0 than those in the control group. Taken together, the results suggest that the brief alcohol intervention positively affected the transitions of all four attributes by either alleviating existing mental health symptoms or preventing them from developing. However, the beneficial effects may have to be examined for men and women more carefully.

Table 22.7 The corrected and updated attribute prevalence of the mental health symptoms at pretest and posttest and the marginal transition probabilities

	Attribute			
	AP	AN	HO	DE
Attribute prevalence (%)				
Pretest	27.63	20.70	25.49	34.16
Posttest	27.93	20.80	24.74	33.62
Transition probability (total sample)				
From 0 to 1	0.10	0.14	0.15	0.22
From 1 to 0	0.25	0.52	0.47	0.44
Transition probability (by intervention status)				
From 0 to 1 (control group)	0.11	0.15	0.16	0.20
From 0 to 1 (intervention group)	0.09	0.12	0.14	0.24
From 1 to 0 (control group)	0.26	0.46	0.42	0.39
From 1 to 0 (intervention group)	0.24	0.57	0.51	0.50

Note. *AP* alcohol-related problems, *AN* anxiety, *HO* hostility, *DE* depression. For transition probabilities from 0 to 0 and from 1 to 1, please refer to Table 22.1

22.5 Discussion

This study demonstrated how to analyze the transition of mental health symptoms over time with a model that accounts for the covariates' effects on both the initial state and transition probabilities within a CDM analytic framework. We demonstrated the use of a three-step latent transition CDM with covariates (Liang et al., 2023b) in this data application, which extends the work by Tan et al. (2023) to transitions and the effects of covariates on transitions. We walked through the three-step estimation procedure, where CDM parameters (measurement model) and regression coefficients (structural model) can be flexibly estimated while maintaining classification accuracy through correction. The current study showcases the first application of CDMs for longitudinal clinical data.

The three-step approach in this study disentangled the relationships between covariates and each mental health symptom. This approach can answer the question of which covariates are associated with the transition of which constructs at the attribute level. It allows researchers to understand the mental health symptoms at the most granular level of information and enables evaluation of intervention effects. This fine-grained approach may assist with developing better intervention designs. Specifically, by testing intervention group membership as a covariate, we can probe whether respondents in the intervention group exhibited a higher transition probability from one state to another. For example, in the present study, the intervention did not have any significant effects on respondents' transition from presence to absence with respect to alcohol-related problems and hostility. However, it had "spillover" effects on anxiety and depression, in the sense that the intervention was not designed to improve anxiety and depression. This suggests that the brief alcohol intervention tested may have a broader range of health effects, while its targeted effect may need to be strengthened.

Note that we utilized the three-step approach to latent transition CDM. Because each step is separately estimated, it is possible to examine an additional set of covariates or different sets of covariates without fitting the entire model again. With the advent of powerful computing capacity, it may be helpful to assess clinical and nonclinical endpoints more broadly to assess the health benefits of brief alcohol interventions across time.

The current study adopted Bayes' theorem to integrate the classification information obtained from the first step of CDM estimation and information from the latent logistic regressions to update the posterior probability. This procedure was demonstrated by Sun and de la Torre (2020), which incorporated covariates into CDMs and updated posterior probabilities at a single time point. This procedure may be useful to improve the classification of respondents even if covariates are not related to attributes. Future studies could investigate to what extent using covariates and updating posterior probabilities in longitudinal CDMs can improve the classification of individuals at multiple time points.

Despite the utility of the three-step latent transition CDM approach, this study has several potential limitations worth mentioning. First, this study dichotomized polytomous responses for model simplicity. Dichotomization may lead to information loss and affect structural and item parameter estimation. Based on the pretest mental health data from the same project, Tan et al. (2023) conducted a sensitivity analysis of the polytomous data using a sequential G-DINA model (Ma & de la Torre, 2016). They showed that the patterns of results obtained with polytomous data were similar to those with dichotomized data. Therefore, it is quite likely that the current results would be similar even with polytomous response data. In the existing literature, most methodological developments (e.g., the three-step approach) and analysis procedures (e.g., absolute model fit evaluation and classification accuracy evaluation) are developed based on dichotomous CDMs. Future research should consider extending the dichotomous CDM methodologies to polytomous CDMs in longitudinal studies and proposing appropriate procedures for polytomous CDMs. For instance, future large sample studies can examine the three-step approach of latent transition CDM with covariates in polytomous response data.

Second, we analyzed the complete response data in this study, consistent with Tan et al. (2023). However, complete response data are not necessary for CDM analysis, and response data with some missing values can also be analyzed using the approach demonstrated in this study. The presence of missing data could lead to biased parameter estimation in psychometric models (e.g., De Ayala et al., 2001; Pohl et al., 2014). However, in the CDM literature, it remains an outstanding task to systematically examine how different patterns of missing data may affect model estimation and classification (Pan & Zhan, 2020; Shan & Wang, 2020; Xu & von Davier, 2006). Only a few studies have empirically evaluated various approaches for handling missing data in CDM applications (e.g., Dai et al., 2018; Ömür Sünbül, 2018). Because respondents with missing data do not receive any diagnostic feedback when a complete case analysis is employed, researchers may want to utilize the response data with missing data. In such cases, we suggest that researchers identify the missing data mechanisms and employ proper analytic approaches to missing data.

Third, only two binary covariates were used in this study, and they were expected to be unrelated due to randomization. However, latent transition CDM can accommodate continuous covariates (e.g., see Liang et al., 2023b). When continuous covariates are used, they may be associated. When the number of covariates in the model increases, however, multicollinearity may become an issue in empirical studies, just as it is a common concern in generalized linear mixed models. In such situations, variable selection and regularization may be considered. Iaconangelo and de la Torre (2017) incorporated the correction weights with the latent-class lasso as a variable selection procedure for CDMs at a single time point, which led to a more helpful subset of nonredundant variables. This variable selection may result in more helpful structural parameter estimation. However, developing variable selection methods for the three-step latent transition CDM with covariates remains a future challenge to address.

Lastly, the present study used the same measurement model with item parameter constraints at different time points to ensure measurement invariance—this may degrade assessment flexibility. Most longitudinal CDM methodologies were developed under such constraints. However, theoretically, using identical test forms and fixing item parameters across time points are not necessary. For different test forms with common items at different time points, multiple-group CDMs (e.g., multiple-group GDINA; Ma et al., 2021) can be used to detect whether items function differentially across different time points and determine item parameters for each time point separately. Moreover, measurement models could be different across time points. We suggest that the feasibility of using different measurement models at multiple time points (Asparouhov & Muthén, 2014) be explored, as this allows researchers to select the most appropriate CDMs for each time point. Di Mari et al. (2016) also noted that the three-step approach for LTA can be adapted to allow time-specific measurement models by performing the first and second steps separately at each time point.

In closing, the current paper demonstrated the utility of CDM for classifying and understanding transition in attributes in longitudinal item-response data. Individual-focused profiles can serve as the basis for tailored or personalized learning or treatment. For example, Wu (2019) developed an online personalized tutor program based on the CDM diagnostic profiles and found that the personalized program outperformed traditional remedial instruction. In addition, the program benefited all students by improving their learning outcomes. The same promise may be possible in clinical settings with more appropriate and specific intervention programs that can be tailored to individuals' diagnostic profiles. To this end, novel analytical techniques, such as CDM, in conjunction with technology, are promising in facilitating the successful implementation of tailored interventions.

Appendix

Table 22.8 The CEP matrices of the other three mental health symptoms at pretest and posttest

	Pretest		Posttest	
	Estimated			
True	0	1	0	1
AN				
0	0.98	0.03	0.97	0.03
1	0.19	0.81	0.19	0.81
HO				
0	0.95	0.06	0.94	0.06
1	0.15	0.84	0.15	0.86
DE				
0	0.97	0.03	0.97	0.03
1	0.06	0.93	0.06	0.94

Note. *AN* anxiety, *HO* hostility, *DE* depression. Row totals of each CEP matrix may not add up to 1.00 due to rounding

Table 22.9 Coefficients of the latent logistic regression obtained in Step 3 when Gender = 1 for females

	AP			AN			HO			DE		
	β	*OR*	*d*	β	*OR*	*d*	β	*OR*	*d*	β	*OR*	*d*
Initial state												
Intercept (β_0)	−0.82	0.44	−0.45	−1.42	0.24	−0.78	−0.60	0.55	−0.33	−0.58	0.56	−0.32
Gender (β_1)	0.00	1.00	0.00	0.58**	1.79	0.32	−0.20*	0.82	−0.11	−0.03	0.97	−0.02
Transition from absence (0) to presence (1)												
Intercept (β_0)	−1.49	0.22	−0.82	−1.43	0.24	−0.79	−1.17	0.31	−0.64	−1.41	0.24	−0.78
Gender (β_1)	−0.62**	0.54	−0.34	−0.01	0.99	−0.01	−0.01	0.99	−0.01	0.30	1.35	0.16
Intervention (β_2)	−0.74**	0.48	−0.41	−0.37	0.69	−0.20	−0.01	0.99	0.00	0.38*	1.47	0.21
Gender × Intervention (β_3)	0.83**	2.28	0.46	0.25	1.28	0.14	−0.30	0.74	−0.17	−0.41	0.67	−0.22
Transition from presence (1) to absence (0)												
Intercept (β_0)	−0.97	0.38	−0.53	0.21	1.23	0.11	−0.04	0.96	−0.02	−0.37	0.69	−0.20
Gender (β_1)	0.20	1.22	0.11	−0.59**	0.56	−0.32	−0.20	0.82	−0.11	−0.13	0.88	−0.07
Intervention (β_2)	0.13	1.14	0.07	0.07	1.08	0.04	−0.18	0.84	−0.10	0.62**	1.86	0.34
Gender × Intervention (β_3)	−0.32	0.73	−0.18	0.37*	1.45	0.21	0.55**	1.74	0.31	−0.17	0.85	−0.09

Note. Gender = 0 for males, and = 1 for females; Intervention = 0 for control group, and = 1 for intervention group; *AP* alcohol-related problems, *AN* anxiety, *HO* hostility, *DE* depression; d = Cohen's $d = \beta \times \frac{\sqrt{3}}{\pi}$ (Sanchez-Meca et al., 2003); *OR* = Odds Ratio = exp (β); $^*p < 0.05$, $^{**}p < 0.01$

References

Asparouhov, T., & Muthén, B. (2014). Auxiliary variables in mixture modeling: Three-step approaches using Mplus. *Structural Equation Modeling, 21*(3), 329–341. https://doi.org/10.1080/10705511.2014.915181

Baum, L. E., & Petrie, T. (1966). Statistical inference for probabilistic functions of finite state Markov chains. *The Annals of Mathematical Statistics, 37*(6), 1554–1563. http://www.jstor.org/stable/2238772

Bolck, A., Croon, M., & Hagenaars, J. (2004). Estimating latent structure models with categorical variables: One-step versus three-step estimators. *Political Analysis, 12*(1), 3–27. https://doi.org/10.1093/pan/mph001

Byrd, R. H., Lu, P., Nocedal, J., & Zhu, C. (1995). A limited memory algorithm for bound constrained optimization. *SIAM Journal on Scientific Computing, 16*(5), 1190–1208. https://doi.org/10.1137/0916069

Chen, J., de la Torre, J., & Zhang, Z. (2013). Relative and absolute fit evaluation in cognitive diagnosis modeling. *Journal of Educational Measurement, 50*(2), 123–140. https://doi.org/10.1111/j.1745-3984.2012.00185.x

Chen, H., Cai, Y., & de la Torre, J. (2022). Investigating second language (L2) reading subskill associations: A cognitive diagnosis approach. *Language Assessment Quarterly*, 1–24. https://doi.org/10.1080/15434303.2022.2140050

Clogg, C. C. (1981). New developments in latent structure analysis. In D. J. Jackson & E. F. Borgatta (Eds.), *Factor analysis and measurement in sociological research* (pp. 215–246). Sage Publications.

Collins, L. M., & Lanza, S. T. (2010). *Latent class and latent transition analysis: With applications in the social, behavioral, and health sciences*. Wiley. https://doi.org/10.1002/9780470567333

Dai, S., Svetina, D., & Chen, C. (2018). Investigation of missing responses in Q-Matrix validation. *Applied Psychological Measurement, 42*(8), 660–676. https://doi.org/10.1177/0146621618762742

Dayton, C. M., & Macready, G. B. (1988). Concomitant-variable latent-class models. *Journal of the American Statistical Association, 83*(401), 173–178. https://doi.org/10.2307/2288938

De Ayala, R. J., Plake, B. S., & Impara, J. C. (2001). The impact of omitted responses on the accuracy of ability estimation in item response theory. *Journal of Educational Measurement, 38*(3), 213–234. https://doi.org/10.1111/j.1745-3984.2001.tb01124.x

de la Torre, J. (2011). The generalized DINA model framework. *Psychometrika, 76*(2), 179–199. https://doi.org/10.1007/s11336-011-9207-7

de la Torre, J., & Chiu, C. Y. (2016). A general method of empirical Q-matrix validation. *Psychometrika, 81*(2), 253–273. https://doi.org/10.1007/s11336-015-9467-8

de la Torre, J., & Minchen, N. (2014). Cognitively diagnostic assessments and the cognitive diagnosis model framework. *Psicologia Educativa, 20*(2), 89–97. https://doi.org/10.1016/j.pse.2014.11.001

de la Torre, J., van der Ark, L. A., & Rossi, G. (2018). Analysis of clinical data from a cognitive diagnosis modeling framework. *Measurement and Evaluation in Counseling and Development, 51*(4), 281–296. https://doi.org/10.1080/07481756.2017.1327286

Derogatis, L. R. (1975). *Brief symptom inventory*. Clinical Psychometric Research.

Di Mari, R., Oberski, D. L., & Vermunt, J. K. (2016). Bias-adjusted three-step latent Markov modeling with covariates. *Structural Equation Modeling, 23*(5), 649–660. https://doi.org/10.1080/10705511.2016.1191015

Hansen, M., Cai, L., Monroe, S., & Li, Z. (2016). Limited-information goodness-of-fit testing of diagnostic classification item response models. *The British Journal of Mathematical and Statistical Psychology, 69*(3), 225–252. https://doi.org/10.1111/bmsp.12074

Hartz, S. M. (2002). *A Bayesian framework for the unified model for assessing cognitive abilities: Blending theory with practicality* (Doctoral dissertation). University of Illinois at Urbana-Champaign.

Henson, R. A., Templin, J. L., & Willse, J. T. (2009). Defining a family of cognitive diagnosis models using log-linear models with latent variables. *Psychometrika, 74*(2), 191–210. https://doi.org/10.1007/s11336-008-9089-5

Huebner, A., & Wang, C. (2011). A note on comparing examinee classification methods for cognitive diagnosis models. *Educational and Psychological Measurement, 71*(2), 407–419. https://doi.org/10.1177/0013164410388832

Humphreys, K., & Janson, H. (2000). Latent transition analysis with covariates, nonresponse, summary statistics and diagnostics: Modelling children's drawing development. *Multivariate Behavioral Research, 35*(1), 89–118. https://doi.org/10.1207/S15327906MBR3501_4

Iaconangelo, C. (2017). *Uses of classification error probabilities in the three-step approach to estimating cognitive diagnosis models* (Doctoral dissertation). The State University of New Jersey, Rutgers. https://doi.org/10.7282/T3W95D95

Iaconangelo, C., & de la Torre, J. (2016, July 11–15). *Three-step estimation of cognitive diagnosis models with covariate* (Paper presentation). 81st International Meeting of the Psychometric Society, Asheville, NC, United States.

Junker, B. W., & Sijtsma, K. (2001). Cognitive assessment models with few assumptions, and connections with nonparametric item response theory. *Applied Psychological Measurement, 25*(3), 258–272. https://doi.org/10.1177/01466210122032064

Lanza, S. T., Patrick, M. E., & Maggs, J. L. (2010). Latent transition analysis: Benefits of a latent variable approach to modeling transitions in substance use. *Journal of Drug Issues, 40*(1), 93–120. https://doi.org/10.1177/002204261004000106

Lanza, S. T., Tan, X., & Bray, B. C. (2013). Latent class analysis with distal outcomes: A flexible model-based approach. *Structural Equation Modeling, 20*(1), 1–26. https://doi.org/10.1080/10705511.2013.742377

Larimer, M. E., Lee, C. M., Kilmer, J. R., Fabiano, P. M., Stark, C. B., Geisner, I. M., Mallett, K. A., Lostutter, T. W., Cronce, J. M., Feeney, M., & Neighbors, C. (2007). Personalized mailed feedback for college drinking prevention: A randomized clinical trial. *Journal of Consulting and Clinical Psychology, 75*(2), 285–293. https://doi.org/10.1037/0022-006X.75.2.285

Liang, Q., de la Torre, J., & Law, N. (2021). Do background characteristics matter in Children's mastery of digital literacy? A cognitive diagnosis model analysis. *Computers in Human Behavior, 122*, 106850. https://doi.org/10.1016/j.chb.2021.106850

Liang, Q., de la Torre, J., Larimer, M. E., & Mun, E.-Y. (2023a). Mental health symptom profiles over time: A three-step latent transition cognitive diagnosis modeling analysis with covariates. *Mendeley Data*. https://doi.org/10.17632/kpjp3gnwbt.1

Liang, Q., de la Torre, J., & Law, N. (2023b). Latent transition cognitive diagnosis model with covariates: A three-step approach. *Journal of Educational and Behavioral Statistics. (in press)*. https://doi.org/10.3102/10769986231163320

Liu, Y., Tian, W., & Xin, T. (2016). An application of M2 statistic to evaluate the fit of cognitive diagnostic models. *Journal of Educational and Behavioral Statistics, 41*(1), 3–26. https://doi.org/10.3102/1076998615621293

Lu, I. R. R., & Thomas, D. R. (2008). Avoiding and correcting bias in score-based latent variable regression with discrete manifest items. *Structural Equation Modeling, 15*(3), 462–490. https://doi.org/10.1080/10705510802154323

Ma, W. (2020). Evaluating the fit of sequential G-DINA model using limited-information measures. *Applied Psychological Measurement, 44*(3), 167–181. https://doi.org/10.1177/0146621619843829

Ma, W., & de la Torre, J. (2016). A sequential cognitive diagnosis model for polytomous responses. *The British Journal of Mathematical and Statistical Psychology, 69*(3), 253–275. https://doi.org/10.1111/bmsp.12070

Ma, W., Iaconangelo, C., & de la Torre, J. (2016). Model similarity, model selection, and attribute classification. *Applied Psychological Measurement, 40*(3), 200–217. https://doi.org/10.1177/0146621615621717

Ma, W., Minchen, N., & de la Torre, J. (2020). Choosing between CDM and unidimensional IRT: The proportional reasoning test case. *Measurement: Interdisciplinary Research and Perspectives, 18*(2), 87–96. https://doi.org/10.1080/15366367.2019.1697122

Ma, W., Terzi, R., & de la Torre, J. (2021). Detecting differential item functioning using multiple-group cognitive diagnosis models. *Applied Psychological Measurement, 45*(1), 37–53. https://doi.org/10.1177/0146621620965745

Maris, E. (1999). Estimating multiple classification latent class models. *Psychometrika, 64*(2), 187–212. https://doi.org/10.1007/BF02294535

Maydeu-Olivares, A., & Joe, H. (2014). Assessing approximate fit in categorical data analysis. *Multivariate Behavioral Research, 49*(4), 305–328. https://doi.org/10.1080/00273171.2014.911075

Mun, E., de la Torre, J., Atkins, D. C., White, H. R., Ray, A. E., Kim, S., Jiao, Y., Clarke, N., Huo, Y., Larimer, M. E., Huh, D., & The Project INTEGRATE Team. (2015). Project INTEGRATE: An integrative study of brief alcohol interventions for college students. *Psychology of Addictive Behaviors, 29*(1), 34–48. https://doi.org/10.1037/adb0000047

Nash, J. C. (1990). *Compact numerical methods for computers: Linear algebra and function minimisation* (2nd ed.). Adam Hilger.

Ömür Sünbül, S. (2018). The impact of different missing data handling methods on DINA model. *International Journal of Evaluation and Research in Education (IJERE), 7*(1), 77. https://doi.org/10.11591/ijere.v7i1.11682

Pan, Y., & Zhan, P. (2020). The impact of sample attrition on longitudinal learning diagnosis: A prolog. *Frontiers in Psychology, 11*, 1051. https://doi.org/10.3389/fpsyg.2020.01051

Pohl, S., Gräfe, L., & Rose, N. (2014). Dealing with omitted and not-reached items in competence tests: Evaluating approaches accounting for missing responses in item response theory models. *Educational and Psychological Measurement, 74*(3), 423–452. https://doi.org/10.1177/0013164413504926

Reboussin, B. A., Reboussin, D. M., Liang, K.-Y., & Anthony, J. C. (1998). Latent transition modeling of progression of health-risk behavior. *Multivariate Behavioral Research, 33*(4), 457–478. https://doi.org/10.1207/s15327906mbr3304_2

Sánchez-Meca, J., Marín-Martínez, F., & Chacón-Moscoso, S. (2003). Effect-size indices for dichotomized outcomes in meta-analysis. *Psychological Methods, 8*(4), 448–467. https://doi.org/10.1037/1082-989X.8.4.448

Shan, N., & Wang, X. (2020). Cognitive diagnosis modeling incorporating item-level missing data mechanism. *Frontiers in Psychology, 11*, 564707. https://doi.org/10.3389/fpsyg.2020.564707

Sun, Y., & de la Torre, J. (2020). Improving attribute classification accuracy in high dimensional data: A four-step latent regression approach. In H. Jiao & R. W. Lissitz (Eds.), *Innovative psychometric modeling and methods* (pp. 17–44). Information Age.

Tan, Z., de la Torre, J., Ma, W., Huh, D., Larimer, M. E., & Mun, E.-Y. (2023). A tutorial on cognitive diagnosis modeling for characterizing mental health symptom profiles using existing item responses. *Prevention Science, 24*, 480–492. https://doi.org/10.1007/s11121-022-01346-8

Tatsuoka, K. K. (1983). Rule space: An approach for dealing with misconceptions based on item response theory. *Journal of Educational Measurement, 20*(4), 345–354. https://doi.org/10.1111/j.1745-3984.1983.tb00212.x

Tatsuoka, K. K. (2009). *Cognitive assessment: An introduction to the rule space method.* Routledge. https://doi.org/10.4324/9780203883372

Templin, J. L., & Henson, R. A. (2006). Measurement of psychological disorders using cognitive diagnosis models. *Psychological Methods, 11*(3), 287. https://doi.org/10.1037/1082-989X.11.3.287

Tjoe, H., & de la Torre, J. (2014). The identification and validation process of proportional reasoning attributes: An application of a cognitive diagnosis modeling framework. *Mathematics Education Research Journal, 26*(2), 237–255. https://doi.org/10.1007/s13394-013-0090-7

Vermunt, J. K. (2010). Latent class modeling with covariates: Two improved three-step approaches. *Political Analysis, 18*(4), 450–469. https://doi.org/10.1093/pan/mpq025

Vermunt, J. K., Langeheine, R., & Bockenholt, U. (1999). Discrete-time discrete-state latent Markov models with time-constant and time-varying covariates. *Journal of Educational and Behavioral Statistics, 24*(2), 179–207. https://doi.org/10.3102/10769986024002179

von Davier, M., & Lee, Y.-S. (Eds.). (2019). *Handbook of diagnostic classification models* (1st ed.). Springer. https://doi.org/10.1007/978-3-030-05584-4

Wang, W., Song, L., Chen, P., Meng, Y., & Ding, S. (2015). Attribute-level and pattern-level classification consistency and accuracy indices for cognitive diagnostic assessment. *Journal of Educational Measurement, 52*(4), 457–476. https://doi.org/10.1111/jedm.12096

White, R. H., & Labouvie, E. W. (1989). Towards the assessment of adolescent problem drinking. *Journal of Studies on Alcohol, 50*(1), 30–37. https://doi.org/10.15288/jsa.1989.50.30

Wu, H. M. (2019). Online individualised tutor for improving mathematics learning: A cognitive diagnostic model approach. *Educational Psychology, 39*(10), 1218–1232. https://doi.org/10.1080/01443410.2018.1494819

Xu, X., & von Davier, M. (2006). Cognitive diagnosis for NAEP proficiency data. *ETS Research Report Series, 2006*(1), i–25. https://doi.org/10.1002/j.2333-8504.2006.tb02014.x

Part V
Longitudinal Modeling and Estimation of Missing Data

Chapter 23
Multiple Imputation of Longitudinal Data: A Comparison of Robust Imputation Methods Regarding Sample Size Requirements, with an Application to Corporal Punishment Data

Kristian Kleinke, Markus Fritsch, Mark Stemmler, and Friedrich Lösel

23.1 Introduction

Multiple imputation (MI) based on Rubin's (1987) theory is one of the state-of-the-art approaches to analyse incomplete datasets (cf Schafer & Graham, 2002). For complex multilevel data structures such as clustered or panel data, MI methods are available (for an overview, see, e.g., Spiess et al., 2021) but depend on the assumptions that the covariance matrix of the error terms is spherical and the error terms are normally distributed. The quality of statistical inferences of these methods usually depends on the extent to which the modeling assumptions are met (see, e.g., the simulations in Kleinke, 2021). Kleinke et al. (2021) have evaluated the performance of robust semi- and nonparametric imputation methods and found that robust alternatives yielded more precise statistical inferences compared to fully parametric methods when the assumptions that the covariance matrix of

K. Kleinke (✉)
Department of Psychology, University of Siegen, Siegen, Germany
e-mail: kristian.kleinke@uni-siegen.de

M. Fritsch
University of Passau, School of Business, Economics and Information Systems, Passau, Germany
e-mail: markus.fritsch@uni-passau.de

M. Stemmler
Institute of Psychology, Friedrich-Alexander-University of Erlangen-Nuremberg (FAU), Erlangen, Germany
e-mail: mark.stemmler@fau.de

F. Lösel
Friedrich-Alexander-University (FAU) Erlangen-Nuremberg, Institute of Psychology, Erlangen, Germany

Cambridge University, Institute of Criminology, Cambridge, UK
e-mail: friedrich.loesel@fau.de

M. Stemmler et al. (eds.), *Dependent Data in Social Sciences Research*,
https://doi.org/10.1007/978-3-031-56318-8_23

the error terms is spherical and the error terms are normally distributed are violated. Unfortunately, semi- and nonparametric imputation methods have not been generalized to multilevel data yet, and only single-level imputation methods are available for such data structures. When facing missing values in multilevel data, applied researchers need to choose between parametric multilevel imputation methods and semi- and nonparametric single-level imputation methods. Choosing an imputation method for 'non-normal' panel data thus involves trading off the assumptions that the covariance matrix of the error terms is spherical and the error terms are normally distributed with the assumption that the errors are independent.

Kleinke et al. (2011) have compared single-level imputation methods applied to wide-format panel data (i.e., when repeated measurements of a certain variable are stored as separate variables in the dataset) to multilevel imputation methods. They found that imputation in wide format using an appropriate imputation method yields statistical inferences that are widely unbiased, when cluster effects are not too large. Repeated measurements are usually highly correlated, and missing information from some panel waves can often be predicted by observed information from neighbouring panel waves. Based on these findings, we currently suggest to use a robust single-level imputation method such as quantile regression (QR)-based imputation also for the imputation of 'non-normal' clustered or panel data—if sample size is sufficiently large (for overviews, see Wilcox, 2017 on robust statistical methods and the precision of estimators when using robust methods, and Koenker, 2005 for a comprehensive treatment of quantile regression). Unfortunately, it is currently unclear what exactly a sufficiently large sample size is—especially in the context of robust imputation of missing values.

The purpose of the present chapter is to shed more light on the sample size requirements of QR-based multiple imputation. First, we extend the simulations in Kleinke et al. (2021) by systematically varying the available sample size from $N = 50$ to $N = 600$ and contrast the results against further robust imputation methods such as predictive mean matching, which is robust towards outliers and milder forms of 'non-normality', and random forest-based imputation, which is robust towards the assumptions that the covariance matrix of the error terms is spherical and that the error terms are normally distributed. Second, we apply the imputation methods to an empirical five-wave panel dataset on corporal punishment of children by their fathers and investigate the relationship between corporal punishment and the development of children's social behaviour problems.

23.2 Theoretical Background

Missing values are an omnipresent problem in empirical research. If the percentage of missing values in the dataset is not negligible, and the values are not missing completely at random in the sense of Rubin (1976), then state-of-the art imputation techniques should be applied to obtain parameter estimates and unbiased standard errors (cf. Schafer & Graham, 2002). One of these state-of-the-art techniques is

(Bayesian) multiple imputation (MI) based on Rubin's (1987) theory. The following subsections detail the theoretical background of MI and three different techniques to apply MI: predictive mean matching, quantile regression-based MI and random forest-based MI.

Missing Values and Multiple Imputation

Application of MI in the context of frequentist inference is a three-step process: In the first step, each missing value is imputed $M > 1$ times by *plausible* values (see Bodner, 2008; Graham, 2009, for recommendations on the number of imputations). This step produces M complete datasets that differ only in the formerly missing part. *Plausible* values in this context refer to any values that lead to unbiased statistical inferences.[1] In the second step, these M complete datasets are analysed separately by any technique that yields point estimates and standard errors for the parameters of interest. In the final step, the M sets of modeling results are combined into an overall set of results based on combination rules such as the large-sample formula given in Rubin (1987) or adaptations for small samples (Barnard & Rubin, 1999). For a detailed discussion, see (Kleinke et al., 2020, Chap. 4). The combination rules yield a combined point estimate and a combined standard error for each model parameter: The MI point estimate is simply the average of the M estimates. Corresponding standard errors are obtained by combining a within-imputation and a between-imputation variance component. The latter component reflects the amount of missing information with respect to the respective parameter: Fewer information to predict the missing values leads to higher uncertainty in the imputations and a larger between variance.

For a comprehensive introduction to missing data and MI, an overview on recent developments in this field and various applications, see, for example, Kleinke et al. (2020) or Enders (2022). A general introduction to the problem of missing data, missing data patterns and mechanisms and a taxonomy of missing data methods, as well as a discussion of the rationale of MI in comparison to other methods to account for uncertainty in parameter estimates due to missing data using resampling methods (such as the bootstrap or jackknife), may also be found in (Little and Rubin, 2020, ; see especially Chapters 1, 4 and 5). A further good resource for applied researchers is Raghunathan et al. (2018), who discuss applications of MI in various contexts including linear and generalised linear models, categorical data analysis, survival analysis, structural equation modeling, longitudinal data analysis and complex survey settings.

Assumptions Approaches for MI typically assume that the *mechanism generating the missing values* in a vector of realisations of a variable $\boldsymbol{y}$ of length n is *ignorable*

[1] See Rubin (1996) for a discussion of criteria that are required to create *proper* imputations.

(Rubin, 1976, 1987; see also Kleinke et al., 2020, Chap. 2): Observing or not observing the i−th entry of $\boldsymbol{y}$ can be seen as the result of a dichotomous process generating a binary response r_i, which obtains the value '1' if the value of the corresponding variable is observed and '0' otherwise. Transitioning from the sample of realisations to the subsample of observed realisations can be interpreted as an additional sampling step. The underlying process, however, is usually unknown during data collection and assumed to be a random process. When the mechanism generating the missing values is ignorable, datasets can be analysed without the need to explicitly model the mechanism. An ignorable mechanism can be assumed when the missing values are generated at random, given all observed information, or, stated differently, the probability of not observing certain realisations of a variable depends on observed information only. To ensure the credibility of this (at least partially untestable) assumption, applied researchers typically specify *rich* imputation models and include all variables that are related to both the variables of the model of scientific interest and the missing data indicators (the so-called auxiliary variables) into the imputation model—even if these additional variables are omitted from the analysis model later on (cf Collins et al., 2001).

How imputations are generated MI based on Rubin's (1987) theory can be considered a Bayesian approach: Imputations are generated by drawing from the posterior predictive distribution of the variables whose values are missing given all observed values. From a mathematical point of view, the posterior predictive distribution is usually difficult to approximate, because in most empirical datasets, different types of variables are present, e.g., continuous, truncated, categorical or count, and missing value patterns are usually not monotone.

A popular approach to handle these situations is the `mice` (i.e., multiple imputation by chained equations) framework (Van Buuren & Groothuis-Oudshoorn, 2011; Van Buuren, 2018; Van Buuren et al., 2006), where users can specify separate conditional imputation models for each variable with missing values. The models can be specifically tailored to the distribution and type of each variable. The software then cycles iteratively through all variables with missing values and imputes them step by step using predictions from a pre-specified regression model. Values of all other (observed and imputed) variables are used by the procedure. The approach is inter alia implemented in R package `mice` (for details, see Van Buuren & Groothuis-Oudshoorn, 2011; Van Buuren, 2018). Available imputation methods in `mice` include fully parametric ones such as Bayesian linear regression (`norm`), Bayesian logistic regression (`logreg`), semi-parametric ones such as predictive mean matching (`pmm`) and 'tree'-based methods such as `cart` (imputation by classification and regression trees), or `rf` (random forest-based imputation). Extensions of the functionality of `mice` are also available from other packages such as `Qtools` (Geraci, 2016a,b) for quantile regression-based MI.

Robust Imputation by Predictive Mean Matching

Predictive mean matching (pmm) was first proposed by Rubin (1986) and Little (1988) as a single-imputation approach and later extended to the context of MI. There are many different variants and implementations of the pmm method in the different software packages, and some type of pmm is available in all current versions of the major statistical packages including SAS, Stata and SPSS. Predictive mean matching is also the default imputation method for incomplete numeric vectors in R package `mice` (Van Buuren & Groothuis-Oudshoorn, 2011), which is described in detail in (Van Buuren, 2018, Algorithm 3.3) and summarised in the following paragraph: The illustrations are based on the general regression model:

$$\boldsymbol{y} = f(\boldsymbol{X}) + \boldsymbol{\varepsilon}, \tag{23.1}$$

where $\boldsymbol{y}$ denotes the vector of response variables, $\boldsymbol{X}$ is a matrix of explanatory variables, $f(\cdot)$ is a function of the explanatory variables, and $\boldsymbol{\varepsilon}$ is a vector of errors. Note that we assume an additive decomposition of Eq. (23.1) into the systematic model component $f(\boldsymbol{X})$ and the unsystematic model component $\boldsymbol{\varepsilon}$. For illustration, we assume that there are only missing values in the vector of response variables $\boldsymbol{y}$.[2] When carrying out pmm, first assume that the systematic model component can be approximated reasonably well by the linear in parameters functional $\boldsymbol{X\beta}$, with coefficient vector $\boldsymbol{\beta}$. Then, estimate the coefficient vector based on the n_1 elements of the vector of observed response variables $\boldsymbol{y}_{\text{obs}}$ and matrix of corresponding explanatory variables $\boldsymbol{X}_{\text{obs}}$ by the OLS estimator $\hat{\boldsymbol{\beta}} = (\boldsymbol{X}'_{\text{obs}}\boldsymbol{X}_{\text{obs}})^{-1}\boldsymbol{X}'_{\text{obs}}\boldsymbol{y}_{\text{obs}}$. Second, compute $\boldsymbol{\beta}^*$ according to Algorithm 3.1 in Van Buuren (2018) by drawing randomly from the posterior distribution of $\boldsymbol{\beta}$ to incorporate the underlying sampling variability when predicting the missing values. Third, obtain the vector of fitted values corresponding to the vector of observed response variables by $\hat{\boldsymbol{y}}_{\text{obs}} = \boldsymbol{X}_{\text{obs}}\hat{\boldsymbol{\beta}}$ and the vector of predicted values for the n_2 missing values by $\hat{\boldsymbol{y}}_{\text{mis}} = \boldsymbol{X}_{\text{mis}}\boldsymbol{\beta}^*$. This yields exactly one predicted value for each missing value. Fourth, for each of the $j = 1, \ldots, n_2$ elements of the vector of predicted values $\hat{\boldsymbol{y}}_{\text{mis}}$, compute the distance to each of the $i = 1, \ldots, n_1$ elements in $\hat{\boldsymbol{y}}_{\text{obs}}$ according to a pre-defined distance metric (e.g., absolute distance $|\hat{y}_{\text{mis},j} - \hat{y}_{\text{obs},i}|$). Fifth, generate the set of potential donor cases or *donor pool* for each missing value j by selecting the K elements from the vector of actually observed values $\boldsymbol{y}_{\text{obs}}$, which correspond to the K elements from the vector of fitted values $\hat{\boldsymbol{y}}_{\text{obs}}$ that fulfil a pre-specified criterion based on the distance metric (e.g., the elements with the K smallest absolute distances). Note that this yields one donor pool with K elements for each missing value. Finally, to

[2] If values are also missing from some of the predictors, these variables would become the response variables in their respective own imputation model. Note again that `mice` iteratively cycles through each incomplete variable and fills the incomplete values based on user-specified methods and models for the respective incomplete variable at hand.

select the j-th imputed value, sample one observation (the so-called donor) from the corresponding donor pool according to a given sampling scheme.

Since the distance metric, the number of potential donors K and the sampling scheme need to be specified, there are a number of proposals for donor selection. One approach implemented in the R package `mice` samples from the $K = 5$ potential donors with the smallest absolute distance by default. Simulations by Schenker and Taylor (1996) suggest that this should work well, when sample size is sufficiently large. In case of a small sample size, the size of the donor pool usually has to be adjusted (see, e.g., the simulations in Kleinke, 2018), and nearest neighbour imputation (i.e., setting $K = 1$) often produces better results than using the defaults. Another commonly employed approach was proposed by (Siddique and Belin, 2008) and is available from the SAS macro MIDAS (Siddique & Harel, 2009). The approach was adapted for R via the method `midastouch` in the package `mice` by Gaffert et al. (2016). Here, for each missing value j, the pool of potential donors is set to all observed realisations $K = n_1$, and the donor is sampled with selection probabilities depending on the absolute distance between the missing value and the actually observed realisations $|\hat{y}_{\text{mis},j} - \hat{y}_{\text{obs},i}|$, where cases with smaller distances receive higher selection probabilities and vice versa. Especially when the sample size is small, this is a good alternative to classical pmm, since the user does not have to specify the size of the donor pool (see the discussions and simulations in Kleinke, 2018).

Van Buuren (2018, Section 3.4.5) summarises the advantages and disadvantages of pmm as follows: By imputing actually observed realisations of the variable, the method 'provides imputations that possess many characteristics of the complete data. Predictive mean matching cannot be used to extrapolate beyond the range of the data, or to interpolate within the range of the data if the data at the interior are sparse. Also, it may not perform well with small datasets. Bearing these points in mind, predictive mean matching is a great all-around method'. The main problem with small sample sizes is that there might not be any or many suitable donor cases in the dataset to produce unbiased inferences. On the other hand, if suitable donor cases are available, the main advantage of pmm is that by imputing actually observed realisations of the variable, many characteristics of the empirical data are preserved. This makes the method quite robust against violations of assumptions imposed by the underlying linear regression model such as a spherical covariance matrix of the errors (see the simulation results in Kleinke et al., 2011) and normally distributed errors (Kleinke, 2017).

Quantile Regression-Based Multiple Imputation

While standard parametric regression-based MI approaches create imputations by modeling the conditional mean of the response variable given certain explanatory variables assuming that the covariance matrix of the error terms is spherical and the error terms are normally distributed, quantile regression (QR) models the conditional *median* or another quantile τ ($\tau \in]0, 1[$) of the response variable given the explanatory variables (cf. Koenker & Hallock, 2001; Koenker, 2005).

The conditional τ-quantile function is

$$Q_{Y|X}(\tau) = \boldsymbol{X}\boldsymbol{\beta}_\tau, \tag{23.2}$$

where Y is the response, $\boldsymbol{X}$ is a matrix of explanatory variables, and $\boldsymbol{\beta}_\tau$ denotes the vector of regression coefficients for the conditional τ-quantile. Parameter estimates are obtained by minimising a criterion based on asymmetrically weighted absolute errors. Assuming spherical and normally distributed errors is not required and replaced by the identifying condition that the τ-quantile of the errors is zero (for more details, see Koenker, 2005).

By estimating separate coefficients for each conditional quantile, QR provides a more nuanced picture of the conditional distribution of the response variable and the explanatory variables compared to considering only the conditional mean. In many applications, higher, average and lower intensities of the response variable are generated by different structures of explanatory variables (for an example, see Haupt et al., 2014). Together with the weak assumptions about the distribution of the errors employed by QR, this may lead to improved imputations and related inferences compared to fully parametric imputation methods. This is especially the case, when parametric modeling assumptions are violated. Geraci (2016b) has implemented quantile regression-based MI in R package `Qtools` (see also Bottai & Zhen, 2013; Geraci, 2016a). Geraci's functions work as an add-on to package `mice` and are fully integrated into the `mice` architecture. The main imputation function is called `mice.impute.rq()` and works as follows:

Let again $\boldsymbol{y}$ denote the vector of response variables with missing values, and let $\boldsymbol{y}_{obs}$ and $\boldsymbol{y}_{mis}$ denote the vectors of observed and missing values with $i = 1, \ldots, n_1$ and $j = 1, \ldots, n_2$ elements, respectively. Let $\boldsymbol{X}$ be a matrix where all elements are observed, with $\boldsymbol{X}_{\text{obs}}$ being the subset of explanatory variables from the imputation model that correspond to $\boldsymbol{y}_{obs}$, and $\boldsymbol{X}_{\text{mis}}$ the subset of explanatory variables from the imputation model that correspond to the missing values. Variability in the $m = 1, \ldots, M$ imputations is introduced by generating an independent uniform random draw for the imputation of each missing value in each data set: In the first step, for each missing value y_j, a random uniform number u_j is generated from $u_j \sim \mathcal{U}(\epsilon, 1-\epsilon)$, where ϵ is small (e.g., $\epsilon = 0.001$), and model parameters $\boldsymbol{\beta}_u$ are estimated based on $\boldsymbol{y}_{\text{obs}}$ and $\boldsymbol{X}_{\text{obs}}$. In the second step, for each missing value y_j, the vector of imputations $\hat{\boldsymbol{y}}_j$ with M elements is obtained by $\hat{\boldsymbol{y}}_j = \boldsymbol{x}_{j,\text{mis}}\boldsymbol{\beta}_u$. Note that $\boldsymbol{x}_{\text{mis},j}$ is the j-th column of matrix $\boldsymbol{X}_{\text{mis}}$. Kleinke et al. (2021) have demonstrated that QR-based imputation works reasonably well in a range of practically relevant scenarios, when sample size is sufficiently large ($N > 600$).

Random Forest-Based Multiple Imputation

Random forest (RF) is a machine learning method that can be employed for classification and regression problems. The RF algorithm approximates the function $f(\boldsymbol{X})$ in Eq. (23.1) by fitting multiple decision trees to a dataset of response variables $\boldsymbol{y}$ and explanatory variables $\boldsymbol{X}$ via recursive binary partitioning (for algorithms for fitting regression trees and RFs, see Sections 2.1 and 2.2 in Fritsch et al., 2019). Recursive binary partitioning partitions the space spanned by the explanatory variables into several non-overlapping regions according to a pre-defined splitting criterion. At each split (or node), one explanatory variable is used to split the space spanned by the explanatory variables into two regions until some stopping criteria are reached (e.g., a minimum number of observations in a node, a maximum number of splits or a minimum improvement in the splitting criterion). The first split in the tree is referred to as root node, all other splits are referred to as internal nodes, and the regions where the splitting is stopped are called terminal nodes (or leafs). For more details, see the vignette of R package `rpart` (Therneau & Atkinson, 2022). As individual trees exhibit substantial instabilities when varying only a small fraction of the dataset, the idea behind the RF algorithm is to resolve the instability by combining multiple de-correlated decision trees (Hastie et al., 2009). This is achieved by drawing bootstrap samples from the original data and fitting one decision tree to each bootstrap sample while using different subsets of the explanatory variables for deciding about the different splits of the individual trees (Breiman, 2001). The RF algorithm has also been adapted for MI of missing values and is implemented in the R package `mice` via the function `mice.impute.rf()` (see Algorithm 1 in Doove et al., 2014). The method can be summarised as follows:

For all missing values in the vector of response variables $\boldsymbol{y}$, draw a value from the vector of observed response variables $\boldsymbol{y}_{\text{obs}}$ as initial imputation $\hat{\boldsymbol{y}}_0$. Then, draw B bootstrap samples from the dataset without missing values (i.e., $\boldsymbol{y}_{\text{obs}}$ and $\boldsymbol{X}_{\text{obs}}$) and fit one regression tree to each bootstrap sample while considering only a fraction s, with $s \in]0, 1]$ of explanatory variables for partitioning the space spanned by the explanatory variables at each node of the tree. Each of the B trees is then used for imputation by determining the terminal node of each tree in which each of the $j = 1, \ldots, n_2$ missing values in $\boldsymbol{y}_{\text{mis}}$ is located. All observed responses that are located in the same terminal node as the j-th missing value are the corresponding potential donors. In order to obtain the imputed value $\hat{y}_j$ for the j-th missing value $y_{j,\text{mis}}$, one of the potential donors is then sampled randomly. This produces the first of the M imputed datasets. For the current default settings, see `help("mice.impute.rf")`.

Imputation by RF is particularly appealing for large datasets, since no imputation model and auxiliary variables need to be specified. The latter are variables related to the employed response and explanatory variables and correlated with the missing data indicators, but which are not of substantive interest in the data analysis (cf. Collins et al., 2001). A further advantage of random forest-based imputation is that no functional form needs to be specified by the researcher as the underlying functional is approximated in data-driven fashion. For an in-depth discussion regarding the problems that can arise in the imputation of nonlinear terms in multilevel models, see Kleinke (2021).

23.3 Monte Carlo Simulation

The present simulation extends the scenarios reported in Kleinke et al. (2021) based on the empirical data from the Erlangen-Nuremberg Development and Prevention Study (Lösel et al., 2009). Our main focus is to explore to what extent the respective imputation approaches can handle small sample sizes. The study has a 3×8 (imputation method $\times$ sample size) factorial design. We created complete datasets with sample sizes $\mathcal{N} \in \{50, 100, 150, 200, 300, 400, 500, 600\}$. Each of the 24 conditions was replicated 1000 times. From the complete datasets, we then deleted 40% of values from the response and from one of the explanatory variables following a missing at random (MAR) missing data pattern. Missing values were imputed by QR- and random forest-based MI, as well as by the pmm variant 'midastouch'. In the following subsections, we detail the evaluation criteria, the stylized facts of the empirical dataset, the missing value patterns and imputation schemes and the simulation results regarding the bias and related inferences of the different imputation techniques.

Evaluation Criteria

We evaluated the different imputation techniques based on the goal to yield unbiased statistical inferences: *Bias* in point estimates was defined as the difference between the respective true population parameter θ and the average estimate $\widehat{\theta}$ across the imputations. We report bias relative to the true population parameter as

$$\%\text{Bias} = \frac{|\text{ Bias }|}{\theta} \cdot 100\%. \tag{23.3}$$

In determining if relative biases are substantial, we follow Forero and Maydeu-Olivares (2009), who deem absolute values of less than 10% parameter bias as acceptable. *Coverage rate* (CR) is defined as the percentage of 95% confidence intervals that included the true population parameter. Obviously, CR should be close to the nominal level of 95%. Schafer and Graham (2002) deem CR below 90% as serious undercoverage. Undercoverage may result from biased point estimates and/or an underestimated variance. The confidence interval either could be too narrow to include the true parameter or could exhibit too large a shift to include the true population parameter. A CR close to 100% indicates that the standard errors are too large, which increases the risk of Type II errors. Together, bias and CR give a good impression of the accuracy of statistical inferences.

Sample and Measures

Our simulation is based on empirical data from the Erlangen-Nuremberg Development and Prevention Study (Lösel et al., 2009), which examines the origins and development of social behaviour problems in childhood. It combines a prospective longitudinal investigation of preschool children and their families with an experimental evaluation of a parent and a child training programme. For this simulation, we used data from 609 families (675 preschool children). The simulation builds upon the empirical analyses reported in Haupt et al. (2014), who identified risk factors for *corporal punishment* of children by their fathers (CP): parenting behaviour, socio-economic status and the gender of the children. The response in our model CP, as well as two risk factors *inconsistent discipline* (ID) and *other discipline practices* (OD), was assessed using a German version of the Alabama Parenting Questionnaire (APQ, Lösel et al., 2003). Items were rated on five-point scales ranging from 1 ('never') to 5 ('always'). Scale scores were obtained by calculating the means of the respective items. The CP scale was based on three items such as 'You spank your child with your hand when he/she has done something wrong'. To avoid estimation problems due to the discrete nature of the CP index, we mildly jittered the data by adding random numbers drawn from $\mathcal{N}(0, 0.001)$. Inconsistent discipline (six items) summarises parenting behaviours such as 'The punishment you give your child depends on your mood'. Other discipline practices (seven items) refer to actions such as 'You take away privileges or money from your child as a punishment'. The computation of the *socio-economic status* (SES) index was based on Geißler (1994). FEMALE is a binary gender indicator. Detailed information about the distributions of all model variables can be found in (Haupt et al., 2014, Table 1 and Figure 1). For summary statistics of the corporal punishment index, see also Table 23.1a in this chapter.

Procedure

For each condition, we simulated 1000 bootstrap samples from the original $N = 485$ complete cases. We then introduced MAR patterns (Rubin, 1976): We deleted 40% of the values of the response CP and additionally from the explanatory variable OD. Values were deleted using the following rule: For each observation i ($i = 1, \ldots, N$), we computed the probability of a missing value with a logit model, where the probability of the event that a value is missing for individual i, $Pr(\text{Mis}_i)$, is given by

$$Pr(\text{Mis}_i) = \text{invlogit}(\alpha_0 + \alpha_1 \cdot \text{SES}_i). \tag{23.4}$$

Table 23.1 Descriptive statistics

(a) Corporal punishment						(b) Social behaviour problems				
	t1	*t*2	*t*3	*t*4	*t*5	*t*1	*t*2	*t*3	*t*4	*t*5
M	1.58	1.53	1.48	1.32	1.28	13.47	11.76	11.37	10.92	10.69
SD	0.45	0.43	0.40	0.36	0.32	7.29	7.21	7.08	7.35	7.46
Minimum	1.00	1.00	1.00	1.00	1.00	1.00	0.00	0.00	0.00	0.00
First quartile	1.33	1.33	1.00	1.00	1.00	8.00	6.00	6.00	5.00	5.00
Median	1.67	1.33	1.33	1.33	1.17	13.00	11.00	10.00	10.00	9.00
Third quartile	1.67	1.67	1.67	1.67	1.50	18.00	16.00	15.00	15.00	15.00
Maximum	4.00	3.00	3.00	2.67	2.33	44.00	41.00	43.00	55.00	43.00
Skewness	0.94	0.64	0.77	0.92	0.92	0.81	0.98	1.02	1.16	1.04
Excess	2.26	0.03	0.60	0.15	−0.04	0.79	1.19	1.36	2.62	1.15
%NA	4.89	11.73	18.24	23.94	26.38	4.56	11.56	18.24	23.78	26.55

Coefficients in (23.4) were set to $\alpha_0 = 5$ and $\alpha_1 = -2$ to ensure that the probabilities of missing values range from around 0.2 to 0.8. This implies that participants from lower social layers were assigned larger probabilities of having missing values in our simulation. The necessary number of cases were then sampled with selection probabilities $Pr(\text{Mis}_i)$ and their respective values were deleted from the bootstrap samples.

Missing values were then imputed with QR- and RF-based MI, as well as with the pmm variant 'midastouch' using functions implemented in packages `mice`, and `Qtools`. We employed the default settings of all functions. When predicting/imputing the missing values in a particular column, all other columns of the dataset were employed. As common practice (e.g., Kleinke, 2018) suggests, the number of imputations was set to the percentage of missing values.

We then compared coefficient estimates based on the imputed datasets with the means of the 1000 parameter estimates obtained from the complete cases. We took the latter as proxies for the unknown true population parameters. Coefficient estimates were obtained from a QR model, in which we regressed the response CP on SES, FEMALE, ID and OD regarding quantiles $\tau \in \{0.2, 0.3, 0.4, 0.5, 0.6, 0.7, 0.8\}$. Heteroscedasticity robust standard errors were computed as in Haupt et al. (2014).

Results

Results regarding bias in point estimates are summarized in Fig. 23.1. Overall, QR-based MI yielded the lowest bias across all sample sizes. On average, the bias was 9% across all sample sizes, coefficient estimates and quantiles, whereas

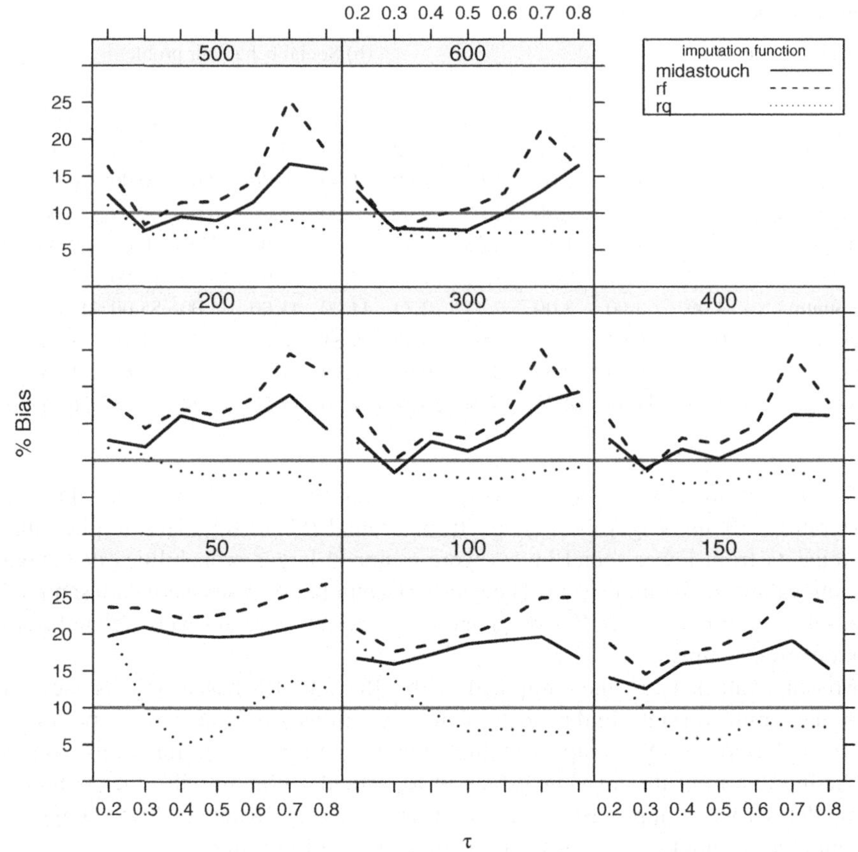

Fig. 23.1 Average parameter bias (in %) produced by pmm, QR-based MI and RF-based MI for eight different sample sizes. *Note.* Results were averaged across all coefficient estimates; midastouch is a pmm variant using distance-aided selection of donors; rq is QR-based MI; rf is RF-based MI; black horizontal line indicates parameter bias of 10% relative to true population parameter

pmm and RF-based MI yielded average parameter biases of 15% and 18%, respectively.

When sample size was small ($N = 50$), QR-based MI yielded acceptable estimates only regarding the quantiles in the centre (τ between 0.4 and 0.5)—with the exception of some coefficient estimates: Biases ranged between 0.3 and 16%. The coefficient estimate for the intercept was generally unbiased, as were the coefficient estimates of explanatory variables FEMALE, ID and OD. The coefficient estimate of the SES score (i.e., the 'cause' of missingness in our model) produced the largest biases ranging between 9 and 16%.

In comparison, imputation based on pmm and RF yielded biases of around 20 and 25% of the coefficient estimates, respectively, across all quantiles. Similar patterns were also found for sample sizes of $N = 100$ and $N = 150$. With increasing sample size, relative biases generally decreased—also regarding the off-centre parts of the conditional distribution. When $N \geq 200$, QR-based MI yielded acceptable results for all, but the extreme quantiles, where bias was still slightly above 10%. The largest observed bias was 18% for the coefficient estimate corresponding to OD, when $\tau = 0.3$. For pmm, the bias in coefficient estimates was generally acceptable in the centre parts of the conditional distribution (τ between 0.3 and 0.6) when the sample size was $N \geq 500$. Overall, RF-based MI produced the worst results with relative biases of up to 53%. We conclude that the examined methods for MI require a sufficiently large sample size. This is especially true, when applied researchers want to look beyond the centre part of the conditional distribution.

Figure 23.2 displays 95% confidence interval CR. Undercoverage was not observed, but with decreasing sample size, CR generally became unacceptably

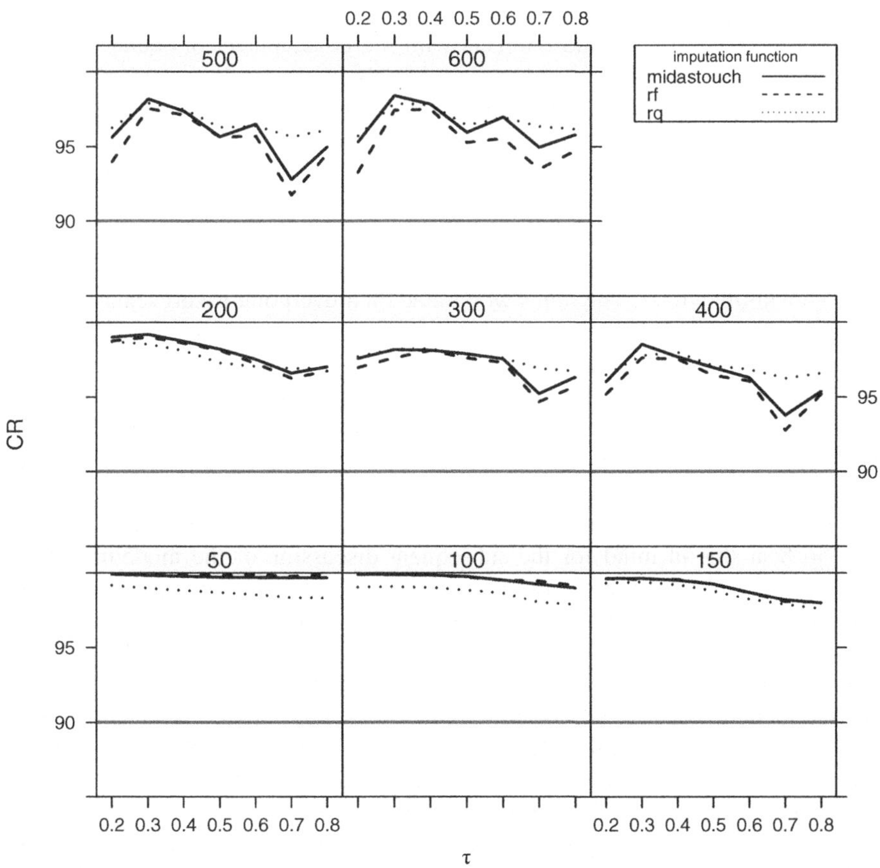

Fig. 23.2 Average coverage rates of 95% confidence intervals produced by pmm, QR-based MI and RF-based MI for eight different sample sizes. *Note.* Results were averaged across all parameter estimates; midastouch is a pmm variant using distance-aided selection of donors; rq is QR-based MI; rf is RF-based MI; black horizontal line indicates CR of 90%

large. Especially pmm and RF-based MI produced CR of very close to 100%, which dramatically increases the risk for Type II errors. Standard errors were inadequately large. With increasing sample size, CR improved and were more closely centred around 95%.

Overall, our results suggest that point estimates obtained after QR-based MI are reasonable even when they are based on limited sample sizes. Reliable standard errors and statistical inferences, however, require sufficiently large sample sizes. Our simulation results suggest sample sizes of at least $N \geq 500$, preferably more.

23.4 Application: Development of Corporal Punishment by Fathers over Time

In the second part of this chapter, we investigate how CP by fathers develops over time and how CP affects and interacts with the development of social behaviour problems in their children. The empirical basis of the analyses reported in this section is data from $N = 614$ children and their fathers across five panel waves of the Erlangen-Nuremberg Development and Prevention Study (Lösel et al., 2009). Families where the mother was a single parent were excluded from the analysis. As already stated above, CP was assessed by self-report and was measured by three items from the Alabama Parenting Questionnaire regarding to what extent fathers slapped or spanked their child either with the hand, or a belt, when the child has done something wrong. The items were rated on a five-point Likert scale (from 1 = 'never' to 5 = 'always'), and the scale score was computed as the mean of the item scores. Descriptive statistics are given in Table 23.1a. As can be seen, the maximum observed score decreases over time from 4.0 to 2.33, while the percentage of missing values increases from around 5–26%. The mean of CP decreases from 1.58 to 1.28 over time. If the decrease in CP was predominantly driven by fathers who exhibited more severe forms of CP dropping out of the study, missing values would be missing not at random (MNAR) and the means of CP could be biased downwards. We need to bear this in mind for the subsequent discussion of the modeling results. Additionally note that CP scores are skewed (see Table 23.1a), which means that applying some sort of robust imputation and analysis method seems to be in order. Missing value patterns of CP are displayed in Table 23.2: 399 out of the 614 cases were complete cases, while 23 fathers refused to provide information about CP at all measurement timepoints. We can also see that wave nonresponse is rather negligible and that this dataset is mainly affected by panel attrition with 25 fathers quitting after wave 1, 27 after wave 2, 40 after wave 3 and another 28 after wave 4.

Children's social behaviour problems were rated by their fathers at each timepoint using the Social Behavior Questionnaire (SBQ, Tremblay et al., 1987, 1991, 1992), in the German adaptation by Lösel et al. (2002). The total SBQ score is computed as the sum of all items that ask about problematic behaviours such as 'he/she kicks, bites and hits other children', which are rated on a three-point scale:

Table 23.2 Corporal punishment: Missing value patterns

	t1	t2	t3	t4	t5	Freq	Freq%
1	1	1	1	1	1	399	64.98
2	1	1	1	0	0	40	6.51
3	1	1	1	1	0	28	4.56
4	1	1	0	0	0	27	4.40
5	1	0	0	0	0	25	4.07
6	0	0	0	0	0	23	3.75
7	1	1	1	0	1	17	2.77
8	1	1	0	1	1	16	2.61
9	1	0	1	1	1	8	1.30

Note. The table displays absolute and relative frequencies (in %) of most frequent missing value patterns (>1%); 1 = observed value; 0 = missing value

'never or not true' (0), 'sometimes or somewhat true' (1) or 'often or very true' (2). Descriptive statistics are given in Table 23.1b. Again, we see that the mean of the social behaviour (SB) score decreases over time. Similar to before, the decrease in the mean SB score could be biased downwards, if the dropout mechanism is MNAR and it is predominantly those dyads where the child exhibits more problems, who drop out of the study. The missing value patterns are highly similar to the ones reported in Table 23.2, since the SB score was rated by the father and the missing data problem appears to be mainly a problem of panel attrition.

A Parallel Latent Growth Curve Model of Corporal Punishment and Children's Social Behaviour Problems

Parallel growth curve models (for an introduction to growth curve models, see Bollen & Curran, 2006) are used to model change in two (or more) response variables simultaneously over time. In our application, we are interested in the parallel development of CP by fathers and the development of social behaviour problems in their children. In addition to the average effects—i.e., the average starting levels of CP and the social behaviour problems at the beginning of the study, as well as the average levels of linear and non-linear change across time—these models also allow to explain how and why the individual fathers and their children vary in this regard. The general measurement model equation for the growth curve model is

$$\boldsymbol{y} = \boldsymbol{\Lambda}\boldsymbol{\eta} + \boldsymbol{\epsilon}, \tag{23.5}$$

where $\boldsymbol{y}$ is the $p \times 1$ vector of the repeatedly measured manifest indicators, $\boldsymbol{\eta}$ is the $q \times 1$ vector of latent growth factors and $\boldsymbol{\epsilon}$ is the $p \times 1$ vector of time-specific measurement errors. $\mathbf{\Lambda}$ is the $p \times q$ matrix of factor loadings. Latent growth factors $\boldsymbol{\eta}$ are described as a function of their vector of means $\boldsymbol{\alpha}$, their relationship to exogenous variables $\boldsymbol{\xi}$ and residuals $\boldsymbol{\zeta}$:

$$\boldsymbol{\eta} = \boldsymbol{\alpha} + \mathbf{\Gamma}\boldsymbol{\xi} + \boldsymbol{\zeta}, \tag{23.6}$$

where $\mathbf{\Gamma}$ is a $q \times r$ matrix of regression coefficients describing the effects of r exogenous variables $\boldsymbol{\xi}$ on the q growth factors. $\mathbf{\Psi}$ is the variance-covariance matrix of residuals $\boldsymbol{\zeta}$. Factor loadings are constrained to represent the constant, linear and—if necessary—non-linear development. The estimated means $\boldsymbol{\alpha}$ of latent variables $\boldsymbol{\eta}$ represent the mean starting level and mean levels of change of the response variable. The variances of the growth factors represent the degree of individual variation around the respective means. Covariances of the growth factors describe their interrelationships, for example, if individuals with lower starting levels tend to have steeper linear and quadratic slopes and vice versa. Parallel growth curve models allow to examine if and to what extent the latent growth factors of two or more response variables are correlated.

Figure 23.3 displays the development of children's SB and CP by their fathers over time. The estimates are based on full information maximum likelihood estimation using a robust estimator. More detailed results, as well as a comparison of different robust imputation methods, can be found in Table 23.3. Note that both the SB score and the intensity of CP decrease over time (assuming that data are MAR). The latent intercept and slope of SB is furthermore negatively correlated, which means that children with an already low SB score at the first measurement timepoint tend to exhibit a less steep decline over time than children with high SB scores. Also, the intercepts of SB and CP are positively correlated, meaning that

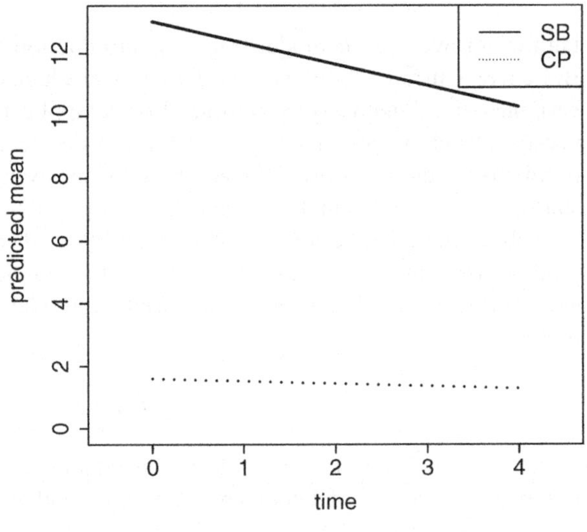

Fig. 23.3 Parallel latent growth curve model of children's social behaviour problems and corporal punishment by their fathers over five measurement timepoints. *Note.* SB denotes the extent of children's social behaviour problems, and CP is the corporal punishment index

Table 23.3 Parameter estimates of the parallel latent growth curve model based on different missing value imputation methods

Parameter	fiml	rf	rq	pmm
	Means			
ISB	13.02	12.95	12.94	12.92
SSB	−0.68	−0.66	−0.58	−0.64
ICP	1.60	1.59	1.60	1.60
SCP	−0.08	−0.08	−0.08	−0.08
	Variances			
ISB	37.52	36.07	37.90	36.78
SSB	1.71	1.56	1.89	1.78
ICP	0.15	0.14	0.15	0.14
SCP	0.01	0.01	0.01	0.01
	Covariances			
ISB WITH SSB	−2.87	−3.23	−3.01	−3.59
ISB WITH ICP	0.89	0.88	0.92	0.85
ISB WITH SCP	−0.05	−0.06	−0.06	−0.06
SSB WITH ICP	−0.03	−0.05	−0.02	−0.04
SSB WITH SCP	0	0	0	0
ICP WITH SCP	−0.02	−0.02	−0.02	−0.02

Note. The table displays unstandardized coefficient estimates; fiml denote full information maximum likelihood estimates; rf is RF-based MI (`mice.impute.rf()`); rq is QR-based MI (`mice.impute.rq()`); pmm is predictive mean matching using distance-aided selection of donors (`mice.impute.midastouch()`); SB denotes social behaviour problem scores; CP is the corporal punishment index at respective panel wave (1–5); I denotes the latent intercept and S the latent slope

children with higher SB scores also tend to receive more severe CP at the beginning of the study. Finally, the intercepts and slopes of CP are negatively correlated, indicating that children who receive more corporal punishment at the beginning of the study tend to experience a steeper reduction in CP over time in comparison to children, who already receive little or no CP at the beginning of the study. Note that all imputation methods produce highly similar estimates and lead to the same substantive conclusions. One noticeable difference is the estimate of the mean slope of SB. In comparison to the methods pmm and RF-based MI (−0.64 and −0.66), QR-based MI produced a less steep slope (−0.58) and also a somewhat larger variance of the slopes between individuals (1.89 in comparison to 1.78 and 1.56). Since the underlying dataset is a real-world dataset, the true population parameters are unknown, and it might be a good idea to report a range of results based on several plausible imputation methods. This gives an impression of the stability or variability of model results given the respective assumptions and properties of the applied methods. In the given application, the results are very similar across the employed methods.

A Cross-Lagged Panel Model of Corporal Punishment and Children's Social Behaviour Problems

To further disentangle the relationship between CP and SB, we fit a cross-lagged panel model, which can be enumerated among the model class of (parallel) autoregressive models: Autoregressive models are popular models to analyse stability and change in one or more response variables over time (cf. Kessler & Greenberg, 1981). The most simple model predicts each repeated observation y_t by its predecessor y_{t-1}:

$$y_t = a_t + b_t y_{t-1} + \varepsilon_t, \tag{23.7}$$

where a is an intercept, b is the autoregressive coefficient which indicates the strength of the relationship and the stability of the response construct across time and ε is the respective error term. The usual assumptions of regression models apply. The model for the first measurement timepoint t_0 is

$$y_0 = a_0 + \varepsilon_0. \tag{23.8}$$

If more than one response variable is observed multiple times, the so-called cross-lagged paths are usually of interest, to examine if and to what extent earlier measurements of one construct predict later measurements of the other construct(s) and vice versa.

Figure 23.4 displays a cross-lagged panel model of SB and CP and gives a more detailed picture regarding how children's social behaviour problems and corporal punishment affect each other over time. The coefficient estimates are based on full information maximum likelihood estimation using a robust estimator. A comparison with the robust imputation methods can be found in Table 23.4. First of all, it can be seen that the stability of both SB and CP over time is large. Standardized coefficient estimates (based on FIML estimation using a robust estimator) range between 0.65 and 0.76 for SB and between 0.88 and 0.92 for CP. Furthermore, SB and CP are correlated. The cross-lagged coefficient estimates indicate a reciprocal

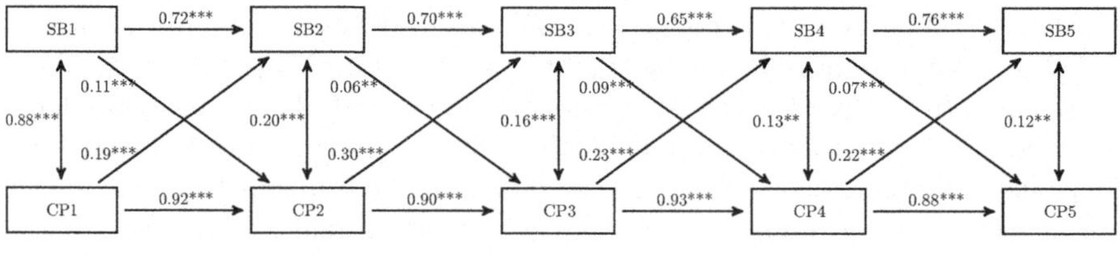

Fig. 23.4 Cross-lagged panel model of children's social behaviour problems and corporal punishment by their fathers over five measurement timepoints. *Note.* SB denotes social behaviour problem score and CP corporal punishment index

Table 23.4 Parameter estimates of cross-lagged panel model based on different missing data strategies

Parameter	fiml	rf	rq	pmm
	Stability			
SB5 ON SB4	0.72	0.63	0.70	0.66
SB4 ON SB3	0.70	0.61	0.70	0.63
SB3 ON SB2	0.65	0.59	0.63	0.60
SB2 ON SB1	0.76	0.72	0.74	0.73
CP5 ON CP4	0.92	0.91	0.92	0.72
CP4 ON CP3	0.90	0.88	0.89	0.89
CP3 ON CP2	0.93	0.91	0.92	0.91
CP2 ON CP1	0.88	0.87	0.88	0.87
	Cross-lagged paths			
SB2 ON CP1	0.19	0.22	0.20	0.21
SB3 ON CP2	0.30	0.35	0.31	0.34
SB4 ON CP3	0.23	0.30	0.22	0.28
SB5 ON CP4	0.22	0.30	0.24	0.27
CP2 ON SB1	0.11	0.12	0.11	0.11
CP3 ON SB2	0.06	0.07	0.06	0.07
CP4 ON SB3	0.09	0.10	0.10	0.09
CP5 ON SB4	0.07	0.08	0.07	0.07

Note. Table displays standardized coefficient estimates; fiml denote full information maximum likelihood coefficient estimates; rf is RF-based MI (`mice.impute.rf()`); rq is QR-based MI (`mice.impute.rq()`); and pmm is predictive mean matching using distance-aided selection of donors (`mice.impute.midastouch()`); SB denotes social behaviour problem scores, and CP is corporal punishment index at respective panel wave (1–5)

relationship, meaning that the intensity of CP seems to be a slightly stronger predictor for later behaviour problems than vice versa. Cross-lagged paths from CP to later SB range between 0.19 and 0.30, whereas cross-lagged paths from SB to later CP range only between 0.06 and 0.11. The results indicate that children's social behaviour problems are more a consequence and less so a cause of corporal punishment.

When comparing the three imputation methods, we see that QR-based MI yields the largest stability coefficient estimates overall. The corresponding coefficient estimates obtained after RF-based MI and pmm are somewhat smaller. Regarding the cross-lagged paths, coefficient estimates are highly similar, with QR-based MI producing somewhat lower point estimates in comparison to both RF-based MI and pmm. Overall, all methods lead to the same substantive conclusions that CP is a stronger predictor for later social behaviour problems than vice versa.

23.5 Discussion

The main purpose of this paper was to evaluate different techniques for MI regarding their ability to produce unbiased point estimates and standard errors when the empirical data violate distributional assumptions of standard parametric models. We built upon and extended the scenarios in Kleinke et al. (2021) and focused on sample size requirements for MI methods that are robust in some regard: predictive mean matching, RF-based MI and QR-based MI.

Predictive Mean Matching

First of all, pmm—the default MI method in `mice` for continuous variables—imputes actually observed values, thereby preserving many properties of the (remaining) empirical data. This is an advantage as it leads to some robustness but unfortunately also a drawback when the sample size is small: With smaller sample size, it becomes less likely that suitable observed donor cases are available in the dataset for the respective incomplete cases. Recall that the method can neither be used to make extrapolations beyond the range of the (remaining) observed data nor can it make intrapolations in regions where data are sparse. In our simulation, for example, we were interested in predictors for corporal punishment of children by their fathers. The QR model that we fitted allowed us to get different coefficient estimates for low, medium and high intensities of CP. If certain regions, for example, high intensities of CP are selectively thinned out, then pmm cannot produce imputations in these regions with the desired statistical properties. While estimates regarding the centre part of CP might be fine, results regarding the upper quantiles might be biased in this case. Generally speaking, if the remaining observed cases are no longer representative of the population one wants to make inferences about, results will be biased to some extent. We have seen in our simulation that precision in point estimates produced by pmm generally increased with an increasing sample size and that results were not too heavily biased when sample size exceeded $N = 500$, at least with respect to the estimates regarding the centre part of CP. Note that the empirical distribution of CP was skewed to the right and that many fathers reported no or very low intensities of CP. Of scientific interest, however, were also those fathers who report higher intensities of CP and pmm did not appear to be well suited for that task.

Limitations In the present simulation, we have only examined the pmm variant based on distance-aided selection of donors (`midastouch`), which has proven to yield acceptable statistical inferences in previous research regarding the applicability of pmm, when sample size was small (Kleinke, 2018). This pmm variant, however, is based on a 'normal' linear regression model to match the donors to the donees. Future research should establish if more robust (e.g., nonparametric) pmm variants might be better suited when it comes to violations of parametric modeling assumptions such as normally distributed errors with a spherical covariance matrix. Our guess is that these variants would need a sufficiently large sample to produce acceptable statistical inferences, as well.

Random Forest-Based Multiple Imputation

Secondly, RF-based MI also requires a substantial sample size to be able to produce predictions with low bias and CR close to the nominal value. Simplistically speaking, RF consists of multiple steps—the relevant ones for this discussion being (a) the bootstrap resampling and (b) the training of the decision trees based on the bootstrap samples. First, it needs to be stated that bootstrapping is typically not a cure for small samples. If the sample size is small, the same cases tend to get selected again and again. If there are only $N = 50$ cases in the data file, all the bootstrap samples will consist of a maximum of $N = 50$ different cases. So, when sample size is small, we have a basis for the training part of the RF algorithm that does not have too much variability. Additionally, the decision trees are trained by splitting the data based on the defined explanatory variables (i.e., the imputation model) to identify subsamples that have the best discriminatory ability. Depending on the number of splits, these subsamples become quite small, and in consequence, estimates of the conditional mean are based on very few cases. In our simulation, RF-based MI produced the largest biases overall, with biases generally getting smaller with increasing sample size. As the RF-based MI method that we used for the present simulation focused on the conditional mean, estimates regarding the off-centre parts of CP were still biased, even when sample size was $N = 600$.

Limitations and future research A very fruitful avenue for future research would be to extend RF-based MI to more robust quantile forest regression models (Meinshausen, 2006). First ideas already exist in this regard (Yadav & Ravi, 2018). This however could solve only the biased inferences in the off-centre regions of the dependent variable, not the sample size requirements. RF is a method that requires a sufficiently large sample size and is therefore not well suited to impute missing values when sample size is small.

Quantile Regression-Based Multiple Imputation

Across all scenarios, QR-based MI produced the least biased point estimates. Again, performance generally increased with an increasing sample size. Our results here are in line with previous research regarding sample size requirements for QR models: Tarr (2012), for example, has examined the performance of different methods to directly estimate upper and lower bounds of confidence intervals by QR (direct estimation, rank test inversion and resampling), when sample size was very small.

He examined scenarios, were sample size ranged between $N = 50$ and $N = 200$ and how certain model traits such as heavy-tailed errors or skewed covariates affected confidence intervals. He found that (a) all three approaches did not perform well when estimating the intercept for high and low quantiles in the presence of heavy-tailed errors; (b) undercoverage was serious regarding high and low quantiles, when there was strong heteroskedasticity; and (c) generally extreme caution is in order, when the independence of errors is violated.

Based on these, as well as on our findings, we recommend an absolute minimum of 150–200 cases to apply QR-based MI for descriptive and exploratory analyses, and of 500 and more for statistical inferences, where also the off-centre quantiles are of interest. Future research should extend QR-based imputation to linear quantile mixed models to be able to explicitly model the non-independence of errors due to clustering.

References

Barnard, J., & Rubin, D. B. (1999). Small-sample degrees of freedom with multiple imputation. *Biometrika, 86*(4), 948–955. https://doi.org/10.1093/biomet/86.4.948

Bodner, T. E. (2008). What improves with increased missing data imputations? *Structural Equation Modeling, 15*(4), 651–675. https://doi.org/https://doi.org/10.1080/10705510802339072

Bollen, K. A., & Curran, P. J. (2006). *Latent curve models: A structural equation perspective*. Wiley.

Bottai, M., & Zhen, H. (2013). Multiple imputation based on conditional quantile estimation. *Epidemiology, Biostatistics and Public Health, 10*(1), Article e8758. https://doi.org/https://doi.org/10.2427/8758

Breiman, L. (2001). Random forests. *Machine Learning, 45*(1), 5–32. https://doi.org/10.1023/A:1010933404324

Collins, L. M., Schafer, J. L., & Kam, C. M. (2001). A comparison of inclusive and restrictive missing-data strategies in modern missing-data procedures. *Psychological Methods, 6*(4), 330–351. https://doi.org/10.1037/1082-989X.6.4.330

Doove, L. L., Van Buuren, S., & Dusseldorp, E. (2014). Recursive partitioning for missing data imputation in the presence of interaction effects. *Computational Statistics & Data Analysis, 72*, 92–104. https://doi.org/https://doi.org/10.1016/j.csda.2013.10.025

Enders, C. K. (2022). *Applied missing data analysis* (2nd ed.). Guilford.

Forero, C. G., & Maydeu-Olivares, A. (2009). Estimation of IRT graded response models: Limited versus full information methods. *Psychological Methods, 14*(3), 275–299. https://doi.org/10.1037/a0015825

Fritsch, M., Haupt, H., Lösel, F., & Stemmler, M. (2019). Regression trees and random forests as alternatives to classical regression modeling: Investigating the risk factors for corporal punishment. *Psychological Test and Assessment Modeling, 61*(4), 389–417. https://www.psychologie-aktuell.com/fileadmin/Redaktion/Journale/ptam-2019-4/02_Fritsch.pdf

Gaffert, P., Meinfelder, F., & Bosch, V. (2016). *Midastouch: Towards an MI-proper predictive mean matching*. https://citeseerx.ist.psu.edu/document?repid=rep1&type=pdf&doi=5f13dd8d367c9b0ebc127c24487aa06773a2f00f

Geißler, R. (Ed.). (1994). *Soziale Schichtung und Lebenschance in Deutschland* [Social stratification and life chances in Germany]. Enke.

Geraci, M. (2016a). Estimation of regression quantiles in complex surveys with data missing at random: An application to birthweight determinants. *Statistical Methods in Medical Research, 25*(4), 1393–1421. https://doi.org/10.1177/0962280213484401

Geraci, M. (2016b). Qtools: A collection of models and tools for quantile inference. *The R Journal*, *8*(2), 117–138. https://doi.org/10.32614/RJ-2016-037

Graham, J. W. (2009). Missing data analysis: Making it work in the real world. *Annual Review of Psychology*, *60*, 549–576. https://doi.org/10.1146/annurev.psych.58.110405.085530

Hastie, T., Tibshirani, R., & Friedman, J. (2009). *The elements of statistical learning* (2nd ed.). Springer. https://doi.org/10.1007/978-0-387-84858-7

Haupt, H., Lösel, F., & Stemmler, M. (2014). Quantile regression analysis and other alternatives to ordinary least squares regression: A methodological comparison on corporal punishment. *Methodology*, *10*(3), 81–91. https://doi.org/10.1027/1614-2241/a000077

Kessler, R. C., & Greenberg, D. F. (1981). *Linear panel analysis: Models of quantitative change*. Academic Press.

Kleinke, K. (2017). Multiple imputation under violated distributional assumptions – a systematic evaluation of the assumed robustness of predictive mean matching. *Journal of Educational and Behavioral Statistics*, *42*(4), 371–404. https://doi.org/10.3102/1076998616687084

Kleinke, K. (2018). Multiple imputation by predictive mean matching when sample size is small. *Methodology*, *14*(1), 3–15. https://doi.org/10.1027/1614-2241/a000141

Kleinke, K. (2021). Estimation of partially observed non-linear terms in a multilevel model: An evaluation of the robustness of ad hoc and state-of-the-art missing data methods. *Psychological Test and Assessment Modeling*, *63*(3), 432–455. https://www.psychologie-aktuell.com/fileadmin/Redaktion/Journale/ptam-2021-3/PTAM__3-2021_6_kor.pdf

Kleinke, K., Fritsch, M., Stemmler, M., Reinecke, J., & Lösel, F. (2021). Quantile regression-based multiple imputation of missing values – an evaluation and application to corporal punishment data. *Methodology*, *17*(3), 205–230. https://doi.org/10.5964/meth.2317

Kleinke, K., Reinecke, J., Salfrán, D., & Spiess, M. (2020). *Applied multiple imputation. Advantages, pitfalls, new developments and applications in R*. Springer Nature.

Kleinke, K., Stemmler, M., Reinecke, J., & Lösel, F. (2011). Efficient ways to impute incomplete panel data. *Advances in Statistical Analysis*, *95*(4), 351–373. https://doi.org/10.1007/s10182-011-0179-9

Koenker, R. (2005). *Quantile regression*. Cambridge University Press.

Koenker, R., & Hallock, K. F. (2001). Quantile regression. *Journal of Economic Perspectives*, *15*(4), 143–156. https://doi.org/10.1257/jep.15.4.143

Little, R. J. A. (1988). Missing-data adjustments in large surveys. *Journal of Business & Economic Statistics*, *6*(3), 287–296. https://doi.org/10.1080/07350015.1988.10509663

Little, R. J. A., & Rubin, D. B. (2020). *Statistical analysis with missing data*. Wiley.

Lösel, F., Beelmann, A., Jaursch, S., Scherer, S., Stemmler, M., & Wallner, S. (2003). *Skalen zur Messung elterlichen Erziehungsverhaltens bei Vorschul- und Grundschulkindern. Die deutsche Version des Alabama Parenting Questionnaires (APQ)* (unpublished manuscript).

Lösel, F., Beelmann, A., & Stemmler, M. (2002). *Skalen zur Messung sozialen Problemverhaltens bei Vorschul- und Grundschulkindern. Die deutschen Versionen des Eyberg Child Behavior Inventory (ECBI) und des Social Behavior Questionnaire (SBQ)* (unpublished manuscript).

Lösel, F., Stemmler, M., Jaursch, S., & Beelmann, A. (2009). Universal prevention of antisocial development: Short-and long-term effects of a child-and parent-oriented program. *Monatsschrift für Kriminologie und Strafrechtsreform*, *92*(2–3), 289–307. https://doi.org/10.1515/mks-2009-922-314

Meinshausen, N. (2006). Quantile regression forests. *Journal of Machine Learning Research*, *7*, 983–999. https://jmlr.org/papers/v7/meinshausen06a.html

Raghunathan, T., Berglund, P. A., & Solenberger, P. W. (2018). *Multiple imputation in practice: With examples using IVEware*. CRC Press.

Rubin, D. B. (1976). Inference and missing data. *Biometrika*, *63*(3), 581–592. https://doi.org/10.1093/biomet/63.3.581

Rubin, D. B. (1987). *Multiple imputation for nonresponse in surveys*. Wiley.

Rubin, D. B. (1996). Multiple imputation after 18+ years. *Journal of the American Statistical Association*, *91*(434), 473–489. https://doi.org/10.1080/01621459.1996.10476908

Rubin, D. B. (1986). Statistical matching using file concatenation with adjusted weights and multiple imputations. *Journal of Business & Economic Statistics*, *4*(1), 87–94. https://doi.org/10.1080/07350015.1986.10509497

Schafer, J. L., & Graham, J. W. (2002). Missing data: Our view of the state of the art. *Psychological Methods*, *7*(2), 147–177. https://doi.org/10.1037/1082-989X.7.2.147

Schenker, N., & Taylor, J. M. (1996). Partially parametric techniques for multiple imputation. *Computational Statistics & Data Analysis*, *22*(4), 425–446. https://doi.org/10.1016/0167-9473(95)00057-7

Siddique, J., & Belin, T. R. (2008). Multiple imputation using an iterative hot-deck with distance-based donor selection. *Statistics in Medicine*, *27*(1), 83–102. https://doi.org/10.1002/sim.3001

Siddique, J., & Harel, O. (2009). MIDAS: A SAS macro for multiple imputation using distance-aided selection of donors. *Journal of Statistical Software*, *29*(9), 1–18. https://doi.org/10.18637/jss.v029.i09

Spiess, M., Kleinke, K., & Reinecke, J. (2021). Proper multiple imputation of clustered or panel data. In P. Lynn (Ed.), *Advances in longitudinal survey methodology* (pp. 424–446). Wiley.

Tarr, G. (2012). Small sample performance of quantile regression confidence intervals. *Journal of Statistical Computation and Simulation*, *82*(1), 81–94. https://doi.org/10.1080/00949655.2010.527844

Therneau, T., & Atkinson, B. (2022). *Rpart: Recursive partitioning and regression trees* [R package version 4.1.16]. https://CRAN.R-project.org/package=rpart

Tremblay, R. E., Desmarais-Gervais, L., Gagnon, C., & Charlebois, P. (1987). The Preschool Behavior Questionnaire. Stability of its factor structure between cultures, sexes, ages and socioeconomic classes. *International Journal of Behavioral Development*, *10*, 467–484. https://doi.org/10.1177/016502548701000

Tremblay, R. E., Loeber, R., Gagnon, C., Charlebois, R., Larive, S., & LeBlanc, M. (1991). Disruptive boys with stable and unstable high fighting behavior patterns during junior elementary school. *Journal of Abnormal Child Psychology*, *19*(3), 285–300. https://doi.org/10.1007/BF00911232

Tremblay, R. E., Vitaro, F., Gagnon, C., Piché, C., & Royer, N. (1992). A prosocial scale for the preschool behavior questionnaire: concurrent and predictive correlates. *International Journal of Behavior Development*, *15*(2), 227–245. https://doi.org/10.1177/016502549201500204

Van Buuren, S. (2018). *Flexible imputation of missing data* (2nd ed.). CRC Press. https://stefvanbuuren.name/fimd/

Van Buuren, S., Brand, J. P. L., Groothuis-Oudshoorn, C. G. M., & Rubin, D. B. (2006). Fully conditional specification in multivariate imputation. *Journal of Statistical Computation and Simulation*, *76*(12), 1049–1064. https://doi.org/10.1080/10629360600810434

Van Buuren, S., & Groothuis-Oudshoorn, K. (2011). MICE: Multivariate imputation by chained equations in R. *Journal of Statistical Software*, *45*(3), 1–67. https://doi.org/10.18637/jss.v045.i03

Wilcox, R. R. (2017). *Introduction to robust estimation and hypothesis testing* (4th ed.). Academic Press.

Yadav, M., & Ravi, V. (2018). Quantile regression random forest hybrids based data imputation. In Y. Wang, S. Kwong, J. Feldman, N. Howard, P. Sheu, & B. Widrow (Eds.), *Proceedings of the 17th International Conference on Cognitive Informatics & Cognitive Computing* (pp. 195–201). IEEE. https://ieeexplore.ieee.org/abstract/document/8482040

Chapter 24
Multiple Imputation of Incomplete Panel Data Based on a Piecewise Growth Curve Model: An Evaluation and Application to Juvenile Delinquency Data

Kristian Kleinke and Jost Reinecke

24.1 Introduction and Overview

Missing data are a nuisance. Complex multilevel data—and panel data can be seen as a special type of multilevel data where repeated measurements are clustered in persons—can be affected by various 'sources' of nonresponse including missingness in the target variables and in predictors at level 1, level 2, or higher hierarchical levels or even missingness in the cluster identifiers. Depending on the type and pattern of missing data, unobserved information poses a threat to the validity of statistical inferences (Schafer & Graham, 2002; Van Buuren, 2011). Modern imputation approaches therefore try to recreate the sample in a way as if ideally no information had gone missing. To this end, the respective model that is used to create the imputations usually needs to reflect the assumed (and typically unknown) true data-generating process (DGP) and, if necessary, the mechanism that created the missing data patterns well. If the imputation model is somehow misspecified, bias is to be expected (e.g. Kleinke, 2017, 2021; Kleinke & Reinecke, 2013; Yu et al., 2007). This includes that the imputation model typically has to reflect the clustered structure of the data to obtain unbiased statistical inferences (see, e.g. the discussions in Drechsler, 2015, and the references therein). The amount of bias inter alia depends on the type and 'strength' of the missing data processes, the degree of model misspecifications and the number and patterns of missing data.

K. Kleinke (✉)
Department of Psychology, University of Siegen, Siegen, Germany
e-mail: kristian.kleinke@uni-siegen.de

J. Reinecke
Faculty of Sociology, University of Bielefeld, Bielefeld, Germany
e-mail: jost.reinecke@uni-bielefeld.de

M. Stemmler et al. (eds.), *Dependent Data in Social Sciences Research*,
https://doi.org/10.1007/978-3-031-56318-8_24

Furthermore, if the development of the target variable over time is non-linear, suitable model specifications that cater for the non-linear relationships in the data should be used in both the imputation and the analysis process. Simple curvilinear functions such as u-shaped or inverted u-shaped relationships can be modeled by a quadratic growth curve model (see the next section). More complex nonlinear trajectories can be approximated by piecewise growth curve models (which are also known as linear piecewise trajectory models). Here, the nonlinear time function is split into multiple parts and approximated by two or more linear piecewise splines (e.g. Bollen & Curran, 2006).

We propose to use piecewise growth curve models for model-based missing data imputation since they are (a) relatively easy to specify either as a structural equation model or as a multilevel regression model and (b) application is straightforward with already existing MI software such as the popular `mice` package in R (Van Buuren & Groothuis-Oudshoorn, 2011; Van Buuren, 2018).

In the following, we (a) introduce piecewise linear trajectory models as a general-purpose imputation model for incomplete non-linear panel data that can, for example, be specified using any of the two-level imputation functions from package `mice` or its add-on packages (e.g. `countimp`, Kleinke & Reinecke, 2019), (b) demonstrate in a tutorial-like section how imputations can be obtained using `mice`, (c) evaluate the proposed approach by means of a Monte Carlo simulation and (d) apply the method to an empirical 12-wave panel data set focusing on the development of juvenile delinquency—contrasting the results with related imputation approaches that might also be deemed useful for the situation at hand. We end with a general discussion of our findings.

24.2 Theoretical Background

In this section, we give a brief introduction to (latent) growth curve modeling, introduce the general model class and then focus on piecewise growth curve models. Next, we give a brief general introduction to multiple imputation based on Rubin's (1987) theory and the chained equations multiple imputation framework (Van Buuren, 2018). Finally, we demonstrate how multiple imputation based on a piecewise growth curve model works.

Growth Modeling in a Nutshell

Growth curve models (e.g. Bollen & Curran, 2006) are highly useful to model both linear and non-linear changes in a target variable of interest across time. In addition to average effects—i.e. the average starting level of the target variable and average levels of linear and non-linear change—these models also allow to explain how and why individuals vary in this regard.

The general measurement model equation for the growth curve model is

$$y_{it} = \Lambda \eta_i + e_{it} \tag{24.1}$$

y is a $T \times 1$ vector of the repeatedly measured manifest indicators for observational units i, $i = 1, ..., n$ at timepoint t, $t = 1, ..., T$. η is an unknown $Q \times 1$ vector of latent growth factors, and e is the $T \times 1$ vector of time-specific measurement errors. Λ is a user-defined $T \times Q$ matrix of factor loadings.

For example, the model equation of an unconditional growth curve model over 12 measurement time points that describe a curvilinear development of the target construct y across time would be

$$\begin{pmatrix} y_{i1} \\ y_{i2} \\ y_{i3} \\ y_{i4} \\ \vdots \\ y_{i12} \end{pmatrix} = \begin{pmatrix} 1 & 0 & 0 \\ 1 & 1 & 1 \\ 1 & 2 & 4 \\ 1 & 3 & 9 \\ \vdots & \vdots & \vdots \\ 1 & 11 & 121 \end{pmatrix} \cdot \begin{pmatrix} \eta_{1i} \\ \eta_{2i} \\ \eta_{3i} \end{pmatrix} + \begin{pmatrix} e_{i1} \\ e_{i2} \\ e_{i3} \\ e_{i4} \\ \vdots \\ e_{i12} \end{pmatrix}$$

Factor loadings are constrained in a way to represent the constant (η_1), linear (η_2) and quadratic (η_3) development.

Latent growth factors η are furthermore described as a function of their means α, their relationship to exogenous variables ξ (if applicable) and residuals ζ:

$$\eta_i = \alpha + \Gamma \xi_i + \zeta_i \tag{24.2}$$

Γ is a $Q \times R$ matrix of regression coefficients describing the effects of R exogenous variables ξ on the Q growth factors. The variances and covariances of ζ are labelled ψ_{kl}, where subscripts k and l denote the respective growth factor, with $k, l = 1, ..., Q$.

The estimated means α of latent variables η represent the average starting level and average levels of change of the target variable. The variances of the growth factors represent the degree of individual variation around the respective averages. Covariances of the growth factors describe their interrelationships, for example, if individuals with lower starting levels tend to have steeper linear and quadratic slopes and vice versa.

A graphical representation of this model is given in Fig. 24.1. If no constraints are set, this model estimates three means of growth factors η_1–η_3, three variances as well as three covariances of growth factors η_1–η_3 and 12 measurement error variances, one for each of the 12 measurement timepoints.

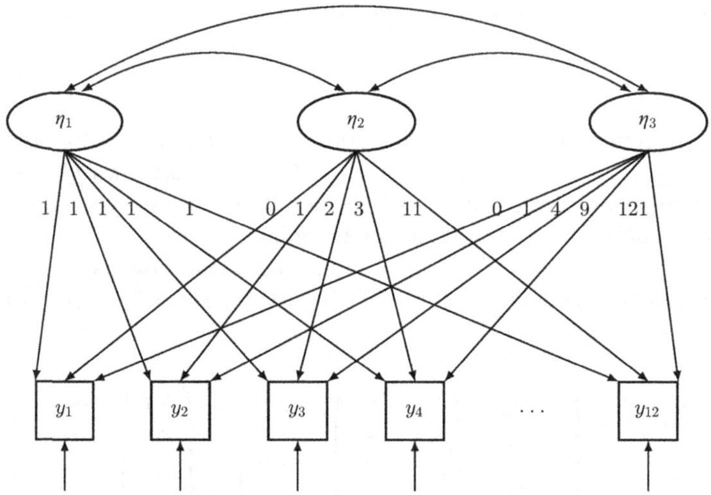

Fig. 24.1 Unconditional latent growth curve model over 12 measurement timepoints that describes a curvilinear development of the target construct across time. *Note.* y_1–y_{12} denote the repeated measurements of the target variable, η_1–η_3 the latent growth factors representing constant (η_1), linear (η_2), and quadratic (η_3) change across time; double-headed arrows represent covariances of the latent growth factors

Piecewise Growth Models

More complex trajectories across time can be described by piecewise growth models, also known as (latent) piecewise linear trajectory models (Bollen & Curran, 2006, Chap. 4.3). The general idea here is to split a complex non-linear trajectory into multiple parts that can be described by simple linear functions. To this end, one first needs to identify one or more transition points—either theory driven or data driven, for example, by inspecting the mean development across time—and to fit linear trajectories from the current transition point to the next one accordingly.

The model equations are the same as in (24.1) and in (24.2). However, η now have a different meaning and denote the starting level and the linear change in the user-defined linear pieces. Again, we can obtain means, variances and covariances of the growth factors and include predictors why individuals vary in these factors.

To get an idea how exactly the loadings of the linear pieces need to be constrained, let us assume we want to analyse data from 12 panel waves showing a non-linear average trajectory, where we first observe a steep increase in the mean of the target variable y, followed by a steep decrease, and finally a slighter decrease later on. The average trajectory for this example is displayed in Fig. 24.2. The first piece comprises panel waves 1 and 2, where we observe an increase in the mean of the target construct. The second piece comprises waves 3–8. Here, the mean of the

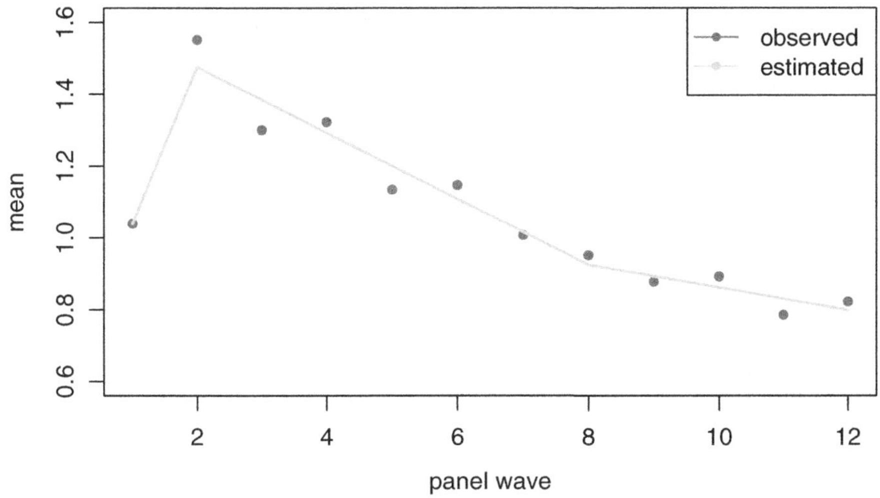

Fig. 24.2 Example: A non-linear average trajectory that can be described by three linear pieces

target variable decreases. Finally, in the last part (waves 9–12), we still see a slight decrease over time, but less steep than before.

To model these three linear parts, loading matrix Λ needs to have four columns, one for the starting level and three for the slopes in the three linear pieces:

$$\Lambda = \begin{pmatrix} 1 & 0 & 0 & 0 \\ 1 & \mathbf{1} & 0 & 0 \\ 1 & 1 & 1 & 0 \\ 1 & 1 & 2 & 0 \\ 1 & 1 & 3 & 0 \\ 1 & 1 & 4 & 0 \\ 1 & 1 & 5 & 0 \\ 1 & 1 & \mathbf{6} & 0 \\ 1 & 1 & 6 & 1 \\ 1 & 1 & 6 & 2 \\ 1 & 1 & 6 & 3 \\ 1 & 1 & 6 & 4 \end{pmatrix}$$

The first column of 1s represents the starting level, the second column defines the linear trajectory up to the first transition point (here: wave 2), the third column models the linear development in the next piece (transition point: wave 8) and the final column models the development from the last transition point onwards. After each transition point, the loadings in the columns describing the pieces before remain constant, and the loadings for the respective next piece 'take over' to describe the linear development in the respective section. Up to their respective transition

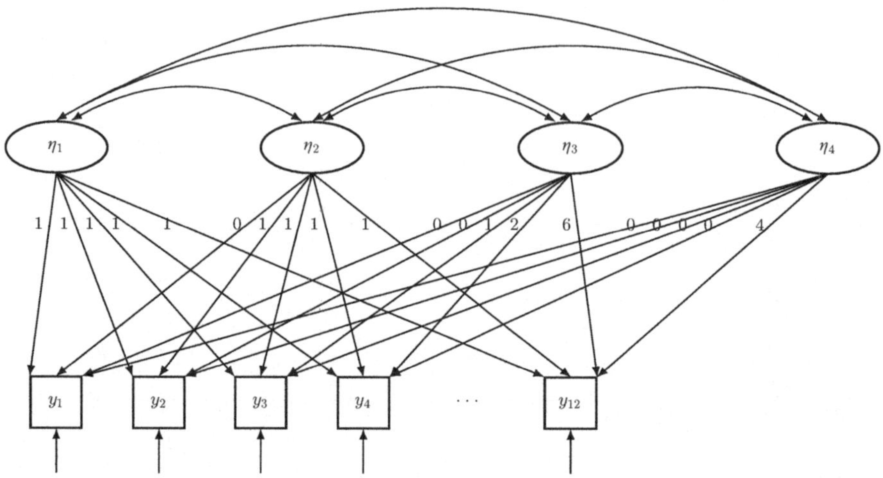

Fig. 24.3 Unconditional piecewise growth curve model with three linear pieces and transition points at panel waves 2 and 8. *Note.* y_1–y_{12} are the repeated observations of the target variable. η_1 represents the starting level of the target variable, and η_2, η_3 and η_4 are the slopes representing the linear development before and after the transition points, respectively. Factor loadings λ are constrained to represent the assumed piecewise linear development. Double-headed arrows represent covariances of the latent growth factors

points, the later pieces are coded zero, similar to the concept of dummy coding in classical regression analysis. A graphical representation of this model is given in Fig. 24.3.

Missing Data and Multiple Imputation

Unfortunately, most empirical data sets are affected by missing values. A missing value is defined as a value that has not been observed for *whatever* reason. This includes both unplanned and planned missingness, where some items were never intended to be observed for some units to reduce the burden on the surveyed units. One of the standard approaches to handle item-level missingness that avoids exclusion of cases and consequently loss of sample size and statistical power is to create multiple replacements of each missing value, i.e. multiple imputation (MI). Bayesian MI based on Rubin's (1987) theory is a three-step process: First, each missing value is filled-in $M > 1$ times by *plausible* values, so that researchers end up with M complete data sets. Plausible in this context refers to any value that leads to unbiased statistical inferences (for a discussion of criteria that are required to create *proper* imputations, see Rubin, 1996). Typically, the imputation process is model-based, i.e. imputations are generated from $P(y_{\text{mis}}|X)$—the conditional distribution of the incomplete data y_{mis} given predictors X under some statistical model that is deemed appropriate for the situation at hand. In the second step, the M completed data sets are analysed separately by any complete data technique that yields point estimates and standard errors for the parameters of interest. In the final step, the M sets of modeling results are combined into an overall result based on combination rules such as the large-sample formula given in Rubin (1987) or adaptations for small samples (Barnard & Rubin, 1999). For a detailed discussion, see Kleinke et al. (2020, Chap. 4). The combination rules yield pooled point estimates and standard errors for each model parameter. The combined point estimate is simply the average of the M estimates. Corresponding standard errors are obtained by combining a within-imputation and a between-imputation variance component. The latter component reflects the extra amount of uncertainty regarding the respective parameter due to missing information.

Assumptions

Multiple imputations are often generated under the assumption that the *mechanism* that leads to values being unobserved is *ignorable* and that missingness only depends on observed information in the data file and not additionally on unobserved information—i.e. missing data are 'missing at random' (MAR) in the sense of Rubin (1976); see also Kleinke et al. (2020, Chap. 2). In our Monte Carlo simulation, we also evaluate how the proposed method can handle non-ignorable missingness (NMAR), i.e. missingness that is not missing at random in Rubin's (1976) sense.

24.3 Multiple Imputation by a Piecewise Growth Model in `mice`

In model-based Bayesian imputation, replacements of each missing value are generated by drawing from the posterior predictive distribution of the variables whose values are missing given some predictors X. So-called joint modeling approaches specify one multivariate model for all variables in the data file and approximate the posterior predictive distribution by iterative Markov chain Monte Carlo (MCMC) algorithms, whereas conditional modeling approaches specify separate conditional models for each incomplete variable in the data set. The `mice` software in R—i.e. multiple imputation by chained equations (Van Buuren & Groothuis-Oudshoorn, 2011; Van Buuren, 2018)—can be enumerated among the latter approaches. The advantage here is that individual models can be specifically tailored to the respective distribution and type of each variable at hand. The software then cycles iteratively through all variables with missing values and imputes them step by step using predictions from the pre-specified regression models (see Van Buuren, 2018, Chap. 4.5 for details).

Imputations are generated by calling the main `mice` function. Its most important arguments are `method`, `predictorMatrix`, `m`, `maxit` and `seed`: Argument `method` is used to specify the type of the respective imputation method (e.g. "`logreg`" for a Bayesian logistic regression model). `predictorMatrix` specifies the corresponding imputation model—the set of complete or previously completed predictor variables that is used in the respective prediction model for each incomplete target variable. `m` sets the number of imputations to be generated and `maxit` the iterations of the MCMC algorithm. For reproducible results, it is advisable to set a `seed` of the random number generator. For comprehensive information about the `mice` framework, see Van Buuren (2018). For a brief overview on the most important aspects, see VanBuuren and Groothuis-Oudshoorn (2011).

Imputation by a Piecewise Growth Curve Model

In this tutorial-like section, we want to demonstrate how the `mice` package in R can be used to impute incomplete panel data based on a piecewise growth curve model and how to automate the repeated data analysis process and pooling of results. We assume that readers are at least somewhat familiar with R packages `lavaan` (Rosseel, 2012) and `lme4` (Bates et al., 2015).

First, we simulate some 'toy data' using the `simulateData`-function from R package `lavaan` for latent variable analysis. The required input is a `lavaan` model with user-defined population values and the desired sample size. We simulate data for the model in Fig. 24.3 ($n = 500$) setting the means of the growth factors to $\alpha_1 = 1$, $\alpha_2 = 0.5$, $\alpha_3 = -0.1$, $\alpha_4 = -0.02$, the variance of η_1 to $\psi_{11} = .1$ and all other variances and covariances of the growth factors to 0. Loadings in Λ are constrained according to the assumed transition points at panel waves 2 and 8.

```
population.model =
"

# Growth factor loadings
# eta 1
i =~ 1*y1 +1*y2 + 1*y3 + 1*y4 + 1*y5 +
     1*y6 + 1*y7 + 1*y8 + 1*y9 + 1*y10 +
     1*y11 + 1*y12

# eta 2
s1 =~ 0*y1 + 1*y2 + 1*y3 + 1*y4 + 1*y5 +
      1*y6 + 1*y7 + 1*y8 + 1*y9 + 1*y10 +
      1*y11 + 1*y12

# eta 3
s2 =~ 0*y1 + 0*y2 + 1*y3 + 2*y4 + 3*y5 +
      4*y6 + 5*y7 + 6*y8 + 6*y9 + 6*y10 +
```

```
      6*y11 + 6*y12

# eta 4
s3 =~ 0*y1 + 0*y2 + 0*y3 + 0*y4 + 0*y5 +
      0*y6 + 0*y7 + 0*y8 + 1*y9 + 2*y10 +
      3*y11 + 4*y12

# Population Parameters
i  ~ 1*1    # mean of eta 1
s1 ~ .5*1   # mean of eta 2
s2 ~ -.1*1  # mean of eta 3
s3 ~ -.02*1 # mean of eta 4

# Only intercept (eta 1) varies across persons
# variances and covariances of the (other)
# growth factors are fixed to zero in this illustration

i ~~ .1*i # variance of eta 1
i ~~ 0*s1
i ~~ 0*s2
i ~~ 0*s3
s1 ~~ 0*s1
s1 ~~ 0*s2
s1 ~~ 0*s3
s2 ~~ 0*s2
s2 ~~ 0*s3
s3 ~~ 0*s3
"
set.seed(1234)
library(lavaan)
n = 500
dat = simulateData(population.model, sample.nobs = n)
save(dat, file = "piecewise.rda")
```

Next, we introduce some random panel attrition. In this illustration, about 3% of the sample drop out at each panel wave:

```
p = .03 # drop out rate per panel wave
set.seed(12334)
mdat = dat
for (j in 2:ncol(dat)){
sel = sample(1:n, round(p*n), replace=FALSE)
mdat[sel,j:ncol(dat)] = NA
}
```

Note that the currently available multilevel imputation functions in `mice` or its add-on packages require a view of the latent growth curve model from a multilevel regression modeling perspective (e.g. Raudenbush & Bryk, 2002), which require data in long format. That means that the repeated measurements of our target variable y_1–y_{12} have to be stacked upon another to yield a single data vector y. So, in the first step, we need to bring our incomplete data file into long format, for example, using R's `reshape` function:

```
dat.long = reshape(
  data = mdat,
  varying = paste0("y", 1:12),
  direction = "long",
  v.names = "y",
  times = 0:11,
  timevar = "time",
  idvar = "id")
```

Next, we need to create time variables according to our assumed transition points (panel waves 2 and 8). Since the piecewise linear trajectory model is specified as a multilevel regression model, we do not fix factor loadings λ (like in Fig. 24.3), but create three separate time variables (also in long format) for the three linear pieces and attach them to our data frame in long format:

```
dat.long$time1 = rep(c(0,1,1,1,1,1,1,1,1,1,1,1),
                     each = n)
dat.long$time2 = rep(c(0,0,1,2,3,4,5,6,6,6,6,6),
                     each = n)
dat.long$time3 = rep(c(0,0,0,0,0,0,0,0,1,2,3,4),
                     each = n)
```

`time1` represents the linear development in the first piece, `time2` the linear change in the second one and `time3` the linear development in the third piece. Note that the coding is the same as in Fig. 24.3.

Now we are ready to choose an appropriate two-level imputation function from package `mice` or its add-on packages. For this illustration, we use `mice.impute.2l.pan()`, which is the chained equation variant of the method proposed by Schafer and Yucel (2002) to generate imputations based on a linear mixed effects model assuming normal homoscedastic errors.

```
library(mice)

# set up method and predictorMatrix
ini = mice(dat.long, maxit = 0)
meth = ini$method
pred = ini$predictorMatrix
meth["y"] = "2l.pan"
pred["y", ] = c(0, 0, -2, 1, 1, 1)
```

```
imp_2l_pan = mice(
  data = dat.long,
  m = 100,
  method = meth,
  predictorMatrix = pred,
  print = FALSE,
  seed = 1234)
save(imp_2l_pan, file = "imp_2l_pan.imp")
```

The first line in the code above loads the `mice` package—assuming that it has been already installed. Next, we let `mice` set up `method` and `predictorMatrix` using its defaults settings, so that we only need to change some aspects and do not have to build everything from scratch.

We set the imputation method for the incomplete variable y in long format to "`2l.pan`" and specify our two-level imputation model via the `predictorMatrix`. The coding scheme is as follows: A "0" entry in the `predictorMatrix` means that the corresponding variable is *not* included in this imputation model; a "1" would define a fixed effect. Fixed effects in multilevel regression are variables for which only a mean α (see Eqs. (24.1) and (24.2)) is estimated, but no variances or covariances. "2" defines a fixed and random effect, i.e. in addition to the average value, also a variance ψ, and if applicable a covariance is estimated. "-2" defines the class variable. Note that in two-level models, only one class variable is allowed.

In the code above, we define the three time variables as fixed effects. The intercept by default is random. This is in accordance with our data-generating model.

In the final line, we save the output with the multiple imputations of each missing value to disk for further use.

Now, we have several options for repeated data analysis and pooling of results: We could, for example, bring the completed data sets back into wide format and use package `semTools` (Jorgensen et al., 2022) to repeatedly fit the latent piecewise trajectory model with `lavaan` and apply Rubin's formula. An example how to apply functions from package `semTools` to data imputed in `mice` is given in Kleinke et al. (2020, Chap. 5). However, since the imputed data are already in long format, we can also use `lmer()` from package `lme4` to fit the piecewise trajectory model as a multilevel regression model:

```
library(mitml)
implist = mids2mitml.list(imp_2l_pan)
library(lme4)
fit = with(implist, lmer(y ~ 1 + time1 + time2 + time3 + (1|id)))
testEstimates(fit, extra.pars = TRUE)
```

```
Final parameter estimates and inferences obtained from 100 imputed data sets.

             Estimate Std.Error   t.value        df   P(>|t|)       RIV       FMI
(Intercept)     0.947     0.046    20.463 3.055e+50     0.000     0.000     0.000
time1           0.500     0.053     9.381 1.559e+05     0.000     0.026     0.025
time2          -0.088     0.008   -10.882 5.829e+03     0.000     0.150     0.131
```

```
time3            -0.020     0.014    -1.452 1.193e+03     0.147     0.405     0.289

                           Estimate
Intercept~~Intercept|id     0.113
Residual~~Residual          0.957
ICC|id                      0.105

Unadjusted hypothesis test as appropriate in larger samples.
```

Package `mitml` (Grund et al., 2023) provides useful helper functions that facilitate the process: `mids2mitml.list()` extracts a list of completed data sets from the `mice` output. The generic `with`-function can then be used to repeatedly fit the linear piecewise trajectory model to the list of $M = 100$ completed data sets. Finally, `testEstimates()` applies Rubin's (1987) rules for MI inference and reads out classical MI output. RIV is the relative increase in variance due to nonresponse. FMI is the estimated fraction of missing information, a statistic that indicates to what extent the estimate is affected by missing information. If the statistic is large (say above 0.4), then the trustworthiness of the obtained inferences largely depends upon the correct specification of the imputation model and upon whether the (partially) untestable assumptions of the MI approach are met.

The argument `extra.pars = TRUE` tells the function to also pool the random part of the model. The last three lines of the output show the intercept variance, the level 1 residual variance and the ICC—the intraclass correlation coefficient.

24.4 Monte Carlo Simulation

Overview, Rationale and Aims of the Simulation

To evaluate the proposed approach of imputing incomplete non-linear panel data by a piecewise growth curve model, we simulated a panel data set comprising 12 repeated measurements of one target variable of interest. Population parameters were defined to resemble existing empirical panel data from the *Crime in the Modern City* (CrimoC) project, focussing on the emergence and development of juvenile delinquency during adolescence and young adulthood in a 'normal' population (see also section 'Application'). The main dependent variable here is the count of deviant and delinquent behaviours reported for the respective previous year. In our simulation, we generated the counts resembling this delinquency index based on a zero-inflated Poisson process, since most participants reported zero delinquent behaviours in the respective previous year (see also Table 24.6): Zero-inflation models (Lambert, 1992) are mixture models and assume two data-generating processes. The excessive number of zero counts is usually expressed by a binomial generalized linear model (GLM) with a logit link determining if a certain case belongs to the zero inflation or to the count process. A count model (typically a Poisson or negative binomial model with a log link) then predicts the respective count for each participant who is not in the zero-inflation group. Different sets of predictors can be specified for the zero and the count parts of the model. A brief introduction to zero-inflation models related to the CrimoC data can be found in Reinecke and Weins (2013) and Kleinke et al. (2020, Chap. 6).

The data-generating model in our simulation as well as in the subsequent analysis model was a piecewise growth curve model (assuming a zero-inflated Poisson data-generating process) consisting of two linear pieces—which shall resemble the typical age-crime relationship, i.e. first a rise in delinquency in early adolescence, followed by a general decline later on (e.g. Moffitt, 1993). The curvilinear relationship between age and crime has been observed regularly in criminological studies: A rapid increase of delinquency occurs in the mid- to late teens, followed by a rapid decline in the early to mid-20s. A further gradual decline is observed for increasingly older age groups (Britt, 2019). The total sample size is $N = 3943$. Data were simulated and analysed with M*plus* (Muthén & Muthén, 1998–2017). We examined five missing data scenarios. Panel attrition is a serious issue and an ubiquitous problem in longitudinal research. In the first three conditions, we simulated a monotone MAR pattern with constant dropout rates over time leading to a total of 20%, 30%, or 40% missing data, respectively, after 12 panel waves. The first panel wave was always completely observed and was defined as the 'cause of missingness'. We used `mice`'s `ampute` function to delete the values according to the pre-specified missing data patterns and frequencies. The 12 dropout patterns of the first three conditions are displayed in Table 24.1. The `weights` matrix of `mice`'s `ampute` function—a matrix with dimension number of patterns by number of variables, which defines the type and strength of the missing data mechanisms—contained a column of 1s for the first timepoint, and columns of 0s for the other timepoints, defining wave 1 as the cause of missingness (for details, see `?mice::ampute`). In conditions 4 and 5, we simulated a more realistic missing data pattern comprising both panel attrition and arbitrary item missingness. The difference between conditions 4 and 5 is that in condition 4, missing data

Table 24.1 Simulated dropout patterns—conditions 1–3 (20%, 30%, 40% missing data)

t_1	t_2	t_3	t_4	t_5	t_6	t_7	t_8	t_9	t_{10}	t_{11}	t_{12}
1	1	1	1	1	1	1	1	1	1	1	1
1	1	1	1	1	1	1	1	1	1	1	0
1	1	1	1	1	1	1	1	1	1	0	0
1	1	1	1	1	1	1	1	1	0	0	0
1	1	1	1	1	1	1	1	0	0	0	0
1	1	1	1	1	1	1	0	0	0	0	0
1	1	1	1	1	1	0	0	0	0	0	0
1	1	1	1	1	0	0	0	0	0	0	0
1	1	1	1	0	0	0	0	0	0	0	0
1	1	1	0	0	0	0	0	0	0	0	0
1	1	0	0	0	0	0	0	0	0	0	0
1	0	0	0	0	0	0	0	0	0	0	0

Note. 1 = observed value; 0 = missing value

Table 24.2 Simulated dropout patterns—conditions 4–5 (MAR, partially NMAR)

	t_1	t_2	t_3	t_4	t_5	t_6	t_7	t_8	t_9	t_{10}	t_{11}	t_{12}	freq.	freq. %
1	1	1	1	1	1	1	1	1	1	1	1	1	585	14.84
2	1	1	0	0	0	0	0	0	0	0	0	0	272	6.90
3	1	1	1	1	1	0	0	0	0	0	0	0	155	3.93
4	1	1	1	0	0	0	0	0	0	0	0	0	143	3.63
5	0	1	1	1	1	1	1	1	1	1	1	1	123	3.12
6	1	1	1	1	0	0	0	0	0	0	0	0	101	2.56
7	1	0	1	1	1	1	1	1	1	1	1	1	79	2.00
8	0	0	1	1	1	1	1	1	1	1	1	1	68	1.72
9	0	0	0	0	1	1	1	1	1	1	1	1	67	1.70
10	1	1	1	1	1	1	0	0	0	0	0	0	58	1.47
11	0	1	1	1	0	0	0	0	0	0	0	0	56	1.42
12	0	1	1	1	1	0	0	0	0	0	0	0	51	1.29
13	0	1	1	0	0	0	0	0	0	0	0	0	50	1.27
14	0	0	1	1	1	0	0	0	0	0	0	0	48	1.22
15	1	1	1	0	1	0	0	0	0	0	0	0	46	1.17
16	1	1	0	1	1	1	1	1	1	1	1	1	40	1.01

Note. 1 = observed value; 0 = missing value; the last two columns give the observed absolute and relative frequency in percent of the 16 most frequently observed missing data patterns (i.e. $\geq 1\%$)

were merely MAR, whereas in condition 5, the pattern was partially not missing at random (NMAR). Missingness here depended on the partially unobserved values of the first panel wave itself. In condition 4, the `weights` matrix was identical to the simulated missing data patterns, which are displayed in Table 24.2. In condition 5, again we used a column of 1s for the first variable and a column of 0s for all other variables. Item-level missingness in these latter two scenarios ranged between around 10% and 35%. Each scenario was replicated 1000 times. The resulting incomplete data files were then imputed with R package `countimp` (Kleinke & Reinecke, 2019) by an imputation model that was fully compatible to the data-generating model i.e. a zero-inflated Poisson piecewise growth curve model reflecting the defined two linear pieces.

We also contrasted results against a quadratic growth curve imputation model, and against a method, which is the default setting in many imputation software packages—predictive mean matching (pmm, Van Buuren, 2018, Chap. 3), which is a quite robust all-round imputation method: In real-life applications, the true data-generating process (DGP) is typically unknown, and researchers often have the choice between more than one 'plausible' method or model with an acceptable fit to the empirical data. Like in our case, for example, a rise in delinquency early on in adolescence and a decline later on could be approximated by both a quadratic growth curve model and a piecewise model. Our purpose here is to elucidate to what extent minor model misspecifications that could easily occur in everyday empirical research affect substantive model results. In other words, if the true DGP is a piecewise growth curve model, and the imputer falsely assumes a quadratic growth curve model being the data-generating model, does this influence the substantive interpretation of the developmental process?

In addition, we intend to corroborate the research regarding the robustness of the all-round imputation method pmm. We want to elucidate how this affects substantive model results, if no explicit non-linear panel model is specified. Predictive mean matching samples one case from a pool of observed cases that are similar to the incomplete case. The observed value from this *donor* is then used to fill in the corresponding missing value of the *donee*. The matching is done by similarity in predicted means for the observed and unobserved part of the variable based on a linear regression model—thus the name. The approach is typically enumerated among the semi-parametric imputation procedures since the underlying parametric linear model is used only as a means for matching the incomplete case to a potential donor. Then, actually observed values are imputed. This typically preserves many properties of the observed part of the data including cluster effects. Based on research by Kleinke et al. (2011), we apply pmm to the wide format data (i.e. the repeated measurements are stored as separate variables in the data file). We let `mice`'s `quickpred`-function automatically identify appropriate predictors using its default settings. While the imputation of actual observed cases makes pmm robust against mild to moderate violations of distributional assumptions of the normal linear regression model (cf. Kleinke, 2017), 'normal' model-based pmm could yield biased inferences regarding an extremely skewed zero-inflated variable (Kleinke & Reinecke, 2022).

Hypotheses

We expect best results, when both the imputation model and the model of the data analyst correctly reflect the true DGP. We furthermore expect that minor model misspecifications (like when the imputer falsely assumes the true DGP to be a quadratic growth model while the true DGP is a piecewise model with two linear pieces) can be buffered quite well, since model misspecifications of the imputation model affect only the formerly missing part. Effects thus shall become more pronounced the more data have to be imputed. Finally, previous missing data research has shown that 'normal' model-based pmm can handle minor violations of distributional assumptions quite well, but not severe violations. Again, the magnitude of bias is to be expected to depend on the missing data percentage and might be negligible, when the missing data problem is only minor. Regarding the imputation of a zero-inflated target variable, we expect normal model-based pmm to yield the overall worst inferences in comparison to the other two solutions in the scenarios considered here.

Evaluation Criteria

We evaluated each imputation method by its capability to produce unbiased point estimates and unbiased measures of uncertainty. Bias in point estimates was defined as the absolute difference between the true parameter Q and the average estimate $\widehat{Q}$ across the replications:

$$\text{Bias} = |\widehat{Q} - Q| \tag{24.3}$$

Bias obviously shall be close to zero. We deemed bias 'significant' if it leads to a serious drop in coverage rates—the percentage of 95% confidence intervals that include the true parameter. CR should be close to 95%. Following Schafer and Graham (2002), we define CR $< 90\%$ as seriously low. CR in fact is a hybrid measure that reflects both large biases and inadequate standard errors: Undercoverage may result from biased point estimates and/or an underestimated variance. The confidence interval could either be too narrow to include the true parameter, and/or it can be shifted too much to the left or to the right to include the true parameter. Overcoverage, i.e. coverage close to 100%, is an indicator of too large standard errors and thus overly conservative statistical test decisions—increasing the probability for type II errors. Together, bias and CR give a good impression of the accuracy of statistical inferences.

Results and Discussion

Information about combined model fit statistics regarding the zero-inflated Poisson piecewise growth curve model can be obtained from Table 24.3. The table displays means and standard deviations across the imputations and replications of the log likelihood (LL), the Akaike information criterion (AIC) and the Bayesian information criterion (BIC), respectively. One noteworthy finding here is that—regardless which imputation method has been used—with an increasing severity and complexity of the missing data processes, the variance of the model fit statistics increases reflecting a greater uncertainty and variability in the imputed data across the M sets of imputations, respectively. While the piecewise and the quadratic growth curve imputation models produce highly similar model fit statistics, pmm produces substantially larger variance in model fit. We will discuss this phenomenon later on.

Bias in point estimates is summarized in Table 24.4. The table displays bias regarding the means of the growth factors for the count part and the inflation part of the model, where I denotes the latent intercept and S1 and S2 denote the two slopes in the count part of the model and S1I and S2I the two linear slopes in the zero-inflation part of the model. Variances and covariances of the growth factors were not considered in this simulation.

Table 24.3 Model fits of the simulation conditions

Condition	LL_Mean	LL_SD	AIC_Mean	AIC_SD	BIC_Mean	BIC_SD
	Piecewise model					
1. Monotone 20%	−24605.52	59.71	49223.05	119.41	49260.73	119.41
2. Monotone 30%	−24609.29	74.71	49230.58	149.41	49268.26	149.41
3. Monotone 40%	−24610.16	89.23	49232.31	178.46	49269.99	178.46
4. Realistic MAR	−24611.78	122.33	49235.56	244.66	49273.24	244.66
5. Realistic NMAR	−24266.28	120.92	48544.56	241.85	48582.24	241.85
	Quadratic model					
1. Monotone 20%	−24612.74	59.72	49237.47	119.44	49275.15	119.44
2. Monotone 30%	−24620.01	74.23	49252.01	148.46	49289.69	148.46
3. Monotone 40%	−24623.74	89.12	49259.48	178.24	49297.15	178.24
4. Realistic MAR	−24768.34	124.30	49548.68	248.61	49586.36	248.61
5. Realistic NMAR	−24456.69	123.09	48925.37	246.19	48963.05	246.19
	Predictive mean matching (pmm)					
1. Monotone 20%	−24591.27	328.10	49194.55	656.19	49232.23	656.19
2. Monotone 30%	−24583.57	500.71	49179.13	1001.42	49216.81	1001.42
2. Monotone 40%	−24573.52	691.70	49159.05	1383.41	49196.73	1383.41
4. Realistic MAR	−24507.97	681.82	49027.94	1363.63	49065.62	1363.63
5. Realistic NMAR	−24170.09	1134.97	48352.19	2269.95	48389.87	2269.95

Note. LL is the log likelihood, AIC the Akaike information criterion and BIC the Bayesian information criterion. The table displays the average means and standard deviations of the statistics across the M sets of imputations and 1000 replications

The piecewise model was the only imputation model that was identical to the data-generating model and produced the overall lowest biases. The other two imputation methods, however, also produced point estimates that were accurate most of the time. If the mechanism—however—was partially NMAR, bias in some parameters was substantial regardless of which imputation method had been used. In our classification of 'significant' parameter bias, we also follow Forero and Maydeu-Olivares (2009), who deem bias significant, if it exceeds more than 10% of the true parameter.

Substantial bias is also reflected in a drop in coverage rates (see Table 24.5). Imputation based on a piecewise growth curve model produced serious undercoverage regarding the coefficients for the first linear slope in the count part and the zero-inflation part of the model, respectively, as well as regarding the intercept in the count part, when data were NMAR. The (slightly) misspecified quadratic growth curve imputation model produced low coverage rates not only in the NMAR but also in the arbitrary MAR condition. In all other conditions, coverage rates were acceptable and sufficiently close to 95%. Neither serious overcoverage nor undercoverage was observed. Predictive mean matching produced coverage rates of 100% or close to 100% in many scenarios. This means that standard errors were not adequate and far too large.

Table 24.4 Average absolute bias across the replications

Condition	I	S1	S2	S1I	S2I
	Piecewise model				
1. Monotone 20%	0.001	0.002	0.000	0.001	0.000
2. Monotone 30%	0.001	0.002	0.000	0.002	0.000
3. Monotone 40%	0.001	0.002	0.000	0.001	0.000
4. Realistic MAR	0.002	0.002	0.000	0.001	0.000
5. Realistic NMAR	0.049	**0.051**	0.000	**0.154**	0.000
	Quadratic model				
1. Monotone 20%	0.001	0.001	0.000	0.001	0.000
2. Monotone 30%	0.001	0.002	0.000	0.002	0.000
3. Monotone 40%	0.001	0.002	0.000	0.002	0.000
4. Realistic MAR	0.026	**0.031**	0.002	0.044	0.002
5. Realistic NMAR	0.012	0.008	0.002	**0.096**	0.002
	Predictive mean matching (pmm)				
1. Monotone 20%	0.001	0.003	0.001	0.002	0.000
2. Monotone 30%	0.001	0.004	0.001	0.003	0.000
3. Monotone 40%	0.001	0.006	0.002	0.004	0.000
4. Realistic MAR	0.010	0.013	0.002	0.026	0.000
5. Realistic NMAR	0.058	**0.066**	0.004	**0.157**	0.000

Note. I is the latent intercept, S1 and S2 are the two slopes in the count part of the model and S1I and S2I are the two linear slopes in the zero-inflation part of the model. Bias is defined as $|\widehat{Q} - Q|$, where Q is the 'true' population parameter and $\widehat{Q}$ the average estimate across the replications; use of boldface type indicates bias that exceeds 10% of the true parameter

We have shown that the proposed method of using a piecewise growth curve model to impute incomplete panel data works well with monotone and arbitrary missing data patterns with up to 40% missingness, when all modeling assumptions were met (i.e. neither the imputation nor the analysis model was misspecified, parametric modeling assumptions were met and assumptions about the missing data mechanisms were correct). Typically, imputations are created under the assumption of ignorable MAR missingness, so that having to make strong (and at least partially untestable) assumptions about the mechanisms that created the missing data can be avoided. Our findings showed that violations of MAR can lead to biased statistical inferences.

Future research should give more insight into how and to what extent model misspecifications of either the imputation or the analysis model affect results. Statistical models are only an approximation, and usually also a simplification of 'reality', and also, empirical data seldom perfectly follow exactly a zero-inflated Poisson process (as in this simulation). A conceptual replication of this study based on empirical data can shed more light in this regard, how deviations from mathematically convenient distributions can affect results.

Table 24.5 Coverage rates of the 95% confidence intervals across the replications

Condition	I	S1	S2	S1I	S2I
	Piecewise model				
1. Monotone 20%	93.80	94.60	93.70	94.50	95.10
2. Monotone 30%	93.80	94.90	94.90	94.40	95.60
3. Monotone 40%	93.80	94.70	94.10	95.10	94.40
4. Realistic MAR	94.40	94.90	93.40	94.50	94.60
5. Realistic NMAR	**63.40**	**75.90**	93.80	**23.60**	94.80
	Quadratic model				
1. Monotone 20%	93.80	94.60	93.30	94.80	95.00
2. Monotone 30%	93.80	95.50	95.00	94.20	95.20
3. Monotone 40%	93.80	94.40	94.80	94.60	94.50
4. Realistic MAR	**86.00**	**88.10**	92.30	**88.90**	94.50
5. Realistic NMAR	95.20	96.60	92.80	**61.90**	93.30
	Predictive mean matching (pmm)				
1. Monotone 20%	93.80	98.40	99.90	98.10	100.00
2. Monotone 30%	93.80	99.30	100.00	99.90	100.00
3. Monotone 40%	93.80	99.90	100.00	100.00	100.00
5. Realistic MAR	99.80	99.80	100.00	99.80	100.00
4. Realistic MMAR	100.00	100.00	100.00	100.00	100.00

Note. I is the latent intercept, S1 and S2 are the two slopes in the count part of the model and S1I and S2I are the two linear slopes in the zero-inflation part of the model. Coverage rate is defined as the percentage of 95% confidence intervals that include the true parameter; use of boldface type indicates seriously low coverage below 90%

Furthermore, in this simulation study, we have focused on the 'fixed' part of the model, i.e. the means of the growth factors alone. We disregarded their variances and covariances, since estimating these parameters under the assumption of a zero-inflation process requires numerical integration, which takes a tremendous amount of time to compute (100 imputations per scenario per replication). We will address this issue in future research.

24.5 Application

In the final section, we apply the proposed approach of creating imputations based on a piecewise growth curve model to an empirical data set and contrast results against related and also plausible imputation approaches. Since in 'real-life' applications the true data-generating process is unknown, we discuss the appropriateness of the imputation approaches by similarity and differences in substantive modeling results.

Sample and Data

The empirical basis for the analysis stems from the longitudinal research project *Crime in the Modern City* (CrimoC), an interdisciplinary ongoing research project located at the Universities of Münster and Bielefeld (see www.crimoc.org for details). The main focus of the study lies on the emergence and the development of deviant and delinquent behaviour of juveniles and the social control surrounding it—both formal control, meaning the police and the judiciary, and informal control, referring to school, family and peers. The annual survey started in 2002 with a sample of 3411 seventh grade students from all school types in the industrial town of Duisburg. Their mean age was 13. Up to 2009, annual panel waves were conducted. From 2009 to 2019, the data collection was biannual. The present data set allows us to analyse the development of deviant and delinquent behaviours from ages 13 to 28. In the first period of the study, questionnaires were completed in class (self-report data). When participants left school, the mode of data collection changed to a mail survey. To obtain reasonable response rates, an incentive of up to 25 Euros was offered for the participation at each panel wave. Detailed information about the panel design and main results of the first 11 panel waves are presented and discussed in Boers and Reinecke (2019).

The target variable of interest in the present analysis is the count of delinquent behaviours. At each panel wave, participants answered 16 questions about various types of offences (property-related crimes, as well as violent crimes, fraud and drug-related crimes—for details, see Table 8 in Reinecke & Weins, 2013). Participants indicated for each offence, if they committed the particular behaviour in the 12 months prior to the interview (0 = no; 1 = yes). The individual delinquency index was computed as the sum of these 16 binary items, reflecting the versatility of delinquent activities. The higher the index value (with the maximum of 16), the more versatile are the individual criminal activities.

Table 24.6 gives the means and standard deviations (SD) of the delinquency index for each panel wave as well as observed minimum (Min) and maximum (Max) values.

Mean delinquency rates are always below one, meaning that on average, participants reported less than one different delinquent activity in the respective past year. In accordance with the expected trajectory of the age-crime curve, the mean delinquency rate first increases with a maximum at the second panel wave. From the third wave on, the rate constantly decreases until the last panel wave. The same can be observed for the variability (SD) of the delinquency index in each panel wave.

Because the sample of the CrimoC study is not stratified to vulnerable groups of students with a higher probability to become delinquent, the distributions of the delinquency scores contain a large amount of zeros, i.e. no delinquency reported in that particular year at all. Even at the maximum at wave 2, the amount of zeros is about 63% (cf. the right column in Table 24.6).

Table 24.6 Overview of the observed delinquency scores over 12 panel waves

Delinquency	N	Mean	SD	Min	Max	% zeros
t_1	2652	0.717	1.560	0	15	69.65
t_2	2869	0.980	1.894	0	15	63.02
t_3	2922	0.808	1.746	0	15	66.56
t_4	2952	0.623	1.466	0	15	72.90
t_5	2811	0.417	1.071	0	12	78.62
t_6	2489	0.233	0.753	0	15	86.30
t_7	2294	0.137	0.526	0	8	90.76
t_8	2213	0.096	0.441	0	8	93.31
t_9	2050	0.077	0.341	0	4	93.95
t_{10}	1847	0.056	0.276	0	3	95.40
t_{11}	1764	0.056	0.310	0	6	95.58
t_{12}	1734	0.046	0.300	0	7	96.42

Data Imputation and Analysis

For these reasons, we opted for a model that can explicitly address the excess zeros in the data: a zero-inflated Poisson model. The age-crime relationship was expressed by two linear piecewise splines. The model was used both for imputation and analysis of the empirical data. Similar to the model specifications in the simulation study, the nonlinear age-crime relationship was also approximated by a quadratic model. To see if an explicit model-based imputation strategy leads to different results than the default in `mice`, comparisons are made to model results based on semi-parametric predictive mean matching, where no explicit panel imputation model for zero-inflated count data is specified (pmm rather implicitly preserves the properties of the empirical data by imputing actual observed values). We applied pmm to the wide format data. Donors and donees are matched by means of similarity in predicted means based on a standard linear regression model. All other panel waves were included in the prediction models for an unobserved panel wave.

Results and Discussion

Model fit statistics are given in Table 24.7. We approximated the development of delinquency over time by a piecewise growth curve model assuming a zero-inflated Poisson data-generating process. Not surprisingly, using the same model for both imputation and data analysis resulted in the best model fit statistics. However, using the quadratic term in the growth curve model instead of a second linear piece or relying on semi-parametric pmm produced similar results both regarding the average model fit across the set of imputations and regarding substantive results. Note again that wide format pmm produced a far greater variability in model fit across the imputations in comparison to the other two approaches.

Table 24.7 Comparison of different imputation models and methods regarding their model fit

Model	Method	LL_Mean	LL_SD	AIC_Mean	AIC_SD	BIC_Mean	BIC_SD
Piecewise	zip	−30777.11	157.69	61566.22	315.38	61618.81	315.38
Quadratic	zip	−31156.31	196.34	62424.61	392.69	62377.20	392.69
Linear	pmm	−34045.96	640.85	68103.92	1281.71	68156.51	1281.71

Note. The first two left columns denote the imputation model and the imputation method. In all cases, the results are based on the zero-inflated piecewise growth curve model. For the abbreviations of the log likelihood and information criteria, see Table 24.3.

Table 24.8 Comparison of different imputation models and methods regarding parameter estimates and standard errors

Model	Method	I	S1	S2	II	S1I	S2I
		Point estimates					
Piecewise	zip	0.723	0.179	−0.162	0.627	−0.217	0.205
Quadratic	zip	0.796	0.087	−0.155	0.514	−0.045	0.190
Linear	pmm	0.730	0.181	−0.164	0.553	−0.171	0.170
		Standard errors					
Piecewise	zip	0.027	0.031	0.008	0.047	0.055	0.009
Quadratic	zip	0.026	0.030	0.007	0.045	0.052	0.009
Linear	pmm	0.063	0.068	0.010	0.117	0.130	0.024
		Estimated fraction of missing information					
Piecewise	zip	0.307	0.291	0.491	0.324	0.335	0.482
Quadratic	zip	0.404	0.317	0.442	0.320	0.332	0.441
Linear	pmm	0.884	0.865	0.797	0.895	0.885	0.934

Note. The first two left columns indicate the imputation model and method. In all cases, the results are based on a zero-inflated piecewise growth curve model. I is the intercept, S1 and S2 are the two slopes in the count part of the model and II, S1I and S2I are the counterparts in the zero-inflation part of the model

Point estimates and standard errors are given in Table 24.8. Average trajectories for the zero-inflation and count parts of the model are also displayed in Figs. 24.4 and 24.5. The positive first slope (labelled S1 in the table) represents the average increase in delinquency from the first to the second panel wave. The model with the quadratic term produced a noticeably less steep slope parameter, whereas pmm leads to a comparable estimate of the first slope in comparison to the result using the piecewise model. The negative coefficient of the second slope (labelled S2 in the table) represents the constant average decrease in delinquency from the second wave onwards. Again using a quadratic model for the imputation leads to a slightly less pronounced decrease, whereas pmm leads to a virtually identical decrease. Similar patterns of results were found in the zero-inflation part of the model: The negative first slope (labelled S1I in the table) represents the decreasing chance of belonging to the structural zero class from the first to the second wave, whereas the positive coefficient for the second slope (labelled S2I in the table) represents a constant increase in the probability of belonging to the certain zero category.

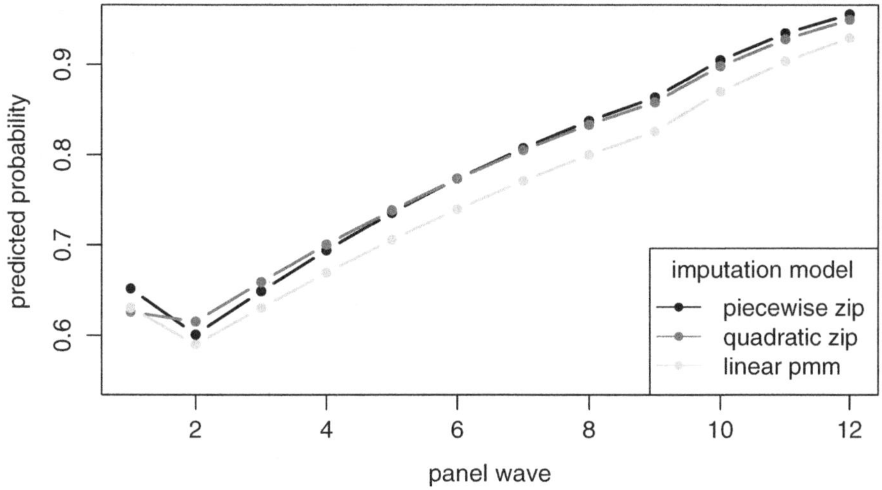

Fig. 24.4 Predicted probabilities for belonging to the certain zero group per panel wave and imputation method

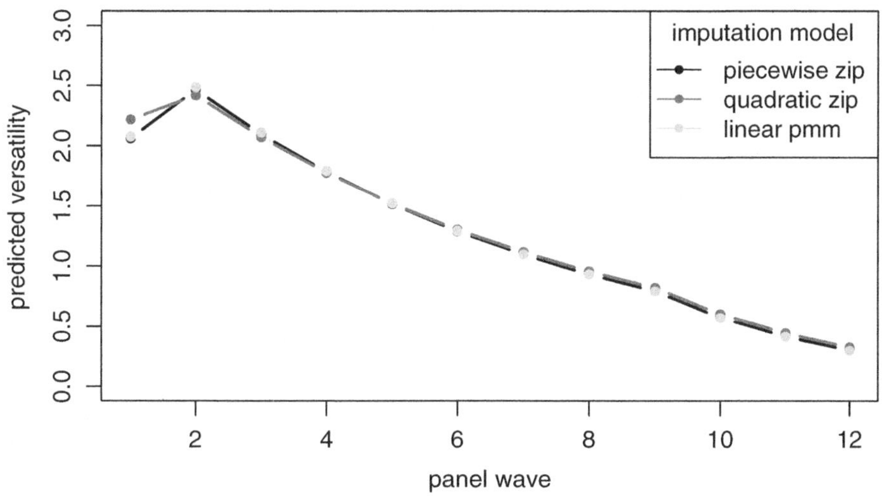

Fig. 24.5 Predicted versatility of delinquent behaviours per panel wave and imputation method

The parameter estimate of S1I using pmm was less steep in comparison to the piecewise model, while the parameter estimate produced by the quadratic model is far less steep. Likewise, the slopes S2I of the quadratic model are less steep in comparison to the piecewise model. Despite these nuances, the general patterns of results are going in the same substantive direction reflecting the expected age-crime relationship. Interestingly, imputation based on semi-parametric matching with a (here inappropriate) underlying normal linear regression model (pmm) produced by far the largest (combined) standard errors, expressing more uncertainty regarding the unobserved part of the data and thus resulting in more conservative test decisions in comparison to the other two approaches. This greater uncertainty is also reflected in the estimated fractions of missing information (see the bottom part of Table 24.8). Fraction of missing information quantifies to what extent the respected estimate is affected by missing information and in consequence gives an idea about how certain or uncertain one can be about the obtained inferences.

24.6 Summary and Conclusions

A crucial step when analysing complex incomplete panel data is to think about appropriate statistical models that reflect the assumed (and usually unknown) true data-generating process well. Multiple imputation functions based on either a univariate or multivariate linear mixed effects model under the assumption of both normal homoscedastic and heteroscedastic errors are available from several software packages nowadays, and extensions to the generalized linear mixed effects modeling framework have also become available more recently (for an overview, see Spiess et al., 2021; Kleinke et al., 2020). More robust semi- and nonparametric approaches have not yet been generalized to multilevel data (see the Chapter by Kleinke, Fritsch, Stemmler and Lösel in this book). In the present simulation and application, we have used functions from R package `countimp` (Kleinke & Reinecke, 2019) for the imputation of longitudinal (discrete) count data with an excessive number of zero observations. The imputation functions from this package can be used to fit piecewise growth curve models under various count models (e.g. Poisson, negative binomial and zero-inflated Poisson). In a tutorial-like section, we have demonstrated that non-linear trajectories over time can be easily modeled by fitting linear piecewise growth curve models.

Our Monte Carlo simulation has shown that data imputation based on these models works reasonably well, when models are correctly specified and when assumptions about the missing data processes are correct. Furthermore, our results have shown that bias is to be expected, when assumptions about the mechanism that created the missing data patterns are violated: Falsely assuming an ignorable MAR mechanism has led to biased statistical inferences, when the data were in fact NMAR.

In our application using data from 12 waves of the CrimoC study, we have applied model-based imputation using a piecewise growth curve model assuming a zero-inflated Poisson data-generating process. Our choice of imputation method and model was guided by the long-term research regarding the age-crime curve (Britt, 2019), which can also be found in the panel data of the CrimoC study (see earlier analyses reported in Reinecke & Weins, 2013; Kleinke et al., 2021). Since in empirical applications the true DGP and missing data processes are unknown, we have compared how different imputation methods and models affect substantive modeling results: Imputation based on a quadratic growth curve model worked similarly well (see also the results reported in Kleinke et al., 2021). The relatively robust imputation technique pmm leads to some minor differences in substantive model results compared to the previous ones. In the scenario considered here, pmm-based imputation leads to too large standard errors leading to overly conservative statistical test decisions.

Implications for Applied Researchers

In real-life applications, the true DGP is typically unknown, and applied researchers often have to make the choice between several plausible imputation and analysis models. Our simulation has shown that minor model misspecification did not dramatically affect statistical results. What is more: Sometimes the imputation and analysis methods with a perfect fit are simply not feasible from a practical point of view (e.g., when computation time is of the essence, or when software is not yet available; see also the chapter by Kleinke, Fritsch, Stemmler and Lösel in this book). Thus, from a pragmatic point of view, it might be necessary and acceptable to make some simplifying assumptions. Monte Carlo simulations are a useful method in this regard to get an idea when simplifying the imputation and analysis process is acceptable and when it is not. Future research, for example, should establish what amount of bias is to be expected in the 'random' part of the model (i.e. regarding the variances and covariances of the growth factors), if the imputation model considers fixed effects (i.e. the means of the growth factors) only.

References

Barnard, J., & Rubin, D. B. (1999). Small-sample degrees of freedom with multiple imputation. *Biometrika, 86*(4), 948–955. https://doi.org/10.1093/biomet/86.4.948

Bates, D., Mächler, M., Bolker, B., & Walker, S. (2015). Fitting linear mixed-effects models using lme4. *Journal of Statistical Software, 67*(1), 1–48. https://doi.org/10.18637/jss.v067.i01

Boers, K., & Reinecke, J. (2019). *Delinquenz im Altersverlauf: Erkenntnisse der Langzeitstudie Kriminalität in der modernen Stadt*. Waxmann.

Bollen, K. A., & Curran, P. J. (2006). *Latent curve models: A structural equation perspective*. Wiley.

Britt, C. L. (2019). Age and crime. In D. P. Farrington, L. Kazemian, & A. R. Piquero (Eds.), *The Oxford handbook of developmental and life-course criminology* (pp. 13–33). Oxford University Press.

Drechsler, J. (2015). Multiple imputation of missing multilevel data—rigor versus simplicity. *Journal of Educational and Behavioral Statistics, 40*(1), 69–95. https://doi.org/10.3102/1076998614563393

Forero, C. G., & Maydeu-Olivares, A. (2009). Estimation of IRT graded response models: Limited versus full information methods. *Psychological Methods, 14*(3), 275–299. https://doi.org/10.1037/a0015825

Grund, S., Robitzsch, A., & Luedtke, O. (2023). *Mitml: Tools for multiple imputation in multilevel modeling* [R package version 0.4-5]. https://cran.r-project.org/package=mitml

Jorgensen, T. D., Pornprasertmanit, S., Schoemann, A. M., & Rosseel, Y. (2022). *semTools: Useful tools for structural equation modeling* [R package version 0.5-6]. https://CRAN.R-project.org/package=semTools

Kleinke, K. (2017). Multiple imputation under violated distributional assumptions—a systematic evaluation of the assumed robustness of predictive mean matching. *Journal of Educational and Behavioral Statistics, 42*(4), 371–404. https://doi.org/10.3102/1076998616687084

Kleinke, K. (2021). Estimation of partially observed non-linear terms in a multilevel model: An evaluation of the robustness of ad hoc and state-of-the-art missing data methods. *Psychological Test and Assessment Modeling, 63*(3), 432–455. https://www.psychologie-aktuell.com/fileadmin/Redaktion/Journale/ptam-2021-3/PTAM__3-2021_6_kor.pdf

Kleinke, K., & Reinecke, J. (2013). Multiple imputation of incomplete zero-inflated count data. *Statistica Neerlandica, 67*(3), 311–336. https://doi.org/10.1111/stan.12009

Kleinke, K., & Reinecke, J. (2019). *countimp version 2—A multiple imputation package for incomplete count data* (Technical Report). University of Siegen, Department of Education Studies and Psychology. Siegen, Germany. https://www.kkleinke.de/countimp

Kleinke, K., & Reinecke, J. (2022). How to and how not to impute incomplete count data. In A. Hernández & I. Tomás (Eds.), *Proceedings from the 9th European Congress of Methodology* (pp. 86–92). Universitat de València.

Kleinke, K., Reinecke, J., Salfrán, D., & Spiess, M. (2020). *Applied multiple imputation. Advantages, pitfalls, new developments and applications in R*. Springer Nature.

Kleinke, K., Reinecke, J., & Weins, C. (2021). The development of delinquency during adolescence: A comparison of missing data techniques revisited. *Quality & Quantity, 55*(3), 877–895.

Kleinke, K., Stemmler, M., Reinecke, J., & Lösel, F. (2011). Efficient ways to impute incomplete panel data. *Advances in Statistical Analysis, 95*(4), 351–373. https://doi.org/10.1007/s10182-011-0179-9

Lambert, D. (1992). Zero-inflated poisson regression. With an application to defects in manufacturing. *Technometrics, 34*(1), 1–14.

Moffitt, T. E. (1993). Adolescence-limited and life-course-persistent antisocial behavior: A developmental taxonomy. *Psychological Review, 100*, 674–701.

Muthén, L. K., & Muthén, B. O. (1998–2017). *Mplus user's guide* (8th ed.). Muthén & Muthén.

Raudenbush, S. W., & Bryk, A. S. (2002). *Hierarchical linear models: Applications and data analysis methods*. Sage.

Reinecke, J., & Weins, C. (2013). The development of delinquency during adolescence: A comparison of missing data techniques. *Quality & Quantity, 47*(6), 3319–3334.

Rosseel, Y. (2012). lavaan: An R package for structural equation modeling. *Journal of Statistical Software, 48*(2), 1–36. https://doi.org/10.18637/jss.v048.i02

Rubin, D. B. (1976). Inference and missing data. *Biometrika, 63*(3), 581–592. https://doi.org/10.1093/biomet/63.3.581

Rubin, D. B. (1987). *Multiple imputation for nonresponse in surveys*. Wiley.

Rubin, D. B. (1996). Multiple imputation after 18+ years. *Journal of the American Statistical Association, 91*(434), 473–489. https://doi.org/10.1080/01621459.1996.10476908

Schafer, J. L., & Graham, J. W. (2002). Missing data: Our view of the state of the art. *Psychological Methods, 7*(2), 147–177. https://doi.org/10.1037/1082-989X.7.2.147

Schafer, J. L., & Yucel, R. M. (2002). Computational strategies for multivariate linear mixed-effects models with missing values. *Journal of Computational and Graphical Statistics, 11*(2), 437–457.

Spiess, M., Kleinke, K., & Reinecke, J. (2021). Proper multiple imputation of clustered or panel data. In P. Lynn (Ed.), *Advances in longitudinal survey methodology* (pp. 424–446). Wiley.

Van Buuren, S. (2011). Multiple imputation of multilevel data. In J. J. Hox & J. K. Roberts (Eds.), *Handbook of advanced multilevel analysis* (pp. 173–196). Taylor & Francis.

Van Buuren, S. (2018). *Flexible imputation of missing data* (2nd ed.). CRC press. https://stefvanbuuren.name/fimd/

Van Buuren, S., & Groothuis-Oudshoorn, K. (2011). MICE: Multivariate imputation by chained equations in R. *Journal of Statistical Software, 45*(3), 1–67. https://doi.org/10.18637/jss.v045.i03

Yu, L. M., Burton, A., & Rivero-Arias, O. (2007). Evaluation of software for multiple imputation of semi-continuous data. *Statistical Methods in Medical Research, 16*, 243–258. https://doi.org/10.1177/0962280206074464

Chapter 25
Impact of Inconsistent Imputation Models in Mediation Analysis with Clustered Data

Bo Ye, Recai Yucel, and Donna L. Coffman

25.1 Introduction

Missing data can easily complicate most analyses in many disciplines. Since the seminal work by Dempster et al. (1977), missing data has been the subject of much research not only in statistics but also in many subject-matter areas. In particular, inference by multiple imputation (MI) has increasingly become the norm in the analysis of incomplete data (Rubin, 1987). Another approach makes use of expectation-maximization (EM)-type solutions leading to maximum likelihood estimation (MLE) in problem-specific settings. In this work, we focus on how the choice of imputation model influences multilevel mediation analysis.

Underlying missingness mechanism assumptions as defined by Rubin (1976) have been extensively discussed in the missing data literature. We consider three missingness mechanisms: missing completely at random (MCAR), which means that missingness probabilities are independent of any observed data; missing at random (MAR), which means that missingness probabilities may depend on observed data but not on variables subject to missing data; and missing not at random (MNAR), which means that probabilities of missingness depend on the missing values and hence must be modeled. In this paper, we focus on investigating performance under MCAR and MAR.

B. Ye
Department of Epidemiology and Biostatistics, School of Public Health, State University of New York, Albany, NY, USA

R. Yucel (✉)
Department of Epidemiology and Biostatistics, College of Public Health, Temple University, Philadelphia, PA, USA
e-mail: recai.yucel@temple.edu

D. L. Coffman
Department of Psychology, University of South Carolina, Columbia, SC, USA

M. Stemmler et al. (eds.), *Dependent Data in Social Sciences Research*,
https://doi.org/10.1007/978-3-031-56318-8_25

In general, there are two methods available for MI: joint modeling (JM; Carpenter and Kenward 2013; Schafer and Yucel 2002) and fully conditional (or sequential or variable-by-variable) specification (FCS) modeling (Van Buuren, 2011).

Multilevel JM draws imputations from a multivariate mixed model (two-level hierarchical model). Specifically, JM employs the Bayesian theory to generate imputations from two distributions. The missing values are drawn from a multivariate distribution that conditions on the complete variable, random effects, and fixed effects. The random effects are drawn from a multivariate distribution with zero means and corresponding variances.

Multilevel FCS imputation employs multiple draws for one incomplete variable at a time based on univariate mixed model. Specifically, the FCS imputation procedure employs a similar Bayesian procedure as JM. For each variable with missing data, imputations are drawn from a univariate distribution that conditions on all the other variables, fixed effects, and a random effect. The random effect is drawn from a univariate distribution with zero mean and corresponding variance.

There is a relatively short history of applying JM and FCS to multilevel data, so less is known about their performance. Van Buuren (2011) extended the application of FCS to multilevel data based on univariate linear mixed models. Carpenter and Kenward (2013) incorporated cluster-level variable means as covariates in the FCS imputation model. Schafer and Yucel (2002) applied JM to multilevel missing data problems based on a multivariate linear mixed model.

Multilevel imputation methods incorporate the clustering effects into the imputation process. Imputation models that account for random effects can appropriately handle multilevel data with missingness. Grund et al. (2018) applied MI to multilevel data and suggested that MI provides an effective treatment of missing data in multilevel research. Mistler and Enders (2017) examined four multilevel MI approaches and concluded that their analytic work and computer simulations showed that FCS is more restrictive by imposing implicit equality constraints on functions of the within- and between-cluster covariance matrices.

Multilevel mediation analysis is also impacted by missing data. Krull and MacKinnon (2001) and other researchers introduced multilevel mediation to appropriately assess mediation effects of a mediator in mediating the relationship between the independent variable and response variable for clustered data. However, limited research has been conducted to analyze the impact of imputation models on multilevel mediation analysis. In a single-level mediation analysis model, Zhang and Wang (2013) introduced and compared four approaches to deal with missing data including listwise deletion, pairwise deletion, MI, and a two-stage maximum likelihood (TS-ML) under various missingness mechanisms through simulation studies. They showed that MI performed well when the missingness mechanism is independent of any of the observed variables (i.e., MCAR) or when the missingness mechanism depends only on the observed variables but not on the missing variables (i.e., MAR). Milletich II (2018) introduced a Bayesian bootstrap (BB) method to generate posterior inferences for the mediation effect and impute missing data using linear and logistic regression models. They incorporated a BB method with MI and claimed it performed well. However, they did not extend their research to multilevel mediation. Ye and Yucel (2020) compared the performance of imputation models (JM and FCS) for single-level mediation analysis and found that JM and FCS perform similarly in estimating the mediation effect. However, the same result is not guaranteed in the multilevel mediation analysis model because it is constructed by multiple univariate mixed-model equations.

Many of the developments so far on MI for multilevel data have been made available to the practitioner. A JM approach for clustered data was developed by Schafer and Yucel (2002) for the linear mixed-effect model (LMM) in the R package, `pan`. Multilevel JM draws imputations from a multivariate mixed model. van Buuren (2012) developed the R package `mice` that implements a wide variety of algorithms under FCS imputation and provides additional options to draw from the potential conditional predictive distributions. Multilevel FCS imputation employs multiple imputation draws for one incomplete variable at a time based on a univariate mixed model. There are also programs available in other platforms. Raghunathan et al. (2001) developed a SAS macro, "IveWare," for the application of FCS imputation in SAS. In STATA, Royston and White (2011) developed the procedure "ICE" to implement MI.

We focus on analytically comparing JM and FCS imputation methods in multilevel mediation analysis with missing data. First, we introduce how to estimate multilevel mediation effects, followed by MI methods for this model. Then, the analytical comparison of the two MI methods is provided. Finally, we conduct a simulation study with finite samples to assess and compare the performance of the methods under varying scenarios (different missingness rate, mediation effect size, missingness mechanism, and intracluster correlation coefficient (ICC) size).

25.2 Methods

Notation

Suppose K random variables $D = (D_1, \ldots, D_K)^T$ are intended to be observed on N subjects. Subscripts p and q are used to index subjects and variables, respectively, $(p = 1, \ldots, N; q = 1, \ldots, K)$. Let d_{pq} denote the pth row and qth column element in an $(N * K)$ matrix. We denote the column q of matrix d by $d_q = (d_{1q}, \ldots, d_{Nq})^T$. We denote the complete data matrix as $U = (u_{pq})$, the missing data indicator matrix as $R = (R_{pq})$, and ϕ as unknown parameters. In the multilevel mediation analysis model, Y is denoted as the response variable, M is denoted as the mediator, and X is denoted as the independent variable. Let y_{ij} denote a value of the random variable Y for subject $j = 1, 2, \ldots, n_i$ in cluster $i = 1, 2, \ldots, m$. For example, the independent variable X_{ij} indicates the value for jth subject at the ith hospital.

Three variables (X_{ij}, M_{ij}, and Y_{ij}) are included in the data matrix D. η is the matrix that includes M and Y, δ represents the mediation effect, and β is the matrix of regression coefficients for the mediator. All three variables, X_{ij}, M_{ij}, and Y_{ij}, are measured at the individual level. For MI, all variables, including covariates and the response variable, need to be included in the imputation model. Finally, we can partition Y_{ij} into $Y_{ij(obs)}$, the portion of the outcome variable that is observed and $Y_{ij(mis)}$, the portion of the outcome variable that is unobserved.

1-1-1 Multilevel Mediation Analysis Model

Mediation analysis examines the indirect effect (mediation effect) of an independent variable on the response variable by a mediator. In contrast to the single-level mediation analysis model, the multilevel mediation model considers interventions that occur in clusters, such as hospitals or schools (Ellickson et al., 2003). In this study, we focus on the two-level 1-1-1 mediation model (Krull & MacKinnon, 2001), in which level 1 is nested within level 2. In the 1-1-1 multilevel mediation model, the independent variable (X_{ij}), mediator (M_{ij}), and response variable (Y_{ij}) are all evaluated at level 1 (individual level). The ij subscript on each variable refers to an individual j within cluster i.

Equations (25.1)–(25.3) present the 1-1-1 multilevel mediation model:

$$Y_{ij} = \alpha_{1i} + cX_{ij} + \epsilon_{1ij}, \tag{25.1}$$

$$M_{ij} = \alpha_{2i} + aX_{ij} + \epsilon_{2ij}, \tag{25.2}$$

$$Y_{ij} = \alpha_{3i} + c'X_{ij} + bM_{ij} + \epsilon_{3ij}, \tag{25.3}$$

where Y_{ij} represents the outcome (or response) variable, which is a continuous variable in this study. M_{ij} represents the mediator, which is also a continuous variable. X_{ij} refers to the independent variable, which could be either a discrete or continuous variable. c, a, c', and b are fixed effects. ϵ_{1ij}, ϵ_{2ij}, and ϵ_{3ij} are error terms, and α_{1i}, α_{2i}, and α_{3i} are random effects. We assume normal distributions for the error terms and random effects: $\epsilon_{1ij} \sim N(0, \sigma^2_{\epsilon_{1ij}})$, $\epsilon_{2ij} \sim N(0, \sigma^2_{\epsilon_{2ij}})$, $\epsilon_{3ij} \sim N(0, \sigma^2_{\epsilon_{3ij}})$, $\alpha_{1i} \sim N(0, \sigma^2_{\alpha_{1i}})$, $\alpha_{2i} \sim N(0, \sigma^2_{\alpha_{2i}})$, and $\alpha_{3i} \sim N(0, \sigma^2_{\alpha_{3i}})$ for $i = 1, 2, \ldots, m$ and $j = 1, 2, \ldots, n_i$. We assume that ϵ_{1ij}, ϵ_{2ij}, and ϵ_{3ij} are independent of the corresponding dependent and independent variables. c is the total effect of X on Y. c' is the direct effect of X on Y that does not go through M. $a \times b$ is the indirect effect (i.e., mediation effect) of X on Y that goes through variable M.

Multiple Imputation

MI is a very useful method to impute missing values by replacing them with values that are typically multiple independent draws from a posterior predictive distribution of missing data to generate m complete datasets. The missing values are drawn from their posterior predictive distribution as follows:

$$P(Y_{ij(mis)}|Y_{ij(obs)}) = \int P(Y_{ij(mis)}|Y_{ij(obs)}, \phi)P(\phi|Y_{ij(obs)})d\phi \tag{25.4}$$

After imputation, there are m complete datasets, which can be pooled using the standard MI combination rules (Rubin, 1987). Because $P(\phi|Y_{ij(obs)})$ is difficult to estimate, we use a Markov chain Monte Carlo (MCMC) method. Specifically, we use a Gibbs sampler, which is a Bayesian simulation technique that samples from the conditional distributions (Van Buuren et al., 2006).

First, we draw ϕ from the conditional distribution of $P(\phi|\alpha, Y)$. This process includes three steps.

(1) Draw $\alpha^{(t+1)}$ from the following conditional distribution:

$$\alpha^{(t+1)} \sim P(\alpha|Y_{ij(obs)}, Y_{ij(mis)}^{(t)}, \phi^{(t)}) \tag{25.5}$$

(2) Draw $\phi^{(t+1)}$ as follows:

$$\phi^{(t+1)} \sim P(\phi|Y_{ij(obs)}, Y_{ij(mis)}^{(t)}, \alpha^{(t+1)}) \tag{25.6}$$

(3) Draw $Y_{ij(mis)}^{(t+1)}$ as follows:

$$Y_{ij(mis)}^{(t+1)} \sim P(Y_{ij(mis)}|Y_{ij(obs)}, \phi^{(t+1)}, \alpha^{(t+1)}) \tag{25.7}$$

Based on the initial values of $(\alpha, \phi, Y_{ij(mis)})^{(0)}$, we can use the above steps to finalize a full cycle of the Gibbs sampler. The cycle will keep generating sequences for ϕ and $Y_{ij(mis)}$ (i.e., $\{\phi^{(1)}, \phi^{(2)}, \phi^{(3)}, \ldots\}$ and $\{Y_{ij(mis)}^{(1)}, Y_{ij(mis)}^{(2)}, Y_{ij(mis)}^{(3)}, \ldots\}$) that follow a Markov chain.

25.3 Analytic Comparison of JM Versus FCS Imputation

Population Joint Distribution

We consider a multivariate normal distribution as the imputation model, which is one of the most commonly used joint parametric structures, for the three variables

Y, M, and X:

$$\begin{bmatrix} Y \\ M \\ X \end{bmatrix} \sim N_3(\boldsymbol{\mu}, \boldsymbol{\Sigma}) \tag{25.8}$$

where the mean vector $\boldsymbol{\mu}$ and covariance matrix $\boldsymbol{\Sigma}$ are

$$\boldsymbol{\mu} = \begin{bmatrix} \mu_Y \\ \mu_M \\ \mu_X \end{bmatrix}$$

$$\boldsymbol{\Sigma} = \boldsymbol{\Sigma_W} + \boldsymbol{\Sigma_B} = \begin{bmatrix} \sigma^2_{Y(W)} + \sigma^2_{Y(B)} & \sigma_{YM(W)} + \sigma_{YM(B)} & \sigma_{YX(W)} + \sigma_{YX(B)} \\ \sigma_{MY(W)} + \sigma_{MY(B)} & \sigma^2_{M(W)} + \sigma^2_{M(B)} & \sigma_{MX(W)} + \sigma_{MX(B)} \\ \sigma_{XY(W)} + \sigma_{XY(B)} & \sigma_{MX(W)} + \sigma_{MX(B)} & \sigma^2_{X(W)} + \sigma^2_{X(B)} \end{bmatrix},$$

where μ_Y, μ_M, and μ_X are expectations of Y, M, and X, respectively. The variance-covariance matrix, $\boldsymbol{\Sigma}$, is partitioned into within- and between-cluster components ($\boldsymbol{\Sigma_W}$ and $\boldsymbol{\Sigma_B}$) as shown above. For example, σ^2_Y is the variance of Y and σ_{YX} is the covariance of Y and X. σ^2_Y and σ_{YX} are partitioned into within-cluster variance and covariance (i.e., $\sigma^2_{Y(W)}$ and $\sigma_{YX(W)}$) and between-cluster variance and covariance (i.e., $\sigma^2_{Y(B)}$ and $\sigma_{YX(B)}$).

The between-cluster variance of Y (i.e., $\sigma^2_{Y(B)}$) is assumed to be equal to the variance of the random effect in the model for Y (i.e., α_{Y_i}). The between-cluster variance of M (i.e., $\sigma^2_{M(B)}$) is assumed to be equal to the random effect in the model for M (i.e., α_{M_i}). Assuming that α_{Y_i} and α_{M_i} are independent, we have the joint distribution of Y, M, and X and the random effects (α_Y and α_M) as follows:

$$\begin{bmatrix} Y_{ij} \\ M_{ij} \\ X_{ij} \\ \alpha_{Y_i} \\ \alpha_{M_i} \end{bmatrix} \sim N(\boldsymbol{\mu}, \boldsymbol{\Sigma})$$

where the mean vector and covariance matrix are

$$\boldsymbol{\mu} = \begin{bmatrix} \mu_Y \\ \mu_M \\ \mu_X \\ 0 \\ 0 \end{bmatrix}$$

$$\boldsymbol{\Sigma}=\begin{bmatrix} \sigma^2_{Y(W)}+\sigma^2_{Y(B)} & \sigma_{YM(W)}+\sigma_{YM(B)} & \sigma_{YX(W)}+\sigma_{YX(B)} & \sigma^2_{Y(B)} & \sigma_{YM(B)} \\ \sigma_{YM(W)}+\sigma_{YM(B)} & \sigma^2_{M(W)}+\sigma^2_{M(B)} & \sigma_{MX(W)}+\sigma_{MX(B)} & \sigma_{MY(B)} & \sigma^2_{M(B)} \\ \sigma_{XY(W)}+\sigma_{XY(B)} & \sigma_{MX(W)}+\sigma_{MX(B)} & \sigma^2_{X(W)}+\sigma^2_{X(B)} & \sigma_{XY(B)} & \sigma_{MX(B)} \\ \sigma^2_{Y(B)} & \sigma_{MY(B)} & \sigma_{XY(B)} & \sigma^2_{Y(B)} & 0 \\ \sigma_{YM(B)} & \sigma^2_{M(B)} & \sigma_{MX(B)} & 0 & \sigma^2_{M(B)} \end{bmatrix}.$$

In the following sections, the process of constructing conditional distributions underlying the computation and drawing the multiple imputations using JM and FCS is based on this joint distribution of $Y_{ij}, M_{ij}, X_{ij}, \alpha_{Y_i}, \alpha_{M_i}$.

Implied Conditional Distributions Under JM

We first formulate the population joint distribution as a conditional distribution in terms of mixed-model parameters. The LMM as given in Eqs. (25.1)–(25.3) can be rewritten as a two-level hierarchical model:

$$\begin{bmatrix} Y_{ij} \\ M_{ij} \end{bmatrix} | X_{ij}, \alpha_Y, \alpha_M \sim N\left(\begin{bmatrix} \alpha_{Y_i} + \beta_{Y|X} X_{ij} \\ \alpha_{M_i} + \beta_{M|X} X_{ij} \end{bmatrix}, \boldsymbol{\Sigma}_{Y,M|X,\alpha_Y,\alpha_M}\right) \tag{25.9}$$

$$\begin{bmatrix} \alpha_{Y_i} \\ \alpha_{M_i} \end{bmatrix} \sim N\left(\begin{bmatrix} 0 \\ 0 \end{bmatrix}, \begin{bmatrix} \sigma^2_{Y(B)} & 0 \\ 0 & \sigma^2_{M(B)} \end{bmatrix}\right) \tag{25.10}$$

where α_{Y_i} and α_{M_i} are random effects. $\beta_{Y|X}$ and $\beta_{M|X}$ are fixed effects. The prior distribution for the random effects, $\boldsymbol{\alpha_i}$, is multivariate normal with zero means and variances ($\sigma^2_{\alpha_{Y_i}}$ and $\sigma^2_{\alpha_{M_i}}$). $\boldsymbol{\Sigma}_{Y,M|X,\alpha_Y,\alpha_M}$ is the variance-covariance matrix for the conditional distribution of $Y, M|X, \alpha_Y, \alpha_M$, and it is calculated as

$$\boldsymbol{\Sigma}_{Y,M|X,\alpha_Y,\alpha_M} = \boldsymbol{\Sigma}_{11} - \boldsymbol{\Sigma}_{12}\boldsymbol{\Sigma}_{22}^{-1}\boldsymbol{\Sigma}_{21} \tag{25.11}$$

where $\boldsymbol{\Sigma}_{11}$, $\boldsymbol{\Sigma}_{12}$, and $\boldsymbol{\Sigma}_{22}$ are given as

$$\boldsymbol{\Sigma}_{11} = \begin{bmatrix} \sigma^2_{Y(W)}+\sigma^2_{Y(B)} & \sigma_{YM(W)}+\sigma_{YM(B)} \\ \sigma_{YM(W)}+\sigma_{YM(B)} & \sigma^2_{M(W)}+\sigma^2_{M(B)} \end{bmatrix},$$

$$\boldsymbol{\Sigma}_{12} = \begin{bmatrix} \sigma_{YX(W)}+\sigma_{YX(B)} & \sigma^2_{Y(B)} & \sigma_{YM(B)} \\ \sigma_{MX(W)}+\sigma_{MX(B)} & \sigma_{YM(B)} & \sigma^2_{M(B)} \end{bmatrix},$$

$$\boldsymbol{\Sigma}_{22} = \begin{bmatrix} \sigma^2_{X(W)}+\sigma^2_{X(B)} & \sigma_{XY(B)} & \sigma_{MX(B)} \\ \sigma_{XY(B)} & \sigma^2_{Y(B)} & 0 \\ \sigma_{MX(B)} & 0 & \sigma^2_{M(B)} \end{bmatrix},$$

where $\mathbf{\Sigma}_{21}$ is the transpose of $\mathbf{\Sigma}_{12}$. Details on the elements of $\mathbf{\Sigma}_{22}^{-1}$ and $\mathbf{\Sigma}_{Y,M|X,\alpha_Y,\alpha_M}$ are given in Appendix section "Derivation for $\mathbf{\Sigma}_{22}^{-1}$, $\mathbf{\Sigma}_{Y_{ij},M_{ij}|X_{ij},\alpha_Y,\alpha_M}$".

Individual Distributions Under FCS

Parametric FCS employs a set of mixed models to approximate the joint predictive posterior distribution of missing data for Y and M. Specifically,

$$Y_{ij} \mid M_{ij}, X_{ij}, \alpha_{Y_i} \sim N(\alpha_{Y_i} + \beta_{Y|X} X_{ij} + \beta_{Y|M} M_{ij}, \sigma^2_{Y|M,X}) \tag{25.12}$$

$$\alpha_{Y_i} \sim N(0, \sigma^2_{Y(B)}). \tag{25.13}$$

This specification forms the basis for drawing from a posterior predictive distribution of Y, typically using an MCMC-type algorithm incorporating uncertainties due to model parameters in Eqs. (25.12) and (25.13). The procedure of imputing missing values of M proceeds in a same manner under the following model:

$$M_{ij} \mid Y_{ij}, X_{ij}, \alpha_{M_i} \sim N(\alpha_{M_i} + \beta_{M|X} X_{ij} + \beta_{M|Y} Y_{ij}, \sigma^2_{M|Y,X}) \tag{25.14}$$

$$\alpha_{M_i} \sim N(0, \sigma^2_{M(B)}). \tag{25.15}$$

Bias Under JM and FCS

In this section, we assess the impact of imputation models on the multilevel mediation effect, denoted as δ. We first formulate the MLE of δ and then analytically estimate the bias generated by imputation models using the JM and FCS approaches as outlined in sections "Population joint distribution", "Implied Conditional Distributions Under JM", and "Individual Distributions Under FCS". Under the multilevel mediation analysis models given in Eqs. (25.1)–(25.3), the direct effect is estimated by the coefficient of X_{ij}, c', in Eq. (25.3). The mediation effect, δ, is estimated by the multiplication of the coefficient of X_{ij} on M_{ij} (a in Eq. (25.2)) and the coefficient of M_{ij} on Y_{ij} (b in Eq. (25.3)). The total effect is estimated by the coefficient of X_{ij} on Y_{ij}, c, in Eq. (25.1). Mackinnon et al. (1995) have shown the algebraic equivalence of the $a \times b$ and $c - c'$ estimates of the mediation effect by normal theory ordinary least squares and MLE.

We rewrite each LMM in Eqs. (25.1), (25.2), and (25.3) as a two-level model. Then, using MLE or the weighted least squares estimator (WLS), we can compute $\hat{c}$ in Eq. (25.1) as

$$\hat{c} = (X^T V_1^{-1} X)^{-1} X^T V_1^{-1} Y$$

where

$$Y = \begin{pmatrix} Y_1 \\ \vdots \\ Y_n \end{pmatrix} \in \Re^{n\times 1}, \text{ where } n = \sum_{i=1}^{m} n_i, \; X = \begin{pmatrix} X_1 \\ \vdots \\ X_n \end{pmatrix} \in \Re^{n\times 1}, \; V_1 = I_n G_{\alpha_{1i}} I_n^T + R_1$$

$G_{\alpha_{1i}}$ is a diagonal matrix of the variance of α_{1i} ($\sigma^2_{\alpha_{1i}}$), I_n is an identity matrix of size n, and R_1 is a $n \times n$ variance matrix of residuals for the regression model in Eq. (25.1). Similarly, the MLE or WLS of $\hat{a}$ in Eq. (25.2) is as follows:

$$\hat{a} = (X^T V_2^{-1} X)^{-1} X^T V_2^{-1} M$$

where

$$M = \begin{pmatrix} M_1 \\ \vdots \\ M_n \end{pmatrix} \in \Re^{n\times 1}, \; V_2 = I_n G_{\alpha_{2i}} I_n^T + R_2$$

$G_{\alpha_{2i}}$ is a diagonal matrix of the variance of α_{2i}, and R_2 is a $n \times n$ variance matrix of residuals for the regression model in Eq. (25.2). The MLE or WLS of $\hat{c}'$ and $\hat{b}$ is as follows:

$$(\hat{c}', \hat{b})^T = (X_{[M]}^T V_3^{-1} X_{[M]})^{-1} X_{[M]}^T V_3^{-1} Y$$

where

$$X_{[M]} = \begin{pmatrix} X_1 & M_1 \\ \vdots & \vdots \\ X_n & M_n \end{pmatrix} \in \Re^{n\times 2}, \; V_3 = I_n G_{\alpha_{3i}} I_n^T + R_3$$

$G_{\alpha_{3i}}$ is a diagonal matrix of the variance of α_{3i}, and R_3 is a $n \times n$ variance matrix of residuals for the LMM in Eq. (25.3).

Using the invariance property of the ML theory, the MLE of $\hat{\delta}$ can be obtained as

$$\hat{\delta} = \hat{a}\hat{b} = Z(X^T V_2^{-1} X)^{-1} X^T V_2^{-1} M (X_{[M]}^T V_3^{-1} X_{[M]})^{-1} X_{[M]}^T V_3^{-1} Y \qquad (25.16)$$

where Z is a 1×2 matrix, i.e., $Z = (0, 1)$. Computational details of deriving this MLE of $\hat{a}$, $\hat{b}$, $\hat{c}$, $\hat{c}'$, and $\hat{\delta}$ are provided in Appendix section "Derivation of MLE Estimator for $\hat{c}$, $\hat{a}$, $\hat{b}$, and $\hat{\delta}$".

Mediation Effect Estimate Under JM and FCS

When Y is subject to missing values and M and X are fully observed in a dataset and MI proceeds under JM or FCS (Eqs. (25.12) and (25.13)), the expression for $\hat{\delta}$ is consistent for both JM and FCS imputation methods in estimating the mediation effect under the multilevel mediation analysis model. When there are two incomplete variables (Y and M) and one fully observed variable (X) in the dataset, the bias of the MLE estimate of the mediation effect using JM and FCS can be found as follows.

First, consider FCS as the imputation model. FCS employs the conditional distributions of Y_{ij} given M_{ij}, X_{ij}, and α_{Yi} with the prior distribution of α_{Y_i}, and M_{ij} given Y_{ij}, X_{ij}, and α_{Mi} with the prior distribution of α_{M_i} to impute incomplete values for Y_{ij} and M_{ij}, respectively. The conditional distribution of Y_{ij} given M_{ij}, X_{ij}, and α_{Yi}, and prior distribution of α_{Y_i} are given in Eqs. (25.12) and (25.13). We multiply the conditional distribution of Y_{ij} given M_{ij}, X_{ij}, and α_{Y_i} with the prior distribution of α_{Y_i}. The conditional distribution of Y_{ij} and α_{Y_i} given M_{ij} and X_{ij} is proportional to (details are shown in Appendix section "Derivation of $f(Y|M, X)$ and $E(Y|M, X)$ for FCS")

$$f(Y_{ij}, \alpha_{Y_i}|M_{ij}, X_{ij}) \propto f(Y_{ij} \mid M_{ij}, X_{ij}, \alpha_{Y_i}) \times f(\alpha_{Y_i})$$

Integrating the conditional distribution of Y_{ij} and α_{Y_i} given M_{ij} and X_{ij} with respect to α_{Y_i}, the conditional distribution of Y_{ij} given M_{ij} and X_{ij} is free of the random effect, α_{Y_i}. Details of this derivation of the distribution of Y_{ij} given M_{ij} and X_{ij} are provided in the Appendix section "Derivation of $f(Y|M, X)$ and $E(Y|M, X)$ for FCS". The distribution of Y_{ij} given M_{ij} and X_{ij} is normal with mean, $\beta_{Y|X}X_{ij} + \beta_{Y|M}M_{ij}$, and variance, $\sigma^2_{Y|M,X} + \sigma^2_{\alpha_Y}$. The conditional expectation of Y_{ij} given M_{ij} and X_{ij} is $\beta_{Y|X}X + \beta_{Y|M}M$, and the estimate of $\hat{\beta}_{Y|M}$ ($\hat{b}$ in mediation analysis model) is

$$\hat{b} = \hat{\beta}_{Y|M} = \frac{\hat{\sigma}^2_M \hat{\sigma}_{YX} - \hat{\sigma}_{MX}\hat{\sigma}_{MY}}{\hat{\sigma}^2_M \hat{\sigma}^2_X - \hat{\sigma}^2_{MX}}$$

Next, multiplying the conditional distribution of M_{ij} given Y_{ij}, X_{ij}, and α_{M_i} with the prior distribution of α_{M_i} in Eqs. (25.14) and (25.15) results in the conditional distribution of M_{ij}, α_{M_i} given Y_{ij} and X_{ij} (details are shown in Appendix section "Derivation of $f(Y|M, X)$ and $E(Y|M, X)$ for FCS"):

$$f(M_{ij}, \alpha_{M_i}|Y_{ij}, X_{ij}) = f(M_{ij} \mid Y_{ij}, X_{ij}, \alpha_{M_i}) \times f(\alpha_{M_i})$$

The conditional distribution of M_{ij} given Y_{ij} and X_{ij} is obtained in a similar fashion to the previous derivation (see Appendix section "Derivation of $f(M|Y, X)$ for FCS" for details) and is normally distributed with mean, $\beta_{M|X}X_{ij} + \beta_{M|Y}Y_{ij}$, and variance, $\sigma^2_{M|Y,X} + \sigma^2_{\alpha_M}$. Then, we find the conditional distribution of M_{ij} and

Y_{ij} given X_{ij} by multiplying the conditional distribution of M_{ij} given Y_{ij} and X_{ij} with the distribution of Y_{ij}. Y_{ij} is distributed as $N(\mu_Y, \sigma^2_{Y(W)} + \sigma^2_{Y(B)})$. Integrating the conditional distribution of M_{ij} given Y_{ij} and X_{ij} with respect to Y_{ij} gives the distribution of M_{ij} given X_{ij}:

$$\begin{aligned} f(M_{ij}|X_{ij}) &= \int f(M_{ij}|Y_{ij}, X_{ij}) f(Y_{ij}) dY \\ &= \int f(M_{ij}, Y_{ij}|X_{ij}) dY \end{aligned}$$

As we show in Appendix section "Derivation of $f(M|X)$ and $E(M|X)$ for FCS", the conditional distribution of M_{ij} given X_{ij} is a shifted normal distribution:

$$f(M_{ij}|X_{ij}) = \frac{1}{\sqrt{\pi(\beta^2_{M|Y}\sigma^2_Y + \sigma^2_{M|Y,X} + \sigma^2_{\alpha_M})}} e^{A_2 M^2 + B_2 M + C_2}$$

where A_2, B_2, and C_2 are functions of σ^2_Y, $\sigma^2_{\alpha_M}$, $\beta_{M|Y}$, etc. The expressions for A_2, B_2, and C_2 are provided in Appendix section "Derivation of $f(M|X)$ and $E(M|X)$ for FCS". By calculating the expectation of M_{ij} given X_{ij} (see Appendix section "Derivation of $f(M|X)$ and $E(M|X)$ for FCS" for details), the estimate of $\hat{a}$ is given as

$$\hat{a} = [\frac{\hat{\sigma}^2_Y \hat{\beta}^2_{M|Y} \hat{\beta}_{M|X}}{(\hat{\beta}^2_{M|Y}\hat{\sigma}^2_Y + \hat{\sigma}^2_{M|Y,X} + \hat{\sigma}^2_\alpha)(\hat{\sigma}^2_{M|X,Y} + \hat{\sigma}^2_\alpha)} - \frac{\hat{\beta}_{M|X}}{\hat{\sigma}^2_{M|X,Y} + \hat{\sigma}^2_\alpha}] \times \frac{1}{2\pi \hat{A}\sqrt{2\hat{A}(\hat{\beta}^2_{M|X}\hat{\sigma}^2_Y + \hat{\sigma}^2_{M|Y,X} + \hat{\sigma}^2_\alpha)}}$$

Thus, the estimate of the mediation effect after applying FCS to data with missing values is

$$\hat{\delta}_{FCS} = \hat{a}\hat{b} = [\frac{\hat{\sigma}^2_Y \hat{\beta}^2_{M|Y} \hat{\beta}_{M|X}}{(\hat{\beta}^2_{M|Y}\hat{\sigma}^2_Y + \hat{\sigma}^2_{M|Y,X} + \hat{\sigma}^2_\alpha)(\hat{\sigma}^2_{M|X,Y} + \hat{\sigma}^2_\alpha)} - \frac{\hat{\beta}_{M|X}}{\hat{\sigma}^2_{M|X,Y} + \hat{\sigma}^2_\alpha}] \times \frac{\hat{\beta}_{M|X}}{2\pi \hat{A}\sqrt{2\hat{A}(\hat{\beta}^2_{M|X}\hat{\sigma}^2_Y + \hat{\sigma}^2_{M|Y,X} + \hat{\sigma}^2_\alpha)}} \tag{25.17}$$

We implement a similar strategy to estimate bias for JM as we did for FCS. JM draws incomplete variables, Y_{ij} and M_{ij}, together from the joint conditional distribution of Y_{ij} and M_{ij} given X_{ij}, α_Y, and α_M with the prior joint distribution of α_Y and α_M in Eqs. (25.9) and (25.10). However, in contrast to FCS, a and b are not independently generated, so δ cannot be directly estimated by multiplying a and b. Therefore, we use the equivalent $\hat{c}-\hat{c}'$ to estimate $\hat{\delta}$. The details for estimating $\hat{c}$ and $\hat{c}'$ are given in Appendices section "Derivation of $f(Y, M|X)$ for JM", "Derivation of $f(Y|X)$, $E(Y|X)$, and $f(M|X)$ for JM", and "Derivation of $f(Y|M, X)$ and $E(Y|M, X)$ for JM". The estimate of the mediation effect after applying JM to data with missing values is

$$\begin{aligned}\hat{\delta}_{JM} &= \hat{c} - \hat{c}' \\ &= -\frac{2\sqrt{2(\hat{f}\hat{k}-\hat{g}\hat{h})\hat{\sigma}_Y^2 H}}{\sqrt{(\hat{k}\hat{\sigma}_Y^2+\hat{f}\hat{k}-\hat{g}\hat{h})\hat{\Sigma}_C\hat{\Sigma}_\alpha(\hat{P}^2-4\hat{Q}\hat{H})}} \times \frac{\hat{L}\hat{R}}{\hat{P}^2-4\hat{H}\hat{Q}} - \\ &\quad \frac{\hat{E}\sqrt{(\hat{f}\hat{k}-\hat{g}\hat{h})\hat{\sigma}_Y^2(\hat{\beta}_{M|Y}^2\hat{\sigma}_Y^2+\hat{\sigma}_{M|Y,X}^2+\hat{\sigma}_\alpha^2)}}{\sqrt{(\hat{k}\hat{\sigma}_Y^2+\hat{f}\hat{k}-\hat{g}\hat{h})\hat{\Sigma}_C\hat{\Sigma}_\alpha\hat{A}_4}}\left(\frac{\hat{F}}{\hat{E}}\hat{\beta}_{M|X}-\hat{\beta}_{Y|X}\right)\end{aligned} \quad (25.18)$$

where the expressions for $\hat{f}$, $\hat{g}$, $\hat{h}$, and $\hat{k}$ are given in Appendix section "Derivation for $\mathbf{\Sigma}_{22}^{-1}$, $\mathbf{\Sigma}_{Y_{ij},M_{ij}|X_{ij},\alpha_Y,\alpha_M}$"; the expressions for $\hat{A}_4$, $\hat{H}$, and K are given in Appendix section "Derivation of $f(Y, M|X)$ for JM"; the expressions for $\hat{P}$, $\hat{Q}$, and $\hat{R}$ are given in Appendix section "Derivation of $f(Y|X)$, $E(Y|X)$, and $f(M|X)$ for JM"; and the expressions for $\hat{E}$, $\hat{F}$, and $\hat{L}$ are given in Appendix section "Derivation of $f(Y|M, X)$ and $E(Y|M, X)$ for JM".

θ is denoted as the bias (difference between the expected value of the estimator and true value) of the estimate of the mediation effect after applying MI. θ_{FCS} is the difference between the expectation of δ after implementing FCS and its corresponding true value:

$$\hat{\theta}_{FCS} = E(\hat{\delta}_{FCS}) - \hat{\delta} \quad (25.19)$$

where $\hat{\delta}_{FCS}$ is given in Eq. (25.17). θ_{JM} is the difference between the expectation of δ after implementing JM and its corresponding true value:

$$\hat{\theta}_{JM} = E(\hat{\delta}_{JM}) - \hat{\delta} \quad (25.20)$$

where $\hat{\delta}_{JM}$ is given in Eq. (25.18). The true value of $\hat{\delta}$ in Eqs. (25.19) and (25.20) is estimated based on the MLE or WLS estimate of δ in Eq. (25.16) before applying any imputation models. MLE has the consistency property that it converges to the true parameter in probability as $n \to \infty$. Applying Slutsky's theorem to find the expectation of $E(\hat{\delta}_{FCS})$ and $E(\hat{\delta}_{JM})$, $Y_n X_n \to aX$ in distribution, if $X_n \to X$ in distribution and $Y_n \to a$ in probability, where a is a constant.

These formulas can be further simplified by factoring out $\beta_{M|X}$ (the coefficient of $\mu_{M|X,Y,\alpha_M}$ for X) in the bias estimate for FCS. The relationship between M and

X is directly related to the expected bias generated by FCS. Unfortunately, factoring out $\beta_{M|X}$ is not possible for JM.

25.4 Simulation Study

We conduct a simulation study to assess our derivations as well as the performance of different MI methods as implemented in the R `pan` and `mice` packages in multilevel mediation analysis. The `pan` package uses a joint multivariate LMM as the imputation model, while the `mice` package imputes continuous two-level data based on univariate LMMs.

Data Generation

The data generation process first simulates the independent variable, X_{ij}, from an independent normal distribution, i.e., $X_{ij} \sim N(0, \sigma_X^2)$. Next, the mediator M_{ij} is simulated from the following mixed-effects formulation:

$$M_{ij} = aX_{ij} + \alpha_{1i} + \epsilon_{1ij},$$

where α_{1i} is a cluster random effect assumed to follow a normal distribution, i.e., $\alpha_{1i} \sim N(0, \sigma_{\alpha_1}^2)$ for any i. ϵ_{ij} refers to the residual error term and is assumed to be generated from a normal distribution, i.e., $\epsilon_{1ij} \sim N(0, \sigma_{\epsilon_1}^2)$ for any i and j. α_{1i} and ϵ_{1ij} are assumed to be independent. Further, α_{1i} and X as well as ϵ_{1ij} and X are also assumed to be independent. There is an independence assumption for $\{X, M\}$ and α_{3i} as well as $\{X, M\}$ and ϵ_{3ij}.

Next, we simulated Y_{ij} conditional on M_{ij} and X_{ij} using the following mixed-effects formulation:

$$Y_{ij} = c'X_{ij} + bM_{ij} + \alpha_{3i} + \epsilon_{3ij}$$

where α_{3i} and ϵ_{3ij} refer to the random effect and residual error, respectively. α_{3i} and ϵ_{3ij} are assumed to be generated from independent normal distributions, i.e., $\alpha_{3i} \sim N(0, \sigma_{\alpha_3}^2)$ and $\epsilon_{3ij} \sim N(0, \sigma_{\epsilon_3}^2)$. We use this framework to simulate a population of 1000 observations under 100 clusters. We repeat this process for 100 times to study the performance criteria stated in section "Estimation and Evaluation Criteria".

In the first simulation experiment, we set δ to 0.01 (i.e., $a = b = 0.1$, $\delta = 0.01$, and $c' = 0.5$) to test the performance of imputation methods with a small mediation effect. In the second simulation experiment, we set δ to 0.1521 (i.e., $a = b = 0.39$, $\delta = 0.1521$, and $c' = 0.5$) to study the impact of imputation methods with a medium mediation effect. In the third simulation experiment, we set δ to 0.3481 (i.e., $a =$

$b = 0.59$, $\delta = 0.3481$, and $c' = 0.5$) to examine the performance of imputation models with a large mediation effect.

The next step of data generation is imposing missing values under MCAR, MAR, and MNAR mechanisms (Rubin, 1976). We impose the missing values on the response variable, Y, and the mediator, M, under different missingness mechanisms. Let r_Y and r_M denote missingness indicators for Y and M, respectively. Under MCAR, both r_Y and r_M are simulated from a Bernoulli distribution. We set the success probability to 10%, 20%, and 30% to study the behavior under these rates of missingness for Y and M. Under MAR, the probability of missingness of Y and M depends on the fully observed variable X:

$$logit(P(r_Y = 1|x)) = \gamma_0^Y + \gamma_1^Y X,$$

$$logit(P(r_M = 1|x)) = \gamma_0^M + \gamma_1^M X,$$

where the γ (γ_0 and γ_1) for both Y and M are chosen to lead to approximately 10%, 20%, and 30% rates of missingness. Under MNAR, missingness in Y and M depends on Y and M themselves, respectively. If Y is smaller than its sth percentile, Y is missing, and if M is smaller than its sth percentile, M is missing. We set s to be 10, 20, and 30 to control the missingness rates of Y and M at 10%, 20%, and 30%, respectively.

We repeat this simulation process three times to investigate the impact of different ICC sizes. We vary σ_{α_1} from 0.5 to 2, and σ_{ϵ_1} from 2 to 4, where ICC=$\frac{\sigma_\alpha^2}{\sigma_\alpha^2 + \sigma_\epsilon^2}$. We report results only for the scenarios where ICC equals 0.05, 0.1, and 0.2.

Estimation and Evaluation Criteria

For each simulation scenario, we create a total of five imputations under each MI method. Literature on MI suggests that 3–40 imputations yield excellent results. Our choice of five imputations is based on the empirical results that indicate the fraction of missing information is stabilized mostly after five imputations. The general choice for the number of imputations is discussed in Graham et al. (2007); Royston (2005), and Bodner (2008). The estimates and their standard errors are combined using Rubin's rules (Rubin, 1976).

Comparisons of different imputation methods are conducted based on different effect sizes of mediation effects, which are always focal to mediation analysis. We use the prespecified three different mediation effects (δ) as the "true values" to test the accuracy of different imputation methods. We use four different criteria to assess the consistency and accuracy for each imputation method. Let $\hat{l}_r$ and $\hat{u}_r$ denote the lower and upper limits of the 95% confidence interval (CI) for the mediation effect in the rth imputed data. We also denote $\hat{\delta}_r = \hat{a}_r\hat{b}_r$ as the mediation effect estimate for the rth replication, where $r = 1, \ldots, 100$.

The first criterion to assess the performance is the coverage rate (CR):

$$CR = \frac{\sum_{r=1}^{100} I_r(\hat{l}_r < \delta < \hat{u}_r)}{100},$$

where I_r is an indicator function such that

$$I_r = \begin{cases} 1 & \delta \in (\hat{l}_r, \hat{u}_r) \\ 0 & \delta \notin (\hat{l}_r, \hat{u}_r). \end{cases}$$

CR evaluates the proportion of CIs that contain the true mediation effect value. The actual rate should be close to the nominal rate (i.e., 95%). Collins et al. (2001) suggest that the performance of an imputation method is poor if the CR is below 90% for a 95% nominal rate. Second, we estimate the standardized bias (SB) for imputation models, calculated as

$$SB = 100 \times \frac{\bar{\hat{\delta}} - \delta}{SE},$$

where SE is the standard error of the estimates and $\bar{\hat{\delta}}$ is the average estimate of the mediation effect. Collins et al. (2001) comment that once the SB exceeds 40%, the bias may have a noticeable negative impact on the imputed parameter. In this study, we consider SB as substantially different when its absolute value is greater than 40%.

The third criterion is mean square error (MSE), a measure of accuracy and precision, computed as follows:

$$MSE = \frac{1}{100}\sum_{r=1}^{100}(\hat{\delta} - \delta)^2,$$

We also calculated the average width (AW) as

$$AW = \frac{\sum_{r=1}^{100}(\hat{u}_r - \hat{l}_r)}{100}.$$

Among the four different criteria to assess the consistency and accuracy for each method under all conditions, CR and SB are the most important criteria to show the performance of imputation methods under the multilevel mediation analysis model.

The same evaluation criteria will also be assessed for complete-case-only (CC) analysis. For CC analysis, we exclude all units for which the response variable and mediator (Y and M) are missing and keep the remaining complete observations.

Then, we apply the mediation analysis model to obtain the estimate of mediation effect and compare the results with the MI methods.

Summary of the Results

We compare the performance of the FCS and JM methods for incomplete data under the multilevel mediation analysis model. Tables 25.1, 25.2, and 25.3 show CR, MSE, AW, and SB for the different scenarios for the small mediation effect condition

Table 25.1 Simulation results for small mediation effect (**δ=0.01, ICC=0.05**)

MM	Imputation model	JMR	CR	MSE	AW	SB
mcar	FCS	0.190	0.94	0.00013	0.033	0.142
	JM		0.90	0.00011	0.022	0.326
	CC		0.96	0.00005	0.033	0.171
	FCS	0.361	0.92	0.00009	0.036	0.366
	JM		0.96	0.00011	0.032	0.267
	CC		0.92	0.00007	0.036	0.151
	FCS	0.510	0.93	0.00009	0.037	0.369
	JM		0.93	0.00013	0.051	0.100
	CC		0.91	0.00011	0.040	0.266
mar	FCS	0.100	0.91	0.00013	0.054	0.111
	JM		0.92	0.00008	0.035	0.281
	CC		0.95	0.00007	0.036	0.391
	FCS	0.200	0.91	0.00014	0.061	0.070
	JM		0.96	0.00012	0.055	0.213
	CC		0.92	0.00009	0.042	0.292
	FCS	0.300	0.90	0.00010	0.063	0.264
	JM		0.93	0.00015	0.072	0.094
	CC		0.96	0.00014	0.050	**0.460**
mnar	FCS	0.193	**0.54**	0.00008	0.020	**0.505**
	JM		**0.86**	0.00006	0.014	0.220
	CC		**0.74**	0.00006	0.023	0.255
	FCS	0.368	**0.43**	0.00008	0.019	**0.428**
	JM		**0.73**	0.00008	0.020	**0.429**
	CC		**0.63**	0.00007	0.022	**0.594**
	FCS	0.522	**0.43**	0.00009	0.020	**0.526**
	JM		**0.50**	0.00010	0.025	**0.424**
	CC		**0.53**	0.00008	0.021	**0.417**

Note. MM, missingness mechanism; MR, missingness rate; JMR, joint missingness rate; CR, coverage rate; MSE, mean square error; AW, average width; SB, standardized bias; FCS, fully conditional specification; JM, joint modeling; CC, complete-case-only analysis; mcar, missing completely at random; mar, missing at random; mnar, missing not at random

Table 25.2 Simulation results for small mediation effect (**δ=0.01, ICC=0.1**)

MM	Imputation model	JMR	CR	MSE	AW	SB
mcar	FCS	0.190	0.91	0.00068	0.068	0.129
	JM		0.94	0.00128	0.085	0.225
	CC		0.93	0.00133	0.144	0.300
	FCS	0.360	0.91	0.00034	0.083	0.316
	JM		0.97	0.00111	0.138	0.271
	CC		0.94	0.00165	0.162	0.343
	FCS	0.510	0.95	0.00027	0.088	0.189
	JM		0.97	0.00160	0.182	0.273
	CC		0.94	0.00208	0.193	0.333
mar	FCS	0.100	0.97	0.00063	0.133	0.268
	JM		0.95	0.00184	0.166	0.222
	CC		0.95	0.00174	0.159	0.225
	FCS	0.200	0.99	0.00043	0.147	0.381
	JM		0.96	0.00263	0.233	0.286
	CC		0.91	0.00258	0.185	0.381
	FCS	0.300	0.96	0.00024	0.140	0.259
	JM		0.97	0.00271	0.312	**0.669**
	CC		**0.87**	0.00266	0.213	**0.664**
mnar	FCS	0.193	**0.79**	0.00020	0.045	0.144
	JM		**0.83**	0.00033	0.053	0.343
	CC		**0.85**	0.00037	0.088	0.356
	FCS	0.368	**0.81**	0.00013	0.042	**0.454**
	JM		**0.88**	0.00032	0.073	**0.685**
	CC		**0.85**	0.00040	0.090	**0.500**
	FCS	0.522	**0.79**	0.00010	0.043	**0.666**
	JM		**0.87**	0.00030	0.091	**0.553**
	CC		**0.85**	0.00044	0.097	**0.425**

Note. MM, missingness mechanism; MR, missingness rate; JMR, joint missingness rate; CR, coverage rate; MSE, mean square error; AW, average width; SB, standardized bias; FCS, fully conditional specification; JM, joint modeling; CC, complete-case-only analysis; mcar, missing completely at random; mar, missing at random; mnar, missing not at random

($a = b = 0.1$, $\delta = ab = 0.01$, $c' = 0.5$) with ICC fixed at 0.05, 0.1, and 0.2. Tables 25.4, 25.5, and 25.6 show the results for the large mediation effect ($a = b = 0.59$, $\delta = ab = 0.3481$, $c' = 0.5$) condition and ICC fixed at 0.05, 0.1, and 0.2. Figures 25.1, 25.2, and 25.3 show the SB verses the JMR for each of the mediation effect size conditions with ICC fixed at 0.05, 0.1, and 0.2, respectively. For each, we compare CRs, MSEs, AWs, and SBs of the estimated mediation effect under CC analysis and the imputation methods. Different missingness mechanisms (MCAR,

Table 25.3 Simulation results for small mediation effect (**δ=0.01, ICC=0.2**)

MM	Imputation model	JMR	CR	MSE	AW	SB
mcar	FCS	0.190	0.94	0.00230	0.129	0.204
	JM		0.93	0.00309	0.141	0.106
	CC		0.94	0.00352	0.243	0.196
	FCS	0.359	0.93	0.00136	0.152	0.244
	JM		0.93	0.00394	0.214	0.280
	CC		0.95	0.00444	0.284	0.304
	FCS	0.510	0.96	0.00078	0.147	0.341
	JM		0.96	0.00357	0.287	0.300
	CC		0.94	0.00613	0.314	0.218
mar	FCS	0.100	0.95	0.00164	0.259	0.265
	JM		0.93	0.00686	0.327	0.282
	CC		0.91	0.00621	0.294	0.288
	FCS	0.200	0.97	0.00100	0.254	0.329
	JM		0.97	0.00768	0.449	0.286
	CC		0.94	0.00695	0.345	0.384
	FCS	0.300	0.98	0.00070	0.241	0.383
	JM		0.98	0.00922	0.586	0.393
	CC		0.95	0.01025	0.390	**0.511**
mnar	FCS	0.193	**0.76**	0.00086	0.070	**0.477**
	JM		**0.88**	0.00170	0.104	**0.479**
	CC		**0.85**	0.00162	0.178	0.320
	FCS	0.367	0.93	0.00066	0.087	0.346
	JM		**0.87**	0.00125	0.136	**0.502**
	CC		**0.86**	0.00127	0.165	**0.577**
	FCS	0.521	**0.89**	0.00021	0.098	**0.728**
	JM		**0.88**	0.00118	0.161	**0.507**
	CC		**0.86**	0.00159	0.176	**0.513**

Note. MM, missingness mechanism; MR, missingness rate; JMR, joint missingness rate; CR, coverage rate; MSE, mean square error; AW, average width; SB, standardized bias; FCS, fully conditional specification; JM, joint modeling; CC, complete-case-only analysis; mcar, missing completely at random; mar, missing at random; mnar, missing not at random

MAR, and MNAR) and variations in the missingness rates (10%, 20%, and 30%) are considered for a sample size of 1000. In the figures, we focus on comparing SBs with respect to varying joint missingness rates (JMRs).

Table 25.1 presents the results for the small mediation effect scenario. The missingness rates for M and Y are fixed at 10%, 20%, and 30% marginally for each variable, and the JMR varies from 10% to 52.2%. CRs and SBs that are above acceptable limits are shown in bold. We find that FCS and JM are consistent

Table 25.4 Simulation results for large mediation effect (**δ=0.3481, ICC=0.05**)

MM	Imputation model	JMR	CR	MSE	AW	SB
mcar	FCS	0.190	0.93	0.00011	0.034	0.280
	JM		0.97	0.00007	0.021	**0.485**
	CC		0.93	0.00007	0.033	0.268
	FCS	0.360	**0.88**	0.00007	0.040	0.332
	JM		0.94	0.00009	0.034	0.391
	CC		0.96	0.00010	0.038	0.346
	FCS	0.510	0.94	0.00008	0.043	**0.433**
	JM		0.95	0.00010	0.044	0.352
	CC		0.92	0.00011	0.042	**0.454**
mar	FCS	0.100	0.91	0.00014	0.055	0.068
	JM		0.90	0.00010	0.040	0.352
	CC		**0.89**	0.00008	0.037	0.366
	FCS	0.200	0.94	0.00012	0.062	0.118
	JM		0.95	0.00013	0.057	0.181
	CC		**0.88**	0.00010	0.042	0.302
	FCS	0.300	0.96	0.00010	0.064	0.271
	JM		0.97	0.00019	0.070	0.083
	CC		**0.88**	0.00013	0.088	**0.467**
mnar	FCS	0.193	**0.55**	0.00010	0.031	0.264
	JM		**0.59**	0.00006	0.055	0.377
	CC		**0.74**	0.00006	0.053	**0.423**
	FCS	0.367	**0.45**	0.00010	0.070	0.243
	JM		**0.56**	0.00007	0.089	**0.582**
	CC		**0.64**	0.00007	0.071	**0.556**
	FCS	0.523	**0.45**	0.00010	0.051	**0.693**
	JM		**0.55**	0.00008	0.085	**0.449**
	CC		**0.64**	0.00008	0.092	**0.487**

Note. MM, missingness mechanism; MR, missingness rate; JMR, joint missingness rate; CR, coverage rate; MSE, mean square error; AW, average width; SB, standardized bias; FCS, fully conditional specification; JM, joint modeling; CC, complete-case-only analysis; mcar, missing completely at random; mar, missing at random; mnar, missing not at random

in estimating the mediation effect under MCAR and MAR mechanisms. Under MCAR, CRs ranged from 92% to 94% for FCS and from 92% to 96% for JM. Under MAR, CRs range from 90% to 91% for FCS and from 92% to 96% for JM. However, FCS, JM, and CC do not perform well under the MNAR mechanism, and we observe low CR values (below 90%). We find that SBs are below 40% when the missingness mechanisms are MCAR and MAR (except for the case when we apply CC and have a 30% JMR).

Table 25.5 Simulation results for large mediation effect (**δ=0.3481, ICC=0.1**)

MM	Imputation model	JMR	CR	MSE	AW	SB
mcar	FCS	0.190	0.94	0.00045	0.066	0.160
	JM		0.92	0.00084	0.071	0.179
	CC		0.92	0.00092	0.130	0.139
	FCS	0.360	0.95	0.00036	0.084	0.164
	JM		0.94	0.00128	0.120	0.210
	CC		0.92	0.00157	0.150	0.257
	FCS	0.510	0.97	0.00019	0.073	0.054
	JM		0.98	0.00111	0.161	0.257
	CC		0.97	0.00153	0.173	0.327
mar	FCS	0.100	0.96	0.00045	0.121	0.237
	JM		0.96	0.00138	0.165	0.232
	CC		0.94	0.00137	0.152	0.319
	FCS	0.200	0.98	0.00044	0.142	0.292
	JM		0.99	0.00205	0.241	0.335
	CC		0.93	0.00229	0.180	0.324
	FCS	0.300	0.99	0.00026	0.138	0.276
	JM		0.98	0.00310	0.316	**0.695**
	CC		0.91	0.00286	0.211	**0.662**
mnar	FCS	0.193	**0.84**	0.00023	0.051	0.190
	JM		**0.85**	0.00045	0.060	0.352
	CC		**0.76**	0.00046	0.096	0.381
	FCS	0.368	**0.81**	0.00013	0.041	0.369
	JM		0.93	0.00029	0.068	0.347
	CC		**0.78**	0.00032	0.087	**0.475**
	FCS	0.522	**0.89**	0.00013	0.042	0.360
	JM		**0.87**	0.00035	0.085	**0.490**
	CC		**0.75**	0.00043	0.091	**0.549**

Note. MM, missingness mechanism; MR, missingness rate; JMR, joint missingness rate; CR, coverage rate; MSE, mean square error; AW, average width; SB, standardized bias; FCS, fully conditional specification; JM, joint modeling; CC, complete-case-only analysis; mcar, missing completely at random; mar, missing at random; mnar, missing not at random

Tables 25.4, 25.5, and 25.6 show the results for the large mediation effect for ICC = 0.05, 0.1, and 0.2, respectively. FCS and JM perform well under MCAR and MAR. The compatibility between FCS and JM remains consistent when the JMR increases. However, none of the methods perform well under MNAR.

Figure 25.4 shows the comparison between exact bias and estimated bias by the imputation method. The bias generated by the imputation methods is very close to

Table 25.6 Simulation results for large mediation effect (**δ=0.3481, ICC=0.2**)

MM	Imputation model	JMR	CR	MSE	AW	SB
mcar	FCS	0.190	0.95	0.00169	0.124	0.279
	JM		0.98	0.00375	0.143	0.278
	CC		0.96	0.00461	0.260	0.194
	FCS	0.360	0.98	0.00107	0.134	0.293
	JM		0.92	0.00446	0.244	0.223
	CC		0.95	0.00495	0.287	0.297
	FCS	0.509	0.97	0.00052	0.147	0.466
	JM		0.96	0.00687	0.301	0.305
	CC		0.93	0.00803	0.343	0.372
mar	FCS	0.100	0.95	0.00168	0.239	0.153
	JM		**0.86**	0.00560	0.297	0.121
	CC		0.91	0.00613	0.286	0.214
	FCS	0.200	0.98	0.00115	0.244	0.218
	JM		0.93	0.00879	0.424	0.347
	CC		0.93	0.00813	0.331	0.389
	FCS	0.300	0.99	0.00082	0.222	0.393
	JM		0.91	0.00990	0.605	0.290
	CC		0.94	0.00936	0.387	**0.489**
mnar	FCS	0.193	**0.87**	0.00083	0.079	0.253
	JM		**0.79**	0.00170	0.090	0.269
	CC		**0.65**	0.00183	0.169	0.302
	FCS	0.368	**0.87**	0.00030	0.088	0.260
	JM		**0.85**	0.00110	0.424	**0.529**
	CC		**0.77**	0.00116	0.556	**0.543**
	FCS	0.522	**0.83**	0.00023	0.679	**0.669**
	JM		**0.88**	0.00093	0.551	**0.528**
	CC		**0.79**	0.00098	0.564	**0.529**

Note. MM, missingness mechanism; MR, missingness rate; JMR, joint missingness rate; CR, coverage rate; MSE, mean square error; AW, average width; SB, standardized bias; FCS, fully conditional specification; JM, joint modeling; CC, complete-case-only analysis; mcar, missing completely at random; mar, missing at random; mnar, missing not at random

the exact bias especially when the missingness mechanism is MCAR. When the missingness mechanism is MAR or MNAR, the estimated bias generated by the imputation methods is larger than the exact bias.

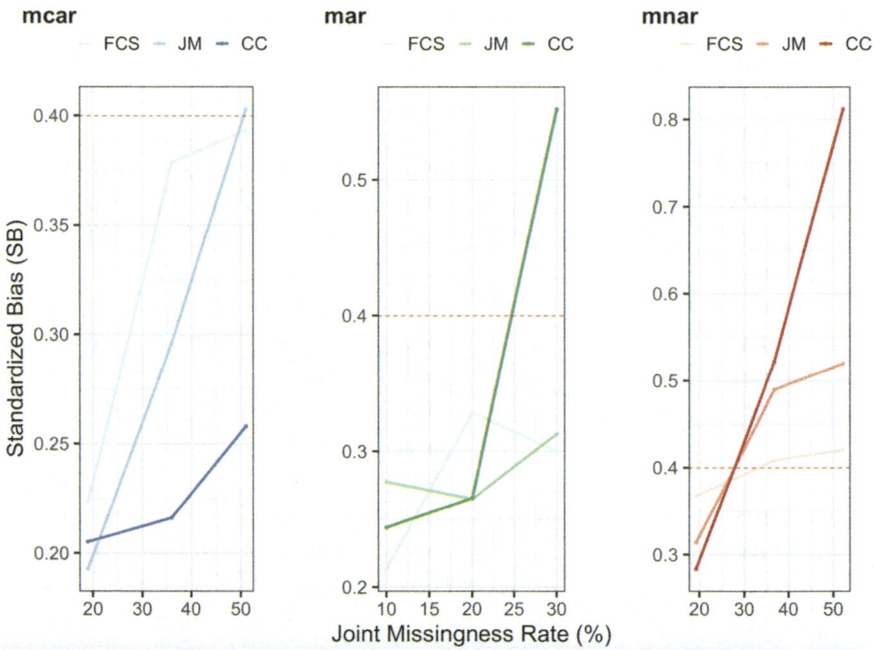

Fig. 25.1 Estimated standardized bias versus joint missingness rate under different missingness mechanisms (ICC=0.05, small mediation effect size). FCS, fully conditional specification; JM, joint modeling; CC, complete-case-only analysis; mcar, missing completely at random; mar, missing at random; mnar, missing not at random

25.5 Discussion

The objective of the simulation study was to verify the analytically derived bias formulas and determine the performance of FCS and JM imputation models in the multilevel mediation analysis model. We find that the choice of imputation model did not have an impact on the multilevel mediation analysis model and the bias formulas were verified.

We recommend that researchers apply imputation methods to assess the mediation effect for incomplete data when the missingness mechanism is assumed to be MCAR or MAR. When we apply imputation methods to incomplete data with varying mediation effect sizes (small, medium, and large), we do not find a substantial SB when the missingness mechanism is MCAR with varying ICC (from 0.05 to 0.2). Although JM and FCS are performing as well as CC analysis for MCAR regarding the SB, MSE, CR, etc., we lose sample size, efficiency, and power with CC analysis. Mediation analysis is mostly dealing with multivariate conditions; therefore, losing sample size may have a negative impact on the model.

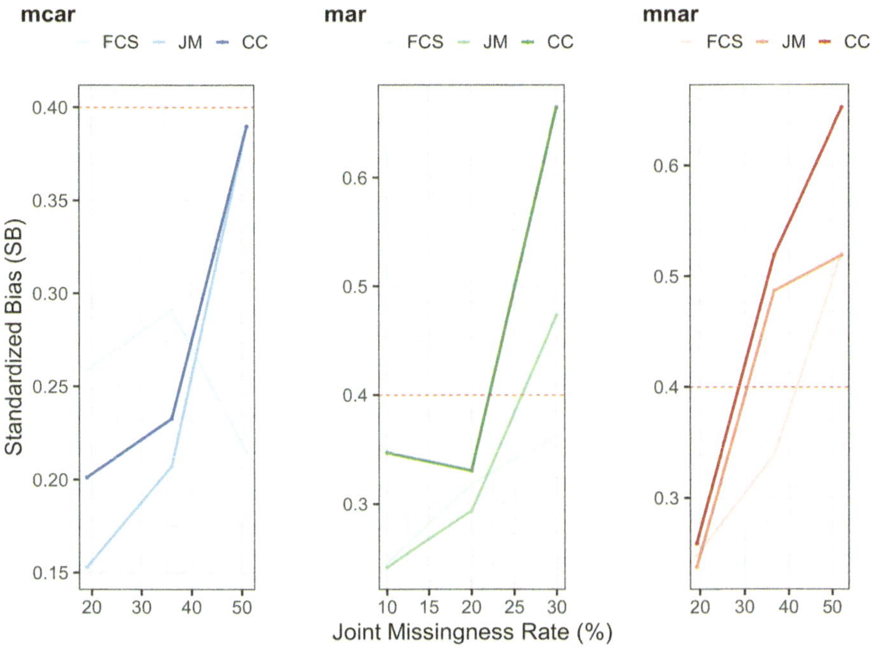

Fig. 25.2 Estimated standardized bias versus joint missingness rate under different missingness mechanisms (ICC=0.1, medium mediation effect size). FCS, fully conditional specification; JM, joint modeling; CC, complete-case-only analysis; mcar, missing completely at random; mar, missing at random; mnar, missing not at random

As expected, when the missingness mechanism is MAR, MI inference outperforms the CC analysis regardless of the imputation method and ICC. CC will bias the whole sample because the missing value depends on the covariates. None of the methods performed well under MNAR; in particular, CC led to substantial biases. Our simulation study found that across all simulation scenarios, MI outperformed CC under MCAR as well as MAR. We could not judge the performance under MNAR as MI models assume MAR.

This study has several limitations. First, comparison of JM and FCS is based on a pseudorandom simulation study, aimed to mimic a real-world problem. We supplemented this simulation with analytical assessment of the bias induced by each imputation method. Second, we only considered one mediator in the mediation analysis model. We did not investigate different types of response variables or covariates/mediators (e.g., binary). These aspects will be the subject of future research.

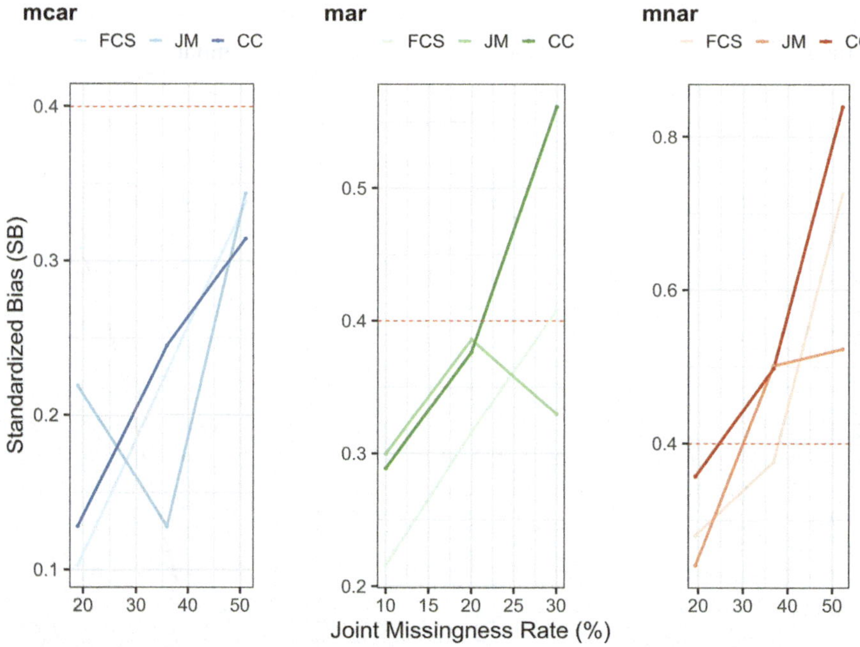

Fig. 25.3 Estimated standardized bias versus joint missingness rate under different missingness mechanisms (ICC=0.2, large mediation effect size). FCS, fully conditional specification; JM, joint modeling; CC, complete-case-only analysis; mcar, missing completely at random; mar, missing at random; mnar, missing not at random

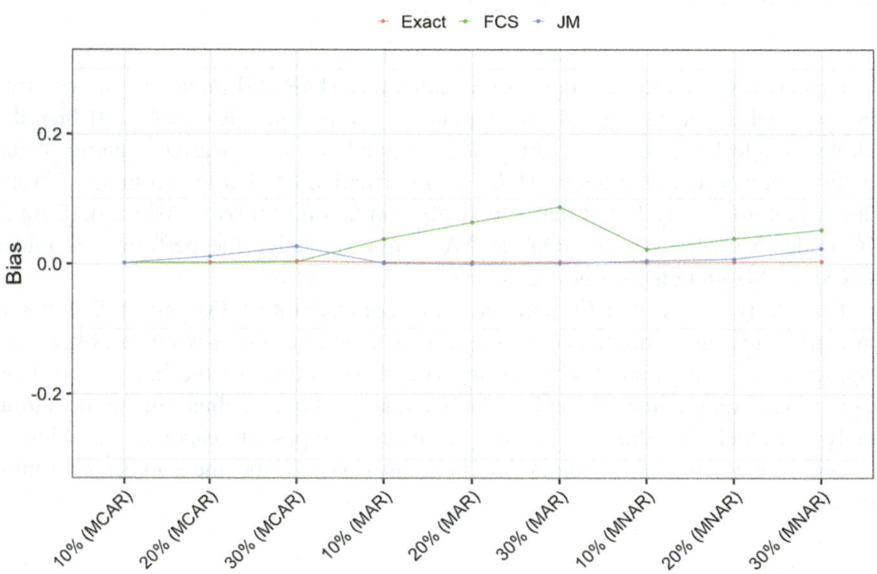

Fig. 25.4 Comparison between exact bias and estimated bias by imputation method. FCS, fully conditional specification; JM, joint modeling; mcar, missing completely at random; mar, missing at random; mnar, missing not at random

Appendix

Derivation for $\boldsymbol{\Sigma}_{22}^{-1}$, $\boldsymbol{\Sigma}_{Y_{ij},M_{ij}|X_{ij},\alpha_Y,\alpha_M}$

We take four steps to calculate $\boldsymbol{\Sigma}_{22}^{-1}$. We restate $\boldsymbol{\Sigma}_{22}$ as follows:

$$\boldsymbol{\Sigma}_{22} = \begin{bmatrix} \sigma^2_{X(W)} + \sigma^2_{X(B)} & \sigma_{XY(B)} & \sigma_{MX(B)} \\ \sigma_{XY(B)} & \sigma^2_{Y(B)} & 0 \\ \sigma_{MX(B)} & 0 & \sigma^2_{M(B)} \end{bmatrix} \tag{25.21}$$

In the first step, we calculate the matrix of minors of $\boldsymbol{\Sigma}_{22}$, and we have the matrix of minors for each cell as follows:

$$\begin{bmatrix} \sigma^2_{Y(B)}\sigma^2_{M(B)} & \sigma_{XY(B)}\sigma^2_{M(B)} & -\sigma^2_{Y(B)}\sigma_{MX(B)} \\ \sigma_{XY(B)}\sigma^2_{M(B)} & (\sigma^2_{X(B)} + \sigma^2_{X(W)})\sigma^2_{M(B)} - \sigma^2_{MX(B)} & -\sigma_{XY(B)}\sigma_{MX(B)} \\ -\sigma^2_{Y(B)}\sigma_{MX(B)} & -\sigma_{XY(B)}\sigma_{MX(B)} & \begin{pmatrix} (\sigma^2_{X(B)} + \sigma^2_{X(W)})\sigma^2_{Y(B)} \\ -\sigma^2_{XY(B)} \end{pmatrix} \end{bmatrix}$$

Then, we employ the second step to turn the matrix of minors into the matrix of cofactors as follows:

$$\begin{bmatrix} \sigma^2_{Y(B)}\sigma^2_{M(B)} & -\sigma_{XY(B)}\sigma^2_{M(B)} & -\sigma^2_{Y(B)}\sigma_{MX(B)} \\ -\sigma_{XY(B)}\sigma^2_{M(B)} & (\sigma^2_{X(B)} + \sigma^2_{X(W)})\sigma^2_{M(B)} - \sigma^2_{MX(B)} & \sigma_{XY(B)}\sigma_{MX(B)} \\ -\sigma^2_{Y(B)}\sigma_{MX(B)} & \sigma_{XY(B)}\sigma_{MX(B)} & \begin{pmatrix} (\sigma^2_{X(B)} + \sigma^2_{X(W)})\sigma^2_{Y(B)} \\ -\sigma^2_{XY(B)} \end{pmatrix} \end{bmatrix}$$

The third step is to transpose the matrix. We have a symmetric matrix and it remains the same. The fourth step is to multiply the matrix by corresponding $\frac{1}{|\Sigma|}$. The determinant $|\Sigma|$ is calculated as

$$|\Sigma| = -\sigma^2_{MX(B)}\sigma^2_{Y(B)} + \sigma^2_{M(B)}(\sigma^2_{X(W)}\sigma^2_{Y(B)} + \sigma^2_{X(B)}\sigma^2_{Y(B)} - \sigma^2_{XY(B)}) \tag{25.22}$$

Therefore, we have the $|\Sigma_{22}^{-1}|$ as follows:

$$\Sigma_{22}^{-1} = \frac{1}{|\Sigma|}\times$$

$$\begin{bmatrix} \sigma^2_{Y(B)}\sigma^2_{M(B)} & -\sigma_{XY(B)}\sigma^2_{M(B)} & -\sigma^2_{Y(B)}\sigma_{MX(B)} \\ -\sigma_{XY(B)}\sigma^2_{M(B)} & (\sigma^2_{X(B)} + \sigma^2_{X(W)})\sigma^2_{M(B)} - \sigma^2_{MX(B)} & \sigma_{XY(B)}\sigma_{MX(B)} \\ -\sigma^2_{Y(B)}\sigma_{MX(B)} & \sigma_{XY(B)}\sigma_{MX(B)} & \begin{pmatrix} (\sigma^2_{X(B)} + \sigma^2_{X(W)})\sigma^2_{Y(B)} \\ -\sigma^2_{XY(B)} \end{pmatrix} \end{bmatrix}$$

After we calculate the Σ_{22}^{-1}, we employ the following formula of calculating conditional variance:

$$\boldsymbol{\Sigma}_{Y,M|X,\alpha_Y,\alpha_M} = \boldsymbol{\Sigma}_{11} - \boldsymbol{\Sigma}_{12}\boldsymbol{\Sigma}_{22}^{-1}\boldsymbol{\Sigma}_{21} \tag{25.23}$$

As we have shown Σ_{11}, Σ_{12}, Σ_{22}^{-1}, and Σ_{21}, we have $\boldsymbol{\Sigma}_{Y,M|X,\alpha_Y,\alpha_M}$ as follows:

$$\boldsymbol{\Sigma}_{Y,M|X,\alpha_Y,\alpha_M} = \begin{bmatrix} f & g \\ h & k \end{bmatrix} \tag{25.24}$$

where

$$\begin{aligned}
f = {} & \sigma^2_{Y(W)} + \sigma^2_{Y(B)} - \frac{1}{|\Sigma|}[\sigma_{YX(W)}\sigma^2_{Y(B)}\sigma^2_{M(B)}\sigma_{XY(W)} \\
& + \sigma^2_{Y(B)}\sigma^2_{M(B)}(\sigma^2_{X(W)} + \sigma^2_{X(B)}) \\
& - 2\sigma_{YM(B)}\sigma_{MX(B)}\sigma^2_{Y(B)}\sigma_{XY(W)} - \sigma^4_{Y(B)}\sigma^2_{MX(B)} - \sigma^2_{YX(B)}\sigma^2_{M(B)}\sigma^2_{Y(B)} \\
& + \sigma^2_{YM(B)}\sigma^2_{Y(B)}(\sigma^2_{X(W)} + \sigma^2_{X(B)}] - \sigma^2_{YM(B)}\sigma^2_{XY(B)} \\
g = {} & \sigma_{YM(W)} + \sigma_{YM(B)} + \sigma_{YX(W)}\sigma^2_{Y(B)}\sigma^2_{M(B)}\sigma_{MX(W)} \\
& - \sigma_{YM(B)}\sigma_{MX(B)}\sigma^2_{Y(B)}\sigma_{MX(W)} \\
& - \sigma_{YM(B)}\sigma^2_{MX(B)}\sigma^2_{Y(B)} - \sigma_{YX(W)}\sigma_{XY(B)}\sigma^2_{M(B)}\sigma_{MY(B)} - \sigma^2_{YX(B)}\sigma^2_{M(B)}\sigma_{MY(B)} \\
& + \sigma^2_{MY(B)}\sigma_{MX(B)}\sigma_{XY(B)} + \sigma^2_{M(B)}\sigma_{YM(B)}\sigma^2_{Y(B)}\sigma^2_{X(W)} + \sigma^2_{M(B)}\sigma_{YM(B)}\sigma^2_{Y(B)}\sigma^2_{X(B)} \\
& - \sigma^2_{M(B)}\sigma_{YM(B)}\sigma^2_{XY(B)} \\
h = {} & \sigma_{YM(W)} + \sigma_{YM(B)} + \sigma_{MX(W)}\sigma^2_{Y(B)}\sigma^2_{M(B)}\sigma_{XY(W)} \\
& - \sigma_{YM(B)}\sigma_{XY(B)}\sigma^2_{M(B)}\sigma_{XY(W)} \\
& - \sigma_{YM(B)}\sigma_{XY(B)}\sigma^2_{M(B)}\sigma_{XY(B)} + \sigma^2_{Y(B)}\sigma_{YM(B)}\sigma^2_{X(W)}\sigma^2_{M(B)} \\
& + \sigma^2_{Y(B)}\sigma_{YM(B)}\sigma^2_{X(B)}\sigma^2_{M(B)} \\
& - \sigma^2_{Y(B)}\sigma_{YM(B)}\sigma^2_{MX(B)} - \sigma_{YM(B)}\sigma_{MX(W)}\sigma^2_{Y(B)}\sigma_{MX(B)} \\
& - \sigma_{YM(B)}\sigma_{MX(B)}\sigma^2_{Y(B)}\sigma_{MX(B)} \\
& + \sigma^2_{YM(B)}\sigma_{XY(B)}\sigma_{MX(B)} + \sigma_{YM(B)}\sigma^2_{M(B)}\sigma^2_{Y(B)} + \sigma_{YM(B)}\sigma^2_{X(W)} + \sigma_{YM(B)}\sigma^2_{X(B)} \\
& - \sigma_{YM(B)}\sigma^2_{XY(B)}\sigma^2_{M(B)} \\
k = {} & \sigma^2_{M(W)} + \sigma^2_{M(B)} + \sigma_{MX(W)}\sigma^2_{Y(B)}\sigma^2_{M(B)}\sigma_{MX(W)} - \sigma_{YM(B)}\sigma_{XY(B)}\sigma^2_{M(B)}\sigma_{MX(W)}
\end{aligned}$$

$$
\begin{aligned}
&- \sigma_{MY(B)}\sigma_{MX(W)}\sigma_{XY(B)}\sigma_{M(B)} + \sigma^2_{MY(B)}\sigma^2_{X(W)}\sigma^2_{M(B)} + \sigma^2_{MY(B)}\sigma^2_{X(B)}\sigma^2_{M(B)} \\
&- \sigma^2_{MY(B)}\sigma^2_{MX(B)} - \sigma^2_{M(B)}\sigma^2_{MX(B)}\sigma^2_{Y(B)} + \sigma^4_{M(B)}\sigma^2_{Y(B)} + \sigma^2_{M(B)}\sigma^2_{X(W)} \\
&+ \sigma^2_{M(B)}\sigma^2_{X(B)} \\
&- \sigma^2_{M(B)}\sigma^2_{XY(B)}\sigma^2_{M(B)} \qquad (25.25)
\end{aligned}
$$

Derivation of MLE Estimator for $\hat{c}$, $\hat{a}$, $\hat{b}$, and $\hat{\delta}$

The linear mixed model is as follows:

$$
\boldsymbol{Y} = \boldsymbol{X\beta}_c + \boldsymbol{\alpha}_1 + \boldsymbol{\epsilon}_1 \qquad (25.26)
$$

We rewrite the mixed model as $\boldsymbol{Y} = \boldsymbol{X\beta}_c + \boldsymbol{\epsilon}_1^*$, where $\boldsymbol{\epsilon}_1^* = \boldsymbol{\alpha}_1 + \boldsymbol{\epsilon}_1$, and $\boldsymbol{\epsilon}_1^* \sim N(\boldsymbol{0}, \boldsymbol{V})$, with $\boldsymbol{V} = \boldsymbol{G} + \boldsymbol{R}$. We have $\boldsymbol{V} = \boldsymbol{V}^{\frac{1}{2}}(\boldsymbol{V}^{\frac{1}{2}})^T$. We multiply $\boldsymbol{V}^{-\frac{1}{2}}$ on both sides of $\boldsymbol{Y} = \boldsymbol{X\beta}_c + \boldsymbol{\epsilon}_1^*$, and we have

$$
\boldsymbol{V}^{-\frac{1}{2}}\boldsymbol{Y} = \boldsymbol{V}^{-\frac{1}{2}}\boldsymbol{X\beta} + \boldsymbol{V}^{-\frac{1}{2}}\boldsymbol{\epsilon}_1^* \qquad (25.27)
$$

where

$$
\boldsymbol{V}^{-\frac{1}{2}}\boldsymbol{\epsilon}_1^* \sim N(\boldsymbol{0}, \boldsymbol{I}_n) \qquad (25.28)
$$

Therefore, we generate the equation for $\hat{\boldsymbol{\beta}}$ as follows:

$$
\boldsymbol{V}^{-\frac{1}{2}}\boldsymbol{Y} = \boldsymbol{V}^{-\frac{1}{2}}\boldsymbol{X}\hat{\boldsymbol{\beta}} \qquad (25.29)
$$

We multiply $\boldsymbol{X}^T(\boldsymbol{V}^{-\frac{1}{2}})^T$ on both sides of the equation above, and we have

$$
\boldsymbol{X}^T(\boldsymbol{V}^{-\frac{1}{2}})^T\boldsymbol{V}^{-\frac{1}{2}}\boldsymbol{Y} = \boldsymbol{X}^T(\boldsymbol{V}^{-\frac{1}{2}})^T\boldsymbol{V}^{-\frac{1}{2}}\boldsymbol{X}\hat{\boldsymbol{\beta}} \qquad (25.30)
$$

Thus, we find the MLE or WLS of $\hat{\beta}$ by multiplying $\boldsymbol{X}^T(\boldsymbol{V}^{-\frac{1}{2}})^T\boldsymbol{V}^{-\frac{1}{2}}\boldsymbol{X}$ on both sides of the equation above, and we have

$$
\hat{\boldsymbol{\beta}} = (\boldsymbol{X}^T\boldsymbol{V}^{-1}\boldsymbol{X})^{-1}\boldsymbol{X}^T\boldsymbol{V}^{-1}\boldsymbol{Y} \qquad (25.31)
$$

$\boldsymbol{X}$ is a $n \times 1$ matrix, and $\hat{\boldsymbol{\beta}}$ in this case is a scalar, and we have $\hat{c} = \hat{\boldsymbol{\beta}}$. Similar procedures are employed to estimate $\hat{a}$ and $\hat{b}$ based on different mixed models as follows:

$$M = X\boldsymbol{\beta}_a + \boldsymbol{\alpha}_2 + \boldsymbol{\epsilon}_2 \tag{25.32}$$

$$Y = X'\boldsymbol{\beta}_{bc'} + \boldsymbol{\alpha}_3 + \boldsymbol{\epsilon}_3 \tag{25.33}$$

Derivation of $f(Y|M, X)$ and $E(Y|M, X)$ for FCS

We have the distribution of Y_{ij} given α_{Y_i}, M_{ij}, and X_{ij}, and the corresponding prior distribution of α_{Y_i}. We find the distribution of Y_{ij}, α_{Y_i} given M_{ij}, X_{ij} as follows:

$$\begin{aligned} f(Y_{ij}, \alpha_{Y_i}|M_{ij}, X_{ij}) &= f(Y_{ij} \mid M_{ij}, X_{ij}, \alpha_{Y_i}) \times f(\alpha_{Y_i}) \\ &= \frac{1}{\sqrt{2\pi}\sigma_{Y|M,X}} \times e^{\dfrac{-(Y_{ij} - \alpha_{Y_i} - \beta_{Y|X}X_{ij} - \beta_{Y|M}M_{ij})^2}{2\sigma^2_{Y|M,X}}} \times \\ &\quad \frac{1}{\sqrt{2\pi}\sigma_{\alpha Y_i}} \times e^{\dfrac{-\alpha^2_{Y_i}}{2\sigma^2_{\alpha Y_i}}} \end{aligned} \tag{25.34}$$

We reformulate the distribution of Y_{ij}, given α_{Y_i}, M_{ij}, X_{ij} into a function of α_{Y_i}, and we have

$$\begin{aligned} f(Y_{ij}, \alpha_{Y_i}|M_{ij}, X_{ij}) &= \frac{e^{-(\frac{1}{2\sigma^2_{Y|M,X}}+\frac{1}{2\sigma^2_{\alpha Y_i}})(\alpha^2_{Y_i}-\dfrac{\beta_{Y|X}X_{ij}+\beta_{Y|M}M_{ij}-Y_{ij}}{\frac{1}{2}+\frac{\sigma^2_{Y|M,X}}{2\sigma^2_{\alpha Y_i}}}\alpha_{Y_i})}}{2\pi\sigma_{Y|M,X}\sigma_{\alpha Y_i}} \times \\ &\quad \frac{1}{2\pi\sigma_{Y|M,X}\sigma_{\alpha Y_i}} e^{-\dfrac{(\beta_{Y|X}X_{ij}+\beta_{Y|M}M_{ij}-Y_{ij})^2}{2\sigma^2_{Y|M,X}}} \\ &= \frac{1}{2\pi\sigma_{Y|M,X}\sigma_{\alpha Y_i}} e^{-(\frac{1}{2\sigma^2_{Y|M,X}}+\frac{1}{2\sigma^2_{\alpha Y_i}})[\alpha_{Y_i}-\frac{\sigma^2_{\alpha Y_i}(\beta_{Y|X}X+\beta_{Y|M}M-Y)}{\sigma^2_{\alpha Y_i}+\sigma^2_{Y|M,X}}]^2+G} \end{aligned} \tag{25.35}$$

where

$$G = \frac{(\beta_{Y|X}X + \beta_{Y|M}M - Y)^2}{2\sigma^2_{Y|M,X} + \dfrac{2\sigma^4_{Y|M,X}}{\sigma^2_{\alpha Y_i}}} - \frac{(\beta_{Y|X}X + \beta_{Y|M}M - Y)^2}{2\sigma^2_{Y|M,X}} \tag{25.36}$$

Then, we develop the conditional distribution of Y given M, X by integrating the distribution $f(Y_{ij}, \alpha_{Y_i}|M_{ij}, X_{ij})$ with regard to α_{Y_i}, and we have $f(Y|M, X)$ as follows:

$$
\begin{aligned}
f(Y|M,X) &= \int f(Y_{ij}, \alpha_{Y_i}|M_{ij}, X_{ij}) d\alpha_{Y_i} \\
&= \frac{1}{\sqrt{2\pi}\sigma_{Y|M,X}\sigma_{\alpha_{Y_i}}} \times \sqrt{\frac{\sigma^2_{Y|M,X}\sigma^2_{\alpha_{Y_i}}}{\sigma^2_{Y|M,X}+\sigma^2_{\alpha_{Y_i}}}} e^{\frac{(\beta_{Y|X}X+\beta_{Y|M}M-Y)^2}{2\sigma^2_{Y|M,X}+\frac{2\sigma^4_{Y|M,X}}{\sigma^2_{\alpha_{Y_i}}}}} \times \\
&\quad e^{-\frac{(\beta_{Y|X}X+\beta_{Y|M}M-Y)^2}{2\sigma^2_{Y|M,X}}} \\
&= \frac{1}{\sqrt{2\pi(\sigma^2_{Y|M,X}+\sigma^2_{\alpha_{Y_i}})}} e^{-\frac{(Y-\beta_{Y|X}X-\beta_{Y|M}M)^2}{2(\sigma^2_{\alpha_{Y_i}}+\sigma^2_{Y|M,X})}}
\end{aligned}
\tag{25.37}
$$

Therefore, we find that $f(Y|M, X)$ is a normal distribution with mean $\beta_{Y|X}X + \beta_{Y|M}M$ (i.e., $E(Y|M, X) = \beta_{Y|X}X + \beta_{Y|M}M$) and variance $\sigma^2_{\alpha_{Y_i}} + \sigma^2_{Y|M,X}$.

Derivation of $f(M|Y, X)$ for FCS

We have the distribution of M_{ij} given α_{M_i}, Y_{ij}, and X_{ij} and the corresponding prior distribution of α_{M_i}. We find the distribution of M_{ij}, α_{M_i} given Y_{ij} and X_{ij} as follows:

$$
\begin{aligned}
f(M_{ij}, \alpha_{M_i}|Y_{ij}, X_{ij}) &= f(M_{ij} \mid Y_{ij}, X_{ij}, \alpha_{M_i}) \times f(\alpha_{M_i}) \\
&= \frac{1}{\sqrt{2\pi}\sigma_{M|Y,X}} \times e^{\frac{-(M_{ij}-\alpha_{M_i}-\beta_{M|X}X_{ij}-\beta_{M|Y}Y_{ij})^2}{2\sigma^2_{M|Y,X}}} \\
&\quad \times \frac{1}{\sqrt{2\pi}\sigma_{\alpha_{M_i}}} \times e^{\frac{-\alpha^2_{M_i}}{2\sigma^2_{\alpha_{M_i}}}}
\end{aligned}
\tag{25.38}
$$

Similarly as we did for the distribution of M_{ij} given α_{M_i}, Y_{ij}, and X_{ij}, we reconstruct the distribution of M_{ij}, α_{M_i} given Y_{ij} and X_{ij} into a function of α_{M_i}, and we have

$$
\begin{aligned}
& f(M_{ij}, \alpha_{M_i} | Y_{ij}, X_{ij}) \\
&= \frac{e^{-(\frac{1}{2\sigma^2_{M|Y,X}} + \frac{1}{2\sigma^2_{\alpha M_i}})(\alpha^2_{M_i} - \frac{\beta_{MX} X_{ij} + \beta_{M|Y} Y_{ij} - M_{ij}}{\frac{1}{2} + \frac{\sigma^2_{M|Y,X}}{2\sigma^2_{\alpha M_i}}} \alpha_{M_i})}}{2\pi \sigma_{M|Y,X} \sigma_{\alpha M_i}} \times \\
& \frac{1}{2\pi \sigma_{M|Y,X} \sigma_{\alpha M_i}} e^{-\frac{(\beta_{M|X} X_{ij} + \beta_{M|Y} Y_{ij} - M_{ij})^2}{2\sigma^2_{M|Y,X}}} \\
&= \frac{1}{2\pi \sigma_{M|Y,X} \sigma_{\alpha M_i}} e^{-(\frac{1}{2\sigma^2_{M|Y,X}} + \frac{1}{2\sigma^2_{\alpha M_i}})[\alpha_{M_i} - \frac{\sigma^2_{\alpha M_i} (\beta_{M|X} X + \beta_{M|Y} Y - M)}{\sigma^2_{\alpha M_i} + \sigma^2_{M|Y,X}}]^2 + G_1}
\end{aligned}
\tag{25.39}
$$

where

$$
G_1 = \frac{(\beta_{M|X} X + \beta_{M|Y} Y - M)^2}{2\sigma^2_{M|Y,X} + \frac{2\sigma^4_{M|Y,X}}{\sigma^2_{\alpha M_i}}} - \frac{(\beta_{M|X} X + \beta_{M|Y} Y - M)^2}{2\sigma^2_{M|Y,X}}
\tag{25.40}
$$

Then, we develop the conditional distribution of M given Y, X by integrating the distribution of $f(M_{ij}, \alpha_{M_i} | Y_{ij}, X_{ij})$ with respect to α_{M_i}, and we have $f(M|Y, X)$ as follows:

$$
\begin{aligned}
f(M|Y, X) &= \int f(M_{ij}, \alpha_{M_i} | Y_{ij}, X_{ij}) d\alpha_{M_i} \\
&= \frac{1}{\sqrt{2\pi} \sigma_{M|Y,X} \sigma_{\alpha M_i}} \times \sqrt{\frac{\sigma^2_{M|Y,X} \sigma^2_{\alpha M_i}}{\sigma^2_{M|Y,X} + \sigma^2_{\alpha M_i}}} e^{\frac{(\beta_{M|X} X + \beta_{M|Y} Y - M)^2}{2\sigma^2_{M|Y,X} + \frac{2\sigma^4_{M|Y,X}}{\sigma^2_{\alpha M_i}}}} \times \\
& e^{-\frac{(\beta_{M|X} X + \beta_{M|Y} Y - M)^2}{2\sigma^2_{M|Y,X}}} \\
&= \frac{1}{\sqrt{2\pi (\sigma^2_{M|Y,X} + \sigma^2_{\alpha M_i})}} e^{-\frac{(M - \beta_{M|X} X - \beta_{M|Y} Y)^2}{2(\sigma^2_{\alpha M_i} + \sigma^2_{M|Y,X})}}
\end{aligned}
\tag{25.41}
$$

Therefore, we find that $f(M|Y, X)$ is a normal distribution with mean $\beta_{M|X}X + \beta_{M|Y}Y$ (i.e., $E(M|Y, X) = \beta_{M|X}X + \beta_{M|Y}Y$) and variance $\sigma^2_{\alpha M_i} + \sigma^2_{M|Y,X}$.

Derivation of $f(M|X)$ and $E(M|X)$ for FCS

First, we find the expression of the conditional distribution of M given Y and X as follows:

$$\begin{aligned} f(M, Y|X) &= f(M|Y, X)f(Y) \\ &= \frac{1}{\sqrt{2\pi(\sigma^2_{M|Y,X} + \sigma^2_{\alpha M})}} e^{-\dfrac{(M - \beta_{M|X}X - \beta_{M|Y}Y)^2}{2(\sigma^2_{M|Y,X} + \sigma^2_{\alpha})}} \times \\ &\quad \frac{1}{\sqrt{2\pi}\sigma_Y} e^{-\dfrac{(Y - \mu_Y)^2}{2\sigma^2_Y}} \end{aligned} \tag{25.42}$$

Then, we reformulate the distribution of Y, M given X as a function of Y, and we have

$$f(M, Y|X) = \frac{1}{2\pi\sigma_Y\sqrt{\sigma^2_{M|Y,X} + \sigma^2_{\alpha M}}} e^{-A_1(Y-B_1)^2} + C_1 \tag{25.43}$$

where

$$\begin{aligned} A_1 &= \frac{\beta^2_{M|Y}}{2\sigma^2_{M|Y,X} + 2\sigma^2_{\alpha M}} + \frac{1}{2\sigma^2_Y} \\ B_1 &= \frac{\mu_Y(\sigma^2_{M|Y,X} + \sigma^2_{\alpha}) - \beta_{M|Y}(\beta_{M|X}X - M)\sigma^2_Y}{\beta^2_{M|Y}\sigma^2_Y + \sigma^2_{M|Y,X} + \sigma^2_{\alpha M}} \\ C_1 &= \frac{\sigma^2_Y(\sigma^2_{M|Y,X} + \sigma^2_{\alpha M})}{2(\beta^2_{M|Y}\sigma^2_Y + \sigma^2_{M|Y,X} + \sigma^2_{\alpha M})} (\frac{\mu_Y}{\sigma^2_Y} - \frac{\beta_{M|X}(\beta_{M|X} - M)}{\sigma^2_{M|Y,X} + \sigma^2_{\alpha M}})^2 - \\ &\quad \frac{(\beta_{M|X}X - M)^2}{2(\sigma^2_{M|Y,X} + \sigma^2_{\alpha M})} - \frac{\mu^2_Y}{2\sigma^2_Y} \end{aligned} \tag{25.44}$$

Therefore, we integrate $f(M, Y|X)$ with respect to Y, and we have the distribution of M given X as follows:

$$
\begin{aligned}
f(M|X) &= \int f(M, Y|X)dY \\
&= \frac{1}{\sqrt{\pi(\beta_{M|Y}^2\sigma_Y^2 + \sigma_{M|Y,X}^2 + \sigma_{\alpha_M}^2)}} e^{A_2M^2+B_2M+C_2}
\end{aligned}
\tag{25.45}
$$

where

$$
A_2 = \frac{\beta_{M|Y}^2\sigma_Y^2}{2(\beta_{M|Y}^2\sigma_Y^2 + \sigma_{M|Y,X}^2 + \sigma_{\alpha_M}^2)(\sigma_{M|Y,X}^2 + \sigma_{\alpha_M}^2)} - \frac{1}{2(\sigma_{M|Y,X}^2 + \sigma_{\alpha_M}^2)}
$$

$$
B_2 = \frac{\mu_Y\beta_{M|Y}}{\beta_{M|Y}^2\sigma_Y^2 + \sigma_{M|Y,X}^2 + \sigma_{\alpha_M}^2} - \frac{\sigma_Y^2\beta_{M|Y}^2\beta_{M|X}X}{(\beta_{M|Y}^2\sigma_Y^2 + \sigma_{M|Y,X}^2 + \sigma_{\alpha_M}^2)(\sigma_{M|Y,X}^2 + \sigma_{\alpha_M}^2)} + \frac{\beta_{M|X}X}{\sigma_{M|Y,X}^2 + \sigma_{\alpha_M}^2}
$$

$$
C_2 = -\frac{\mu_Y^2}{2\sigma_Y^2} - \frac{\beta_{M|X}^2X^2}{2(\sigma_{M|Y,X}^2 + \sigma_{\alpha_M}^2)} + \frac{\mu_Y^2(\sigma_{M|Y,X}^2 + \sigma_{\alpha_M}^2)}{2\sigma_Y^2(\beta_{M|Y}^2\sigma_Y^2 + \sigma_{M|Y,X}^2 + \sigma_{\alpha_M}^2)} - \frac{\mu_Y\beta_{M|Y}\beta_{M|X}X}{\beta_{M|X}^2\sigma_Y^2 + \sigma_{M|Y,X}^2 + \sigma_{\alpha_M}^2} + \frac{\sigma_Y^2\beta_{M|Y}^2\beta_{M|X}^2X^2}{2(\beta_{M|Y}\sigma_Y^2 + \sigma_{\alpha_M}^2 + \sigma_{M|Y,X}^2)(\sigma_{M|Y,X}^2 + \sigma_{\alpha_M}^2)}
\tag{25.46}
$$

Derivation of $f(Y, M|X)$ for JM

We find the distribution of $Y_{ij}, M_{ij}, \alpha_{Y_i}, \alpha_{M_i}$ given X_{ij} by multiplying the conditional distribution of Y_{ij}, M_{ij} given $X_{ij}, \alpha_{Y_i}, \alpha_{M_i}$ and the prior distribution of $\alpha_{Y_i}, \alpha_{M_i}$ as follows:

$$
\begin{aligned}
f(Y_{ij}, M_{ij}, \alpha_{Y_i}, \alpha_{M_i}|X_{ij}) &= f(Y_{ij}, M_{ij}|X_{ij}, \alpha_{Y_i}, \alpha_{M_i}) \times f(\alpha_{Y_i}, \alpha_{M_i}) \\
&= \frac{1}{4\pi^2\sqrt{\boldsymbol{\Sigma}_C\boldsymbol{\Sigma}_\alpha}} e^{-\frac{k(Y-\alpha_Y-\beta_{Y|X}X)^2+f(M-\alpha_M-\beta_{M|X}X)^2}{2(fk-gh)}} \times \\
&\quad e^{-\frac{(g+h)(M-\alpha_M-\beta_{M|X}X)(Y-\alpha_Y-\beta_{Y|X}X)}{2(fk-gh)}} \times \\
&\quad e^{\frac{1}{2\sigma_Y^2\sigma_M^2}(\sigma_M^2\alpha_Y^2+\sigma_Y^2\alpha_M^2)}
\end{aligned}
\tag{25.47}
$$

$$f(Y_{ij}, M_{ij}, \alpha_{Y_i}, \alpha_{M_i}|X_{ij}) = f(Y_{ij}, M_{ij}|X_{ij}, \alpha_{Y_i}, \alpha_{M_i}) \times f(\alpha_{Y_i}, \alpha_{M_i})$$

$$= \frac{exp(-\frac{1}{2}(\boldsymbol{x}-\boldsymbol{\mu})^T \boldsymbol{\Sigma_c}^{-1}(\boldsymbol{x}-\boldsymbol{\mu}))}{\sqrt{(2\pi)^2|\boldsymbol{\Sigma_c}|}} \times \frac{exp(\boldsymbol{\alpha}^T \boldsymbol{\Sigma_\alpha}^{-1}\boldsymbol{\alpha})}{\sqrt{(2\pi)^2 \boldsymbol{\Sigma_\alpha}}}$$

In the above derivation, we have the expression of $\boldsymbol{\Sigma}_C$ ($\boldsymbol{\Sigma}_{Y_{ij},M_{ij}|\alpha_{Y_i},\alpha_{M_i},X_{ij}}$ in Eq. (25.11), we use $\boldsymbol{\Sigma}_C$ to simplify the notation), and the expression of $\boldsymbol{\Sigma}_\alpha$ is shown in Eq. (25.10). We find the conditional distribution of Y, M given X by integrating the conditional distribution of Y_{ij}, M_{ij}, α_{Y_i}, α_{M_i} given X_{ij} regarding α_{Y_i} and α_{M_i} to obtain the conditional distribution of Y, M given X (i.e., $f(Y, M|X)$) as follows:

$$f(Y_{ij}, M_{ij}|X_{ij}) = \iint f(Y_{ij}, M_{ij}, \alpha_{Y_i}, \alpha_{M_i}|X_{ij}) d\alpha_{Y_i} d\alpha_{M_i} \tag{25.48}$$

We reformulate $f(Y_{ij}, M_{ij}, \alpha_{Y_i}, \alpha_{M_i}|X_{ij})$ into a function of α_Y; therefore, we have

$$f(Y_{ij}, M_{ij}, \alpha_{Y_i}, \alpha_{M_i}|X_{ij}) = \frac{1}{4\pi^2\sqrt{\boldsymbol{\Sigma}_C \boldsymbol{\Sigma}_\alpha}} e^{A_3\alpha_Y^2 + B_3\alpha_Y + C_3} \tag{25.49}$$

where

$$\begin{aligned}
A_3 &= -\frac{k}{2(fk-gh)} - \frac{1}{2\sigma_Y^2} \\
B_3 &= -\frac{k(\beta_{Y|X}X - Y) + (g+h)(M - \alpha_M - \alpha_{M|X}X)}{fk - gh} \\
C_3 &= -\frac{k(\beta_{Y|X}X - Y)^2 - (g+h)(M - \alpha_M - \beta_{M|X}X)(Y - \beta_{Y|X}X)}{2(fk-gh)} - \\
&\quad \frac{f(M - \alpha_M - \beta_{M|X}X)^2}{2(fk-gh)}
\end{aligned} \tag{25.50}$$

Then we find the expression of $f(Y_{ij}, M_{ij}, \alpha_M|X_{ij})$ and reformulate it into a function of α_M as follows:

$$f(Y_{ij}, M_{ij}, \alpha_M|X) = \frac{\sqrt{(fk-gh)\sigma_Y^2}}{2\pi\sqrt{\pi(k\sigma_Y^2 + fk - gh)\boldsymbol{\Sigma}_C\boldsymbol{\Sigma}_\alpha}} e^{A_4\alpha_M^2 + B_4\alpha_M + C_4} \tag{25.51}$$

where

$$A_4 = \frac{(g+h)^2(fk-gh)\sigma_Y^2}{2(k\sigma_Y^2 + fk - gh)(fk-gh)^2}$$

$$B_4 = \frac{(g+h)[k(\beta_{Y|X}X - Y) + (g+h)(M - \beta_{M|X}X)]^2\sigma_Y^2(fk-gh)}{(k\sigma_Y^2 + fk - gh)(fk-gh)^2} +$$

$$\frac{(g+h)(Y - \beta_{Y|X}X) - 2a(M - \beta_{M|X}X)}{2(fk-gh)}$$

$$C_4 = \frac{[k(\beta_{Y|X}X - Y) + (g+h)(M - \beta_{M|X}X)]^2\sigma_Y^2(fk-gh)}{2(k\sigma_Y^2 + fk - gh)(fk-gh)} -$$

$$\frac{k[\beta_{Y|X}X - Y]^2 - (g+h)(M - \beta_{M|X}X)(Y - \beta_{Y|X}X) + f(M - \beta_{M|X}X)^2}{2(fk-gh)} \tag{25.52}$$

Then, we find the expression of $f(Y, M|X)$ by integrating the conditional distribution $f(Y, M, \alpha_M|X)$ with respect to α_M. We show $f(Y, M|X)$ as a function of Y as follows:

$$\begin{aligned} f(Y_{ij}, M_{ij}|X_{ij}) &= \int f(Y_{ij}, M_{ij}, \alpha_M|X_{ij})d\alpha_M \\ &= \frac{\sqrt{(fk-gh)\sigma_Y^2}}{\pi\sqrt{2(k\sigma_Y^2 + fk - gh)\Sigma_C\Sigma_\alpha A_4}} e^{-HY^2+JY+K} \end{aligned} \tag{25.53}$$

where

$$H = [\frac{(g+h)k\sigma_Y^2}{(k\sigma_Y^2 + fk - gh)(fk-gh)} - \frac{g+h}{2(fk-gh)}]^2 - \frac{k^2\sigma_Y^2}{2k\sigma_Y^2 + 2(fk-gh)} +$$

$$\frac{k}{2(fk-gh)}$$

$$J = [\frac{(g+h)k\sigma_Y^2}{(k\sigma_Y^2 + fk - gh)(fk-gh)} - \frac{g+h}{2(fk-gh)}]^2\beta_{Y|X}X - \frac{f\sigma_Y^2\beta_{Y|X}X}{k\sigma_Y^2 + (fk-gh)}$$

$$+ \frac{2\beta_{Y|X}Xd}{2(fk-gh)} - [\frac{k(g+h)\sigma_Y^2}{k\sigma_Y^2 + 2(fk-gh)} + \frac{g+h}{2(fk-gh)}](M - \beta_{M|X}X) +$$

$$2(\frac{(g+h)k\sigma_Y^2}{(k\sigma_Y^2 + fk - gh)(fk-gh)} - \frac{g+h}{2(fk-gh)})$$

$$(\frac{(g+h)^2\sigma_Y^2}{(k\sigma_Y^2 + fk - gh)(fk-gh)} - \frac{f}{(fk-gh)}) \times (M - \beta_{M|X}X)$$

$$
\begin{aligned}
K =&[\frac{k(g+h)\sigma_Y^2}{k\sigma_Y^2+2(fk-gh)}+\frac{g+h}{2(fk-gh)}](M-\beta_{M|X}X)\beta_{Y|X}X- \\
&2(\frac{(g+h)k\sigma_Y^2}{(k\sigma_Y^2+fk-gh)(fk-gh)}-\frac{g+h}{2(fk-gh)}) \\
&(\frac{(g+h)^2\sigma_Y^2}{(k\sigma_Y^2+fk-gh)(fk-gh)}-\frac{f}{fk-gh})\times \\
&(M-\beta_{M|X}X)\beta_{Y|X}X-(M-\beta_{M|X}X)^2\times \\
&[(\frac{(g+h)^2\sigma_Y^2}{(k\sigma_Y^2+fk-gh)(fk-gh)}-\frac{f}{fk-gh})^2-\frac{2(g+h)^2\sigma_Y^2}{k\sigma_Y^2+fk-gh}+\frac{f}{2(fk-gh)}]
\end{aligned}
\tag{25.54}
$$

Derivation of $f(Y|X)$, $E(Y|X)$, and $f(M|X)$ for JM

In order to find the expression of $f(Y|X)$, we reformulate $f(Y, M|X)$ as a function of M as follows:

$$
f(Y_{ij}, M_{ij}|X_{ij}) = \frac{\sqrt{(fk-gh)\sigma_Y^2}}{\pi\sqrt{2(k\sigma_Y^2+fk-gh)\Sigma_C\Sigma_\alpha A_4}}e^{-H_1M^2+J_1M+K_1} \tag{25.55}
$$

where

$$
\begin{aligned}
H_1 &= \frac{(g+h)\sigma_Y^2}{2(k\sigma_Y^2+fk-gh)}+\frac{f}{2(fk-gh)} \\
J_1 &= \frac{(g+h)[k(\beta_{Y|X}X-Y)-(g+h)\beta_{M|X}X]^2}{(k\sigma_Y^2+fk-gh)} \\
&\quad -\frac{(g+h)(Y-\beta_{Y|X}X)}{2(fk-gh)}-\frac{f\beta_{Y|X}X}{fk-gh} \\
K_1 &= \frac{[k(\beta_{Y|X}X-Y)-(g+h)\beta_{M|X}X]^2\sigma_Y^2}{2(k\sigma_Y^2+fk-gh)} \\
&\quad -\frac{k(\beta_{Y|X}-Y)^2+(g+h)\beta_{M|X}X(Y-\beta_{Y|X}X)}{2(fk-gh)}
\end{aligned}
\tag{25.56}
$$

We find the expression of $f(Y_{ij}|X_{ij})$ by integrating $f(Y_{ij}, M_{ij}|X_{ij})$ with respect to M_{ij} and obtain the pdf of Y_{ij} given X_{ij} as follows:

$$
\begin{aligned}
f(Y_{ij}|X_{ij}) &= \int f(Y_{ij}, M_{ij}|X_{ij})dM \\
&= \frac{\sqrt{2\sigma_Y^2(fk-gh)}}{\sqrt{\pi(k\sigma_Y^2+fk-gh)\Sigma_C\Sigma_\alpha A_4 H_1}} e^{K_1 - \frac{J_1^2}{4H_1}}
\end{aligned} \tag{25.57}
$$

We find the expression of $f(M_{ij}|X_{ij})$ by integrating $f(Y_{ij}, M_{ij}|X_{ij})$ with respect to Y_{ij}, and we show $f(M_{ij}|X_{ij})$ as a function of M_{ij} as follows:

$$
\begin{aligned}
f(M_{ij}|X_{ij}) &= \int f(Y_{ij}, M_{ij}|X_{ij})dY \\
&= \frac{\sqrt{\sigma_Y^2(fk-gh)}}{\sqrt{\pi(k\sigma_Y^2+fk-gh)\Sigma_C\Sigma_\alpha A_4 H}} e^{UM^2+VM+W} \\
&= \frac{\sqrt{\sigma_Y^2(fk-gh)}}{\sqrt{\pi(k\sigma_Y^2+fk-gh)\Sigma_C\Sigma_\alpha A_4 H}} e^{U(M+\frac{V}{2U})^2+W-\frac{V^2}{4U}}
\end{aligned} \tag{25.58}
$$

where

$$
\begin{aligned}
P =& [\frac{(g+h)k\sigma_Y^2}{(k\sigma_Y^2+fk-gh)(fk-gh)} - \frac{g+h}{2(fk-gh)}]^2 \\
&- \frac{2k^2\sigma_Y^2}{2k\sigma_Y^2+2(fk-gh)} + \frac{k}{2(fk-gh)} \\
Q =& [2(\frac{(g+h)k\sigma_Y^2}{(k\sigma_Y^2+fk-gh)(fk-gh)} - \frac{g+h}{2(fk-gh)}) \\
&(\frac{(g+h)^2\sigma_Y^2}{(k\sigma_Y^2+fk-gh)(fk-gh)} - \frac{f}{fk-gh}) - \frac{k(g+h)\sigma_Y^2}{k\sigma_Y^2+fk-gh} - \frac{g+h}{2(fk-gh)}]^2 \\
R =& [\frac{(g+h)^2\sigma_Y^2}{(k\sigma_Y^2+fk-gh)(afk-gh)} + \frac{f}{(fk-gh)}]^2 \\
&- \frac{f}{2(fk-gh)} - \frac{2(g+h)^2\sigma_Y^2}{d\sigma_Y^2+fk-gh} \\
U =& Q^2 - R \\
V =& (2\beta_{Y|X}P - Q\beta_{M|X})QX - Q\beta_{Y|X}X + 2R\beta_{M|X}X \\
W =& [(2\beta_{Y|X}P - Q\beta_{M|X})X]^2 + Q\beta_{M|X}X^2\beta_{Y|X} - R\beta_{M|X}^2X^2
\end{aligned} \tag{25.59}
$$

Therefore, we find M given X is a shifted normal distribution, and we find the expression for $E(M|X)$ as follows:

$$E(M|X) = \frac{-V\sqrt{\sigma_Y^2(fk-gh)}}{\sqrt{2U(k\sigma_Y^2+fk-gh)\Sigma_C\Sigma_\alpha A_4 H}} e^{W-\frac{V^2}{2U}} \tag{25.60}$$

Derivation of $f(Y|M,X)$ and $E(Y|M,X)$ for JM

We find the conditional distribution of Y_{ij} given M_{ij}, X_{ij} using the Bayes theorem to generate the formula of $f(Y|M,X)$ as follows:

$$\begin{aligned} f(Y_{ij}|M_{ij},X_{ij}) &= \frac{f(Y_{ij},M_{ij}|X_{ij})}{f(M_{ij}|X_{ij})} \\ &= \frac{1}{\sqrt{2\pi H}} e^{-H(Y-\frac{J}{2H})^2+\frac{J^2}{4H}+K-UM^2-VM-W} \end{aligned} \tag{25.61}$$

Therefore, the conditional distribution of Y given M, X is a shifted normal distribution, and we have the expression of $E(Y|M,X)$ as follows:

$$E(Y|M,X) = \frac{J}{2H} e^{\frac{J^2}{4H}+K-UM^2-VM-W} \tag{25.62}$$

We find the expression of $\hat{c}'$ by assessing the coefficient of X_{ij} on Y_{ij} within the expression of $E(Y|M,X)$. We find Y_{ij} given M_{ij} and X_{ij} is a shifted normal distribution; therefore, the formula of $E(Y|M,X)$ is as follows:

$$\begin{aligned} E(Y|M,X) = &\frac{E\sqrt{(fk-gh)\sigma_Y^2(\beta_{M|Y}^2\sigma_Y^2+\sigma_{M|Y,X}^2+\sigma_\alpha^2)}}{\sqrt{(k\sigma_Y^2+fk-gh)\Sigma_C\Sigma_\alpha A_4}} \\ &\times[(\frac{F}{E}\beta_{M|X}-\beta_{Y|X})X-\frac{FM}{E}] \end{aligned}$$

And the expression of $\hat{c}'$ is as follows:

$$\hat{c}' = \frac{\hat{E}\sqrt{(\hat{f}\hat{k}-\hat{g}\hat{h})\hat{\sigma}_Y^2(\hat{\beta}_{M|Y}^2\hat{\sigma}_Y^2+\hat{\sigma}_{M|Y,X}^2+\hat{\sigma}_\alpha^2)}}{\sqrt{(\hat{k}\hat{\sigma}_Y^2+\hat{f}\hat{k}-\hat{g}\hat{h})\hat{\Sigma}_C\hat{\Sigma}_\alpha\hat{A}_4}}(\frac{\hat{F}}{\hat{E}}\hat{\beta}_{M|X}-\hat{\beta}_{Y|X})$$

where

$$
\begin{aligned}
F &= \frac{(g+h)^2\sigma_Y^2}{(f\sigma_Y^2+kf-gh)(fk-gh)} - \frac{k}{fk-gh} \\
E &= \frac{(g+h)f\sigma_Y^2}{2(fk-gh)(f\sigma_Y^2+fk-gh)} + \frac{g+h}{2(fk-gh)}
\end{aligned} \tag{25.63}
$$

We determine $\hat{c}$ by calculating $E(Y|X)$ and assessing the association between Y and X. We estimate $\hat{c}'$ by assessing the coefficient of X on Y within the expression of $E(Y|M, X)$.

The expression of $f(Y_{ij}|X_{ij})$ is as follows (details of development are provided in Appendix section "Derivation of $f(Y|X)$, $E(Y|X)$, and $f(M|X)$ for JM"):

$$
f(Y_{ij}|X_{ij}) = \frac{\sqrt{\sigma_Y^2(fk-gh)}}{\sqrt{\pi(k\sigma_Y^2+fk-gh)\Sigma_C\Sigma_\alpha A_4 H}} e^{UM^2+VM+W}
$$

where U, V, and W are constant to X_{ij}. The conditional distribution of Y_{ij} given X_{ij} is a shifted normal distribution, and the expression of E($Y|X$), by the property of a shifted normal distribution (details of development are provided in Appendix section "Derivation of $f(Y|X)$, $E(Y|X)$, and $f(M|X)$ for JM"), is as follows:

$$
E(Y|X) = \frac{2\sqrt{2(fk-gh)\sigma_Y^2 H}}{\sqrt{(k\sigma_Y^2+fk-gh)\Sigma_C\Sigma_\alpha(P^2-4QH)}} \times \frac{2HR-LRX}{P^2-4HQ}
$$

where

$$
\begin{aligned}
t &= \frac{f}{2(fk-gh)} + \frac{1}{2\sigma_Y^2} \\
L &= 2\beta_{M|X}\left(F^2 - \frac{(g+h)^2}{4t(fk-gh)} + \frac{k}{2(fk-gh)}\right) + \left(2EF - \frac{f(g+h)}{2t(fk-gh)}\right. \\
&\quad \left. + \frac{g+h}{2(fk-gh)}\right)\beta_{Y|X} \\
P &= \left[\frac{(g+h)k\sigma_Y^2}{(k\sigma_Y^2+fk-gh)(fk-gh)} - \frac{g+h}{2(fk-gh)}\right]^2 - \frac{2k^2\sigma_Y^2}{2k\sigma_Y^2+2(fk-gh)} \\
&\quad + \frac{k}{2(fk-gh)}
\end{aligned} \tag{25.64}
$$

c equals the coefficient of X in the expression of $E(Y|X)$. We find the expression of $\hat{c}$ as follows:

$$\hat{c} = -\frac{2\sqrt{2(\hat{f}\hat{k} - \hat{g}\hat{h})\hat{\sigma}_Y^2 H}}{\sqrt{(\hat{k}\hat{\sigma}_Y^2 + \hat{f}\hat{k} - \hat{g}\hat{h})\hat{\Sigma}_C\hat{\Sigma}_\alpha(\hat{P}^2 - 4\hat{Q}\hat{H})}} \times \frac{\hat{L}\hat{R}}{\hat{P}^2 - 4\hat{H}\hat{Q}}$$

We show the details of deriving $\hat{c}$ in Appendix section "Derivation of $f(Y|X)$, $E(Y|X)$, and $f(M|X)$ for JM".

References

Bodner, T. E. (2008). What improves with increased missing data imputations?. *Structural Equation Modeling, 15*(4), 651–675.

Carpenter, J. R., & Kenward, M. G. (2013). Multiple imputation and its application. Wiley.

Collins, L. M., Schafer, J. L., & Kam, C. M. (2001). A comparison of inclusive and restrictive strategies in modern missing data procedures. *Psychological Methods, 6*(3), 330–351.

Dempster, A. P., Laird, N. M., & Rubin, D. B. (1977). Maximum likelihood from incomplete data via the EM algorithm. *Journal of the Royal Statistical Society: Series B (Methodological), 39*(1), 1–22.

Ellickson, P. L., McCaffrey, D. F., Ghosh-Dastidar, B., & Longshore, D. L. (2003). New inroads in preventing adolescent drug use: Results from a large-scale trial of project ALERT in middle schools. *American Journal of Public Health, 93*(11), 1830–1836.

Graham, J. W., Olchowski, A. E., & Gilreath, T. D. (2007). How many imputations are really needed? Some practical clarifications of multiple imputation theory. *Prevention Science, 8*(3), 206–213.

Grund, S., Lüdtke, O., & Robitzsch, A. (2018). Multiple imputation of missing data for multilevel models: Simulations and recommendations. *Organizational Research Methods, 21*(1), 111–149.

Krull, J., & MacKinnon, D. (2001). Multilevel modeling of individual and group level mediated. *Multivariate Behavioral Research, 36*(2), 227–248.

Mackinnon, D. P., Warsi, G., & Dwyer, J. (1995). A simulation study of mediated effect measures. *Multivariate Behavioral Research, 30*(1), 41.

Milletich II, R. J. (2018). Multiple imputation of missing data in structural equation models with mediators and moderators using gradient boosted machine learning. *Dissertation Abstracts International: Section B: The Sciences and Engineering, 78*(7-B(E)).

Mistler, S. A., & Enders, C. K. (2017). A comparison of joint model and fully conditional specification imputation for multilevel missing data. *Journal of Educational and Behavioral Statistics, 42*(4), 432–466.

Raghunathan, T., Lepkowski, J., Van Hoewyk, J., & Solenberger, P. (2001). A multivariate technique for multiply imputing missing values using a sequence of regression models. *Survey Methodology, 27*(1), 85–96.

Royston, P. (2005). Multiple imputation of missing values: Update of ice. *Stata Journal, 5*(4), 527–536.

Rubin, D. B. (1976). Inference and missing data. *Biometrika, 63*(3), 581–592.

Rubin, D. B. (1987). Multiple imputation for nonresponse in surveys. *Journal of Marketing Research, 26*(4), 485.

Schafer, J. L., & Yucel, R. M. (2002). Computational strategies for multivariate linear mixed-effects models with missing values. *Journal of Computational and Graphical Statistics, 11*(2), 437–457.

Van Buuren, S. (2011). *Multiple imputation of multilevel data* (Vol. 11237).

van Buuren, S. (2012). *Flexible imputation of missing data.*

Van Buuren, S., Brand, J. P., Groothuis-Oudshoorn, C. G., & Rubin, D. B. (2006). Fully conditional specification in multivariate imputation. *Journal of Statistical Computation and Simulation, 76*(12), 1049–1064.

Ye, B., & Yucel, R. M. (2020). *Impact of inconsistent imputation models in mediation analysis.* American Statistical Association, Alexandria, VA.

Zhang, Z., & Wang, L. (2013). Methods for mediation analysis with missing data. *Psychometrika, 78*(1), 154–184.

Chapter 26
Ecological Momentary Assessment (EMA) Designs with Planned Missingness

Diane Losardo, Sy-Miin Chow, A. T. Panter, Melissa Burkley, and Edward Burkley

26.1 Ecological Momentary Assessment (EMA) Designs with Planned Missingness

Ecological momentary assessment (EMA) studies provide rich and nuanced information concerning individuals' momentary experience in everyday lives, and yet are data-intensive and consequently involve considerable respondent burden (Bolger et al., 2003; Iida et al., In press). In a typical EMA study, respondents provide assessments on relevant constructs over a specified period (e.g., 14 days, 21 days) at a specified time of the day (interval-contingent), after an experimenter-invoked signal (signal-contingent), or through the trigger of an event in everyday life (event-contingent). Responses are then transmitted back to the investigators via some chosen medium such as the Internet, regular postal mail, or in-person visits to the laboratory (Bolger et al., 2003; Reis & Gable, 2000). Typically, researchers hope to

The bulk of this research was completed while Dr. Losardo was at the University of North Carolina at Chapel Hill. Research reported in this publication was supported by awards from the National Institutes of Health (NIH) Intensive Longitudinal Health Behavior Cooperative Agreement Program under U24AA027684, the National Center for Advancing Translational Sciences (UL1TR002538; 5TL1TR002540), and National Science Foundation grants IGE-1806874 and BCS-0826844

D. Losardo (✉) · A. T. Panter
The University of North Carolina at Chapel Hill, Chapel Hill, NC, USA

S.-M. Chow
The Pennsylvania State University, University Park, PA, USA
e-mail: symiin@psu.edu

M. Burkley · E. Burkley
Oklahoma State University, Stillwater, OK, USA

M. Stemmler et al. (eds.), *Dependent Data in Social Sciences Research*,
https://doi.org/10.1007/978-3-031-56318-8_26

obtain assessments for all respondents on all days and offer various incentives and training mechanisms to help minimize the incomplete data and attrition arising from respondent burden. Such sampling methods are sometimes referred to as ecological momentary assessment (EMA) methods or experience sampling methods (ESM; Stone and Shiffman, 2002).

Given the intensive nature of EMA designs in terms of the required time commitment and number of assessments, a major drawback is high levels of respondent burden (Bolger et al., 2003; Enders & Gottschall, 2011; Iida et al., In press). High respondent burden may in turn lead to attrition and other missing data issues, including possible presence of non-ignorable missingness (Little & Rubin, 1987). Thus, it is important to understand the conditions under which participant burden can be reduced while also preserving high-quality study data. In fact, the success of an EMA design is contingent on participants staying committed to the study, following the rules of the study protocol, being unaffected by testing threats, and being unaffected by study burden (Stone et al., 1991; Stone & Shiffman, 2002; Wheeler & Reis, 1991). Furthermore, Enders and Gottschall (2011) discussed how the overall ethical quality of a study may be enhanced if costs and participant burden can be reduced without compromising the integrity of study results.

The designs examined in this investigation reflect different ways that researchers can reduce respondent burden by strategically reducing the number of occasions. Even though we focus on one specific variation of EMA commonly known as daily diary designs, wherein the data are equally spaced and the smallest time interval is a day, similar ideas can be extended to other EMA designs with smaller time intervals. We propose five planned missingness designs and investigate the statistical properties of the point and standard error estimates in the context of four longitudinal models with different degrees of planned missingness.

Planned Missingness Designs in the Literature

The idea of implementing designs incorporating planned missingness has a long history in the psychometric literature. In the structural equation modeling (SEM) framework, McArdle (1994) and McArdle et al. (2002) argue that incorporating missing data into study designs can be beneficial and should be considered more routinely. Planned missingness is also contained in certain randomized block designs, such as Latin square designs (Winer, 1962). Such designs randomly allocate a treatment effect once to each row and once to each column in a given factorial design. Thus, not every level of each factor receives the same treatment, but this allows the researcher to conduct a much smaller experiment and focus on important main effects. The "missingness" is then confined to parts of the experimental design which are not important to the hypothesis at hand.

In the multilevel modeling (MLM) literature, Raudenbush and Liu (2000) discussed optimal allocation of resources in designing a repeated measures multilevel study. This work showed analytically and by example that a researcher's study planning for multilevel statistical power in a repeated measures case with complete data and a continuous outcome is informed by consideration of the three main data collection features: study duration, the frequency of observation, and sample size. This research provided an excellent and relevant framework for EMA researchers in the design planning stages. For example, when study duration is kept constant (e.g., the typical 14-day EMA study), statistical power for detecting the treatment-related polynomial effect is substantially increased by halving the frequency of observation and doubling the sample size (see Table 4; Raudenbush & Liu, 2000). This specific statistical power trade-off balances adding assessments for a given participant versus enrolling new participants who complete less frequent (fewer) assessments. These study parameters are aspects within a researcher's control, and thus the relative costs of alternative designs can be weighed.

Graham et al. (2001) described a method for planned missingness using random sampling techniques to maximize statistical power while reducing respondent burden. In their approach each participant received only a subset of the items, with some degree of overlap of items received across participants. Thus, each participant received fewer questions, but the overall number of questions issued was increased. This allowed for the collection of more data without compromising the quality of the data by asking too many questions to a given participant. Also, the overlap ensured that participants may be compared to each other.

The use of planned missingness designs has instigated theoretical and methodological advancements in many areas of psychology, particularly developmental psychology. One design put forth by Nessleroade and Baltes (1979), termed the cohort sequential design (*CSD*) or accelerated longitudinal design, was initially discussed by Bell (1953) under the name of convergence method. In those designs, the sample is divided into different age cohorts such that the age cohorts are staggered and overlap to some degree. Thus, a shorter developmental trajectory is obtained for individual groups and then combined into one developmental trajectory that spans the whole time frame. This type of design was later applied within the SEM framework. Horn and McArdle (1980) discussed this design in the context of an autoregressive process, while others (McArdle & Hamagami, 1991, 1992; Duncan et al., 1996; Graham et al., 2001) described and evaluated using such a procedure with growth curve models.

Variations of CSDs have been considered. For example, Duncan et al. (1996) investigated a design with six occasions of measurement and four cohorts that each received three consecutive assessments. Overall, there were 2 days of overlap between consecutive cohorts. Both McArdle and Hamagami (1991) and McArdle and Hamagami (1992) considered a design where data arose from a latent curve model (LCM) with five occasions of measurement and four cohorts. Each cohort was assessed on two consecutive occasions, with 1 day of overlap between consecutive cohorts. Graham et al. (2001) considered a different kind of design with five occasions of measurement. Their first design contained a cohort with complete case data and five other cohorts representing all possible combinations with one time point missing. They also considered designs that included all possible combinations of one or two missing time points, and all possible combinations of only two missing time points. Across studies, although designs differed in overall number of occasions of measurement, structure and number of cohorts, and degree of overlap, similar results were found indicating that the planned missingness designs were able to suitably recover parameter estimates.

While the previous studies provide psychologists with information that helps enhance the effectiveness of planned missingness designs, a few gaps in the literature remain. Previous studies have focused on examining panel data, where relatively few repeated measures are assessed across many individuals. The current investigation adds to the existing literature by incorporating more repeated measures, i.e., 14, as is more typical in an EMA study. In addition, previous literature has only examined the effects of the planned missingness designs in the context of models that posit more systematic change patterns (e.g., linear and quadratic trends as captured by growth curve models). In contrast, the constructs of interest in many EMA studies may show less systematic, "noise-like" fluctuations (see, e.g., Diener et al., 1995; Shifren et al., 1997) that are not well represented by growth curve models. As an example, consider the trajectories illustrated in Fig. 26.1

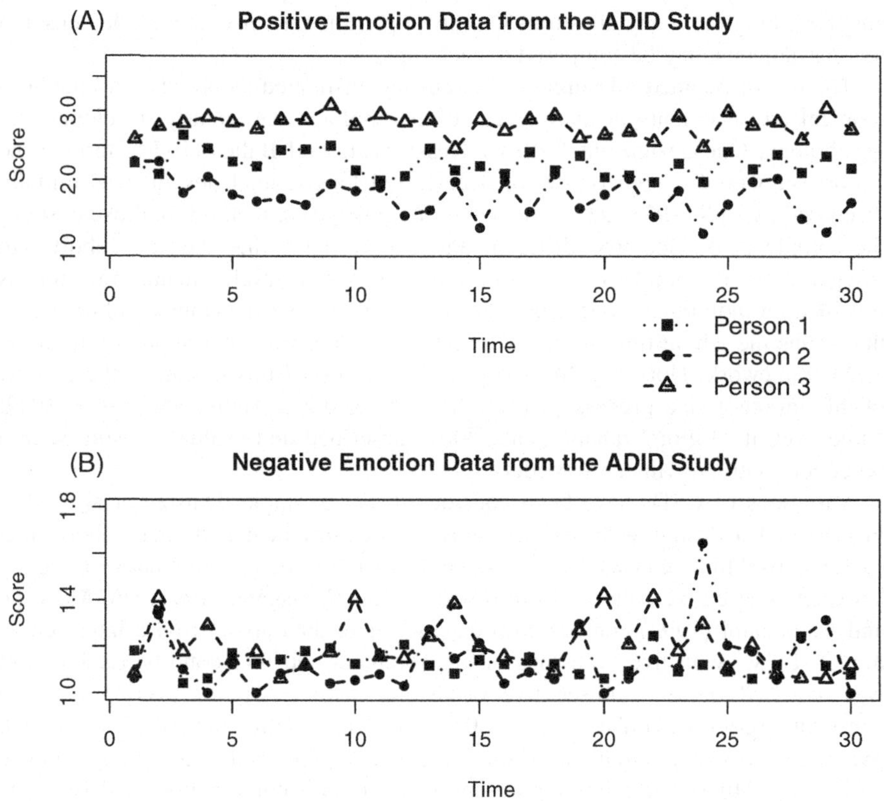

Fig. 26.1 Sample trajectories from the ADID study for (**a**) positive emotion, (**b**) negative emotion

adopted from data collected in the Affective Dynamics and Individual Differences (ADID; Emotions and Dynamic Systems Laboratory, 2010) study. In this EMA study, participants were asked to respond to questions regarding both positive and negative emotions multiple times a day for 30 days. Figure 26.1 displays three individual's composite scores for positive emotion (A) and negative emotion (B) over the course of 15 days with two averaged scores per day. Such data show ongoing fluctuations that reflect a person's emotional ebbs and flows over time. In the current investigation, we consider time series models in addition to latent growth models to better capture such change trajectories.

Deterministic vs. Stochastic Change Models

One possible way of parameterizing less systematic changes is to consider models for stochastic as opposed to deterministic processes. A process is considered to be deterministic when the current time point can be perfectly predicted given knowledge of the previous state of the process. A process is considered to be stochastic, however, when the current time point is a function of the previous state plus some added uncertainties, known as "random shocks" or process noise (Browne & Nesselroade, 2005). For stochastic change, the true score for a given time point cannot be perfectly predicted even if previous true scores are known because of the process noise. As we will illustrate, deciding which type of change is present is integral to the design of effective experience sampling studies with planned missingness.

In our investigation, we consider models that are both deterministic (i.e., a latent growth model and multilevel model) and stochastic (i.e., time series models). Traditional growth curve models such as LCMs are deterministic models. These models allow for the estimation of inter-individual differences in intraindividual (i.e., change within an individual over time) patterns of change that are perfectly predictable over time (Bollen & Curran, 2006; Singer & Willett, 2003). Any within-person deviations from the hypothesized growth patterns are regarded as measurement errors that affect one time point only.

Time series models incorporate stochastic change by allowing for process noise components to affect the current as well as future time points. Through the inclusion of process noise, within-person deviations are allowed. Two common time series models, known as the autoregressive (AR) models and moving average (MA) models, allow for previous process noise to structurally affect current and future time points in two different ways (Chatfield, 2004; Hamilton, 1994; Shumway & Stoffer, 2006; Wei, 1990). That is, in an MA process, previous process noise affects the future state of a process directly through the MA weights, whereas in an AR process, the influences of previous process noise continue to persist over time through autoregressive relationships among past and future states. In the present chapter, we consider AR and MA processes structured using the raw data likelihood

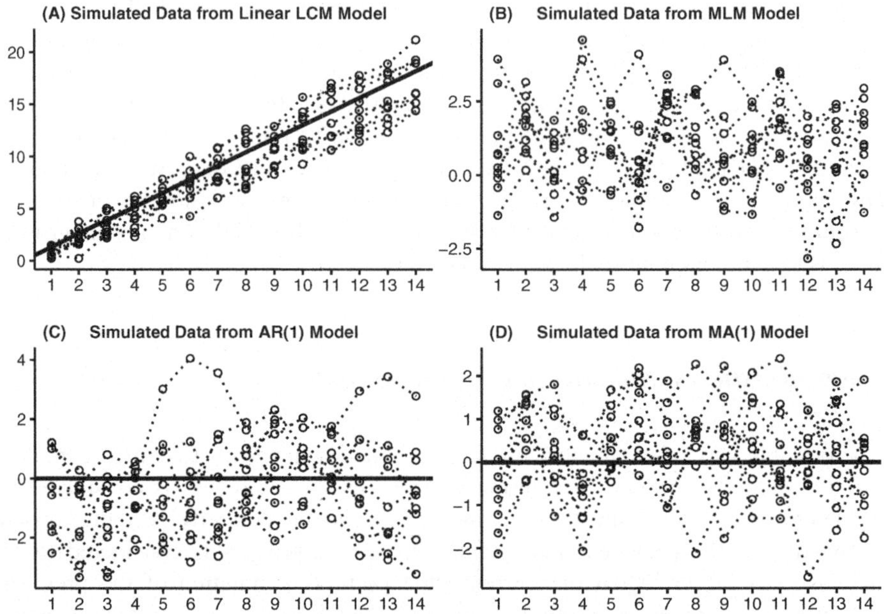

Fig. 26.2 Examples of change trajectories generated using (**a**) the LCM model, (**b**) the MLM model, (**c**) the AR(1) model, (**d**) the MA(1) model. The solid line indicates the mean change over time while the dotted lines represent ten individual cases drawn from the hypothesized model

approach in SEM (for details see Du Toit and Browne, 2001; Du Toit & Browne, 2007).

We now describe each model in more detail and provide substantive examples of data that may have been generated by each model. Figure 26.2 displays hypothetical change trajectories simulated using the four models considered in the present study. To illustrate model differences, we ask the reader to consider an example of the repeated measures representing a daily report of anxiety for an individual. The assumptions imposed by each model will be discussed within this hypothetical context.

Linear LCM We first considered a linear LCM (Bollen & Curran, 2006; McArdle & Epstein, 1987; Meredith & Tisak, 1990). This model assumes there is linear growth over time and each individual's intercept and slope remain constant across time points. However, the intercept and slope are allowed to vary over individuals. This model is expressed as

$$y_{it} = (\mu_\alpha + \zeta_{\alpha_i}) + (\mu_\beta + \zeta_{\beta_i})\lambda_{it} + \epsilon_{it} \tag{26.1}$$

where y_{it} represents the manifest observation of person i at time t, ϵ_{it} is person i's residual at time t, and λ_{it} is person i's time-specific loading for occasion t, constrained in the context of our linear growth curve model to be a linear function of time. μ_α and ζ_{α_i} are the group intercept and person-specific deviation in intercept, respectively, and μ_β and ζ_{β_i} are the group slope and individual-specific deviation in slope, respectively. The residual variance of y_{it}, ϵ_{it}, is assumed to be normally distribution with mean zero and variance, $var(\epsilon_{it})$, which is held constant over time.

Within the context of modeling daily anxiety, the LCM depicts individuals' levels of anxiety as growing in a linear fashion with respect to time. The average change in anxiety (i.e., the group slope), μ_β, does not change with respect to time; however, there may be variability around this slope, represented by ζ_{β_i}, allowing each individual to have his or her own specific linear slope. The LCM is deterministic: an individual's level of anxiety grows by the exact same amount from day to day. Thus, having knowledge concerning the individual's (true) level of anxiety on the previous day, as well as other true parameters, allows us to predict the individual's (true) anxiety level on the next day perfectly.

MLM with Time-Varying Covariate The second model considered is a MLM (Raudenbush & Bryk, 2002) with a time-varying covariate. The model is defined as

$$y_{it} = (\gamma_0 + u_{0i}) + (\gamma_1 + u_{1i})x_{it} + \epsilon_{it} \tag{26.2}$$

where y_{it} represents the manifest repeated measures over time, ϵ_{it} represents the residuals, and x_{it} is a time-varying covariate that varies in value for each individual at each time point. γ_0 and u_{0i} are the group intercept and person-specific deviation in intercept, respectively, together denoted as b_{0i}, and γ_1 and u_{1i} are the group regression slope for the time-varying covariate and person-specific deviation in regression slope, respectively, together denoted as b_{1i}. The residual variance of y_{it} does not vary over time. This model is identical to the LCM, except that the "time-varying covariate" in the LCM consists of time loadings, whereas in this particular model, the time-varying covariates may be variables (e.g., stress) thought to be associated with concurrent changes in the dependent variable (i.e., anxiety).

The MLM depicts anxiety as fluctuating with respect to another time-varying covariate, such as stress on a particular day. Unlike the LCM, anxiety from the previous day does not help to predict anxiety today. The MLM is still deterministic, however, because perfect prediction of an individual's true anxiety level (i.e., the true score) can be obtained provided that the true parameters and values of the time-varying covariates are known. The time-varying covariates in the MLM may assume a more diverse range of values than the time loadings in the LCM. In the context of our simulation study, the MLM provides a helpful platform for assessing whether spurious covariates may emerge as a consequence of planned missingness.

Autoregressive Model of Order 1 (AR(1)) Model The third model considered is a first-order autoregressive AR(1) time series process (Box & Jenkins, 1976; Hamilton, 1994; Wei, 1990). The repeated measures are a function of the previous time point plus some random shock or process noise. The lag of one indicates that only the previous time point $y_{i,t-1}$ influences the current time point, y_{it}, *directly*.

The lingering effects of earlier time points are exerted indirectly through $y_{i,t-1}$. The model is expressed as

$$y_{it} = \alpha y_{i,t-1} + z_{it} \tag{26.3}$$

where y_{it} represents the manifest repeated measure over time, α represents the AR(1) weight, and z_{it} represents the random shock or process noise variable. The process noise variable is typically specified to be independently and identically distributed from a normal distribution with a mean of zero and variance $\sigma^2_{z_{it}}$. We are assuming that the AR(1) model is weakly stationary, or in other words, the mean and variance-covariance structures of the process remain constant across time points. Thus, the covariance between two observed time points does not depend on time but, rather, the time lag between the two measurement occasions[1] (Chatfield, 2004; Hamilton, 1994; Shumway & Stoffer, 2006; Wei, 1990). By assumptions of stationarity, a process is by definition stable over time because the influences of past process noises become less pronounced (i.e., dissipate), as opposed to being intensified over time. An AR(1) model is stationary when $|\alpha| < 1$.

The AR model is a stochastic model because of the added uncertainties imposed by the process noise variable, z_{it}. In the context of our anxiety example, the process noise variable may represent environmental stress that is specific to each day (e.g., crowded classroom, strained family relationship). In this regard, the AR model describes an individual's current state of anxiety as directly affected by an individual's previous state of anxiety plus some "random shocks" stemming from environmental stress. The environmental stress from a previous day, however, is uncorrelated with the environmental stress of the current day and has only an indirect effect on the anxiety of the current day.

Moving Average Model of Order 1 (MA(1)) Model The final model considered is a moving average process of order 1 (MA(1); Chatfield, 2004; Hamilton, 1994; Shumway & Stoffer, 2006; Wei, 1990). The manifest observations are a function of the previous process noise variable and the current process noise variable. This is formally expressed as

$$y_{it} = \beta z_{i,t-1} + z_{it} \tag{26.4}$$

where β represents the MA weight. The process noise variable is specified to be normally distributed with mean zero and variance $\sigma^2_{z_{it}}$. All MA processes are inherently stationary, but one property of importance is whether an MA process is *invertible* or has parameters that can be uniquely identified. An MA(1) process is identified, or invertible, when $|\beta| < 1$ (Chatfield, 2004; Hamilton, 1994; Shumway

[1] This is in contrast to strong stationarity, which assumes that all statistical properties (including higher moments) of every set of observations remain the same when there is an arbitrary shift in time.

& Stoffer, 2006; Wei, 1990). In the present study, we only consider MA processes that are invertible.

The MA process describes an individual's current state of anxiety as affected by the current day's environmental stress but also as a function of the previous day's environmental stress. However, this model dictates that the environmental stress for a given day is uncorrelated with the previous day's environmental stress.

In general, higher-order lags may be incorporated as researchers desire to yield an AR(p) model with time dependency up to lag p and an MA(q) process of order q. In some cases, AR(p) and MA(q) processes may be combined to yield an ARMA(p,q) model, which allows for more complicated change trajectories without having to substantially increase the lag order p or q (Chatfield, 2004; Shumway & Stoffer, 2006). For didactic purposes, we chose the simpler cases of AR(1) and MA(1) processes.

Current Investigation

In the current investigation, we evaluate potentially optimal EMA designs that balance the need for adequate statistical power and efficiency with the need to reduce participant burden and study costs. We focus on five EMA designs that involve planned incomplete data that arise from a researcher's study design and intervention. This non-exhaustive set of designs varies on assessment dimensions relating to assessment interval, blocking, and reference dates. These are dimensions that we consider to be plausible for EMA researchers who are interested in reducing the number of assessments made by their respondents. These designs also reflect characteristics of designs proposed from previous research (McArdle, 1994). We evaluate the effectiveness of each design by conducting a Monte Carlo simulation study and generating data according to change trajectories arising from the LCM, MLM, AR, and MA models. For each design and change trajectory, we evaluate how well parameter estimates and standard errors are recovered and the relative reduction in respondent burden. Our aim is to identify reduced designs that still produce results with acceptable statistical properties, in which case we will conclude that such designs may be preferred over other alternative designs given the potential reduction in participant burden.

26.2 Method

Building from previous work (e.g., McArdle, 1994) that incorporates missing data features as part of a planned design, we evaluated five alternative designs that involve the collection of fewer assessments by subgroups of study participants, along with a complete case design, for a total of six different designs. As mentioned, the proposed designs were not exhaustive, but they were strongly motivated by previous work in the optimal design literature (McArdle, 1994; McArdle & Hamagami, 1991, 1992; Duncan et al., 1996; Graham et al., 2001), and were intended to stimulate interest in possible research designs that would be suited to capturing changes evidenced by more volatile processes, such as those typically studied in experience sampling studies.

Our selection of design features was motivated by typical EMA studies in the psychological literature. We performed a search of EMA studies between 1997 and 2010 using PsycINFO from the following journals: *Journal of Personality and Social Psychology, Psychology and Aging, Developmental Psychology, and Child Development.* Based on over 50 studies identified, we chose 14 days as our hypothetical study period. Some designs include a regular or irregular interval between assessments. Some involve respondent assessment blocks that are either completely separated (no overlap), partially overlapping (some overlap), or interwoven or staggered. Some designs involve certain "reference" assessment days to denote special status of the assessment times, such as Saturday and Sunday for a given week.

A summary of the six designs considered in the present study is shown in Table 26.1. To aid comparisons across designs, the percentages of data reduction and the required number of assessment days are also summarized in Table 26.1. Key characteristics of each design are next highlighted.

Design 1: Complete Case Design The first design is the complete case design, in which all respondents complete all assessments on the 14 days of the study. This control condition reflects the ideal case of EMA investigators, as currently implemented in practice in the literature.

Designs 2–3: Variations of Cohort Sequential Designs (CSDs) The second and third designs are variations of CSDs, as motivated by previous literature including McArdle (1994, see Figure 3a), McArdle and Hamagami (1992, see Figure 6) and Duncan et al. (1996, see Figure 1). The second design is a CSD with 2 days of overlap between subgroups of participants. Individuals are randomly assigned to one of the four groups, with each group being assessed for five consecutive days. There is a 64% reduction of required assessments as compared to the complete case design. The third design is a variation on the previous CSD and has fewer subgroups of participants (3) and only one day overlap between groups. Two groups are assessed for five consecutive days, while one group is assessed for six consecutive days. This design has a 60% reduction of required assessments as compared to the complete case design.

Design 4: CSD with Reference Day The fourth design is also a CSD that is very similar to the second design, with the following exception: two reference days on which all participants are assessed. Reference days were placed at the 1st and 14th days; however, this is an arbitrary placement, and other ways of reference day placement would be equally viable. There is a 54% reduction of required assessments as compared to the complete case design. This model was motivated

Table 26.1 EMA designs considered

	Complete case design													
	Week 1							Week 2						
	M_1	T_2	W_3	Th_4	F_5	S_6	Su_7	M_8	T_9	W_{10}	Th_{11}	F_{12}	S_{13}	Su_{14}
P_1	x	x	x	x	x	x	x	x	x	x	x	x	x	x
P_2	x	x	x	x	x	x	x	x	x	x	x	x	x	x
P_3	x	x	x	x	x	x	x	x	x	x	x	x	x	x
P_4	x	x	x	x	x	x	x	x	x	x	x	x	x	x
P_5	x	x	x	x	x	x	x	x	x	x	x	x	x	x
P_6	x	x	x	x	x	x	x	x	x	x	x	x	x	x
P_7	x	x	x	x	x	x	x	x	x	x	x	x	x	x
P_8	x	x	x	x	x	x	x	x	x	x	x	x	x	x
P_9	x	x	x	x	x	x	x	x	x	x	x	x	x	x
P_{10}	x	x	x	x	x	x	x	x	x	x	x	x	x	x

	Cohort sequential design with one-day overlap													
	Week 1							Week 2						
	M_1	T_2	W_3	Th_4	F_5	S_6	Su_7	M_8	T_9	W_{10}	Th_{11}	F_{12}	S_{13}	Su_{14}
P_1	x	x	x	x	x									
P_2	x	x	x	x	x									
P_3	x	x	x	x	x									
P_4	x	x	x	x	x									
P_5					x	x	x	x	x	x				
P_6					x	x	x	x	x	x				
P_7					x	x	x	x	x	x				
P_8										x	x	x	x	x
P_9										x	x	x	x	x
P_{10}										x	x	x	x	x

	Cohort sequential design with two-days overlap													
	Week 1							Week 2						
	M_1	T_2	W_3	Th_4	F_5	S_6	Su_7	M_8	T_9	W_{10}	Th_{11}	F_{12}	S_{13}	Su_{14}
P_1	x	x	x	x	x									
P_2	x	x	x	x	x									
P_3	x	x	x	x	x									
P_4				x	x	x	x	x						
P_5				x	x	x	x	x						
P_6							x	x	x	x	x			
P_7							x	x	x	x	x			
P_8										x	x	x	x	x
P_9										x	x	x	x	x
P_{10}										x	x	x	x	x

	Cohort sequential design with reference days													
	Week 1							Week 2						
	M_1	T_2	W_3	Th_4	F_5	S_6	Su_7	M_8	T_9	W_{10}	Th_{11}	F_{12}	S_{13}	Su_{14}
P_1	x	x	x	x	x									x
P_2	x	x	x	x	x									x
P_3	x	x	x	x	x									x
P_4	x			x	x	x	x	x						x
P_5	x			x	x	x	x	x						x
P_6	x						x	x	x	x	x			x
P_7	x						x	x	x	x	x			x
P_8	x									x	x	x	x	x
P_9	x									x	x	x	x	x
P_{10}	x									x	x	x	x	x

(continued)

Table 26.1 (continued)

	Two consecutive days max design													
	Week 1							Week 2						
	M_1	T_2	W_3	Th_4	F_5	S_6	Su_7	M_8	T_9	W_{10}	Th_{11}	F_{12}	S_{13}	Su_{14}
P_1	x		x		x		x		x		x		x	x
P_2	x		x		x		x		x		x		x	x
P_3	x		x		x		x		x		x		x	x
P_4	x		x		x		x		x		x		x	x
P_5	x		x		x		x		x		x		x	x
P_6	x	x		x		x		x		x		x		x
P_7	x	x		x		x		x		x		x		x
P_8	x	x		x		x		x		x		x		x
P_9	x	x		x		x		x		x		x		x
P_{10}	x	x		x		x		x		x		x		x

	Signal-contingent design													
	Week 1							Week 2						
	M_1	T_2	W_3	Th_4	F_5	S_6	Su_7	M_8	T_9	W_{10}	Th_{11}	F_{12}	S_{13}	Su_{14}
P_1	x	x		x			x			x				
P_2			x		x	x							x	
P_3			x				x		x		x	x		
P_4	x			x	x		x			x				
P_5		x		x			x		x	x				
P_6		x	x					x	x	x				
P_7			x				x		x				x	x
P_8		x		x			x	x			x			
P_9		x					x			x		x	x	
P_{10}	x				x	x			x		x			

Name	% Reduction	Max # of days per participant	Specific features
Complete case	0	14	Every participant assessed at all times
Cohort sequential design with two-day overlap	64%	5	Two days overlap
Cohort sequential design with one-day overlap	60%	5–6	One day overlap
Cohort sequential design with reference days	54%	5–6	Two days overlap; everyone assessed first and last days
Two consecutive days max	43%	8	Everyone assessed first and last days; no other overlap
Partially randomized design	64%	5	At least two consecutive days per participant

by the reference day model proposed by McArdle (1994, see Figure 4a), which captures specific effects that only arise on certain days of the week. For example, if a researcher is studying the effects of drinking alcohol in a sample of college students, having everyone assessed on the weekends may be helpful as it is known that college students typically drink during this time frame. This rationale is consistent with the optimal design discussion by McClelland (1997). Specifically, McClelland (1997) suggested using a larger sample size for categories of an independent variable that are at the extremes (i.e., the two categories that are hypothesized to most likely show a significant difference to each other) with the hope of maximizing the chances of detecting an effect. In our case, we are maximizing the chances of detecting an effect by using a larger sample size on days when previous theory suggests an effect may be especially pronounced.

Design 5: Two Consecutive Days Max Design The fifth design was motivated by the fact that obtaining assessments on two consecutive observations (e.g., two consecutive days) would be needed to capture inherent lagged effects in certain time series models. Thus, this design is structured such that each participant has a maximum of only two consecutive assessment days. We call this design the two consecutive days max (TCDM) design. This design requires assessments for all participants on the first and last days. In addition, two participant subgroups are assessed on alternating days, with one group having a gap between the first and third day and one group having a gap between the second to last and last day (see Table 26.1). In such a design, the day-to-day dynamics may not be easily captured, making the main focus the performance of the time series models. This design has a 43% reduction of required assessments as compared to the complete case design.

Design 6: Partially Randomized Design The sixth and final design is a partially randomized design. In this design, each participant is assessed on at least two consecutive assessment points with the same total number of time points, but the pattern of response intervals for each participant is random. This design was adapted from popular designs often utilized in the ESM and EMA studies, wherein participants are prompted to respond at random times throughout the day (e.g., using a beeper or pager; see Heron & Smyth, 2010). It is also of interest to determine how such a controlled randomized design performs as compared to other more structured planned missingness designs. To make more direct comparisons to the first CSD (Design 2), this design also has a 64% reduction of required assessments as compared to the complete case design, yielding a total of five assessment points per individual.

Simulation Design Characteristics

Simulation I: Recovering Parameters from the True Models with Planned Missingness Data were simulated according to each of the four models described earlier: LCM, MLM, AR(1), and MA(1). For the LCM and MLM models, we considered two sets of parameter values. One set described a more "idealized" scenario with a small residual variance and a large ratio of explained variance relative to total observed variance. Specifically, for the LCM the explained variance ranged from 0.92 to over 0.999 going from time point 1 to time point 14, and for the MLM the explained variance displayed a range of 0.92 to 0.95 across all time points, depending on the value of x_{it}. The other parameter value set described a set that more accurately reflects parameter ranges commonly seen in empirical EMA studies (Birditt et al., 2005; Flook & Fuligni, 2008; Hankin et al., 2005; Ong et al., 2009; Stawski et al., 2008). Specifically, for the LCM, the explained variance ranged from 0.2 to 0.94, going from time point 1 to time point 14, and for the MLM the explained variance displayed a range of 0.2 to 0.3 across all time points, depending on the value of x_{it}. The parameters for the LCM were set to

$$\text{"Idealized" Case: } [\alpha_i, \beta_i]' \sim MVN\left[\begin{pmatrix} 3 \\ 1.2 \end{pmatrix}, \begin{pmatrix} 1.2 & .3 \\ .3 & .4 \end{pmatrix}\right] \quad \epsilon_{it} \sim N(0, .1) \tag{26.5}$$

$$\text{Empirical Case: } [\alpha_i, \beta_i]' \sim MVN\left[\begin{pmatrix} 1 \\ .02 \end{pmatrix}, \begin{pmatrix} .3 & .05 \\ .05 & .1 \end{pmatrix}\right] \quad \epsilon_{it} \sim N(0, 1.2) \tag{26.6}$$

We also assumed linear growth and set the latent intercept coefficients to one across all time points and the latent slope coefficients to 0 through 13, incrementing by 1 unit at each time point. Thus, each person displays linear growth and the intercept reflects the level at the first time point.

The parameters for the MLM were set to

"Idealized" Case:

$$[b_0, b_1]' \sim MVN\left[\begin{pmatrix} 3 \\ 1.2 \end{pmatrix}, \begin{pmatrix} 1.2 & .3 \\ .3 & .4 \end{pmatrix}\right] \quad x_{it} \sim Unif[0, 1] \quad \epsilon_{it} \sim N(0, .1) \tag{26.7}$$

Empirical Case:

$$[b_0, b_1]' \sim MVN\left[\begin{pmatrix} 1 \\ .02 \end{pmatrix}, \begin{pmatrix} .3 & .05 \\ .05 & .1 \end{pmatrix}\right] \quad x_{it} \sim Unif[0, 1] \quad \epsilon_{it} \sim N(0, 1.2) \tag{26.8}$$

For the AR and MA models, the parameters are selected to be stationary and invertible (Chatfield, 2004; Hamilton, 1994; Shumway & Stoffer, 2006; Wei, 1990). Specifically, the true value for the AR parameter α was set to .7, and the true value for the MA parameter β was set to .7. The true value for the variance of the process noise variables was set to be 1 across all time points. Consistent with the convention in the time series literature, we assume that the researcher has previously removed the mean and other systematic (e.g., linear) trends from the data by differencing or other detrending procedures (Shumway & Stoffer, 2006). Such "demeaned" and detrended processes by definition have an intercept value of zero, and we thus did not include an intercept term in either the AR(1) or the MA(1) model.

The complete case design maintained its form, while the data for the remaining designs were modified by deleting cases to correspond to each of the five EMA designs: CSD, CSD with fewer days overlap, CSD with reference days, two consecutive days max, and partially random. The missingness was generated using a missing completely at random (MCAR; Shafer & Graham, 2002; Little & Rubin, 1987) mechanism. For each of these cases, three sample sizes were evaluated: $N = 50$, $N = 100$, and $N = 300$. That is, within each sample size and design condition, data were simulated using a total of 50, 100, or 300 participants, each participant having up to a maximum of 14 time points (in the complete case design). Due to differences in the planned missingness designs, different percentages of incomplete data (and hence, complete observations) were present in each design condition. The incomplete observations were specified to be equally scattered throughout the span of each design, subject to the constraints inherent to each design. The percentages of missing observations reported in Table 26.1 reflect the percentages of missing observations present in each design out of all the data from all subjects and time points. The simulation design was 4(model) × 6(design) × 3(sample size), for a total of 72 cells. Each cell had a total of 5,000 Monte Carlo replications. All data were simulated using the R statistical package (R Development Core Team, 2023).

Simulation II: Detecting Model Misspecification with Planned Missingness The above simulation study assumes that a researcher knows the underlying model that gave rise to the process. Since this is virtually never true in applied research, we also conducted a simulation study where we generated data from the four models described previously, induced missingness according to several EMA designs, and subsequently fit the "wrong" models to each data set. Overall, the goal of this study is to establish whether the EMA designs we considered would be able to detect if an incorrect model is being estimated. Specifically, we wanted to determine whether fitting a "wrong" model to EMA data with planned missingness will result in inflated Type I error rates wherein the model incorrectly identifies a parameter estimate as significant. In addition, we also sought to determine whether, in the presence of planned missingness, model fit statistics such as the root-mean-square error of approximation (RMSEA; Steiger & Lind, 1980), the Akaike information criterion (AIC; Akaike, 1973), and Bayesian information criterion (BIC; Raftery, 1993; Schwarz, 1978) would be able to detect model misspecification and identify the true data generating model as more appropriate for the data.

The models used as the true data generation models in the population included the LCM (condition with large residual variance and small explained variance), the AR, and the MA models with parameters fixed at the values described previously. In the present simulation, we did not use the MLM as one of the true data generating models because the critical element that drives the within- and between-person differences in the MLM is a time-varying covariate that is naturally omitted when other candidate models (the LCM, AR, and MA models) are fit to the corresponding data. The question of whether the parameters associated with the time-varying covariate could be correctly recovered in the presence of planned missingness was already addressed by our first simulation and, in part, by simulation results involving the LCM. Of greater interest to us was whether spurious covariates that were not present in the other three models at the population level would be misidentified as statistically significant when a misspecified model, namely, the MLM, was fit to data with planned missingness. When the incorrect MLM model was fit to data generated from each of the other models, we included a spurious variable with time-varying values drawn from a uniform(0,1) distribution as a covariate. We were interested in evaluating the Type I error rates across conditions: that is, instances where the parameters associated with the spurious covariate would be misidentified as significantly different from zero even though they are not in the true data generating models. In addition, rather than considering all five of the planned missingness designs, we focused in the present study on using selected designs that were concluded to be optimal based on our first simulation study.

Data Analysis and Outcome Measures

All model formulation and model fitting were performed within the SEM or MLM framework using Mplus version 5.0 (Muthén & Muthén, 1998–2007). Missing data were handled using the standard full information maximum likelihood (FIML) procedure. As the time series models we used may not be familiar to applied researchers, we provided sample Mplus code for estimating the AR(1) and MA(1) models in the appendix. This SEM formulation is identical to that proposed by Du Toit and Browne (2001).

Measures used to assess estimation results included absolute bias, relative bias, root-mean-square error (RMSE), confidence interval coverage and width, and power for detecting nonzero parameters. Absolute bias and relative bias provide information concerning the difference between the average estimate of the parameter and the population parameter and they were calculated as

$$\text{Absolute Bias} = \frac{\sum_{r=1}^{R} |\hat{\theta}_r - \theta|}{R} \tag{26.9}$$

$$\text{Relative Bias} = \frac{\frac{\sum_{r=1}^{R} \hat{\theta}_r}{R} - \theta}{\theta} \tag{26.10}$$

where r represents a particular Monte Carlo run, R is the total number of replications, and θ is the true parameter value in the population. Lower values indicated greater parameter accuracy. MSE was calculated as

$$\mathrm{MSE} = \frac{\sum_{r=1}^{R} (\hat{\theta}_r - \theta)^2}{R} \tag{26.11}$$

Lower values indicated greater parameter precision. Confidence interval coverage was calculated as the percentage of replications in which the population parameter falls between the lower confidence limit and the upper confidence limit with an alpha of 0.05. Specifically, we calculated the upper and lower confidence limits as

$$\text{Lower Confidence Limit} = \hat{\theta} - 1.96 * S\hat{E}_\theta \tag{26.12}$$

$$\text{Upper Confidence Limit} = \hat{\theta} + 1.96 * S\hat{E}_\theta \tag{26.13}$$

where $\hat{\theta}$ is the estimated value of a given parameter and $S\hat{E}_\theta$ the estimated standard error for that parameter. Power was calculated as the percentage of replications in which the null hypothesis of $\theta = 0$ is rejected. The estimated standard errors, i.e., the mean of the estimated standard errors ($\hat{SE}$), were compared against the empirical or "true" standard errors, i.e., the standard deviation of the parameter estimates (SE_θ). To evaluate efficiency, we compared the estimated standard errors across designs, with lower estimated standard errors indicating better parameter efficiency. To evaluate the consistency of the standard error estimation, we compared the estimated standard errors against the empirical standard errors by computing the difference $SEDIFF = SE_\theta - \hat{SE}$. Lower values indicated better consistency of standard error estimation.

To make more direct comparisons among the various designs, we computed averages of some summary statistics across parameters. Specifically, across parameters within a given model, we computed the average of the absolute value of relative biases (ARB), the average of the RMSE values (ARMSE), the average of the estimated standard error values $A\hat{SE}$, and the average of the absolute difference between the estimated and true standard errors (ASEDIFF). For the LCM and MLM models, this computation involved averaging across six parameters, while for the AR and MA models, this involved averaging across two parameters. For the LCM and MLM models, we also computed an average of the parameters associated with the fixed effects (i.e., the group average intercept and slope) and the parameters associated with the random effects in the models (i.e., variances of slope and intercept, covariance of slope and intercept, and residual variance). This allowed for a direct comparison of the recovery of fixed versus random effects components.

Due to the large amount of summary results, we opted to use a series of meta-models to more formally evaluate our simulation results, as suggested by Skrondal (2000). We chose to use analysis of variance (ANOVA) models as our meta-models and estimated a series of ANOVAs using the simulation outcome measures listed above as dependent variables. Thus, we ran separate meta-models for all outcome measures and used the following sets of independent variables: 1. sample size and design, 2. parameter type (fixed effect vs. random effects) and design, and 3. parameter type (fixed effect vs. random effects) and sample size. Formally, these models are expressed as

$$\text{ANOVA Model 1: SimOutcome}_i = \text{SampleSize}_i + \text{Design}_i + \text{SampleSize}_i * \text{Design}_i + \epsilon_i \quad (26.14)$$

$$\text{ANOVA Model 2: SimOutcome}_i = \text{ParType}_i + \text{Design}_i + \text{ParType}_i * \text{Design}_i + \epsilon_i \quad (26.15)$$

$$\text{ANOVA Model 3: SimOutcome}_i = \text{SampleSize}_i + \text{ParType}_i + \text{SampleSize}_i * \text{ParType}_i + \epsilon_i \quad (26.16)$$

For ANOVA model 26.14, we estimated separate models: (1) aggregating across all parameters and (2) for each individual parameter. Given the large number of replications, most effects are found to be significant. Thus, we consider ANOVA main effect and interaction results that have an R^2 value of at least 0.1 to be relatively strong effects, and report R^2 values when discussing our results. Due to space constraints, we only report effects that showed approximately a medium or larger effect size (i.e., $R^2 \geq 0.1$). [2]

26.3 Results

Given the large amount of results, we chose to highlight our main findings graphically. The figures capture interaction and main effects of the ANOVA models that can help shed light on the relative merits of the different planned missingness designs under different model and sample size conditions.

Before discussing the specific results from each model, a few general observations can be noted across all models. First, as expected, as sample size increased, biases and MSEs of all parameters tended to decrease in magnitude, coverage rates approached 0.95, and power tended to increase in magnitude. Second, as sample size increased, estimated standard errors and empirical standard errors tended to decrease in magnitude, and the difference between estimated standard errors and empirical standard errors also tended to decrease in magnitude. In addition, the complete case design displayed the lowest average MSE values and larger or equivalent values of power as compared to the other designs. We next discuss specific results associated with each model.

[2] We acknowledge that this cut-off value is arbitrary, however, it coincides with Cohen's 1992 popular effect size guidelines.

LCM and MLM Results

"Idealized" Parameter Condition with Small Residual Variance and Large Explained Variance

Overall, estimation results were satisfactory across all designs, and differences among designs were not very salient. Given that all of the planned missingness designs were able to recover the point and SE estimates well, the key factor that determined the relative effectiveness of the different designs was the number of non-missing occasions (see Figs. 26.3 and 26.4a, b, respectively, for the LCM and MLM). For both LCM and MLM, the effect sizes associated with sample size were relatively large for most simulation outcomes, e.g., for confidence interval width (LCM, $R^2 = 0.17$; MLM, $R^2 = 0.23$), and for coverage rates (LCM,$R^2 = 0.24$; MLM, $R^2 = 0.27$). For both models, when compared to the fixed effect parameters, the variance-covariance components of the random parameters (i.e., the variance of the intercept, the variance of the slope, and the covariance between intercept and slope) tended to have higher relative biases (LCM, $R^2 = 0.13$; MLM, R^2 associated with parameter type in ANOVA = 0.12) and displayed poorer coverage rates (LCM, $R^2 = 0.10$; MLM, R^2 for parameter type = 0.15). There was also no substantial interaction effect between design and sample size, i.e., all R^2 values for this interaction effect of ANOVA model 26.14 were less than 0.01.

Table 26.2 displays the percentage of reduction in power for all planned missingness designs as compared to the complete case design. Only parameters that displayed a reduction in power are shown. The planned missingness designs were ordered across the columns by percentage of reduction in required assessments, from largest to smallest. For both the LCM and MLM, the reduction in power was negligible when designs with fewer assessment occasions were used. Even among the parameters that evidenced the greatest reduction in power (namely, the random effects parameters), the effect sizes associated with design condition for ANOVA models 26.14 and 26.15 were small ($R^2 < 0.01$ in all cases). Reduction in power was most consistently observed for the CSD with 2-day overlap, the CSD with 1-day overlap, and for the MLM, the partially randomized design. The two consecutive days max design consistently showed the least and, in most cases, only negligible reduction in power.

Empirically Based Parameter Condition with Large Residual Variance and Small Explained Variance

The simulation outcome measures, on average, displayed somewhat less satisfactory results in this condition than in the "idealized" parameter condition. In particular, the difference between designs was magnified when larger residual or error variance was present (see the plots in Figs. 26.3 and 26.4a as compared to those in (C)). Correspondingly, the R^2 values for the design main effect using the ANOVA

Table 26.2 Percentage reduction in power for LCM "idealized" condition, MLM "idealized" condition, and MA(1)

		θ	CSD two-days overlap ($T = 5$)	Partially random ($T = 5$)	CSD one-day overlap ($T = 5 - 6$)	CSD reference days ($T = 5 - 6$)	Two consecutive days max ($T = 8$)
$N = 50$	LCM: "idealized case"	$\psi_{\alpha\beta}$	13.8%	3.9%	10.6%	2.7%	0%
	MLM "idealized" case	ψ_{b1}	12.1%	11.3%	9%	3.3%	0.7%
		ψ_{b0b1}	23.6%	21.2%	21.1%	15.9%	8.6%
	MA	β	2.7%	18.5%	2%	2.7%	62.6%
		σ_y^2	2.5%	15.7%	1.8%	2.3%	35.2%
$N = 100$	LCM "idealized" case	$\psi_{\alpha\beta}$	2.9%	0.2%	1.9%	0%	0%
	MLM "idealized" case	ψ_{b1}	0.4%	0.3%	0.1%	0%	0%
		ψ_{b0b1}	10.1%	9.9%	9.2%	6.6%	2.7%
	MA	β	0.7%	10.9%	0.4%	0.4%	30.4%
		σ_y^2	0.6%	9.8%	0.4%	0.4%	24.2%
$N = 300$	MA	β	0%	3.4%	0%	0%	14.3%
		σ_y^2	0%	3.1%	0%	0%	13%

Columns are ordered by percentage of reduction in required assessments, starting with the largest percentage reduction to the smallest

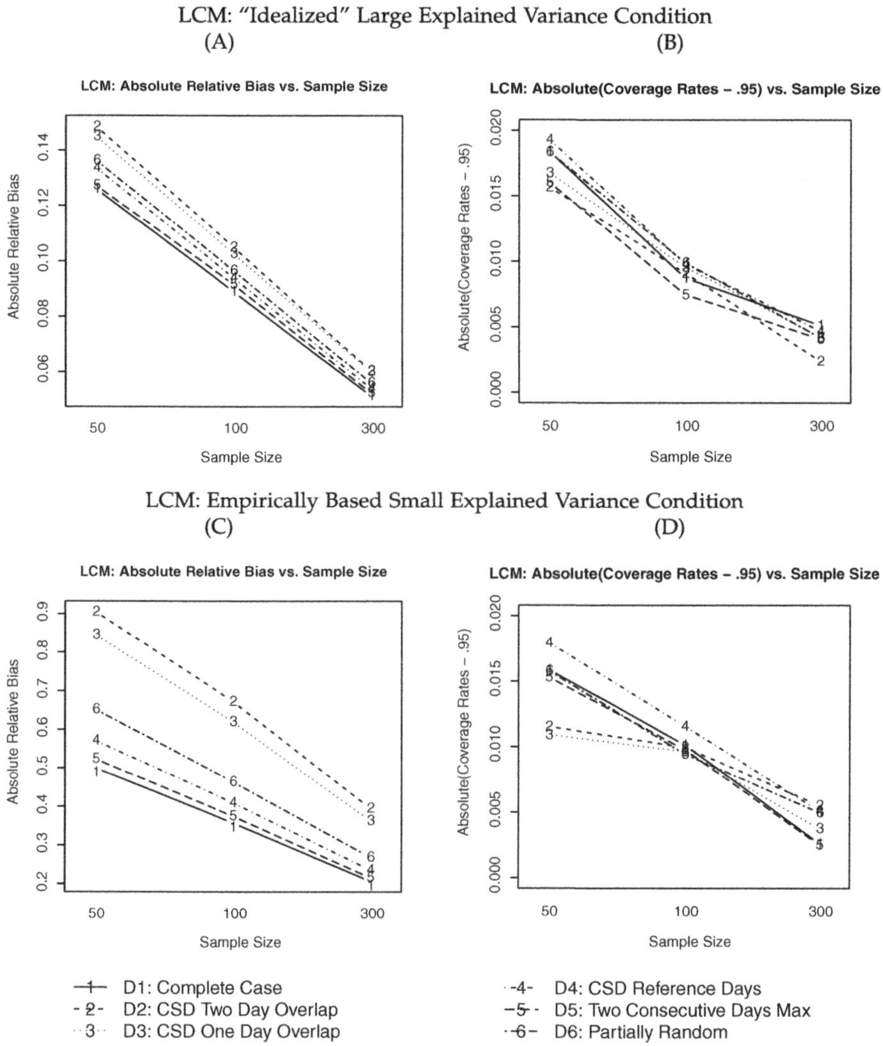

Fig. 26.3 Absolute bias and coverage rates, calculated as the absolute difference from 0.95, for the average of all parameters of the LCM comparing (**a**) and (**b**) the "idealized" parameter condition to (**c**) and (**d**) the empirically based parameter condition

model of Eq. 26.14 were all larger in magnitude in the empirically based condition as compared to the idealized condition. As illustrated when comparing results in Tables 26.2 and 26.3, differences in power between designs were also more noticeable.

Inspection of Fig. 26.5a and c in terms of absolute biases suggested that for the LCM, the CSDs with both 1-day and 2-day overlap displayed the largest relative biases and CI width, except for the residual of the variance parameter,

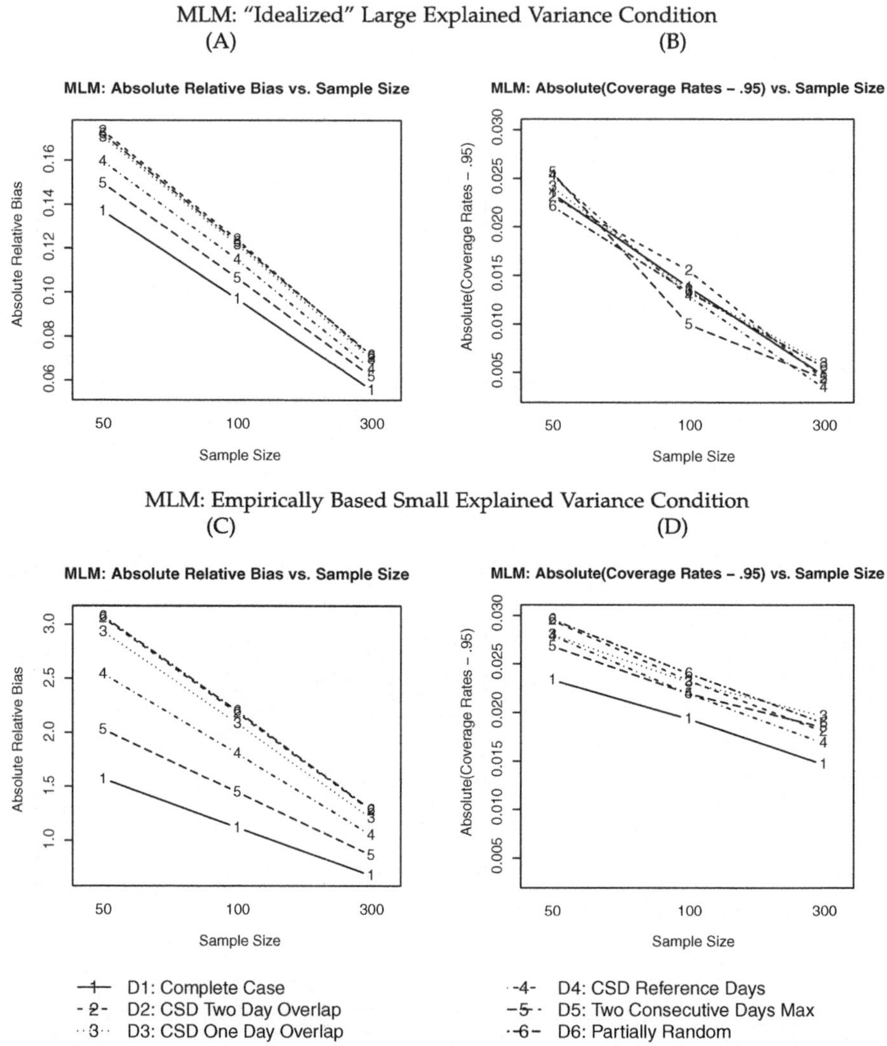

Fig. 26.4 Absolute bias and coverage rates, calculated as the absolute difference from 0.95, for the average of all parameters of the MLM comparing (**a**) and (**b**) the "idealized" parameter condition to (**c**) and (**d**) the empirically based parameter condition

which displayed a larger CI width for the partially randomized design. Similarly, Table 26.3, highlighting results for parameters that showed a relatively large decrement in power, indicated that for the LCM, the CSD variations without reference day yield less satisfactory power estimates specifically for the mean of the slope, variance of the intercept, and covariance between intercept and slope,

Consonant with results in the "idealized" parameter condition, reduction in estimation performance in the case of the MLM appeared to be due primarily to

Table 26.3 Percentage reduction in power for LCM empirical case and MLM empirical case

		θ	CSD two-days overlap ($T = 5$)	Partially Random ($T = 5$)	CSD one-day overlap ($T = 5 - 6$)	CSD reference days ($T = 5 - 6$)	Two consecutive Days Max ($T = 8$)
$N = 50$	LCM empirical case	μ_β	13.9%	2.5%	7.7%	5.1%	0%
		$\psi_{\alpha\alpha}$	98.4%	86.8%	96.6%	65.4%	30.9%
		$\psi_{\alpha\beta}$	92.9%	39.3%	85.4%	24.4%	11%
	MLM empirical case	ψ_{b0}	79.1%	78.2%	76.9%	67.8%	45.1%
		ψ_{b1}	52%	48%	48%	48%	44%
$N = 100$	LCM empirical case	μ_β	16.3%	8.2%	15.3%	0%	1%
		$\psi_{\alpha\alpha}$	93.3%	68.5%	87.8%	35.3%	8.4%
		$\psi_{\alpha\beta}$	82%	46%	75%	31.1%	11.3%
	MLM empirical case	ψ_{b0}	65.7%	64.3%	62.2%	46.9%	20.1%
		ψ_{b1}	60.9%	56.5%	58.7%	52.8%	32.6%
$N = 300$	LCM empirical case	μ_β	33.5%	9.8%	30.4%	7.7%	9.8%
		$\psi_{\alpha\alpha}$	59%	15.4%	44.3%	0.6%	0%
		$\psi_{\alpha\beta}$	75.2%	36.3%	72%	18.6%	4.9%
	MLM empirical case	ψ_{b0}	8.5%	8.8%	5.4%	1.2%	0%
		ψ_{b1}	75%	72.1%	75.7%	69.3%	53.6%

Columns are ordered by percentage of reduction in required assessments, starting with the largest percentage reduction to the smallest

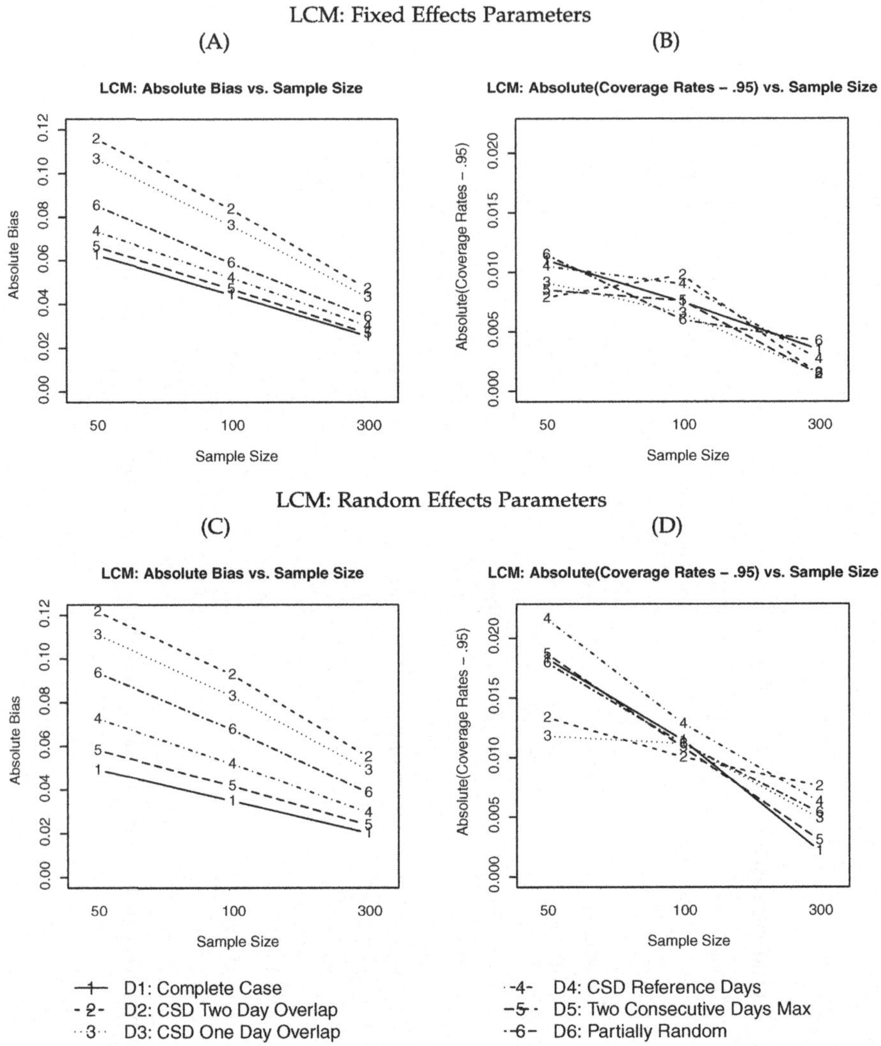

Fig. 26.5 Absolute bias and coverage rates, calculated as the absolute difference from 0.95, for the average of (**a**) and (**b**) fixed effects parameters versus (**c**) and (**d**) random effects parameters for the LCM with large residual variance and small explained variance

reduction in T, particularly in terms of relative biases and coverage rates for the random effects components compared to the fixed effects components (see panel (B) as compared to panel (D) in Fig. 26.6). The key difference was just that the effects due to reduction in T were magnified in the empirically based condition.

It is of particular interest to compare the CSD with 2-day overlap and the partially randomized design, both of which were characterized by the same reduction in required assessments. For the LCM model, the absolute values of the relative

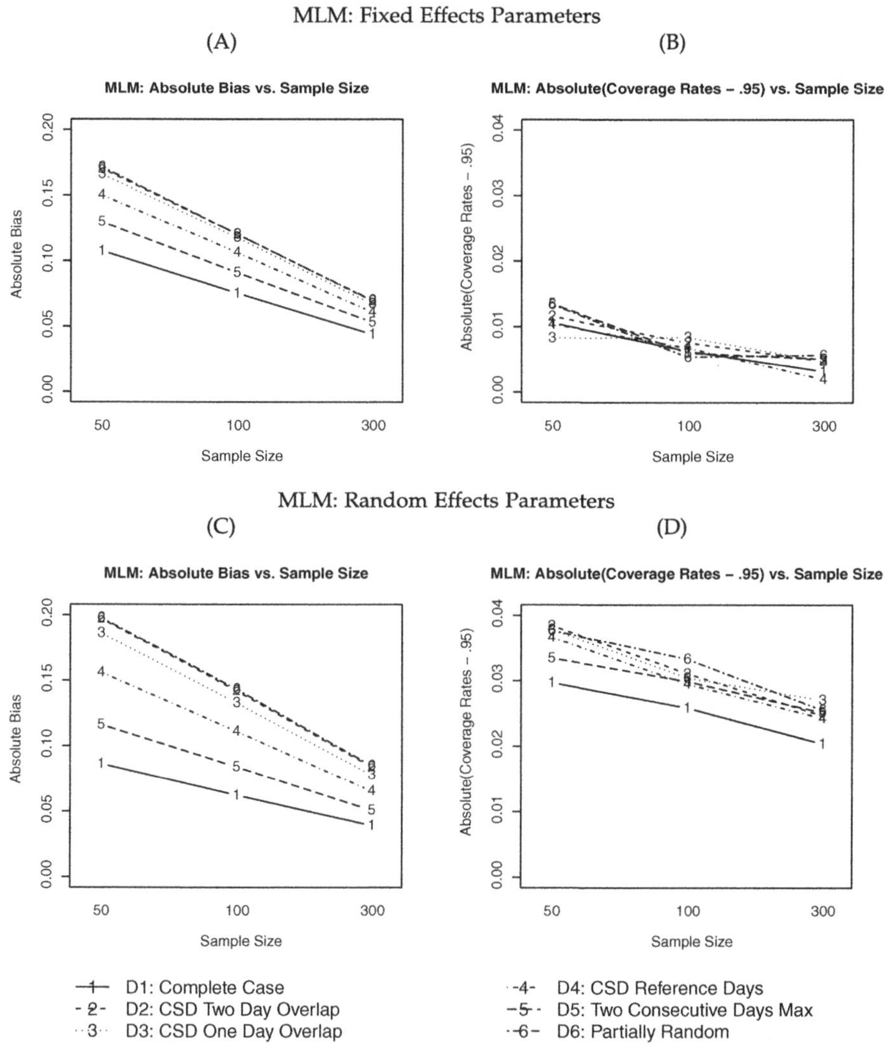

Fig. 26.6 Absolute bias and coverage rates, calculated as the absolute difference from 0.95, for the average of (**a**) and (**b**) fixed effects parameters versus (**c**) and (**d**) random effects parameters for the MLM with large residual variance and thus, smaller explained variance

biases were similar across the two designs, with the partially randomized design displaying slightly lower absolute value of the relative biases, as illustrated in Fig. 26.3 panel (A). The MSE values and estimated standard errors tended to be smaller for the partially randomized design than the CSD with 2-day overlap, and this was especially true for the variance of the intercept parameter and the mean of the intercept parameter. Also, the partially randomized design yielded a smaller reduction in power for the covariance between intercept and slope parameter at sample sizes of 50 and 100. For the MLM model, however, the performance of the CSD with 2-day overlap and the partially randomized design were very similar in simulation outcomes. In fact, power increased almost solely as a function of the reduction in required assessments. This was not the case for the LCM, wherein factors such as including a reference day for all participants on the first measurement occasion and having a partially randomized design all helped to increase power without notable differences in the number of complete assessments.

AR(1) and MA(1) Results

Plots of the absolute biases and confidence interval coverage rates of the parameters in the AR(1) and MA(1) models are shown in Figs. 26.7 and 26.8, respectively. Across all designs, the AR(1) model displayed low values of absolute biases, relative biases, low MSE values, low estimated standard errors, adequate coverage rates, small CI widths, high power, and a small difference between estimated standard errors and empirical standard errors. Thus, the AR(1) model performed well with respect to parameter accuracy, precision, efficiency, and consistency of standard error estimation, which is nicely illustrated in Fig. 26.7. The R^2 values for the main effect of sample size ranged from 0.05 to 0.36, indicating that all simulation outcomes significantly improved with increasing sample size. For the complete case design, the MA(1) model also performed well with respect to all simulation outcomes. For the planned missingness designs, however, the MA(1) model did not perform as well.

For the AR(1) model, the complete case design performed well with respect to all simulation outcomes as compared with the other planned missingness designs. Properties of the point and SE estimates of other planned missingness designs were close to those observed in the complete data case. Coverage rates were also adequate and power was high across all designs. Overall, the two consecutive days max design showed slight improvements in coverage rates ($R^2 = 0.09$ design effect when estimating ANOVA model 26.14) and confidence interval width ($R^2 = 0.10$ design effect when estimating ANOVA model 26.14) as compared to the other planned missingness designs. As illustrated when comparing panel (A) and panel (C) in Fig. 26.7, a more notable design by sample size interaction effect was observed for the process noise parameter than the absolute biases of the AR(1) parameter than ($R^2 = 0.13$). The design by sample size interaction effect on coverage rates was relatively large ($R^2 = 0.554$ for AR(1) parameter, $R^2 = 0.632$ for process noise parameter) and is illustrated in panels (B) and (D) of Fig. 26.7. However, all coverage rates are very close to the nominal rate of 0.95 with the largest discrepancy from 0.95 being equal to only 0.007 units.

The two designs with the same percentage of reduction in required assessments, i.e., the CSD with 2-day overlap and the partially randomized design, performed similarly with respect to all simulation outcomes. The former showed slightly smaller MSE values for the variance of the process noise parameter, indicating that

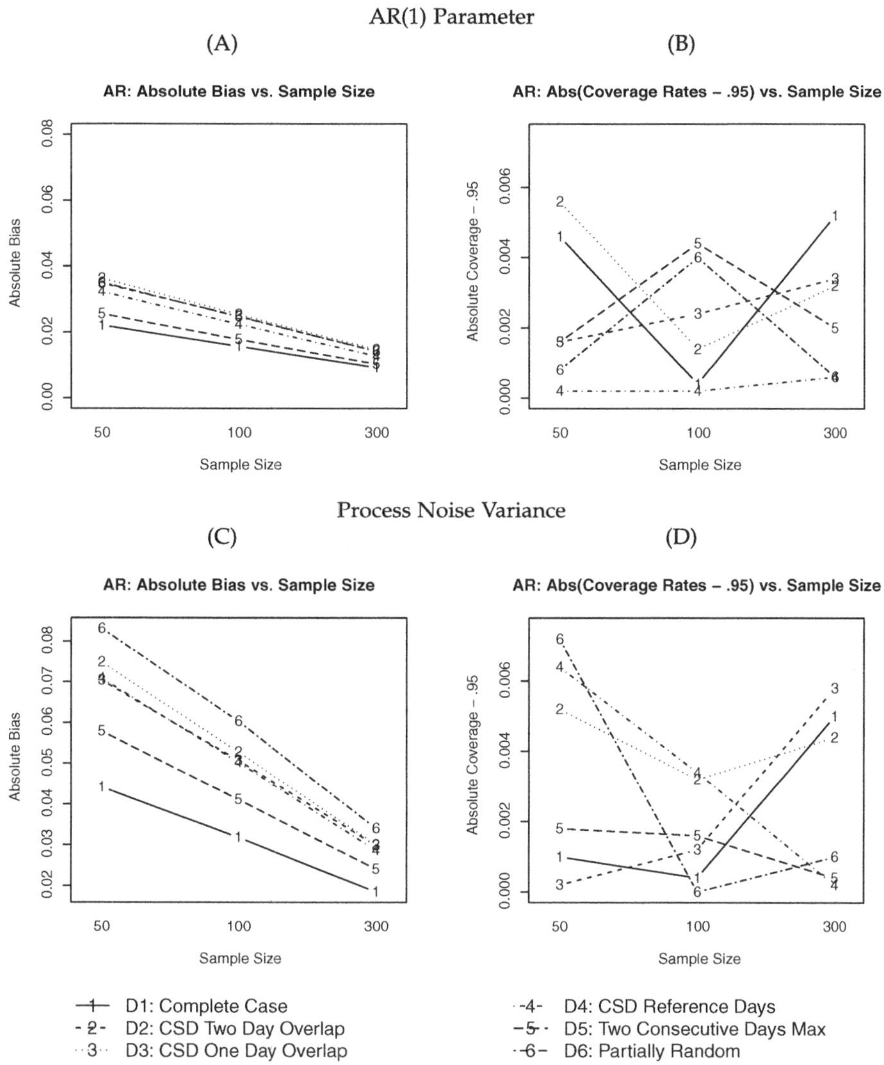

Fig. 26.7 Absolute bias and coverage rates, calculated as the absolute difference from 0.95, for the AR(1) parameter and process noise variance of the AR(1) model

having a longer stretch of consecutive measurements from each participant may help improve the precision of the process noise variance parameter in this particular model. Also supporting this idea is the finding that the absolute bias, relative bias, and MSE values for the process noise variance in the partially randomized design were largest when compared to all other designs, even across all sample size conditions.

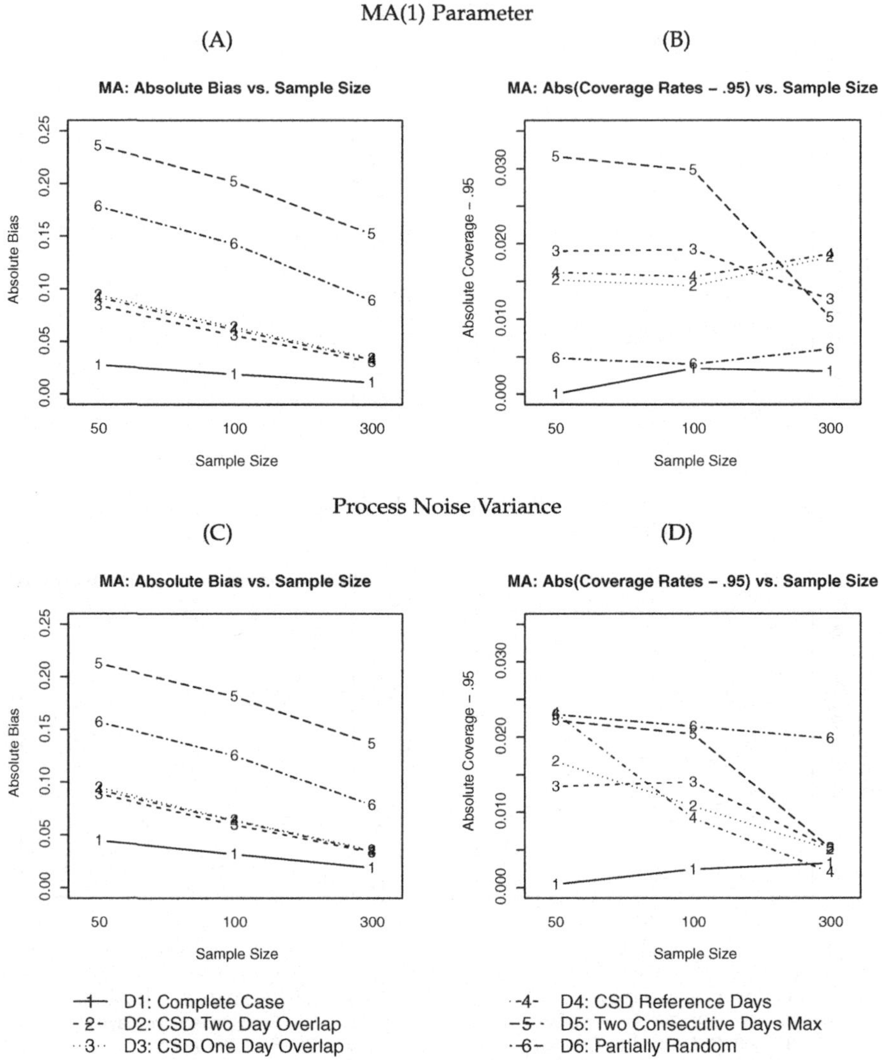

Fig. 26.8 Absolute bias and coverage rates, calculated as the absolute difference from 0.95, for the MA(1) parameter and process noise variance of the MA(1) model

For the MA(1) model, the complete case design substantially outperformed the other planned missingness designs as it displayed distinctly smaller absolute biases (see plots (A) and (C) in Fig. 26.8) and relative biases, smaller MSE values, smaller estimated standard errors, smaller differences between estimated standard errors and empirical standard errors, smaller confidence interval widths, and higher power. In fact, the effect sizes for the main effect of design when estimating ANOVA model 26.14 were the largest in magnitude when compared to the other modeling conditions (ranging from $R^2 = 0.17$ to $R^2 = 0.34$), and it is clear from Fig. 26.8 that the performance of the complete case design is distinguishable from the performance of other designs. The difference in simulation outcome results was most noticeable between the complete case design and both the two consecutive days max design and partially randomized design. The CSD with 2-day overlap, CSD with 1-day overlap, and CSD with reference days all produced similar results and performed closest to the complete case design.

When averaging across the MA parameter and process noise variance parameter in the MA(1) model, Fig. 26.8a and c illustrates that the partially randomized design displayed the largest absolute value of the relative biases for sample sizes of 50 and 100, while the two consecutive days max design displayed the largest absolute value of the relative bias for a sample size of 300 and the largest MSE values for all sample sizes. Figure 26.8b and d illustrates the optimal performance of the complete case design with respect to confidence interval coverage rates and an interesting design by sample interaction effect ($R^2 = 0.13$). For the partially randomized design, coverage rates for the variance of the process noise were overestimated by a larger margin and did not show as much improvement with increasing sample size compared to other designs. For the MA(1) parameter, coverage rates for the two consecutive days max design were overestimated by a larger margin at sample sizes of 50 and 100 when compared to other designs, but revealed comparable coverage rates to other designs at a sample size of 300. Taking both relative biases and coverage rates into consideration, there is a design by sample size interaction effect such that the partially random and two consecutive days max designs yielded less satisfactory estimation results compared to all other designs, particularly at small sample sizes. When comparing the two designs with the same percentage of reduction in required assessments, i.e., the CSD with 2-day overlap and the partially randomized design, the CSD with 2-day overlap performed better with regard to all simulation outcomes.

For the MA(1) model, Table 26.2 illustrates that both the MA parameter and variance of the process noise parameter displayed a small reduction in power for the CSD with 2-day overlap, CSD with 1-day overlap, and CSD with reference days when compared to the complete case design. For the partially randomized design, there was a larger reduction in power for both parameters. Of all planned missingness designs, the two consecutive days max design displayed the largest reduction in power, especially for the MA parameter. For example, at a sample size of 50, there was a 62.6% reduction in power, at a sample size of 100, there was a 24.2% reduction in power, and at a sample size of 300, there was a 14.3% reduction in power.

The suboptimal performances of the partially randomized design and the two consecutive days max design is to be expected because, for an MA(1) model, the covariance between nonadjacent time points is zero. This can be seen by examining the autocovariance function, which describes the covariation between two observations of the same variable across specific time lags. For an invertible MA(1) process, the autocovariance function for a time lag of 1 is $\sigma^2_{z_{it}}\beta$, while the autocovariance for a time lag greater than 1 is zero (Shumway & Stoffer, 2006). Thus, for the two consecutive days max and partially random designs, relatively few adjacent time points were available in the simulated data sets to allow the MA and process noise variance parameters to be recovered accurately.

Fitting Misspecified Models

Based on results from the first simulation study, we focused on three EMA designs: CSD with 2-day overlap, two consecutive days max, and partially randomized design. These three designs were found to work particularly well with most models, so we were interested in determining whether they would be robust to identifying when a model is fit that is *not* the true data generating model. Also, we considered the CSD with 2-day overlap to make direct comparisons to the partially randomized design as these two designs have the same number of required assessments.

Table 26.4 illustrates the results for AIC, BIC, sample size-adjusted BIC, and RMSEA. These results suggested that, when generating data from the LCM, AR, and MA models, the CSD with 2-day overlap, two consecutive days max, and partially randomized designs were able to detect when an incorrect model was fitted to the data. Several conditions do not display RMSEA results for one of two reasons. First, when fitting the MLM in Mplus, the output does not provide an RMSEA as it is fitting this model as a mixed model. We chose to fit the model in the MLM as opposed to the SEM framework to mimic the typical methods researchers use when analyzing EMA data (see, e.g., Flook & Fuligni, 2008; Kianget et al., 2006; Silk et al., 2003; Birditt et al., 2005; Campbell et al., 2010). Second, Mplus, by default, does not estimate the saturated model and provide RMSEA value when all variables and pairs of variables do not have data for at least ten percent of the sample (Muthén & Muthén, 1998–2007). When using the partially randomized design, due to the large amount of missingness between subsequent stretches of data, this default was not satisfied and Mplus did not output an RMSEA estimate.

When available, the RMSEA provided evidence suggesting that the data fit with the correct data generating model was superior across all conditions. This was true for all models and all three designs, which is promising evidence that this test statistic will be able to successfully choose the correct data generating model. The AIC, BIC, and sample size-adjusted BIC would select the correct model for both the two consecutive days max and partially randomized designs, across all conditions. However, the results were mixed when examining the CSD with 2-day overlap. For example, in the condition involving the AR(1) model as the true data generating model, the AIC, BIC, and sample size-adjusted BIC were frequently found to favor the LCM and MA model over the correct data generating model. At a sample size of 100 and 300, the AIC would choose the MLM model over the correct AR model, and at a sample size of 300, the sample size-adjusted BIC would choose the MLM model over the correct AR model. Also, the AIC, BIC, and sample size-adjusted

Table 26.4 Fit statistics obtained from fitting misspecified models to data under different planned missingness conditions

		Fitted	$N = 50$				$N = 100$				$N = 300$			
	DGM[a]	model	AIC	BIC	BIC N-adj	RMSEA	AIC	BIC	BIC N-adj	RMSEA	AIC	BIC	BIC N-adj	RMSEA
D2:	**LCM**	**LCM**	912.13	923.60	904.77	0.07	1822.95	1838.58	1819.63	0.03	5457.46	5479.68	5460.66	0.01
CSD	LCM	MLM	971.60	992.76	973.74		1935.45	1960.76	1941.72		5756.98	5788.86	5769.80	
two days	LCM	AR	965.90	973.55	960.99	0.15	1934.21	1944.63	1931.99	0.14	5796.94	5811.75	5799.07	0.14
overlap	LCM	MA	1122.31	1129.96	1117.40	0.27	2253.04	2263.46	2250.82	0.27	6754.26	6769.07	6756.39	0.27
($T = 5$)	**AR**	**AR**	793.11	800.76	788.20	0.06	1583.57	1593.99	1581.36	0.03	4745.82	4760.63	4747.95	0.01
	AR	LCM	784.16	795.63	776.80	0.13	1562.02	1577.66	1558.71	0.11	4676.28	4698.51	4679.48	0.11
	AR	MLM	793.75	814.88	795.83		1580.51	1605.79	1586.75		4731.23	4763.11	4744.05	
	AR	MA	786.05	793.70	781.14	0.13	1567.20	1577.62	1564.99	0.12	4694.62	4709.43	4696.74	0.11
	MA	**MA**	791.47	799.12	786.57	0.06	1583.14	1593.56	1580.93	0.03	4743.23	4758.05	4745.36	0.01
	MA	LCM	800.03	811.50	792.66	0.15	1595.21	1610.84	1591.89	0.14	4773.23	4795.45	4776.43	0.13
	MA	MLM	801.76	822.89	803.87		1597.20	1622.49	1603.44		4781.04	4812.92	4793.86	
	MA	AR	763.55	771.20	758.65	0.10	1523.15	1533.57	1520.94	0.08	4560.81	4575.63	4562.94	0.07
D5:	**LCM**	**LCM**	1588.02	1599.49	1580.66	0.06	3170.65	3186.28	3167.33	0.03	9501.14	9523.36	9504.34	0.01
two consecutive	LCM	MLM	1978.56	2003.212	1984.17		3948.30	3977.11	3958.06		11855.84	11891.25	11872.18	
days max	LCM	AR	1710.07	1717.72	1705.16	0.18	3416.67	3427.10	3414.46	0.17	10246.15	10260.96	10248.28	0.17
($T = 8$)	LCM	MA	2147.18	2154.82	2142.27	0.36	4304.05	4314.47	4301.84	0.36	12919.98	12934.80	12922.11	0.36
	AR	**AR**	1414.24	1421.89	1409.34	0.06	2824.74	2835.16	2822.53	0.03	8466.44	8481.26	8468.57	0.01
	AR	LCM	1481.29	1492.76	1473.92	0.14	2956.43	2972.06	2953.11	0.12	8858.59	8880.81	8861.78	0.12
	AR	MLM	1534.29	1558.95	1539.91		3061.04	3089.85	3070.80		9171.92	9207.33	9188.26	
	AR	MA	1505.69	1513.34	1500.79	0.15	3006.83	3017.25	3004.62	0.14	9009.40	9024.21	9011.53	0.14
	MA	**MA**	1286.56	1294.21	1281.65	0.05	2569.92	2580.34	2567.70	0.02	7696.29	7711.11	7698.42	0.01
	MA	LCM	1453.32	1464.79	1445.96	0.11	2900.89	2916.52	2897.57	0.09	8693.06	8715.28	8696.25	0.09
	MA	MLM	1461.30	1485.96	1466.92		2915.29	2944.11	2925.05		8732.853	8768.26	8749.20	
	MA	AR	1430.85	1438.50	1425.94	0.08	2857.08	2867.50	2854.87	0.06	8562.38	8577.19	8564.51	0.05

(continued)

Table 26.4 (continued)

	DGM[a]	Fitted model	$N = 50$ AIC	BIC	BIC N-adj	RMSEA	$N = 100$ AIC	BIC	BIC N-adj	RMSEA	$N = 300$ AIC	BIC	BIC N-adj	RMSEA
D6:	**LCM**	**LCM**	936.76	948.23	929.40		1867.03	1882.66	1863.71		5589.74	5611.96	5592.93	
partially	LCM	MLM	1094.40	1115.54	1096.52		2183.30	2208.59	2189.55		6531.95	6563.83	6544.77	
random	LCM	AR	997.87	1005.52	992.96	0.17	1991.97	2002.39	1989.76	0.13	5971.38	5986.20	5973.51	0.11
($T = 5$)	LCM	MA	1180.39	1188.04	1175.49	0.25	2368.82	2379.25	2366.61	0.22	7115.22	7130.04	7117.35	0.21
	AR	**AR**	803.61	811.26	798.70		1603.86	1614.28	1601.65		4801.17	4815.98	4803.30	
	AR	LCM	833.02	844.50	825.66		1659.77	1675.40	1656.45		4968.03	4990.25	4971.22	
	AR	MLM	852.68	873.81	854.79		1698.65	1723.93	1704.89		5084.58	5116.46	5097.40	
	AR	MA	832.21	839.85	827.30	0.15	1659.56	1669.98	1657.35	0.10	4968.22	4983.03	4970.35	0.07
	MA	**MA**	786.97	794.62	782.06		1569.13	1579.55	1566.91		4702.16	4716.97	4704.29	
	MA	LCM	808.40	819.87	801.04		1611.99	1627.62	1608.67		4823.76	4845.98	4826.95	
	MA	MLM	812.56	833.69	814.67		1617.95	1643.24	1624.20		4843.39	4875.27	4856.21	
	MA	AR	791.74	799.39	786.83	0.14	1578.72	1589.14	1576.51	0.07	4727.81	4742.63	4729.94	0.03

Models in bold indicate the correctly specified model fit to itself. Thus, the values for these rows should be more optimal, i.e., smaller in magnitude, than for the subsequent rows where a misspecified model is fitted

[a] DGM = data generating model

BIC would choose the AR model over the correct MA model across all sample sizes.

We were also interested in examining whether fitting the MLM model with a spurious time-varying covariate to data generated from the LCM, AR, and MA models would lead to false detection of the covariate as having a statistically significant effect. This did not occur in any of the modeling conditions. The MLM was consistently identified as less favorable compared to the true models by all three of the information criterion measures. Across all the data generating conditions, the fixed effect slope associated with the spurious covariate in the MLM, γ_1, was not statistically significantly different from zero when averaged across all the Monte Carlo replications. The average estimated parameter for γ_1 divided by estimated standard error was 0.0055, −0.0013, and 0.0047 when the true data generating model was the LCM, AR, and MA model, respectively. This parameter was only significantly different from zero in 5.6% of replications when the MA model was the data generating model, 5.7% of the replications when the AR model was the data generating model, and 6.4% of the replications when the LCM was the data generating model. Thus, the Type I error rates for false detection of a spurious covariate were close to the nominal rate of 5%.

26.4 Discussion

Our study results illustrate how EMA designs requiring fewer planned assessments perform with various types of growth trajectories. Overall, we found that our estimation results involving the LCM and MLM were consistent with previous findings of incomplete data designs using such types of growth models (Duncan et al., 1996; Graham et al., 2001; McArdle & Hamagami, 1991, 1992). This study illustrated that well-known planned incomplete designs such as cohort sequential designs can be effectively used in the EMA context to reduce participants' response burden. We next discuss key factors that influence design and modeling choices, which are summarized in Fig. 26.9.

For the LCM and MLM, the design with two consecutive days max performed the best, yielding results with the closest match to the complete case design, although only marginally so. This design had the lowest reduction of required assessments (43%), which may be the key reason behind its optimal performance across these models. For the MLM model, the placement of data points does not appear to be as important as the number of data points. However, for the LCM model, this was not the case: the partially randomized design evidenced slightly better performance than did the CSD with 2-day overlap, despite both designs having the same number of non-missing assessment points. The partially randomized design has the possibility of capturing data points that are more representative of the entire 14-day time span. For example, a given person may have assessments for time points 1, 4, 8, 9, and 13, whereas for the CSD with 2-day overlap, the only possible data structures involve assessments on days 1 through 5, days 4 through 9, days 7 through 11, or days

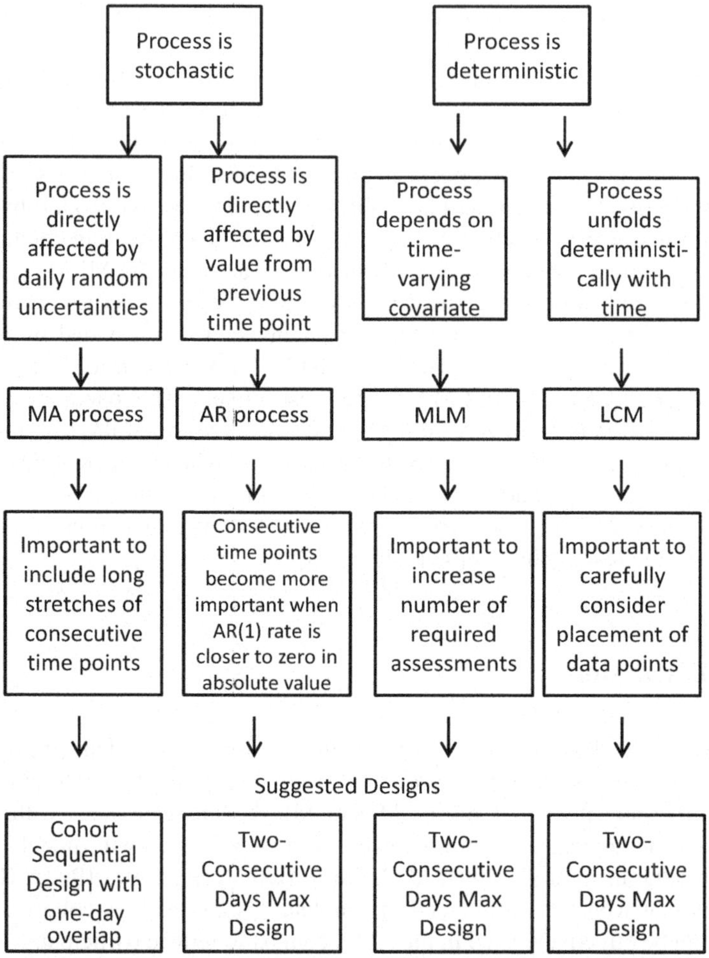

Fig. 26.9 Flowchart to help make decisions when designing a daily diary study

10 through 14. This assessment schedule may explain why the partially random design performed better than the CSD with 2-day overlap. In addition, including a reference day for all participants on the first measurement occasion was found to increase power without notable differences in the number of assessments. For the MLM model, because time does not explicitly appear in the model, the specific placement of the assessment occasions may not matter as much.

Satisfactory point and SE estimates were also obtained from the AR(1) model, which accommodates stochastic change and is suited for describing constructs that show less systematic fluctuations around a baseline (e.g., emotions, stress). We found that the other stochastic change model, the MA(1) model, did not perform as well as compared to the other hypothesized models. The statistical reason for MA(1) model's suboptimal reason has been discussed earlier. Some added insights concerning the differences between the AR(1) and MA(1) model may be helpful here. In a stationary AR(1) model, the covariance between any two observations within an individual is attenuated exponentially as the two observations are further apart in time, but the covariance does not diminish (i.e., become zero) completely. This can be understood mathematically by examining the autocovariance function of an AR(1) model for a given lag l, given by $\alpha^l \sigma^2_{z_{it}}$. At the moderately high AR(1) weight of 0.7, the decay rate is relatively slow. Thus, the AR(1) model considered in the present study did not require many consecutive time points to recover the parameters accurately. We did find evidence that having a longer stretch of consecutive measurements from each participant helps improve the precision of the process noise variance parameter in the AR(1) model. In particular the CSD with 2-day overlap yielded greater accuracy in this parameter when compared with the partially randomized design even though both designs were characterized by the same number of non-missing observations. Differences were relatively small, however. The need for measurement occasions that are closer in time would generally be higher as the AR(1) parameter is closer to zero (i.e., the decay rate is higher).

It was also interesting to note that the two consecutive days max and partially randomized designs were robust to detecting the fit of a "wrong" model. The CSD with 2-day overlap did not detect this in every condition when the AR and MA models were used as the true data generating model. This may suggest that using the former two designs would provide the researcher with more confidence that he or she will be able to determine which model best approximates the true underlying trajectory.

Recommendations for Applied Researchers

A flowchart summarizing some of the key issues to consider in designing EMA studies is shown in Fig. 26.9. Based on our findings, we have a number of specific recommendations for researchers interested in adopting planned missingness designs in collecting EMA data:

1. We recommend that at the very earliest points in designing an EMA study, researchers carefully consider the characteristics of the hypothesized change trajectory(ies) of interest. As illustrated, certain EMA designs are better equipped at capturing different change processes. Thus, if one hypothesizes a change trajectory that follows an MA(1) model, it would be wise to allow for multiple consecutive assessment points within each participant. Such a design, however, would not be as important to consider if one is examining change resulting from an LCM or MLM model. Further, a change trajectory hypothesized in psychological research may be a combination of an AR and MA model, namely, an ARMA(p,q) model (Shumway & Stoffer, 2006). In such cases, the performance of MA parameter must be explicitly considered.

2. When the change trajectory is not known or cannot be reasonably conjectured prior to data collection, we recommend a design with multiple consecutive assessment points per each individual and fewer missing time points.
3. We also recommend, based on our findings, that researchers consider the importance of particular parameters in answering hypothesized research questions. We found in these simulation studies that the variance components of the LCM and the MLM models were not as accurately recovered when compared to the fixed effects. These findings are consistent with McArdle et al. (2002) who found that, when using a growth model in the MLM context, fixed effects were easily recoverable, but random effects, such as the covariance between intercept and slope, were much more difficult to recover as the number of required assessments was reduced. Our results indicate a similar pattern wherein the covariance parameters for the LCM and MLM models were the most difficult to recover accurately.
4. If there is a phenomenon that is likely to appear at a certain day of the week, such as on the weekends, we recommend researchers use a reference group design. This recommendation follows McClelland's (1997) notion of "loading up" extreme values of an independent variable. However, we are now "loading" up participants on certain days of the week when an effect is more likely to be pronounced.
5. The variations of cohort sequential designs considered offer a promising alternative to the commonly adopted partially randomized design. This is true despite the fact that cohort effects are not the focus of daily dairy designs, unlike developmental studies wherein the cohort sequential design is a vastly popular design. The partially randomized design, which is very popular among EMA researchers (e.g., Heron & Smyth, 2010; Hawkley et al., 2003), may not be the best choice unless a sufficient number of time points can be obtained from each participant to get at different types of trends and change processes.

Limitations and Future Directions

The designs considered in the present chapter are by no means exhaustive. Our results are limited to the five planned missingness designs and the four models considered. For instance, each model was univariate in nature. To represent change accurately, it is important to collect and analyze multivariate data (Browne & Nesselroade, 2005). Furthermore, models with covariates, more complicated functional forms, differing parameter values, and larger degrees of lagged effects (e.g., an AR model of order 2) were not considered. We also did not examine models incorporating both AR and MA processes, i.e., ARMA models. Thus, it is unclear whether and to what extent results will generalize to these more complicated cases.

Another limitation is that we focused only on daily dairy designs wherein the time units were equally spaced and measured in days, with a maximum of 14 "days." Similar ideas can be adapted to yield designs based on denser units of time.

For instance, to capture hourly fluctuations in a process of interest, researchers or practitioners could adopt variations of the two consecutive "days" max design so each participant would provide data for two consecutive hours during a targeted time window during the day. Doing so of course requires that researchers/practitioners have some preconceived ideas on the time units of interest—a requirement that may be challenging to fulfill in many empirical applications. Thus, future work may examine a broader collection of EMA designs, possibly as enriched by alternative intensive longitudinal data, (e.g., passive location and physical activity data collected using smartphones and smartwatches), to allow more effective use of planned missingness in intensive longitudinal designs.

We did not consider other mechanisms of missingness, particularly data that are missing at random. Finally, we only considered time series processes that are stationary. A nonstationary process requires the proper specification of an initial condition distribution, the choice of which can affect parameter estimates (De Jong, 1991; De Jong & Chu-Chun Lin, 1994), especially given a relatively small number of time points (Oud et al., 1990). Also, given a high degree of non-stationarity, i.e., a mean and/or variance that changes drastically with time, it may be important to create a design that ensures that data are collected at the appropriate times to capture such changes. Researchers should thus consider examining nonstationary processes in the context of planned missingness EMA studies in future studies, especially given that such processes are common in psychological data.

Our results are also limited to the particular estimation procedures we chose to implement. For example, we chose to use FIML to deal with the planned missingness. We did not consider using other approaches such as a multiple imputation (Little & Rubin, 1987) method. While using FIML and multiple imputation produce similar results asymptotically, there is some evidence that multiple imputation performs better in smaller samples (Shafer & Graham, 2002). Thus, it may be worthwhile for future studies to compare the two methods when implementing planned missingness EMA studies. We also did not implement a reduced information estimation procedure, such as restricted maximum likelihood (REML), which may have produced more accurate results in the conditions with smaller sample sizes.

Consequences of practice effects, a well-known confound in longitudinal research, may be especially pronounced in experience sampling studies. Subtle changes may arise when participants are required to contribute an intensive amount of information over a short period of time. It is not clear whether an individual may display a different pattern of results solely due to the fact that he or she is required to respond in a different pattern. For example, if a participant knows he or she needs to respond every day for 14 days, the participant's response pattern may be different than if he or she is anticipating only responding every day for 5 days. Also, unplanned missingness may not be reduced but actually may increase with our diary designs, especially with the partially randomized design, given the sporadic nature of this design. For example, a participant may respond more often given a diary schedule that requires him or her to respond every day, thus allowing the person to develop a habit of responding. However, we argue that, given other benefits of partially randomized designs, such as obtaining a response that more accurately reflects how a person feels "in the moment," and given that researchers are already hypothesizing such designs, it is still worthwhile to consider these designs. The current study did not address this concern with actual participants, but future studies may be able to do so using randomized groups where one group is required to respond daily and another given a partially randomized design.

Overall, the planned missingness EMA designs investigated here have the potential to reduce participant burden substantially while preserving the statistical properties of the change processes of interest using the reduced data sets. The designs proposed should be utilized depending upon the researcher's hypothesized change trajectory. Future studies should examine how well differing AR, MA, ARMA, and other multivariate change models perform as a function of the number of consecutive days for a given subgroup of respondents.

Appendix: Sample AR(1) and MA(1) Mplus Code

```
!!!!!!!!!!!!!!!!!!!!!!!!!!!!!!!!!!!!!!!!!!!!!!!!!!!!!!!!!!!!!!!!!!!!!!!!!!!!!!!!
!Mplus code for fitting the AR(1) model in Equation 3
!!!!!!!!!!!!!!!!!!!!!!!!!!!!!!!!!!!!!!!!!!!!!!!!!!!!!!!!!!!!!!!!!!!!!!!!!!!!!!!!
title: AR(1) sample code;
data:
file is simAR1.txt; !Simualted data

variable:
names = y1-y14; !variable names
USEVARIABLES =  y1-y14; !variables used
missing = .; !missing values
model:

!Initial latent variable
inity by y1@0;
y1 on inity@1;
!Free variance of initial condition
inity*;
[inity@0]; !Mean of initial condition fixed at zero

!Auto-regression weights
y2 on y1(1);y3 on y2(1);y4 on y3(1);y5 on y4(1);
y6 on y5(1);y7 on y6(1);y8 on y7(1);y9 on y8(1);
y10 on y9(1); y11 on y10(1); y12 on y11(1);
y13 on y12(1); y14 on y13(1);

!Random shocks
z1 by y1@1; z2 by y2@1; z3 by y3@1; z4 by y4@1;
z5 by y5@1; z6 by y6@1; z7 by y7@1; z8 by y8@1;
z9 by y9@1; z10 by y10@1 z11 by y11@1;
z12 by y12@1; z13 by y13@1; z14 by y14@1;

!no corrs among random shocks, and between random shocks and initial variable;
z1 with z3-z14@0;
z2 with z3-z14@0;
z3 with z5-z14@0;
z4 with z5-z14@0;
z5 with z7-z14@0;
```

```
z6 with z7-z14@0;
z7 with z9-z14@0;
z8 with z9-z14@0;
z10 with z10-z14@0;
z11 with z12-z14@0;
z12 with z13-z14@0;
z13 with z14@0;

z1-z14 with inity@0;

!equal resid vars over time;
z1-z14 (2);

!Intercepts of y1-y14 set to zero
[y1-y14@0];
!No residual variance for y1-y14.
!The random shocks serve as the "measurement errors"
y1-y14@0;

!saving the parameter estimates
savedata: results = "simEstimates.dat";


!!!!!!!!!!!!!!!!!!!!!!!!!!!!!!!!!!!!!!!!!!!!!!!!!!!!!!!!!!!!!!!!!!!!!!!!!!!!!!!!
!Mplus code for fitting the MA(1) model in Equation 4
!!!!!!!!!!!!!!!!!!!!!!!!!!!!!!!!!!!!!!!!!!!!!!!!!!!!!!!!!!!!!!!!!!!!!!!!!!!!!!!!

title: MA(1) sample code;

data:
file is simMA1.txt; !Simualted data

variable:
names = y1-y14; !variable names
USEVARIABLES =  y1-y14; !variables used
missing = .; !missing values
model:

!Initial latent variable
inity by y1@0;
y1 on inity@1;
!Free variance of initial condition
inity*;
[inity@0]; !Mean of initial condition fixed at zero

!Moving average weights
y2 on z1(1);  y3 on z2(1);  y4 on z5(1);  y5 on z7(1);
y6 on z2(1);  y7 on z6(1);  y8 on z7(1);  y9 on z8(1);
y10 on z9(1); y11 on z10(1); y12 on z11(1); y13 on z12(1);
y14 on z13(1); y15 on z14(1);


!Random shocks
z1 by y1@1; z2 by y2@1; z3 by y3@1; z4 by y4@1;
z5 by y5@1; z6 by y6@1; z7 by y7@1; z8 by y8@1;
z9 by y9@1; z10 by y10@1 z11 by y11@1;
z12 by y12@1; z13 by y13@1; z14 by y14@1;

!no corrs among random shocks, and between random shocks and initial variable;
z1 with z3-z14@0;
z2 with z3-z14@0;
z3 with z5-z14@0;
z4 with z5-z14@0;
```

```
z5 with z7-z14@0;
z6 with z7-z14@0;
z7 with z9-z14@0;
z8 with z9-z14@0;
z10 with z10-z14@0;
z11 with z12-z14@0;
z12 with z13-z14@0;
z13 with z14@0;

z1-z14 with inity@0;

!equal resid vars over time;
z1-z14 (2);

!Intercepts of y1-y14 set to zero
[y1-y14@0];
!No residual variance for y1-y14.
!The random shocks serve as the "measurement errors"
y1-y14@0;

!saving the parameter estimates
savedata: results = "simEstimates.dat";
```

References

Akaike, H. (1973). Information theory and an extension of the maximum likelihood principle. In B. N. Petrov & F. Csáki (Eds.), *Second international symposium in information theory* (pp. 267–281). Akadémiai Kiadó.

Bell, R. Q. (1953). Convergence: An accelerated longitudinal approach. *Child Development, 24*, 145–152.

Birditt, K. S., Fingerman, K. L., & Almeida, D. M. (2005). Age differences in exposure and reactions to interpersonal tensions: A daily diary study. *Psychology and Aging, 20*(2), 330–340.

Bolger, N., Davis, A., Olchowski, A. E., & Rafaeli, E. (2003). Diary methods: Capturing life as it is lived. *Annual Review of Psychology, 54*, 579–616.

Bollen, K. A., & Curran, P. J. (2006). *Latent curve models: A structural equation perspective*. Wiley.

Box, G. E. P., & Jenkins, G. M. (1976). *Time series analysis: Forecasting and control (Rev. ed.)*. Holden–Day.

Browne, M. W., & Nesselroade, J. R. (2005). Representing psychological processes with dynamic factor models: Some promising uses and extensions of arma time series models. In A. Maydeu-Olivares & J. J. McArdle (Eds.), *Advances in psychometrics: A festschrift for roderick p. mcdonald* (pp. 415–452). Erlbaum.

Campbell, L., Simpson, J. A., Boldry, J. G., & Rubin, H. (2010). Trust, variability in relationship evaluations, and relationship processes. *Journal of Personality and Social Psychology, 99*(1), 14–31.

Chatfield, C. (2004). *The analysis of time series: An introduction* (6th edn.). Chapman & Hall/CRC.

Cohen, J. (1992). A power primer. *Psychological Bulletin, 112*(1), 155–159.

De Jong, P. (1991). The diffuse kalman filter. *The Annals of Statistics, 19*(2), 1073–1083.

De Jong, P., & Chu-Chun Lin, S. (1994). Stationary and non-stationary state space models. *Journal of Time Series Analysis, 15*, 151–166.

Diener, E., Fujita, F., & Smith, H. (1995). The personality structure of affect. *Journal of Personality and Social Psychology, 69*(1), 130–141.

Duncan, S. C., Duncan, T. E., & Hops, H. (1996). Analysis of longitudinal data within accelerated longitudinal designs. *Psychological Methods*, *1*, 236–248.

Du Toit, S. H. C., & Browne, M. W. (2001). The covariance structure of a vector time series. In R. Cudeck, S. H. C. D. Toit, & D. Sorbom (Eds.), (pp. 279–314). Scientific Software International Inc.

Du Toit, S. H. C., & Browne, M. W. (2007). Structural equation modeling of multivariate time series. *Multivariate Behavioral Research*, *42*(1), 67–101.

Emotions and Dynamic Systems Laboratory. (2010). *The Affective Dynamics and Individual Differences (ADID) study: Developing non-stationary and network-based methods for modeling the perception and physiology of emotions*. Unpublished manual, University of North Carolina at Chapel Hill.

Enders, C. K., & Gottschall, A. C. (2011). The impact of missing data on the ethical quality of a research study. In A. T. Panter & S. K. Sterba (Eds.), *Handbook of ethics in quantitative methodology* (pp. 357–381). Routledge.

Flook, L., & Fuligni, A. J. (2008). Family and school spillover in adolescents' daily lives. *Child Development*, *79*(3), 776–778.

Graham, J. W., Taylor, B. J., & Cumsille, P. E. (2001). Planned missing data designs in analysis of change. In L. Collins & A. Sayer (Eds.), *New methods for the analysis of change* (pp. 503–541). American Psychological Association.

Hamilton, J. D. (1994). *Time series analysis*. Princeton University Press.

Hankin, B. L., Fraley, R. C., & Abela, J. R. Z. (2005). Daily depression and cognitions about stress: Evidence for a traitlike depressogenic cognitive style and the prediction of depressive symptoms in a prospective daily diary study. *Journal of Personality and Social Psychology*, *88*(4), 673–685.

Hawkley, L. C., Burleson, M. H., Berntson, G. G., & Cacioppo, J. T. (2003). Loneliness in everyday life: Cardiovascular activity, psychosocial context, and health behaviors. *Journal of Personality and Social Psychology*, *85*, 105–120.

Heron, K. E., & Smyth, J. M. (2010). Ecologial momentary interventions: Incorporating mobile technology into psychosocial and health behaviour treatments. *British Journal of Health Psychology*, *15*, 1–39.

Horn, J. L., & McArdle, J. J. (1980). Perspectives on mathematical/statistical model building (masmob) in research on aging. In L. W. Poon (Ed.), *Aging in the 1980's: Selected contemporary issues in the psychology of aging* (pp. 503–541). American Psychological Association.

Iida, M., Shrout, P. E., Laurenceau, J., & Bolger, N. (In press). Using intensive longitudinal methods in psychological research (3 volumes). In H. Cooper, P. Coutanche, L. McMullin, A. T. Panter, D. Rindskopf, & K. S. K. (Eds.), *APA handbooks in psychology: APA handbook of research methods in psychology*. American Psychological Association.

Kiang, L., Yip, T., Gonzales-Backen, M., Witkow, M., & Fuligni, A. J. (2006). Ethnic identity and the daily psychological well-being of adolescents from mexican and chinese backgrounds. *Child Development*, *77*(5), 1338–1350.

Little, R. J. A., & Rubin, D. B. (1987). *Statistical analysis with missing data*. Wiley.

McArdle, J. J. (1994). Structural factor analysis experiments with incomplete data. *Multivariate Behavioral Research*, *29*, 409–454.

McArdle, J. J., & Epstein, D. (1987). Latent growth curves within developmental structural equation models. *Child Development*, *58*(1), 110–133.

McArdle, J. J., Ferrer-Caja, E., Hamagami, F., & Woodcock, R. W. (2002). Comparative longitudinal structural analyses of the growth and decline of multiple intellectual abilities over the life span. *Developmental Psychology*, *38*(1), 115–142.

McArdle, J. J., & Hamagami, F. (1991). Modeling incomplete longitudinal and cross-sectional data using latent growth structural models. In L. Collins & J. Horn (Eds.), *Best methods for the analysis of change* (pp. 275–303). American Psychological Association.

McArdle, J. J., & Hamagami, F. (1992). Modeling incomplete longitudinal and cross-sectional data using latent growth structural models. *Experimental Aging Research*, *18*, 145–166.

McClelland, G. H. (1997). Optimal design in psychological research. *Psychological Methods, 2*, 3–19.
Meredith, W., & Tisak, J. (1990). Latent curve analysis. *Psychometrika, 55*, 107–122.
Muthén, L. K., & Muthén, B. O. (1998–2007). *Mplus user's guide* (5th edn.). Muthén & Muthén.
Nessleroade, J. R., & Baltes, P. B. E. (1979). *Longitudinal research in the study of behavior and development*. Academic Press.
Ong, A. D., Fuller-Rowell, T., & Burrow, A. L. (2009). Racial discrimination and the stress process. *Journal of Personality and Social Psychology, 96*(6), 1259–1271.
Oud, J. H. L., van den Bercken, J. H., & Essers, R. J. (1990). Longitudinal factor score estimation using the kalman filter. *Applied psychological measurement, 14*(4), 395–418.
Raftery, A. E. (1993). Bayesian model selection in structural equation models. In K. A. Bollen & J. S. Long (Eds.), *Testing structural equation models* (pp. 163–180). Sage.
Raudenbush, S. W., & Bryk, A. S. (2002). *Hierarchical linear models: Applications and data analysis methods*. Sage.
Raudenbush, S. W., & Liu, X. (2000). Statistical power and optimal design for multisite randomized trials. *Psychological Methods, 5*(2), 199–213.
R Development Core Team. (2023). R: A language and environment for statistical computing [Computer software manual]. Vienna, Austria. Retrieved from http://www.R-project.org (ISBN 3-900051-07-0).
Reis, H. T., & Gable, S. L. (2000). Event sampling and other methods for studying daily experience. In H. T. Reis & C. M. Judd (Eds.), *Handbook of research methods in social and personality psychology* (pp. 190–222). Cambridge University.
Schwarz, G. (1978). Estimating the dimensions of a model. *Annals of Statistics, 6*, 461–464.
Shafer, J. L., & Graham, J. W. (2002). Missing data: Our view of the state of the art. *Psychological Methods, 7*, 147–177.
Shifren, K., Hooker, K. A., Wood, P. K., & Nesselroade, J. R. (1997). Structure and variation of mood in individuals with Parkinson's disease: A dynamic factor analysis. *Psychology and Aging, 12*, 328–339. https://doi.org/10.1037/0882-7974.12.2.328.
Shumway, R. H., & Stoffer, D. S. (2006). *Time series analysis and its applications*. Springer Texts in Statistics.
Silk, J. S., Steinberg, L., & Morris, A. S. (2003). Adolescents' emotion regulation in daily fife: Links to depressive symptoms and problem behavior. *Child Development, 74*(6), 1869–1880.
Singer, J. D., & Willett, J. B. (2003). *Applied longitudinal data analysis: Modeling change and event occurrence*. Oxford University Press.
Skrondal, A. (2000). Design and analysis of monte carlo experiments: Attacking the conventional wisdom. *Multivariate Behavioral Research, 35*(2), 137–167.
Stawski, R. S., Sliwinski, M. J., Almeida, D. M., & Smyth, J. M. (2008). Reported exposure and emotional reactivity to daily stressors: The roles of adult age and global perceived stress. *Psychology and Aging, 23*(1), 52–61.
Steiger, J. H., & Lind, J. (1980). *Statistically based tests for the number of common factors*. Paper presented at the annual meeting of the Psychometric Society, Iowa City.
Stone, A. A., Kessler, R. C., & Haythornthwaite, J. A. (1991). Measuring daily events and experiences: Decisions for the researcher. *Journal of Personality, 59*(3), 575–607.
Stone, A. A., & Shiffman, S. (2002). Capturing momentary, self-report data: A proposal for reporting guidelines. *Annals of Behavioral Medicine, 24*(3), 236–243.
Wei, W. W. S. (1990). *Time series analysis*. Addison–Wesley.
Wheeler, L., & Reis, H. T. (1991). Self recording of every day life events: Origins, types, and uses. *Journal of Personality, 59*, 339–354.
Winer, B. J. (1962). Latin squares and related designs. In B. J. Winer (Ed.), *Statistical principles in experimental design* (pp. 514–577). McGraw-Hill Book Company.

Part VI
Item-Response-Modeling for Dependent Data

Chapter 27
Variants of Estimating an IRT-Based Actor-Partner Interdependence Model (APIM) with R

Rainer W. Alexandrowicz, Linda Maurer, Anna Schultz, and Marcus Mund

27.1 Introduction

Dyadic data structures are essential for research in various fields of psychology and arise in contexts, in which researchers are interested in interdependencies and mutual influences between two individuals (e.g., Kenny et al., 2006). Most prominently, dyadic data are crucial in research involving partner relationships; in this case, the dyad (i.e., the couple) is the natural unit of analysis. Classical questions in this field are concerned with the partners' mutual influence on their relationship satisfaction (Bradbury et al., 2000). Specifically, in this case, *both* partners would provide information on the perceived quality of their relationship, which allows researchers to then compare the perspectives of both partners. Of course, this basic example of a correlation between both partners' relationship satisfaction can be extended to accommodate much more complex research questions, many of which are described in Kenny et al. (2006). For example, it can be examined to what extent one partner's relationship satisfaction is influenced by characteristics of the other partner—which would not be possible if only one partner was considered. The primary workhorse employed to address such research questions is the Actor-Partner Interdependence Model (APIM; Kenny et al., 2006).

R. W. Alexandrowicz (✉)
Abteilung für Methodenlehre, Institut für Psychologie, Universität Klagenfurt, Klagenfurt, Austria
e-mail: rainer.alexandrowicz@aau.at

L. Maurer · M. Mund
Universität Klagenfurt, Abteilung für Persönlichkeitspsychologie und Psychologische Diagnostik, Klagenfurt, Austria

A. Schultz
Medizinische Universtität Graz, Institut für Sozialmedizin und Epidemiologie, Graz, Austria

M. Stemmler et al. (eds.), *Dependent Data in Social Sciences Research*,
https://doi.org/10.1007/978-3-031-56318-8_27

Although the APIM is an established tool and used very widely across several research fields, its implementation typically hinges on assumptions that may not always be met in research contexts. In the present chapter, we discuss those assumptions and the pitfalls of implementing the APIM in a "cookbookery"-like fashion. Furthermore, we demonstrate and compare several alternative implementations using cutting-edge IRT-based models.

Self-Esteem and Relationship Satisfaction

To illustrate our methodological expositions, we build on the example of self-esteem and relationship satisfaction. Self-esteem is one of the most widely examined constructs in psychology and describes an individual's self-regard as a person of worth (Donnellan et al., 2011). It has been found to be prospectively associated with a wide range of important life outcomes including aspects of health and well-being, occupational status, and interpersonal satisfaction (Harris & Orth, 2020; Arsandaux et al., 2020; Orth et al., 2012). Relationship satisfaction is typically considered a crucial aspect of broader relationship quality (Fletcher et al., 2000) and has been the key focus of countless empirical studies (for reviews, see Karney & Bradbury, 1995; Bradbury et al., 2000). Thus, research on these two constructs is widely established and ideally suited to demonstrate the added value of the analytic approach presented in this chapter over more classic analyses.

Across several studies, it has been found that self-esteem is robustly and positively related to relationship satisfaction in partner relationships (for a meta-analysis, see Harris & Orth, 2020). These studies have been conducted in both cross-sectional and longitudinal dyadic settings employing both path analysis with manifest variables and structural equation modeling with latent variables (Mund et al., 2015; Erol & Orth, 2013, 2014; Schaffhuser et al., 2014). The general pattern that has emerged from those studies and that is, thus, to be expected to occur in the present analysis as well is that self-esteem shows positive, though small, actor and partner effects on relationship satisfaction (Fig. 27.1).

Variants of Modeling an APIM

Basically, the APIM is a sophisticated regression model, taking into account four central regression equations (see Fig. 27.1): the effect of $X \rightarrow Y$ of actors one and two (paths a_1 and a_2) along with the cross-effects of the X of actor one upon the Y of actor two (path p_{21}) and vice versa (path p_{12}). Additionally, the covariance of both actors' X (c_x) and the residual correlation of both actors' Y (c_y) are taken into account along with all four variables' variances (vx_1, vx_2, ve_1, ve_2). Several approaches to determine these elements are available, most prominently

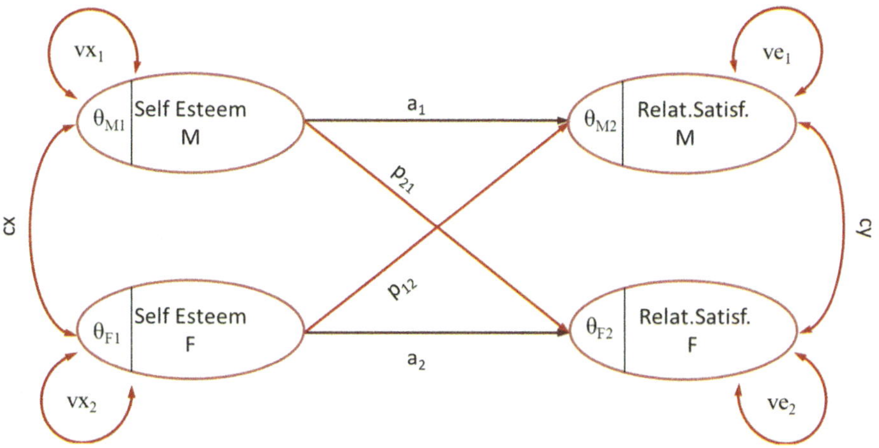

Fig. 27.1 The Actor-Partner Interdependence Model (APIM). *Notes:* a_1, male actor effect; a_2, female actor effect; p_{12}, partner effect female to male; ap_{21}, partner effect male to female; vx_1/vx_2 predictor variance male/female; ve_1/ve_2 outcome variance male/female; c_x, residual covariance predictors; c_y, residual covariance outcomes

path analysis (PA; Wright, 1934) and structural equation models (SEMs; Hoyle, 2023).

Path Analysis Path analysis is a technique to determine a complex structure of dependencies among several variables. It is based on regression analysis of scores obtaining path coefficients following rules proposed by Wright (1934) (see also Duncan, 1966; Lleras, 2005; Hollar, 2018, Ch. 5; Pedhazur, 1982). Path models can be regarded a predecessor of structural equation models, but restricted to "recursive" structures, i.e., models without feedback loops. Due to their history, they do not involve latent constructs but evaluate manifest scores.

Considering that most psychological constructs evade direct measurability, the usual assessment involves a set of questions reflecting facets of such a latent construct. This raises the question of how to generate an appropriate numeric quantification of the construct itself. The most simple yet still applied approach of taking the sum ignores the fact that such a procedure implicitly involves assumptions, which if not met may render the results inaccurate. Jöreskog (2004) unmistakably states that "*Ordinal variables are not continuous variables and should not be treated as if they are. It is common practice to treat scores 1, 2, 3, . . . assigned to categories as if they have metric properties but this is wrong. Ordinal variables do not have origins or units of measurements. Means, variances, and covariances of ordinal variables have no meaning. The only information we have are counts of cases in each cell of a multiway contingency table*" (p. 1).

Aside from these fundamental problems, another obviously problematic assumption is that all responses enter the sum unweightedly, i.e., they contribute equally to the latent construct. This contradicts the desideratum of "Objective Measurement," which "*is the repetition of a unit amount that maintains its size, within an allowable range of error, no matter which instrument, intended to measure the variable of interest, is used and no matter who or what relevant person or thing is measured*" (The Program Committee of the Institute for Objective Measurement, 2000). McNeish and Wolf (2020) work out in detail the intricacies resulting from inadequate scoring (see also McNeish, 2022; Widaman & Revelle, 2023b; Widaman & Revelle, 2023a).

A third issue with the unreflected use of the simple sum is that the measurement error remains undetected, i.e., the stochastic element of each single response enters the model without correction. Mislevy (1993) clearly works out that what we actually observe is a fallible indicator $x \sim N(\theta, \sigma_e^2)$, rather than the true θ (p. 82; the symbol θ referring to a latent variable, which is detailed below; σ_e^2 denotes the error variance of θ). It is generally known, even for decades, that ignoring the measurement error may produce biased estimates (e.g., Fuller, 1987; Stefanski, 1985). Schofield (2015) further notes that "*The use of noisy measures of latent variables can cause researchers to* mis-estimate *the effects of the latent variables on the outcome of interest and the effects of other correlated covariates in the analysis*" (p. 2134; emphasis in the original). Also (Zidek et al., 1996) show that interaction effects of the measurement error and explanatory variables may cause that "*[a] causal variable measured with error may be overlooked and its significance transferred to a surrogate*" (p. 441). For a comprehensive treatment, see, e.g., Yi et al. (2021). Hence, Junker et al. (2012), argue that "*Estimates of θ_i should be used in place of total score X_i whenever possible*" (p. 5; θ_i again indicating a latent variable).

Structural Equation Modeling (SEM) The SEM approach possesses two advantages over PA on simple sums. First, it contains a measurement part, which models the strength of association of the manifest indicators (the questions/items) on the one hand and the latent construct, which is hypothesized to cause the manifest responses. That way, we tackle the measurement error by extracting the information regarding the common latent cause based on the responses depending on this construct. Second, it offers a structural part, which allows for modeling simultaneously complex relationships between the latent constructs. Moreover, SEMs allow for inspecting the adequacy of the hypothesized model in the light of the data by means of a plethora of fit measures and indices (for an overview see, e.g., West et al., 2023).

In their original form, SEMs estimate all parameters (i.e., both the loadings of the measurement model and the latent regression coefficients) from the Pearson-type covariance matrix of all manifest indicators, which requires interval scaled manifest indicators. However, many psychological applications use ordered categorical response formats (frequently termed "Likert-type items"[1]), which requires to determine the covariance matrix from the polychoric correlation coefficient (Drasgow, 1988; Jöreskog, 1994a,b; Olsson, 1979; Olsson et al., 1982). This coefficient

[1] See Uebersax (1999), for a clarification of terms, and Bernstein (2006) for some fundamental problems.

estimates the correlation of two ordered categorical variables by introducing a further distributional assumption, that is, they follow a bivariate normal distribution, each of which is cut at unknown thresholds (frequently termed τ) forming the observed categories. While plausible from a psychological point of view, Kolbe et al. (2021) found that "*the bivariate normal and skew-normal distributions rarely hold in the empirical datasets*" (p. 562) and that estimates assuming an inadequate distribution may be substantially biased (cf. Foldnes & Grønneberg, 2020). Foldnes and Grønneberg (2019) propose a bootstrap test outperforming the test of Maydeu-Olivares (2006). Moreover, Grønneberg and Foldnes (2022) show that not only the marginal and bivariate normality assumption may fail with severe consequences but that we require substantive arguments regarding the marginals. They even work out, under which circumstances polychoric correlations systematically overestimate the size of the true response correlation (p. 75).

Therefore, other techniques for determining the covariance matrix are indicated (e.g., Choi et al., 2010; Foldnes & Grønneberg, 2019, 2022). However, such procedures may require separate checks of the adequacy of their assumptions (cf. Kampen and Weeren, 2016), but these are rarely applied. Moreover, the estimation routines have to be adapted as the polychoric correlation matrix lacks the requirements of the maximum likelihood (ML) method. Here the diagonally weighted least squares method (WLSMV; Brown, 2006), a limited information estimation method, has become established as standard, but challenged by the robust ML method (MLR; Li, 2016).

Another sometimes overseen aspect is that invariance testing (cf. Meade & Lautenschlager, 2004) involves additional steps regarding the invariance of the thresholds τ (Bowen & Masa, 2015). Moreover, the model fit indices RMSEA, CFI, and TLI require special treatment when using WLSMV (Xia & Yang, 2019). However, the "polychoric + WLSMV"–ritual seems most prominent for SEM with categorical data, despite these issues. It somehow may be regarded as *cookbookery* in the sense of Box (1976, p. 797).

Item Response Theory (IRT) A different approach would be to treat the Likert-type responses for what they are, ordered categories, which brings us ultimately to Item Response Theory (IRT; cf. de Ayala, 2022). In contrast to SEM, the latent continuous measure (usually termed[2] θ) is "directly" estimated, i.e., without assuming an intermediate "categorization" (τ) to reconstruct a latent covariance matrix, which in turn serves as the foundation for parameter estimation. Another difference of SEM and IRT is that parameter estimation in SEMs is based on the matrix of second moments, whereas IRT models employ the entire response matrix (Baker & Kim, 2004). Therefore, the estimation principle is sometimes termed "full information" (cf. Croon, 2002; Cai & Hansen, 2013).[3] Moreover, Lechner et al. (2021) argue that "*IRT is arguably better to handle incomplete block designs,*

[2] Not to be mixed up with the residual covariances termed θ^{δ} and θ^{ϵ} in SEM.

[3] Not to be mixed up with the "Full Information Maximum Likelihood" principle (FIML; cf. Enders, 2010).

population models with background variables, and computer adaptive testing" (p. 4).

When applying IRT, we model the probability of choosing a response category $P(X)$ (rather than X itself) by specifying a specific regression model $P(X|\theta, \Xi)$, with θ indicating a person parameter (termed "ability," "proficiency," "skill," or "propensity to endorse a statement") and Ξ a set of item parameters specific for the chosen model. As $P(X)$ is bounded to the interval (0,1), a linear link function would produce out-of-bound predictions; hence, it must employ a nonlinear function (Rusch et al., 2016). Historically, the probit link was a common choice, now predominantly replaced by the logistic function $(1 + e^{-\vartheta})^{-1}$ for its mathematical advantages, the exponent ϑ substituting the respective model parametrization. Recently, also the third link function known from generalized linear models, the complementary log-log function, has been proposed for IRT models (Shim et al., 2023a,b).

Regarding the item parameters (Ξ), there are several variants available, differing with respect to their assumptions and interpretation. First of all, we distinguish models for dichotomous vs. polytomous response format. The former provide one location parameter per item (the difficulty parameter) and the latter several, depending on the number of response categories. Second, (Thissen & Steinberg, 1986) developed a taxonomy of IRT models distinguishing divide-by-total models from difference models. These differ with regard to the way the category boundaries are interpreted (see also Ostini & Nering, 2006). Third, IRT models can be classified by specific properties of the link functions (i.e., regression lines, here termed item characteristic functions (ICC), in the dichotomous case, and category characteristic functions/threshold characteristic functions, CCC/TCC in the polytomous case; cf. Wright & Masters, 1982). These may either run parallel across items/categories or can intersect. Parallel ICCs/TCCs establish the fundamental feature of specific objectivity (SO; Rasch, 1961, 1966a,b), which is a necessary and sufficient condition for the simple sum of scores to constitute a sufficient statistic for the person parameter. Models containing a discrimination ("steepness") parameter require special weighting procedures for estimating the person parameters. Hence, the basic difference is that models supporting SO allow for easily calculating a person statistic per hand, whereas the others require complex estimation methods (albeit frequently ignored). Fourth, we may apply a unidimensional and a multidimensional model, with the person parameter being a scalar in the former and a vector in the latter case (Reckase, 2009). As IRT models formulate a specific hypothesis regarding the structure of the data, we can also check the adequacy of such a hypothesis empirically.

IRT Models

There are three major model variants usually applied to polytomous scoring, the Partial Credit Model (PCM; Masters, 1982, 2005), the Generalized Partial Credit

Model (GPCM; Muraki, 1992), and the Graded Response Model (GRM; Samejima, 1969), each in a uni- and a multidimensional form.

The Partial Credit Model (PCM) The PCM is a member of the divide-by-total models, i.e., it focuses on passing a threshold between two adjacent categories. It is a member of the "Rasch family of models" (cf. Fischer and Molenaar, 1995), thus supporting specific objectivity (SO; Rasch, 1961, 1966a). As a consequence, the unweighted sum of scores is a sufficient statistic for the person parameter.

The Generalized Partial Credit Model (GPCM) The GPCM can be understood as an extension of the PCM by adding a slope (discrimination) parameter (similar to the relation of the dichotomous Rasch model and the Birnbaum 2PL model). This extension allows, on the one hand, for better model fit due to the additional parameters, but, on the other hand, at the price of losing SO. For estimating the person parameters, we have to weigh items with their discrimination parameters, which requires complex software and can—in contrast to the PCM—no longer be performed "by hand" (which would be useful in clinical applications, for example).

The Graded Response Model (GRM) This model is in contrast to the other two a difference model, i.e., the threshold locations are determined by considering all categories below vs. all above each pair of adjacent categories. It does not belong to the Rasch family of models; hence, the same restrictions apply as with the GPCM. For further details regarding the models, see Keller and Alexandrowicz (2024), in this volume.

The Multidimensional Approach Each of the three IRT models taken into consideration here allows for a multidimensional extension with each partner and measurement occasion forming one dimension (see Fig. 27.1). This may prove advantageous, as we estimate four latent constructs simultaneously. Using the multivariate approach allows for considering the latent covariation of these constructs, which go undetected in a unidimensional approach.

IRT-Based APIM Using R

Alexandrowicz (2015) showed how to estimate the parameters of an APIM using IRT models (an approach later extended by Gin et al. (2020)). The proposed procedure used the proprietary software ConQuest (Adams et al., 2012; Wu et al., 2007), which implements the Multidimensional Random Coefficients Multinomial Logit model (MRCMLM; Adams et al., 1997). A specific feature of this program is that it supports an option to estimate regression models of latent variables, very much like the structural model in the SEM context. However, to make the modeling more accessible to a wider audience, the present chapter shall demonstrate how to implement an APIM using R, an open-source software for statistical programming (R Core Team, 2023; see also Alexandrowicz, 2013 for an introduction). There are numerous packages supporting IRT models, most importantly eRm (Mair et al.,

2020), ltm (Rizopoulos, 2006), mirt, (Chalmers, 2012, 2023), psychotools (Zeileis et al., 2021), and TAM (Robitzsch et al., 2020). However, none of these packages supports a "latent regression" feature comparable to ConQuest, so that a direct implementation of an APIM with R requires further consideration. Therefore, the present contribution explores several means to estimate the parameters of an APIM based on IRT models using R packages.

A simple and straightforward approach would be using an IRT model to establish a measurement model for each scale and use the person parameter point estimates for performing the APIM. Albeit this approach elegantly circumvents the problems associated with the measurement part of the SEM for truly categorical responses, it still bears some disadvantages. For example, Mislevy (1991) carries out that "*[f]amiliar IRT procedures based on point estimates for individual examinees break down in efficient designs that solicit relatively few responses from each*" (p. 195). Basically, the problem is regression dilution or errors-in-variables, i.e., an inconsistent and a downward biased estimation of the regression slope due to random error in the *in*dependent variable. The shorter the scale (test) used for the predictor variable(s), the more pronounced is this effect. Lu et al. (2005) use the term finite item bias effect.

Skrondal and Laake (2001, p. 563) proposed a "bias avoiding" method by applying different estimation methods for obtaining the factor scores: regression factor scores for the explanatory latent variables and Bartlett scores (Bartlett, 1937, 1938; Thomson, 1938) for the response latent variables. Alternatively, Croon (2002) discusses a "bias avoiding" method based on random assignment or a Kernel-based procedure. However, both of these approaches rely on what Lu and Thomas (2008) term "truly continuous measurements" in contrast to "*ordinal-interval* scales" (p. 464; emphasis in the original)—the latter referring to the typical integer-based response formats typically applied in psychological research. In their simulation study, Lu and Thomas (2008) further showed that the bias correcting method is sensitive to the skewness of the underlying distribution of each manifest variable.

Regarding *discretized* variables, Takane and de Leeuw (1987) showed, among other things, the equivalence of discrete SEM (i.e., a latent distribution cut by thresholds τ; see also de Leeuw, 1983) and the Graded Response Model (GRM; Samejima, 1969). Albeit this qualifies as a valid IRT approach including latent structure modeling; our present claim is broader for it seeks incorporating any IRT model, not only the GRM (see Sect. 27.2 for more details). For the actual application, we chose a confirmatory multidimensional approach with each of the four measurements (ellipses in Fig. 27.1) constituting a latent dimension. This results in a between-item-multidimensional model (cf. Reckase, 2009).

Another way to arrive at proper person parameter estimates is the Plausible Values (PV; e.g., OECD, 2009) approach, which basically recurs to missing value reasoning as worked out by Rubin (1987). We treat the latent propensities θ as unknowns, the distribution of which is estimated by means of multiple imputation technique (Little & Rubin, 2002). Several random values are drawn from each respondent's specific posterior distribution, which constitute the PVs. This method "*allow[s] for an accounting of the uncertainties due to the sampling of individuals and to the incomplete information on each sampled individual*" (Beaton & Johnson, 1990, p. 9). For the evaluation of the multiple PVs per respondent and the models built thereupon, we apply Rubin's rules (cf.p Rubin, 1987). This refers, for example, to the PV-based actor and partner effects (i.e., regression coefficients) of the APIM. The PV technique has been successfully applied in large-scale assessments, for example, in the National Assessment of Education Progress (NAEP; Beaton, 1987; Beaton & Johnson, 1990; Mislevy et al., 1992; see also Mislevy, 1991; Mislevy et al., 1992; Mislevy, 1993) survey, for it allows for a more precise estimation of the population variance of the trait under consideration. As the APIM is a complex regression model, this approach constitutes an interesting estimation candidate. Luo and Dimitrov (2018) explored how many PVs are required to obtain person parameter estimates with a precision comparable to those from Marginal Maximum Likelihood or Expected A Posteriori estimation methods (MML/EAP; Baker & Kim, 2004), recommending to draw 20 PVs per person. Concurrently, Wu (2005) notes that PVs may improve the estimation of the population variance even if only one PV is drawn per respondent (p. 116). Moreover, to obtain an individual point estimate, the average of a respondent's PVs may be taken, using either the mean (e.g., OECD, 2009) or the median. We prefer the latter, for it may better reflect each individual's PV distribution in case it is skewed or will yield the same value as the mean, if symmetric.

Research Questions

The present study focuses on the Actor-Partner Interdependence Model for dyadic data with regard to self-esteem and relationship satisfaction. For the reasons outlined above, we want to (a) apply IRT for the measurement models and (b) use free and open R routines for estimation. Considering the various models and estimation variants presented so far, we want to compare their possible combinations with respect to model fit, local misfit detection capabilities, and the resulting APIM parameters (see Sect. 27.2).

27.2 Methods

Sample

For this study, we used data from the Partnership Study (PASST; Mund et al., 2023), a longitudinal online study on personality and partner relationships. The study protocol was approved by the Institutional Review Board of the Friedrich Schiller University Jena (FSV 18/47). Notably, PASST was designed to be a dyadic study, thus directly trying to recruit couples via online social media platforms and online advertising. Furthermore, participants could join the study at any time, that is, new participants could also enter after the first measurement occasion in January 2020.

Couples were asked questions concerning among others their self-esteem and relationship satisfaction every 3 months over the course of 3 years (January 2020–July 2022). Each wave comprised 11 questionnaires and participation lasted approximately 45 minutes. Participation was possible at any time in the study period for individuals aged 18 years and above and being in a relationship, regardless of marriage. Participation was rewarded with €10.00 per couple per wave (€5.00 for each individual). Moreover, after completing the questionnaires, all participants could take part in a ruffle of vouchers with a total value of €1000.00. Finally, continuous participation (three or fours times a year) was rewarded by an additional bonus of €5.00 or €10.00, respectively.

For the present analyses, we used data from heterosexual couples providing data at first-time participation and 1 year thereafter, yielding in total 1358 individuals, i.e., 679 heterosexual couples. On average, women were 25.75 years old (SD = 4.41; range = 18–57) and men 28.06 years (SD = 5.82; range = 19–58). At baseline, mean relationship duration differed slightly between women (M = 47.7 months, SD = 49.51) and men (M = 47.82 months,[4] SD = 54.96).

We will consider two scales for our model here (see Sect. 27.2). As usual, there are missing values in each of these scales. To preclude the specific treatment of missings performed by the software to affect the results, we performed a complete case analysis ("listwise deletion"), thus silently assuming missing completely at random (MCAR; cf. van Buuren, 2018). This selection resulted in a sample of n = 134 couples. Albeit certainly arguable, we consider our decision justifiable for the present purpose, which is to compare the various modeling variants. Table 27.1 lists the available demographics of the original sample and the selection. Comparing the characteristics of these two analysis samples, we find generally fairly similar frequencies, so that regarding the demographics, we consider the selection justifiable.

Measures

From the scales used in PASST, two will be used in the present study, self-esteem and relationship satisfaction. *Self-esteem* was measured using the German ten-item version of the Rosenberg self-esteem scale (RSES; Rosenberg, 1989; von Collani & Herzberg, 2003). Items (e.g., *I like myself just the way I am*; for the full list, see Appendix section "The Rosenberg Self-Esteem Scale (RSES)") were answered using a five-point Likert-type rating scale ranging from 1 (*do not agree*) to 5 (*fully agree*).

[4] In fact, couples differ with respect to their perceived duration of the partnership. However, the correlation is usually high, so that we will not further pursue this issue here.

Table 27.1 Baseline sample characteristics

	Selection $n_d = 134$				Entire sample $n_d = 679$			
	women		men		women		men	
	n	%	n	%	n	%	n	%
Education								
Primary	–	–	–	–	4	*0.59*	1	*0.15*
Secondary general school	–	–	1	*0.75*	–	–	12	*1.77*
Comprehensive school leaving certificate	3	*2.24*	7	*5.22*	30	*4.42*	59	*8.69*
Vocational school	3	*2.24*	7	*5.22*	32	*4.71*	50	*7.36*
University entrance qualification	46	*34.32*	51	*38.06*	243	*35.79*	246	*36.23*
University degree	81	*60.45*	65	*48.51*	360	*53.02*	291	*42.86*
Doctorate	–	–	3	*2.24*	3	*0.44*	15	*2.21*
Other	1	*0.75*	–	–	7	*1.03*	3	*0.44*
Not specified	–	–	–	–	–	–	2	*0.29*
Occupational status								
Trainee	2	*1.49*	7	*5.22*	25	*3.68*	33	*4.86*
Student	74	*55.22*	58	*43.28*	361	*53.17*	279	*41.09*
Employee	35	*26.12*	43	*32.09*	192	*28.28*	265	*39.03*
Official/civil servant	7	*5.22*	9	*6.72*	24	*3.53*	28	*4.12*
Self-employed	3	*2.24*	2	*1.49*	14	*2.06*	18	*2.65*
Disabled/unable to work	1	*0.75*	–	–	3	*0.44*	2	*0.29*
Seeking work	4	*2.99*	4	*2.99*	20	*2.95*	26	*3.83*
Maternal/paternal leave	2	*1.49*	–	–	13	*1.91*	4	*0.59*
Housewife/househusband	2	*1.49*	1	*0.75*	4	*0.59*	2	*0.30*
Other	4	*2.99*	7	*5.22*	22	*3.24*	15	*2.21*
Not specified	–	–	3	*2.24*	1	*0.15*	7	*1.03*
Living with ...								
no one	30	*22.39*	22	*16.42*	125	*18.41*	118	*17.38*
partner	66	*49.25*	72	*53.73*	375	*55.23*	377	*55.52*
children	6	*4.48*	8	*6.97*	41	*6.04*	39	*5.74*
siblings	5	*3.73*	7	*5.22*	26	*3.83*	31	*4.57*
parents	8	*5.97*	13	*9.70*	62	*9.13*	77	*11.34*
other family	–	–	1	*0.75*	9	*1.33*	13	*1.91*
flatmate	25	*18.66*	27	*20.15*	129	*19.00*	118	*17.38*
SES								
up to € 20,000	67	*50.00*	60	*44.78*	312	*45.95*	268	*39.47*
up to € 40,000	21	*15.67*	28	*20.90*	100	*14.73*	134	*19.73*
up to € 60,000	7	*5.22*	20	*14.93*	46	*6.77*	79	*11.63*
up to € 80,000	1	*0.75*	3	*2.23*	4	*0.59*	29	*4.27*
€ 81,000 and more	1	*0.75*	1	*0.75*	2	*0.29*	9	*1.33*
Not specified	5	*3.73*	3	*2.23*	28	*4.12*	25	*3.68*
Not available	32	*23.88*	19	*14.18*	187	*27.55*	137	*19.89*

Note. n_d = number of dyads/couples

Relationship satisfaction was assessed using the German version of the Relationship Assessment Scale (RAS; Sander & Böcker, 1993). The scale contained seven items (e.g., *In general, how satisfied are you with your relationship?*), which the respondents were asked to answer using a five-point Likert-type rating scale with response categories specific to each question (see Appendix section "The Relationship Assessment Scale (RAS)").

Analytic Strategy

IRT Modeling Variants We will compare the results of the three models (PCM, GPCM, and GRM) times two modeling variants each (uni- and multivariate) to the results of using just the simple sum of scores (which would only be admissible from a theoretical point of view if the PCM applies), the SEM approach (which constitutes the current practice in dyadic data analysis), once in the widespread version of regarding the scores as interval and once as ordinal, the IRT-based ML person parameter estimates (which would appear as a straightforward approach for IRT users), and 3 PV variants each (20 PVs, median of 20 PVs, and 1 PV). This results in 27 model variants listed in Table 27.2.

The APIM Parameters The parameters of the 27 APIMs will be estimated with a SEM using the person parameter estimates from each of the models listed in Table 27.2. The rationale is that regression analysis as such is solely based on the second moments, which is what SEMs are based upon. Moreover, SEMs allow for the simultaneous modeling of all four regression models involved in an APIM. A separate SEM will be applied to each variant listed in Table 27.2.

Assessing Model Fit The present study examines various models of different kinds. Each has its specific means of assessing model adequacy. However, as the aim of the present study is to compare them, we chose to use fit measures applicable to as many as possible. For the uni- and multidimensional IRT models, these will be the χ^2-test of Orlando and Thissen (2000), the infit and outfit indices along with their tests (Wright & Stone, 1979), and the AIC, the BIC, the C_2 statistic (Cai & Monroe, 2014) and the Root Mean Squared Error of Approximation (RMSEA; Browne & Cudeck, 1993), the Standardized Root Mean Square Residual (SRMSR; e. g., Hu & Bentler, 1999), the Tucker-Lewis Index (TLI; Tucker & Lewis, 1973), the Comparative Fit Index (CFI; Bentler, 1990) and the χ^2/df ratio statistic (or normed χ^2; NC), the idea of which can be traced back to Jöreskog (1969); see also Tabachnick and Fidell (2014, p. 918) or West et al. (2023, p. 187). Moreover, we use the PIccc diagram to analyze item characteristics and information curves. Note that the PCM would allow for further model checks (e.g., Alexandrowicz, 2022, 2023), but as these do not allow for a direct comparison to the GPCM and the GRM, we did not pursue them further.

Table 27.2 Overview: The 27 model variants

Model ID	Specification	$\hat{\theta}$
(10)	The naïve approach	simple sum
(11)	The popular SEM	SEM/interval
(12)	The ordinal SEM	SEM/ordinal
(21)	Unidimensional PCM per scale	ML
(22)	Unidimensional GPCM per scale	ML
(23)	Unidimensional GRM per scale	ML
(31)	Multidimensional PCM	ML
(32)	Multidimensional GPCM	ML
(33)	Multidimensional GRM	ML
(41)	Unidimensional PCM per scale	10 PV
(42)	Unidimensional GPCM per scale	10 PV
(43)	Unidimensional GRM per scale	10 PV
(51)	Multidimensional PCM	10 PV
(52)	Multidimensional GPCM	10 PV
(53)	Multidimensional GRM	10 PV
(61)	Unidimensional PCM per scale	Md 10 PV
(62)	Unidimensional GPCM per scale	Md 10 PV
(63)	Unidimensional GRM per scale	Md 10 PV
(71)	Multidimensional PCM	Md 10 PV
(72)	Multidimensional GPCM	Md 10 PV
(73)	Multidimensional GRM	Md 10 PV
(81)	Unidimensional PCM per scale	1 PV
(82)	Unidimensional GPCM per scale	1 PV
(83)	Unidimensional GRM per scale	1 PV
(91)	Multidimensional PCM	1 PV
(92)	Multidimensional GPCM	1 PV
(93)	Multidimensional GRM	1 PV

Technical Details All calculations are performed with R (R Core Team, 2023). For the IRT models, we used the mirt package (Chalmers, 2023) and for the SEMs lavaan (Rosseel, 2012). The PIccc diagrams were drawn with the RMX package (Kabic & Alexandrowicz, 2023a,b). For inferential decisions, we used 5 % for the risk of an error of the first kind.

27.3 Results

The reason why we took the present approach was to take advantage of IRT models for building a measurement model. This includes various means to detect weaknesses of the items. Therefore, we will start with inspecting the fit of the IRT models. Table 27.3 shows fit statistics, which apply to all models taken into consideration.

Thereafter, we will inspect some threshold characteristics (exemplarily) and turn finally to the APIM coefficients.

Fit of IRT-Based Measurement Models

Table 27.3 shows the model fit indices comparable across all models. First, regarding the C_2-test, we find significant results for all sub-models except the RAS for women at T2 (i.e., 1 year after first-time participation). However, some of the relative χ^2-values are below the critical threshold of 5 (cf. Schumacker & Lomax, 2016). The residual-based indices RMSEA range from 0.052 to 0.186 and SRMSR from 0.053 to 0.259. All goodness-of-fit indices TLI and CFI of the unidimensional models are above 0.9, whereas those of the multidimensional models are somewhat below 0.9. Generally, all PCM-based models fit worse than the more flexible GPCM and GRM. Hence, the first clear result is that the assumption of equal discrimination across items does not hold for the RSES and the RAS in both the uni- and the multidimensional variants. The choice between GPCM and GRM is not that clear-cut; rather both models' fit has to be regarded weak. The multidimensional models show worse fit, again with the GPCM and the GRM outperforming the PCM. Interestingly, the RAS fits better than the RSES throughout all variants considered here.

Additionally, we also inspect the information-based fit criteria AIC and BIC, which may only compared for models applied to the same data (Table 27.4).

Again, we find little surprisingly for all model variants the PCM showing the worst fit and the GPCM and the GRM quite similar with a slight advantage of the latter.

Item Fit of IRT-Based Measurement Models

To better understand possible reasons for the less-than-convincing model fits, we highlight some characteristic results with PIccc diagrams (Kabic & Alexandrowicz, 2023a). We start with comparing (exemplarily) the PCM and the GPCM results of the RAS for the female respondents (i.e., time point 2).

The PIccc diagrams shown in Fig. 27.2 share some similarity with the conventional Person-Item Map (PI-Map). It utilizes the fact that item and person parameters share a latent scale so that they might be compared with each other. On the top, we find a histogram of the person parameter estimates with the (Gaussian) kernel density line, the test information function (TIF), and the standard error (SE) lines superimposed. Below, we find for each item the category characteristic curves (CCCs), visualizing the probability of choosing a category given a location on the latent continuum. The bullets indicate the threshold locations of adjacent categories,

Table 27.3 Fit statistics of the IRT models

	C_2	df	p	RMSEA	5%	95%	SRMSR	TLI	CFI	NC
UNI PCM RSES M1	165.71	44	< 0.001	0.144	0.121	0.167	0.134	0.908	0.910	3.77
UNI PCM RSES W1	213.78	44	< 0.001	0.170	0.147	0.193	0.106	0.900	0.902	4.86
UNI PCM RAS M2	45.12	20	0.001	0.097	0.059	0.135	0.096	0.962	0.964	2.26
UNI PCM RAS W2	45.23	20	0.001	0.097	0.059	0.135	0.105	0.962	0.964	2.26
UNI GPCM RSES M1	84.06	35	< 0.001	0.103	0.074	0.130	0.083	0.953	0.964	2.40
UNI GPCM RSES W1	162.87	35	< 0.001	0.166	0.140	0.191	0.089	0.905	0.926	4.65
UNI GPCM RAS M2	23.27	14	0.056	0.071	< 0.001	0.119	0.053	0.980	0.987	1.66
UNI GPCM RAS W2	19.09	14	0.162	0.052	< 0.001	0.105	0.064	0.989	0.993	1.36
UNI GRM RSES M1	108.80	35	< 0.001	0.126	0.099	0.153	0.082	0.930	0.945	3.11
UNI GRM RSES W1	195.24	35	< 0.001	0.186	0.160	0.211	0.090	0.882	0.908	5.58
UNI GRM RAS M2	26.38	14	0.023	0.082	0.029	0.128	0.054	0.973	0.982	1.88
UNI GRM RAS W2	20.91	14	0.104	0.061	< 0.001	0.111	0.069	0.985	0.990	1.49
MUL PCM	2462.99	693	< 0.001	0.139	0.132	0.144	0.259	0.809	0.812	3.55
MUL GPCM	1482.86	663	< 0.001	0.096	0.090	0.103	0.245	0.908	0.913	2.24
MUL GRM	1644.75	663	< 0.001	0.106	0.099	0.112	0.245	0.889	0.896	2.48

Table 27.4 Information-based fit statistics of the IRT models

Scale	Sample	Model	AIC	BIC
RSES	M1	PCM	2976.42	3089.43
		GPCM	2917.33	3056.42
		GRM	2904.59	3043.69
	W1	PCM	2950.55	3060.67
		GPCM	2939.34	3075.54
		GRM	2931.39	3067.59
RAS	M2	PCM	1512.19	1581.74
		GPCM	1495.66	1582.60
		GRM	1491.34	1578.28
	W2	PCM	1449.58	1516.23
		GPCM	1443.56	1527.59
		GRM	1433.99	1518.03
MUL		PCM	9970.95	10388.24
MUL		GPCM	9757.70	10261.92
MUL		GRM	9725.19	10229.42

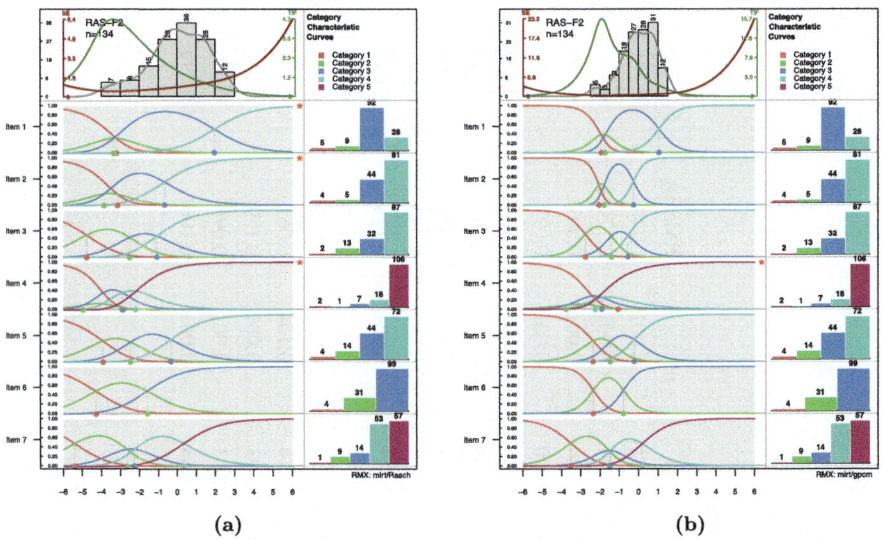

Fig. 27.2 A PIccc showing the category characteristic functions (CCCs), the person parameter distribution, the test information curve, and the standard error of the RAS for the female respondents. These two and Fig. 27.3a were drawn with the `plotPIccc()` function from the RMX package. (**a**) Results of the PCM. (**b**) Results of the GPCM

and the asterisks indicate threshold disordering. On the right side, we find the bar charts of the item category frequencies.

First of all, we find much "wider" CCCs of the PCM compared to the GPCM, which is a consequence of the PCM assuming all discriminations equal with a value of one. This shows the increased flexibility of models allowing for varying discrimination parameters for each item. The discrimination parameters range from $\alpha_4 = 0.823$ to $\alpha_2 = 3.377$ in Fig. 27.2b. In this context, we also find the PCM-based thresholds compared to those of the GPCM spreading more across the latent continuum and generally shifted to the left (i.e., "easier" from a psychometric perspective). As a consequence, we find a broader person parameter distribution for the PCM. Note further that with the PCM, we find three items exhibiting threshold disorder (items 1, 2, and 4), whereas for the GPCM, only item 4 shows this weakness.

One of the strengths of this kind of diagram is that it makes the misalignment of the two kinds of parameters (person and item) visible. It is more pronounced for the PCM than for the GPCM, i.e., from a PCM perspective, the items are very "easy" compared to the locations of the respondents. The heavily skewed item frequency barplots support this conclusion from a descriptive point of view, as most respondents chose the higher response categories. Analogously, we find the TIF far away from the person parameter density estimate. Here, we encounter a situation, which puts the assumptions of the SEM for ordinal data based on the polychoric correlation (bivariate normality) heavily into question. The results of the GRM were similar to those of the GPCM and are therefore not shown. We obtain similar results for the other scales, which makes comprehensible, why the GPCM and the GRM showed generally better fit than the PCM (for more details, including all diagrams, see the online supplement at https://osf.io/e97td/).

As the previous analyses have marked item 4 as problematic, we want to exemplarily explore this item further. Figure 27.3a shows a different kind of PIccc,

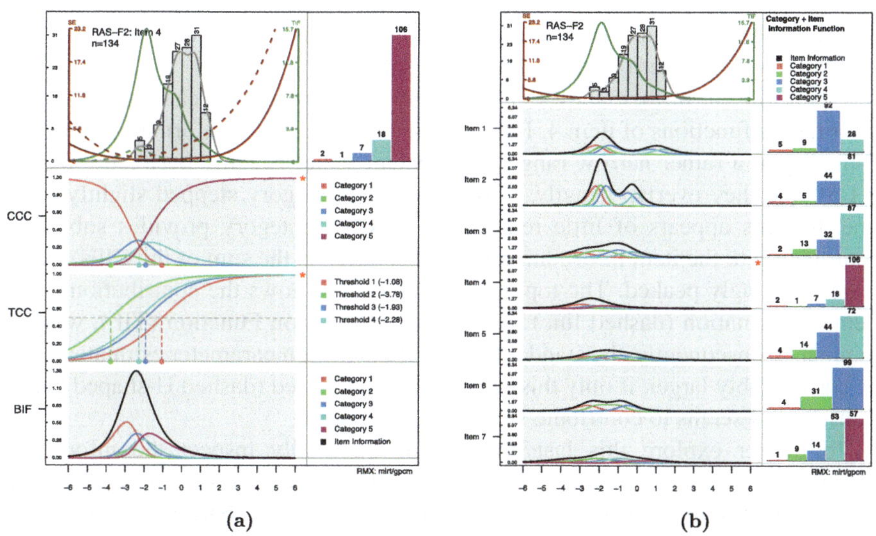

Fig. 27.3 Variants of person-item-CCC (PIccc) diagrams. (**a**) A PIccc showing the CCC, the TCC, the IIF, and the CIF for item 4 of the RAS for the female respondents using the GPCM. Note: "BIF" indicates "Both Information Functions," i.e., item and category information. (**b**) A PIccc showing the category and item information curves of the RAS

Table 27.5 Item parameter estimates and standard errors for item 4

	a	b_1	b_2	b_3	b_4
Estimate	0.82	−1.08	−3.78	−1.93	−2.28
Standard error	0.24	1.56	1.41	0.60	0.67

Notes. a=discrimination parameter; b=threshold parameters

comprising four different kinds of curves for one item, the category characteristic curve (CCC), the threshold characteristic curves (TCCs), the category information curves (CIFs), and the item information curve (IIF). The frequency bar chart has changed to the top, and each diagram has its own legend to its right. The model reveals that item 4 (*How often do you wish you hadn't gotten into this relationship?*) basically made the respondents switch between "often" (the highest response category) and any other one. Therefore, the highest threshold separating categories 4 and 5 can be found at −2.28, which is considerably below the population mean of zero. Hence, it does not require much of relationship discomfort to regret the decision "often." Consequently, this item shows considerable threshold disorder, which can be understood from both a statistical and a substantive perspective: statistically, because the few responses not in category 5 ($28/134 \sim 21\,\%$) split (unevenly) across the remaining four categories, hence there is a large stochastic component in the estimates. Table 27.5 provides the empirical evidence, as the standard errors for thresholds b_1 and b_2 are more than twice as large compared to those of b_3 and b_4.

Second, substantively, the threshold disorders occur, because the respondents did, as concluded before, more or less decide, whether to choose "often" or not—the other categories did not receive that much attention and may have, therefore, been used not in accordance with those of the other items (which is what the model evaluates), hence the disordering. In short, this item functions basically in a dichotomous manner ("often" or not, that's the question).

Figure 27.3a further shows in the bottom segment the category and item information functions of item 4. Here, we find all categories' information functions gathering in a rather narrow range of the latent scale (roughly between −3.5 and −1.5), and they overlap heavily. Maybe, the first category stepped slightly out of line, but this appears of little relevance. Thus, no category provides substantial contribution to the item information function (which is the sum of the CIFs), which itself is strongly peaked. The top diagram now also shows the contribution of this item's information (dashed line) to the Test Information Function (TIF), which is minimal. Consequently, the standard errors of the person parameter estimates would be considerably larger, if only this selected item was used (dashed U-shaped curve). Thus, item 4 seems to contribute very little to the scale.

To further explore this last conclusion, we finally inspect the information functions of all items of this scale, again with a PIccc. Figure 27.3b shows the category and item information for all items. All subplots share the same vertical scale so that we can directly compare each item's contribution to the TIF (which, in turn, is the foundation for the standard error of the person parameter estimates). It becomes obvious that not only item 4 contributes little, but basically all items but the first two. The two IIF peaks of both items 1 and 2 are responsible for the specific shape of the TIF in the upper part of the diagram.

These analyses showed not only some specificities of the scales used here but generally how detailed analyses may give invaluable hints for improvement during scale construction. The PIccc might support such analyses. See Sect. 27.2 of the online supplement at https://osf.io/e97td/ for a full list of PIccc diagrams for all items used in the present analysis.

The APIMs

After having shown how the IRT models allow for identifying the strengths and weaknesses of the used scales, we now turn to the target model, the APIM predicting the partners' relationship satisfaction from their self-esteem. For that purpose, we estimated the SEM parameters of the model shown in Fig. 27.1 for all 27 measurement model candidates, now assuming them to fit sufficiently and ignoring the weaknesses revealed so far. Note that a comparison of fit between these models is not possible for two reasons: first, all are just identified (i.e., $df = 0$), so that χ^2-test and fit indices cannot be computed, and second, a comparison based on AIC, etc. is not possible, because they were applied to different data. Therefore, we will focus on the parameter estimates.

Table 27.6 reveals several interesting patterns: regarding the actor and partner effects, the only clear message is that the partner effect p_{21} is virtually zero in all variants, i.e., the relationship satisfaction of the female partner is not affected by the self-esteem of the male partner (we even find both positive and negative signs, but the value of the coefficients is too small to draw conclusions other than "no effect").

For the other three regression effects, we find considerably varying—and in some instances even contradicting—results. For example, the opposite partner effect p_{12}, i.e., the relationship satisfaction of the male partner regressed on the self-esteem of the female partner, is approximately 0.4 for both the uni- and the multidimensional variants of the PCM, the multidimensional PCM with 10 PV, and the median of the PV, but only approximately 0.25 for the various GPCM and GRM versions. Moreover, if we rely on the significance test of this effect, we will also come to contradicting conclusions: 8 out of the 27 modeling variants yielded a nonsignificant p_{12} (unidimensional, ML PCM+GPCM, 10PV PCM+GPCM+GRM, 1PV GPCM+GRM; multidimensional, 10PV PCM), whereas this coefficient was significant in all other models.

Another contradicting result is that with 10 PVs in a multidimensional GPCM, the partner effect p_{12} is larger than the actor effect $a1$, whereas most other variants state the opposite. Similarly, the variances of all four latent variables (top right and bottom left diagram) are considerably larger for the PCM, and slightly for the covariances (bottom left diagram). Interestingly, all three variants subsumed as "conventional" (i.e., the simple sum and both SEM variants) show the smallest values for most coefficients.

Table 27.6 APIM coefficients

	a1	a2	p12	p21	vx1	vx2	ve1	ve2	cx	cy
(10) Simple Sum/est	0.164	0.288	0.152	0.024	0.521	0.636	0.244	0.241	0.138	0.148
(10) Simple Sum/p-value	0.007	<.001	0.006	0.689	<.001	<.001	<.001	<.001	0.007	<.001
(11) SEM Interval/est	0.151	0.333	0.192	0.013	0.429	0.443	0.186	0.167	0.120	0.121
(11) SEM Interval/p-value	0.029	<.001	0.006	0.843	<.001	<.001	<.001	<.001	0.006	<.001
(12) SEM Ordinal/est	0.224	0.442	0.177	0.019	0.675	0.680	0.377	0.486	0.178	0.315
(12) SEM Ordinal/p-value	<.001	<.001	0.015	0.800	<.001	<.001	<.001	<.001	<.001	<.001
(21) IRT UNI ML PCM/est	0.398	0.441	0.196	−0.016	1.377	2.052	2.605	1.958	0.404	1.264
(21) IRT UNI ML PCM/p-value	0.001	<.001	0.050	0.881	<.001	<.001	<.001	<.001	0.007	<.001
(22) IRT UNI ML GPCM/est	0.251	0.380	0.150	−0.005	0.897	0.911	0.752	0.684	0.225	0.389
(22) IRT UNI ML GPCM/p-value	0.002	<.001	0.065	0.950	<.001	<.001	<.001	<.001	0.005	<.001
(23) IRT UNI ML GRM/est	0.238	0.385	0.164	−0.009	0.905	0.917	0.760	0.701	0.230	0.403
(23) IRT UNI ML GRM/p-value	0.004	<.001	0.044	0.908	<.001	<.001	<.001	<.001	0.005	<.001
(31) IRT MUL ML PCM/est	0.407	0.272	0.219	0.002	1.571	2.068	2.061	0.441	0.604	0.716
(31) IRT MUL ML PCM/p-value	<.001	<.001	0.017	0.969	<.001	<.001	<.001	<.001	<.001	<.001
(32) IRT MUL ML GPCM/est	0.233	0.359	0.258	−0.035	0.980	1.027	0.877	0.461	0.328	0.408
(32) IRT MUL ML GPCM/p-value	0.007	<.001	0.002	.577	<.001	<.001	<.001	<.001	<.001	<.001
(33) IRT MUL ML GRM/est	0.225	0.334	0.234	−0.021	0.977	0.985	0.890	0.456	0.308	0.422
(33) IRT MUL ML GRM/p-value	0.010	<.001	0.007	0.734	<.001	<.001	<.001	<.001	0.001	<.001
(41) IRT UNI 10 PV PCM/est	0.264	0.332	0.154	0.002	1.289	1.709	1.713	1.508	0.275	0.748
(41) IRT UNI 10 PV PCM/p-value	0.027	0.001	0.110	0.982	<.001	<.001	<.001	<.001	0.046	<.001

(42) IRT UNI 10 PV GPCM/est	0.222	0.339	0.162	−0.001	1.025	1.007	0.900	0.889	0.210	0.389
(42) IRT UNI 10 PV GPCM/p-value	0.019	0.002	0.098	0.994	<.001	<.001	<.001	<.001	0.038	<.001
(43) IRT UNI 10 PV GRM/est	0.230	0.374	0.171	< 0.001	0.975	0.984	0.890	0.867	0.208	0.401
(43) IRT UNI 10 PV GRM/p-value	0.015	<.001	0.084	0.997	<.001	<.001	<.001	<.001	0.024	<.001
(51) IRT MUL 10 PV PCM/est	0.353	0.258	0.215	−0.003	1.792	2.366	2.626	0.696	0.626	0.814
(51) IRT MUL 10 PV PCM/p-value	0.006	<.001	0.057	0.961	<.001	<.001	<.001	<.001	0.001	<.001
(52) IRT MUL 10 PV GPCM/est	0.199	0.334	0.259	−0.043	1.085	1.140	1.060	0.616	0.329	0.440
(52) IRT MUL 10 PV GPCM/p-value	0.045	<.001	0.008	0.593	<.001	<.001	<.001	<.001	0.003	<.001
(53) IRT MUL 10 PV GRM/est	0.231	0.299	0.205	−0.020	1.107	1.093	1.040	0.567	0.321	0.428
(53) IRT MUL 10 PV GRM/p-value	0.021	<.001	0.035	0.787	<.001	<.001	<.001	<.001	0.003	<.001
(61) IRT UNI Md 10 PV PCM/est	0.269	0.380	0.191	0.009	1.141	1.546	1.355	1.124	0.292	0.707
(61) IRT UNI Md 10 PV PCM/p-value	0.005	<.001	0.021	0.921	<.001	<.001	<.001	<.001	0.013	<.001
(62) IRT UNI Md 10 PV GPCM/est	0.235	0.366	0.171	0.005	0.935	0.918	0.739	0.694	0.210	0.379
(62) IRT UNI Md 10 PV GPCM/p-value	0.003	<.001	0.032	0.949	<.001	<.001	<.001	<.001	0.011	<.001
(63) IRT UNI Md 10 PV GRM/est	0.253	0.392	0.184	0.015	0.887	0.905	0.777	0.724	0.200	0.399
(63) IRT UNI Md 10 PV GRM/p-value	0.002	<.001	0.025	0.851	<.001	<.001	<.001	<.001	0.012	<.001
(71) IRT MUL Md 10 PV PCM/est	0.383	0.269	0.231	−0.009	1.599	2.128	2.144	0.477	0.643	0.744
(71) IRT MUL Md 10 PV PCM/p-value	<.001	<.001	0.012	0.854	<.001	<.001	<.001	<.001	<.001	<.001

(continued)

Table 27.6 (continued)

	a1	a2	p12	p21	vx1	vx2	ve1	ve2	cx	cy
(72) IRT MUL Md 10 PV GPCM/est	0.209	0.367	0.293	−0.054	0.974	1.025	0.896	0.479	0.328	0.404
(72) IRT MUL Md 10 PV GPCM/p-value	0.017	<.001	0.001	0.399	<.001	<.001	<.001	<.001	<.001	<.001
(73) IRT MUL Md 10 PV GRM/est	0.255	0.324	0.224	−0.021	1.005	0.996	0.902	0.458	0.312	0.403
(73) IRT MUL Md 10 PV GRM/p-value	0.003	<.001	0.009	0.732	<.001	<.001	<.001	<.001	0.001	<.001
(81) IRT UNI 1 PV PCM/est	0.236	0.318	0.245	0.018	1.246	1.713	1.806	1.412	0.268	0.682
(81) IRT UNI 1 PV PCM/p-value	0.025	< 0.001	0.007	0.843	<.001	<.001	<.001	<.001	0.037	<.001
(82) IRT UNI 1 PV GPCM/est	0.286	0.287	0.145	0.014	1.067	0.949	0.909	0.968	0.230	0.419
(82) IRT UNI 1 PV GPCM/p-value	<.001	0.001	0.095	0.870	<.001	<.001	<.001	<.001	0.010	<.001
(83) IRT UNI 1 PV GRM/est	0.201	0.378	0.134	−0.088	0.901	0.957	0.851	0.954	0.271	0.470
(83) IRT UNI 1 PV GRM/p-value	0.022	< 0.001	0.115	0.345	<.001	<.001	<.001	<.001	0.001	<.001
(91) IRT MUL 1 PV PCM/est	0.368	0.313	0.194	−0.002	1.820	2.214	2.510	0.642	0.618	0.840
(91) IRT MUL 1 PV PCM/p-value	0.001	< 0.001	0.045	0.976	<.001	<.001	<.001	<.001	0.001	<.001
(92) IRT MUL 1 PV GPCM/est	0.246	0.308	0.246	−0.052	1.009	1.292	0.933	0.591	0.390	0.426
(92) IRT MUL 1 PV GPCM/p-value	0.005	< 0.001	0.002	0.458	<.001	<.001	<.001	<.001	<.001	<.001
(93) IRT MUL 1 PV GRM/est	0.153	0.303	0.267	−0.051	1.114	1.165	0.928	0.584	0.359	0.422
(93) IRT MUL 1 PV GRM/p-value	0.065	< 0.001	0.001	0.437	<.001	<.001	<.001	<.001	0.001	<.001

27.4 Discussion

In this chapter, we explored several approaches to estimate an APIM: a path analysis based on the simple sum across items of the scales, two "regular" SEM approaches, and a set of IRT-based approaches using the person parameter estimates from a PCM, a GPCM, and a GRM, each in both a uni- and a multidimensional variant. As ML point estimates of person parameters may yield biased and inconsistent population parameter estimates, we also applied the plausible values principle, which has proven successful in large-scale assessments. The combination of all these factors resulted in a total of 27 different models (see Table 27.2). To increase accessibility of the approach, we performed all analyses with the free and open software R. For the SEM-based approaches, we used the lavaan package and for the IRT-based ones mirt, the latter supporting all models taken into consideration.

The IRT model checks revealed that the PCM is unlikely to describe the data adequately. In contrast, we found better and fairly equal fit measures for the GPCM and the GRM. However, the detailed item-level analyses revealed several weaknesses for all models. Interestingly, the actor effects based on the PCM person parameter estimates yielded higher coefficients compared to the other two. Hence, convincing results of the APIM may be based on insufficient measurement models. This can only be explained by influential measurement error components. This demonstrates the utter importance of a solid scale analysis (i.e., the measurement models) before turning to the model of interest (i.e., the APIM).

The most important consequence of a successful application of a PCM is that the simple sum of scores constitutes a sufficient statistic for the person parameter. Simply put, the PCM is a requirement for doing what is frequently done (i.e., summing scores), but without testing. Hence, the fact that the PCM turned out inadequate for all scales used in this analysis renders the path analysis (model 10) invalid from a psychometric point of view. Interestingly, the simple sum approach yielded the smallest APIM coefficients of all models. We discourage generally to follow such a plain approach for complex models, because any measurement error (controlled for when applying a proper measurement model) not only enters the APIM unfiltered, but any kind of item malfunctioning also goes entirely undetected. This will likely obscure (like in the present case) an effect of interest but may as well boost spurious effects, neither of which is desirable. Moreover, we do not learn anything regarding the quality of the items used.

The SEM approach, which does include a measurement model, took an intermediate role in our analyses: the "interval variant" (i.e., evaluating the scores numerically) yielded APIM coefficients not deviating strongly from most of the other models, whereas the more adequate "ordinal variant" yielded the largest female actor effect ($a_2 = 0.442$; the other effects were also comparable to most models). This approach is permissible, for it considers the item scores as category indicators and estimates the threshold locations in between. However, these thresholds are determined by means of the cumulative normal distribution (a fallible assumption, yet seldom checked), which constitutes a similarity to the GRM. Such an approach will—in contrast to the PCM and the GPCM—never reveal threshold disorders and thus possibly disguise irregularities of the item functioning (e.g., due to subgroup non-invariance, ambiguities in wording, overdifferentiated response format, etc.) constituting a model violation (cf. Andrich, 2013). SEMs have indeed the advantage of simultaneously estimating measurement and structural sub-models, but regarding the former, the authors of this chapter are not aware of differentiated analyses of threshold functioning when using the polychoric correlation matrix—or such analyses are maybe just underreported.

The present approach demonstrated how to detect local misfit, i.e., the lack of fit in a specific measurement model only. Separating the estimation of the measurement models and the structural model allows for detecting weaknesses in the former without impairing the estimates of the latter. In a simultaneous setting, measurement errors may propagate to the structural model (i.e., the APIM), which may become difficult to detect. The IRT approach applied here provides various diagnostic tools for detecting local (i.e., scale) misfit. The PIccc diagrams (available in the RMX package of R; see https://osf.io/n9c5r and Kabic & Alexandrowicz, 2023a) additionally proved very helpful, for it allows for visualizing many item details in one diagram. We could easily detect how the threshold disorders rendered certain response categories entirely uninformative, and the information curves further revealed for the RAS-F2 (i.e., the RAS for females at T2) that basically only the first two items contribute substantially to the test information curve, which, in turn, is the foundation of the standard error of the person parameter estimates (see Fig. 27.3b; similar results were also obtained for the other scales; see the online supplement at https://osf.io/e97td).

Applying IRT in the measurement model has further advantages, like "full information" parameter estimation, detailed model diagnostics, or linking and equating of tests (cf. Kolen & Brennan, 2004), just to name a few. In this study, we could, exemplarily, highlight how IRT models allow for detecting item weaknesses, with a special focus on the functioning of the thresholds and their impact upon the scale. We consider the omission of a latent continuum assumed to follow a certain distribution for identifying the thresholds (as is done with the polychoric correlation, even taking skewed distributions into consideration; see Kolbe et al., 2021) the major advantage of our approach. We are, of course, aware that a specific estimation method, the marginal maximum likelihood (MML; cf. Baker & Kim, 2004) method, also assumes a latent distribution, but these may be circumvented with the option `empiricalhist` of mirt. This option derives the latent person parameter distribution from the score pattern frequencies (cf. Bock & Aitkin, 1981), which, in turn, requires the sample to be informative with respect to the latent distribution. The Rasch family of models (i.e., RM, PCM, RSM, and several derivatives) even allows for the conditional ML estimation method (CML; cf. Baker & Kim, 2004; Fischer and Molenaar, 1995), which makes no distributional assumption at all.

After comparing the 27 model variants, what would be the recommendation? First of all, it is indispensable to scrutinize the measurement models. To this end, we consider the IRT-based approaches superior in psychological studies involving rating response scales. Treating the codings as category indicators associated with response probabilities rather than numerical entities is the most favorable approach—the more, as the approaches assuming latent continuous densities cut at thresholds may induce severe adversities. Moreover, IRT models provide many specific model checks, which prove advantageous especially for detecting local misfit. Here, the unidimensional GPCM and GRM performed similarly well (yet not optimal). On that score, we prefer the GPCM, for it is a divide-by-total model allowing for detecting threshold disordering. Next, considering the possible drawbacks of ML point estimates of the person parameters, a defensible approach would be the PV approach along with Rubin's rules. However, taking the median (or any other kind of averaging) of PVs seems little convincing, as we destroy the information gained by the PV's distribution.

The separate univariate IRT approaches share the drawback that the covariance matrix of the latent traits is not estimated properly along with the latent traits. Interestingly, Lu and Thomas (2008) found that univariate scores might serve their purpose in latent variables regression models if "*restricted to a single dependent and a single explanatory variable*" (p. 488). As the APIM considered here has more but still two independent variables, the univariate approaches might be considered yet with great caution. This will, of course, not apply to larger dyadic models, like social relations designs, one-with-many designs, and social network designs (cf. Kenny et al., 2006).

Therefore, we clearly favor the multidimensional IRT approaches, even if they showed inferior fit indices (Table 27.3) compared to their respective unidimensional counterparts. Nevertheless, only these approaches deliver a properly estimated latent covariance matrix, based on which we obtain the latent regression paths of interest. Among these, the favorite candidate would be the PCM, for it provides simple sufficient statistics of the person parameter estimates (i.e., the unweighted sum of scores). However, due to its strong assumptions, it will in many instances be outperformed by the more flexible GPCM and GRM (which was the case here). However, the major task of an APIM analysis is to model the latent relationships (rather than seeking individual assessments); the multidimensional GPCM and GRM remain most favorable for that purpose, with the latter outperforming the former slightly in our analysis.

From a substantive point of view, we have seen that the results of the IRT models largely converge with earlier findings on the association between self-esteem and relationship satisfaction (Erol & Orth, 2013, 2014; Mund et al., 2015; Schaffhuser et al., 2014). However, the results also indicate that using suboptimal measures may disguise or distort important effects. Thus, it will be crucial for future research in the field to take a much closer look at the measurement instruments using the state-of-the-art modeling strategies presented in this chapter. By making the unknown weaknesses of the measures known is a first step toward fixing them and, finally, to arrive at unbiased estimates of the interplay between self-esteem and relationship satisfaction in partner relationships.

One option not applied in the present study is to take a background model (sometimes also termed "latent regression model") into consideration. The background model referred to here allows for correcting the person parameter estimates for further variables, e.g., demographic information (as is done in large-scale assessments). Schofield (2015) has shown that the PV approach may lead to biased estimates if the background model is only applied to the independent variables. This restriction does not apply to the present analyses, because the model structure analyzed here differs from that of Schofield (2015). Comparing the present results to those obtained with the method proposed by Schofield (2015) would be an interesting endeavor to be pursued in further studies.

Scales

These are the wordings of the used scales in English. For the study, a German translation was used.

The Rosenberg Self-Esteem Scale (RSES)

1. On the whole, I am satisfied with myself. *(reverse)*
2. At times, I think I am no good at all.
3. I feel that I have a number of good qualities. *(reverse)*
4. I am able to do things as well as most other people. *(reverse)*
5. I feel I do not have much to be proud of.
6. I certainly feel useless at times.
7. I feel that I'm a person of worth, at least on an equal plane with others. *(reverse)*
8. I wish I could have more respect for myself.
9. All in all, I am inclined to feel that I am a failure.
10. I take a positive attitude toward myself. *(reverse)*

The Relationship Assessment Scale (RAS)

The seven items of the RAS employ a five-category response format with specific labels for the first (1) and the last (5) category, the third (middle) category is always "average" and the even numbered response categories are not labeled:

1. How often does your partner meet your needs?
   1=*poorly* 3=*average*, 5=*extremely well*
2. In general, how satisfied are you with your relationship?
   1=*unsatisfied*, 3=*average*, 5=*excellent*
3. How good is your relationship compared to most?
   1=*poor*, 3=*average*, 5=*excellent*
4. How often do you wish you hadn't gotten into this relationship?
   1=*never*, 3=*average*, 5=very *often*

5. To what extent has your relationship met your original expectations?
   1=*hardly* at all, 3=*average*, 5=*completely*
6. How much do you love your partner?
   1=*not much*, 3=*average*, 5=*very much*
7. How many problems are there in your relationship?
   1=*very few*, 3=*average*, 5=*very many*

References

Adams, R. J., Wilson, M., & Wang, W.-C. (1997). The multidimensional random coefficients multinomial logit model. *Applied Psychological Measurement, 21*, 1–23.

Adams, R. J., Wu, M. L., & Wilson, M. (2012). *Conquest 3.0 [computer software]*. Melbourne: Australian Council for Educational Research (ACER).

Alexandrowicz, R. W. (2013). *R in 10 Schritten. Einführung in die statistische Programmierumgebung [R in 10 Steps. An Introduction into the Statistical Programming Language]*. Vienna: facultas/UTB.

Alexandrowicz, R. W. (2015). Analyzing Dyadic Data with IRT Models. In M. Stemmler, A. von Eye, & W. Wiedermann (Eds.), *Dependent data in social sciences research: Forms, issues, and methods of analysis* (pp. 173–202). Berlin: Springer.

Alexandrowicz, R. W. (2022). GMX: Extended Graphical Model Checks. A Versatile Replacement of the plotGOF() Function of eRm. *Psychological Test and Assessment Modeling, 64*, 215–225.

Alexandrowicz, R. W. (2023). Extending GMX: Conditional likelihood ratio test and extended graphical model checks with psychotools. *Psychological Test and Assessment Modeling, 65*, 259–269.

Andrich, D. (2013). An expanded derivation of the threshold structure of the polytomous Rasch model that dispels any "threshold disorder controversy". *Educational and Psychological Measurement, 73*, 78–124. https://doi.org/10.1177/0013164412450877

Arsandaux, J., Montagni, I., Macalli, M., Bouteloup, V., Tzourio, C., & Galéra, C. (2020). Health risk behaviors and self-esteem among college students: Systematic review of quantitative studies. *International Journal of Behavioral Medicine, 27*(2), 142–159. https://doi.org/10.1007/s12529-020-09857-w

Baker, F. B., & Kim, S.-H. (2004). *Item Response Theory. Parameter Estimation Techniques*. Marcel Dekker.

Bartlett, M. S. (1937). The statistical conception of mental factors. *British Journal of Psychology. General Section, 28*, 97–104. https://doi.org/j.2044-8295.1937.tb00863.x

Bartlett, M. S. (1938). Methods of estimating mental factors. *Nature, 141*, 609–610.

Beaton, A. E. (1987). *The NAEP 1983–1984 Technical Report*. Princeton: Educational Testing Service.

Beaton, A. E., & Johnson, E. G. (1990). The average response method of scaling. *Journal of Educational Statistics, 15*, 9–38. https://doi.org/10.3102/10769986015001009

Bentler, P. M. (1990). Comparative fit indexes in structural models. *Psychological Bulletin, 107*, 238–246. https://doi.org/10.1037/0033-2909.107.2.238

Bernstein, I. H. (2006). Likert scale analysis. In K. Kempf-Leonhard (Ed.), *Encyclopedia of social measurement* (Vols. 2, F–O, pp. 497–504). Elsevier.

Bock, R. D., & Aitkin, M. (1981). Marginal maximum likelihood estimation of item parameters: Application of an EM algorithm. *Psychometrika, 46*, 443–459. https://doi.org/10.1007/BF02293801

Bowen, N. K., & Masa, R. D. (2015). Conducting measurement invariance tests with ordinal data: A guide for social work researchers. *Journal of the Society for Social Work and Research, 6*, 229–249. https://doi.org/10.1086/681607

Box, G. E. P. (1976). Science and statistics. *Journal of the American Statistical Association, 71* , 791–799. https://doi.org/10.1080/01621459.1976.10480949

Bradbury, T. N., Fincham, F. D., & Beach, S. R. H. (2000). Research on the nature and determinants of marital satisfaction: A decade in review. *Journal of Marriage and Family, 62*(4), 964–980. https://doi.org/10.1111/j.1741-3737.2000.00964.x

Brown, T. A. (2006). *Confirmatory factor analysis for applied research*. New York: The Guilford Press.

Browne, M. W., & Cudeck, R. (1993). Alternative ways of assessing model fit. In M. W. Bollen & J. S. Long (Eds.), *Testing structural equation models* (pp. 136–162). Newbury Park, CA: Sage.

Cai, L., & Hansen, M. (2013). Limited-information goodness-of-fit testing of hierarchical item factor models. *British Journal of Mathematical and Statistical Psychology, 66*, 245–276. https://doi.org/10.1111/j.2044-8317.2012.02050.x

Cai, L., & Monroe, S. (2014). *A new statistic for evaluating item response theory models for ordinal data*. Los Angeles, CA: University of California, National Center for Research on Evaluation, Standards, & Student Testing (CRESST), Report 839.

Chalmers, R. P. (2012). mirt: A Multidimensional Item Response Theory Package for the R Environment. *Journal of Statistical Software*, *48*(6), 1–29. https://doi.org/10.18637/jss.v048.i06

Chalmers, R. P. (2023). Multidimensional Item Response Theory [Computer software manual]. Retrieved from https://CRAN.R-project.org/package=mirt (R package version 1.37.10).

Choi, J., Peters, M., & Mueller, R. O. (2010). Correlational analysis of ordinal data: From Pearson's *r* to Bayesian polychoric correlation. *Asia Pacific Educational Revue, 11*, 459–466. https://doi.org/10.1007/s12564-010-9096-y

Croon, M. (2002). Using predicted latent scores in general latent structure models. In G. A. Marcoulides & I. Moustaki (Eds.), *Latent variable and latent structure models* (pp. 195–224). New York: Routledge/Taylor & Francis.

de Ayala, R. J. (2022). *The theory and practice of item response theory* (2nd ed.). New York: Guilford.

de Leeuw, J. (1983). Models and methods for the analysis of correlation coefficients. *Journal of Econometrics, 22*, 113–137. https://doi.org/10.1016/0304-4076(83)90096-9

Donnellan, M. B., Trzesniewski, K. H., & Robins, R. W. (2011). Self-esteem: Enduring issues and controversies. In T. Chamorro-Premuzic, S. v. Stumm, & A. Furnham (Eds.), *The wiley-blackwell handbook of individual differences* (pp. 718–746). Blackwell.

Drasgow, F. (1988). Polychoric and polyserial correlations. In L. Kotz & N. L. Johnson (Eds.), *Encyclopedia of statistical sciences* (Vol. 7, pp. 69–74). New York: Wiley.

Duncan, O. D. (1966). Path analysis: Sociological examples. *The American Journal of Sociology, 72*, 1–16. https://doi.org/10.1086/224256

Enders, C. K. (2010). *Applied missing data analysis*. New York: Guilford.

Erol, R. Y., & Orth, U. (2013). Actor and partner effects of self-esteem on relationship satisfaction and the mediating role of secure attachment between the partners. *Journal of Research in Personality, 47*, 26–35. https://doi.org/10.1016/j.jrp.2012.11.003

Erol, R. Y., & Orth, U. (2014). Development of self-esteem and relationship satisfaction in couples: Two longitudinal studies. *Developmental Psychology, 50*, 2291–2303. https://doi.org/10.1037/a0037370

Fischer, G. H., & Molenaar, I. W. (Eds.). (1995). *Rasch models. Foundations, Recent Developments, and Applications*. New York: Springer.

Fletcher, G. J. O., Simpson, J. A., & Thomas, G. (2000). The measurement of perceived relationship quality components: A confirmatory factor analytic approach. *Personality and Social Psychology Bulletin, 26*, 340–354. https://doi.org/10.1177/0146167200265007

Foldnes, N., & Grønneberg, S. (2019). On identification and non-normal simulation in ordinal covariance and item response models. *Psychometrika, 84*, 1000–1017. https://doi.org/10.1007/s11336-019-09688-z

Foldnes, N., & Grønneberg, S. (2020). Pernicious Polychorics: The impact and detection of underlying non-normality. *Structural Equation Modeling: A Multi- disciplinary Journal,27*, 525–543. https://doi.org/10.1080/10705511.2019.1673168

Foldnes, N., & Grønneberg, S. (2022). The sensitivity of structural equation modeling with ordinal data to underlying non-normality and observed distributional forms. *Psychological Methods, 27*, 541–567. https://doi.org/10.1037/met0000385

Fuller, W. A. (1987). *Measurement error models*. New York: Wiley.

Gin, B., Sim, N., Skrondal, A., & Rabe-Hesketh, S. (2020). A dyadic IRT model. *Psychometrika, 85*, 815–836. https://doi.org/10.1007/s11336-020-09718-1

Grønneberg, S., & Foldnes, N. (2022). Factor analyzing ordinal items requires substantive knowledge of response marginals. *Psychological Methods, 29*, 65–87. https://doi.org/10.1037/met0000495

Harris, M. A., & Orth, U. (2020). The link between self-esteem and social relationships: A meta-analysis of longitudinal studies. *Journal of Personality and Social Psychology, 119*(6), 1459–1477. https://doi.org/10.1037/pspp0000265

Hollar, D. W. (2018). *Trajectory Analysis in Health Care*. Cham, CH: Springer.

Hoyle, R. H. (Ed.). (2023). *Handbook of Structural Equation Modeling* (2nd ed.). New York: The Guilford Press.

Hu, L., & Bentler, P. M. (1999). Cutoff criteria for fit indexes in covariance structure analysis: Conventional criteria versus new alternatives. *Structural Equation Modeling, 6*, 1–55. https://doi.org/10.1080/10705519909540118

Junker, B., Schofield, L. S., & Taylor, L. J. (2012). The use of cognitive ability measures asexplanatory variables in regression analysis. *IZA Journal of Labor Economics, 1*, 1–19. https://doi.org/10.1186/2193-8997-1-4

Jöreskog, K. G. (1969). A general approach to confirmatory maximum likelihood factor analysis. *Psychometrika, 34*, 183–202. https://doi.org/10.1007/BF02289343

Jöreskog, K. G. (1994a). On the estimation of polychoric corrlations and their asymptotic covariance matrix. *Psychometrika, 59*, 381–389. https://doi.org/10.1007/BF02296131

Jöreskog, K. G. (1994b). Structural equation modeling with ordinal variables. In T. W. Anderson, K. T. Fang, & I. Olkin (Eds.), *IMS lecture notes monograph series*(Vol. 24, pp. 297–310). IMS. https://doi.org/10.1214/lnms/1215463803

Jöreskog, K. G. (2004). *Structural equation modeling with ordinal variables using LISREL*. Lincolnwood, IL: Scientific Software International.

Kabic, M., & Alexandrowicz, R. W. (2023a). RMX/PIccc: An Extended Person-Item-Map and a Unified IRT Output for eRm, psychotools, ltm, mirt, and TAM. *psych, 5*, 948–965. https://doi.org/10.3390/psych5030062

Kabic, M., & Alexandrowicz, R. W. (2023b). RMX/PIccc: An Extended Person-Item-Map and a Unified IRT Output for eRm, psychotools, ltm, mirt, and TAM [Computer software manual]. Retrieved from https://osf.io/n9c5r/ (R package version 0.1-3)

Kampen, J. K., & Weeren, A. (2016). A recommendation for applied researchers to substantiate the claim that ordinal variables are the product of underlying bivariate normal distributions. *Quality and Quantity, 51*, 2163–2170. https://doi.org/10.1007/s11135-016-0378-2

Karney, B. R., & Bradbury, T. N. (1995). The longitudinal course of marital quality and stability: A review of theory, methods, and research. *Psychological Bulletin, 118*(1), 3–34. https://doi.org/10.1037/0033-2909.118.1.3

Keller, F., & Alexandrowicz, R. W. (2024). Assessing individual change: A comparison of reliable change indices based on classical test theory and various item response theory models. In M. Stemmler, W. Wiedermann, & F. L. Huang (Eds.), *Dependent data in social sciences research* (2nd ed.). New York: Springer.

Kenny, D. A., Kashy, D. A., & Cook, W. L. (2006). *Dyadic data analysis*. New York: Guilford.

Kolbe, L., Oort, F., & Jak, S. (2021). Bivariate distributions underlying responses to ordinal variables. *mdpi/psych, 3*, 562–578. https://doi.org/10.3390/psych3040037

Kolen, M. J., & Brennan, R. L. (2004). *Test Equating, Scaling, and Linking. Methods and practices* (2nd ed.). Springer.

Lechner, C. M., Bhaktha, N., Groskurth, K., & Bluemke, M. (2021). Why ability point estimates can be pointless: A primer on using skill measures from large-scale assessments in secondary analyses. *Measurement Instruments for the Social Sciences, 3*, 1–16. https://doi.org/10.1186/s42409-020-00020-5

Li, C.-H. (2016). Confirmatory factor analysis with ordinal data: Comparing robust maximum likelihood and diagonally weighted least squares. *Behavior Research Methods, 48*, 936–949. https://doi.org/10.3758/s13428-015-0619-7

Little, R. J., & Rubin, D. B. (2002). *Statistical Analysis with Missing Data* (2. ed.). New York: Wiley.

Lleras, C. (2005). Path Analysis. In K. Kempf-Leonhard (Ed.), *Encyclopedia of Social Measurement* (Vol. 3, pp. 25–30). Elsevier.

Lu, I. R. R., & Thomas, D. R. (2008). Avoiding and correcting bias in score-based latent variable regression with discrete manifest items. *Structural Equation Modeling: A Multidisciplinary Journal, 15*, 462–490. https://doi.org/10.1080/10705510802154323

Lu, I. R. R., Thomas, D. R., & Zumbo, B. D. (2005). Embedding IRT in structural equation models: A comparison with regression based on IRT scores. *Structural Equation Modeling, 12*, 263–277. https://doi.org/10.1207/s15328007sem1202_5

Luo, Y., & Dimitrov, D. M. (2018). A short note on obtaining point estimates of the IRT ability parameter with MCMC estimation in Mplus: How many plausible values are needed? *Educational and Psychological Measurement, 79*, 272–287. https://doi.org/10.1177/0013164418777569

Mair, P., Hatzinger, R., & Maier, M. J. (2020). eRm: Extended Rasch Modeling [Computer software manual]. Retrieved from https://cran.r-project.org/package=eRm (1.0-2)

Masters, G. N. (1982). A Rasch model for partial credit scoring. *Psychometrika, 47*, 149–174. https://doi.org/10.1007/BF02296272

Masters, G. N. (2005). Partial credit model. In K. Kempf-Leonhard (Ed.), *Encyclopedia of social measurement* (Vol. 3, pp. 7–17). Elsevier.

Maydeu-Olivares, A. (2006). Limited information estimation and testing of discretized multivariate normal structural models. *Psychometrika, 71*, 57–77. https://doi.org/10.1007/S11336-005-0773-4

McNeish, D. (2022). Psychometric properties of sum scores and factor scores differ even when their correlation is 0.98: A response to Widaman and Revelle. *Behavior Research Methods*. https://doi.org/10.3758/s13428-022-02016-x

McNeish, D., & Wolf, M. G. (2020). Thinking twice about sum scores. *Behavior Research Methods, 52*, 2287–2305. https://doi.org/10.3758/s13428-020-01398-0

Meade, A. W., & Lautenschlager, G. J. (2004). A comparison of item response theory and confirmatory factor analytic methodologies for establishing measurement equivalence/invariance. *Organizational Research Methods, 7*, 361–388. https://doi.org/10.1177/1094428104268027

Mislevy, R. J. (1991). Randomization-based inference about latent variables from complex samples. *Psychometrika, 56*, 177–196.

Mislevy, R. J. (1993). Should "multiple imputations" be treated as "multiple indicators"? *Psychometrika, 58*, 79–85.

Mislevy, R. J., Beaton, A. E., Kaplan, B., & Sheehan, K. M. (1992). Estimating population characteristics from sparse matrix samples ofitem responses. *Journal of Educational Measurement, 29*, 133–161.

Mislevy, R. J., Johnson, E. G., & Muraki, E. (1992). Scaling Procedures in NAEP. *Journal of Educational Statistics, 17*, 131–154. https://doi.org/10.3102/10769986017002l

Mund, M., Finn, C., Hagemeyer, B., Zimmermann, J., & Neyer, F. J. (2015). The dynamics of self-esteem in partner relationships. *European Journal of Personality, 29*, 235–249. https://doi.org/10.1002/per.1984

Mund, M., Maes, M., Drewke, P. M., Gutzeit, A., Jaki, I., & Qualter, P. (2023). Would the real loneliness please stand up? the validity of loneliness scores and the reliability of single-item scores. *Assessment, 30*(4), 1226–1248. https://doi.org/10.1177/10731911221077227

Muraki, E. (1992). A generalized partial credit model: Application of an EM algorithm. *Applied Psychological Measurement, 16*(2), 159–176. https://doi.org/10.1177/014662169201600206
OECD. (2009). *Plausible values*. Paris: OECD Publishing. (https://www.oecd-ilibrary.org/content/component/9789264056275-7-en; accessed 2023-09-25). https://doi.org/10.1787/9789264056275-7-en
Olsson, U. (1979). Maximum likelihood estimation of the polychoric correlation coefficient. *Psychometrika, 44*, 443–460. https://doi.org/10.1007/BF02296207
Olsson, U., Drasgow, F., & Dorans, N. J. (1982). The polyserial correlation coefficient. *Psychometrika, 47*, 337–347. https://doi.org/10.1007/BF02294164
Orlando, M., & Thissen, D. (2000). Likelihood-based item fit indices for dichotomous item response theory models. *Applied Psychological Measurement, 24*, 50–64. https://doi.org/10.1177/01466216000241003
Orth, U., Robins, R. W., & Widaman, K. F. (2012). Life-span development of self-esteem and its effects on important life outcomes. *Journal of Personality and Social Psychology, 102*, 1271–1288. https://doi.org/10.1037/a0025558
Ostini, R., & Nering, M. L. (2006). *Polytomous item response theory models*. Sage.
Pedhazur, E. J. (1982). *Multiple regression in behavioral research* (2nd ed.). Holt, Rinehard and Winston.
R Core Team. (2023). R: A Language and Environment for Statistical Computing [Computer software manual]. Vienna, Austria. Retrieved from https://www.R-project.org/
Rasch, G. (1961). *On general laws and the meaning of measurement in psychology*. Copenhagen: The Danish Institute of Educational Research.
Rasch, G. (1966a). An individualistic approach to item analysis. In P. F. Lazarsfeld & N. W. Henry (Eds.), *Readings in Mathematical Social Science* (pp. 89–107). Cambridge, MA: The MIT Press.
Rasch, G. (1966b). An informal report on the present state of a theory of objectivity in comparisons. In *Proceedings of the nuffic international summer session in science at "het oude hof", the hague, 14.–28 July*. Retrieved 2022-01-07, from https://rasch.org/memo1966.pdf
Reckase, M. D. (2009). *Multidimensional item response theory*. Springer.
Rizopoulos, D. (2006). ltm: An R package for latent variable modelling and item response theory analyses. *Journal of Statistical Software, 17*(5), 1–25. https://doi.org/10.18637/jss.v017.i05
Robitzsch, A., Kiefer, T., & Wu, M. (2020). TAM: Test Analysis Modules [Computer software manual]. Retrieved from https://CRAN.R-project.org/package=TAM (R package version 3.4-26)
Rosenberg, M. (1989). *Society and the Adolescent Self-Image* (revised ed.). Middletown, CT: Wesleyan University Press.
Rosseel, Y. (2012). lavaan: An R package for structural equation modeling. *Journal of Statistical Software, 48*(2), 1–36. https://doi.org/10.18637/jss.v048.i02
Rubin, D. B. (1987). *Multiple imputation for nonresponse in surveys*. New York: Wiley.
Rusch, T., Lowry, P. B., Mair, P., & Treiblmaier, H. (2016). Breaking free from the limitations of classical TestTheory: Developing and measuring information systems scales using item response theory. *Information & Management, 54*, 189–203. https://doi.org/10.1016/j.im.2016.06.005
Samejima, F. (1969). Estimation of latent ability using a response pattern of graded scores. *PSychometrika, Monograph*, 1–100.
Sander, J., & Böcker, S. (1993). Die Deutsche form der Relationship Assessment Scale (RAS): Eine kurze Skala zur Messung der Zufriedenheit in einer Partnerschaft [The German Form of the Relationship Assessment Scale (RAS): A short scale for assessing satisfaction in close relationships]. *Diagnostica, 39*, 55–62.
Schaffhuser, K., Wagner, J., Lüdtke, O., & Allemand, M. (2014). Dyadic longitudinal interplay between personality and relationship satisfaction: A focus on neuroticism and self-esteem. *Journal of Research in Personality, 53*, 124–133. https://doi.org/10.1016/j.jrp.2014.08.007
Schofield, L. S. (2015). Correcting for measurement error in latent variables used as predictors. *The Annals of Applied Statistics, 9*, 2133–2152. https://doi.org/10.1214/15-AOAS877

Schumacker, R. E., & Lomax, R. G. (2016). *A, beginner's guide to structural equation modeling* (4th ed.). New York: Routledge.

Shim, H., Bonifay, W., & Wiedermann, W. (2023a). Parsimonious asymmetric item response theory modeling with the complementary log-log link. *Behavior Research Methods, 55*, 200–219. https://doi.org/10.3758/s13428-022-01824-5

Shim, H., Bonifay, W., & Wiedermann, W. (2023b). Parsimonious item response theory modeling with the negative log-log link: The role of inflection point shift. *Behavior Research Methods*. https://doi.org/10.3758/s13428-023-02189-z

Skrondal, A., & Laake, P. (2001). Regression among factor scores. *Psychometrika, 66*, 563–575. https://doi.org/10.1007/BF02296196

Stefanski, L. A. (1985). The effects of measurement error on parameter estimation. *Biometrika, 72*, 583–592. https://doi.org/10.2307/2336730

Tabachnick, B. G., & Fidell, L. S. (2014). *Using multivariate statistics* (6th ed.). Boston, MA: Pearson.

Takane, Y., & de Leeuw, J. (1987). On the relationship between item response theory and factor analysis of discretized variables. *Psychometrika, 52*, 393–408. https://doi.org/10.1007/BF02294363

The Program Committee of the Institute for Objective Measurement. (2000). *Definition of objective measurement*. https://www.rasch.org/define.htm. (Accessed: 2023-09-03)

Thissen, D., & Steinberg, L. (1986). A taxonomy of item response models. *Psychometrika, 51*, 567–577. https://doi.org/10.1007/BF02295596

Thomson, G. (1938). Methods of estimating mental factors. *Nature, 141*, 246. https://doi.org/10.1038/141246a0

Tucker, L. R., & Lewis, C. (1973). A reliability coefficient for maximum likelihood factor analysis. *Psychometrika, 38*, 1–10. https://doi.org/10.1007/BF02291170

Uebersax, J. S. (1999). *Likert scales: Dispelling the confusion.* Retrieved 2023-08-17, from https://www.john-uebersax.com/stat/likert.htm

van Buuren, S. (2018). *Flexible imputation of missing data* (2nd ed.). Boca Raton, FL: Chapman & Hall/CRC.

von Collani, G., & Herzberg, P. Y. (2003). Eine revidierte Fassung der deutschsprachigen Skala zum Selbstwertgefühl von Rosenberg [A revised version of the German adaptation of Rosenberg's self-esteem scale]. *Zeitschrift für Differentielle und Diagnostische Psychologie, 24*(1), 3–7. https://doi.org/10.1024//0170-1789.24.1.3

West, S. G., Wu, W., McNeish, D., & Savord, A. (2023). Model fit in structural equation modeling. In R. H. Hoyle (Ed.), *Handbook of structural equation modeling* (2nd ed., pp. 184–205). New York: Guilford.

Widaman, K. F., & Revelle, W. (2023a). Thinking about sum scores yet again, maybe the last time, we don't know, Oh No . . .: A comment on McNeish (2023). *Educational and Psychological Measurement, Online First.* https://doi.org/10.1177/00131644231205310

Widaman, K. F., & Revelle, W. (2023b). Thinking thrice about sum scores, and then some more about measurement and analysis. *Behavior Research Methods, 55*, 788–806. (online first: 2022-04-25). https://doi.org/10.3758/s13428-022-01849-w

Wright, B. D., & Masters, G. N. (1982). *Rating scale analysis*. Chicago: MESA.

Wright, B. D., & Stone, M. H. (1979). *Best test design*. Chicago: MESA.

Wright, S. (1934). The method of path coefficients. *Annals of Mathematical Statistics, 5*, 161–215. https://doi.org/10.1214/aoms/1177732676

Wu, M. L. (2005). The role of plausible values in large-scale surveys. *Studies in Educational Evaluation, 31*, 114–128. https://doi.org/10.1016/j.stueduc.2005.05.005

Wu, M. L., Adams, R. J., Wilson, M. R., & Haldane, S. A. (2007). *ACER ConQuest. Generalised Item Response Modelling Software*. Melbourne: ACER Press.

Xia, Y., & Yang, Y. (2019). RMSEA, CFI, and TLI in structural equation modeling with orderedcategorical data: The story they tell depends on the estimation methods. *Behavior Research Methods, 51*, 409–428. https://doi.org/10.3758/s13428-018-1055-2

Yi, G. Y., Delaigle, A., & Gustafson, P. (Eds.). (2021). *Handbook of measurement error models*. Boca Raton, FL: Chapman & Hall/CRC.

Zeileis, A., Strobl, C., Wickelmaier, F., Komboz, B., Kopf, J., Schneider, L., & Debelak, R. (2021). psychotools: Infrastructure for psychometric modeling [Computer software manual]. Retrieved from https://CRAN.R-project.org/package=psychotools (R package version 0.7-0)

Zidek, J. V., Wong, H., Le, N. D., & Burnett, R. (1996). Causality, measurement error and multicollinearity in epidemiology. *Environmetrics, 7*, 441–451.

Chapter 28
Assessing Individual Change: A Comparison of Reliable Change Indices Based on Classical Test Theory and Various Item Response Theory Models

Ferdinand Keller and Rainer W. Alexandrowicz

28.1 General Introduction and Overview

The effectiveness of an intervention is usually evaluated by randomized controlled trials (RCT) or related types of group comparison designs where group mean values are compared between an intervention group and a control group that receives no or another treatment or placebo. If no control group is available, a single group pre-post design is applicable as well for assessing change. The determination of statistical significance and effect sizes in these study designs yields important information for subsequent application of the intervention in patients and for planning further studies. In contrast, clinicians are often concerned with changes in particular individuals under treatment (Evans et al., 1998), and they often prefer dichotomous outcomes such as improved vs. unchanged (or deteriorated). Thus, their focus is less on the overall group effect (pre-post effect size), but whether an individual patient improves during an intervention. In addition, the change should be meaningful to be considered as improvement. However, the determination of what constitutes a "meaningful change" remains controversial (Trigg et al., 2023) for it involves both substantive and statistical/psychometric aspects; we will concentrate on the latter here. In particular, we focus on differences in evaluating individual change on the basis of classical test theory (CTT) vs. item response theory (IRT). For further details regarding the discussion of the various concepts of meaningful change, see two recent special issues (Reeve & Hays, 2021; Trigg et al., 2023).

F. Keller (✉)
Department of Child and Adolescent Psychiatry and Psychotherapy, University Hospital Ulm, Ulm, Germany
e-mail: Ferdinand.Keller@uniklinik-ulm.de

R. W. Alexandrowicz
Abteilung für Methodenlehre, Institut für Psychologie, Universität Klagenfurt, Klagenfurt, Austria

M. Stemmler et al. (eds.), *Dependent Data in Social Sciences Research*,
https://doi.org/10.1007/978-3-031-56318-8_28

Jacobson and Truax (1991) proposed to summarize change at the level of the individual by using the observed changes in the whole sample. It has mainly been developed in psychotherapy outcome research, but the assessment of significant change for an individual has been applied also in other fields like neuropsychology, neurological, or many other medical diseases (see Duff, 2012 for a comprehensive overview of development and frameworks of change indices). In fact, it is applicable to any clinical problem if the outcome measurement is on a continuous scale (Evans et al., 1998). Recently, some major contributions came from quality of life research (e.g., Reeve & Hays, 2021; Trigg et al., 2023).

The reliable change index (RCI) proposed by Jacobson and Truax (1991) is based on classical test theory; the RCI calculates the difference between pre- and posttest scores of an individual and standardizes this difference by the standard error of the differences (see Sect. 28.2).[1] Thus, the RCI informs about the confidence that the change is beyond what could be attributed to measurement error of the assessment instrument.

However, reliable change alone is considered insufficient to evaluate a patients' progress. An important aim in treatment lies in leaving a dysfunctional level and to reach a level where the patient can be considered as "functional" or below a cutoff for being "symptomatic." Jacobson and Truax (1991) suggest several comparison groups and rules for determining cutoffs to achieve a so-called clinically significant change. Other authors extended their classification (e.g., Tingey et al., 1996). Ronk et al. (2013) emphasized the distinction of in- and outpatient populations with regard to the costs of an intervention and the (realistic) goals for successful treatment. Hence, a clinically significant improvement could also mean to move from in- to outpatient level. Another method would be group-level boundaries previously defined by experts.

Also, norm values could be used to determine cut points, but their estimation relies on the distribution of observed scores in a relevant sample. Thus, these distribution-based indices are *simply a way of expressing the observed change in a standardized metric (. . .) and the reliable change index (. . .) is based on defining change using the statistical convention of exceeding 2 standard errors* (Revicki et al., 2008, p. 106). Complementary to the distribution-based methods, anchor-based approaches have also been proposed. Anchor-based methods use external criteria, such as clinicians' ratings, laboratory measures, or a patient-based global rating of change (Revicki et al., 2008). These methods require further statistical modeling (cf. section "Outlook") and are beyond the scope of this chapter.

A patient should improve to both a reliable and clinically significant amount. Let us consider an example for the BDI-II, where a critical difference of eight raw score points is assumed relevant for reliable change (Hautzinger et al., 2006): a person

[1] Several other approaches to evaluate clinical change on the background of CTT have been suggested (Ferrer & Pardo, 2014), but the RCI of Jacobson and Truax (1991) seems to be most used and discussed. Therefore, the RCI will be used here as an index of statistical evaluation of change that is based on CTT.

should improve by more than 8 points on the BDI-II sum score from pre- to posttest to fulfill the RCI and being labeled as improved; if he/she reaches additionally a sum score of 13 or lower at the posttest (being in the "functional" level, below 14 = cutoff for mild depression), he/she would be labeled as recovered.

There have been several criticisms of the RCI (Wise, 2004; Duff, 2012), including the problems of practice effects or regression to the mean, which in turn have brought up regression-based approaches to evaluate change (e.g., Ferrer & Pardo, 2014). Another major concern is that within CTT the reliability and the standard errors are constant across the entire range of scores. Hence, the standard error for assessing change is the same in the middle of the score distribution and the extremes. Several authors emphasized this shortcoming and proposed solutions resorting to item response theory (IRT) (Brouwer et al., 2013; Jabrayilov et al., 2016). In IRT modeling, the reliability changes over the latent trait (Samejima, 1994; O'Connor, 2018), and standard errors are usually lower in the middle and higher at the extremes of the score distribution.

In this chapter, we would like to elaborate the concept of reliable change in the IRT domain and apply and compare the classical RCI with its IRT-based counterparts on a data set with psychosomatic inpatients who were assessed at the beginning and at the end of inpatient treatment. Our outcome measure is the Beck Depression Inventory-II (Beck et al., 1996). Moreover, we want to compare different IRT models, namely, the PCM, the GPCM, and the GRM (see section "Candidate IRT Models"), with respect to their impact on the results of individual change and their goodness of fit. Finally, we will also discuss the (implicit) postulation of longitudinal invariance that has been assumed in several studies in the empirical literature.

28.2 Assessment of Individual Change

For assessing change with the RCI, two methodological approaches have been proposed: a classical version and an IRT-based variant.

The Classical Reliable Change Index (RCI)

A general approach for assessing individual change is based on the difference of a pretest and a posttest score divided by the standard error of the differences of these two (SE_{diff}). A typical notation of the standard error of measurement (SEM) is

$$SEM(X) = SD(X) \cdot \sqrt{1 - REL(X)}. \tag{28.1}$$

with SD(·) for the standard deviation and *REL*(·) for the reliability. Applying definition (28.1) to the pretest score X_1 and the posttest score X_2, we obtain a generic formula for assessing individual change

$$z = \frac{X_2 - X_1}{\sqrt{SEM(X_1)^2 + SEM(X_2)^2}}, \tag{28.2}$$

which involves $SD(X_1)$, $SD(X_2)$, $REL(X_1)$, and $REL(X_2)$. These four statistics allow for introducing several simplifying assumptions by setting some of them equal.

Assuming equal SD and equal reliability for pre- and posttest (i.e., $SD(X_1) = SD(X_2) = SD(X)$; and likewise for reliability), the denominator of Eq. (28.2) simplifies to

$$\sqrt{SEM(X)^2 + SEM(X)^2} = \sqrt{2 \cdot SEM(X)^2} = \sqrt{2} \cdot SEM(X) = SE_{\text{diff}}, \tag{28.3}$$

arriving at

$$RCI = \frac{X_2 - X_1}{SE_{\text{diff}}} = \frac{X_2 - X_1}{\sqrt{2} \cdot SEM}. \tag{28.4}$$

For applications, we require estimates of the reliability coefficient. Jacobson and Truax (1991) used the reliability estimate based on a test-retest design. However, using the test-retest-reliability estimate seems to have vanished after about 2000 (the most recent reference found was Wise, 2004).[2] Since test-retest reliability is difficult to obtain for an untreated sample and it is generally lower than internal consistency due to additional sources of variation, the application of Cronbach's alpha (α_{CR}) has become a standard coefficient for the reliability in the SEM formula. Note, however, that this can be questioned, and Ferrer and Pardo (2014) found in a simulation study that the rate of false-positive changes was reduced to an acceptable rate if the test-retest reliability was used instead of alpha. In our further analyses, we use the RCI_{CTT} in the sense of Eq. (28.4) using Cronbach's alpha for estimating reliability.

An RCI exceeding a critical z-score of the normal distribution[3] (usually 1.96 for an error of the first kind of 5%) is considered a "reliable change" (improvement or deterioration, depending on the sign and the polarity of the scale). Since the reliability is constant for the entire range of the sum scores, one can also compute a critical difference in raw scores, i.e., a reliable change of an individual has to exceed the interval $\pm 1.96 \cdot \sqrt{2} \cdot SEM$. This coefficient is also known as the minimally detectable change or the coefficient of repeatability (cf. Hays & Peipert, 2021).

---

[2] Interestingly, later on the use of Cronbach's alpha (α_{CR}) has even been seen as standard by some authors. For example, Ferrer and Pardo (2019) state that the test-retest correlation is superior to *the originally proposed internal consistency (Cronbach's alpha)* (p. 98).

[3] Strictly speaking, the index follows a t-distribution, because the standard deviation is obtained from the sample. However, this statistical distinction is of little relevance for practical purposes, as the difference of the critical cutoffs of the t- and the standard normal distribution is marginal for reasonable sample sizes.

Several authors have proposed variants of change indices based on CTT, for example, taking practice effects or regression-based approaches into consideration (see, e.g., Duff, 2012; Ferrer & Pardo, 2014; Hays & Peipert, 2021).

IRT-Based Modeling of Individual Change

Analogously to the classical RCI, an IRT-based approach has been suggested (Brouwer et al., 2013; Jabrayilov et al., 2016), using the difference of the estimated person parameters (θ) of the pre- and the posttest and the pooled standard errors of these parameters in the denominator, i.e.,

$$\mathrm{RCI}_{\mathrm{IRT}} = \frac{\hat{\theta}_2 - \hat{\theta}_1}{\sqrt{SE\left(\hat{\theta}_1\right)^2 + SE\left(\hat{\theta}_2\right)^2}} \tag{28.5}$$

(Jabrayilov et al., 2016). Note the structural similarity to Eq. (28.2). Thus, the $\mathrm{RCI}_{\mathrm{IRT}}$ is handled analogously to the $\mathrm{RCI}_{\mathrm{CTT}}$, i.e., an RCI $\geq$ 1.96 indicates significant change at the 5 % level, again considering the sign to distinguish improvement or deterioration.

Studies Comparing Classical and IRT-Based RCI

Several empirical studies explored how the IRT-based approaches to determine change compare to the classical index. Before summarizing some results thereof, the results of simulations studies comparing CTT-based and IRT-based RCI are presented. Some simulation studies on the behavior of the classical RCI only also are worth mentioning and will be summarized first.

Simulation Studies Within the CTT approach, Ferrer and Pardo (2014) compared seven different coefficients and a regression-based approach (post score lies outside the confidence interval for the individual predicted value) for the evaluation of individual change with respect to their rates of false positives. Although the RCI is not among the two best methods, it (as well as other SEM-based methods) offers an acceptable rate when assessing the reliability according to the test-retest principle rather than the internal consistency. Subsequently, the authors also found increased false-negative rates (Ferrer & Pardo, 2019), when applying five of the best performing methods in their former false-positive evaluation. They conclude that *[...] our results suggest that the widely used distribution-based methods must be applied with caution in a clinical context, because they need huge effect sizes to detect a true change.* (p. 97). In a further simulation study, Estrada et al. (2019) explored the relation of average-based change statistics (e.g., Cohen's *d*) and individual-based change statistics (e.g., the RCI), concluding that—contrary to what previous studies have speculated—these two statistics are closely related. Nonetheless, they recommend to use the individual-based change statistics, because these allow the identification of cases that changed reliably and they facilitate the interpretation and communication of results.

Jabrayilov et al. (2016) compared the RCI_{CTT} and the RCI_{IRT} with regard to the type I error and detection rates of improved persons. In a simulation study, they varied test length (5 items vs. 20 items), the magnitude of the true change, and whether mean item difficulties were homogeneous or heterogeneous. The results did not support the alleged superiority of the IRT over CTT. While IRT was superior to CTT in the long test condition (20 items), they recommended to use CTT for short tests (5 items), since CTT was generally better at correctly detecting change individuals (Jabrayilov et al., 2016). This conclusion was criticized by Hays and Peipert (2021) who state that *CTT should never be better than IRT in this respect* and *CTT raw scores are just an approximation of the IRT model* (p. 789). The authors note, however, that Jabrayilov et al. (2016) used the Fisher information function, while they used standard deviations based on the expected a posterior (EAP) estimates. In their own study, Hays and Peipert (2021) simulated estimated EAP scores based on the graded response model item parameters of two measures and varied the true thetas throughout the continuum (-3 to 3). They found a comparatively low agreement of 67% between CTT and IRT classification as improved in the condition of small change and low thetas (change from -3 to -2). For larger changes in theta and especially with thetas >-2, agreement was good to almost perfect. In a further simulation condition with no true underlying change, the CTT was also consistent with IRT (Hays & Peipert, 2021).

Empirical Studies The few studies that looked at differences in classifications between the RCI_{CTT} and IRT-based RCI come from quite different fields and background. We selected studies from three research domains discussing advantages and limitations of available methods for assessing individual-level change.

Brouwer et al. (2013) assessed change in a sample of clinical outpatients applying for treatment. A calibration sample of 1530 patients was used to estimate the item parameters for a GRM. The outcome was measured with the BDI-II. The estimated GRM item parameters were then used to analyze change in a smaller sample of 104 patients from the same clinic where BDI-II pre- and posttest scores were available. The authors observed that for many individuals, the two RCI methods lead to the same classification, but for eight individuals, there was a different classification, mostly at the cut-off value for the RCI (i.e., eight raw score points of change).

Murray et al. (2021) provide a practical example from a longitudinal cohort study of healthy aging, with a focus on cognitive function. After a review of the RCI and two variants of an IRT-based RCI, they analyzed the performance of 535 participants in a reading test at two occasions being 3 years apart. With regard to the comparison of RCI_{CTT} and RCI_{IRT}, 42 participants showed no change according to the RCI_{CTT}, but were classified as reliable decrease by the RCI_{IRT} (the 42 participants are 7.9%

of the total sample and 8.2% of those classified as no change in RCI_{CTT}). The other mismatching cells were smaller in frequency ($n < 10$).

Hays et al. (2021) used IRT to identify responders to treatment of chronic low back pain and chronic neck pain. Outcome was measured with two short scales (physical function and emotional distress) from the Patient-Reported Outcomes Measurement Information System (PROMIS®). GRM item parameters from PROMIS® were used to estimate standard errors at the two time points (3-month difference). The comparison of RCI_{CTT} and RCI_{IRT} revealed for the physical function scale that 78% were classified as same in CTT versus 91% based on IRT. In the emotional distress scale, 68% stayed the same according to CTT versus 90% based on IRT. Thus, CTT estimates of significant individual change are much more optimistic than IRT-based estimates (Hays et al., 2021). The persons classified as "changed" according to the RCI_{CTT} were located about equally in size in the better or in the worse category. On the group-level analyses, there was no significant change in the emotional distress scale and a significant change in the physical functioning scale, but the effect size was very small. The cross-classification by the two RCI methods revealed for both scales that the percentages of misclassified persons increase considerably if we look at specific rows of the respective cross table (cf. Hays et al., 2021, Tables 4 and 7). For the critical row "better in CTT" with a total of $n = 236$ in the physical functioning scale, we find 147 persons who were classified as same by RCI_{IRT}, i.e., 62.3%, and only 37.7% (89/236) with being consistently better in both RCIs. With regard to the emotional distress scale, the analogue numbers are 67.8% and 32.2%. A similar study with cancer patients (Lee et al., 2023) explored six different PROMIS® subscales with regard to the classifications according to RCI_{CTT} and RCI_{IRT}. Concerning the depression subscale, the agreement was high with 81% and the misclassified patients were about equally distributed in the range of 3 to 6% of the total sample and 0 frequencies in the opposite cells (better/worse and vice versa). With a focus on cross-tabulation and the row "better in CTT," we find, again, larger percentages of misclassification, i.e., from a total of 149 persons classified as better with RCI_{CTT}, 46 persons (30.9%) were classified as same by RCI_{IRT} and 69.1% (103/149) with being better in both RCIs.

Summary and Limitations of Present Studies

There are only few studies that explored empirically the differences in classifications of the RCI_{CTT} and IRT-based RCIs. They found divergent classifications depending on the type of RCI. Murray et al. (2021) found high agreement overall between the RCI_{CTT} and the IRT-based RCI, whereas Hays et al. (2021) found that RCI_{CTT} identified more persons as having changed compared to IRT-based RCI. The percentages of misclassified patients in these studies are in the range of 8–11%

at maximum for some cells of the cross-tables if related to the total sample size. With regard to specific combinations, however, in particular, for the clinically interesting combination "better in CTT and same in IRT," percentages of different classifications go up to 30% or even more than 60% in the PROMIS® studies (see exact numbers in the preceding section). Murray et al. (2021) found a relevant group of persons, in which the RCI_{IRT} indicated worsening in the cognitive function but not the RCI_{CTT}. These heterogeneous findings in both the simulation and the empirical studies indicate that further comparisons are required. Moreover, we want to consider two limitations of the current studies:

Longitudinal Invariance For calculating the classical RCI, we need a reliability estimate. Usually, this is achieved either by using a previously determined value (e.g., from the manual) or by calculating it from the sample at the first measurement occasion. In doing so, one assumes the reliability to remain unchanged by whatever happens between the two time points. However, such an assumption could be overly optimistic, because in a clinical setting, the intervention may likely change the meaning of the items. For example, before a depression therapy, a patient may see worthlessness in a different manner than thereafter. Therefore, the responses at time point two should be regarded differently than before. One issue to consider in this context is to check for Differential Item Functioning (DIF; cf. Penfield & Camilli, 2006), which is readily supported by IRT models. Thus, simply assuming longitudinal invariance as done in the classical context may not be warranted in all cases.

Which IRT Model? Another limitation is the fact that all studies considered so far applied only one specific IRT model, viz., the graded response model (GRM; Samejima, 1969). However, there exist several alternative IRT models, especially in the case of modeling polytomous responses. Some of these models (frequently termed "Rasch Family of Models," e.g., Fischer & Molenaar, 1995) provide stronger means of assessing model fit compared to the GRM (e. g., Alexandrowicz, 2022; Kabic & Alexandrowicz, 2023). Hence, these models should also be taken into consideration when assessing reliable change. Moreover, there are several ways of how to apply each model.

28.3 Extending IRT-Based Approaches

The following overview works out some basic model properties relevant for the present study. For a comprehensive introduction to IRT and further details, see, for example, de Ayala (2009).

Candidate IRT Models

Equation (28.5) is a generic formula applicable to any IRT model. Following the taxonomy of Thissen and Steinberg (1986), IRT models can be split into two basic groups, the so-called divide-by-total models and the difference models (see also Ostini & Nering, 2006). The former focus on modeling the probability of choosing either of two adjacent categories, whereas the latter on the probability of choosing a category k or any category above (both conditional to a proper set of parameters, which are omitted here to point out the core distinction). These two fundamentally different approaches entail several important model characteristics relevant for both parameter estimation and interpretation. We will focus exclusively on models for ordered polytomous response format, which is often applied in the clinical context.

The Partial Credit Model (PCM) The PCM (Masters, 1982) is a member of the divide-by-total family, modeling the thresholds between two adjacent response categories using the logit link function. Moreover, it is a member of the "Rasch family of models," which belong to the exponential family, thus supporting sufficient statistics for both item and person parameters. This feature is especially advantageous in clinical settings, for the simple sum of scored responses is the sufficient statistic for the person parameter thus exploiting all information available from the data. Interestingly, sum scores are typically used in clinical tests (like the BDI, e.g.), but not always backed up by a check, whether the test actually conforms to the assumptions of the PCM.

The Generalized Partial Credit Model (GPCM) The GPCM (Muraki, 1992) can be seen as an extension of the PCM, with an additional discrimination parameter per item. Thus, items may vary with respect to the amount they reflect the common latent trait. However, due to this extension, the GPCM is not a member of the exponential family, so that we cannot use the simple sum of scores to (psychometrically) describe a person. Rather, person parameters have to be determined using the individual response patterns.

The Graded Response Model (GRM) The GRM (Samejima, 1969) is a member of the difference family, i.e., the category thresholds are obtained by subtracting the probabilities of two adjacent categories $P^*(k) = P(k + 1) - P(k)$ (again, conditional on a suitable set of parameters). Originally, the GRM has been introduced using the cumulative normal ogive link function, which can easily be adapted to the logit link. Like the GPCM, it is not a member of the exponential family, so that the simple sum of scores does not constitute a sufficient statistic for the person parameter. The GRM enjoys some popularity in clinical research, and the analyses and simulations described so far were based on this model. Furthermore, the Patient-Reported Outcomes Measurement Information System (PROMIS®; cf. Reeve & Hays, 2021; Hays et al., 2021) is based on GRM analyses.

Uni- Versus Multidimensional Calibration

The PCM, the GPCM, and the GRM are unidimensional models. However, certain tests or research designs may require more than one dimension. In the educational context, for example, the PISA studies modeled student knowledge and skills assuming the four dimensions "Reading," "Mathematics," "Science," and "Problem-Solving" (Hartig & Höhler, 2009). In the clinical context, several studies explored the potentially multidimensional structure of depression as measured with the BDI-II (e.g., Wang & Gorenstein, 2013; Keller et al., 2020) or other depression instruments (e.g., Cole et al., 2011; Stochl et al., 2022).

Wang and Nydick (2020) and Paek et al. (2014) applied multidimensional models to assess change in longitudinal studies. The latter compared three variants of multidimensional longitudinal modeling: separate calibration with linking, multidimensional calibration with one latent dimension per measurement occasion, and multidimensional calibration with one latent dimension per measurement plus one latent factor per item, the latter covering item-specific nuisance factors. This third model follows the idea of a bi-factor analysis (Holzinger & Harman, 1938; Toland et al., 2017). The third model showed the best fit closely followed by the second one, whereas the first variant was least convincing. However, Bonifay et al. (2016) warned that bi-factor models may generally tend to outperform less complex models due to a certain proneness to overfitting. We, therefore, decided to apply the second approach, i.e., estimating one latent dimension per time point. Following the taxonomy of Wang and Nydick (2020), this renders a longitudinal unidimensional IRT (L-UIRT) design.

Research Questions

The present study addresses three research questions:

1. Given that the few empirical studies using the estimates from a previous calibration study for evaluating Eq. (28.5) silently assumed longitudinal invariance, we want to explore the situation, when no calibration is available and the estimates are to be obtained from the sample.
2. As the available studies applied only the GRM, we explore the effects of applying additionally a PCM and a GPCM (along with the GRM).
3. Regarding the pre-/post-design, we include the multidimensional approach with the two measurement occasions constituting a latent dimension each, thus yielding a two-dimensional model. This includes also the question of longitudinal invariance.

We limit our analyses to unidimensional measures in the sense that all items belong to the same latent construct (which seems justified for the BDI-II, cf. Keller et al., 2020). To distinguish the various modeling variants, we term the variant with parameter estimates based on the data of time point one and assuming longitudinal invariance "conventional" and contrast it to the model assuming separate dimensions for the pre and the post measurements, which is termed "two-dimensional."

We inspect the following outcomes:

- Which of the candidate models fits best?
- What are the percentages of improved, unchanged, and deteriorated patients according to the different RCIs, compared across the candidate models considered here?
- Do the models identify the same persons with respect to their classification of change?
- Is the conventional or the two-dimensional modeling approach to be preferred?

28.4 Methods

Sample

The total sample consisted of adult inpatients from a clinic for psychosomatic disorders (N = 10,720) who completed the BDI-II within the routine diagnostic procedure at the beginning of treatment (cf. Keller et al., 2022, for a description and further background). For this study, the subsample of patients completing the BDI-II at two time points (admission and discharge; N = 7276) was used. The mean age of this sample was 40.8 years (SD = 14.5; range 19–75) and 67.4% of the patients were female. The mean BDI-II total score at admission was 28.3 (SD = 11.5) and at discharge 15.8 (SD = 12.3).

The correlation between admission and discharge was $r = .60$ and the correlation of admission score and change score was $r = .39$, indicating a moderate dependency from the initial score. A total of 3590 patients (49.3%) were diagnosed with a primary affective disorder (ICD-10, Chapter F3) as their main diagnosis. When taking F3 as a comorbid diagnosis, another 2401 patients (33.0%) fulfil the criteria of a depression, and 1285 patients (17.7%) revealed no F3 diagnosis.

Instrument

The BDI is one of the most frequently applied questionnaires for measuring depression severity. It consists of 21 items that assess a wide range of depressive symptoms (e.g., sadness, loss of pleasure, suicidal thoughts and wishes, concentration difficulties, loss of energy). Each item has four categories numbered from zero to three that are formulated in a symptom-specific way.

Example: Item 1 "Sadness"
0 = "I do not feel sad."
1 = "I feel sad much of the time."
2 = "I am sad all the time."
3 = "I am so sad or unhappy that I can't stand it."

The total score of these items reflect the severity of depression. In 1996, a minor revision of the BDI was carried out to meet the criteria of the DSM-IV (American Psychiatric Association, 1994) and resulted in the BDI-II (Beck et al., 1996). Reliability and validity of the BDI-II are high (Beck et al., 1996; Wang & Gorenstein, 2013), and these findings could be confirmed for the German version too (Hautzinger et al., 2006; Kuehner et al., 2007).

Symptom scores from 0 to 13 indicate no/minimal depression, 14 to 19 a mild depression, 20 to 28 a moderate, and above 28 a severe depression (Beck et al., 1996). These cut-off scores have been adopted for the German version. Furthermore, a ROC analysis on German patient and community samples found the cut-off score of 14 to be optimal for detecting depressive disorders in adults (Kuehner et al., 2023).

Analytic Plan

The RCI was determined following Eq. (28.4) for the classical variant and Eq. (28.5) for the IRT-based models. The latter include the PCM, the GPCM, and the GRM. First, each model was estimated in the conventional way, i.e., the model parameters were estimated from the admission data and applied to the discharge data, assuming longitudinal invariance. Then, the two-dimensional approach was applied to the three IRT models. That way, we obtain three groups of results: *(i)* the "classical" $\mathrm{RCI}_{\mathrm{CTT}}$, *(ii)* the "conventional" $\mathrm{RCI}_{\mathrm{IRT}}$, and *(iii)* the two-dimensional $\mathrm{RCI}_{\mathrm{IRT}}$.

All analyses were performed with the statistical programming environment R (R Core Team, 2022) using the mirt package (Chalmers, 2012; Version 1.37.1), which applies the marginal maximum likelihood method (MML; cf. Baker & Kim, 2004) for parameter estimation. For the conventional IRT-based RCI, the RCI() function provided by the mirt package was used. For the two-dimensional models, we follow the approach of Olsbjerg and Christensen (2015, p. 1417) and set the category parameters of the first item equal across dimensions (i.e., time points) to establish a common scale[4]. The risk of an error of the first kind was set to $\alpha = 0.05$ for all inferential assessments.

[4] We also estimated a model variant setting equal not only the threshold parameters but also the slope parameter of the first item, which resulted in virtually the same results.

Table 28.1 Fit measures for the IRT-based models

Model	AIC	BIC	M2	df	RMSEA	SRMSR	TLI	CFI
PCM conv.	336,158.7	336,599.8	12,336.24*	167	0.100	0.108	0.871	0.872
GPCM conv.	331,602.7	332,181.7	10,231.21*	147	0.097	0.062	0.878	0.894
GRM conv.	330,834.9	331,413.8	7718.32*	147	0.084	0.062	0.909	0.920
PCM 2-dim.	599,051.8	599,927.1	50,133.28*	776	0.094	0.097	0.886	0.887
GPCM 2-dim.	588,528.5	589,679.5	45,196.49*	736	0.091	0.058	0.892	0.898
GRM 2-dim.	587,089.2	588,240.2	41,433.20*	736	0.087	0.058	0.901	0.906

Notes: conv. = conventional, 2-dim. = two-dimensional; an asterisk (*) marks $p < 0.001$

28.5 Results

The presentation of the results has the following structure: First, we inspect the fit measures of the various IRT models and evaluate two assumptions implicitly made in prior studies, i.e., all IRT models fit the data equally well and longitudinal invariance can be assumed. Then, we turn to the shape and size of the standard errors over the latent trait, and finally, we compare the model-specific frequencies and percentages for change classifications.

Fit of the IRT-Based Models

One of the major advantages of IRT-based modeling is the option to assess model fit. In the following, we compare the overall agreement of the models and evaluate whether the IRT models fit the data equally well and whether assumption of longitudinal invariance holds.

Global Fit Comparison

Table 28.1 shows fit indices for the IRT-based models. Note that we cannot make comparisons of the conventional vs. the two-dimensional fit measures, since the conventional variant of the models was estimated only for time point one, whereas the two-dimensional models were estimated for the entire data set (i.e., both time points simultaneously).

In both the conventional and the two-dimensional designs, the GRM shows the best fit. Comparing the PCM to the GPCM with a Likelihood Ratio Test yields $\chi^2 = 4595.99$ with $df = 20$ for the conventional and $\chi^2 = 10{,}603.31$ with $df = 40$ for the two-dimensional design, both of which are significant.

The GRM cannot be compared that way to the PCM or the GPCM, because these two models are not nested. With regard to the information criteria (AIC, BIC) and the fit indices, the GRM is favored as the best fitting solution in both the conventional and the two-dimensional design—thus constituting the most promising candidate.

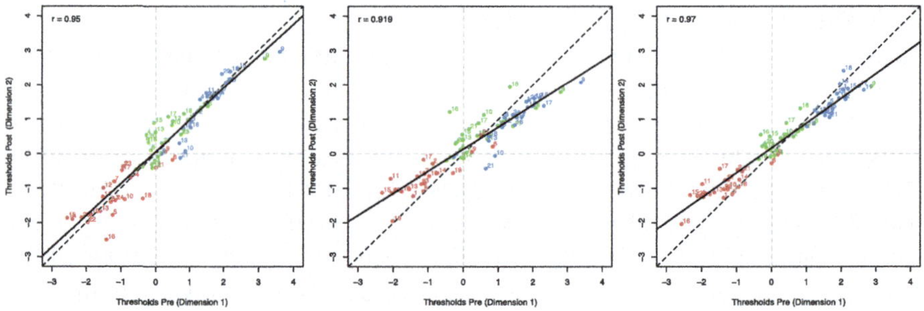

Fig. 28.1 Thresholds pre/post of the three two-dimensional models. Notes: Left: PCM; middle: GPCM; right: GRM; colors indicate thresholds: red = τ_1, green = τ_2, and blue = τ_3; the dashed line is the identity line and the bold line the regression line

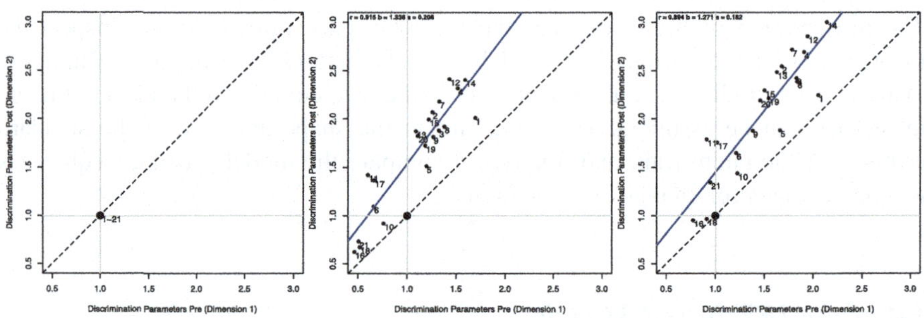

Fig. 28.2 Discrimination parameter estimates of the PCM (left diagram), the GPCM (middle), and the GRM (right) juxtaposed for dimension 1 (pre) and 2 (post) for each model. Gray lines and the fat bullet indicate the value one, the bold line is the regression line

The Longitudinal Invariance Assumption

The conventional IRT-based models assume the parameters of the first measurement to apply to the second measurement. However, this is a strong assumption, which can be checked in the two-dimensional approach by juxtaposing the threshold estimates of the two dimensions.

Figure 28.1 shows the comparison of the estimated thresholds for the two dimensions of each of the three models. We find in general some agreement, but the regression lines indicate a clear tendency for the GPCM and the GRM in that the estimates of dimension 2 (discharge) are smaller in absolute value. Only those of the PCM agree across dimensions—which may constitute both an advantage and a disadvantage: The GPCM and the GRM allow items to vary in their discrimination, thus extracting more information from the data than the PCM does. However, the PCM makes stronger assumptions (which may be confuted by the GPCM and the GRM), but, if it holds, we may use the simple sum of scores as a measure.

Figure 28.2 shows the estimated discrimination parameters of the two-dimensional models.

The left diagram in Fig. 28.2 has only one dot at (1,1), which is the value of the discrimination parameter assumed in the PCM. In contrast, some of the estimates of the GPCM and the GRM deviate strongly from one, most of them showing values larger than one. This indicates that the PCM assumption is questionable for the BDI-II, which is in line with the inferior fit of the PCM shown in Table 28.1.

Visual Inspection of Pre- and Post-Measures with the RCI

To visualize the changes, we plotted the pre- and postscores with colors indicating the RCI status. This was done separately for the classical, the conventional, and the two-dimensional approach of determining the RCI.

The Classical RCI For calculating the classical RCI, we used the coefficients $SD_{pre} = 11.54$ and $\alpha_{\mathrm{CR}} = .911$. From these numbers, we get a $SEM_{\mathrm{pre}} = 3.45$ and $SE_{\mathrm{diff}} = 4.88$. The critical difference for improvement is 9.57, i.e., an individual must improve by at least 10 raw score points for a reliable change.

Figure 28.3 shows the "straight line" structure of the upper and lower significance limits typical for the classical RCI according to Eq. (28.4). The bold vertical lines demonstrate the constant difference of 10 points to achieve a significant change in either direction, and the 45° lines indicate the resulting limits. We added the sometimes used 90% level of significance (dashed lines), with patients in light green or light red who changed according to this reduced level of significance.

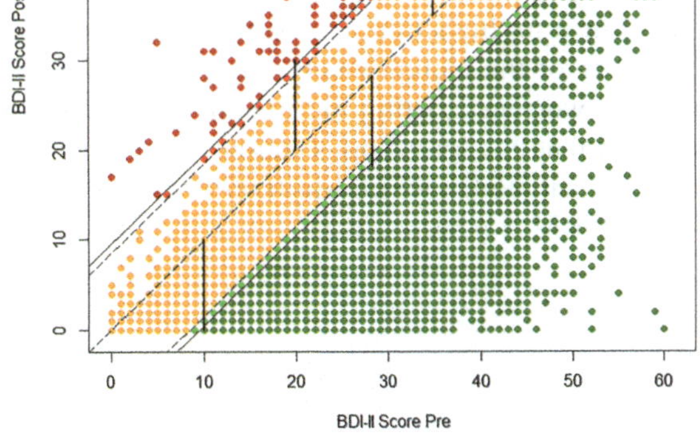

Fig. 28.3 BDI pre- and post-measurements with colors indicating worse (two-sided: dark red; one-sided: light red) unchanged (orange) and better (two-sided: dark green; one-sided: light green) according to Eq. (28.4). The dashed gray lines indicate the limits for one-sided testing and the solid gray lines for two-sided testing

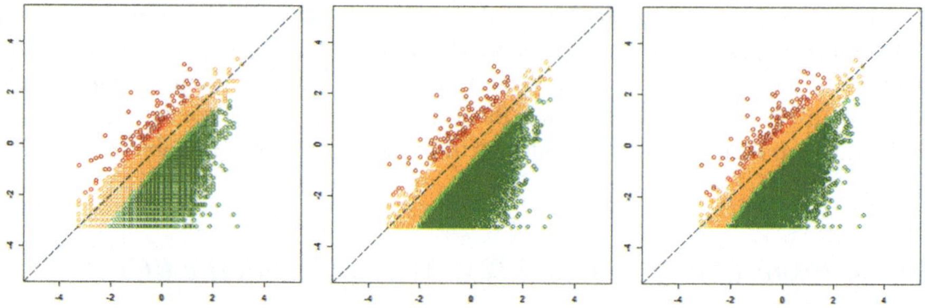

Fig. 28.4 The conventional RCI for the PCM (left diagram), the GPCM (middle diagram), and the GRM (right diagram). Horizontal axes: $\theta^{\wedge}_{pre}$, vertical axes: $\theta^{\wedge}_{post}$; for color coding, see Fig. 28.3

The "Conventional" IRT-Based RCI Figure 28.4 shows the analogue diagram for the conventional IRT-based calculation of the RCI.

Figure 28.4 shows the expected biconcave shape of significance borders due to smaller standard errors in the medium range of the latent dimension rendering smaller differences significant. The left diagram of Fig. 28.4 shows the θ-estimates "lined up" in horizontal and vertical stripes resulting from the sufficient statistics of the PCM: All respondents with the same score will obtain the same person parameter estimate. In contrast, the GPCM and the GRM use the individual response pattern for determining the person parameter thus yielding finer grained estimates; hence these diagrams (middle and right) are more "scattered."

The Two-Dimensional Approach Figure 28.5 shows again the typical biconcave structure for the same reason as in the "conventional" case (Fig. 28.4). We also find the stripes (PCM) and more scattered structure (GPCM and GRM) as before. Additionally, we find the lower bounds of the cloud now concave shaped and points also gathering in stripes. These stripes are formed by the post scores: Take, for example, the lowest stripe representing all respondents scoring zero at discharge, i.e., the prescores are 0, 1, 2,... and the post scores are 0. It bends slightly downwards, i.e., the postscore scale is more expanded (spreaded) for low pre-values than for higher ones. Note that the inverse phenomenon appears at the upper boundary of the cloud, which is not visible here, because (fortunately) only few patients worsened during the treatment. A simulation study (not presented here) asserted that this spreading is caused by the latent correlation estimated in the two-dimensional model, which was 0.661 for the PCM, 0.656 for the GPCM, and 0.653 for the GRM.

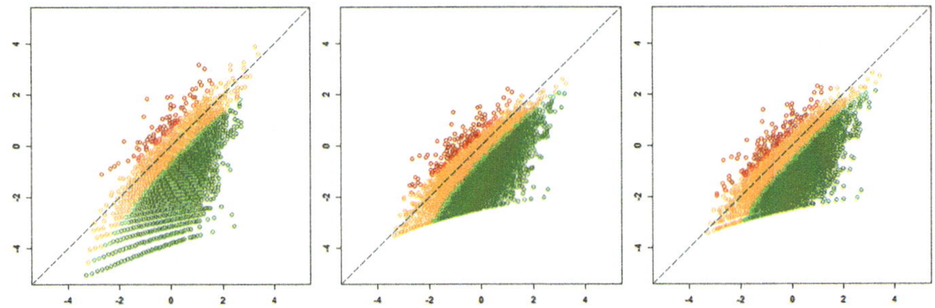

Fig. 28.5 The two-dimensional RCI for the PCM (left diagram), the GPCM (middle diagram), and the GRM (right diagram). For axes and colors, see Fig. 28.4

Table 28.2 Expected marginal frequencies and percentages of individuals classified according to different RCI models

	Better		Unchanged		Worse	
Model	n	%	n	%	n	%
Classical	4322	59.4	2838	39.0	116	1.6
PCM conventional	4359	59.9	2813	38.7	104	1.4
GPCM conventional	4502	61.9	2657	36.5	117	1.6
GRM conventional	4539	62.4	2620	36.0	117	1.6
PCM two-dimensional	4819	66.2	2387	32.8	70	1.0
GPCM two-dimensional	3877	53.3	3270	44.9	129	1.8
GRM two-dimensional	3966	54.5	3189	43.8	121	1.7

Classifications of Change Based on Different RCIs

As the considered models yield different classifications of change, we compare them first by means of the marginal frequencies (all models), followed by a cross-classification of only the classical vs. the IRT-based variants (all 21 possible combinations of cross-classifications would exceed the available space), and finally by the percentages of agreement of all variants.

Marginal Frequencies of Change Classifications

Table 28.2 shows the expected marginal frequencies of change (better/unchanged/worse) according to the RCI determined with the seven approaches under consideration. Interestingly, there is not much difference in the "worse" category, i.e., most of the differences arise from the distinction between the two categories "unchanged" and "better." The classical RCI gives 59.4%, and the three conventional RCI_{IRT} provide slightly higher numbers (59.9–62.4%) for the group "better." The two-dimensional RCI shows larger differences across models with PCM indicating 66.2% as "better," whereas the GPCM and the PCM would classify only about 54% in that category.

Table 28.3 Cross-tabulation of change groups based on the classical RCI (rows) and the various IRT-based RCI (columns)

	Better	Unchanged	Worse	Better	Unchanged	Worse	Total
Classical	**Conventional PCM**			**Two-dimensional PCM**			
Better	4199	123	0	4249	73	0	4322
Unchanged	160	2674	4	570	2268	0	2838
Worse	0	16	100	0	46	70	116
Total	4359	2813	104	4819	2387	70	7276
Classical	**Conventional GPCM**			**Two-dimensional GPCM**			
Better	4203	119	0	3694	628	0	4322
Unchanged	299	2523	16	183	2606	49	2838
Worse	0	15	101	0	36	80	116
Total	4502	2657	117	3877	3270	129	7276
Classical	**Conventional GRM**			**Two-dimensional GRM**			
Better	4204	118	0	3811	511	0	4322
Unchanged	335	2484	19	155	2648	35	2838
Worse	0	18	98	0	30	86	116
Total	4539	2620	117	3966	3189	121	7276

Generally, the various percentages share a similar range across all models. However, this does not necessarily support the conclusion that the same individuals were classified equally. Therefore, we inspect the cross-tabulation of classifications next.

Cross-Tabulation of Change
Table 28.3 shows the cross-tabulation of the classical RCI with the six IRT-based variants (i.e., PCM/GPCM/GRM conventional and two-dimensional).

The comparison of RCI_{CTT} and RCI_{IRT} with the GRM is most frequently applied in the literature, and thus, we first describe the cross-classification between these two models. Of the 2838 patients classified as unchanged according to CTT, 335 (11.8%) were classified as better in the IRT. Conversely, 118 (2.7%) of the patients classified as better in the CTT remained unchanged according to IRT. When taking the two-dimensional GRM, the percentages of misclassifications are similar yet in the opposite direction. Here, 511 patients (11.8%) of the better-group as indicated by the classical RCI are classified as unchanged by the RCI_{IRT}, and the converse number is 155 (5.5%). In the two-dimensional GPCM, these rates of misclassification are similar in structure, but slightly higher (14.5% and 6.4%), whereas the two-dimensional PCM deviates clearly from the structure of the other two-dimensional IRT models. Both opposite misclassification cells (better/worse and vice versa) had zero frequencies in all models.

Total Agreement of Changes
Table 28.4 shows the total agreement of the seven RCI variants under consideration, i.e., the percentage of conforming classifications (better/unchanged/worse), to be

Table 28.4 Agreement of classification of all RCI variants

	Classical	Conventional			Two-dimensional		
		PCM	GPCM	GRM	PCM	GPCM	GRM
Classical	–	95.84	93.83	93.27	90.53	87.69	89.95
PCM conv.	160/123	–	94.91	94.16	93.07	85.56	87.88
GPCM conv.	299/119		–	97.76	92.14	86.05	88.59
GRM conv.	335/118			–	91.93	85.60	88.52
PCM two-dim.	570/ 73				–	79.51	82.11
GPCM two-dim.	183/628					–	96.14
GRM two-dim.	155/511						–

Notes: Upper triangular matrix: Percent agreement; left col: Cell frequencies of "classical unchanged and IRT better"/"classical better and IRT unchanged," taken from Table 28.3

found in the main diagonals of the sub-tables of Table 28.3. Moreover, we here also compare the IRT-based models with one another.

Starting with the classical RCI, the agreement between the classical RCI sums up to 93.3% with the conventional GRM and 90% with the two-dimensional GRM. Within the IRT models, the conventional models agree with 94.2–97.8%, but in the two-dimensional models, the PCM clearly deviates from the GRM and GPCM with only around 80% agreement. GRM and GPCM are very similar in both types. The first column in Table 28.4 repeats (for convenience) the frequencies in the two main disagreement cells from Table 28.3 with regard to the classical RCI. Note the remarkable differences in the two-dimensional case and the opposite classification in PCM vs. GRM/GPCM. In the conventional RCI, the second number ("classical better and IRT unchanged") remains about equal, while the first number ("classical unchanged and IRT better") roughly doubles from PCM to GRM/GPCM. These differences in misclassification rates and especially the sometimes reversed structure are surprising and invite to a closer inspection.

Patterns of Misclassification Visualizing the original scores of only individuals classified differently by the various RCI variants reveals interesting patterns (Figs. 28.6 and 28.7). The PCM assigns lower scoring patients easier as "better," whereas the GPCM and the GRM are more sensitive for higher scoring patients differently compared to the classical RCI. This tendency is even more pronounced for the two-dimensional modeling variant.

28.6 Discussion

In the present chapter, we give an introduction to the assessment of individual change with a focus on modeling change with different IRT models and calibration methods. These methods were applied to a large empirical data set of patients treated

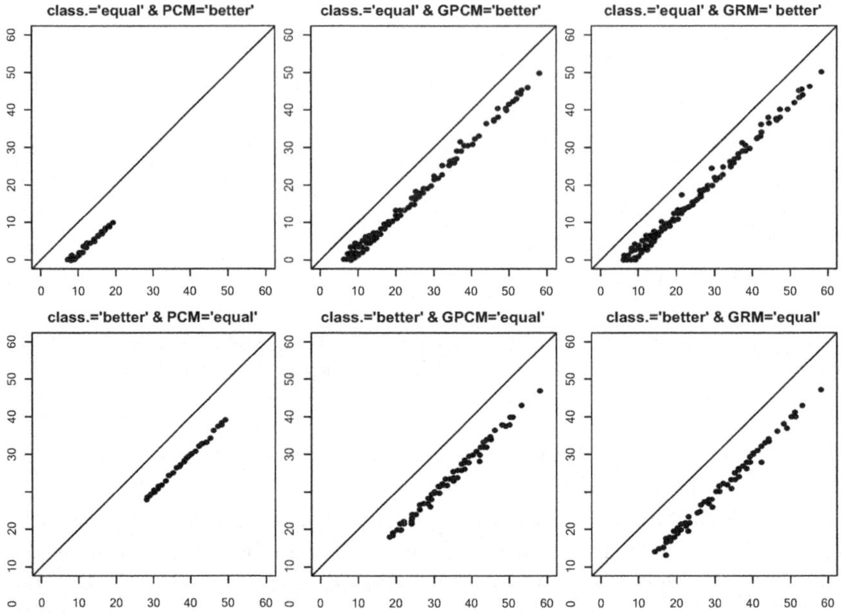

Fig. 28.6 Prescores (horizontal) and postscores (vertical) for individuals with different classification of classical RCI vs. conventional model

at a psychosomatic clinic. With regard to outcome, we used the self-reported severity of depression measured with the BDI-II. Main results were as follows:

- The IRT-based RCIs showed the expected biconcave borders of significant changes, in contrast to the straight lines resulting from the classical RCI.
- About 60% of the patients improved from a statistical point of view and less than 2% deteriorated, indicating successful therapy. These percentages were similar across all models.
- The agreement between classical RCI with conventional IRT comes to $\sim$ 94% and agreement between classical RCI with two-dimensional IRT $\sim$ 90%.
- However, within the (most important) subcategory improved, there were misclassifications up to 10–15% between classical and IRT-based RCI. For example, of the 2838 patients that were classified as unchanged according to CTT, 335 (11.8%) were classified as better in the conventional GRM.
- Regarding the longitudinal invariance assumption, we found differences in both the item difficulty and the discrimination parameter estimates.

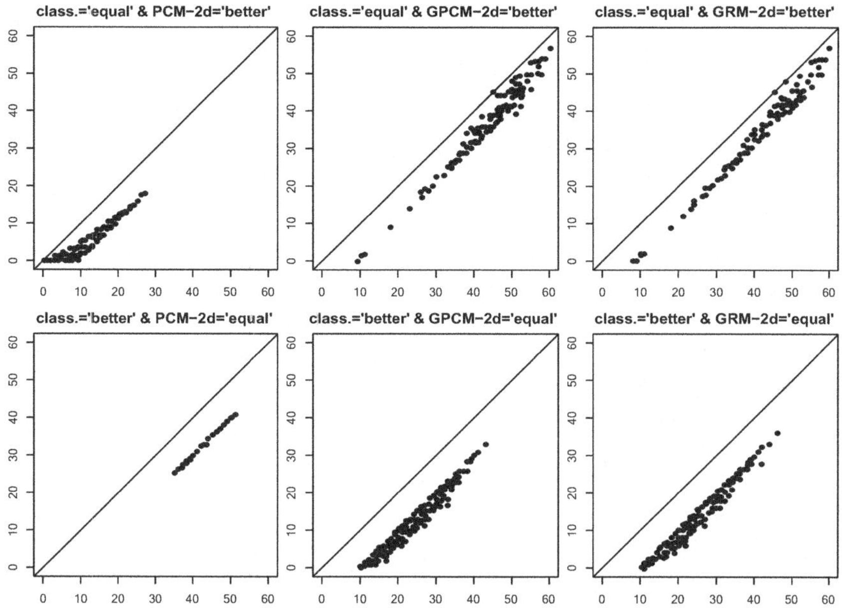

Fig. 28.7 Prescores (horizontal) and postscores (vertical) for individuals with different classification of classical RCI vs. two-dimensional model

Classical RCI Versus IRT-Based RCI

All IRT-based RCIs show the expected biconcave shapes when testing for significance. These result from the fact that the standard errors of the person parameter estimates are (usually) smaller for medium scores than for extreme scores on either end of the latent continuum.

Model Selection Comparing fit, we found the GPCM and the GRM to outperform the PCM. This is, on the one hand, because the additional slope parameters of these two models allow for better approximating differences of the items' discriminations, whereas the PCM assumes strictly parallel threshold characteristic curves. The GRM showed further slightly increased fit compared to the GPCM, which poses the psychometric question, whether the cumulative parametrization of the GRM better reflects the data- generating process compared to the adjacent-category approach of the GPCM. However, the differences in fit of these two models appear too small for a final decision on that point.

However, from a practical point of view, the PCM would still be the optimal choice, for the simple sum of scores constitutes a sufficient statistic for the person parameter. In a clinical context like the one considered here, such a sum allows for a straightforward evaluation of a test result, while the GPCM and the GRM required complex analyses not easily performed by hand to estimate the person parameters.

However, Alexandrowicz et al. (2014) found promising results in a clinical and a student sample when applying the PCM to the BDI-II. Furthermore, Böhnke and Lutz (2014) found no relevant differences in person parameters estimated with the PCM and the GPCM; thus, they question the practical importance, but acknowledge the theoretical difference between the two models and advocate the use of IRT-based person parameter estimates rather than (standardized) scores.

Longitudinal Invariance Regarding the longitudinal invariance assumption, it turned out that almost all of the discrimination parameters increased at the second measurement. This might be taken as an indication that the items were perceived differently after therapy. For example, the item with the largest discrimination parameter was item 14 ("worthlessness"). A likely explanation might be that this symptom plays a fundamental role in depression and has therefore received much attention in the therapeutic process. Confronted again with the response options, the patients improved in choosing the one best reflecting their state of mind at discharge. This, in turn, caused the responses to be much more in line with the individual person parameter estimates based on all items, which, from the "model's view," increased this item's discriminating power. Similar developments likely apply to all items. From a psychometric perspective, such processes contradict the longitudinal invariance assumption made in both the classical as well as the conventional calculation of the RCI. Thus, the proposed two-dimensional approach based on a model estimating the discrimination parameters independently (i.e., GPCM or GRM) appears favorable. However, as the differences appear generally small, longitudinal invariance could be considered to hold to a sufficient extent.

It should be mentioned, too, that our approach did not include strict tests on longitudinal invariance, as provided by the stepwise comparison of nested confirmatory factor models (Millsap, 2011) or by testing for Differential Item Functioning (DIF; cf. Penfield & Camilli, 2006). Furthermore, the presence or absence of longitudinal invariance is not a yes/no decision; in addition, different models may yield different results (Blanchin et al., 2020). Violations of the assumption of longitudinal invariance are discussed in the literature as response shift (e.g., Leitgöb et al., 2023; Sawatzky et al., 2023; Sebille et al., 2021), i.e., patients change their concept of a disorder and the way they view their symptoms during therapy, and may thus get better in assessing their levels of depressive symptoms (Bulteau et al., 2019; Fokkema et al., 2013). Empirical tests of longitudinal invariance for the BDI-II are scarce, but one study found that factor loadings increased at the second time point in a large sample of inpatients (Elhai et al., 2013), which is in good agreement with the increased items' discriminating power found in our study. In a sample of depressed adolescents, there were also signs of response shift (Wu, 2016).

Reasons for Misclassification Rates The improvement rate of about 60% of the patients, together with less than 2% of the patients who deteriorated, reflects a very positive therapy outcome result. In terms of clinical significance (post BDI-II score < 14), 52.3% of the patients are in the category "no depression" at the end of their inpatient treatment. Note, however, that our sample does not include the dropouts, and 10.9% already started in the "no depression" category.

This clear preponderance of improved patients is very different from Hays et al. (2021) and Lee et al. (2023), who analyzed PROMIS® subscales. These authors found not much difference between percentages of improved and deteriorated patients, and on the group level, pre-/post-effect sizes were small and mostly below significance. One plausible reason for misclassification rates differing between CTT- and IRT-based RCI could be due to the higher percentage of change (i.e., many improved patients and a significant decrease in mean values in our study), so that the misclassifications could increase with pre-post score mean differences: For a large difference, at least one of the two scores must be in the upper or lower range, where the standard error (of an IRT-based model) is larger. Thus, the denominator of the RCI increases, which then will tend to not reach significance.

Secondly, Lee et al. (2023) discussed their own differing results to Hays et al. (2021) with regard to test length and sample score distribution. Both studies used short forms which consisted of 4 and 8 (Hays et al., 2021) vs. 8–9 items (Lee et al., 2023), which is still short compared to the BDI-II with 21 items. The score distribution determines how many persons are at the extremes of the scale where the standard error of the IRT-based model is usually larger; however, this pertains to both ends of the scale, i.e., not only the mean pre-/post-change is relevant but also the number of persons with very low or very high values.

Thirdly, we have to consider that the standard errors of the θ estimates depend on the threshold locations: If these cluster in the middle of the scale, we have much information there and small standard errors. Consequently, the standard errors in the margins increase. If, in contrast, the thresholds are scattered across a large range, the standard errors change to a lesser extent (compared from the middle to the margins), so that the biconcavity decreases (cf. Fig. 28.5). In such a case, the IRT-based changes will be more similar to the CTT-based ones, which results in less misclassification of IRT compared to CTT. Correspondingly, Hays et al. (2021) found no differences between classical and GRM-based RCI in a simulation study if there was no true underlying change. From Fig. 28.1, we learn that the thresholds of the BDI-II items are scattered from (approximately) –3 to 3, which seems to be a large interval. In contrast, Hays et al. (2021) report much smaller threshold ranges of the two PROMIS® subscales they analyzed. This could explain the lower misclassification rate in our study compared to Hays et al. (2021) and Lee et al. (2023). Hence, we have to keep in mind that comparing CTT-based and IRT-based RCIs will depend to a substantial extent also on the specific threshold configuration of the used instrument. For our results, at least, the influence of the broad threshold distribution seems to exceed the effect of the higher percentage of patients with reliable change. Furthermore, the practical relevance of discrepancies has to be considered: The classification decision for an individual as having improved is not very important in the low level of symptomatology, for example, for BDI-II scores below 14.

Other Approaches for Evaluating Change

In older treatment studies and in clinical practice, a quite simple method to assess meaningful change has been applied by defining successful change as a 50% improvement against the initial score. In a study on 395 depressive outpatients, Hiller et al. (2012) found slightly higher response rates for the 50% criterion compared to the classical RCI, as assessed with the BDI. The authors considered this criterion as a valuable extension of the established RCI and emphasized that the 50% approach for defining improvement is independent of arbitrarily chosen reliabilities and reference populations. The relation between improvement as subjectively perceived by the patient and the numerical reduction in symptomatology scores has also been studied for the BDI-II. Button et al. (2015) analyzed data from three large depression trials for the minimal clinically important difference (MCID). They found that the MCID was dependent from initial severity, thus percentage improvement was preferable to difference scores in the BDI-II. Note, however, that the MCID is a group-level statistic.

When dealing with a broad clinical sample, some practical problems with the RCI emerge: a significant improvement cannot occur if the pretest score is already low. Specifically, if the BDI-II pretest score < 10, no significant change is possible with an RCI-derived critical difference of 10. In contrast, IRT-based person parameter estimates are not linearly related to scores thus allowing for a more flexible assessment of change. A difference of less than 10 points in the lower score range might indeed reach the critical threshold. Hence, the differences in the cross-tabulations of frequencies of responders according to RCI_{CTT} and RCI_{IRT} might reflect the inability of the CTT-based measure to reflect changes in the low score region. A quite extreme example: A patient in our data set had an initial BDI-II score of 2 and a score of 0 at discharge. Obviously, this patient cannot improve according to the classical RCI, albeit fulfilling the criterion of 50% improvement. Surprisingly, the mentioned studies did not address this problem. This might be due to the fact that the patients in these studies were in the middle or at the upper end of the T-score metric used in PROMIS®, so that the floor effect did not become apparent. In the present study, 6.3% could not show any improvement in the RCI_{CTT} due to a BDI-II score of 10 or below at admission.

A further source of invalid score values comes from persons who produce questionable response patterns. IRT modeling allows to identify (and exclude) inconsistent response patterns, for example, via person-fit statistics (Niessen et al., 2016). However, Conijn et al. (2018) found that inaccurate patient self-reports in the BDI-II are not a major cause of patient-clinician discrepancy in ratings. Self-reported scores may also be influenced by the so-called response style (cf. Böckenholt & Meiser, 2017). Two relevant response styles (RS) are the tendency to prefer the extreme response categories (ERS) and the tendency to choose the middle categories (MRS). Keller and Koller (2015) found response styles to be quite stable from admission to discharge in inpatient treatment, and an association between diagnosis and response style, in that patients with a depression diagnosis tended to be more frequently in the MRS, while other diagnostic groups, for example, substance-related disorders or eating disorders, were overrepresented in the ERS. The addition of response style information, for example, by estimating different person parameters for the latent classes of ERS or MRS, may further improve the classification into improved or unchanged patients (Keller & Koller, 2015).

Limitations

Although the denominator in the classical RCI could be calculated with different terms, we remained for comparability with the simple standard deviation and Cronbach's alpha as reliability, both obtained from the pretest scores. Some studies use two SDs for the two measurement points and some also two reliability coefficients (e.g., Lee et al., 2023). Thus, the comparison to the RCI_{IRT} which uses pre- and posttest information might be not completely fair. However, in our study the SEM at admission (3.45) was close to the SEM at the discharge point (3.01, resulting from SD = 12.3 and alpha = .94) which seems comprehensible since the standard deviation at discharge usually remains about equal or gets somewhat larger and the reliability usually also gets minimally larger. Lee et al. (2023) found that their baseline and follow-up SEMs in six different scales were either very close or equal, and they mention the resulting advantage that a patient can be classified immediately after filling out the follow-up questionnaire, without waiting for the computation of the SEM in the full follow-up sample (which further supports the application of a prescore-based RCI).

Furthermore, relaxed significance criteria have been suggested—in particular, a one-tailed significance level (90%, i.e., $z = 1.645$; cf. Lee et al., 2023). Hays et al. (2021) propose that a considerable but nonsignificant change at $p < .05$ (two-tailed) might still be considered worth noting, and they propose a classification into five categories, with labeling better if passing the two-tailed test or probably better in one-tailed test (and analogously for worse and probably worse), which preserves more information. Even lower criteria (68%, 50%) have been suggested, labeling them as "likely change" (Peipert et al., 2023), but these low levels are only advantageous in conjunction with an anchor item expressing the patient global impression of change. In our additional analyses with one- vs. two-tailed classifications, the "stripes" in Figs. 28.3, 28.4, and 28.5 are rather small, but may be relevant in some cases.

Apart of these rather methodological intricacies, some more fundamental questions should also be addressed—in particular, whether between-subjects information is useful for individual change interpretation (Trigg et al., 2023). Peipert et al. (2023) argue that the measurement error based on group level (SD and reliability) still yields a valid individual-level interpretation and, given the high efforts to collect sufficient data to obtain a purely intraindividual statistic (i.e., variance within a person), is useful in their practicality. Generally speaking, intraindividual statistics and process variables would be preferable. There are indeed examples of promising applications in some (university) outpatient clinics or an initiative for improving access to psychotherapy in England. There, routine outcome monitoring on a session-to-session basis is applied to assess favorable response (see Lutz et al., 2022 for an overview and recent developments on dynamic modeling of change during treatment). However, due to cost and concerns about respondents' burden, in most inpatient clinics and rehabilitation centers as well as during standardized treatment trials, main diagnostic procedures take place at few time points, often at the beginning, after a fixed time interval during treatment and at the end of the stay.

Outlook

IRT-based RCIs have proven to be more flexible and improve the limited approach with the classical RCI. One general advantage of assessing reliable clinical change with IRT models is that these models allow for a more differentiated assessment of model fit, especially regarding malfunctioning items. Also, measurement error is addressed much more convincingly, taking into consideration the varying precision of the person parameters across the latent continuum. The overall agreement between classical RCI and IRT-based RCI seems high at first glance in our sample of inpatients in a psychosomatic clinic, being around 93–96% for the conventional RCI_{IRT} and around 90% for the two-dimensional RCI_{IRT}. However, regarding the classification into improved vs. unchanged, there were differences of around 10% depending on the chosen model. These differences depended also on the PCM vs. GRM/GPCM, and some got reverse in the case of the two-dimensional IRT models. These results are hard to interpret at the moment, and further exploration of different IRT models and calibration applied to other samples are certainly needed. However, Figs. 28.6 and 28.7 showed quite different sensitivity of the various modeling variants.

Finally, we would like to mention that this chapter merely deals with statistical aspects of assessing change. "Meaningfulness" could be inferred from the fact that a systematic factor should be responsible if a reliable difference occurred (cf. Trigg et al., 2023), but a clear content-based operationalization of meaningfulness would be superior. Recent research makes use of anchor items, i.e., mostly a patient-reported global rating of change, and thus increases information about significant and meaningful change in individuals (Lee et al., 2023; Peipert et al., 2023). This added information, however, requires valid anchor items and further statistical estimation procedures, usually based on ROC or predictive modeling. Some drawbacks of these methods are discussed in Bjorner et al. (2023), and the authors suggest a longitudinal IRT model with the inclusion of a transition rating. With regard to the evaluation of within-individual change, valid assessment of intraindividual change needs not only many time points, but the development of sufficiently reliable and valid process variables and good strategies to implement monitoring systems in routine practice (Lutz et al., 2022).

In conclusion, the definition of response is still a challenging task and seems important not only in clinical research and practice, but especially in predictor studies, for example, to predict future outcome from a variety of baseline variables. In our impression, sometimes hundreds of predictors are included in a model and analyzed with highly sophisticated methods, e.g., machine learning, but the left side of the equation, the response variable, receives much less attention and consideration. In light of our results, the decision on what model is used for defining reliable change may become important for such prediction studies.

Acknowledgments We would like to thank Jan R. Böhnke, Mark Stemmler, and Daniela Wetzelhütter for the thoughtful comments on an earlier version of this chapter and Ulrich Voderholzer and Adrian Meule for providing the data set.

References

Alexandrowicz, R. W. (2022). GMX: Extended graphical model checks. A versatile replacement of the plotGOF() function of eRm. *Psychological Test and Assessment Modeling, 64*, 215–225.

Alexandrowicz, R. W., Fritzsche, S., & Keller, F. (2014). Die Anwendbarkeit des BDI-II in klinischen und nichtklinischen Populationen aus psychometrischer Sicht. Eine vergleichende Analyse mit dem Rasch Modell. [Applicability of the BDI-II in clinical and non-clincal populations from a psychometric point of view. A comparative analysis with the Rasch model.]. *Neuropsychiatrie, 28*, 63–73. https://doi.org/10.1007/s40211-014-0104-z

American Psychiatric Association (1994). *Diagnostic and statistical manual of mental disorders (4^{th}ed.) (DSM IV)*. Washington, DC: American Psychiatric Press.

Baker, F. B., & Kim, S.-H. (2004). *Item response theory. Parameter estimation techniques*. Marcel Dekker.

Beck, A., Steer, R., & Brown, G. (1996). *Manual for the Beck depression inventory-II*. Psychological Corporation.

Bjorner, J. B., Terluin, B., Trigg, A., Hu, J., Brady, K. J. S., & Griffiths, P. (2023). Establishing thresholds for meaningful within-individual change using longitudinal item response theory. *Quality of Life Research, 32*, 1267–1276. https://doi.org/10.1007/s11136-022-03172-5

Blanchin, M., Guilleux, A., Hardouin, J.-B., & Sébille, V. (2020). Comparison of structural equation modelling, item response theory and Rasch measurement theory-based methods for response shift detection at item level: A simulation study. *Statistical Methods in Medical Research, 29*, 1015–1029. https://doi.org/10.1177/0962280219884574

Böckenholt, U., & Meiser, T. (2017). Response style analysis with threshold and multiprocess IRT models: A review and tutorial. *British Journal of Mathematical and Statistical Psychology, 70*, 159–181. https://doi.org/10.1111/bmsp.12086

Böhnke, J. R., & Lutz, W. (2014). Using item and test information to optimize targeted assessments of psychological distress. *Assessment, 21*, 679–693. https://doi.org/10.1177/1073191114529152

Bonifay, W., Lane, S. P., & Reise, S. P. (2016). Three concerns with applying a bifactor model as a structure of psychopathology. *Clinical Psychological Science, 5*, 184–186. https://doi.org/10.1177/2167702616657069

Brouwer, D., Meijer, R. R., & Zevalkink, J. (2013). Measureing individual significant change on the Beck Depression Inventory-II through IRT-based statistics. *Psychotherapy Research, 23*, 489–501. https://doi.org/10.1080/10503307.2013.794400

Bulteau, S., Sauvaget, A., Vanier, A., Vanelle, J.-M., Poulet, E., Brunelin, J., & Sebille, V. (2019). Depression reappraisal and treatment effect: Will response shift help improve the estimation of treatment efficacy in trials for mood disorders? *Frontiers in Psychiatry, 10*. https://doi.org/10.3389/fpsyt.2019.00420

Button, K. S., Kounali, D., Thomas, L., Wiles, N. J., Peters, T. J., Welton, N. J., et al. (2015). Minimal clinically important difference on the Beck Depression Inventory – II according to the patient's perspective. *Psychological Medicine, 45*, 3269–3279. https://doi.org/10.1017/S0033291715001270

Chalmers, R. P. (2012). mirt: A multidimensional item response theory package for the R environment. *Journal of Statistical Software, 48*(6), 1–29. https://doi.org/10.18637/jss.v048.i06

Cole, D. A., Cai, L., Martin, N. C., Findling, R. L., Youngstrom, E. A., Garber, J., et al. (2011). Structure and measurement of depression in youths: Applying item response theory to clinical data. *Psychological Assessment, 23*, 819–833. https://doi.org/10.1037/a0023518

Conijn, J. M., Emons, W. H. M., Page, B. F., Sijtsma, K., der Does, W. V., Carlier, I. V. E., & Giltay, E. J. (2018). Response inconsistency of patient-reported symptoms as a predictor of discrepancy between patient and clinician-reported depression severity. *Assessment, 25*(7), 917–928. https://doi.org/10.1177/1073191116666949

de Ayala, R. J. (2009). *The theory and practice of item response theory*. Guilford.

Duff, K. (2012). Evidence-based indicators of neuropsychological change in the individual patient: Relevant concepts and methods. *Archives of Clinical Neuropsychology, 27*, 248–261. https://doi.org/10.1093/arclin/acr120

Elhai, J., Contractor, A., Biehn, T., Allen, J., Oldham, J., Ford, J., et al. (2013). Changes in the Beck Depression Inventory-II's underlying symptom structure over 1 month of inpatient treatment. *Journal of Nervous and Mental Disease, 201*, 371–376. https://doi.org/10.1097/NMD.0b013e31828e1004

Estrada, E., Ferrer, E., & Pardo, A. (2019). Statistics for evaluating pre-post change: Relation between change in the distribution center and change in the individual scores. *Frontiers in Psychology, 9*. https://doi.org/10.3389/fpsyg.2018.02696

Evans, C., Margison, F., & Barkham, M. (1998). The contribution of reliable and clinically significant change methods to evidence-based mental health. *Evidence-Based Mental Health, 1*, 70–72. https://doi.org/10.1136/ebmh.1.3.70

Ferrer, R., & Pardo, A. (2014). Clinically meaningful change: False positives in the estimation of individual change. *Psychological Assessment, 26*, 370–383. https://doi.org/10.1037/a0035419

Ferrer, R., & Pardo, A. (2019). Clinically meaningful change: False negatives in the estimation of individual change. *Methodology, 15*, 97–105. https://doi.org/10.1027/1614-2241/a000168

Fischer, G. H., & Molenaar, I. W. (Eds.). (1995). *Rasch models. Foundations, recent developments, and applications*. Springer.

Fokkema, M., Smits, N., Kelderman, H., & Cuijpers, P. (2013). Response shifts in mental health interventions: An illustration of longitudinal measurement invariance. *Psychological Assessment, 25*, 520–531. https://doi.org/10.1037/a0031669

Hartig, J., & Höhler, J. (2009). Multidimensional IRT models for the assessment of competencies. *Studies in Educational Evaluation, 35*, 57–63. https://doi.org/10.1016/j.stueduc.2009.10.002

Hautzinger, M., Keller, F., & Kühner, C. (2006). *BDI-II. Beck Depressions Inventar Revision – Manual*. Harcourt Test Services.

Hays, R. D., & Peipert, J. D. (2021). Between-group minimally important change versus individual treatment responders. *Quality of Life Research, 30*, 2765–2772.

Hays, R. D., Spritzer, K. L., & Reise, S. P. (2021). Using item response theory to identify responders to treatment: Examples with the patient-reported outcomes measurement information system (PROMIS®) physical function scale and emotional distress composite. *Psychometrika, 86*, 781–792. https://doi.org/10.1007/s11336-021-09774-1

Hiller, W., Schindler, A. C., & Lambert, M. J. (2012). Defining response and remission in psychotherapy research: A comparison of the RCI and the method of percent improvement. *Psychotherapy Research, 22*(1), 1–11. https://doi.org/10.1080/10503307.2011.616237

Holzinger, K. J., & Harman, H. H. (1938). Comparison of two factorial analyses. *Psychometrika, 3*, 45–60. https://doi.org/10.1007/BF02287919

Jabrayilov, R., Emons, W. H., & Sijtsma, K. (2016). Comparison of classical test theory and item response theory in individual change assessment. *Applied Psychological Measurement, 40*, 559–572. https://doi.org/10.1177/0146621616664046

Jacobson, N. S., & Truax, P. (1991). Clinical significance: A statistical approach to defining meaningful change in psychotherapy research. *Journal of Consulting and Clinical Psychology, 59*, 12–19.

Kabic, M., & Alexandrowicz, R. W. (2023). RMX/PIccc: An extended person-item-map and a unified IRT output for eRm, psychotools, ltm, mirt, and TAM. *Psych, 5*, 948–965. https://doi.org/10.3390/psych5030062

Keller, F., & Koller, I. (2015). Mixed Rasch models for analyzing the stability of response styles across time: An illustration with the Beck Depression Inventory (BDI-II). In M. Stemmler, A. von Eye, & W. Wiedermann (Eds.), *Dependent data in social sciences research: Forms, issues, and methods of analysis* (pp. 309–324). Springer. https://doi.org/10.1007/978-3-319-20585-4_13

Keller, F., Kirschbaum-Lesch, I., & Straub, J. (2020). Factor structure and measurement invariance across gender of the Beck Depression Inventory-II in adolescent psychiatric patients. *Frontiers in Psychiatry, 11*. https://doi.org/10.3389/fpsyt.2020.527559

Keller, F., Kühner, C., Alexandrowicz, R. W., Voderholzer, U., Meule, A., Fegert, J. M., et al. (2022). Zur Messqualität des Beck-Depressionsinventars (BDI-II) in unterschiedlichen klinischen Stichproben. Eine Item-Response-Theorie Analyse. [The measurement quality of the Beck Depression-Inventory (BDI-II) in different clinical samples: An item response theory analysis]. *Zeitschrift für Klinische Psychologie und Psychotherapie, 51*, 234–246. https://doi.org/10.1026/1616-3443/a000676

Kuehner, C., Bürger, C., Keller, F., & Hautzinger, M. (2007). Reliabilität und Validität des deutschen Beck Depressions Inventars II (BDI-II). Befunde aus deutschsprachigen Stichproben. [Reliability and validity of the Revised Beck Depression Inventory (BDI-II). Results from German samples]. *Der Nervenarzt, 78*, 651–656. https://doi.org/10.1007/s00115-006-2098-7

Kuehner, C., Keller, F., Schricker, I., Beddig, T., Huffziger, S., Timm, C., et al. (2023). Diagnostische Performanz und Validität des deutschsprachigen BDI-II: Eine Sekundäranalyse mit Daten aus klinischen und nichtklinischen Stichproben [Diagnostic performance and validity of the German version of the BDI-II ? A secondary analysis with data from clinical and nonclinical samples]. *Psychiatrische Praxis, 50*, 36–42. https://doi.org/10.1055/a-1753-2298

Lee, M. K., Peipert, J. D., Cella, D., Yost, K. J., Eton, D. T., Novotny, P. J., et al. (2023). Identifying meaningful change on PROMIS short forms in cancer patients: A comparison of item response theory and classic test theory frameworks. *Quality of Life Research, 32*, 1355–1367. https://doi.org/10.1007/s11136-022-03255-3

Leitgöb, H., Seddig, D., Asparouhov, T., Behr, D., Davidov, E., De Roover, K., et al. (2023). Measurement invariance in the social sciences: Historical development, methodological challenges, state of the art, and future perspectives. *Social Science Research, 110*, 102805. https://doi.org/10.1016/j.ssresearch.2022.102805

Lutz, W., Schwartz, B., & Delgadillo, J. (2022). Measurement-based and data-informed psychological therapy. *Annual Review of Clinical Psychology, 18*(1), 71–98. https://doi.org/10.1146/annurev-clinpsy-071720-014821

Masters, G. N. (1982). A Rasch Model for partial credit scoring. *Psychometrika, 47*(2), 149–174.

Millsap, R. E. (2011). *Statistical approaches to measurement invariance*. Routledge.

Muraki, E. (1992). A generalized partial credit model: Application of an EM algorithm. *Applied Psychological Measurement, 16*(2), 159–176. https://doi.org/10.1177/014662169201600206

Murray, A. L., Vollmer, M., Deary, I. J., Muniz-Terrera, G., & Booth, T. (2021). Assessing individual-level change in dementia research: A review of methodologies. *Alzheimer's Research & Therapy, 13*. https://doi.org/10.1186/s13195-021-00768-w

Niessen, A. S. M., Meijer, R. R., & Tendeiro, J. N. (2016). Detecting careless respondents in web-based questionnaires: Which method to use? *Journal of Research in Personality, 63*, 1–11. https://doi.org/10.1016/j.jrp.2016.04.010

O'Connor, B. P. (2018). An illustration of the effects of fluctuations in test information on measurement error, the attenuation of effect sizes, and diagnostic reliability. *Psychological Assessment, 30*, 991–1003. https://doi.org/10.1037/pas0000471

Olsbjerg, M., & Christensen, K. B. (2015). Modeling local dependence in longitudinal IRT models. *Behavior Research Methods, 47*, 1413–1424. https://doi.org/10.3758/s13428-014-0553-0

Ostini, R., & Nering, M. L. (2006). *Polytomous item response theory models*. Sage.

Paek, I., Park, H.-J., Cai, L., & Chi, E. (2014). A comparison of three IRT approaches to examinee ability change modeling in a single-group anchor test design. *Educational and Psychological Measurement, 74*, 659–676. https://doi.org/10.1177/0013164413507062

Peipert, J., Hays, R., & Cella, D. (2023). Likely change indexes improve estimates of individual change on patient-reported outcomes. *Quality of Life Research, 32*, 1341–1352. https://doi.org/10.1007/s11136-022-03200-4

Penfield, R. D., & Camilli, G. (2006). Differential item functioning and item bias. In C. R. Rao & S. Sinharay (Eds.), *Handbook of statistics* (Vol. 26, pp. 125–167). Elsevier.

R Core Team. (2022). *R: A language and environment for statistical computing [Computer software manual]*, Vienna, Austria. Retrieved from https://www.R-project.org/

Reeve, B. B., & Hays, R. D. (2021). Guest editors' introduction to the invited special section. *Psychometrika, 86*, 671–673. https://doi.org/10.1007/s11336-021-09795-w

Revicki, D., Hays, R. D., Cella, D., & Sloan, J. (2008). Recommended methods for determining responsiveness and minimally important differences for patient-reported outcomes. *Journal of Clinical Epidemiology, 61*, 102–109. https://doi.org/10.1016/j.jclinepi.2007.03.012

Ronk, F. R., Korman, J. R., Hooke, G. R., & Page, A. C. (2013). Assessing clinical significance of treatment outcomes using the DASS-21. *Psychological Assessment, 25*, 1103–1110. https://doi.org/10.1037/a0033100

Samejima, F. (1969). Estimation of latent ability using a response pattern of graded scores. *Psychometrika, 34*, 1–97. https://doi.org/10.1007/BF02290599

Samejima, F. (1994). Estimation of reliability coefficients using the test information function and its modifications. *Applied Psychological Measurement, 18*, 229–244. https://doi.org/10.1177/014662169401800304

Sawatzky, R., Sajobi, T. T., Russell, L., Awosoga, O. A., Ademola, A., Böhnke, J. R., et al. (2023). Response shift results of quantitative research using patient-reported outcome measures: A descriptive systematic review. *Quality of Life Research*. https://doi.org/10.1007/s11136-023-03495-x

Sebille, V., Lix, L. M., Ayilara, O. F., Sajobi, T. T., Janssens, A. C. J. W., Sawatzky, R., et al. (2021). Critical examination of current response shift methods and proposal for advancing new methods. *Quality of Life Research, 30*, 3325–3342. https://doi.org/10.1007/s11136-020-02755-4

Stochl, J., Fried, E., Fritz, J., Croudace, T., Russo, D., Knight, C., et al. (2022). On dimensionality, measurement invariance, and suitability of sum scores for the PHQ-9 and the GAD-7. *Assessment, 29*, 355–366. https://doi.org/10.1177/1073191120976863

Thissen, D., & Steinberg, L. (1986). A taxonomy of item response models. *Psychometrika, 51*, 567–577. https://doi.org/10.1007/BF02295596

Tingey, R., Lambert, M. J., Burlingame, G., & Hansen, N. (1996). Assessing clinical significance: Proposed extensions to method. *Psychotherapy Research, 6*(2), 109–123. https://doi.org/10.1080/10503309612331331638

Toland, M. D., Sulis, I., Giambona, F., Porcu, M., & Campbell, J. M. (2017). Introduction to bifactor polytomous item response theory analysis. *Journal of School Psychology, 60*, 41–63. https://doi.org/10.1016/j.jsp.2016.11.001

Trigg, A., Lenderking, W., & Boehnke, J. (2023). Introduction to the special section: "Methodologies and considerations for meaningful change". *Quality of Life Research, 32*, 1223–1230. https://doi.org/10.1007/s11136-023-03413-1

Wang, Y.-P., & Gorenstein, C. (2013). Psychometric properties of the Beck Depression Inventory-II (BDI-II): A comprehensive review. *Brazilian Journal of Psychiatry, 35*, 416–431. https://doi.org/10.1590/1516-4446-2012-1048

Wang, C., & Nydick, S. W. (2020). On longitudinal item response theory models: A didactic. *Journal of Educational and Behavioral Statistics, 45*, 339–368. https://doi.org/10.3102/1076998619882026

Wise, E. A. (2004). Methods for analyzing psychotherapy outcomes: A review of clinical significance, reliable change, and recommendations for future directions. *Journal of Personality Assessment, 82*, 50–59. https://doi.org/10.1207/s15327752jpa8201_10

Wu, P.-C. (2016). Response shifts in depression intervention for early adolescents. *Journal of Clinical Psychology, 72*(7), 663–675. https://doi.org/10.1002/jclp.22291

Part VII
Other Methods for the Analysis of Dependent Data

Chapter 29
Assessing Unobserved Within-Group Individual Differences

Hoben Thomas

29.1 Easily Observed, Often Difficult to Conceptualize

The recognition that there are often large individual differences among living organisms in skills, abilities, and traits must go back to the dawn of human awareness. Observations from the modern world such as kicking soccer goals, playing cards, spelling words, solving math problems, or playing Haydn string quartets make clear there is no disputing the fact that different individuals, or groups of individuals, display different levels of skill for any task or activity of focus.

Our penchant for quantifying matters means that by some invented procedure, these observations are mapped into numbers. A judge's rating of a gymnast's parallel bar performance, a patient's blood pressure, one's body mass index, a grocer assigning prices to products, and achievement test scores are common examples. Such data are often regarded as realized observations $x_1, x_2, \ldots, x_n$ on random variables $X_1, X_2, \ldots, X_n$. And often, multiple observations on different individuals are regarded as a random sample, which means that all observations are assumed to have been independently sampled from the same probability distribution, typically an assumed normal distribution.

One often arrives at this point without much reflection. And when doing so, this usually ends matters, and standard software procedures are often applied. Occasionally, some feature of the data stream or some unusual observation is noted as the data are gathered, which may suggest some rather different possibility, that perhaps not all observations may be similarly and simply characterized: Namely, some data heterogeneity often unanticipated may have arisen. Unless there is some oddity observed at the time of data collection, or some peculiarity in the observed

H. Thomas (✉)
Pennsylvania State University, University Park, PA, USA
e-mail: hxt@psu.edu

M. Stemmler et al. (eds.), *Dependent Data in Social Sciences Research*,
https://doi.org/10.1007/978-3-031-56318-8_29

data stream, the thought of important unobserved within group heterogeneity usually doesn't occur. And this result is unfortunate and often misleading. The emphasis on effect-size indices, such as d, does not help matters. Such approaches leave no room for conceptualizing individual differences.

The author's first awareness of within-sample heterogeneity, and subsequently a concern in *all* matters of data collection and research considerations, occurred decades ago. Student assistants, while gathering data on children's psychophysical task manipulations to gauge a child's understanding of Piaget's water-level problem, exclaimed that remarkable differences appeared among different children in their task responses. The dramatically different responses of same-aged and same-grade children with similar background profiles clearly implied some within group heterogeneity that had not been expected. Upon reflection, the fact that individual children were required to adjust an apparatus, while directly observed as they did so, was probably key. Such differences are unlikely to be noted when data are collected in group-based assessment settings where an individual's task performance cannot be easily observed, and especially when data are subsequently machine scored.

This chapter is intended as a brief introduction to finite mixtures as an approach to the problem of identifying unobserved within-group heterogeneity. A mixture perspective assumes the distribution of scores follows not a single (often normal) distribution but rather a weighted sum of different probability distributions. Let $f(x)$ be the distribution of observed or manifest scores. Then, a finite mixture distribution has latent unobserved distributions $f_r(x)$, $r = 1, \ldots, R$ with

$$f(x) = \sum_{r=1}^{R} \lambda_r f_r(x) \text{ with } 0 < \lambda_r < 1 = \sum \lambda_r, \tag{29.1}$$

where $f_r(x)$ are called component distributions and λ_r weights or probabilities; (1) is called a R component finite mixture distribution. Most often the components are assumed to be different normal distributions. Each response x is assumed to come from one of the R components with probability λ_r.

$f(x)$ in (1) *is* a probability distribution. So, one may think of a random sample of individuals as "coming from" different components of the mixture distribution. From an "intuitive every-day working research" perspective, (1) seems more realistic than a perspective which assumes all individuals follow the same identical typically normal distribution so often pictured on book covers, book jackets, or commercial ads.

A mixture approach is no panacea for addressing problems of individual differences, as will be made clear, but it does move one closer, in many cases, to reality and what one ultimately desires. That reality, from an individual differences perspective, is that each person has their own perhaps unique characteristic profile, here represented in a person's own probability distribution for the task of focus. But this possibility is not from a probability model parameter estimation perspective typically possible, nor is it scientifically desirable. That is because one goal is to simplify understanding of observations with compact simple models with few unknown model parameters. So a mixture model with some small number of components is likely the only practical possibility achievable, and may be viewed as striking a more appropriate balance.

There are several issues to be addressed in any mixture application. The next section discusses what are regarded as important problems to address in any mixture model application. But before doing so, it seems important to define exactly how individuals are conceptualized as being different:

Individuals are different if and only if their responses follow different probability distributions.

The definition is nothing new but makes explicit how individual differences are viewed throughout. Note it does *not* imply that individuals are different if, say, they have different achievement test scores. Rather, individuals are different if they reside in different probability distributions. Individuals may have identical test scores but reside in different probability distributions. Both boys' and girls' math and reading test score distributions appear well modeled by two component mixture distributions. So, for example, two girls with identical reading scores may well come from different component reading distributions, and similarly for boys (Thomas, 2024).

The two-sample t-test provides a useful framework for thinking about both manifest and latent settings and their contrasting concerns. Think of the two-sample t-test as an estimation, not hypothesis testing methodology, with interest in estimating the two population means. Assume the two population means are different. And suppose that the samples from these two populations were mixed up, so that it is unknown from which population an observation was taken. This is the spirit of a mixture distribution. Thus, from a latent variable mixture distribution perspective, there are $R = 2$ subpopulations, both normal in distribution, with common variance (a two-sample t test assumption) and differing only in mean. Thus, $f_r(x), r = 1, 2$ are these latent components representing the two subpopulations, while observations come from the manifest $f(x)$ distribution.

This perspective can be made real, by considering the data in Table 29.1. Displayed are measured, not self-reported, heights of 59 men and 67 women of college age. Suppose the labels identifying each individual's sex have been lost. The goal is still, like the two-sample t test, to estimate the means of men and women but with heights unidentified as to sex. Before revealing how well a mixture approach retrieves this information, consider the issues any application of mixture models must address. The focus is on the simple univariate setting, but the issues apply most generally to other mixture distribution settings as well.

Table 29.1 Heights of men and women

	Men		Women	
1	65.500	72.000	55.000	66.000
2	65.750	72.250	60.000	66.250
3	66.000	72.250	60.250	66.250
4	66.250	72.250	61.000	66.250
5	66.750	72.375	61.750	66.500
6	67.000	72.500	62.250	66.750
7	67.125	72.500	62.250	66.750
8	67.250	72.875	62.625	66.750
9	68.000	73.000	62.750	66.750
10	68.125	73.000	63.000	67.000
11	68.750	73.000	63.250	67.125
12	68.750	73.250	63.250	67.250
13	69.125	73.250	63.250	67.375
14	69.250	73.625	63.250	67.500
15	69.250	74.000	63.375	67.500
16	69.375	74.625	64.000	67.750
17	69.750	75.000	64.250	67.750
18	70.000	75.000	64.250	67.750
19	70.125	75.050	64.500	68.125
20	70.500	76.000	64.750	68.750
21	70.750	77.000	64.750	69.000
22	70.750	77.125	65.000	69.000
23	71.000	77.500	65.000	69.250
24	71.000	77.500	65.125	69.500
25	71.500	78.500	65.125	69.875
26	71.500		65.167	70.000
27	71.625		65.250	70.000
28	71.750		65.250	70.250
29	71.750		65.250	70.375
30	72.000		65.250	71.000
31	72.000		65.250	71.000
32	72.000		65.500	71.250
33	72.000		65.750	71.750
34	72.000		66.000	

29.2 Central Concerns in Mixture Model Approaches

Are There Subpopulations and If So How Many?

In this height "lost label" example, the answer of course is $R = 2$, the subpopulation of men and the subpopulation of women. In the most interesting problems, R is at best uncertain. How many strategies do different children employ in solving six class inclusion problems (Thomas and Horton, 1997)? Each different strategy is assumed to result in a different probability distribution. It is assumed each child employs the same strategy for each problem, regarded as replicated within-child trials. But of course one can imagine a child employing different strategies on different trials. In fact, specifying and justifying R the number of subpopulations is perhaps the most difficult problem confronted when viewing individual differences within a latent finite mixture setting. While R may be estimated, as will be noted below, the situation when doing so is far more challenging. It is best to have some theoretical or empirical basis for specifying R.

What Is the Shape of Each of the Subpopulations?

This issue concerns the shape of the probability distribution associated with each of the subpopulations, the $f_r(x)$. In most mixture settings, the distribution associated with each subpopulation must be specified. The most popular specification, at least if the data are viewed as continuously distributed, is to assume each subpopulation has a normal distribution. Yet this assumption is often difficult to justify, because there is often a lack of theory or evidence to support the assumption. In the case of heights of men and women, the assumption that the heights of each sex follow, at least roughly, a normal distribution is unlikely to encounter critics. But in psychologically challenging (but interesting) cases, the answer is perhaps *never* obvious. Consider these examples.

In Piaget's water-level task, a common concern is when children understand that still water, within a closed stationary vessel, is invariantly horizontal, and the strategies children employ in route to this understanding. In fact, some adults lack this understanding. In most assessments, children are presented with a test booklet, each page of which pictures a vessel, usually a rectangularly shaped vessel or bottle with cap (e.g., a shampoo bottle, or beverage bottle) with different pages portraying vessels tilted to a different angle, often indexed by its clock-hour position, with the clock face as the reference standard. For example, a 30° tilt to the right is at 1 o'clock. Children are asked to imagine the vessel is partly filled with liquid (such as a beverage). They are asked to draw a line representing their guess as to the position of the surface of liquid in the vessel. If the vessel is upright as if set on a table, or 12 o'clock position, rarely do grade school children draw other than roughly horizontal lines, the correct response. If the vessel is tilted to the 10 o'clock position, a horizontal line is the correct response, but there likely is, with any group of children, a range of drawing errors: Drawn lines may be parallel to vessel's side, i.e., 10 o'clock line position, or parallel the vessel's bottom, the 1 o'clock position. Different vessel orientations elicit different distributions of responses. In all cases of course, a horizontal line is the correct response. How does one assess the cognitive strategy associated with different line drawn responses? And what distributional shapes seem appropriate for these different responses? Thus, the task is to specify the shape of each $f_r(x)$. How does one proceed to do so? An answer to these questions motivates an approach to be considered below, called the cut-point model implemented first in Thomas and Lohaus (1993).

As another example, consider adolescents being asked to specify the number of delinquent acts they committed in the last year. Answers among hundreds of adolescents ranged from 0 to 15 (Wallner et al., 2021). One might initially think that integer-valued responses varying from 0 to 15 in sufficient numbers are sufficiently continuous (although obviously discrete) to assume continuity, and perhaps normality. However, one should be quickly disabused of the correctness this belief after self-reflection: Are not those individuals, in fact the majority, responding zero perhaps rather different adolescents from those with some positive number of responses? A second answer comes from examination of frequency distributions of the scores, which are essentially L-shaped in distribution (Wallner et al., 2021, p. 7) due to the large number of zero responses, thus precluding normal distributions as being plausible distributions. The answer to the shape question is almost always difficult with one's chosen answer more likely reflecting one's faith than empirical or conceptual reality.

How Are Subpopulations Functionally Related to One Another?

This question addresses the heart of how nature or experimental influences exert influence or "work" in the world: That is, what model or idea characterizes, specifically, how the different $f_r(x)$ are functionally related to each other? Nature will never be perfectly modeled, but given its key importance, it is remarkably rare to find even mention of the issue. The showcase model for how influences are typically assumed to be expressed in subpopulations, at least for continuous univariate data settings, is again the two-sample t-test: Both distributions are assumed to be the same normal distribution shape, with (assuming the two distributions are different) one distribution shifted, right or left, relative to the other distribution. Hence, the two-sample t test is a *location-shift* model and similarly for latent distributions.

Nature doesn't, of course, first consult textbooks on experimental design, before acting! And it seems naive to assume location shift is what distinguishes all group differences, although this is the default mentality whether one views the matter from a latent variable perspective or not. That this perspective can be dramatically wrong is illustrated by reading and math test score distributions of boys and girls. Sex differences in test score distributions are not properly modeled by a location-shift model, and have not been since the origins of testing more than 100 years ago (Thomas, 2024). One obvious reason is that the variances for the two sexes are quite different, far more different than the mean differences, thus falsifying the location-shift model. Note that in meta-analysis, effect-size d's model is a scaled location-shift model and only captures the state of the world if it is believed group differences are different only in their means.

One reason for the conceptual dominance of location-shift perspectives, especially in normal mixtures, is that computation is simplified, as will be illustrated shortly. And historically at least, unequal variances have been theoretically difficult for statistical models.

29.3 Two Mixture Model Approaches

The Height "Lost Label" Problem: A Standard Normal Mixture Solution

R is used throughout (R Core Team, 2020), and the R package `mixtools` must be loaded: `library(mixtools)`. It can be obtained from

`https://CRAN.R-project.org/package=mixtools`

Concatenate all observations in Table 29.1 into one vector `xv`. In the present case:

```
xv<-c(maleht,femaleht)
```

Then, with a single line of code, execute:

```
output<-normalmixEM(xv,lambda=c(.5,.5),arbvar=FALSE)
```

This command executes estimates of the model of Eq. (1) where $R = 2$ and the two component distributions $f_r(x), r = 1, 2$ are normal distributions with common standard deviation but with possibly different means. If one is lucky, the following output with perhaps small numerical differences will appear:

```
summary.mixEM(output)
summary of normalmixEM object:
          comp 1    comp 2
lambda  0.551141  0.448859
mu     66.482749 71.078012
sigma   3.453251  3.453251
loglik at estimate:  -357.8088
```

Because there is stochastic variation associated with the starting solution, alternatively one might get:

```
summary of normalmixEM object:
          comp 1     comp 2
lambda  0.0521422  0.947858
mu     64.4450708 68.770933
sigma   4.0278992  4.027899
loglik at estimate:  -357.7881
```

Or perhaps something quite different from either of the above two different solutions given the same identical input. `lambda` is the estimated component weight vector of λ_1 and λ_2, while `mu` is the estimated corresponding mean vector with estimates of μ_1 and μ_2, the population means; `sigma` is the estimate of the assumed common standard deviation σ. A popular criterion for selecting the most appropriate solution is to select the solution with largest loglikelihood, here `loglik`. The second solution above satisfies this criterion, by a small amount. However, this solution is

Table 29.2 Sample values and their corresponding mixture solutions

	$\bar{x}$	$\hat{\mu}$	s	$\hat{\sigma}$	$n/126$	$\hat{\lambda}$	n
Men	71.520	71.078	3.100	3.453	0.468	0.449	59
Women	65.926	66.483	3.069	3.453	0.532	0.551	67

far from the truth because the first component `lambda` estimate suggests that the sample has only about 5% of one sex and is clearly wrong. This reveals a critically important point: Human judgment is required in any solution accepted as plausible. Note, too, one must also recognize that men are taller than women so as to match which column, `comp 1` or `comp 2`, is for men, and which is for women. Mixture solutions are not automatic avenues to "truth."

The first solution above is about as good as one will get, as repeated implementation of the iterative algorithm will reveal. This solution does retrieve parameter estimates which align with the known sample quantities the sample mean, $\bar{x}$ and sample standard deviation, s and n rather well, as Table 29.2 reveals.

In the function `normalmixEM`, `lambda=c(.5,.5)` specifies the component weights, that is, the estimates of $\lambda_r, r = 1, 2$ or probability starting values, and specifies as well that two components in the mixture solution are desired. `arbvar=FALSE` specifies that both component distributions are to have the same variance values.

The starting values can be important. As an example, a solution obtained when `lambda=c(.05,.95)` replaced `lambda=c(.5,.5)` was:

```
summary of normalmixEM object:
          comp 1     comp 2
lambda  0.00727339  0.992727
mu     56.81940970 68.631285
sigma   4.01763861  4.017639
loglik at estimate:   -357.5231
```

This is a seriously misleading solution, but has a larger `loglik`, the loglikelihood estimate, than the best solution. It is important to implement the procedure multiple times, sometimes varying the initial values of the parameter guesses, so that the selected solution is the one which yields the largest loglikelihood—with recognition that the solution selected is the most plausible as well. The function `normalmixEM` can take many other arguments. These are detailed in the manual `mixtools.pdf` and are available at `https://cran.r-project.org`.

Observe the solutions above are location-shift solutions, that is, the population distributions are both normal with the same variance. There is reason to believe the sexes differ in height variance as well. It is well recognized that in many tasks, the variance of males exceeds the variance of females. The sample variances from Table 29.2 are consistent with this finding, but differ modestly by about 0.2, which is perhaps small enough to ignore. But if these data were a real mixture problem, and where knowledge of which subpopulation observations came from was unknown, one might allow for the variances for each subpopulation to be different.

Mixtures of normal distributions where the variances are allowed to be different can be analytically challenging, especially when the subpopulation variances may be very similar, as this case seems to suggest. The result can be unstable and unreliable solutions. The difficulty is revealed by implementing the call `normalmixEM(xv,lambda=c(.5,.5),arbvar=TRUE)` multiple times. Even when the `lambda` proportions of men and women are close to the sample values, quite varying solutions are often the result. A discussion of `normalmixEM` when the variances are allowed to vary with the argument `arbvar=TRUE` is given in the `mixtools` manual.

The Cut-Point Model for Repeated Measures Settings

A key difficulty with finite mixtures being a convincing methodological approach, as noted above, is that simply because latent processes are unobserved, there is often little theory or empirical evidence to guide the selection of appropriate mixture models. In particular, the shape of the unobserved component distributions is a serious concern. And of course, how many such subpopulations or R components are there? The issue can be especially vexing in many settings involving children, with often remarkably different responses of different children which are seemingly quite similar otherwise.

If one is fortunate to have multiple observations, on each individual, often called repeated measures, the component distribution problem can often be addressed in a remarkably simple way. Ideally one would like such multiple responses on each individual to be independent and identically distributed, that is, iid, conditional on any subpopulation that might be present. That is, within a subpopulation the individual's observations are a random sample. Analytical approaches to the problem typically require this structure. However, these structural requirements are rarely achieved, at least where children are involved, especially if multiple measures are obtained during the same assessment setting, on identical or presumably psychologically similar instruments.

It will be assumed, in the following, that a model based on this ideal, at least, holds approximately, and is not egregiously wrong. One often makes this assumption that among different individuals, they all follow the same distribution (most typically a normal distribution), and if multiple observations are made on individuals, they are often treated as if the individual's repeated values are iid, even if they are not. The perspective is hardly new, but infrequently noted in print.

The fundamental idea of the cut-point approach is an elementary analytical fact:

Any nondegenerate (i.e., not a point mass) distribution can be transformed into a Bernoulli distribution.

To illustrate how this can be achieved, consider Fig. 29.1. It displays a density estimation plot based on water-level responses of 39 girls. Their task was to draw a line representing the surface of water in a vessel tilted to the 10'oclock position.

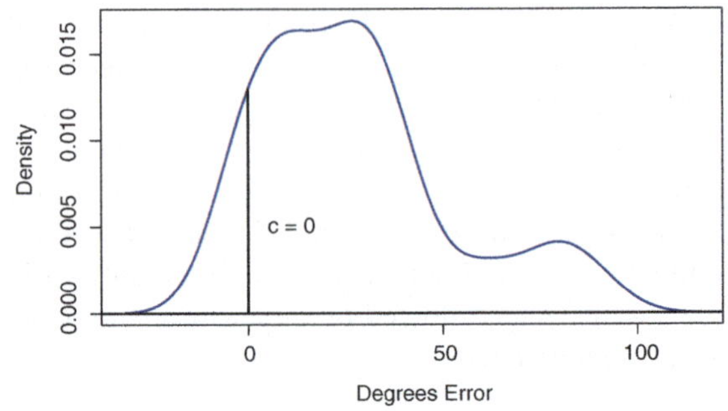

Fig. 29.1 Water-level responses, $n = 39$, girls with 10 o'clock position

The data are drawn line angular deviations from the horizontal, where the horizontal is defined as zero degrees. Here the cut-point $c = 0$ corresponds with a horizontal line, the correct response.

Denote the underlying distribution in Fig. 29.1 as $f(x)$ some unknown distribution where x is a the water-level response. The left tail area or proportion up to the ordinate at $c = 0$ is $\int_{-\infty}^{0} f(x)dx = \theta$ with $0 < \theta < 1$, while the right tail proportion above zero is $\int_{0}^{\infty} f(x)dx = 1 - \theta$ and θ is unknown.

Now define a new random variable Y. Let $Y = 1$ if $x < 0$ while $Y = 0$ if $x > 0$. Then, Y is a binary random variable with $P(Y = 1) = \theta$ and $P(Y = 0) = 1 - \theta$. So Y has a Bernoulli zero or one outcome distribution. Of course the cut point need not be $c = 0$ but could be any value within the range of outcome values.

The idea of creating a defined random variable for data analysis by coding a child's response is of course nothing new. But the cut point creates a known distribution. This general approach was first conceptualized in Thomas and Lohaus (1993) where an interval, not a cut, formed the partition. Suppose $a < b$ and a and b are values typically well within the range of the outcomes observed, but need not be possible outcome values. Then, if a water-level response $x \in (a, b)$, $Y = 1$ while $x \notin (a, b)$ $Y = 0$. Water-level responses were defined in this way with $0 \in (a, b)$. However, subsequently, a cut-point perspective, such as portrayed in Fig. 29.1, has been used. Thus, *any* distribution can be transformed in this way, either discrete or continuous, provided it is not a mass point distribution, that it is not of the form $P(Y = y) = 1$.

Now suppose there are m within-subject repeated measures for a child. Then, if the m responses are iid, the $Y_1, Y_2, \ldots, Y_m$ have a binomial $b(y; \theta, m)$ distribution for a child. Furthermore, an estimate of a child's θ is the $\{\text{number of } x < 0\}/m = \hat{\theta}$. Suppose there are $i = 1, 2, 3, \ldots, n$ individuals, each with m repeated measures. If all n subjects and their m responses follow the same distribution, then the n estimates of $\hat{\theta}$ should all be similar. However, if it is found that the $\hat{\theta}$ substantially vary among different children, this finding suggests that there are $S > 1$ response distributions or subpopulations among the children, with $s = 1, 2, \ldots, S$ components, each with θ_s probability, and thus an S component mixture distribution:

$$g(y) = \sum_{s=1}^{S} \lambda_s b(y, \theta_s, m). \tag{29.2}$$

Equation (2) is an S component binomial mixture distribution, and has the same structure as Eq. (1) with binomial distributions as the component distributions for the new variable Y. Desirable is that S be specified from theory. Otherwise, S must be estimated. To be identifiable, it must be that $m \geq 2S - 1$. The above development has been largely intuitive, but there is a more formal development (Hettmansperger and Thomas, 2000).

The conceptual advantage of the cut-point approach is that it eschews the problematic issue of specifying the shape of $f_r(x)$ latent components. It also eschews the important matter of specifying a model for how the subpopulations differ, such as normal model location-shift difference assumed as the model for height sex differences. But just how these latent components differ will impact the efficiency of the cut-point model.

The application of the cut-point approach is easily implemented with the routines in the `mixtools` R package, as will be seen shortly. But the approach can do more. It can estimate the shape of the observed distribution $f(x)$ at least often reasonably so when x is continuous, as well as the components of $f(x)$, $f_r(x)$. To see how this can be achieved, let F denote the lower tail (cumulative) distribution function of x, and allow c to take different values, then $F(c) = \int_{-\infty}^{c} f(x)dx = \theta_c$. It should be clear that if there were a vector of multiple-cut points c, for example, $c_1 < c_2 < c_3 \ldots$, one could graph-out and estimate of the distribution function $f(x)$ by varying the values of c within the range of the outcomes of x, thus retrieving estimates of the latent subpopulations of x for the task of focus. Examples of the cut-point approach are given in several sources (Benaglia et al., 2010; Elmore et al., 2004; Hettmansperger and Thomas, 2000). These examples all employ the same Piaget water-level data reported in Thomas and Lohaus (1993) which recall did not use the cut-point approach but rather defined intervals in the outcome distribution. These water-level data are included in the `mixtools` package, which, in turn, provides the basis for several examples in the `mixtools` manual. However, because no source provides a "recipe" for implementation which is typically straightforward and relatively easy, what follows is a very simple artificial example of the cut-point approach which hopefully illustrates all aspects of how the cut-point model might be implemented in real-world settings, along with the associated pitfalls of the approach.

So, to begin, let the observed manifest data follow the simple uniform distribution:

$$f(x) = 1/2, 0 < x < 2$$

$f(x)$ is a simple rectangle with height one-half and length two. Histograms of observed data sampled from it will resemble $f(x)$ in appearance. However in the present case, data collected are not of this form. Rather, data arise from an $R = 2$

mixture distribution with component distributions: $f_1(x) = x/2$ a right triangle with hypotenuse slope 1/2, and $f_2(x) = 1 - x/2$ the mirror image of $f_1(x)$ with hypotenuse slope $-1/2$. Set $\lambda_1 = \lambda_2 = 1/2$. Then, the two-component mixture

$$f(x) = \lambda_1 f_1(x) + \lambda_2 f_2(x)$$

gives back $f(x) = 1/2$. Thus, what appears as a rectangle is really the equally weighted sum of two triangular response distributions. The goal is to see how well the cut-point approach retrieves this latent structure.

The first task is to generate $m = 6$ repeated measurements for each of $n = 100$ virtual subjects. m is arbitrary, but it must be that $m \geq 3$ to be identifiable with $S = 2$. Samples from $f_1(x)$ are in vector `xv1` and samples from $f_2(x)$ are in `xv2`. To obtain data, implement:

```
n<-100; m<-6
xv1<-xv2<-numeric(0)
for(j in 1:n) {if(rnorm(1)>0)
xv1<-c(xv1,sqrt(4*runif(m)))else
xv2<-c(xv2,2-2*sqrt(1-runif(m)))
}
```

An observation comes from each component distribution with probability 1/2. This is determined by whether the standard normal random value `rnorm(1)` is positive or negative. Thus, the number of observations in `xv1` and `xv2` will be random, while the total number of observations will be nm. The data generation employs inverse probability transforms (Wikipedia contributors, 2022). Next, combine the observations into an $n = 100$ row by $m = 6$ column matrix `matx`.

```
xv<-c(xv1,xv2)
matx<-matrix(xv,ncol=6,byrow=TRUE)
```

Executing `hist(matx)` will reveal a roughly rectangular histogram of multiple bars resembling $f(x)$. Executing `hist(xv1)` reveals a histogram resembling $f_1(x)$, while `hist(xv2)` reveals bar histograms resembling $f_2(x)$. In the present example, the first 51 rows of `matx` are from `xv1`, while the remaining 49 rows are from `xv2`.

The next task is to assess whether there is evidence in `matx` of a mixture distribution. Different settings will require different strategies to address this matter. One way to do so is using the loglikelihood, but this approach will be postponed until later. Another way to do so is to partition each subject's data vector into counts, reflecting the number of values below a cut point, and the number of counts above the cut point. Because the outcomes are in the interval zero to two, setting a cut point $c = 1$ seems appropriate.

After `mixtools` has been loaded, implement:

```
maty1<-makemultdata(matx,cuts=1)
maty1<-maty1$y
yv1<-maty1[,1]
```

yv1 is a vector of length 100 of the counts of each subject's repeated measures vector which are below the cut point of one. If the values are all from the $f(x)$ uniform population, without subpopulations, they should be binomially distributed with binomial probability $\theta = 1/2$ and have variance approximately the same as the corresponding binomial variance, $m\theta(1 - \theta) = 1.5$. Below are the first ten observations, and the last ten observations from yv1:

```
yv1[1:10]
3 1 2 2 2 3 0 0 1 0
yv1[91:100]
4 5 6 3 4 5 5 6 4 5
```

The first list is sampled from the first component distribution $f_1(x)$; the second list is sampled from the second component $f_2(x)$. The first and last observations look different, and they certainly are. With a focus on the variance of the outcome vector var(yv1), it is 3.67. To be sure this outcome does not reflect sampling variation about 1.5 is a job for the bootstrap (Efron and Tibshirani, 1993). A standard resampling procedure is

```
longv<-numeric(0)
for(i in 1:200)longv<-c(longv,var(sample(yv1,n,replace=TRUE)))
sort(longv)
```

Examining sort(longv) reveals none of 200 bootstrap variance values is as small as 1.5. Thus, there is good support for there being subgroups in the data. It seems safe to proceed with a mixture decomposition. The simplest first analysis is to consider a single cut, $c = 1$. The matrix for this analysis has already been constructed and is maty1 a matrix with 2 columns and 100 rows, where each row sums to 6. Thus:

```
out1<-multmixEM(maty1)
summary(out1)
summary of multmixEM object:
           comp 1   comp 2
lambda 0.476352 0.523648
theta1 0.200681 0.737274
theta2 0.799319 0.262726
loglik at estimate:  -190.796
```

The first row in the output is lambda which estimates λ_1 and λ_2. There are two θ for each component, one value estimates the θ for the interval 0 to 1, while the second value estimates the θ for the interval 1 to 2. For component one, $\int_0^1 f_1(x)dx = 0.250$ while the estimate is 0.201; $\int_1^2 f_1(x)dx = 0.750$ while the estimate is 0.799. For component two, $\int_0^1 f_2(x)dx = 0.750$ with the estimate 0.737, and $\int_1^2 f_2(x)dx = 0.250$ with estimate 0.263. Overall, the estimates seem in-line with the model values.

Now increase the number of cut points in an effort to estimate the shapes $F_1(x) = x^2/4$ and $F_2(x) = x - x^2/4$ the first and second component distribution functions, respectively. Three cut points at 0.5, 1, 1.5, are likely sufficient. Then, implement the following code:

```
maty2<-makemultdata(matx,cuts=c(.5,1,1.5))
out2<-multmixEM(maty2)
summary(out2)
summary of multmixEM object:
          comp 1    comp 2
lambda 0.5207008 0.4792992
theta1 0.0495456 0.4782015
theta2 0.1756397 0.2821013
theta3 0.3174408 0.1802110
theta4 0.4573739 0.0594862
loglik at estimate:  -403.3189
```

The estimates of 0.52 and 0.48 in `lambda` capture the parameter values of $\lambda_1 = \lambda_2 = 1/2$ closely. With three cut points, there are four θ estimates for each component of the 0 to 2 interval: 0 to 1/2, 1/2 to 1, 1 to 3/2, and 3/2 to 2. For example, for $c = 1$ for component one or `comp 1` the estimate $\widehat{F}_1(1) =$ `theta1` + `theta2` $= 0.2252$ while $F_1(1) = 0.25$. Table 29.3 provides the parameter values for $F_r(c)$ and their estimates $\widehat{F}_r(c)$, $r = 1, 2$ for each value of c.

Table 29.3 Component distribution function values with estimates

c	$F_1(c)$	$\widehat{F}_1(c)$	$F_2(c)$	$\widehat{F}_2(c)$
0	0	0	0	0
1/2	0.0625	0.0495	0.4375	0.4782
1	0.2500	0.2252	0.7500	0.7603
3/2	0.5625	0.5462	0.9375	0.9505
2	1	1	1	1

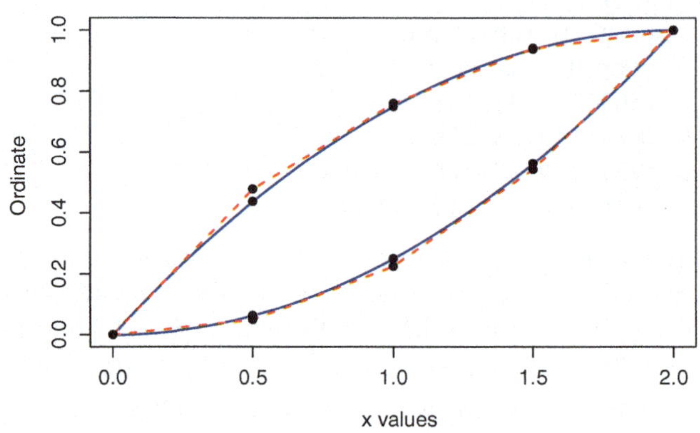

Fig. 29.2 Component distribution functions. Dashed lines are estimates; solid lines are population values

In Fig. 29.2 are graphed both the estimated and true cumulative distributions from Table 29.3. Even with three cut points, the estimated values track the population values well. Estimated unit density histograms for both components can also be obtained from the following lines of code:

```
cuts<-c(.5,1,1.5)
tr<-as.vector(out2$theta[r=2,]) #r=1,2
a<-0;b<-2  #range of data
d<-c(cuts,b)-c(a,cuts)
hts<-tr/sum(tr*d)
xp<-c(a,cuts,b)
plot(xp,seq(0,max(hts),length.out=length(xp)),type="n",
xlab= "x values",ylab="Ordinate")
for(s in 1:length(hts)){
segments(xp[s],hts[s],xp[s+1],hts[s],lwd=2,col="blue")
segments(xp[s],0,xp[s],hts[s],col="blue",lwd=2)
segments(xp[s+1],0,xp[s+1],hts[s],col="blue",lwd=2)}
abline(0,0,lwd=2)
```

The unit histogram for the second triangular component is in Fig. 29.3.

The cut-point approach does retrieve the information about the latent subpopulations rather well. One should not, however, expect the same fidelity with real data where the model assumptions are not likely to be satisfied as they are in this artificial example. In particular, the assumption that the repeated measures on each subject are independent (given a subject's component membership) seems likely to be violated in all real data examples. Still, the approach can be illuminating by providing insights into children's performance on the task (e.g., Benaglia et al. 2010; Thomas and Lohaus 1993).

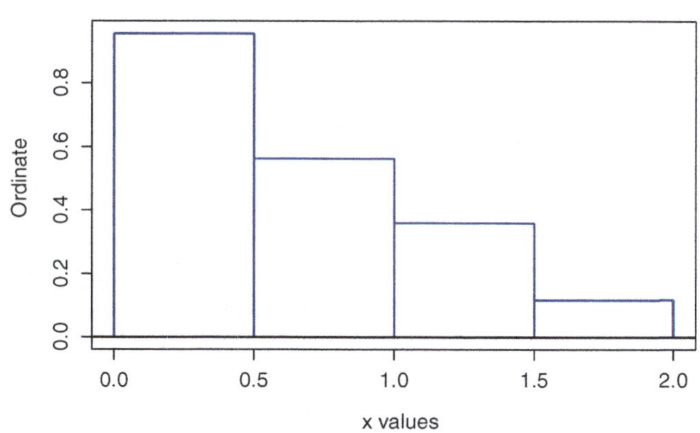

Fig. 29.3 Estimated unit histogram of component 2 triangular density

29.4 Discussion

Implementing mixture solutions, and interpreting the corresponding `summary` outputs without having substantive knowledge regarding the source of the data, and what might be expected is probably not possible. In the height example, as already noted, one would need to know that women are shorter than men to be able to match the two output columns `comp 1` and `comp 2` appropriate for each sex. In the triangles problem, knowing how the data were constructed guided the solution. Setting the single cut point at one given how the data were formed was appropriate. Had other more complex shapes been used as the latent densities, the results might not have been as revealing, and well more than three cut points would likely be needed.

The interpretation problem can be increasingly more difficult with additionally "discovered" components when there is little basis for expecting them. This problem can be especially vexing when normal components are assumed. Extra components are often declared under equal variance assumptions. The equal variance assumption failure may be the primary cause of the "newly found" components.

When subpopulations within the population sampled are suspected, but there is no theory or basis for knowing the number of components, the problem then becomes one of estimation. This problem is usually regarded as a model selection problem. There is a large literature on the matter, but there remains no general agreement as to how best to proceed. Perhaps the most commonly used criteria remain penalized likelihood procedures. A widely used procedure proposed by Schwarz in (1978) and more recently called the Bayesian information criterion or BIC for short, Schwarz defined as

$$\log(M) - (k/2)\log(n).$$

M is the likelihood of a model, and k is the number of parameters estimated in the model, with n sample size. There is a different BIC for each different model which typically vary in the number of k parameters estimated.

Consider the binomial likelihood of the data in vector `yv1`. And suppose that the data were assumed to be independent draws from a binomial with $m = 6$ and $\theta = 1/2$ distribution. Computing `M<-prod(dbinom(yv1,6,.5))` is the estimated likelihood under this model. In this case, there are *no* parameters being estimated from the data, and so BIC $= \log(M) = -231.5274$ or `log(M)`. Recall in `yv1` are the number of responses for each subject which are below one, while a second vector holds the number of responses for each subject above one. BIC is identical for both vectors.

The BIC mixture solution for one cut point with two components estimates three parameters, one θ for each component, and one λ because the within-component probabilities θ must sum to one, and so must the λ weights. Thus, `out1$loglik-1.5*log(100)` equals -197.7038. `out1$loglik` returns -190.796 the loglikelihood value outputted in the solution reported above. Schwarz

(1978) argued that one should select the model with the largest (of what is now called) BIC. Since $-197.7038 > -231.5274$, there is good justification for considering a mixture with two binomial subpopulation components.

So, will the data support a three-component model with one cut point at one? Issuing the command `summary (multmixEM(maty1,k=3))` will typically yield the same or nearly the same loglikelihood, -190.796. (`k` in `mixtools` is the number of components, while k for Schwarz is the number of parameters!) This anomaly is easily explained.

The first clue is that algorithmic matters are not going well. With three components the algorithm usually requires hundreds of iterations to converge, unlike the few iterations required for the two-component solution. More importantly, the only solutions the algorithm seems able to "find" for a third component are with θ estimates coinciding with one of the other component solution estimates or a third component with vanishing near-zero values; neither solution seems sensible. With three components and one cut point, there are five parameters estimated. Thus, BIC is -202.3089 smaller than -197.7038 indicating that the data do not support a three-component solution.

While BIC as defined by Schwarz was intended for model-based methods with likelihoods, it has been adapted to setting where there is no likelihood, such as conventional least squares regression. In these settings a minimum least squares is usually the model selection criterion. BIC for these settings is sometimes defined so that the smallest BIC is desired.

Finally, it might seem quaint to propose simple models with but a handful of parameters and with a simple probability structure. Lurking are artificial intelligence (AI) models with *billions* of parameters, and perhaps the hope that AI and big data can render analytical theory obsolete. Yet, to take one example, it is widely known that any continuous distribution can be approximated by a mixture distribution with a sufficient number of normal components. Assuming a latent structure with normal components is an appropriate or at least useful conceptualization, it remains unclear just how AI might identify which components are psychologically important. The discovery process seems likely to require human cognition for sometime to come, especially when individual differences are of concern.

References

Benaglia, T., Chauveau, D., Hunter, D., & Young, D. (2010). `mixtools`: An R package for analyzing mixture models. *Journal of Statistical Software, 32*(6), 1–29. 10.18637/jss.v032.i06

Efron, B., & Tibshirani, R. (1993). *An introduction to the bookstrap*. Chapman and Hall.

Elmore, R. T., Hettmansperger, T. P., & Thomas, H. (2004). Estimating component distribution functions in finite mixture models. *Communications in Statistics, Theory and Methods, 33*(9), 2075–2086. https://doi.org/10.1081/STA-200026574

Hettmansperger, T. P., & Thomas, H. (2000). Almost nonparametric inference for repeated measures in mixture models. *Journal of the Royal Statistical Society Series B, 62*, 811–825. https://doi.org/10.1111/1467-9868.00266

R Core Team (2020). *R: A language and environment for statistical computing*. R Foundation for Statistical Computing, Vienna, Austria. https://www.R-project.org/

Schwarz, G. (1978). Estimating the dimension of a model. *The Annals of Statistics, 6*(2), 461–464. doi: 10.1214/aos/1176344136

Thomas, H. (2024). *Sex differences in reading and math test score of children.* Springer. https://link.springer.com/book/10.1007/978-3-031-41272-1

Thomas, H., & Horton, J. J. (1997). Competency criteria and the class inclusion task: Modeling judgments and justifications. *Developmental Psychology, 33*, 1060–1073. https://doi.org/10.1037/0012-1649.33.6.1060

Thomas, H., & Lohaus, A. (1993). *Modeling growth and individual differences in spatial tasks.* University of Chicago Press. *Monographs of the society for research in child development* (Vol. 58(9)). Serial 237 (1993). https://doi.org/10.2307/1166121

Wallner, S., Thomas, H., & Stemmler, M. (2021). Change in juvenile offending versatility predicted by individual, familial, and environmental risks. *In ternational Journal of Conflict & Violence, 15*, 1–16. DOI: 10.11576/ijcv-4559

Wikipedia contributors. (2022). Inverse transform sampling. In *Wikipedia, The Free Encyclopedia.* Retrieved 17:10, December 11, 2022, from https://en.wikipedia.org/w/index.php?title=Inversetransformsampling&oldid=1115190568

SPRINGER NATURE

GPSR Compliance

The European Union's (EU) General Product Safety Regulation (GPSR) is a set of rules that requires consumer products to be safe and our obligations to ensure this.

If you have any concerns about our products, you can contact us on ProductSafety@springernature.com

In case Publisher is established outside the EU, the EU authorized representative is:

Springer Nature Customer Service Center GmbH
Europaplatz 3
69115 Heidelberg, Germany

The manufacturer's authorised representative in the EU is Springer Nature Customer Service Centre GmbH, Europaplatz 3, 69115 Heidelberg, Germany. If you have any concerns regarding our products, please contact ProductSafety@springernature.com

Printed and bound by CPI Group (UK) Ltd, Croydon, CR0 4YY
28/09/2026
02224056-0020